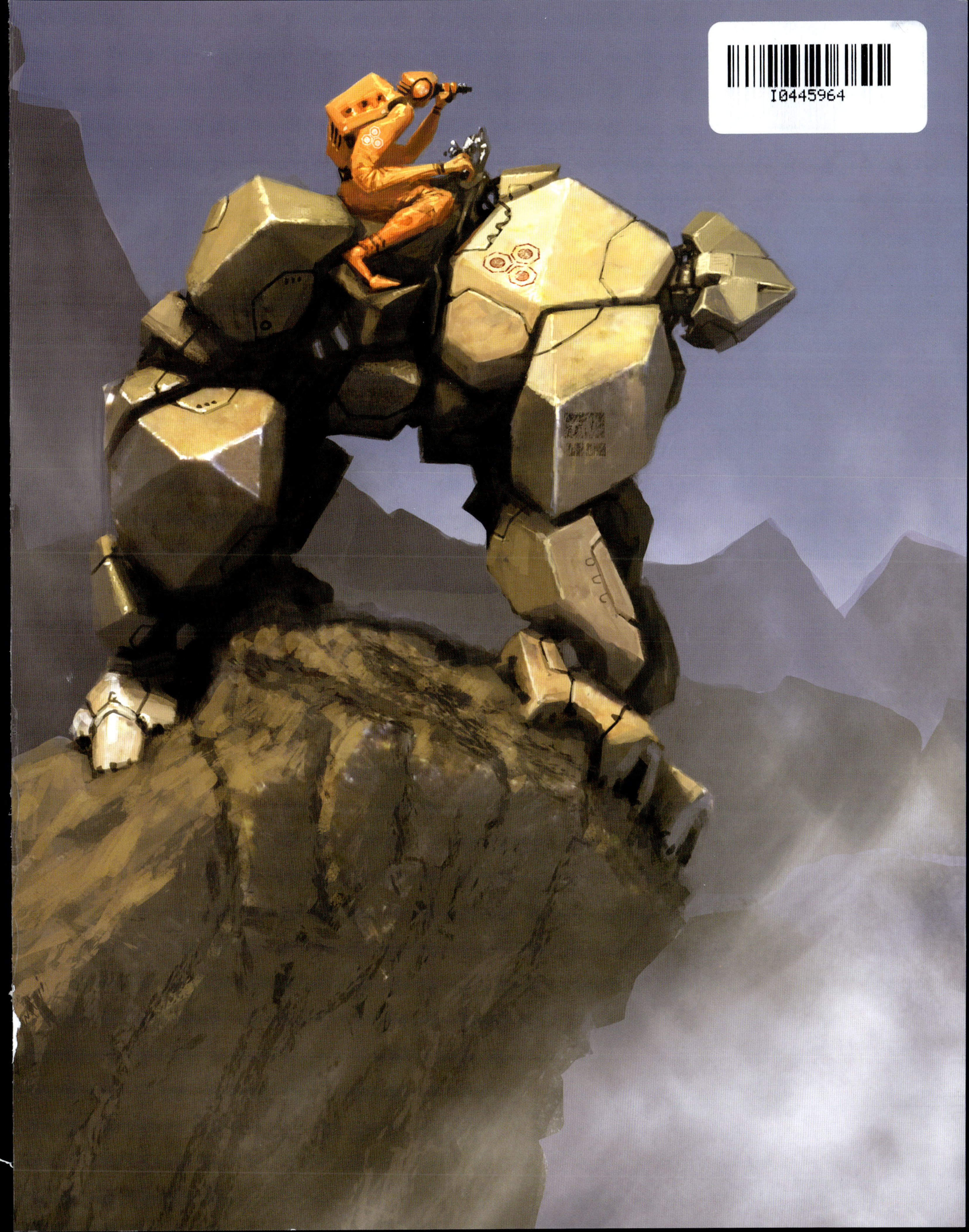

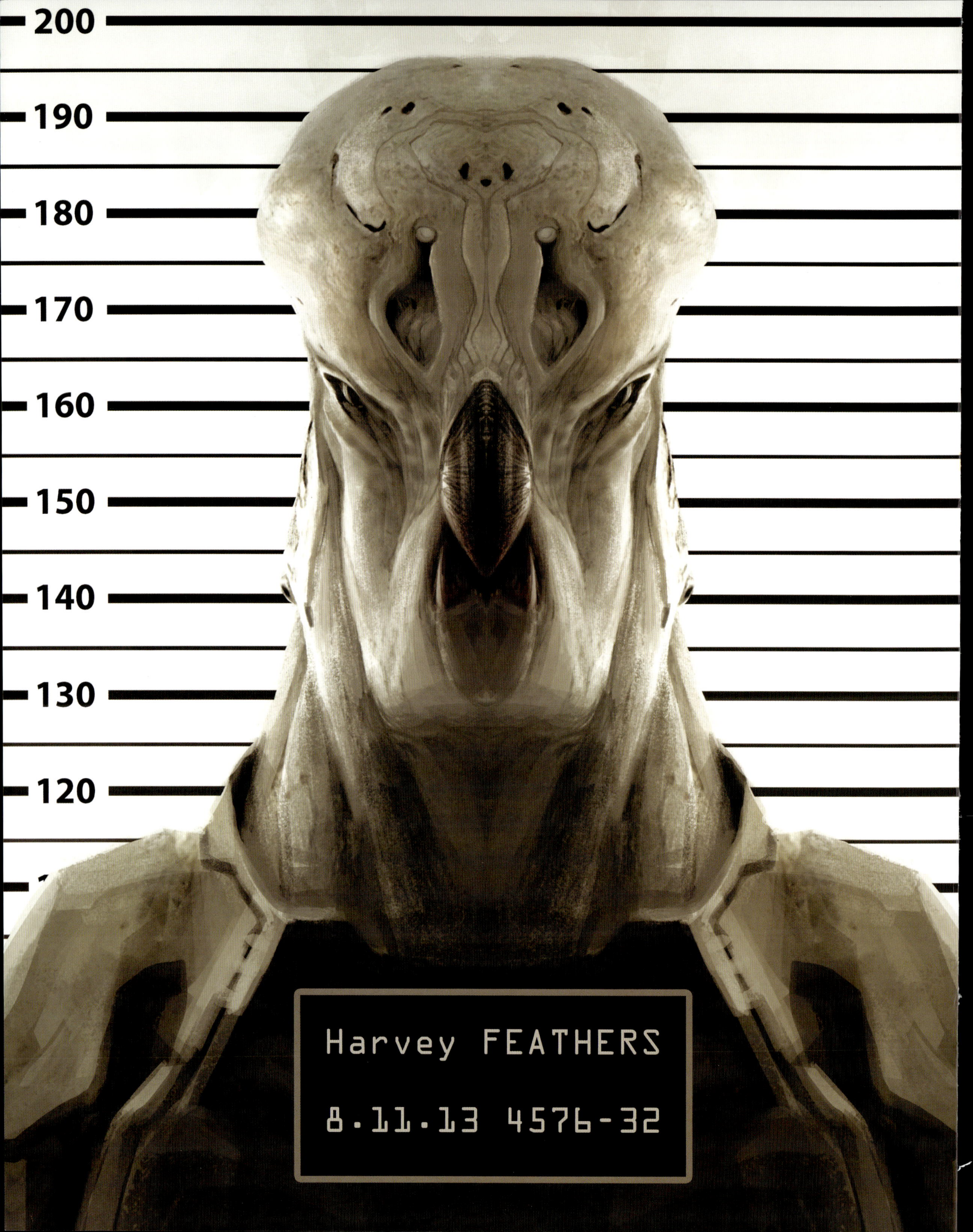
200
190
180
170
160
150
140
130
120
Harvey FEATHERS
8.11.13 4576-32

HOW TO RENDER

the **FUNDAMENTALS** of **LIGHT, SHADOW** and **REFLECTIVITY**

by **Scott Robertson** with **Thomas Bertling**

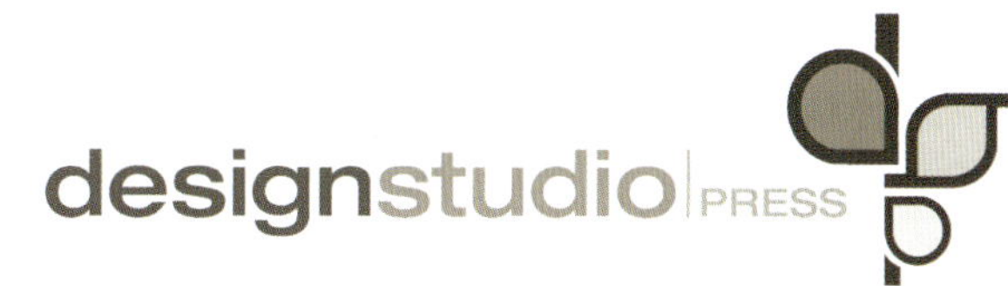

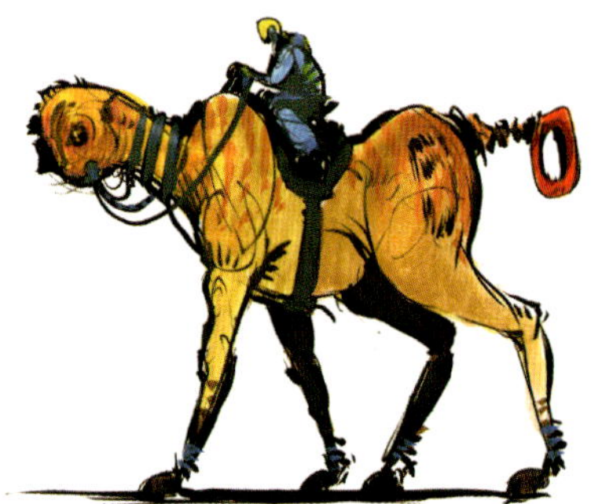

DEDICATION

This book is for those with a passion for rendering and learning. **Never stop!**

BEYOND THIS BOOK:

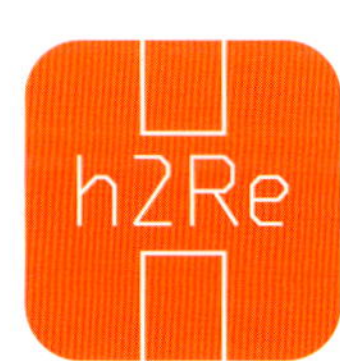

Step-by-step videos are an integral part of the *How To Render* educational experience! Use a smartphone or tablet to open a QR Reader app and scan this QR code. It links to the Design Studio Press image-recognition app needed to play the videos. Download the DSP app, scan Scott's photograph from page 10 and an introductory video will load.

All of the pages in this book that link to educational videos have a play button at the bottom, like this:

No smartphone or tablet? No worries.
Go to page 270 and type in the URL on any computer to gain access to the entire list of links.

Copy Editors: Melissa Kent, Erika G. Bertling, Heather K. Dennis, Teena Apeles | **Graphic Design:** Christopher J. De La Rosa

Published by Design Studio Press
Website: www.designstudiopress.com | **Email:** info@designstudiopress.com

Printed in China | First Edition, November 2014

10 9 8

Library of Congress Control Number | 2014947671 | **Softcover ISBN**: 978-193349296-4 | **Hardcover ISBN**: 978-193349283-4

TABLE OF CONTENTS

INTRODUCTION | PAGE 8

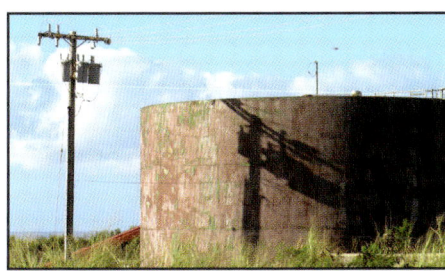

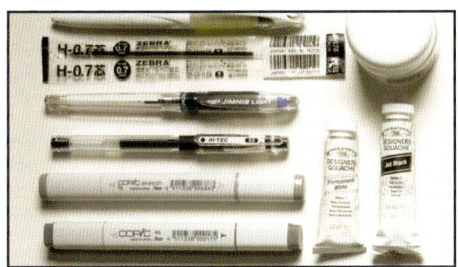

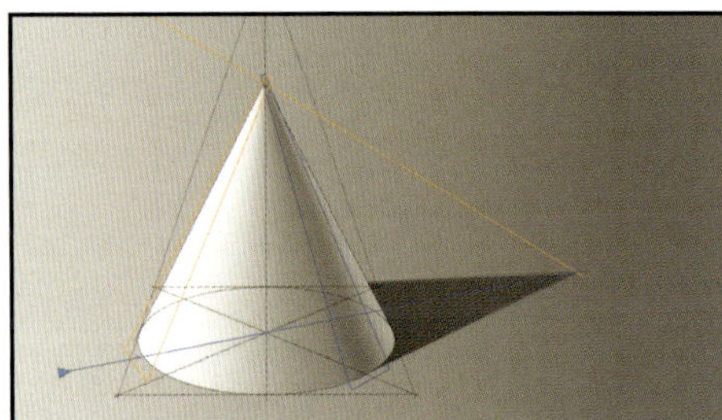

CHAPTER **03 Rendering the Geo Forms** | **continued**

CHAPTER **04 Complex Volumes** | **PAGE 105**

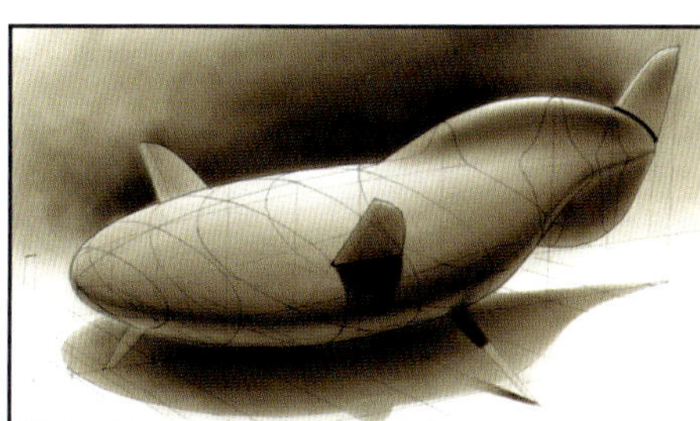

INTRODUCTION

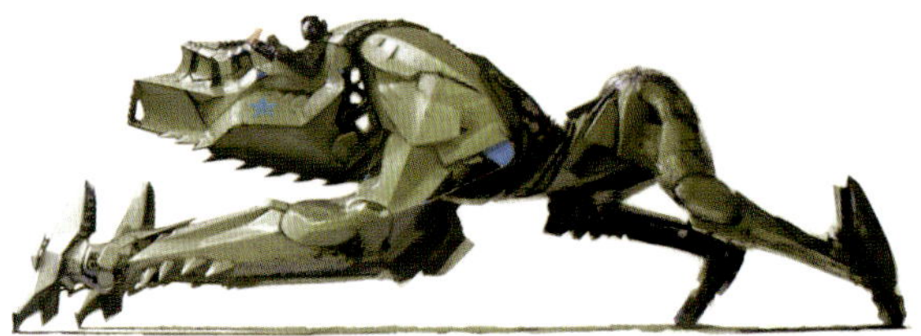

Rendering is the next step after drawing to communicate ideas more clearly. Building on what Thomas Bertling and I wrote about in *How To Draw: Drawing and Sketching Objects and Environments from Your Imagination*, this new book shares almost everything we know about how to render light, shadow and reflective surfaces.

Before getting started, there are a few important things to understand about how this book is organized. It is divided into two major sections: the first explains the physics of light and shadow. You will learn how to construct shadows in perspective and how to apply the correct values to those surfaces. The second section focuses on the physics of reflectivity and how to render a very wide range of materials utilizing this knowledge.

This book is about the fundamentals of light, shadow and reflectivity. It is not about speed-painting, photo-bashing images or working specifically in any one software program. Those are topics for another book. Although many finished rendering examples are included in this book, in order to show what is possible after mastering the fundamentals, the focus is firmly on helping to improve your visual understanding of the world around you and on techniques for representing that world.

Rendering makes volumes look more three-dimensional and realistic. It is much easier than drawing from the imagination, because we can observe nature to learn how light, shadow and reflectivity behave. To illustrate these ideas and to improve your visual awareness and skills of observation, we show a great many photographic examples. This will lead to a deeper understanding of the visual world around you.

Throughout the book, icons appear that indicate "observe" or "act."

Observe 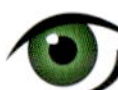

The page or section is about observing reality.

Act 

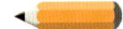

Apply the knowledge and follow the steps to create your own work.

Similar to our previous book, *How To Draw*, this book contains links to free online tutorials that can be accessed via the URL listed on page 270, or through the H2Re app. Page 4 contains more information on accessing and using the apps. It is important to watch these video tutorials in order to get the most out of *How To Render*. In addition, you will also find over forty image files to download using the same URL as the tutorials.

Let's render!

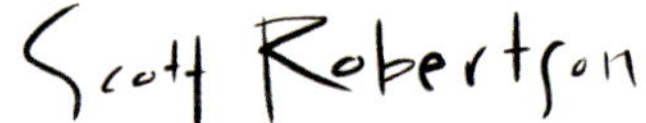

May 31, 2014
Los Angeles, California

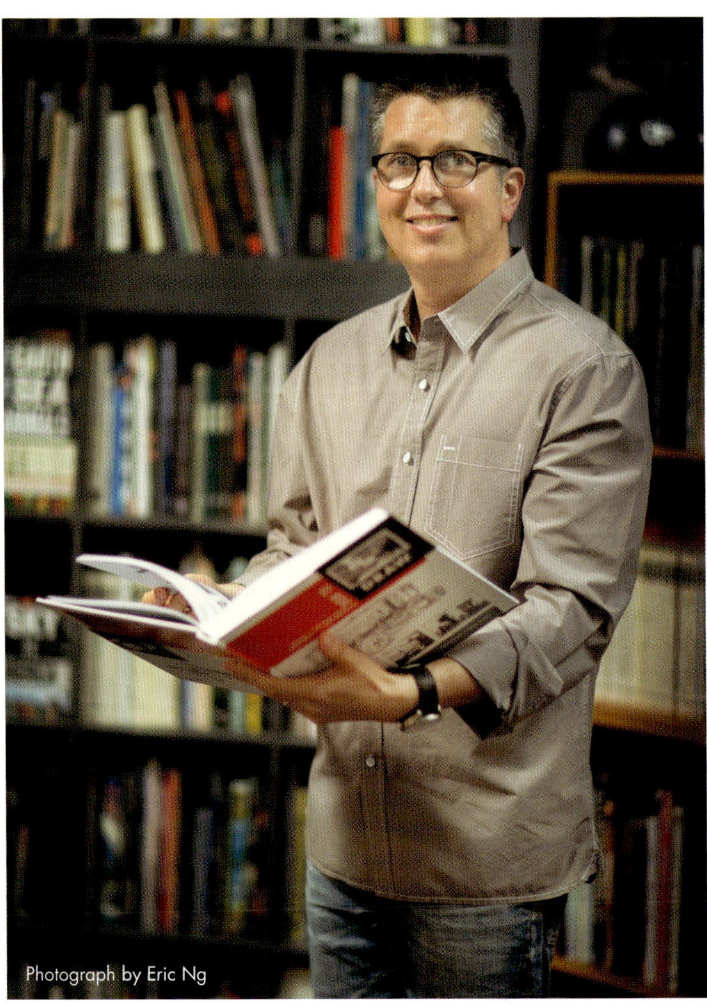

Photograph by Eric Ng

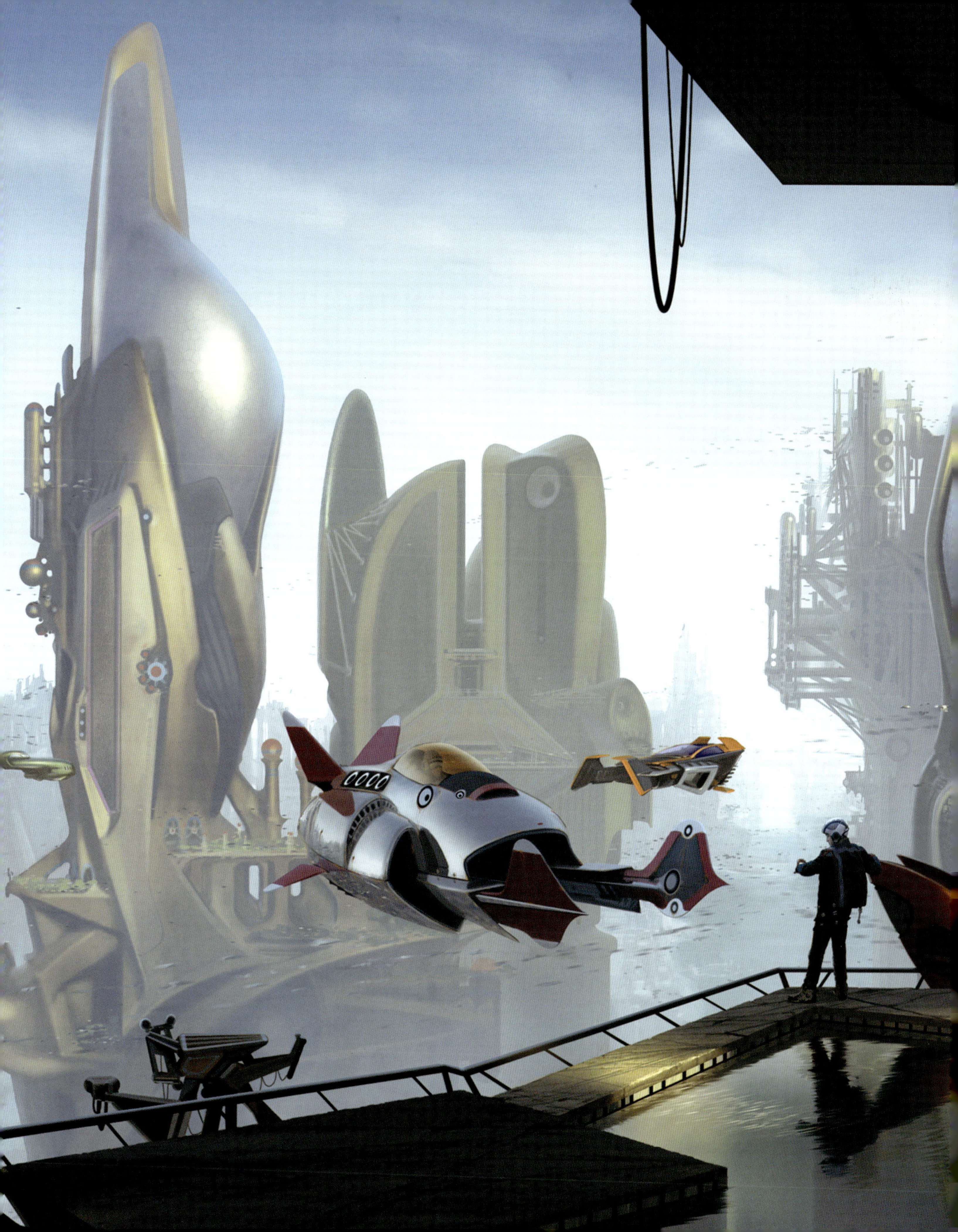

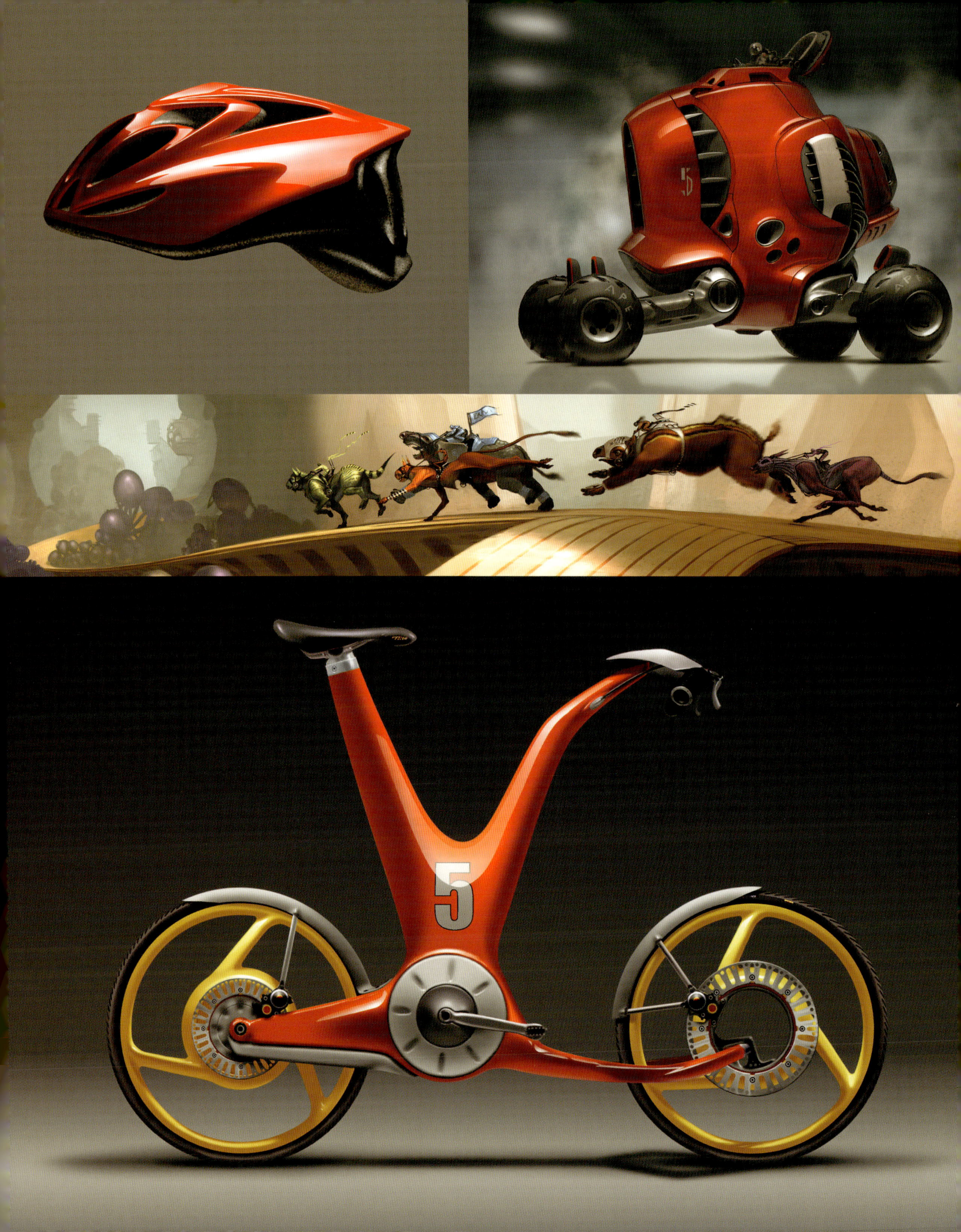
5

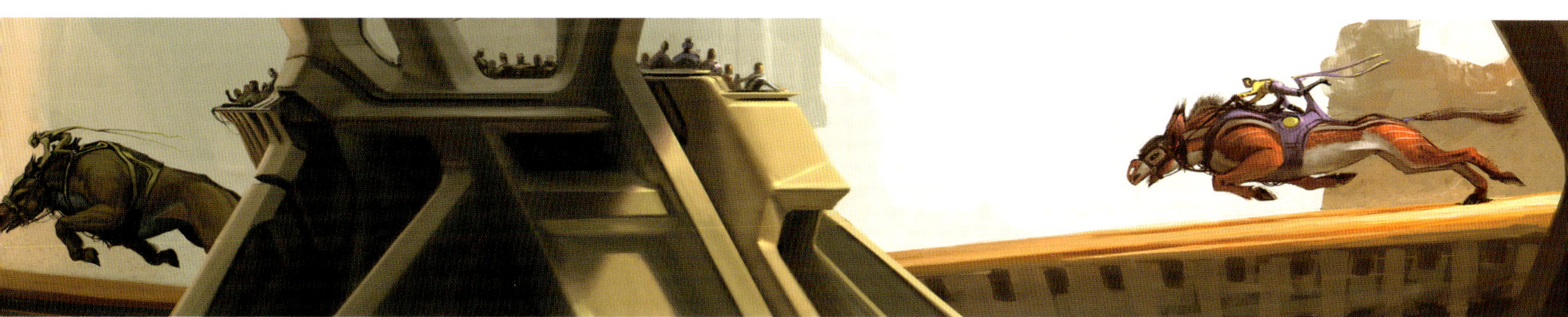

CHAPTER 01

WHAT IS RENDERING? + TOOLS AND MATERIALS

Why is rendering important?

Quickly sketching a concept to share an idea is faster than rendering, and line drawing without value and color can serve to communicate designs, so why learn how to render an original idea to a photo-real level?

Artists and designers often take for granted the ability to understand a concept by looking at a rough line drawing. However, the visual language that most people understand is based on what they see around them every day and in the real world, objects are not surrounded by lines. Silhouettes and surfaces are defined by changes in color and value. That is what the human brain understands best.

To make an imaginary object, character or scene look believable, it needs to be shown in a way that resembles what people see every day. Therefore, artists and designers must learn how to render surfaces and volumes if the goal is to have their designs clearly understood by everyone.

To get started, a few tools and materials are necessary. By no means do all of the items on the following list need to be purchased. The first half of this book's exercises can be practiced and perfected with just a pencil and several pieces of paper.

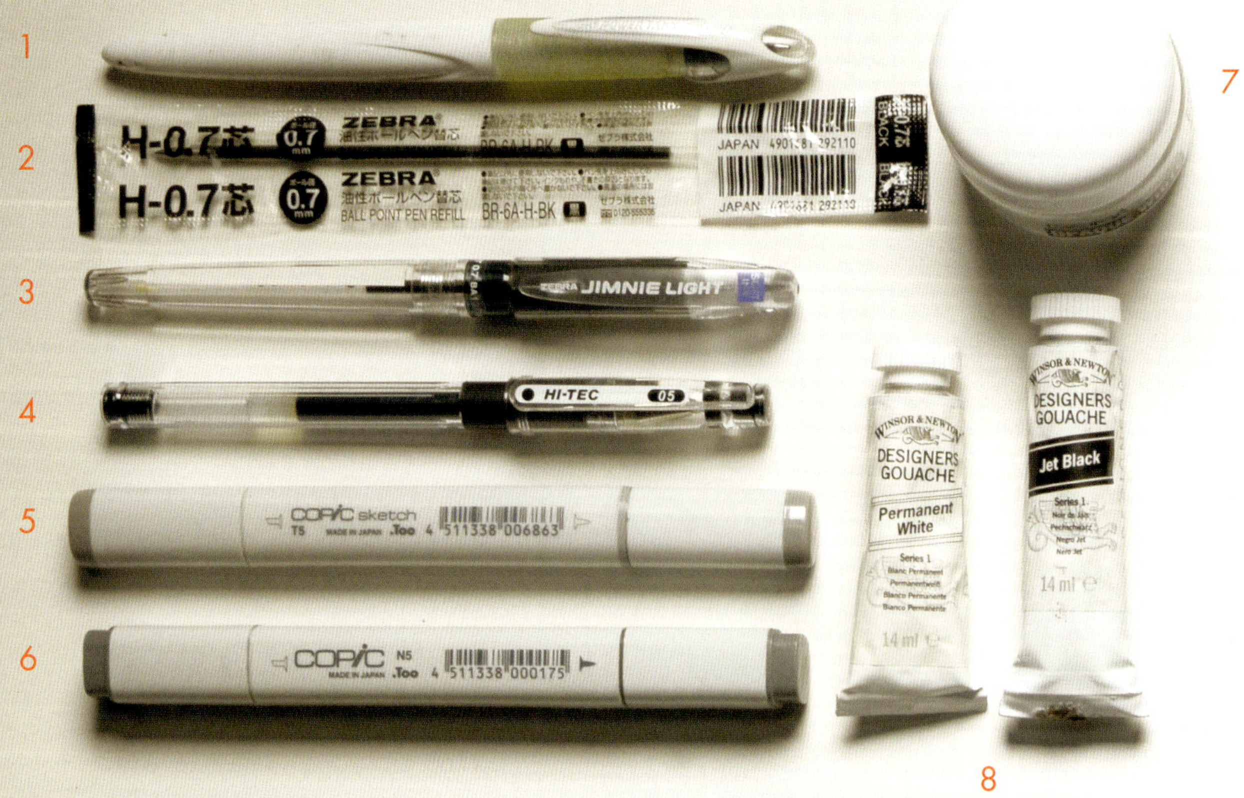

RENDERING TOOLS AND MATERIALS

Here are some excellent rendering tools and materials. Most of these can be found online or at better art supply stores. As you become more proficient you will discover your own favorites that elevate your work. The search for the next best pen or paper seems to never end, so enjoy the journey.

This Page:

1 - Uni POWER TANK 0.7 ballpoint pen

2 - ZEBRA H-0.7 ballpoint pen refill part #BR-6A-H-BK

3 - ZEBRA Jimnie Light (soft ink) ballpoint pen

4 - PILOT HI-TEC pen, available sizes, 0.25, 0.3, 0.4 and 0.5

5 - COPIC - SKETCH marker - chisel and brush tip nibs

6 - COPIC marker - broad chisel and fine point nibs

7 - COPIC - Opaque White paint

8 - Windsor & Newton - Designers Gouache

Opposite Page:

1 - Kneaded Eraser

2 - Tombow - MONO - graphite pencil, various hardness leads

3 - STAEDTLER - Mars Lumograph - graphite pencil

4 - FABER-CASTELL - Polychromos - colored (wax) pencil

5 - Windsor & Newton - series 995, 1/2" flat brush

6 - Windsor & Newton - series 7, number 3 round brush

7 - Pentel - Pocket Brush Pen

8 - Prismacolor - NUPASTEL, chalk

9 - Webril - Handi-Pads, 100% cotton, lintless

10 - Block Eraser, many brands available

11 - Talcum Powder, any brand

12 - Drafting Brush, any brand

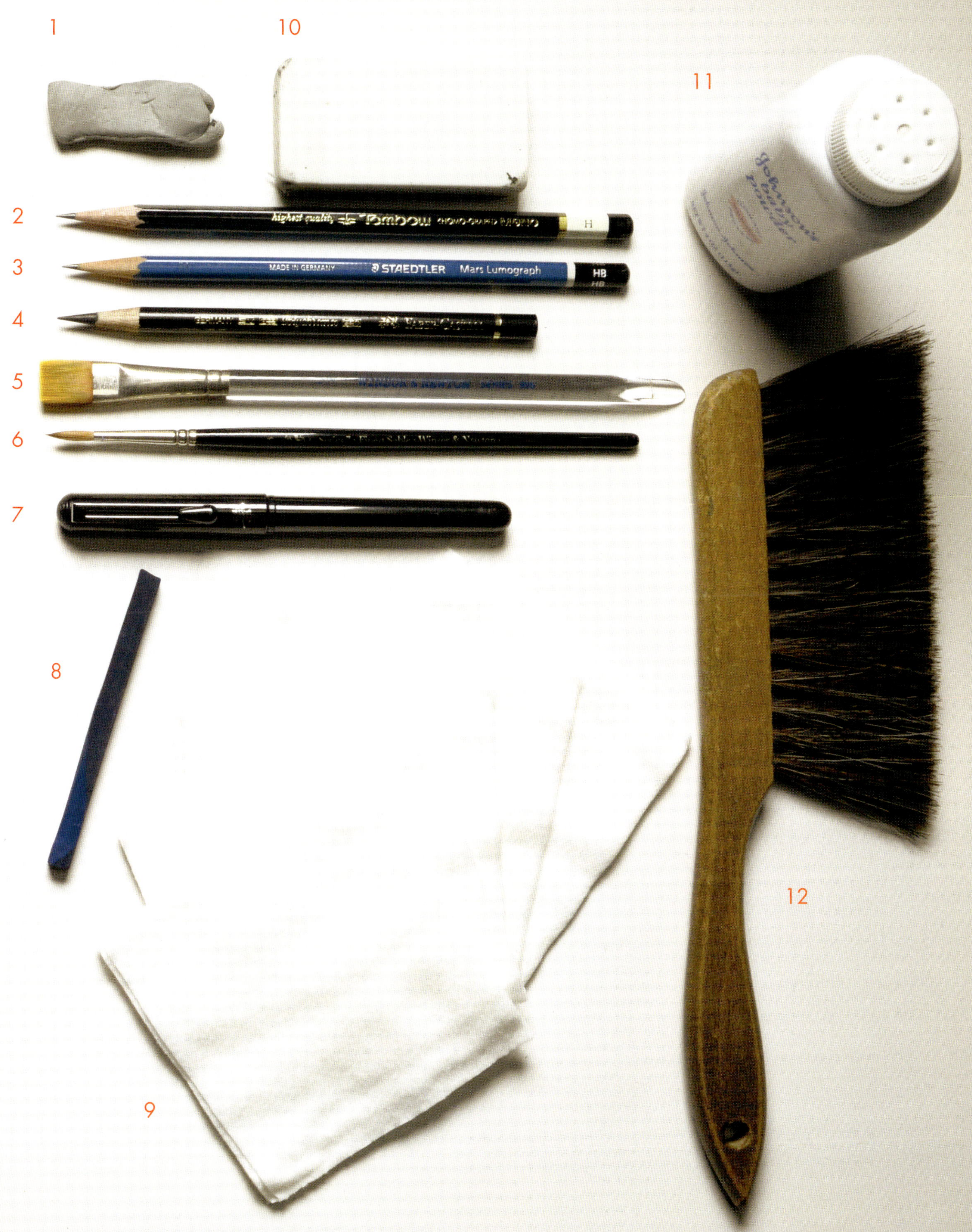
1
10
11
2
highest quality
Tombow
H
3
MADE IN GERMANY
STAEDTLER Mars Lumograph
HB
4
5
6
7
8
9
12

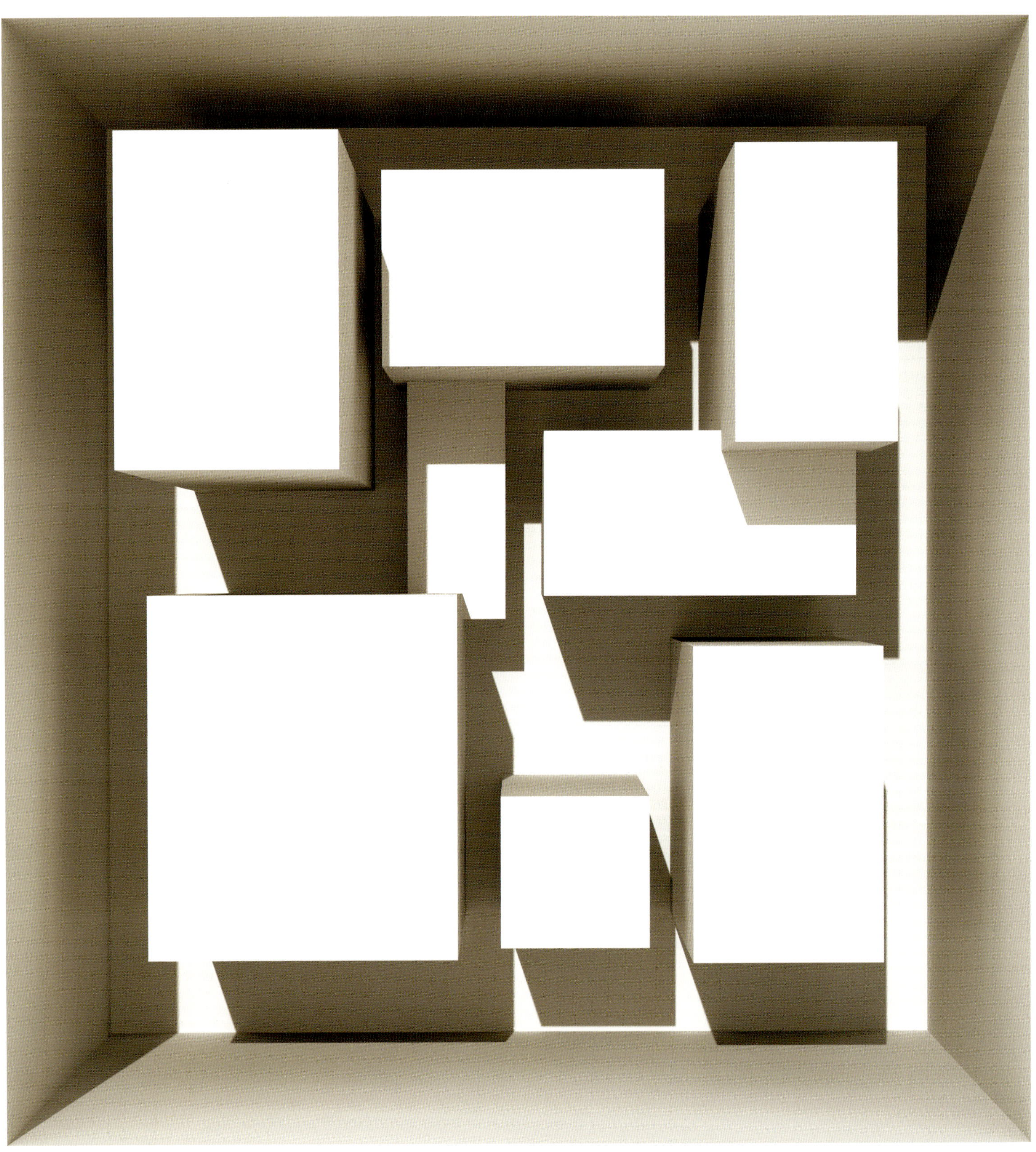

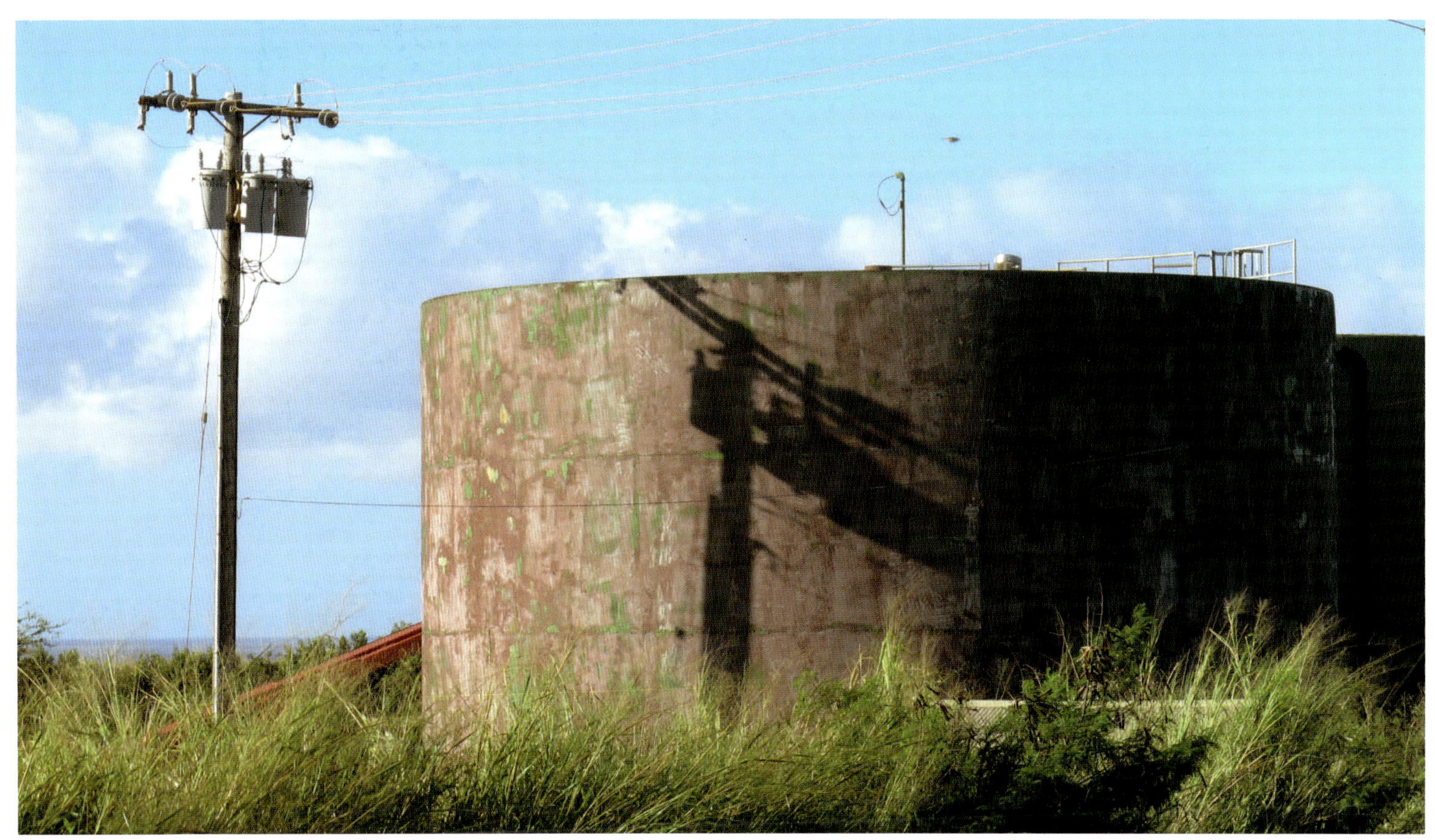

CHAPTER 02
LIGHT TYPES AND CASTING SHADOWS

The simplest and most effective way to start to make a volume look more real is to use light and shadow, creating value changes across the volume instead of using lines to describe it. There is terminology to learn and observations to make before the actual rendering of these volumes can begin.

This chapter begins with a series of pages that provide strong examples of the characteristics of light and shadow. The terms will be defined that describe all of the value renderings and photographs throughout the rest of this book.

Next, the construction of cast shadows will be explained. Cast shadows are integral to creating the illusion of three dimensions and it is important to understand at least the basic concepts of shadow construction before moving on to value assignments and shading in chapter 03.

This chapter, more than any other, relies heavily on the perspective-drawing skills learned from the previous book, *How To Draw*. The techniques and terminology on section drawing will be used throughout this book.

Let's learn about light and casting shadows!

VALUE CHANGE = FORM CHANGE 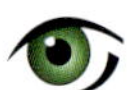

When rays of light illuminate a volume that has variation to its surface, like the bumps on a frog's back (fig. 2.4), or across the corner of a truck (fig. 2.5), the light intersects these varying surface changes at different angles, creating what we see as changes in value. This is most obvious when observing black-and-white images because the brain is not distracted by changes in color across those same surfaces. Even though these images represent a very wide range of subjects, *the form*

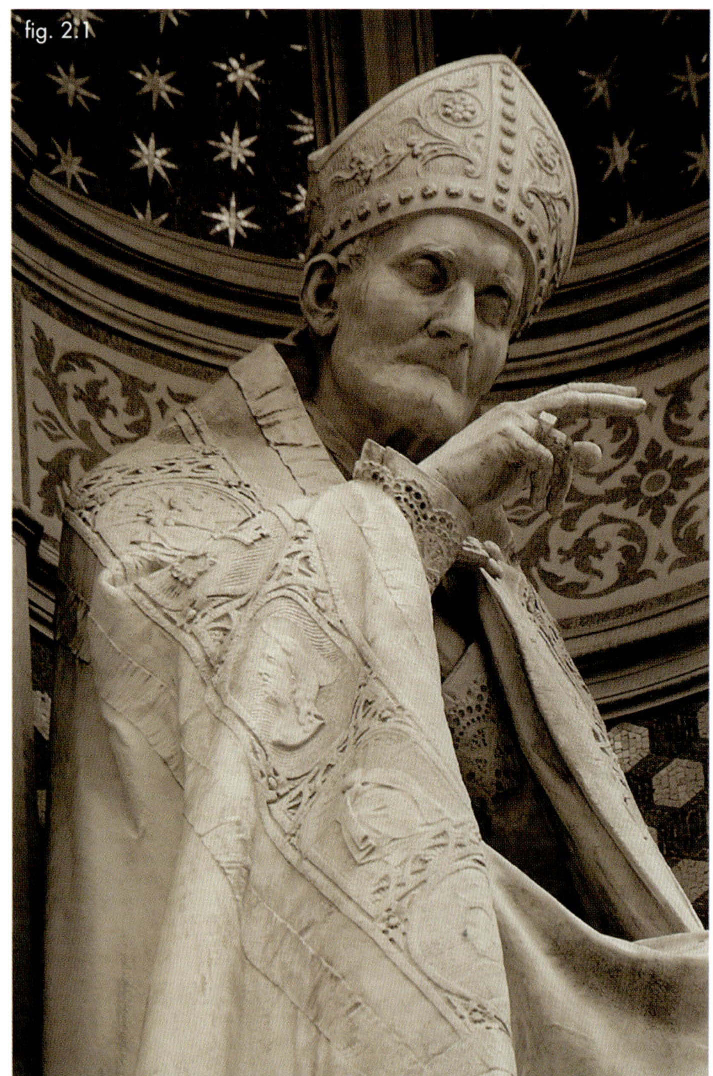
fig. 2.1

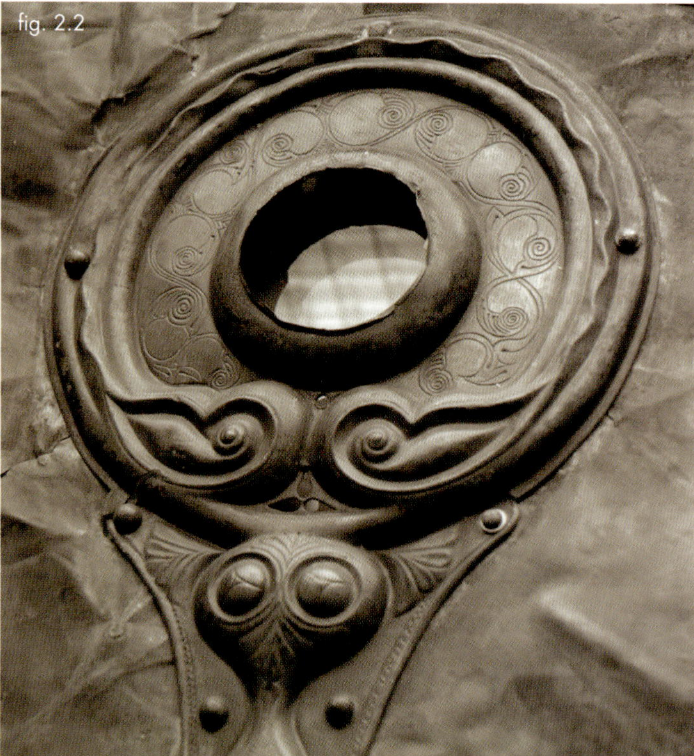
fig. 2.2

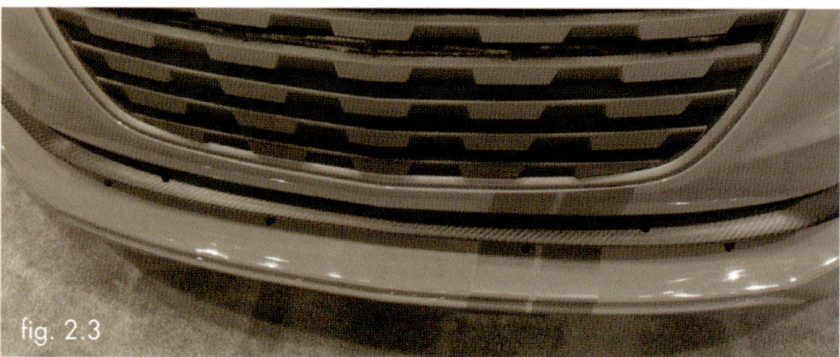
fig. 2.3

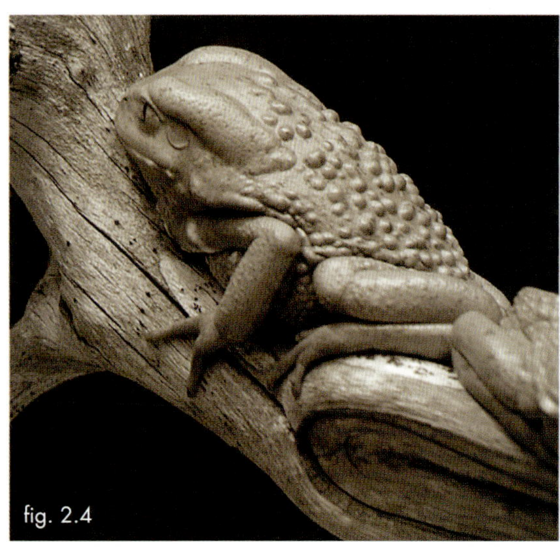
fig. 2.4

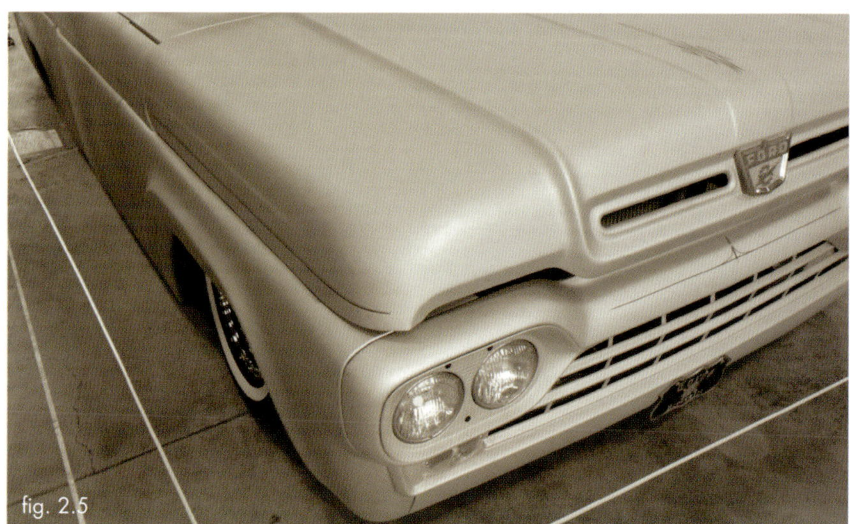

fig. 2.5

changes are all communicated by changes in value. This is the first important lesson to learn about light and shadow. It does not matter whether the object is a sand dune (fig. 2.6), a cloud (fig. 2.7), a car (figs. 2.8 – 2.9), or an airplane (fig. 2.10). As the value changes, the brain registers a change in form. Exploiting this simple fact makes rendering forms so much easier.

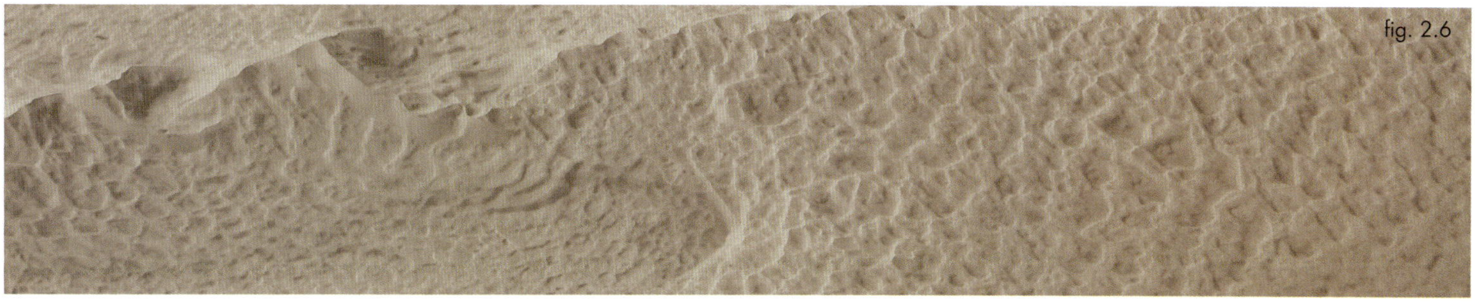

fig. 2.6

fig. 2.7

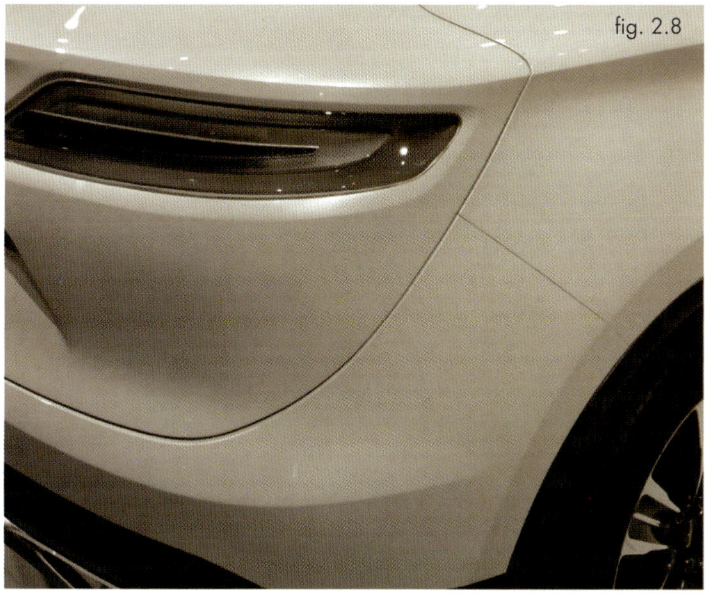

fig. 2.8

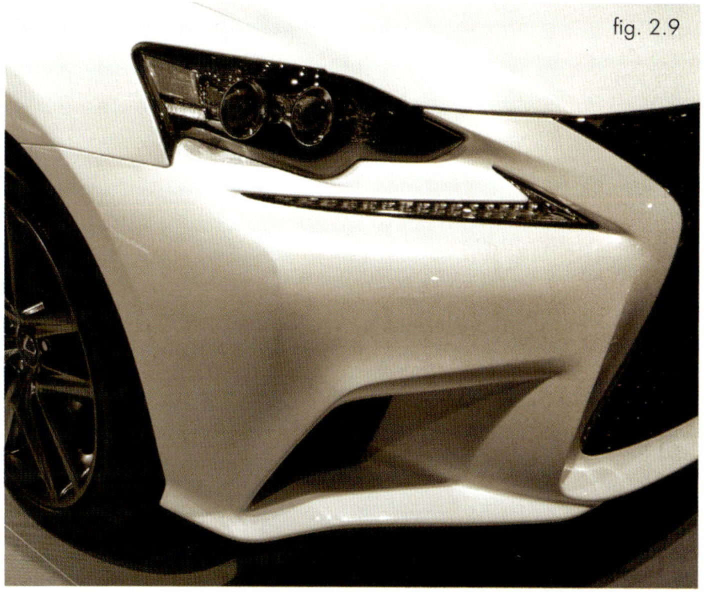

fig. 2.9

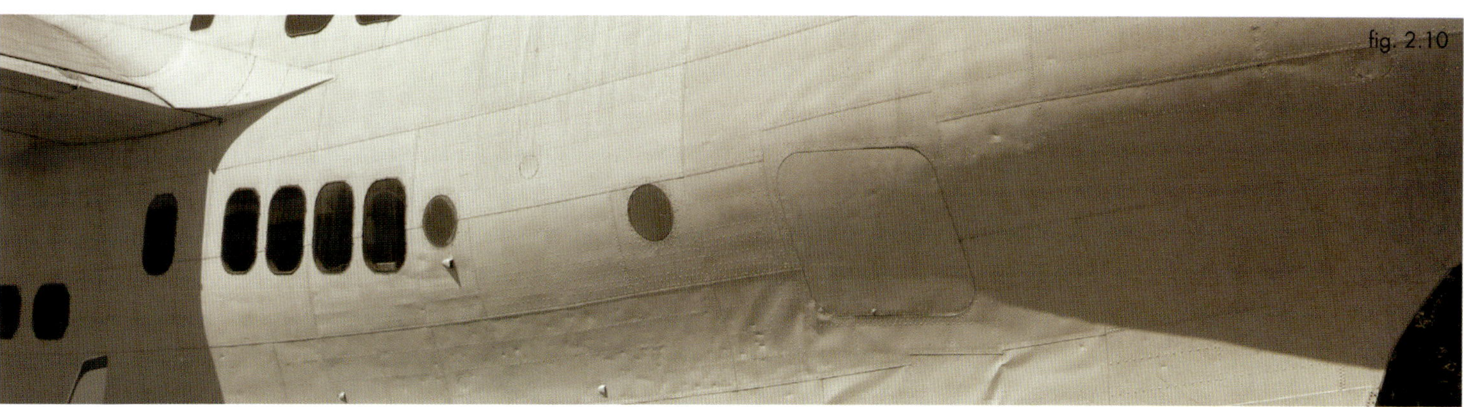
fig. 2.10

DIRECT, HARD LIGHT

When discussing different types of light, it is really about different types of shadows. Direct light casts a shadow with a hard edge, and diffuse light casts a shadow with a soft edge. All rays of light from a particular source are technically exactly the same, but the direction from which they come makes all the difference. The term "hard light" refers to when all of the light rays are very aligned in a singular direction. Hard light can be recognized by the hard edge of the shadow that the illuminated forms cast.

In fig. 2.12 the torso of the stone centaur casts a very hard-edged shadow on the flat, stone surface behind it. The quality of this edge demonstrates that a direct light illuminates the surfaces. Figs. 2.11 and 2.12 show these hard-edged cast shadows with light coming from an indoor artificial source. Hard-edged shadows can also be cast from sunlight (figs. 2.13, 2.14 and 2.15). Even though the sources of light are very different, what matters is that the rays of light are very aligned and traveling from a singular point at the source, whether that source is the sun or a light bulb, the quality of the edge of the cast shadow is the same.

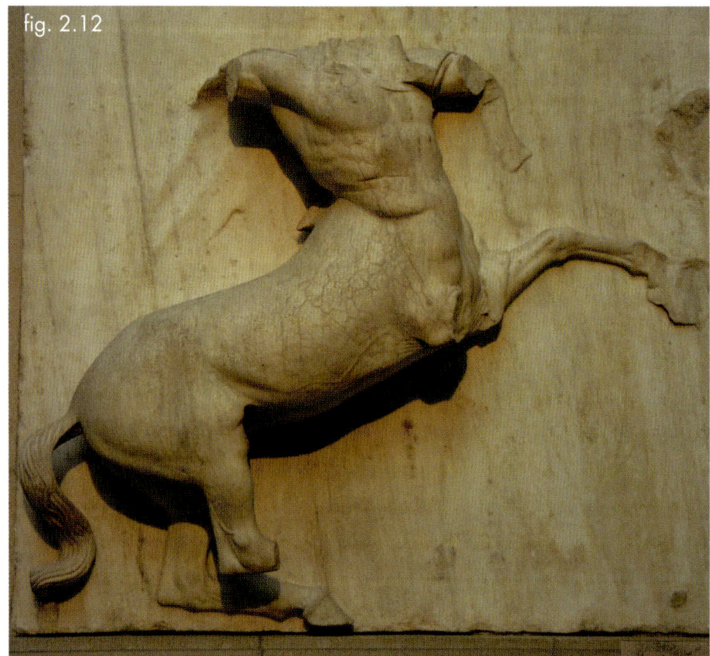
fig. 2.12

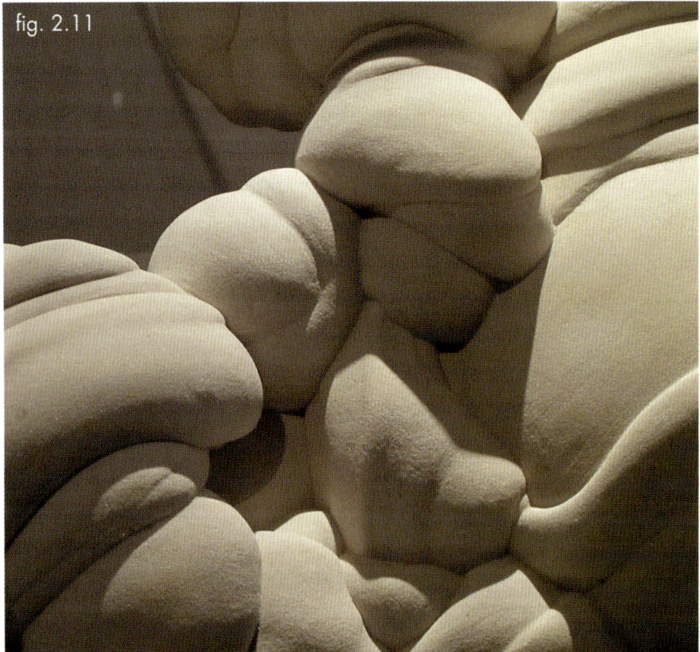
fig. 2.11

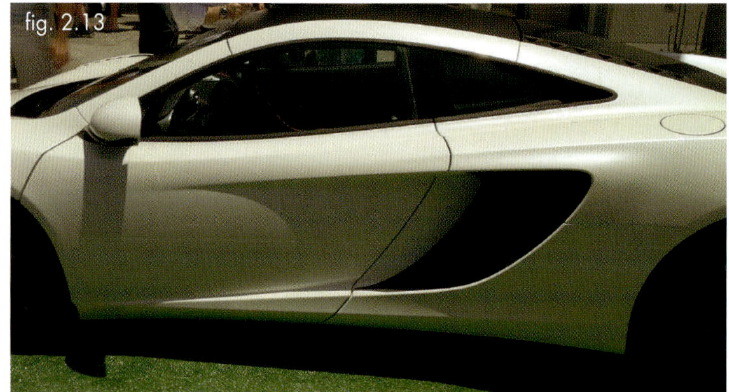
fig. 2.13

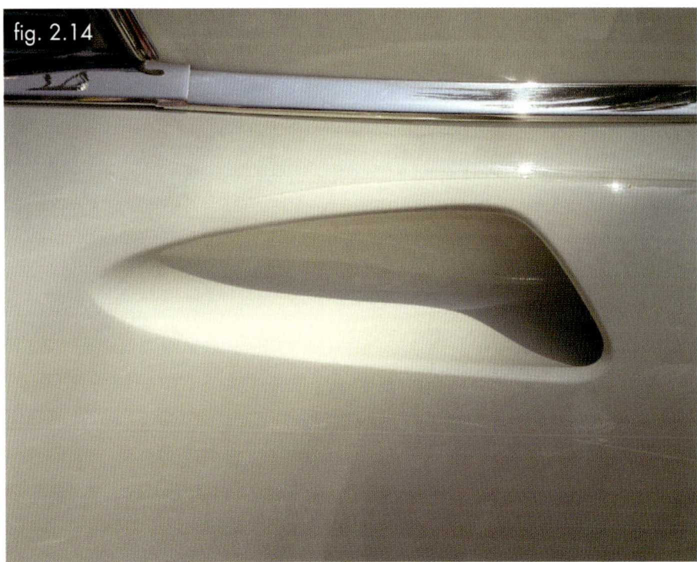
fig. 2.14

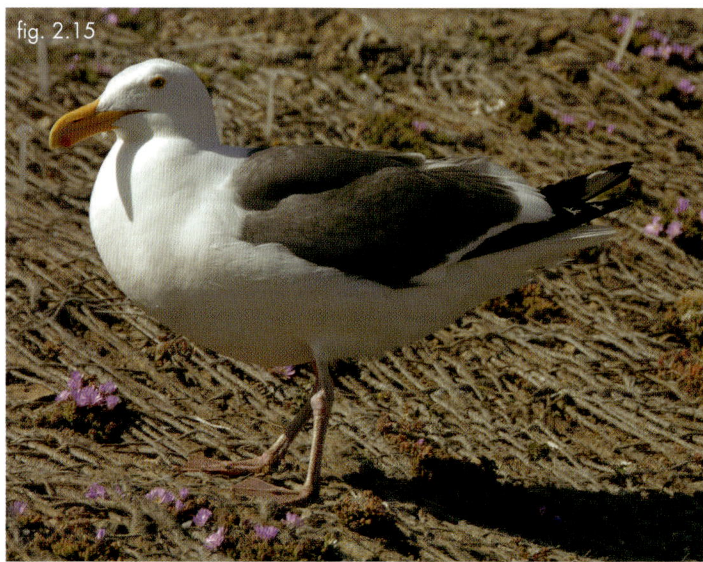
fig. 2.15

The term "soft light" refers to when a surface is illuminated from many different points of light, also called diffuse light, like on a cloudy day. Soft light can be recognized by the soft edge of the shadow that the illuminated forms cast. The diffuse light coming through the skylights in fig. 2.18 casts soft shadows on the vertical surfaces. The three different light sources in the examples on this page are all very diffused. Very strong value changes are still observed across the surfaces of the forms, but the cast shadows have soft edges. Make note of this because when rendering scenes, the cast shadows must correctly match the quality and type of light source. This becomes even more important later when adding reflections.

Sources of soft light can be a light bulb diffused by a lampshade, the sun on an overcast day, or any light source that is obscured, and as a result softened, before it reaches the volume and casts a soft shadow. When starting to observe both types of these cast shadows in the environment, make a mental note of the edge quality of the shadows and the type of light source that is creating them.

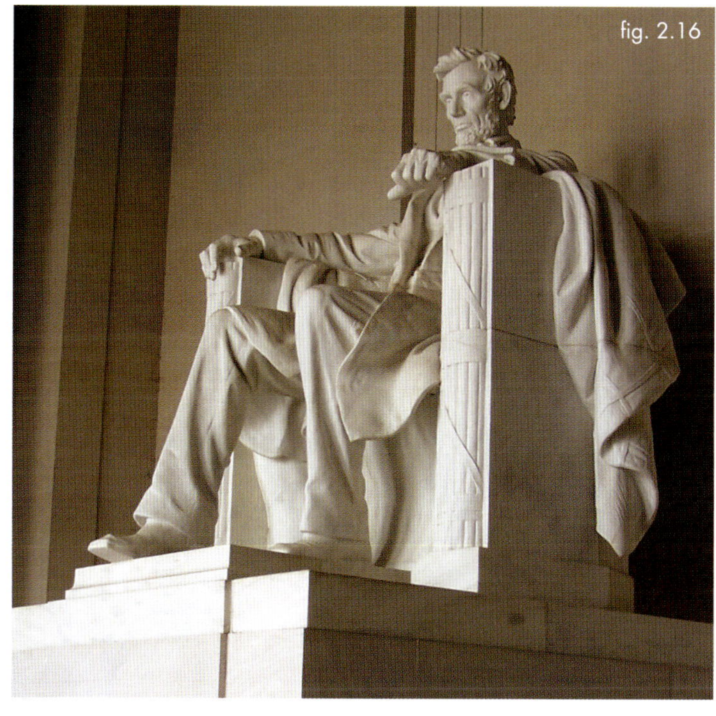

fig. 2.16

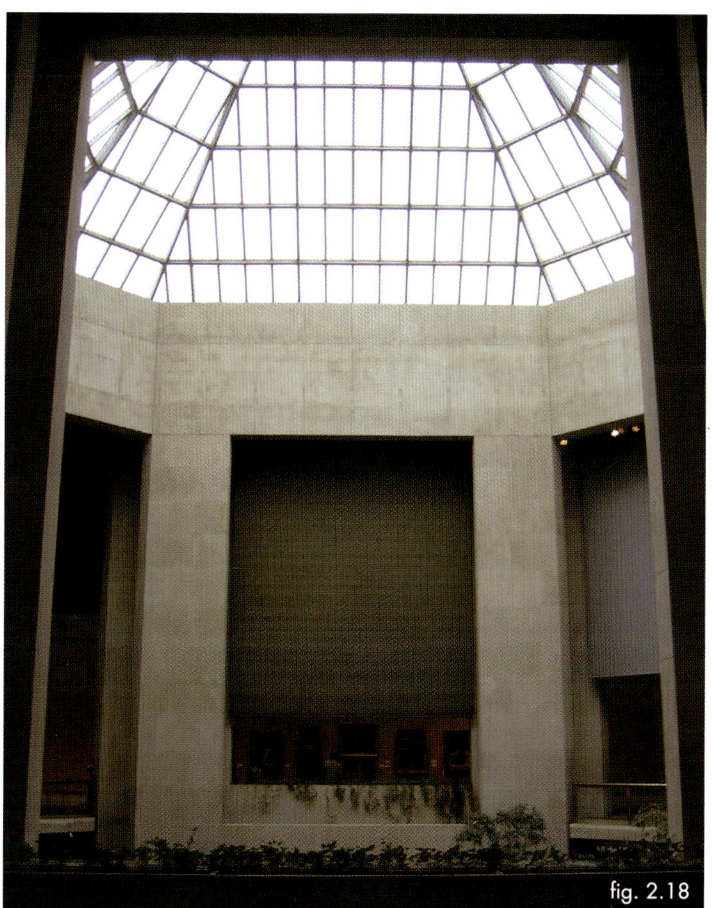

fig. 2.18

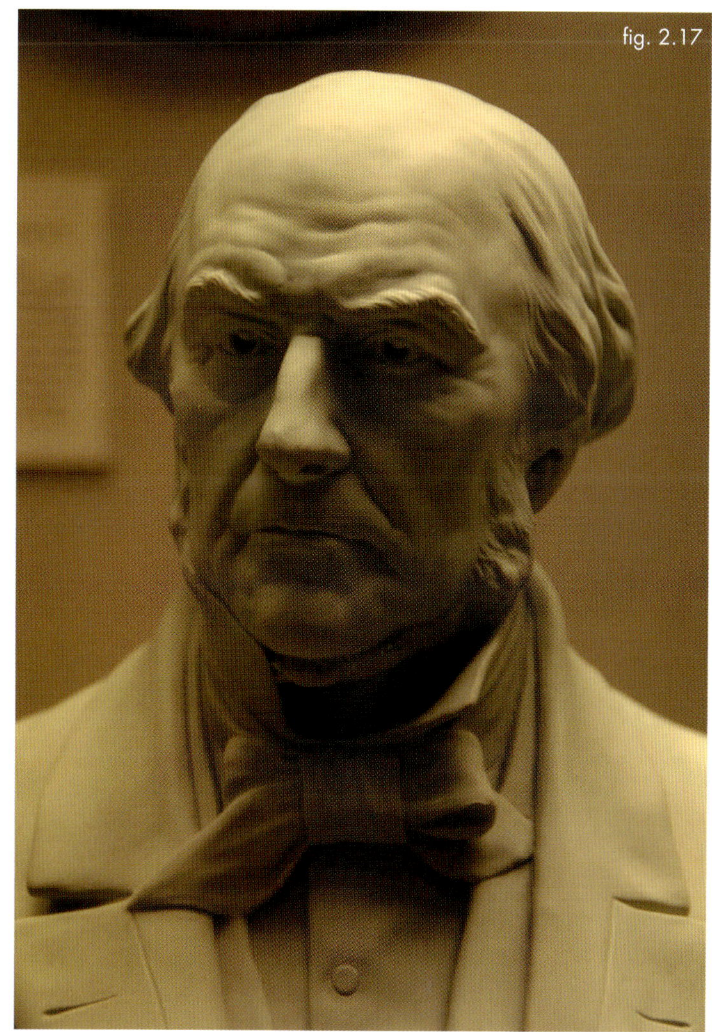

fig. 2.17

LIGHT DECAY, A.K.A. FALLOFF

Light decay, also known as a light's falloff, is the rate at which a light's strength diminishes as the distance from its source increases. This becomes important when rendering the quality of a light as it changes across a surface, because not only does the form change, but the strength of the light actually diminishes. Rendering this effect of light decay goes a long way toward creating realism and providing increased visual interest in scenes. To give an environment a strong sense of mood, render it with a strong light decay.

You can just sense the mood that a strong light falloff creates in fig. 2.19. Keep this observation in mind as you start to create these types of moody, lighting effects in your own illustrations.

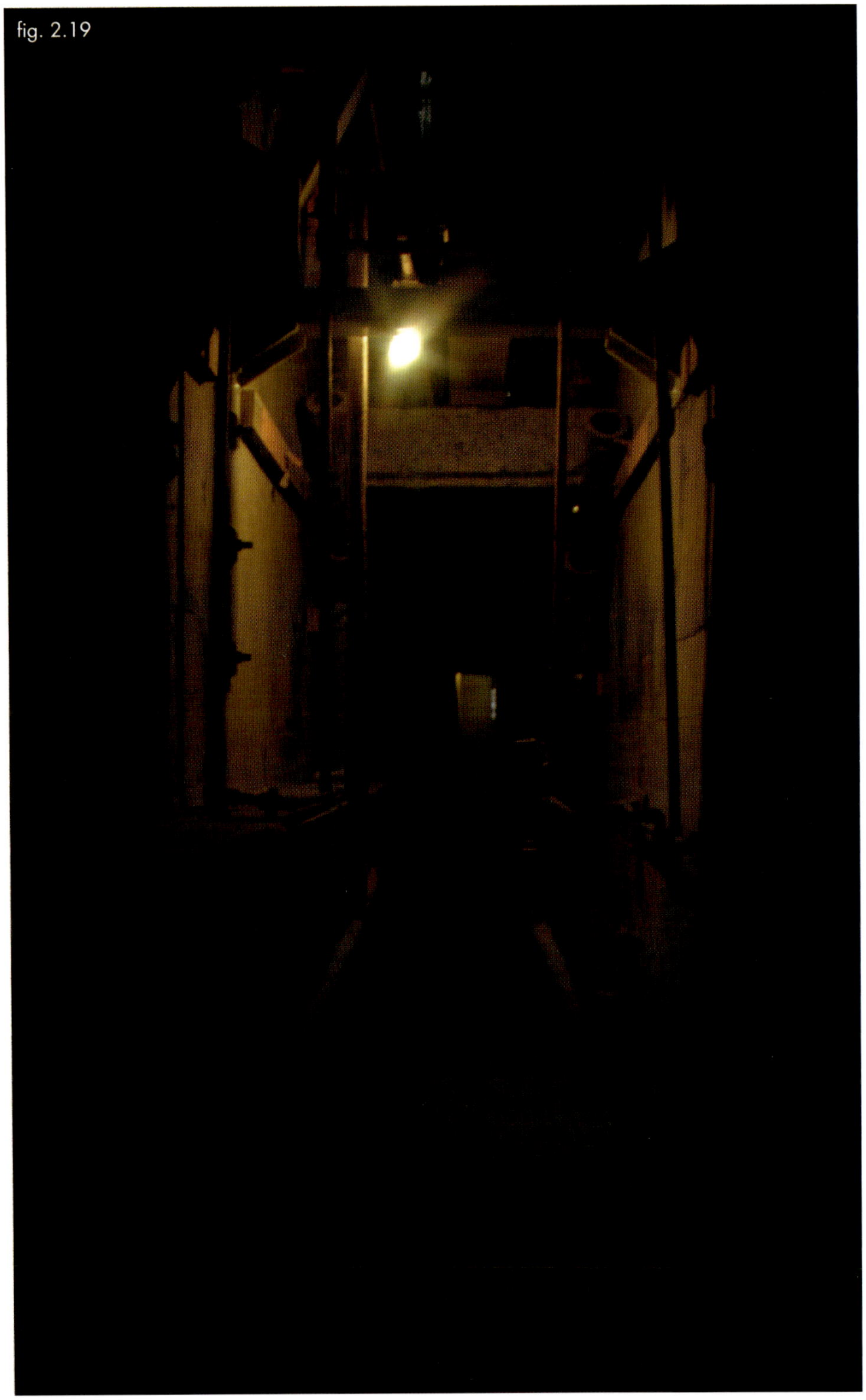

fig. 2.19

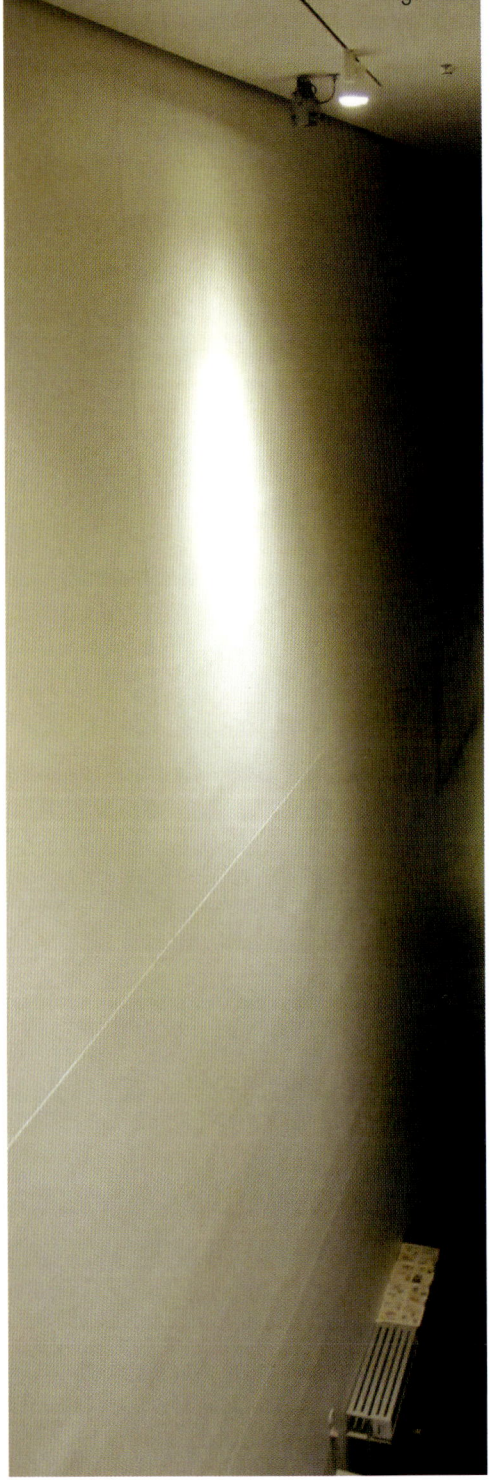

fig. 2.20

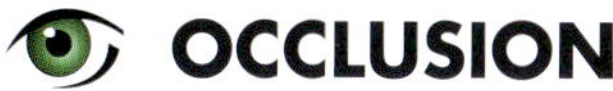

OCCLUSION

Occlusion is a shadow effect of light being blocked by a neighboring surface. This occurs wherever two adjoining surfaces come together to block the light, such as in the corners of walls and ceilings. The arrows point at the strongest examples of occlusion in these images. Adding occlusion to a rendering greatly enhances its realism and visual complexity. It is a shadow quality that is often overlooked by less experienced artists and designers.

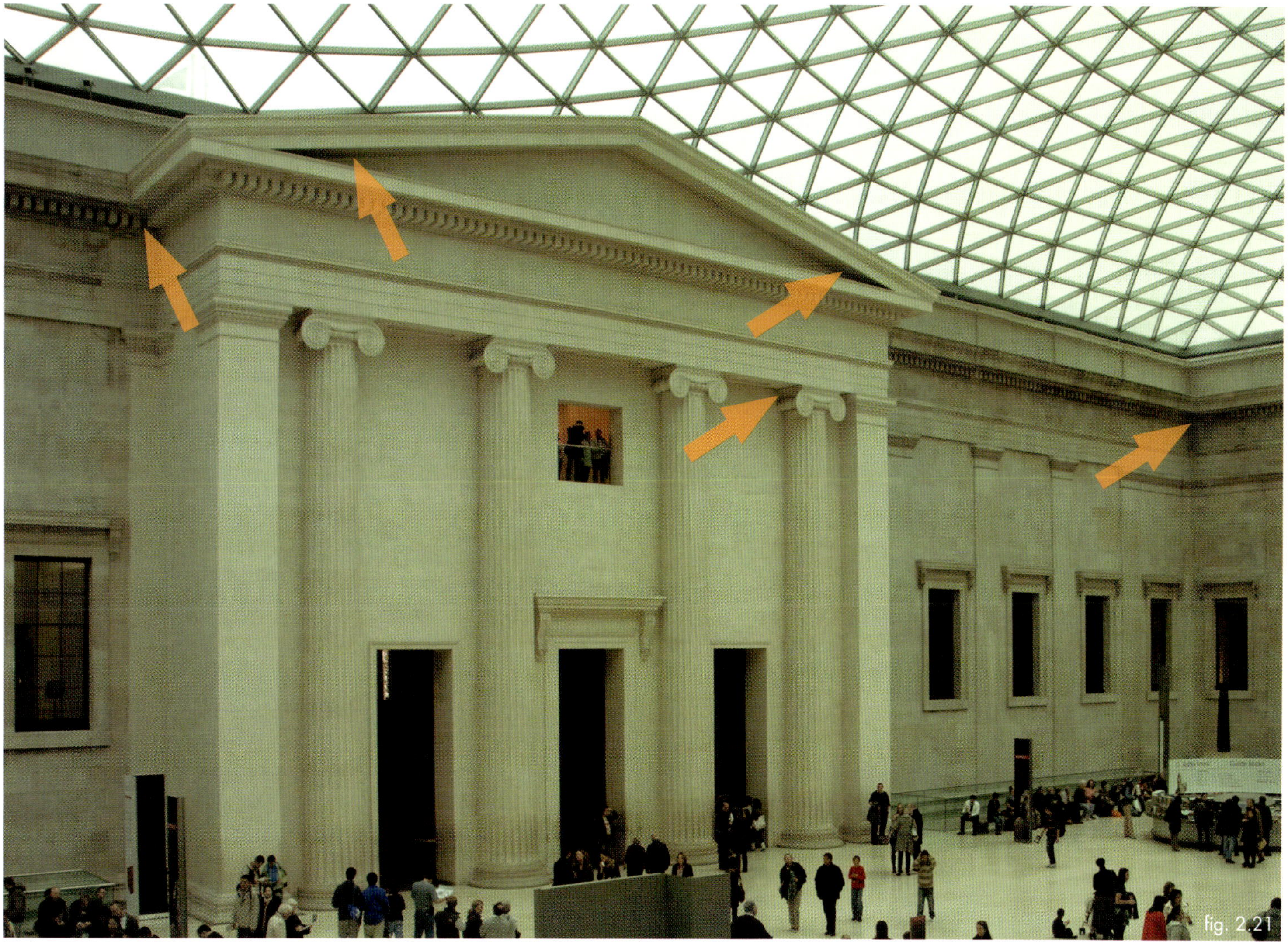

fig. 2.21

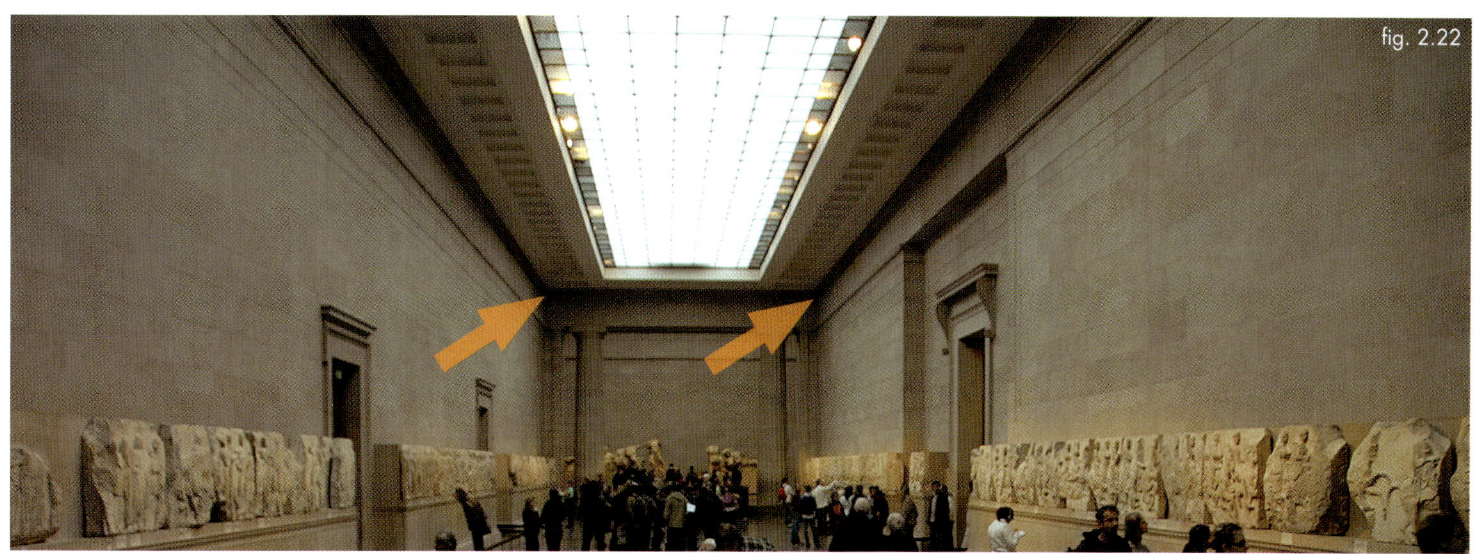

fig. 2.22

EDGE/RIM LIGHT, HALF-LIGHT 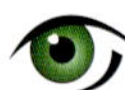

To really emphasize the silhouette of an object without drawing a line around it, which would make it look like a cartoon and kill the realism, try adding edge lighting, also known as rim lighting. This technique accomplishes the same goal as outlining, but with value only. Light shining from behind an object that illuminates just an edge of it, as on the body of the seagull, creates a strong edge-lighting effect that enhances the silhouette. On the statue, the edge lighting is actually the reflection of the bright sky through an archway beside it. The silhouette of the object is visually strengthened. Using this real-world lighting effect in renderings emphasizes silhouettes without resorting to outlining.

Half-light is when an object, or a scene, is only half in the light, creating more visual interest and naturally creating a focal point.

In fig. 2.25, the cast shadow of a large hill behind the camera creates the half-light effect, allowing the sun to illuminate only the top parts of the buildings. A keen observer might wonder, why are the cast shadows of the vertical columns on the brown building hard-edged, when the cast shadow of the hill on the buildings is soft-edged, considering that they are both created by the sun's rays? Jump ahead to find the answer on page 86 .

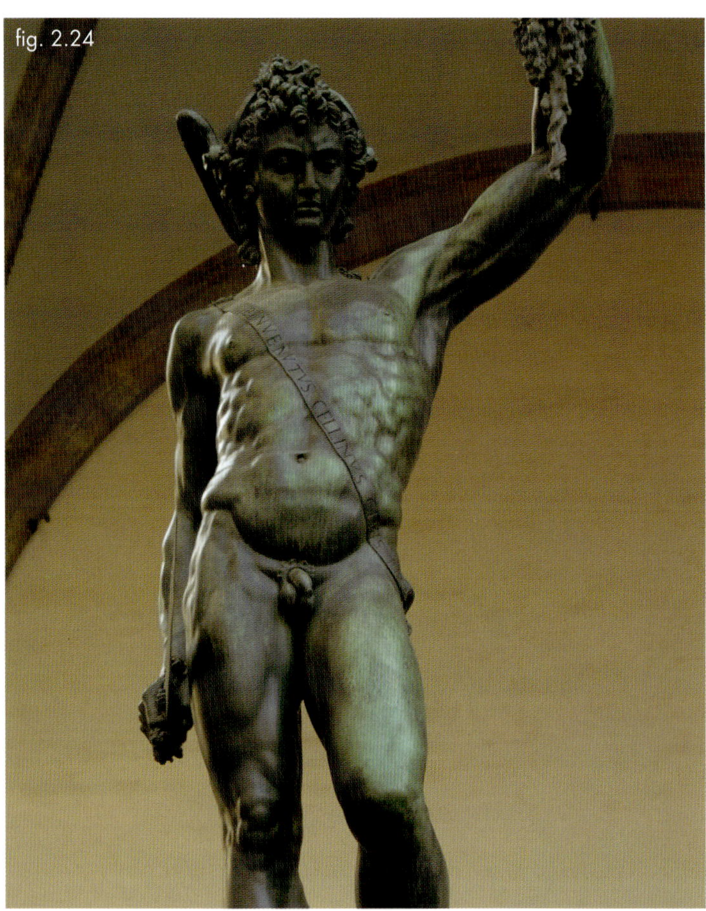
fig. 2.24

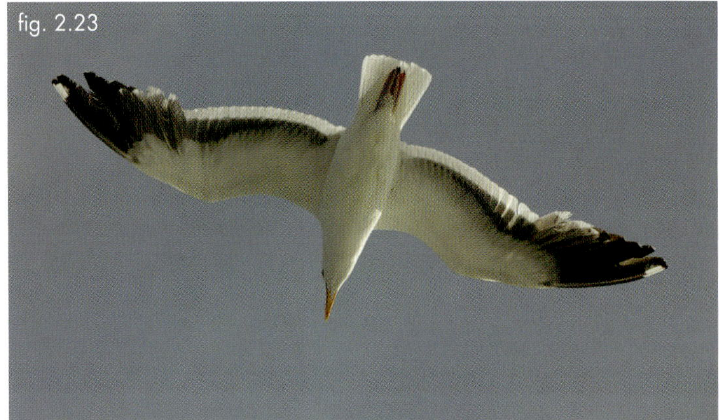
fig. 2.23

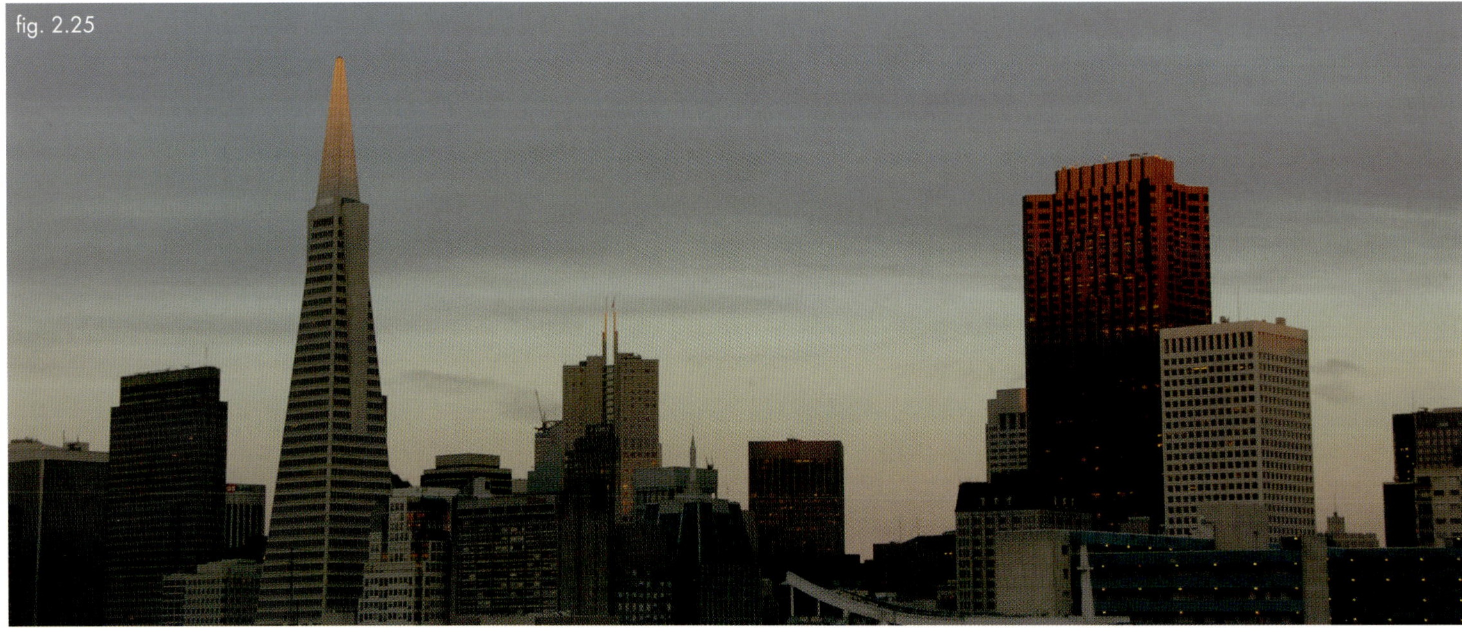
fig. 2.25

MULTIPLE LIGHT SOURCES

When there are multiple light sources illuminating an object, there will be multiple cast shadows. The light sources can be different from each other, so keep this in mind when staging scenes and lighting tests. Knowing that as light sources change, so do the cast-shadow edges, it is quite possible to have multiple types of shadows cast by the same object. Remember, the cast-shadow edge quality is dependent on the type of light. In fig. 2.26, the lights casting the shadows of the two boxes are of the same type and strength. Where the two cast shadows overlap each other they visually multiply in strength, creating a kind of doubling effect.

This is not actually what is happening, but rather the reverse. Each cast shadow is being partially illuminated by the other light source, in effect lighting the parts of the cast shadows that do not overlap. The overlapping part of each cast shadow is no darker than it would be if the second light were turned off. Below, in fig. 2.27, notice this same apparent doubling shadow effect. Incorporating this effect means the construction of two cast shadows instead of one, but the visual interest it creates can be worth the extra effort.

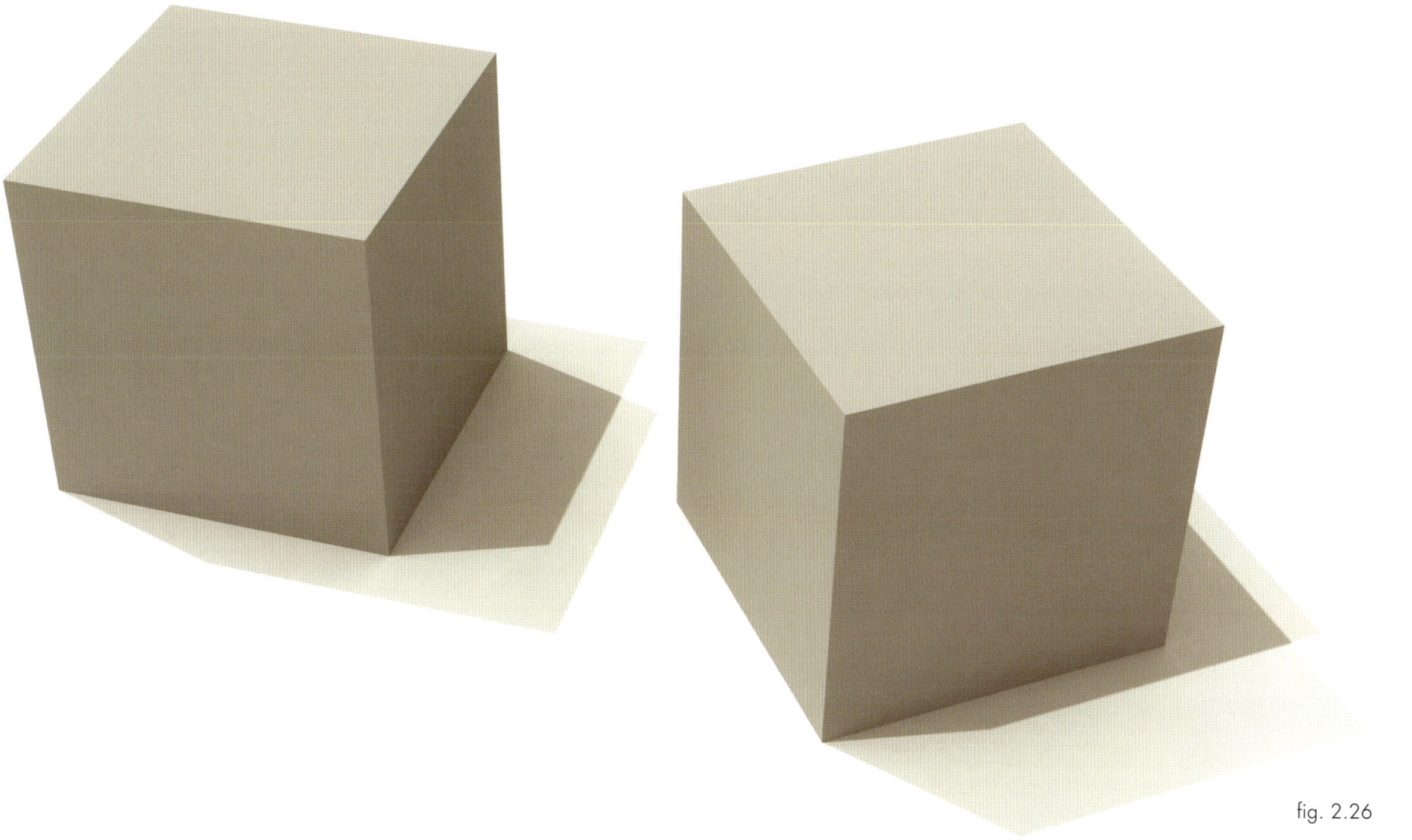

fig. 2.26

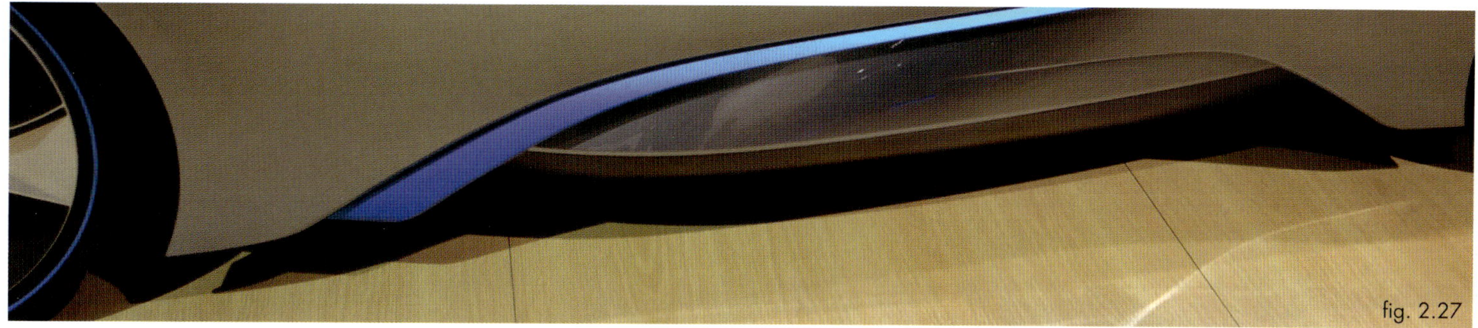

fig. 2.27

REFLECTED LIGHT

Reflected light, commonly known as "bounced" or "fill" light, is light that reflects or bounces off of one surface and illuminates another surface. This often occurs on a different area of the same volume or surface. In fig. 2.28, light reflects off the ground to add some soft light to the underbelly of the cow. Look at the shadow sides of the other examples; the arrows indicate this reflected light occurring. Sometimes the source of the reflected light can be seen within the scene (fig. 2.30), as on the back of the statue's leg with the light reflecting from the adjoining surface. But often it cannot be, as on the shadow side of the forearm.

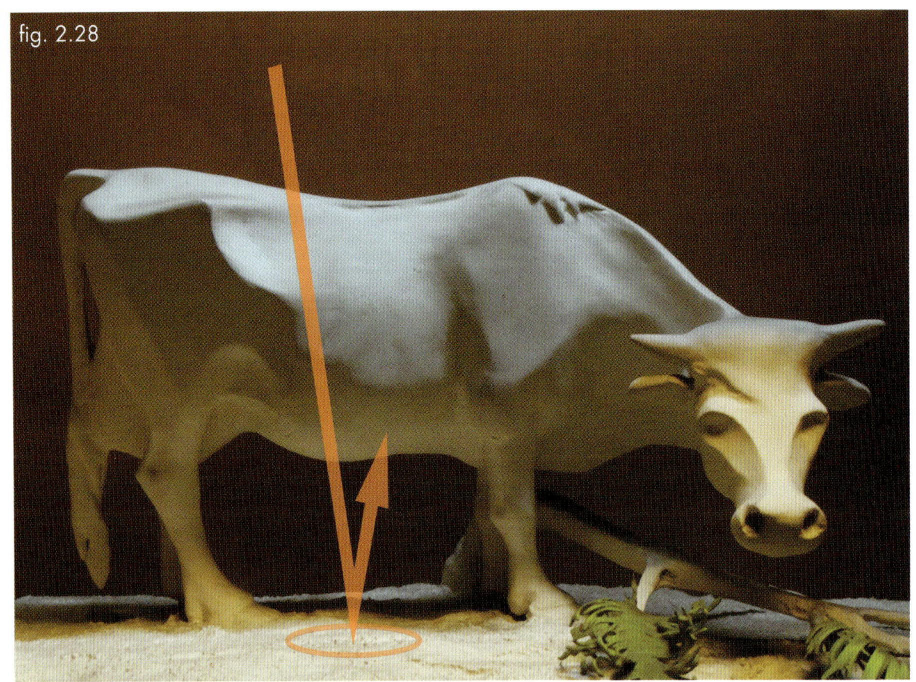
fig. 2.28

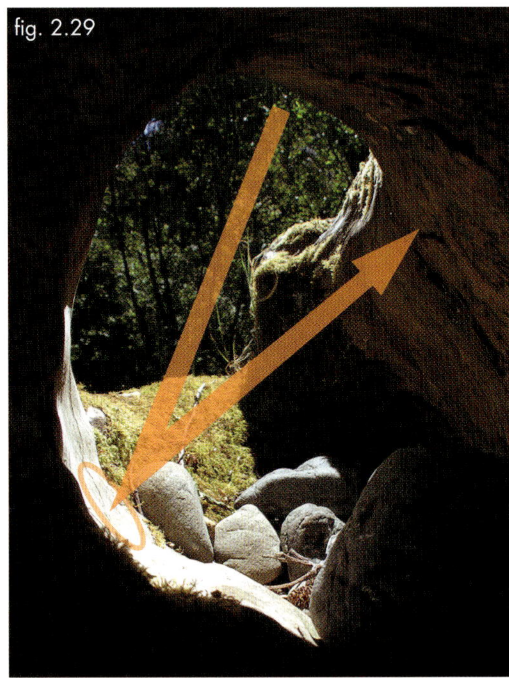
fig. 2.29

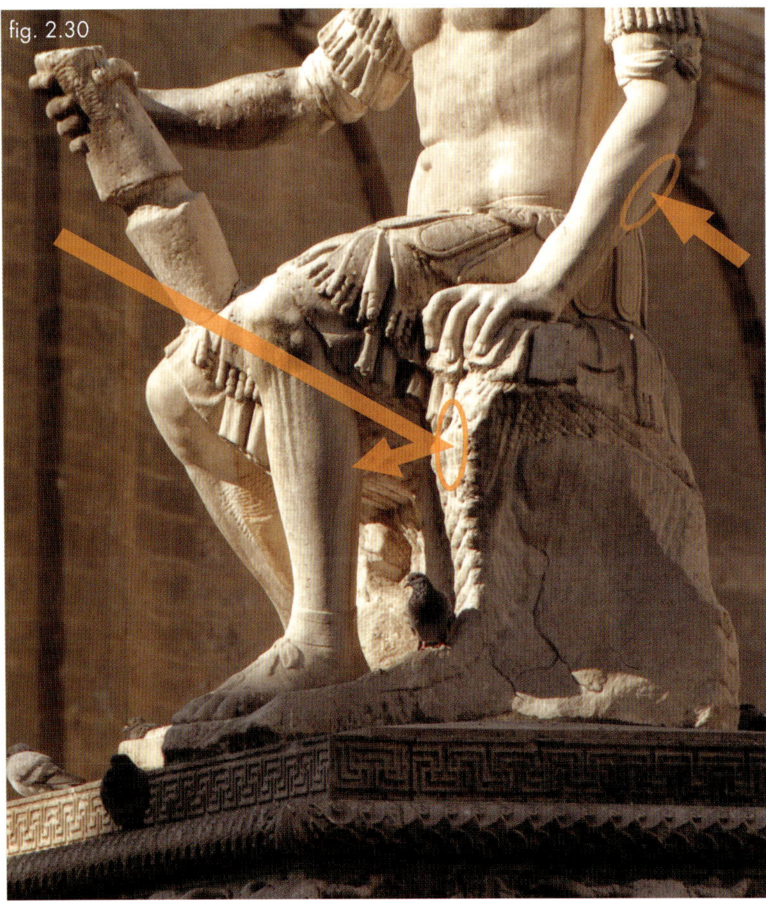
fig. 2.30

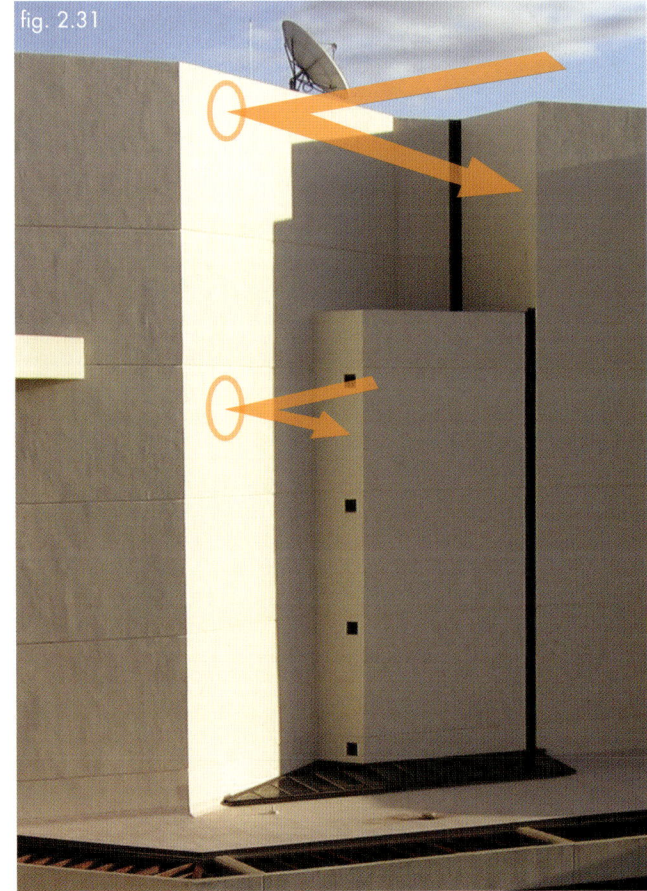
fig. 2.31

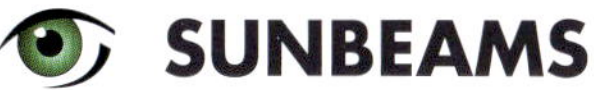

SUNBEAMS

Sunbeams are basically two things at once. They are small areas of atmosphere being illuminated by the sun. They are also the edges of the cast shadows of the clouds. In order for sunbeams to occur, the atmosphere needs quite a bit of moisture, dust or smog. Sunbeams are most visible when the sun is behind the object blocking the light, like the cloud in fig. 2.33. Be aware that illuminated shafts of light in scenes, whether caused by the sun or another light source, tend to draw a lot of interest, so be sure they will enhance the overall rendering before adding them.

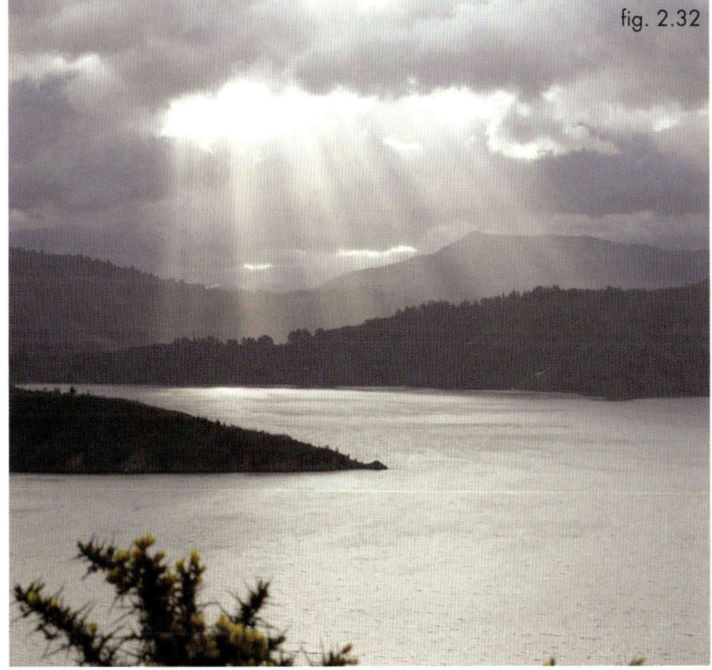

fig. 2.32

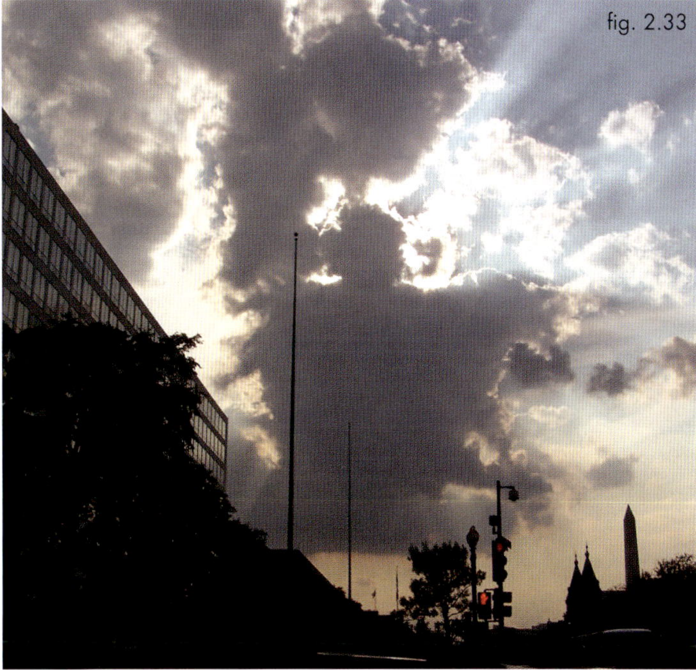

fig. 2.33

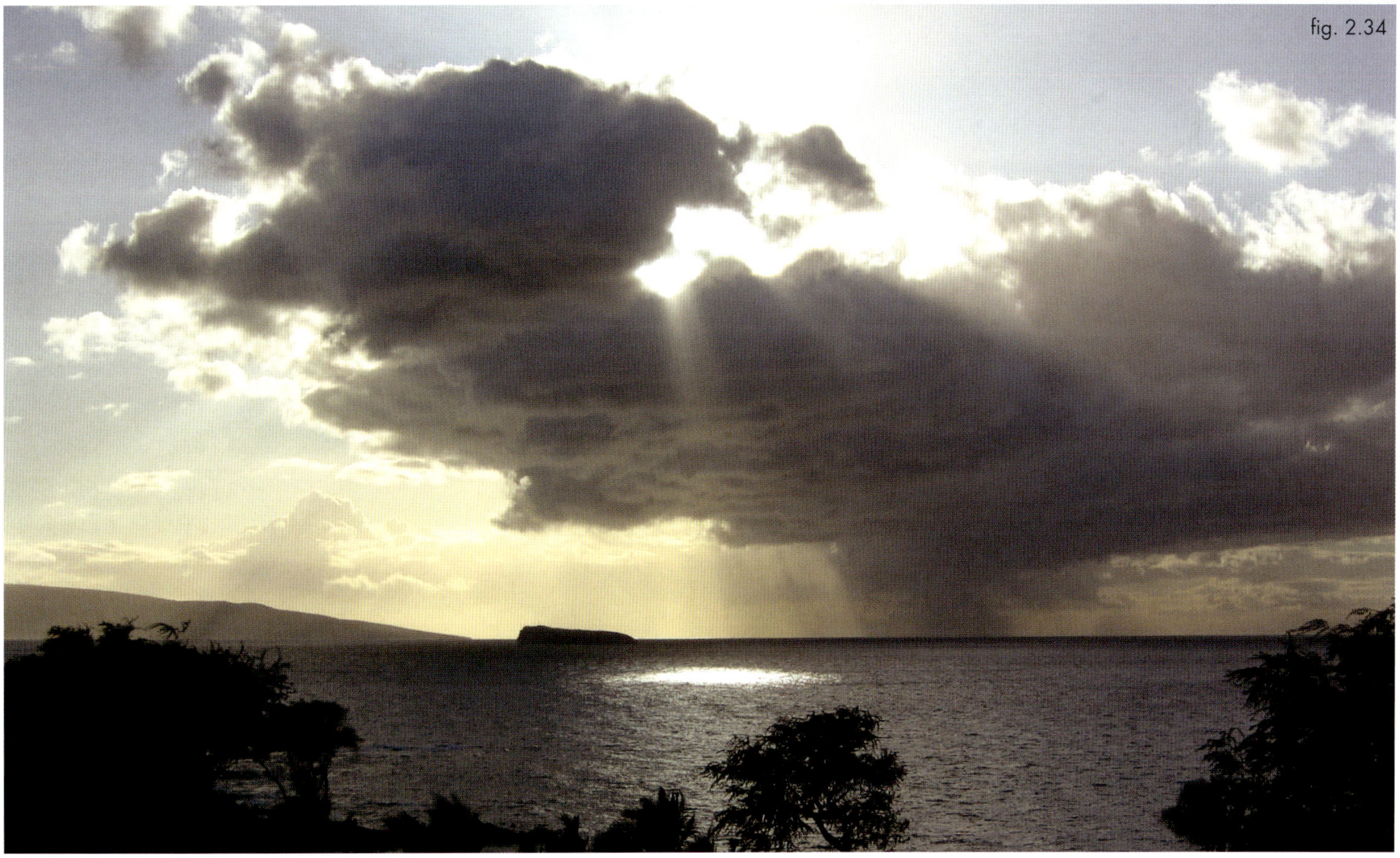

fig. 2.34

ATMOSPHERIC PERSPECTIVE 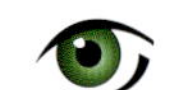

Atmospheric perspective is the effect that surfaces that are farther away appear to have less value contrast (shift between the light and dark values) and less color saturation, than surfaces that are closer to the observer. The effect is strongest at the horizon, because there is more atmosphere through which to look. Again, the more moisture, dust and smog in the air, the stronger the atmospheric perspective. In fig. 2.35 the orange arrows represent the lines of sight straight up in the air and straight out to the horizon. The observer looks through much more atmosphere when facing out toward the distant horizon than when looking straight up. In fig. 2.36 it is easy to see this atmospheric perspective effect between the foreground cliff and the rocks only a short distance away. This is due to the increased moisture in the air from the crashing waves. In fig. 2.37, the effect is less dramatic but still very visible. The buildings in the distance have less value contrast than the ones in the foreground.

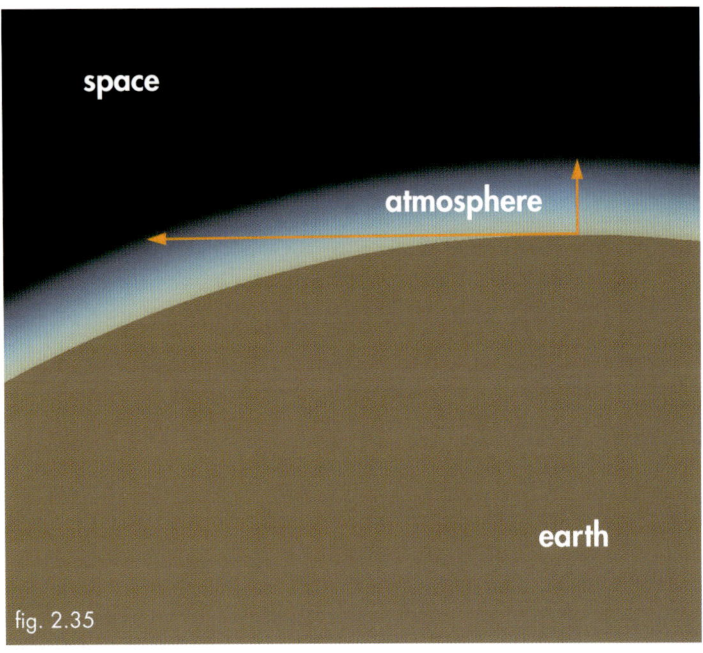

fig. 2.35

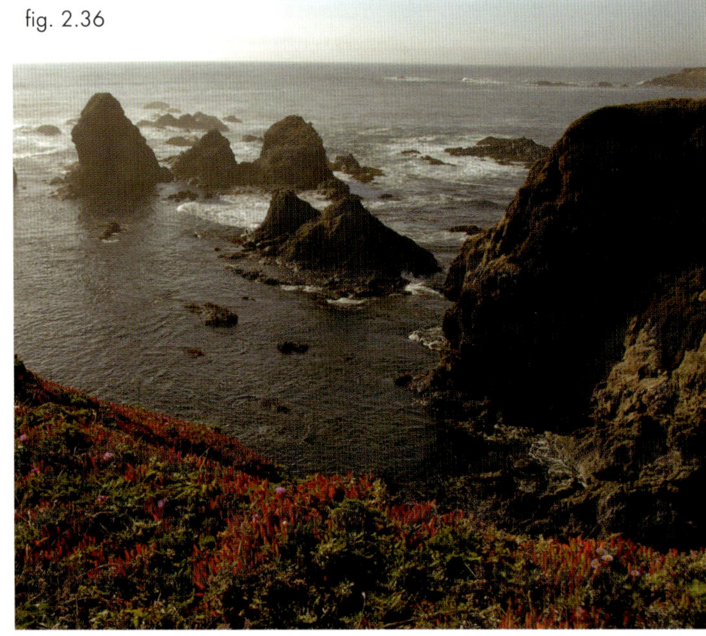
fig. 2.36

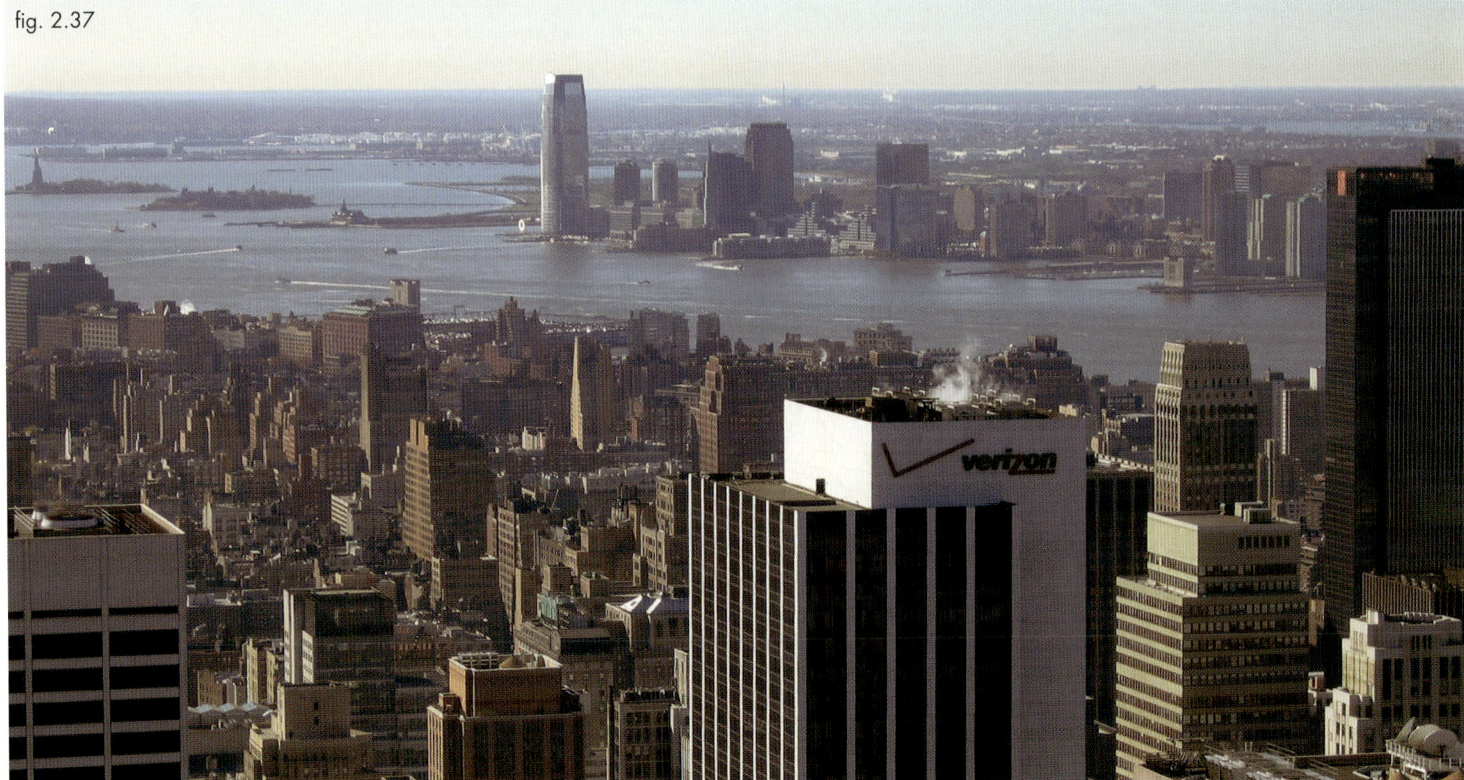

fig. 2.37

In fig. 2.38 compare the value contrast of the buildings in the foreground with that of the container ship in the mid-ground and the shoreline in the background. The atmosphere is very thick here. Fig. 2.39 is a nice example of how color saturation, in addition to value contrast, changes due to the atmosphere. The foreground tower's colors appear much more saturated than those of the large building on the hill in the background, even though they are basically the same local colors and values. In fig. 2.40, the fog really makes this effect easy to observe. Adding atmospheric perspective to renderings is the simplest way to increase perceived depth in scenes.

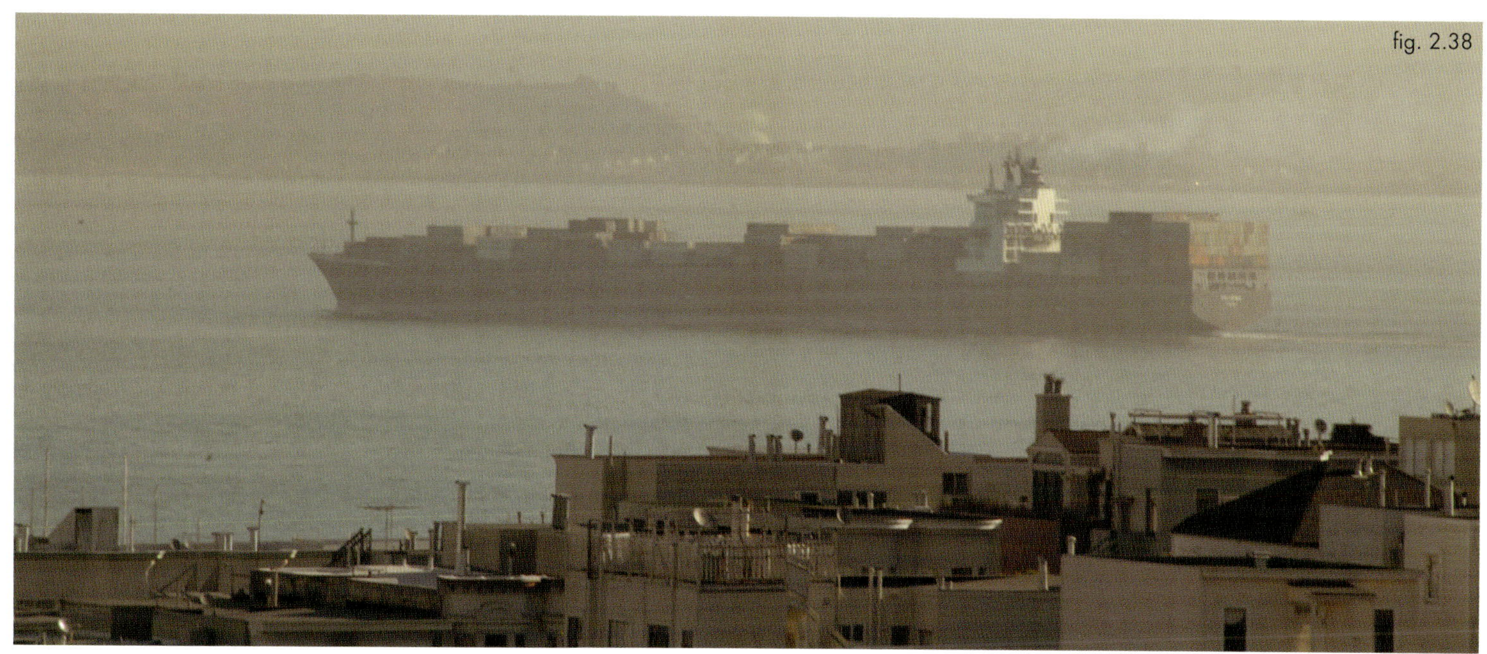
fig. 2.38

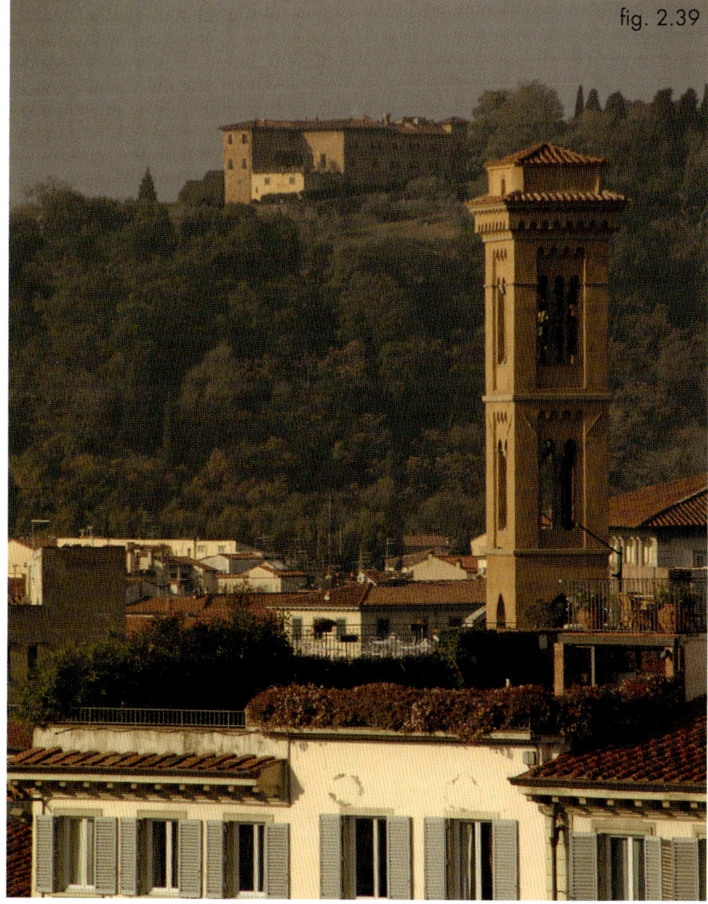
fig. 2.39

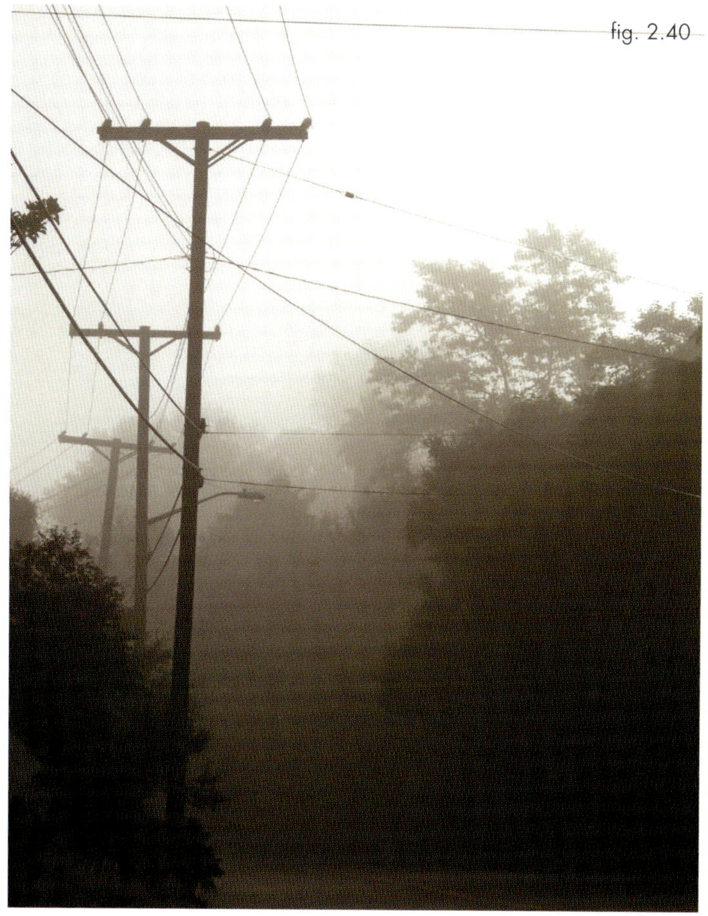
fig. 2.40

CONSTRUCTING CAST SHADOWS

Even when drawing and rendering never-before-seen objects, it is crucial to understand the real-world behavior of objects in light. Whether shadows are cast by sunlight or local light, they follow physical laws that can be learned through practice. A correctly rendered cast shadow can make all the difference in the quality and believability of an image.

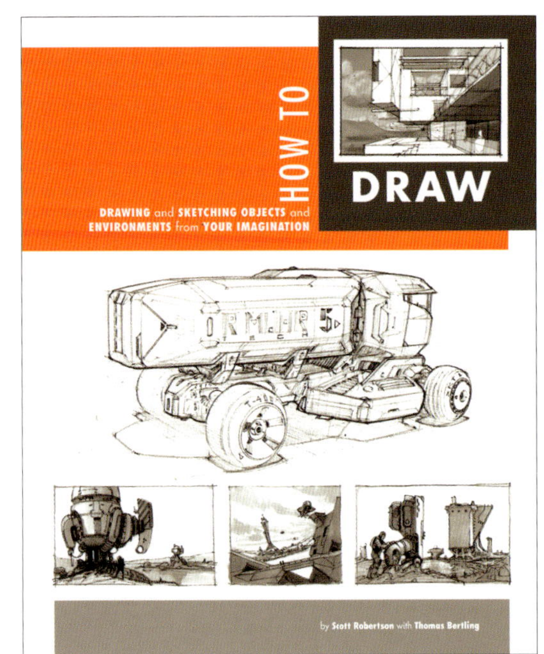

Always use the most efficient media to achieve a set goal, whether that goal is sharpening your own skills, or creating a rendering for a client. Find the most effective combination of working with lines drawn freehand, lines drawn with straight edges and ellipse guides, and computer-aided constructions.

To get the maximum benefit from this chapter on shadow construction, it is essential to have already acquired basic perspective-drawing skills. How to manipulate designs in perspective is covered in the first book of the series, *How To Draw*. Use it to reference the following four skill-sets in order to have a successful experience with constructing cast and core shadows.

Perspective Grid Skills

Before proceeding with the exercises in this chapter, it is absolutely necessary to be proficient in creating perspective grids, including drawing toward Vanishing Points that are off the page, as well as being familiar with Auxiliary Vanishing Points. Mastery of these skills ensures that you are always in control of the scene being drawn, knowing where all points are relative to one another in perspective.

For a reminder on perspective-grid construction, see Chapter 04 in *How To Draw*.

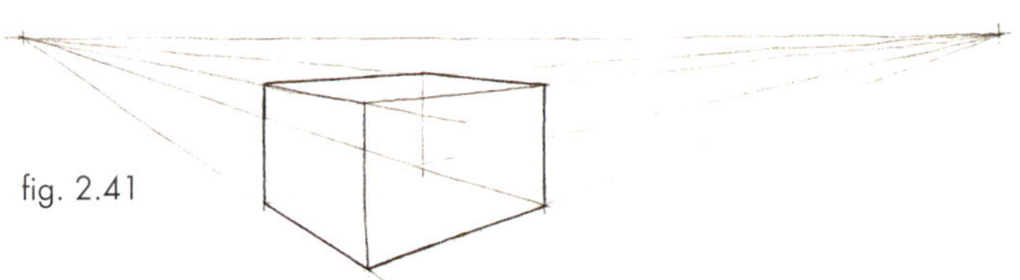
fig. 2.41

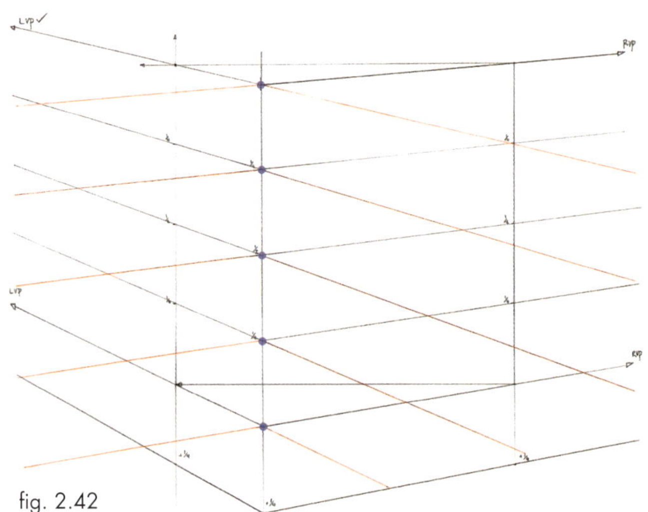
fig. 2.42

Form-Building Skills

Creating X-Y-Z volume constructions helps to understand and communicate a design, and plan for shadow construction. It enables the addition of details to the rendered shape, such as cut lines or other elements. One must have the skills to create these types of constructions for advanced form renderings, since it will be necessary to move back and forth fluidly between plotting construction lines and developing a design with value.

For a refresher on form building, see Chapter 08 in *How To Draw*.

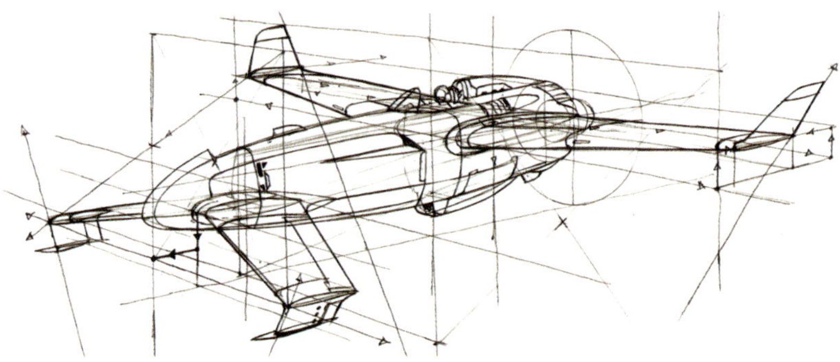
fig. 2.43

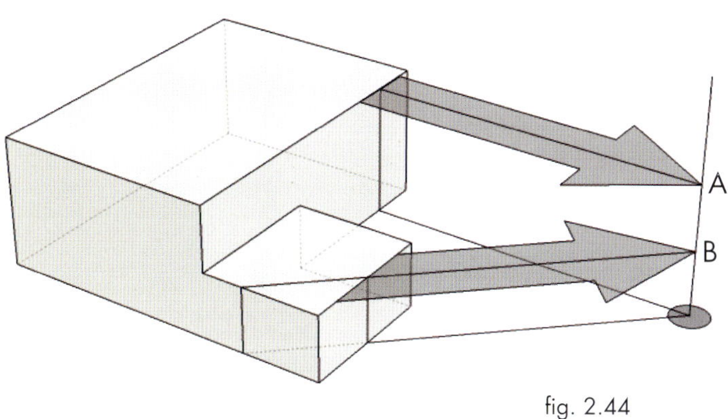

fig. 2.44

Construction Skills

For many of the upcoming shadow-casting constructions, the skills to transfer height points, in perspective, is needed. When objects are under a light source, transferring their height to a vertical line creates what is called a Shadow Origin (A, B). There are multiple ways of doing this, all of which rely on strong perspective construction skills.

To review these construction methods, see Chapter 03 in *How To Draw*.

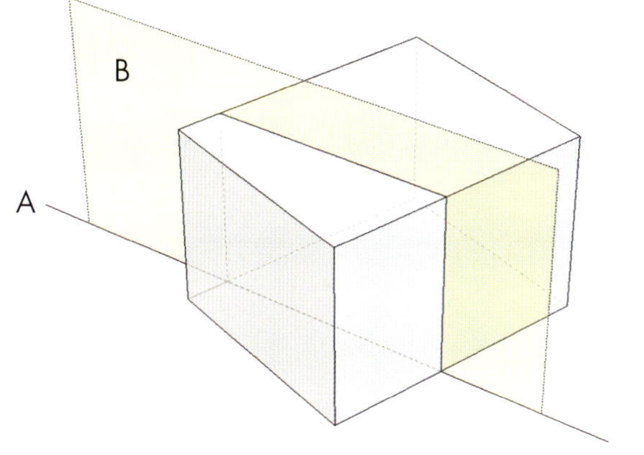

fig. 2.45

Section Drawing Skills

Most constructions require drawing a line on the ground, through the object, in the direction of the cast shadow (A), and then constructing a vertical plane from this line, that cuts through the object (B).

It is important to know how to create section lines, which can range from rectangular sections over complex planar forms, to curved sections over X-Y-Z forms.

For additional practice creating section lines and vertical planes, see Chapter 06 in *How To Draw*.

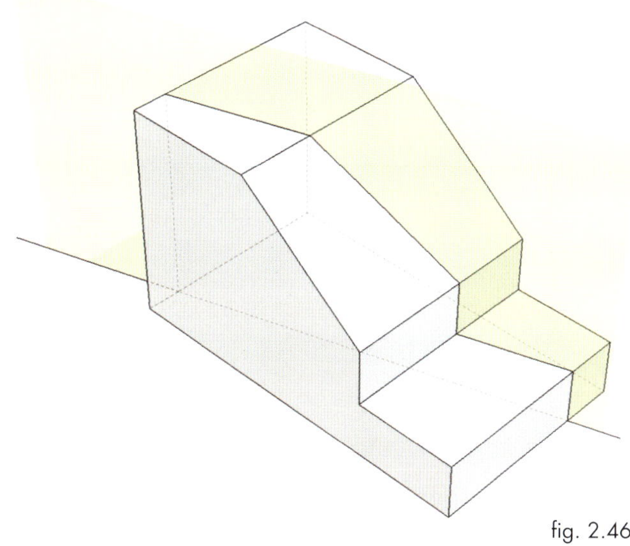
fig. 2.46

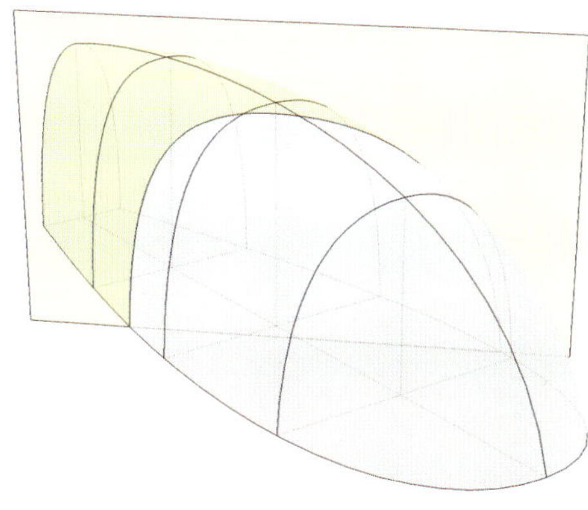
fig. 2.47

A Light Plane is defined by the direction of the light and the direction of the shadow. Both elements must be known to cast a shadow correctly. This applies whether the light is coming from the sun, the moon or a man-made source.

Light/Shadow Direction and Light Planes

Fig. 2.48 is a photograph of a pencil casting a shadow, which is about as simple as a cast shadow can be.

Fig. 2.49: Light Direction (A) is a line that emanates from a light source. Think of the Light Direction line as a ray of light. Shadow Direction (B) is a line originating from a point called the Shadow Origin, which is *always found directly underneath the Light Source on the plane the cast shadow will rest upon,* and in this case is off the page.

The two lines, Light Direction and Shadow Direction, define a plane called the Light Plane (C). The Light Plane sits vertically on the ground plane. Shadows are always constructed on vertical Light Planes. The Light Planes themselves are not always drawn, but this is a structural concept that must be kept in mind.

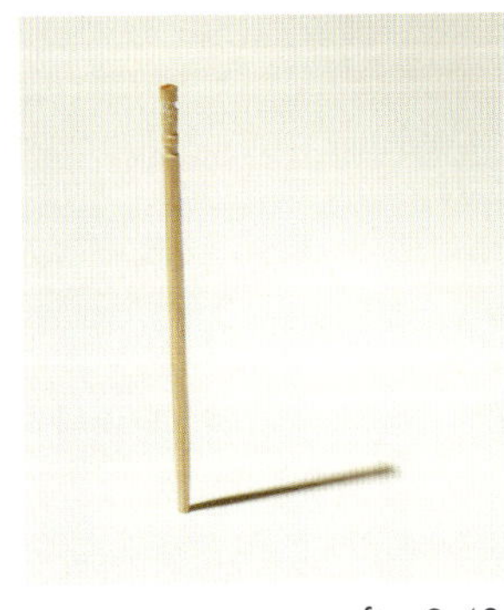

fig. 2.48

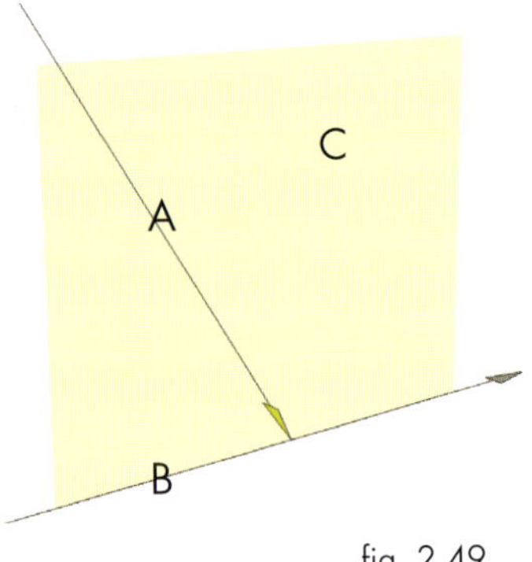

fig. 2.49

Casting the Shadow of a Single Stick in Perspective

In perspective drawing, any line segment is commonly referred to as a "stick." Referring to line segments as "sticks" helps the artist to remember to follow real-world physical laws. To cast the shadow of a stick, one must know three things: Where is the stick in relation to the ground plane? In what direction will the shadow be cast? How long will the shadow be? This is true for both sunlight and local-light situations. As long as there is only one stick, there is no difference between the two constructions.

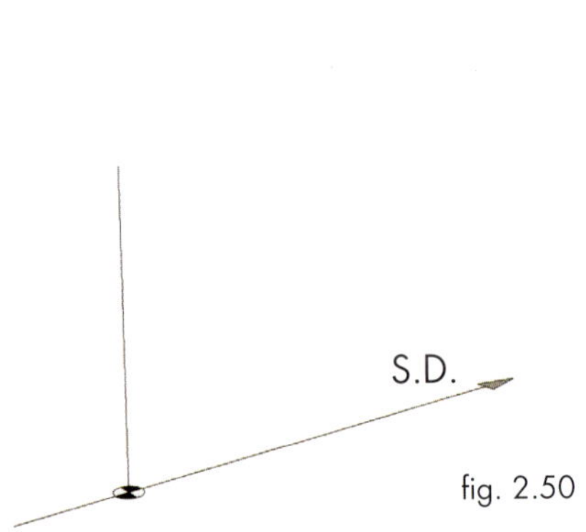

fig. 2.50

STEP 1: Place a stick on the ground, vertically. This will become essential in more complex situations. Run the Shadow Direction (S.D.) through the base of the stick.

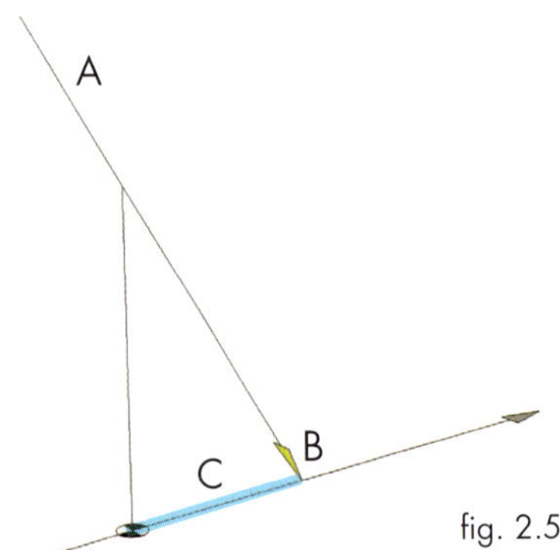

fig. 2.51

STEP 2: Draw a Light Ray (A) that runs through the top of the stick until it intersects with the Shadow Direction (B). The cast shadow (C) starts at the base of the stick and ends where the Light Ray and the Shadow Direction intersect (B).

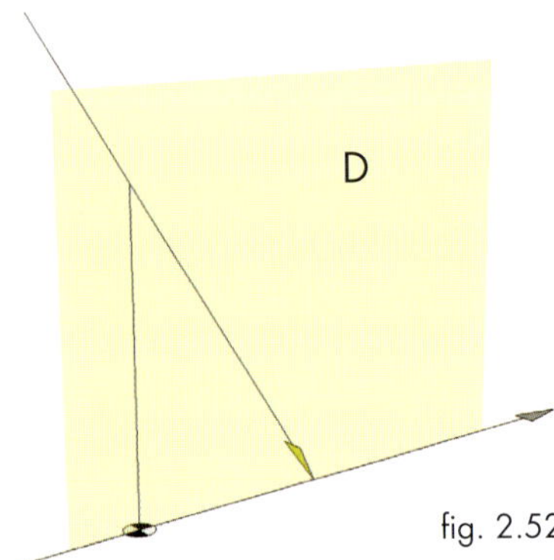

fig. 2.52

Creating a Light Plane, perpendicular to the ground plane, which is the essential first step in constructing cast shadows. The Light Plane itself does not always need to be drawn, but the artist must always remain conscious of its presence (D).

Knowing how to cast shadows of sticks makes it possible to construct the shadow of any point in perspective, leading to more complex shadows. The drawing's perspective must be solid, and the position of the points must be known, relative to the ground plane.

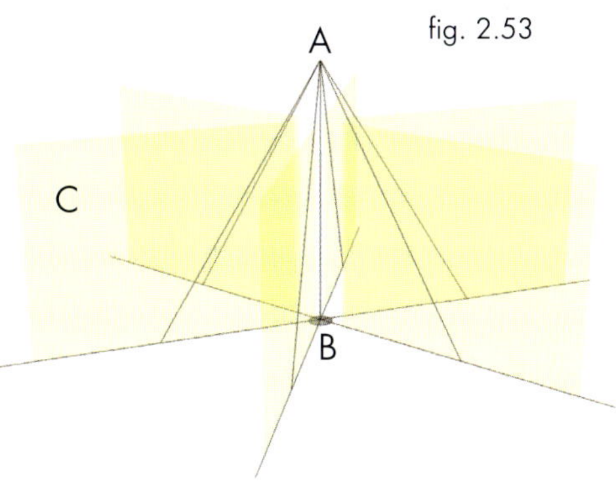

fig. 2.53

SHADOW CONSTRUCTION WITH LOCAL LIGHT

Cast shadow construction with a local light source usually has the advantage of the Light Source and the Shadow Origin being on the page, thus making the construction easier to understand. For the following exercises, the steps can be done in any order. Sticks can be drawn, then the Light Source placed; or the Light Source can be placed before finding the Shadow Origin and drawing sticks.

Fig. 2.53 shows that when using a local Light Source (A), the vertical line that drops down to the ground plane and creates the Shadow Origin (B) forms an axis from which Light Planes (C) radiate out.

fig. 2.54

Local Light Shadow Construction: Multiple Sticks on a Single Level

STEP 1: Draw two vertical sticks (D). Make sure the bases of the sticks are on the same plane, the gray plane in this case (E).

STEP 2: Place the Shadow Origin (B) on the ground and then place the Light Source (A) directly above it.

STEP 3: Draw lines (Light Rays) from the Light Source through the tops of the sticks to the ground plane. Draw lines extending from the Shadow Origin through and beyond the bases of the sticks. Again, the intersections of the Light Direction lines and the Shadow Direction lines determine the lengths of the cast shadows (F).

Local Light Shadow Construction: Multiple Sticks on Multiple Levels

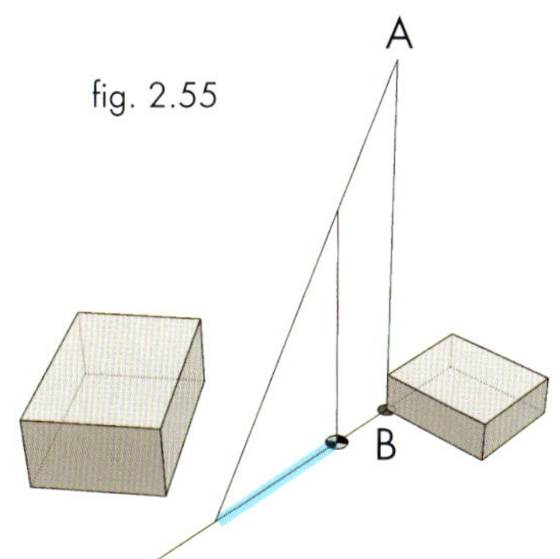

fig. 2.55

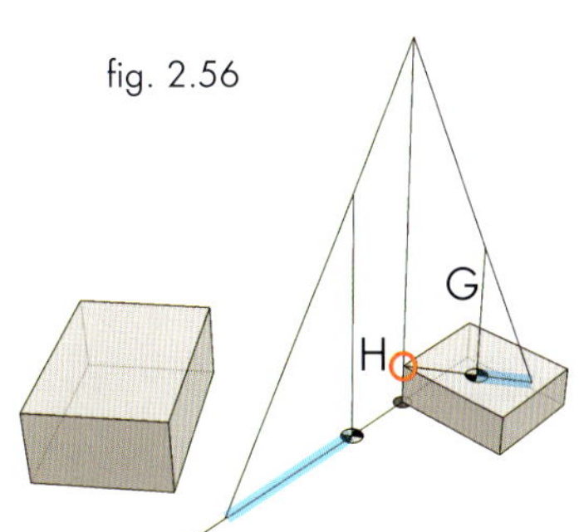

fig. 2.56

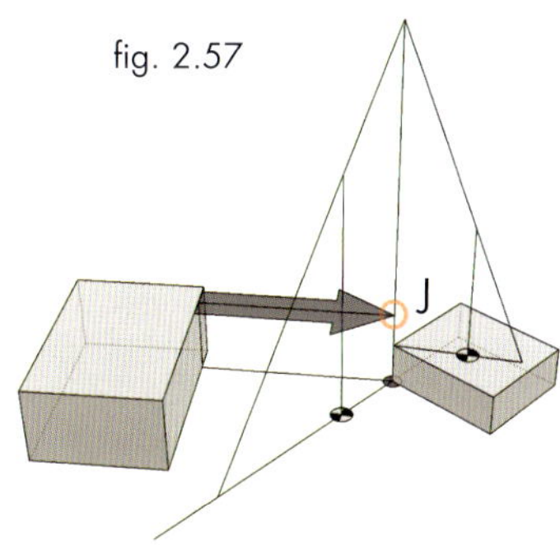

fig. 2.57

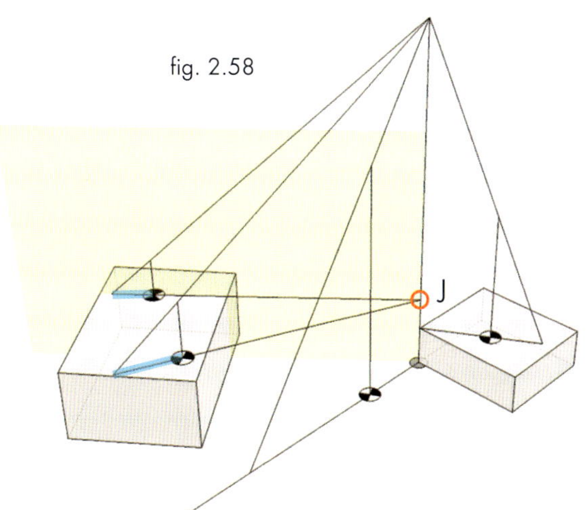

fig. 2.58

STEP 1, fig. 2.55: Define the Light Source (A) and the Shadow Origin (B). Cast the shadow of the stick as usual, since the Shadow Origin and the base of the stick are on the same plane.

STEP 2, fig. 2.56: To cast the shadow of stick G, an additional Shadow Origin (H) must be created. It is located directly below the Light Source, at a level that is equal to the height of the plane on which the base of stick G is located.

STEP 3, fig. 2.57: To cast shadows on the larger block, create an additional Shadow Origin (J) that is level with the top of that box. Use a section line to transfer this height to the vertical line running between the Shadow Origin and the Light Source.

STEP 4, fig. 2.58: Place sticks on the larger box. Use the correct Shadow Origin (J) to plot the cast shadows.

SUNLIGHT CONSTRUCTION

Light Plane Set-Up for Sunlight

Sunlight is the most common lighting situation in which objects are observed and rendered. Since the sun is almost 93 million miles (150 million kilometers) away, only a very small portion of its rays reach Earth. What this means for accurate drawing and rendering is that these light rays are so close to being parallel, they can be considered as such (fig. 2.59). Therefore, in sunlight shadow constructions, the Light Direction lines (A) are parallel; the Shadow Direction (B) lines are parallel, which makes the Light Planes (C) parallel (fig. 2.61).

Parallel lines converge in perspective, to a common Vanishing Point. This means all of the Light Ray lines share a Vanishing Point, and all of the Shadow Direction lines share a Vanishing Point.

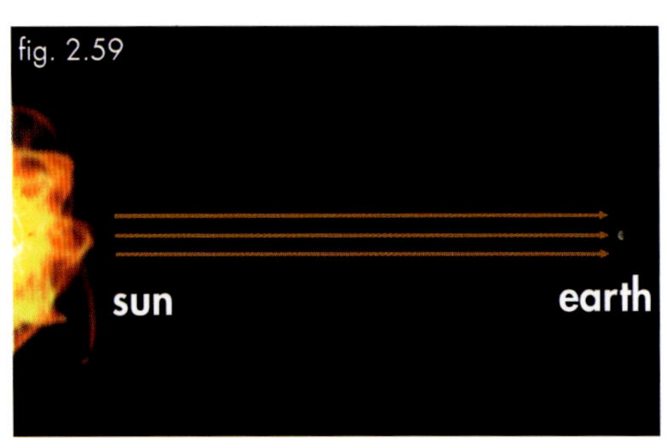

fig. 2.59

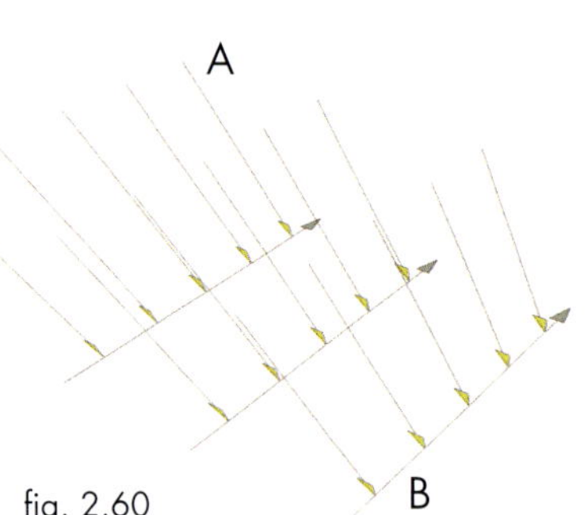

fig. 2.60

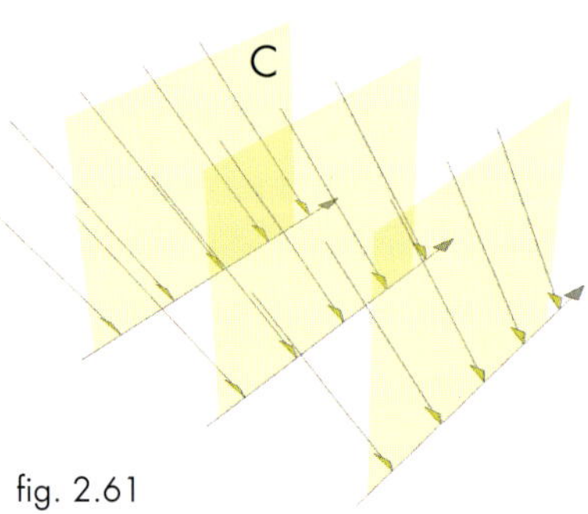

fig. 2.61

Sunlight Construction: Finding the Shadow Origin and Light Source

Fig. 2.62 shows the Shadow Direction lines converging to a Vanishing Point on the Horizon Line (B), which is the sun's Shadow Origin.

In figs. 2.63 and 2.64 Light Rays converge to an Auxiliary Vanishing Point (A) either directly above or below the Shadow Origin. This Auxiliary Vanishing Point is the Light Source (the sun).

There are two possible scenarios: Positive Sunlight when objects are backlit, and Negative Sunlight when objects are front-lit.

In most sunlight constructions, the Shadow Origin and Light Source are not visibly on the page, but it is important to remember that these lines always converge, in perspective, back to the Vanishing Points of the sun and the Shadow Origin.

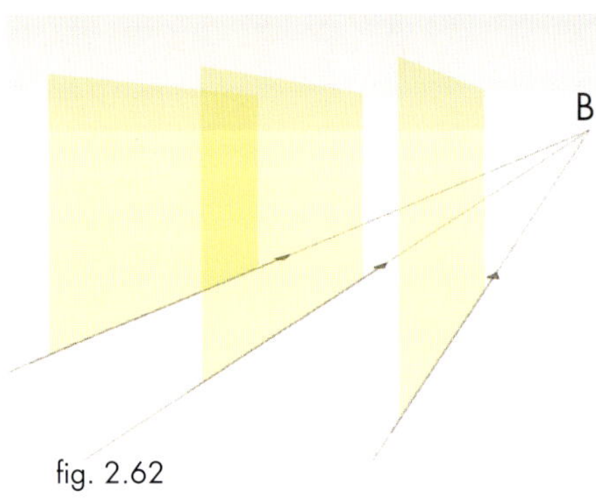

fig. 2.62

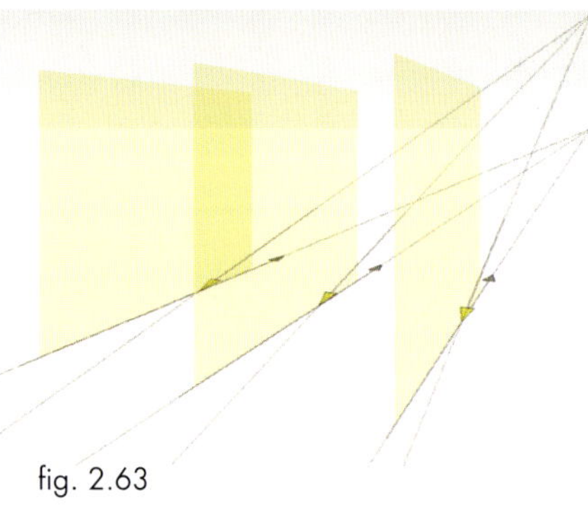

fig. 2.63

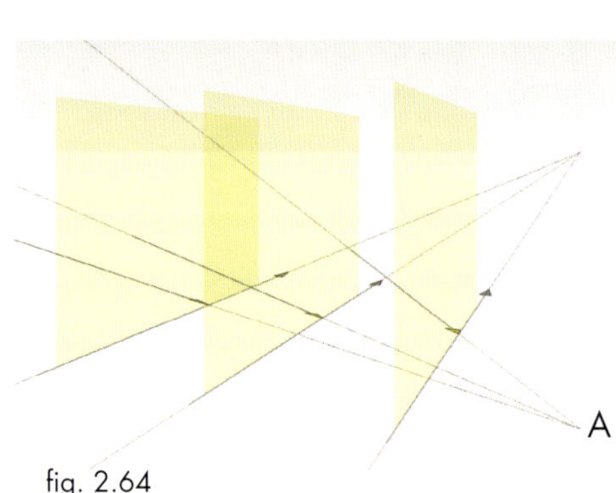

fig. 2.64

Positive Sunlight Construction: Backlight

In a Positive Sunlight construction, the Light Source (A) is above the Horizon Line (B). The sticks are backlit so the shadows are all cast toward the observer. This lighting strategy and construction can be used for dramatic effect, but it can make it difficult to understand the forms of the object, due to the backlighting scenario that makes the silhouette of the object visually dominant.

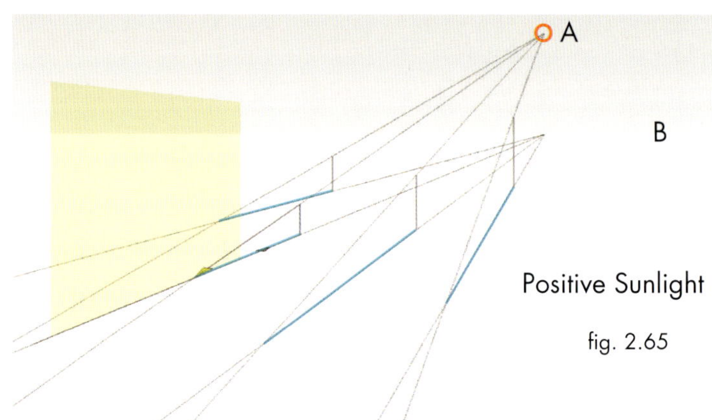

fig. 2.65

Negative Sunlight Construction: Front Light

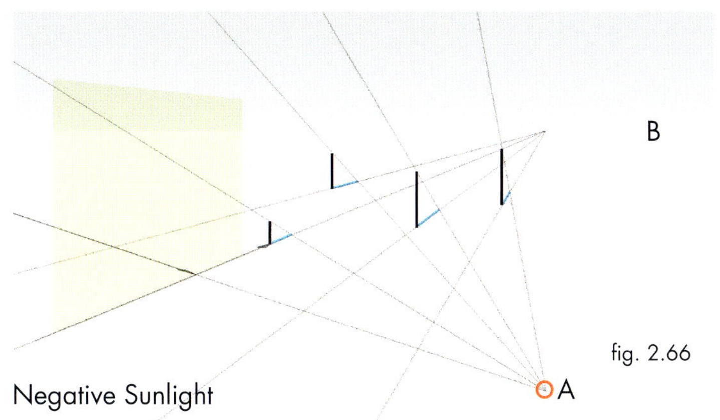

Negative Sunlight

fig. 2.66

In a Negative Sunlight construction, the Light Source (A) is below the Horizon Line (B). The sticks are front-lit, so the shadows are all cast away from the observer. This lighting scenario shows an object's surfaces more clearly than the backlit scene.

In both of these examples the shadows are long, since the Light Source is on the page, close to the Horizon Line. The farther the Light Source moves away from the Horizon Line, the shorter the cast shadows become, reducing the perspective convergence of the Light Rays. This phenomenon can be observed every day, when the shadows are shortest at high noon.

Sunlight Construction: Shadow Origin and Light Source off the Page

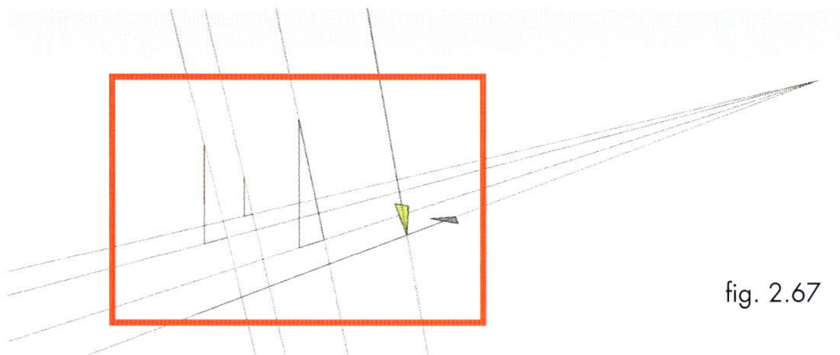
fig. 2.67

Physical drawing area inside orange rectangle with Shadow and Light V.P.s outside this area.

In most constructions, the Shadow Origin and the Light Source are off the page. Just remember how perspective grids work (see Chapter 04 in *How To Draw*) and it will be possible to make a good estimate as to the strength of the perspective convergence. It almost always suffices to estimate the convergence of the Light Rays and Shadow Direction lines without building a perspective grid.

Sunlight Construction without Convergence

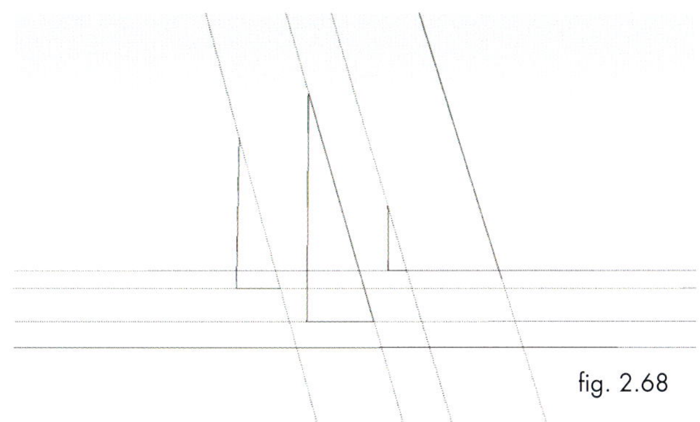
fig. 2.68

To avoid having to work with perspective convergence toward a Shadow Origin or a Light Source, make the Shadow Direction lines parallel to the Horizon Line. Fig. 2.68 shows Light Planes that are perpendicular to the viewer's line of sight, creating a one-point perspective for the cast shadows. The objects casting these shadows can be constructed later in three-quarter view. For more details on perspective grids and layouts, refer to the rotated grids section in Chapter 05 of *How To Draw*.

Sunlight Construction: Casting Shadows on Multiple Levels

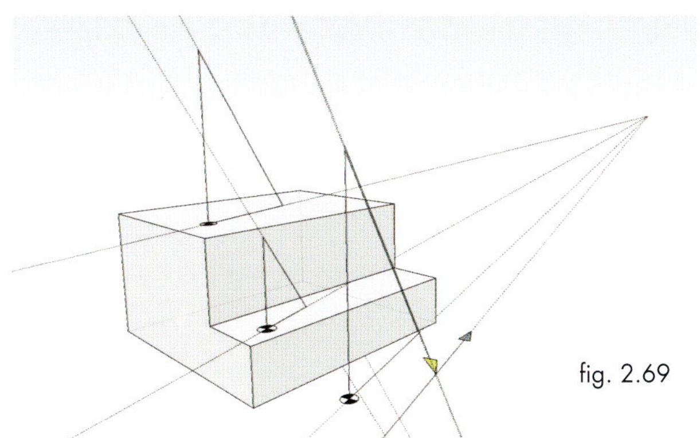
fig. 2.69

Unlike multi-level constructions in local light, multi-level shadows in sunlight do not require the creation of additional Shadow Origins, because all of the Shadow Direction lines are parallel. It does not matter at what height the parallel lines are, since they all converge to a single Vanishing Point.

CASTING SHADOWS OF WALL SHAPES

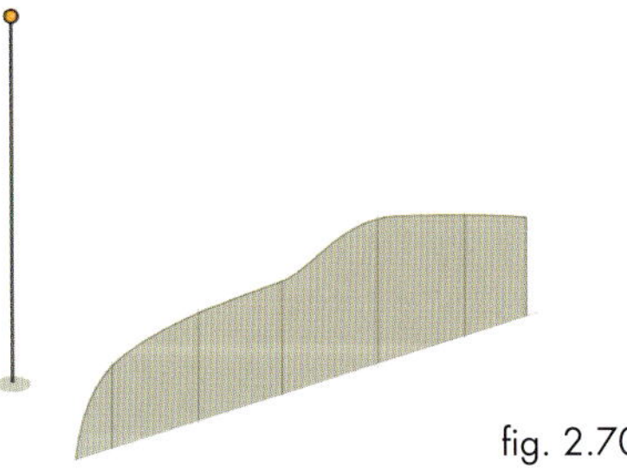

fig. 2.70

To cast the shadow of a more complex shape, add vertical sticks within the shape. Each stick provides an additional point along the shape of the shadow. Use as few sticks as possible. Adding too many sticks may actually hinder accuracy, if one or more of them are less than perfect. This technique works for casting shadows in both sunlight and local light.

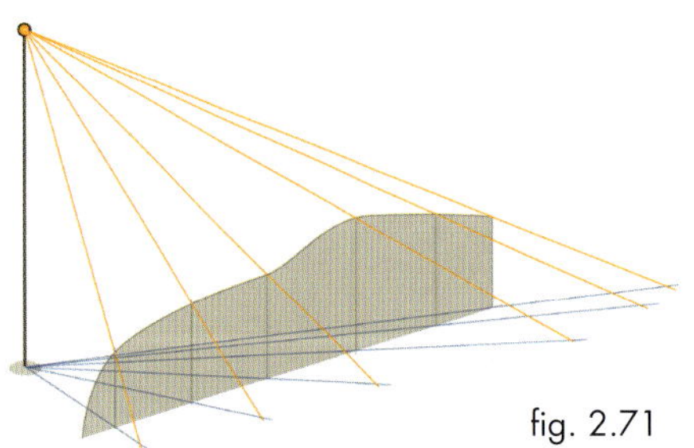

fig. 2.71

Fig. 2.71 is an example in local light. Construct the shadow of each stick by locating the point on the ground plane where the top of the stick casts its shadow.

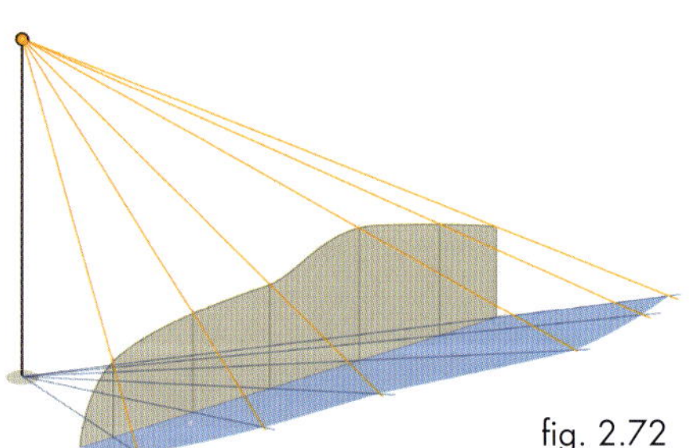

fig. 2.72

In fig. 2.72 these points have been connected to create a smooth cast shadow. Always think about what area or type of shape is casting the shadow. A smooth curve will have a smoothly curving shadow, and a corner will show as a quick turn.

CASTING SHADOWS OVER OBSTACLES

Casting a Stick's Shadow on a Wall

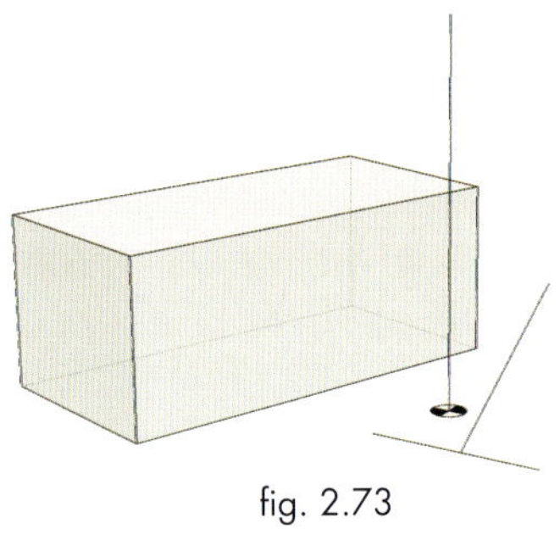

fig. 2.73

In order to cast shadows over obstacles, this scenario involves the sticks, the box, the Light Rays and the Shadow Direction lines.

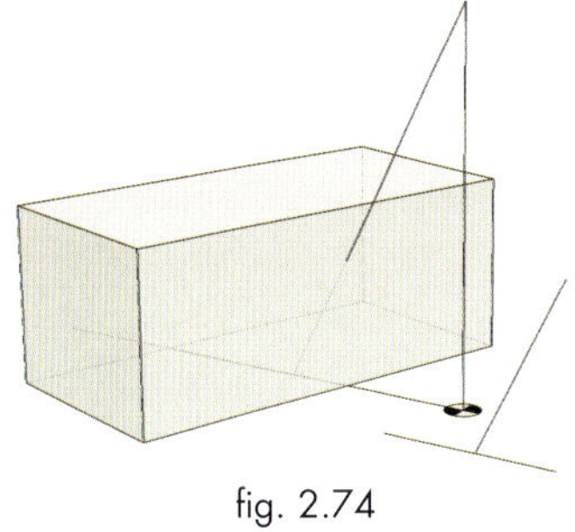

fig. 2.74

First, cast the shadow of the stick on the ground plane, "drawing through" the box. Cast the shadow on one plane at a time, to keep control of the construction.

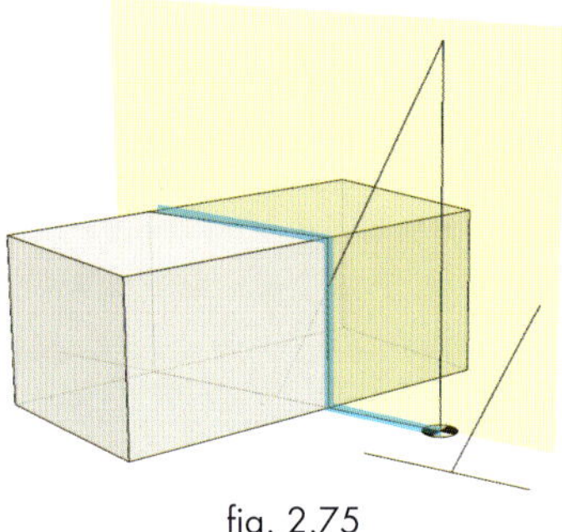

fig. 2.75

Next, create the Light Plane that is defined by the stick and the shadow construction. Use section-drawing skills to define the section line with which the Light Plane cuts through the box.

The shadow of the stick will end where the section line and the Light Ray line intersect (A).

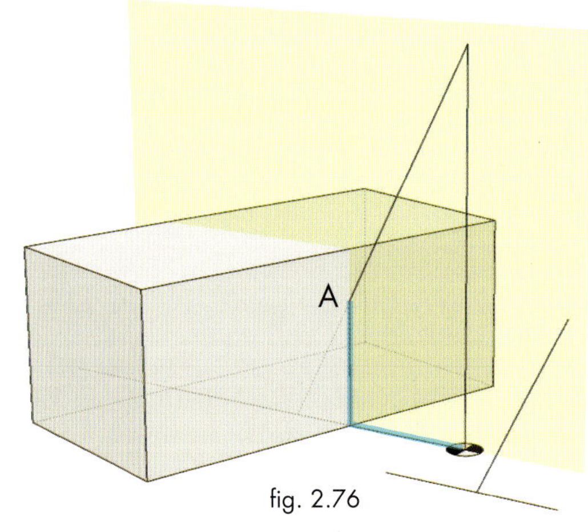

fig. 2.76

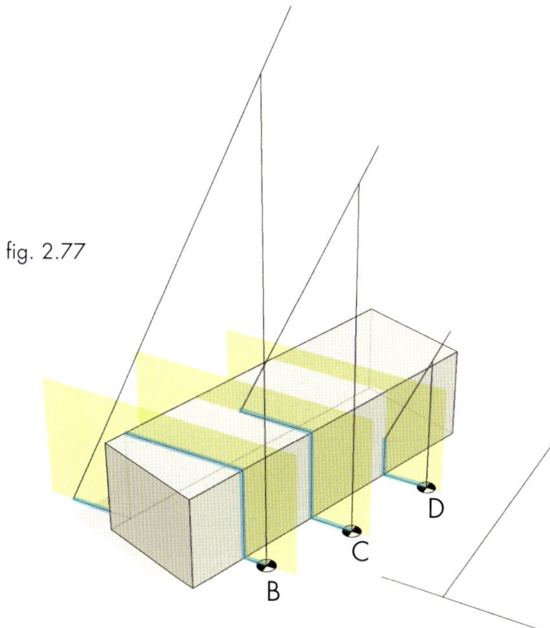

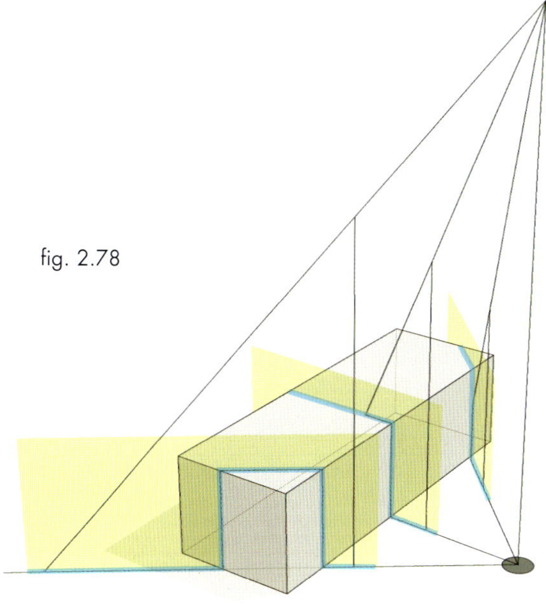

Fig. 2.77: To cast the stick's shadow onthe side of, on top of or over the box (B, C, D respectively), use exactly the same technique of cutting through the box with a Light Plane to find the end point and direction of each shadow. In sunlight, all of these planes are parallel, making the shadows more predictable. With practice it is possible to accurately guess where shadows will fall.

Fig. 2.78: In local light the Light Planes fan out/radiate and it becomes harder to predict where the shadows might fall. To build the shadows, it may be necessary to construct each of the section planes. The Light Plane construction remains the same as in sunlight.

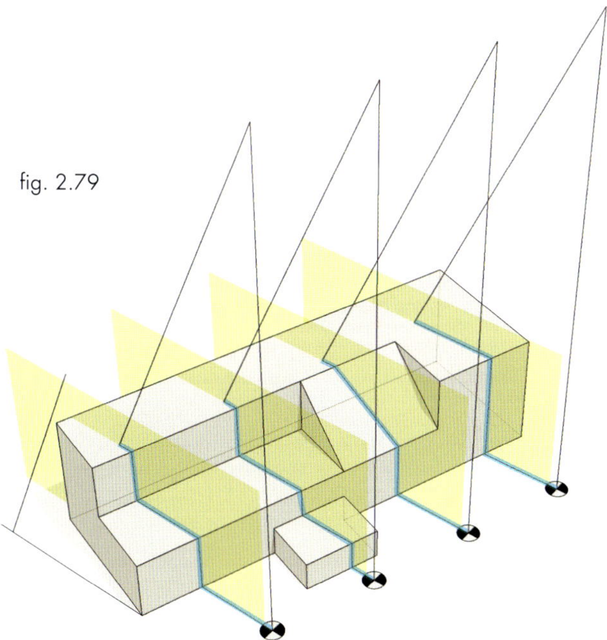

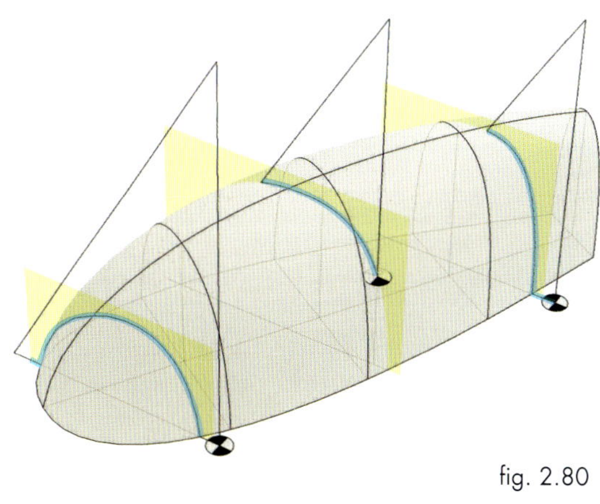

Fig. 2.79: With sloped and/or stepped surfaces, the same Light Plane "cut through" rules apply. To construct cast shadows of more complex objects, it is essential to be in control of the perspective construction. Concentrating on one section at a time makes it possible to resolve even the most complex situations.

Fig. 2.80: This principle applies to X-Y-Z shapes as well. Here it is even more crucial to understand the surface of the object in order to generate the Light Plane section lines properly.

Casting the Shadow of a Vertical Stick onto an Object

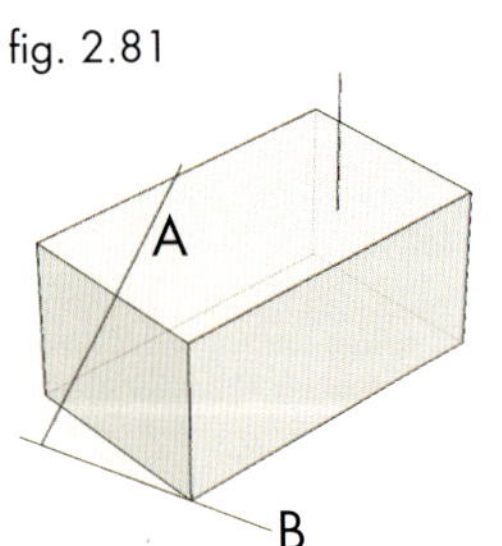

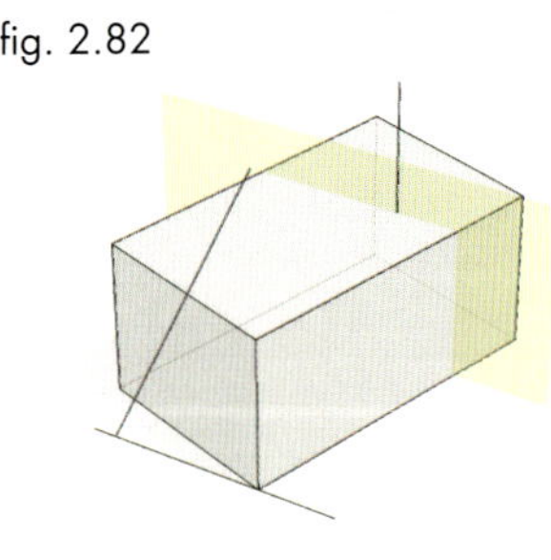

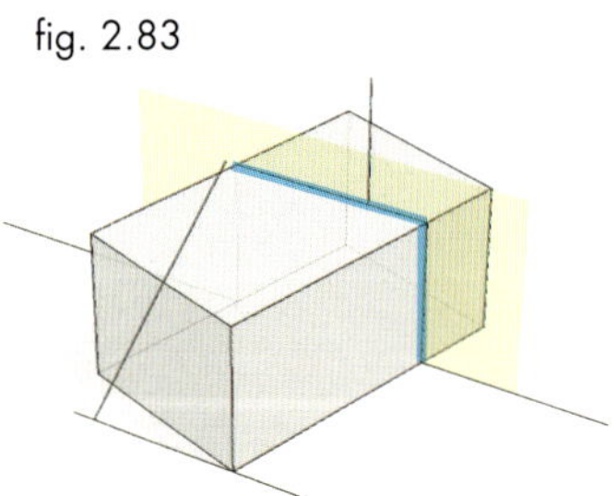

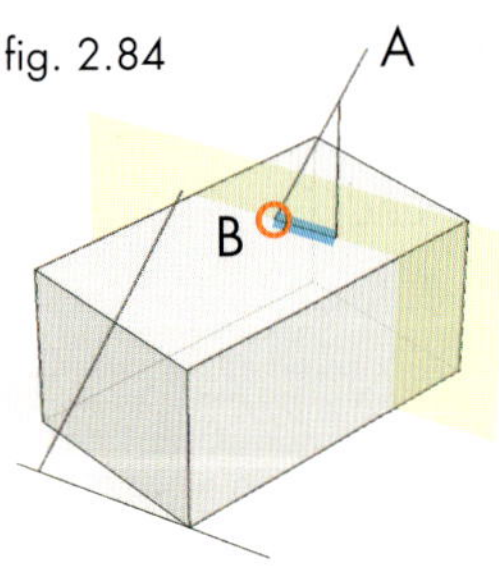

STEP 1, fig. 2.81: In this sunlight scenario, define the Light Direction line (A) and the Shadow Direction line (B).

STEP 2, fig. 2.82: By referencing the Light Ray and Shadow Directions, place a Light Plane through the stick that sits on top of the box.

STEP 3, fig. 2.83: Draw the Light Plane section/cut line on the box. This will be the direction of the cast shadow.

STEP 4, fig. 2.84: Find the length of the shadow by drawing a Light Ray through the top of the stick (A) until it intersects with the Light Plane section/cut line (B). This intersection determines the length of the cast shadow.

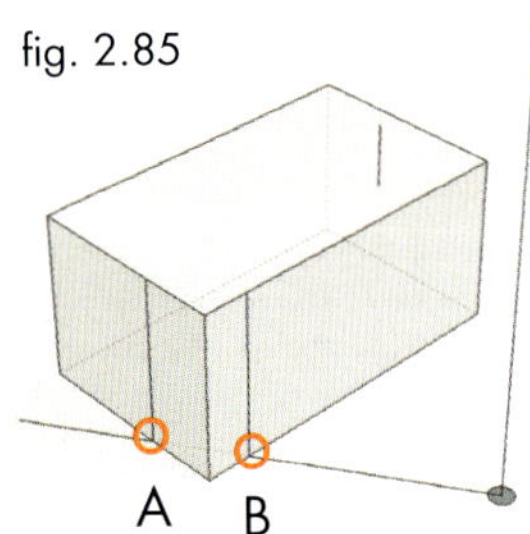

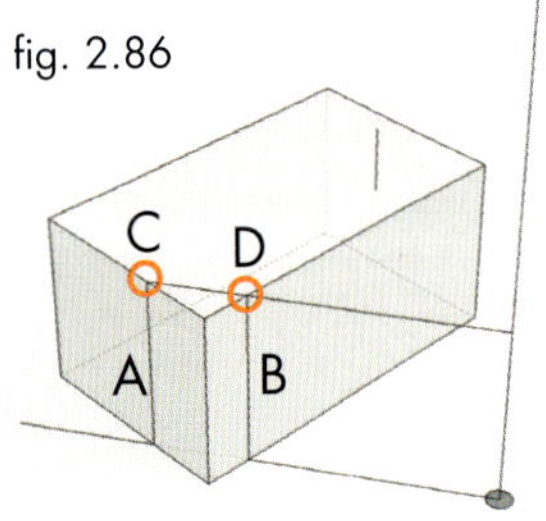

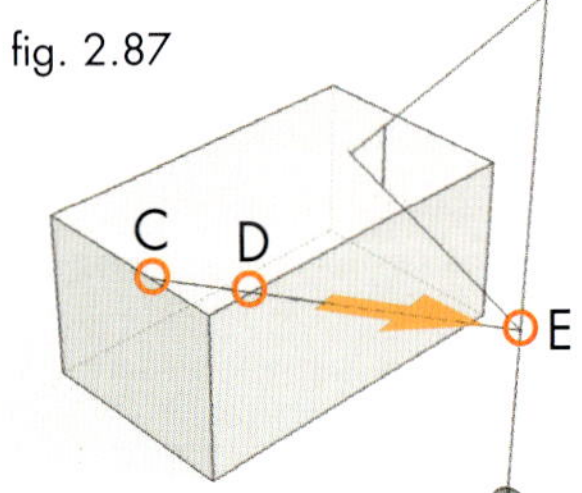

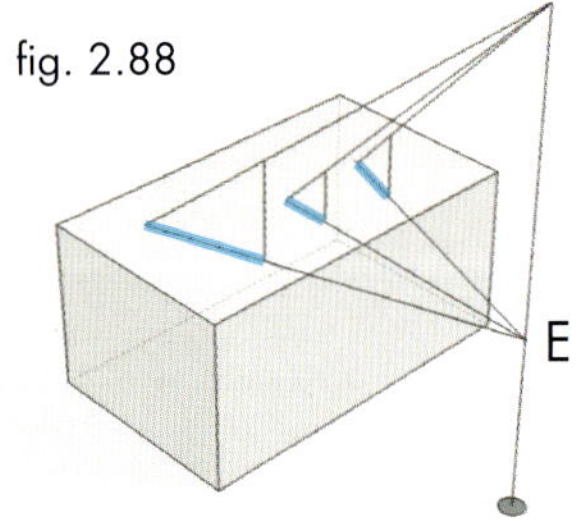

STEP 1, fig. 2.85: In this local light scenario, the Shadow Origin for the stick needs to be level (at the same elevation) with the plane on which the cast shadow is being constructed. First, draw a line on the ground plane from the original Shadow Origin that intersects with the two vertical sides of the box, creating points A and B.

STEP 2, fig. 2.86: Draw vertical lines from points A and B that intersect the top of the box, creating points C and D.

STEP 3, fig. 2.87: Draw a line through points C and D, back to the Shadow Origin/Light Source axis, creating point E.

STEP 4, fig. 2.88: Point E is now the Shadow Origin for any cast shadows drawn on the top surface of the box.

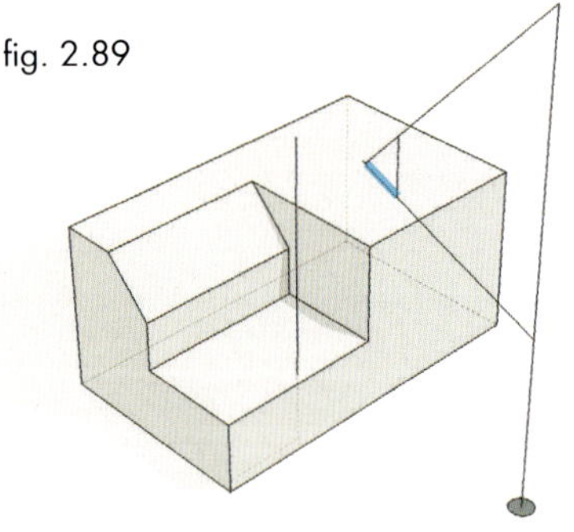

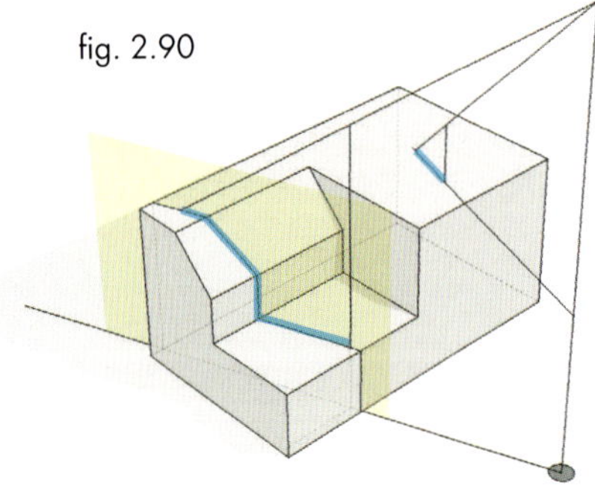

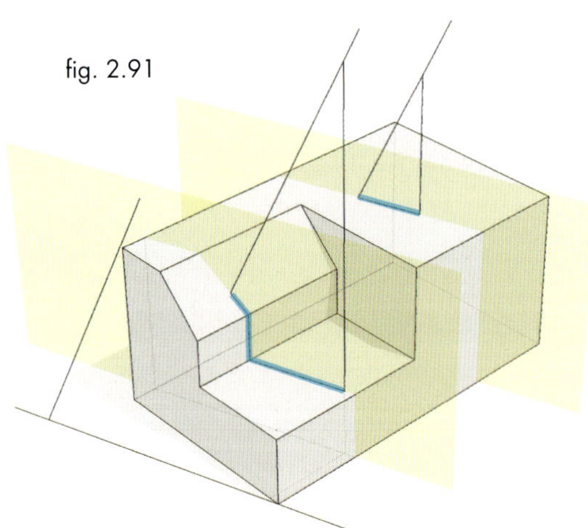

Figs. 2.89, 2.90 and 2.91: Follow the same steps to create multiple shadows on any irregularly shaped surface.

Casting the Shadow of a Horizontal Stick onto a Wall

To cast shadows from horizontal sticks, first create vertical sticks and then focus on casting the top end of the vertical stick on the ground. The points created are the guides for the construction of the horizontal stick's shadow. Let's look at three scenarios in which the shadow will be cast on the wall, on the wall and the ground, and "fall off" the side of the wall. All techniques have one goal: to find an additional point on the wall to determine the Shadow Direction.

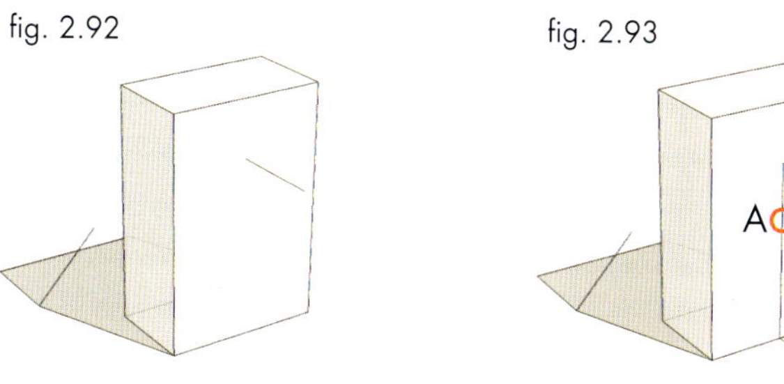

fig. 2.92 fig. 2.93

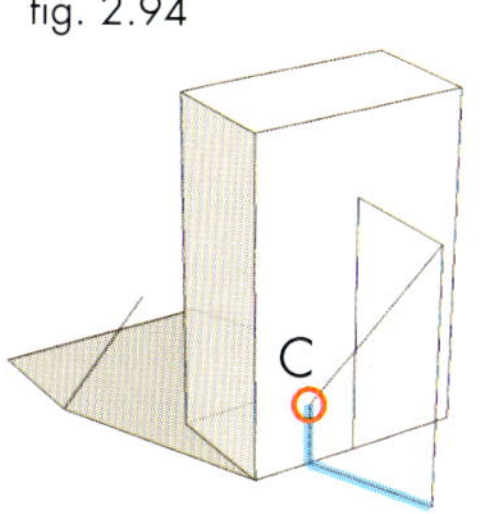

fig. 2.94

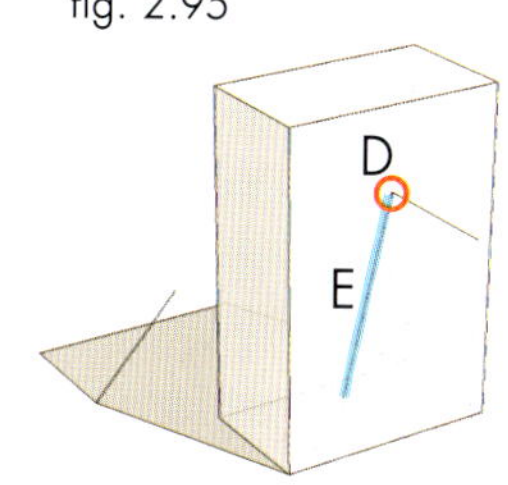

fig. 2.95

STEP 1, fig. 2.92: Place a stick horizontally on the side wall of the box.

STEP 2, fig. 2.93: Draw two vertical sticks (A and B) under the end points of the horizontal stick.

STEP 3, fig. 2.94: Cast the shadow of stick B. This includes casting it onto the wall, creating point C.

STEP 4, fig. 2.95: Connect point C which is the cast shadow's end point, back to the point where the horizontal stick attaches to the wall (point D). Line E is the cast shadow of the horizontal stick on the wall.

Casting the Shadow of a Horizontal Sick onto a Wall and the Ground

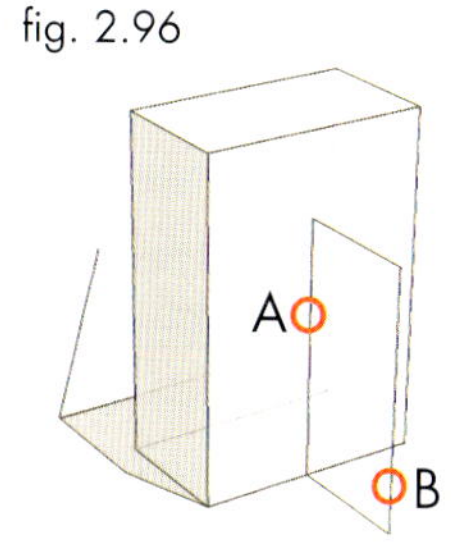

fig. 2.96

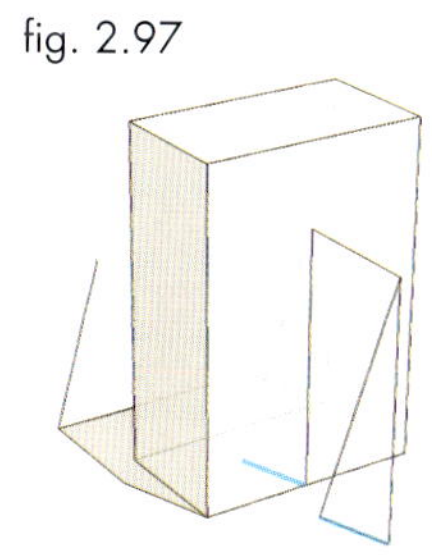

fig. 2.97

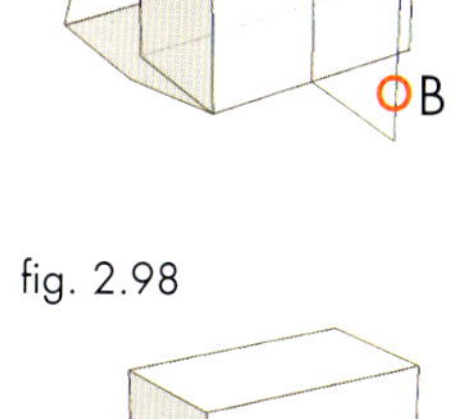

fig. 2.98

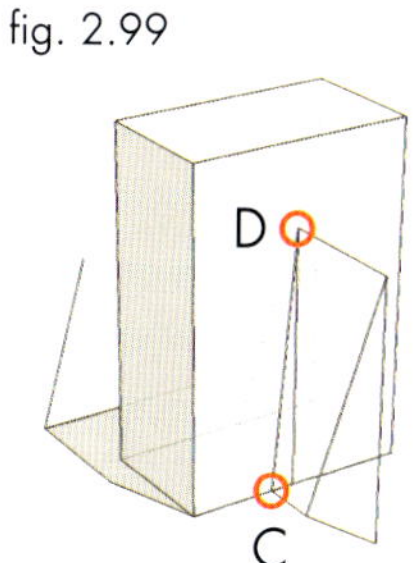

fig. 2.99

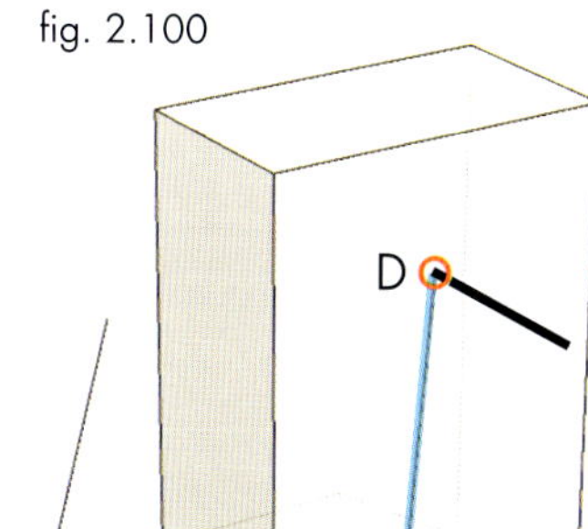

fig. 2.100

The same technique is used when the shadow is on the wall as well as the ground. However, after finding the two vertical sticks, you will construct shadows on two planes: the ground, and the wall.

STEP 1, fig. 2.96: As before, draw two vertical sticks (A and B) under the end points of the horizontal stick.

STEP 2, fig. 2.97: Cast the shadows of the two vertical sticks on the ground plane.

STEP 3, fig. 2.98: Draw through the wall and connect the cast end points of the sticks. Notice that the blue line runs under the wall.

STEP 4, fig. 2.99: Where this shadow intersects the wall (C) it will now extend up the vertical wall and connect to the intersection of the horizontal stick and the wall (D).

STEP 5, fig. 2.100: The line connects straight from point C to the intersection point of the stick and the wall (D) because the shadow being cast is that of a straight line, the stick.

Casting the Shadow of a Horizontal Stick Beyond the Edge of a Wall, onto the Ground

Having a shadow go beyond the edge of a wall requires the creation of the shadow on the ground, at both end points of the horizontal stick. Since this shadow does not intersect the wall at the ground, a different technique is necessary to find the Shadow Direction. Fig. 2.101 shows a horizontal stick on the side of a box, and an established light-source position as evidenced by the box's cast shadow.

Here are three separate ways to find the position where the shadow will cross the vertical edge of the box, continuing on down to the ground.

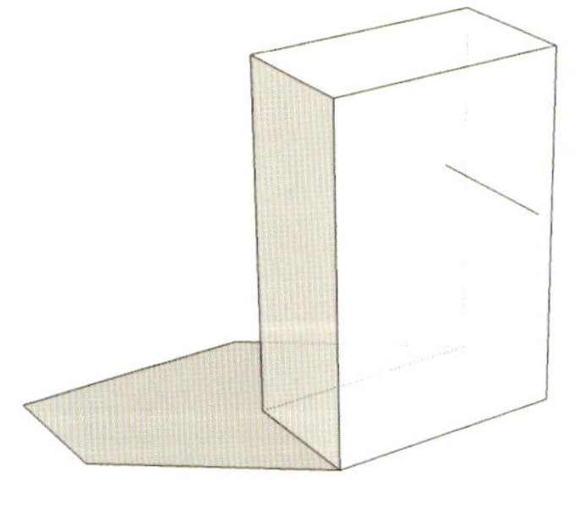

fig. 2.101

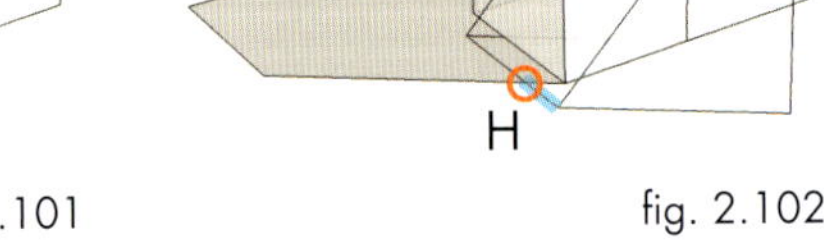

fig. 2.102

Method One:

Fig. 2.103: Building on the process from the previous page, add a vertical stick (A) somewhere along the horizontal stick, and cast its shadow onto the side of the box.

Fig. 2.104: Connect the top of this shadow on the vertical wall (B), to the intersection point of the horizontal stick and the wall (C). This creates the cast shadow of the horizontal stick running down the side, until it intersects the vertical corner of the box (D).

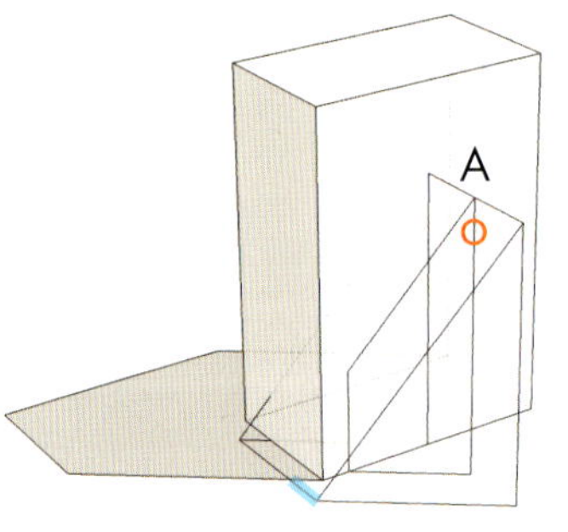

fig. 2.103

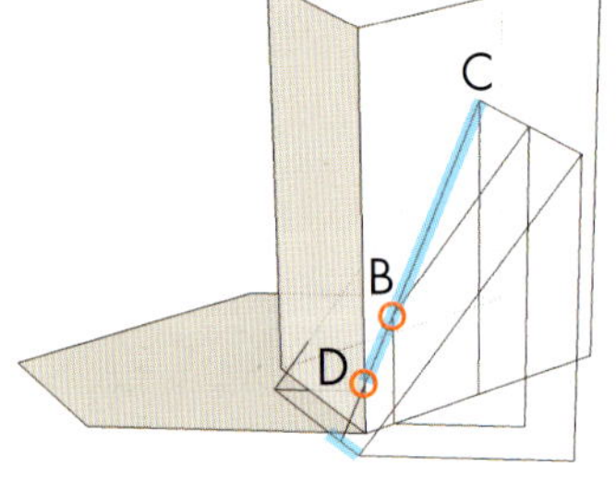

fig. 2.104

Method Two:

Figs. 2.105 and 2.106: Instead of drawing a new vertical stick at a random point along the horizontal stick, draw a vertical stick exactly on the corner of the box (line E). Then, extend the Shadow Direction line (F) along the bottom edge of the box to position a new vertical line (G).

A much more abstract alternative to this method is to observe where the cast shadow of the horizontal stick crosses the cast shadow of the edge of the box on the ground plane. Look back up at fig. 2.102, this would be point H. Simply draw a Light Ray line from point H up to the edge of the box. Connect this point back to the intersection point of the horizontal stick and the wall (point C) to create the cast shadow (fig. 2.104).

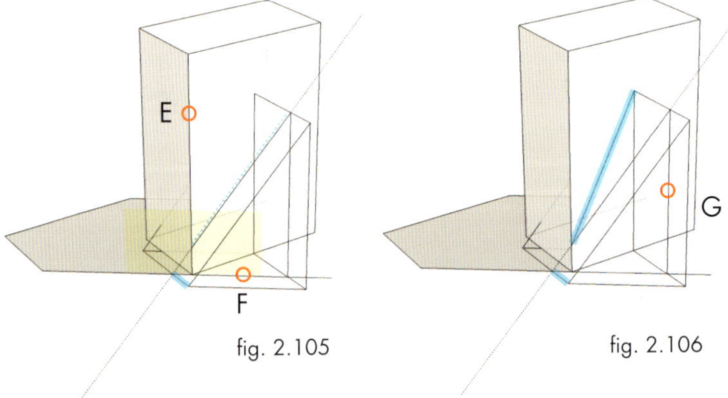

fig. 2.105

fig. 2.106

Method Three:

Figs. 2.107 and 2.108: Extend the side of the box until it intersects with the horizontal stick's cast shadow on the ground, then connect this intersection point, D, back to the intersection point of the horizontal stick (C).

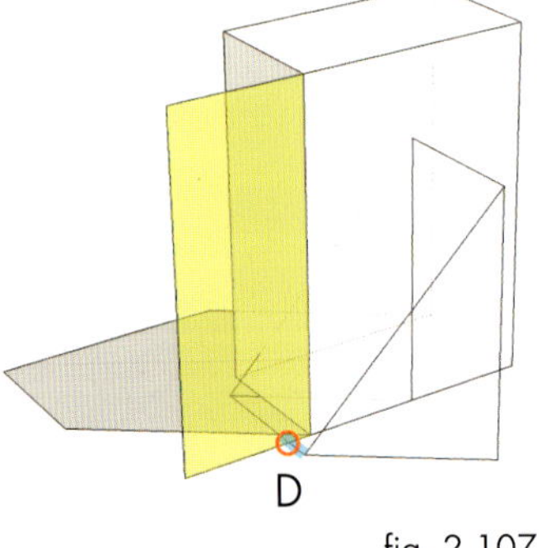

fig. 2.107

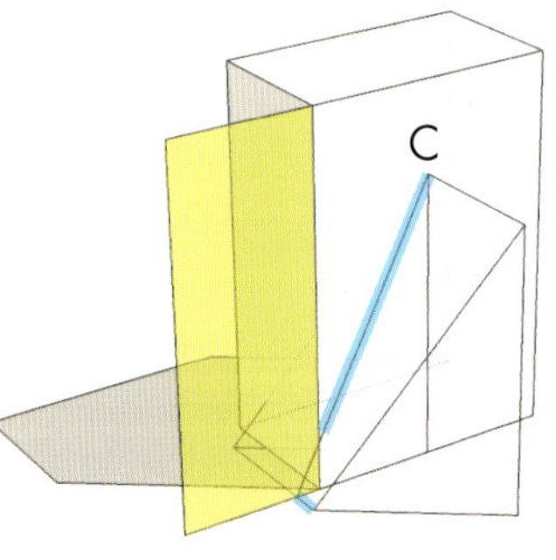

fig. 2.108

Casting the Shadow of a Horizontal Stick over Complex Shapes

Work with one element at a time to cast shadows over complex volumes. Multi-faceted objects can be subdivided into basic forms to construct the shadows and then reassembled for the final result. These sub-constructions can be done with overlays, but most of the time it is simply the most logical thought process to include all the constructions on the same page.

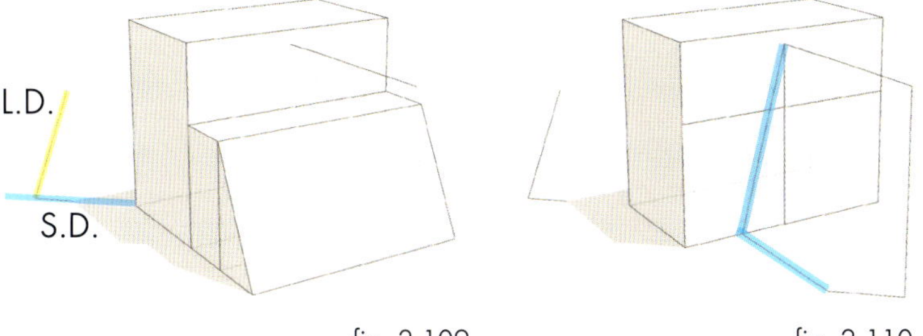

fig. 2.109

fig. 2.110

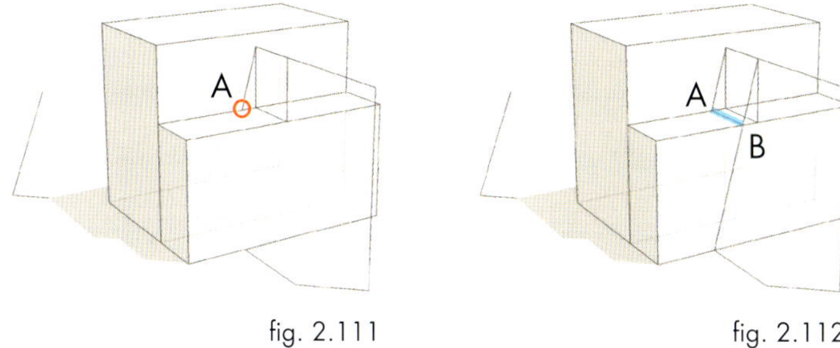

fig. 2.111

fig. 2.112

STEP 1, fig. 2.109: Establish the perspective, Light/Shadow Directions and object. Look for subdivision opportunities of the major volumes.

STEP 2, fig. 2.110: Choose one volume to begin. In this example, it is the largest block. Disregard the other forms for now. Cast the shadow on this large box with the techniques already covered.

STEP 3, fig. 2.111: Add back an additional volume, the smaller box form. Since the horizontal stick is perpendicular to the side of both boxes, a line pointing to the left Vanishing Point can be drawn on top of the small box from point A in fig. 2.112 as one way to establish the direction of the shadow on the top surface of this box.

STEP 4, fig. 2.113: Add the triangular block. Observe where the shadow leaves the top surface of the smaller box (B) and where the shadow on the ground intersects the foot of the sloped surface (C). Connect these two points to complete the construction.

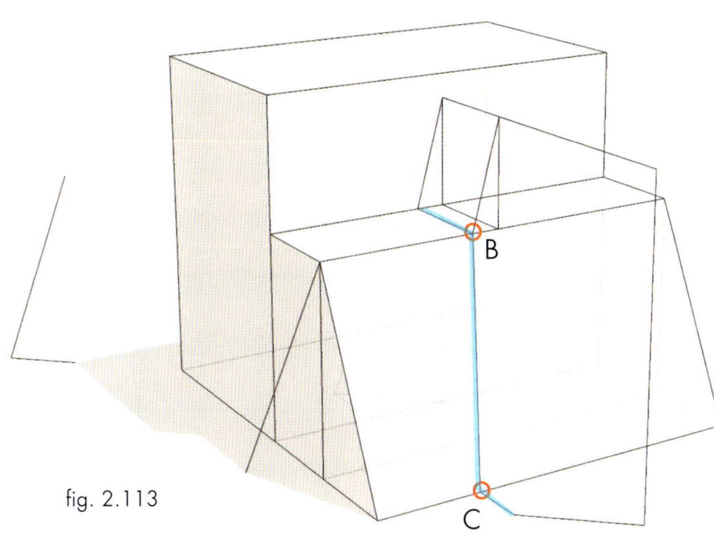

fig. 2.113

Again, notice in fig. 2.113 that the upper and lower parts of the shadow line (in blue) are parallel to the actual stick. Lines that are parallel over level surfaces will cast parallel shadows. This is great information for making quick guesses instead of full constructions.

All four of these lines, A, B, C and D point to the same Vanishing Point.

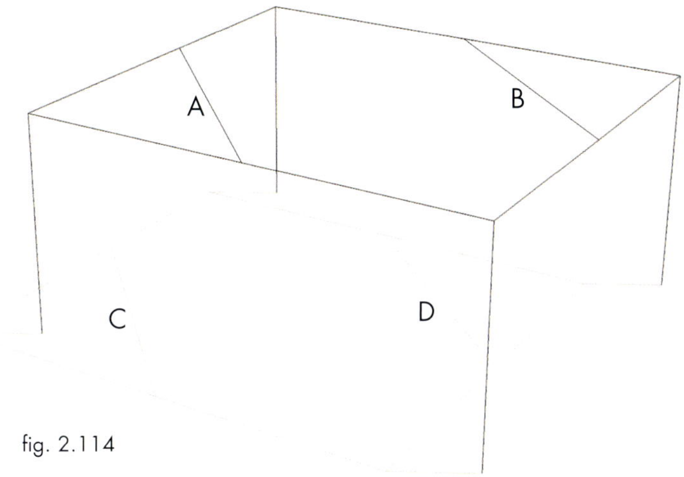

fig. 2.114

Lines that are parallel to a level ground plane will cast parallel shadows. This means that the stick and the shadow of the stick share the same Vanishing Point.

Casting the Shadow of an Angled (Leaning) Stick

Casting the shadow of a stick positioned at an angle to the ground plane and leaning against a box is similar to dealing with horizontal sticks.

STEP 1, fig. 2.115: Define the position of the top of the leaning stick with a supporting vertical line (A).

STEP 2, fig. 2.116: Cast the shadow of the vertical support stick onto the ground plane. Ignore the box being in the way, just draw through it.

STEP 3, fig. 2.117: Create the shadow on the ground by connecting the end of the cast shadow from the previous step with the base of the stick (B).

STEP 4, fig. 2.118: Connect the top point of the angled stick where it is leaning against the box (C) with the intersection point of the ground shadow with the side of the box (D). The cast shadow of the leaning stick is now complete.

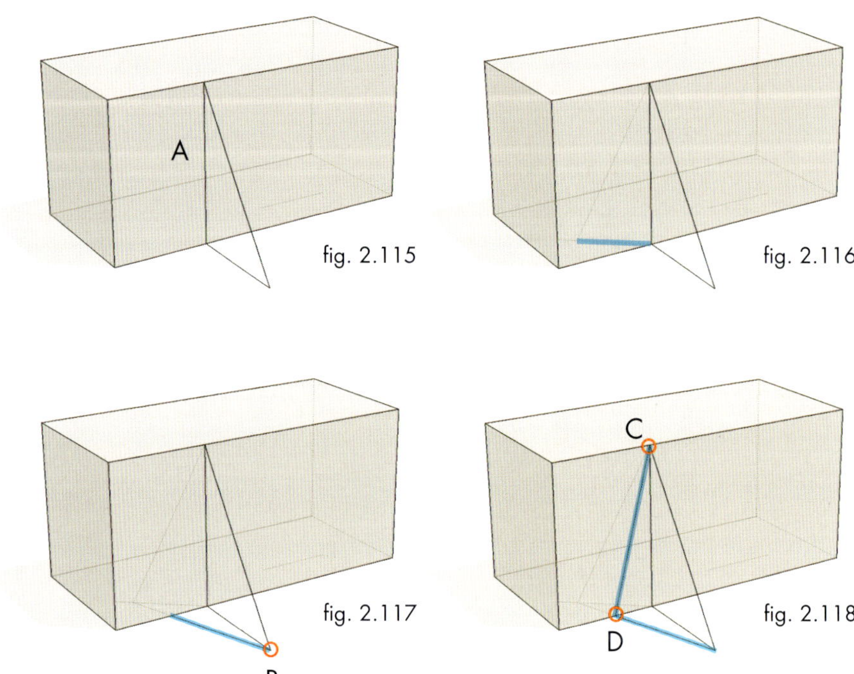

fig. 2.115 fig. 2.116 fig. 2.117 fig. 2.118

Casting the Shadow of a Triangular Shape

Fig. 2.119: The same type of construction works for the yellow triangle extending from the top corner of this box. This is done with a combination of casting a shadow of a stick perpendicular to the box wall and a leaning stick.

STEP 1, fig. 2.120: Draw two vertical lines at points A and B so each corner of the triangle has a vertical stick under it. The corner of the box serves as the vertical stick under point C. Cast the shadows of these three sticks onto the ground plane.

STEP 2, fig. 2.121: Connect the end of each stick's cast shadow on the ground so they create a cast shadow of the triangle. Note where the shadow intersects the box, points D and E.

STEP 3, fig. 2.122: Connect intersection points D and E up to the intersection points of the triangle and the top of the box, points A and C. Point D connects to A, point E connects to C. The blue area (F) is the resulting cast shadow of the triangle on the side of the box and the ground plane.

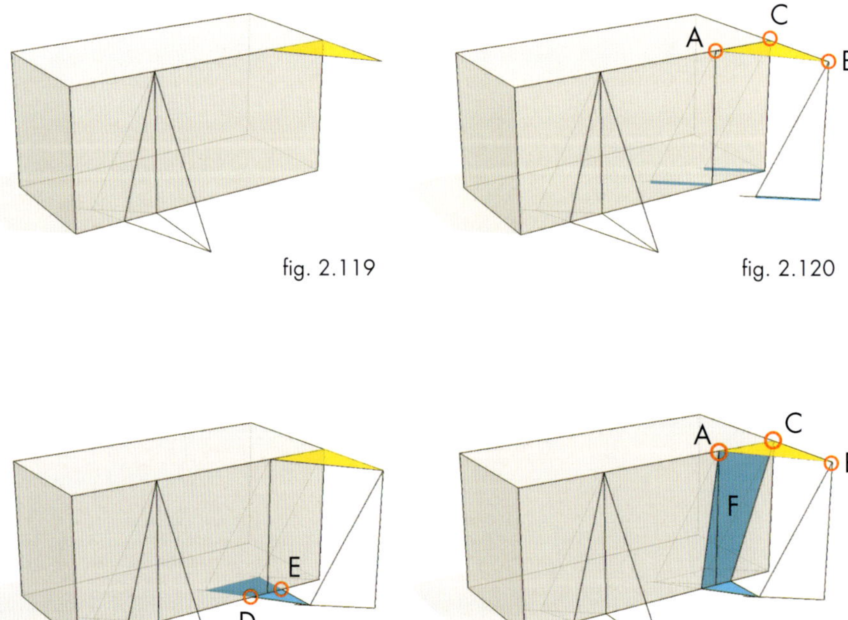

fig. 2.119 fig. 2.120 fig. 2.121 fig. 2.122

With practice, this basic technique of using a stick as a starting point can yield some impressive shadow constructions, as seen in these examples by our former students, Jason Kang and Charles Liu.

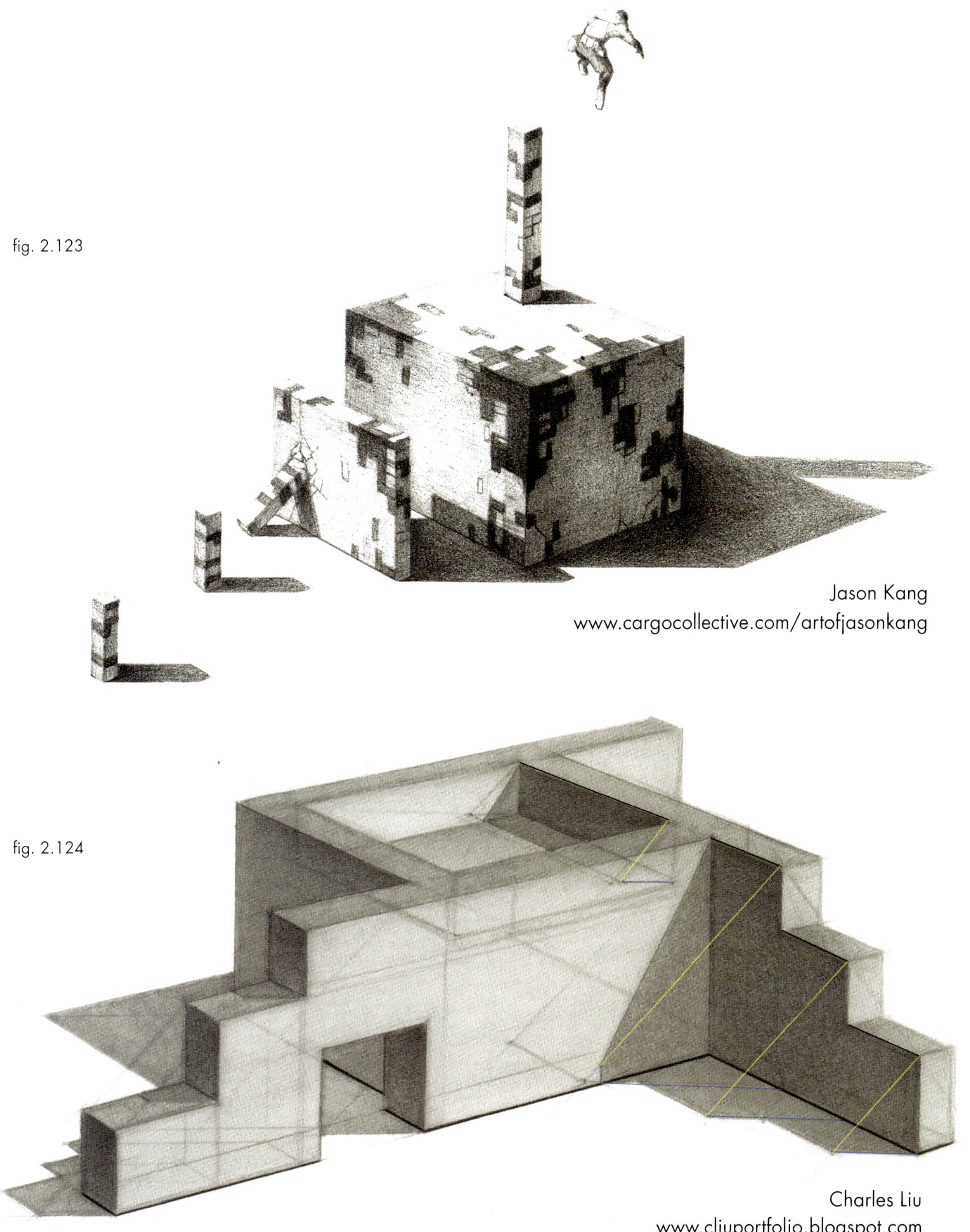

fig. 2.123

Jason Kang
www.cargocollective.com/artofjasonkang

fig. 2.124

Charles Liu
www.cliuportfolio.blogspot.com

CASTING SHADOWS OF SIMPLE VOLUMES 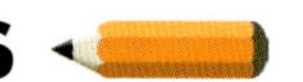

Just as points created by casting the shadows of sticks can combine to form shadow shapes, these techniques can be combined to construct the cast shadows of simple box volumes. The complexity of these constructions can increase exponentially and provides the opportunity to make truly exciting and interesting cast shadows. These types of shadows can be most easily observed in architectural scenes.

Casting Shadows of Simple Volumes: Sunlight

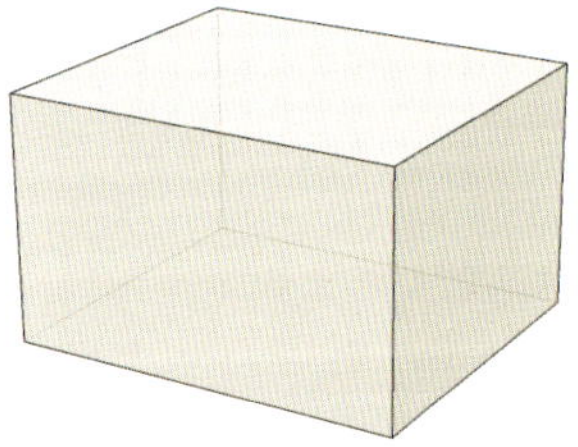

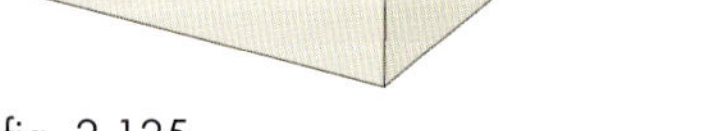

fig. 2.125

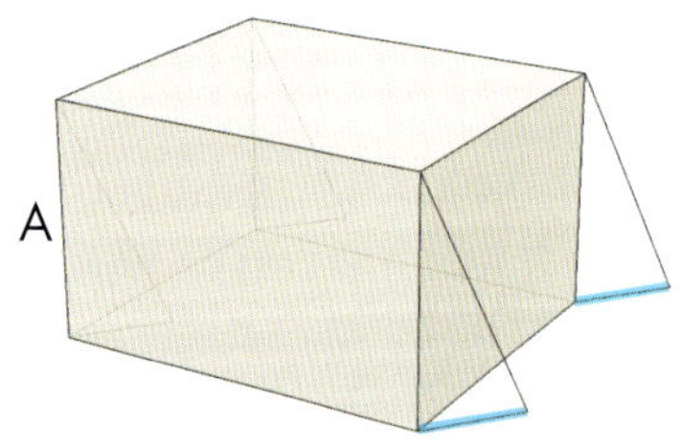

fig. 2.126

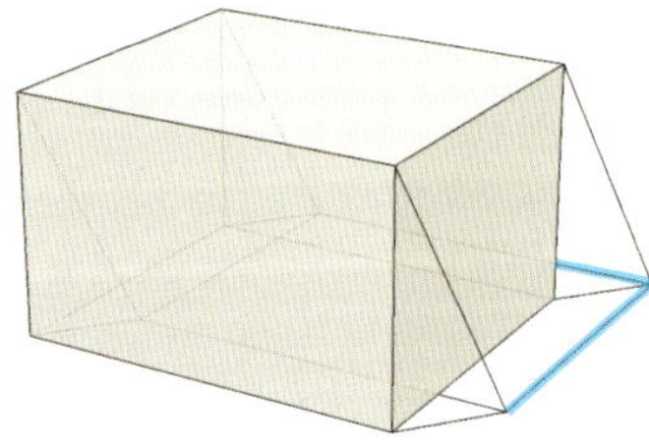

fig. 2.127

STEP 1, fig. 2.125: Establish the object in perspective, drawing through the volume lightly will make it easier to plot the shadow.

STEP 2, fig. 2.126: Cast all of the vertical edge (stick) shadows from each corner of the box onto the ground plane with the exception of the corner (A) because that shadow would be inside the object. Shadows inside solid objects can be ignored, assuming they are not needed to aid in the construction of the rest of the shadow. Learning which sticks are nonessential shortens the construction time.

STEP 3, fig. 2.127: Connect the edges' cast-shadow end points with straight lines to cast the shadow of the "roof" or "top" of the box onto the ground plane.

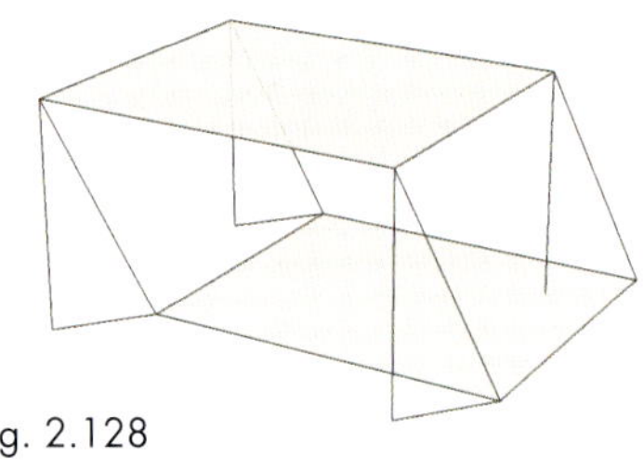

fig. 2.128

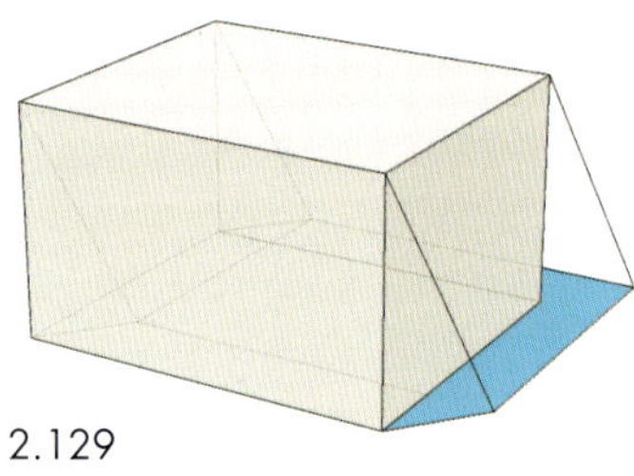

fig. 2.129

STEP 4, fig. 2.128: With the sides of the box now invisible, observe that the interior construction would cast a shadow of the top plane held by four sticks onto the ground.

STEP 5, fig. 2.129: Combine the shadow of the vertical corner of the box with the roof shadow to create the box's full shadow.

Figs. 2.130 and 2.131. Notice the only construction lines used are the essential ones to define the box and the shadow. Gaining knowledge of how to build cast-shadow constructions enables the reduction of construction lines to increase drawing and rendering efficiency.

fig. 2.130

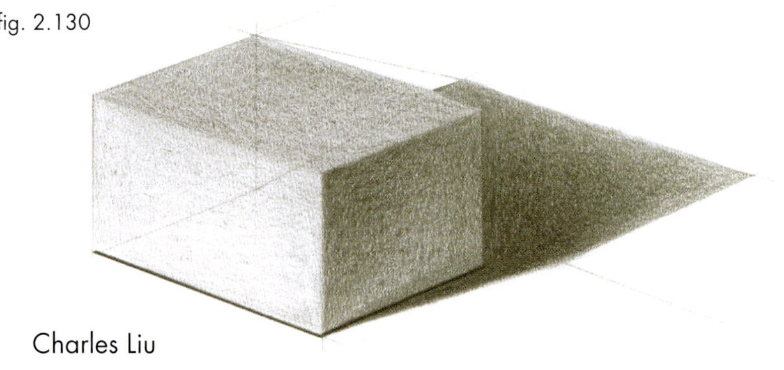

Charles Liu

fig. 2.131

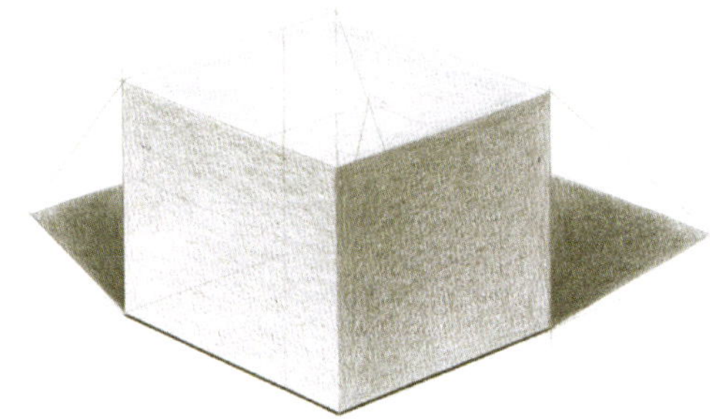

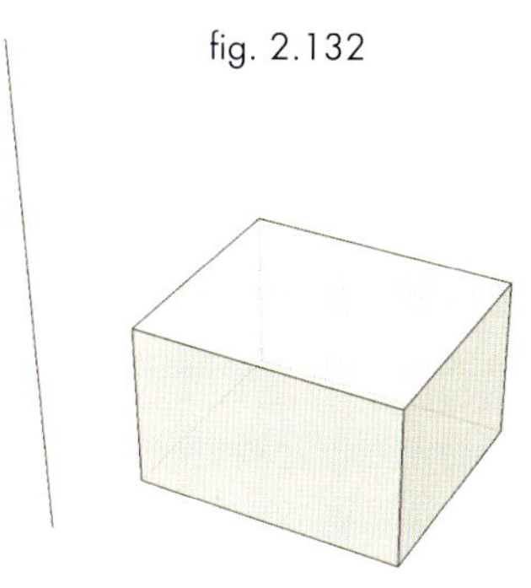
fig. 2.132

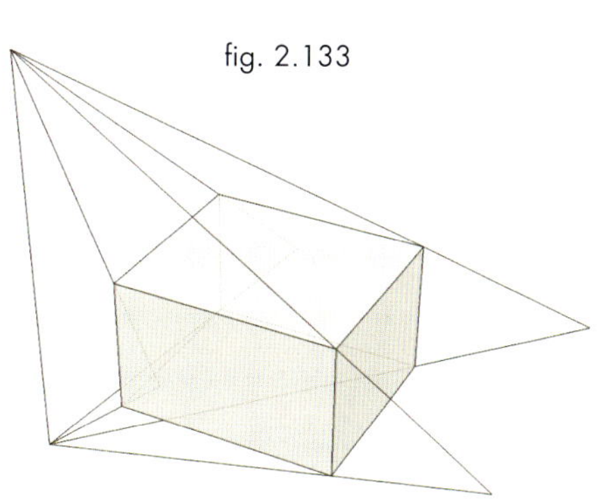
fig. 2.133

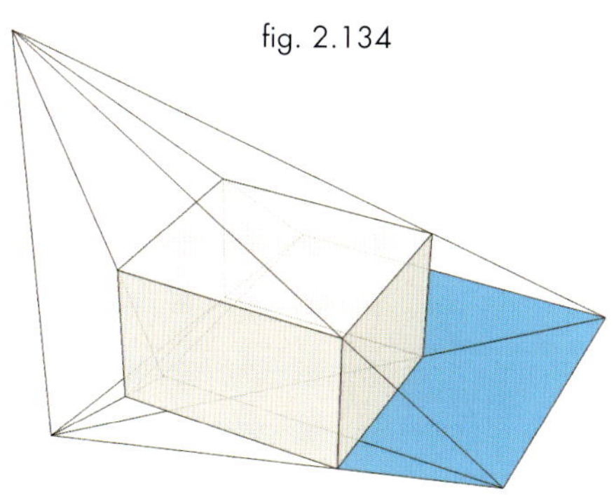
fig. 2.134

The same technique can be applied for local light, but the cast shadow will fan out from the volume, making it harder to guess.

STEP 1, fig. 2.132: Establish the object, the Light Source and Shadow Origin.

STEP 2, fig. 2.133: Cast the shadow of all of the vertical edges onto the ground plane.

STEP 3, fig. 2.134: Connect the end points of the edges' cast shadows with straight lines to cast the shadow of the roof or top of the box onto the ground plane.

fig. 2.135

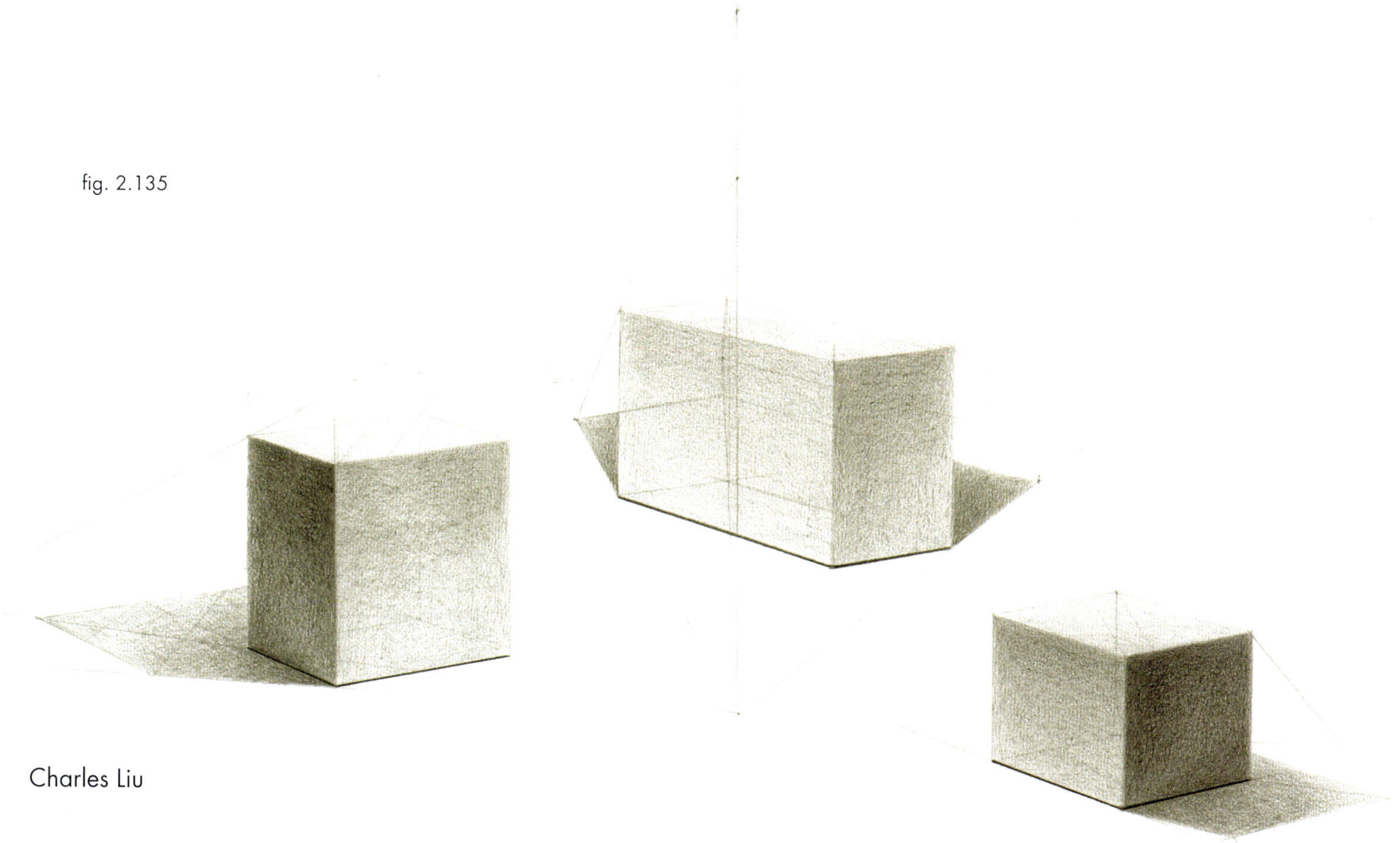

Charles Liu

Although very similar techniques are used in both sunlight and local light constructions, the resulting cast shadows differ quite a lot. Remember that a local light source has an axis that is perpendicular to the ground plane, around which the Light Planes radiate. This means the size of a cast shadow increases as the distance from the local light source increases. In contrast, sunlight Light Planes are parallel to each other so the Shadow Direction for each stick is similarly parallel. This makes sunlight constructions very predictable.

CASTING SHADOWS OF OVERHANGS

Flat surface volumes often include overhangs. Remember, staying focused on constructing cast shadows one stick at a time makes it possible to master these more complex cast shadows. Keep construction lines light to manage the increased number of lines needed. Darken the lines only when the construction is finished. These constructions work in both sunlight and local light.

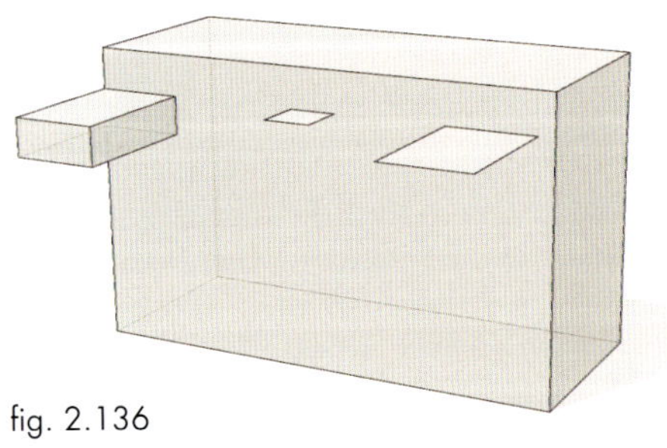

fig. 2.136

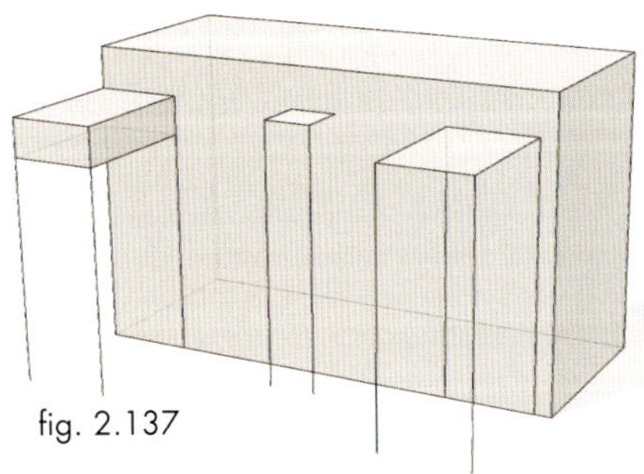

fig. 2.137

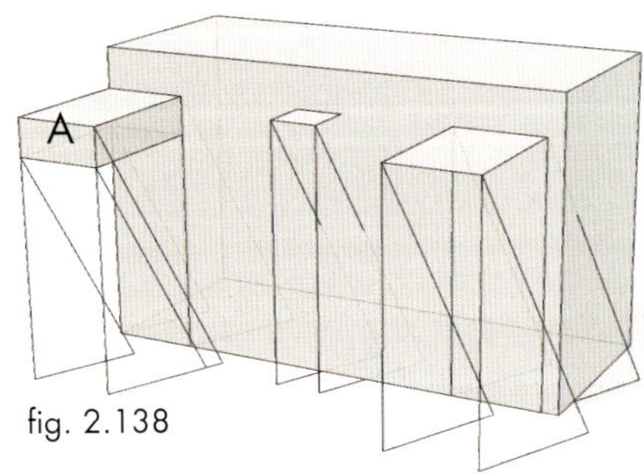

fig. 2.138

STEP 1, fig. 2.136: Establish the object and its overhangs in perspective. Define the Light and Shadow Direction and create the shadow of the box without the overhangs.

STEP 2, fig. 2.137: Place the vertical sticks that will enable the construction of the shadows of the defining edges of the overhangs.

STEP 3, fig. 2.138: Cast the shadows of the end points of each of the vertical sticks onto the ground plane. Project both the top and bottom corners of the thick overhang (A) to accurately reflect this thickness in the cast shadow.

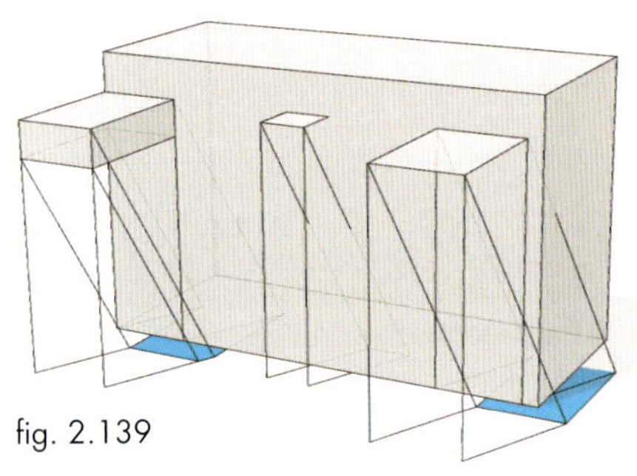

fig. 2.139

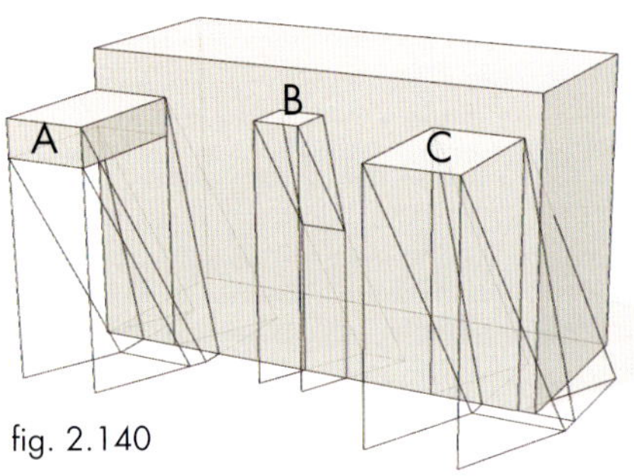

fig. 2.140

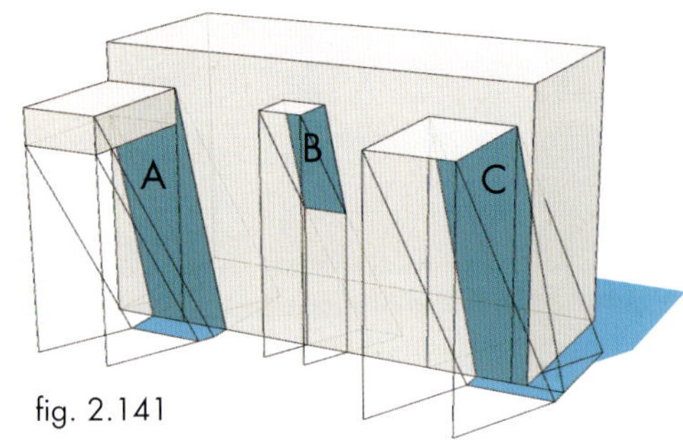

fig. 2.141

STEP 4, fig. 2.139: Define the cast shadows on the ground plane.

STEP 5, figs. 2.140 and 2.141: Observe the different interactions of the shadows with the wall just like the triangle shadow construction on page 42.
A: casts a shadow on the ground and the wall.
B: casts a shadow only on the wall.
C: casts a shadow beyond the edge of the wall and down to the ground.

Finding the correct shadow direction for the right edge of the cast shadow on the wall can be achieved by applying one of the methods on page 40. The shadow on the ground is a combination of the overhang's shadow and the large box's shadow.

Observe these overhang cast shadows in both the real-life and constructed examples below.

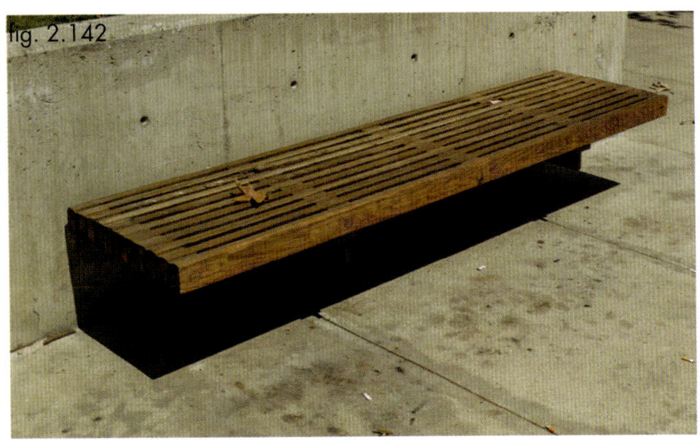

fig. 2.142

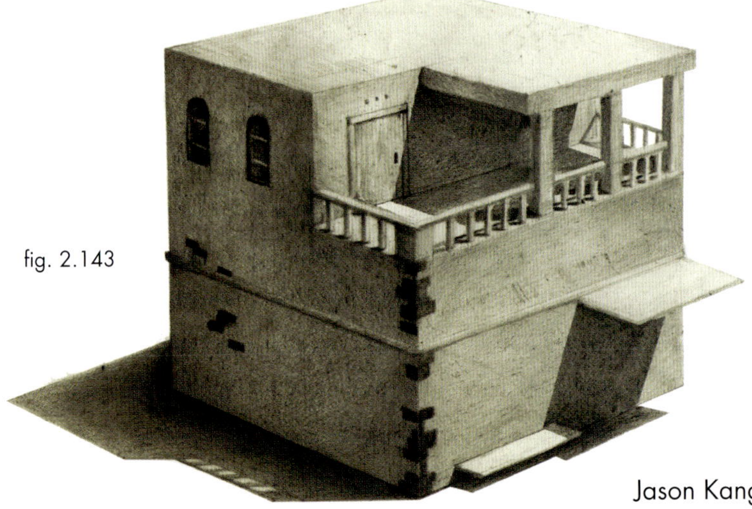

fig. 2.143

Jason Kang

The cast shadows of walls or frames around openings like windows, arches or doorways are best constructed with several well-placed light planes. Taking the time to look around and regularly observe which edges cast these types of shadows will increase the efficiency of creating these constructions. Finding the relevant edges from which to cast the shadows is essential. The following example uses sunlight because it is more common, but the same technique can be used for local light.

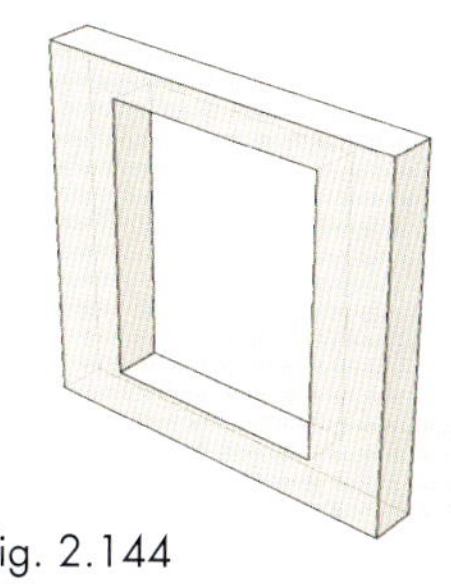

fig. 2.144

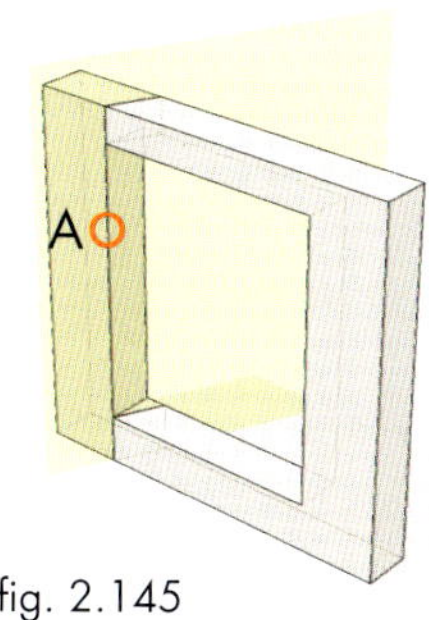

fig. 2.145

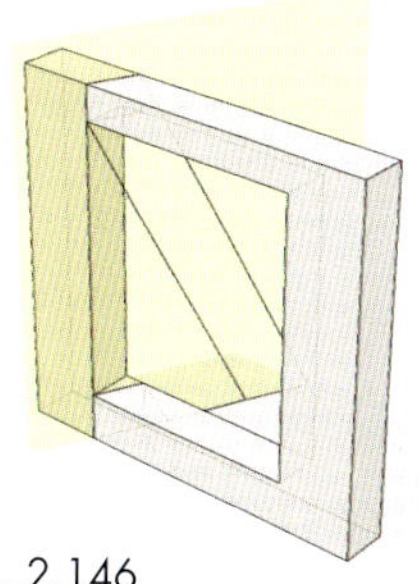

fig. 2.146

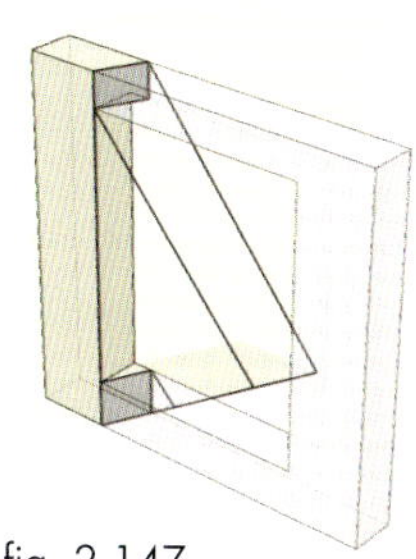

fig. 2.147

STEP 1, fig. 2.144: Construct the cast shadow of the volume while ignoring the opening.

STEP 2, fig. 2.145: Place a Light Plane at the edge (A). This edge will cast a shadow over the lower window frame. Construct and draw the cut-through section lines.

STEP 3, figs. 2.146 and 2.147: Cast the shadow of each of the frame sections created by the frame section cut. First cast the shadow of edge (A) and then transfer the height of the frame sections to this shadow line.

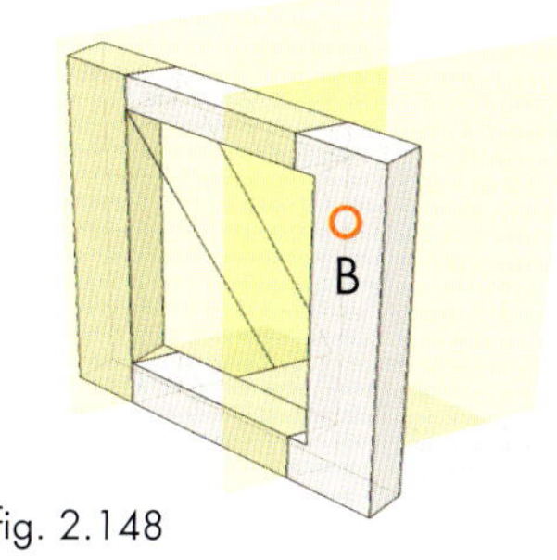

fig. 2.148

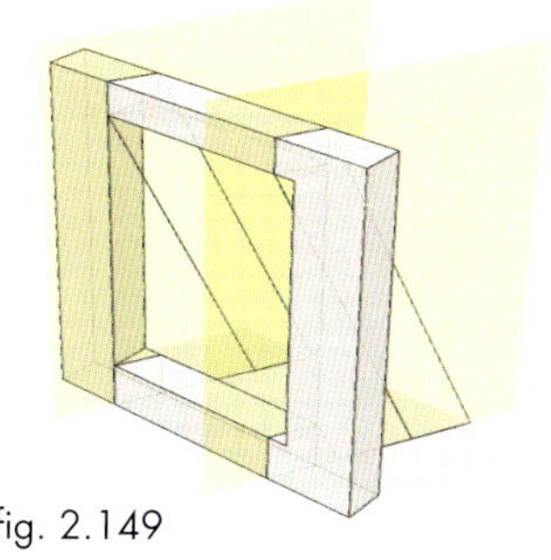

fig. 2.149

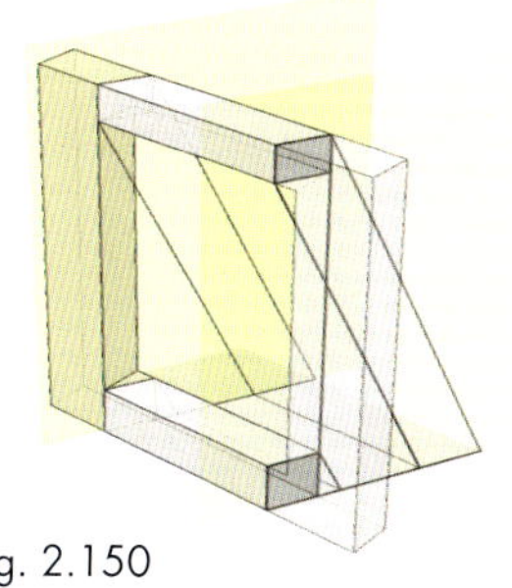

fig. 2.150

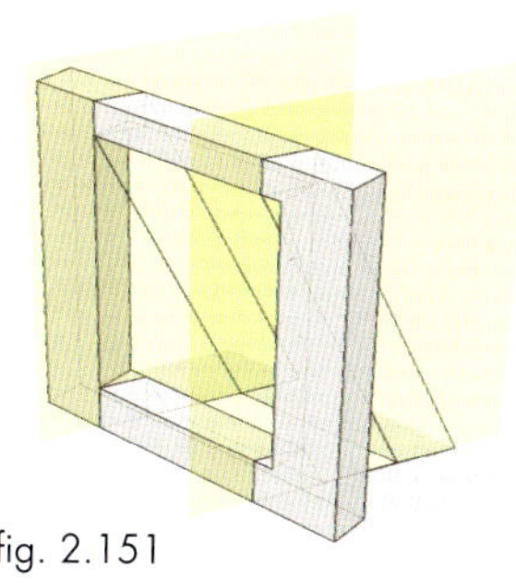

fig. 2.151

STEP 4, fig. 2.148: Place an additional Light Plane that cuts through the hidden back edge (B) and construct the section line shadow the same way as the one for edge (A).

STEP 5, figs. 2.149 and 2.150: Cast the shadow of the frame section profiles onto the ground.

STEP 6, fig. 2.151: Combine the lines on the ground to find the shadow of the frame creating the light area on the ground.

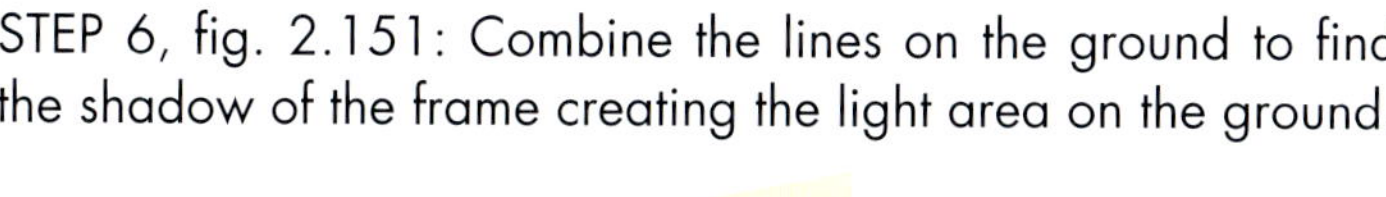

Fig. 2.152: This rotated, reverse view provides an opportunity to study the same construction from a different perspective Look for these types of cast shadows in reality to build a better understanding of this construction.

Below, Charles Liu's example shows his mastery of this type of construction and how interesting the shapes of the cast shadow can become.

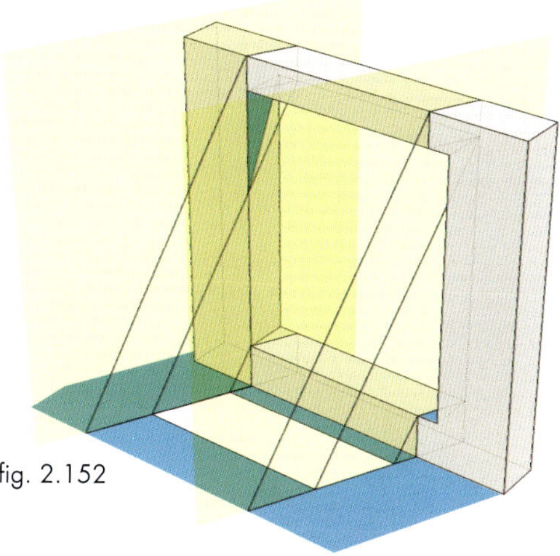

fig. 2.152

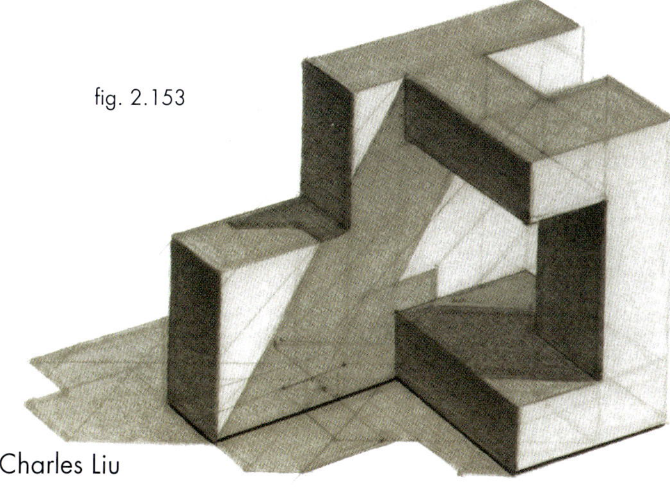

fig. 2.153

Charles Liu

CASTING SHADOWS OF COMPLEX PLANAR VOLUMES

Casting the shadows of complex planar volumes (volumes with flat surfaces) combines all of the prior techniques. Being confident and creative in finding your own solutions becomes essential. Learn to balance the amount of perspective construction with the desired level of accuracy and efficiency. It is totally acceptable to vary the effort within a drawing from full constructions to educated guesses. However, only add cast shadows that enhance a design. Adding wrong or poorly planned shadows will only serve to confuse the viewer.

These examples by Charles Liu show how the casting of shadows can add a lot of dimension and realism to a volume.

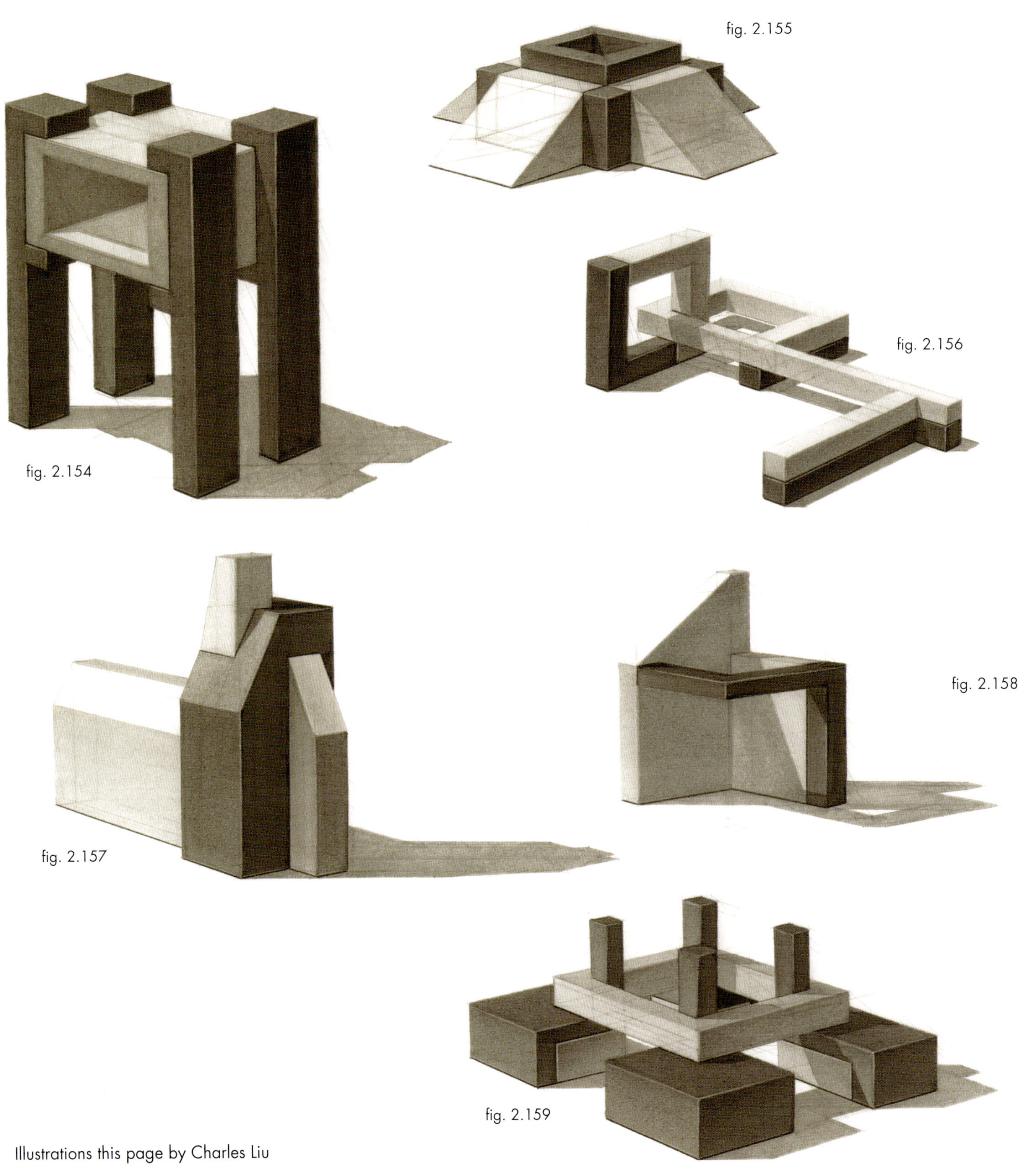

fig. 2.155

fig. 2.154

fig. 2.156

fig. 2.157

fig. 2.158

fig. 2.159

Illustrations this page by Charles Liu

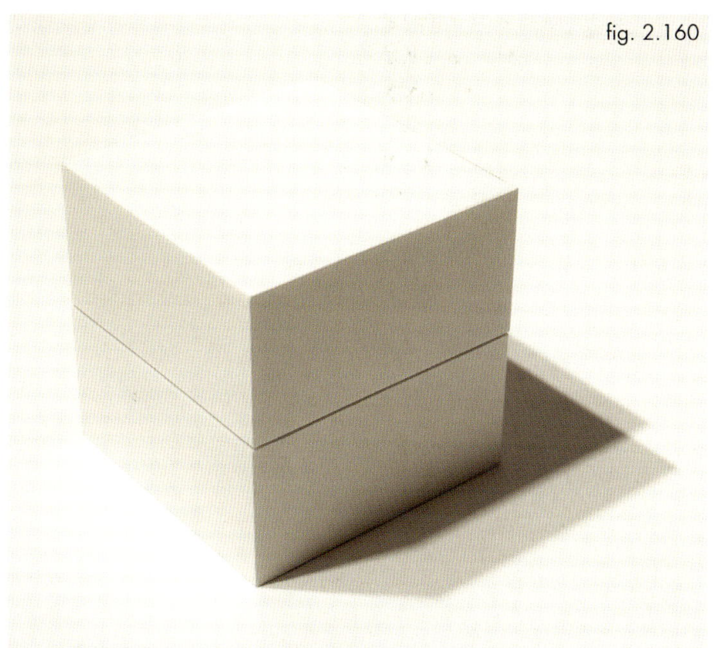
fig. 2.160

Multiple shadows are created when multiple light sources illuminate an object. Casting the shadows in these scenarios requires the construction of multiple cast shadows overlapping to create a final composite outcome. This resulting composite shadow has areas that are not affected by any light source and others that are affected by one or more light sources.

In addition, the shadows assume different values due to the amount of ambient or reflected light they receive. Further details on determining the values of cast shadows and object surfaces will be covered in the next chapter.

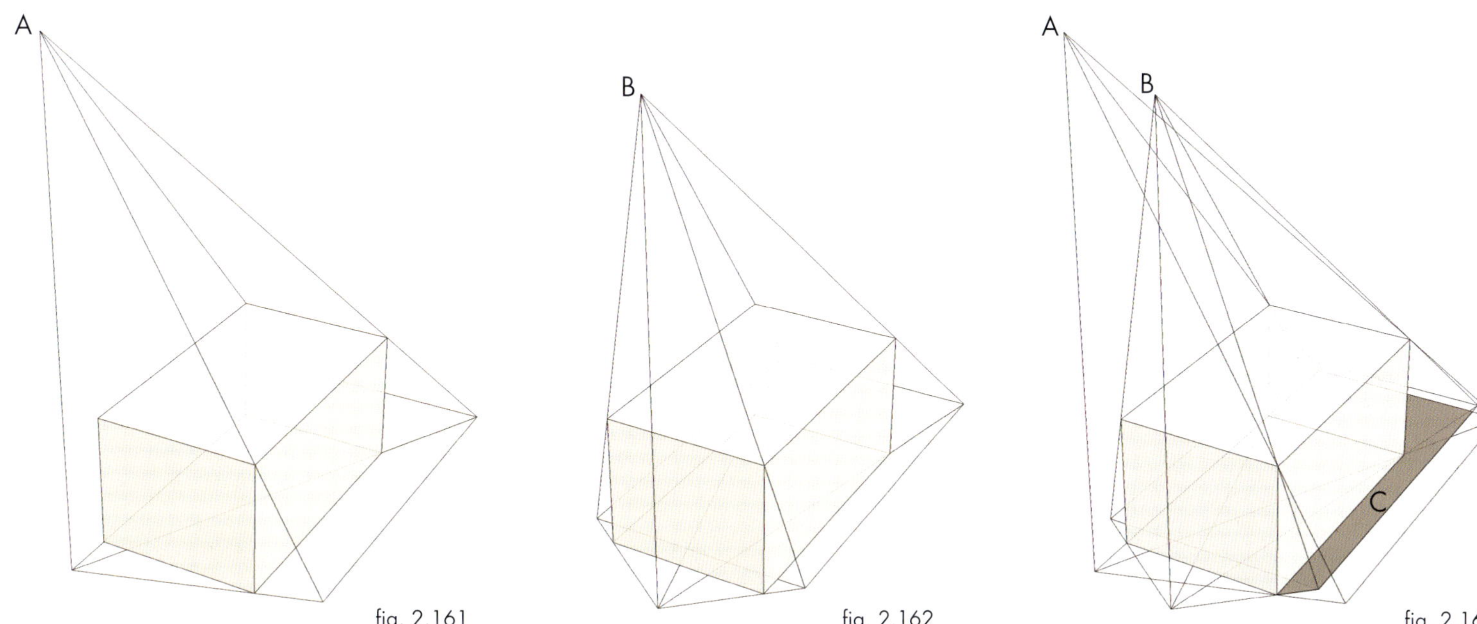

fig. 2.161 fig. 2.162 fig. 2.163

In this example there are two local light sources, A and B. To create the cast shadow of the box, construct the cast shadows from one light source at a time.

STEP 1, fig. 2.161: Cast the shadow of the box from local light source A.

STEP 2, fig. 2.162: Cast the shadow of the box from local light source B.

STEP 3, fig. 2.163: Combine both constructions, either by tracing over one construction, or by creating both constructions on the same page. Note the shadow area that is common for both light sources (C). This area is the darkest since no additional direct light is shining into this shadow. The other shadow areas are lighter depending on the strength of each of the direct light sources.

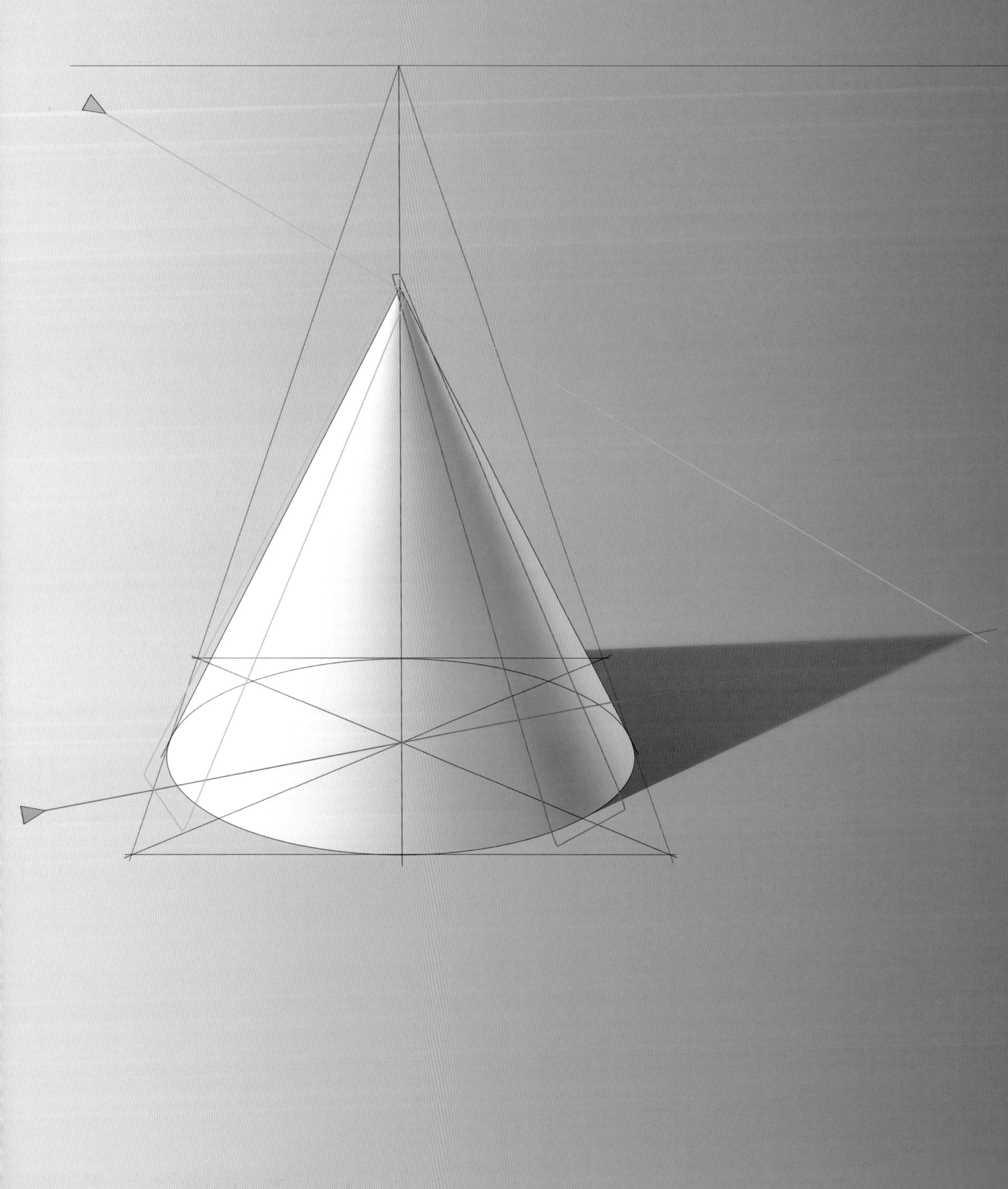

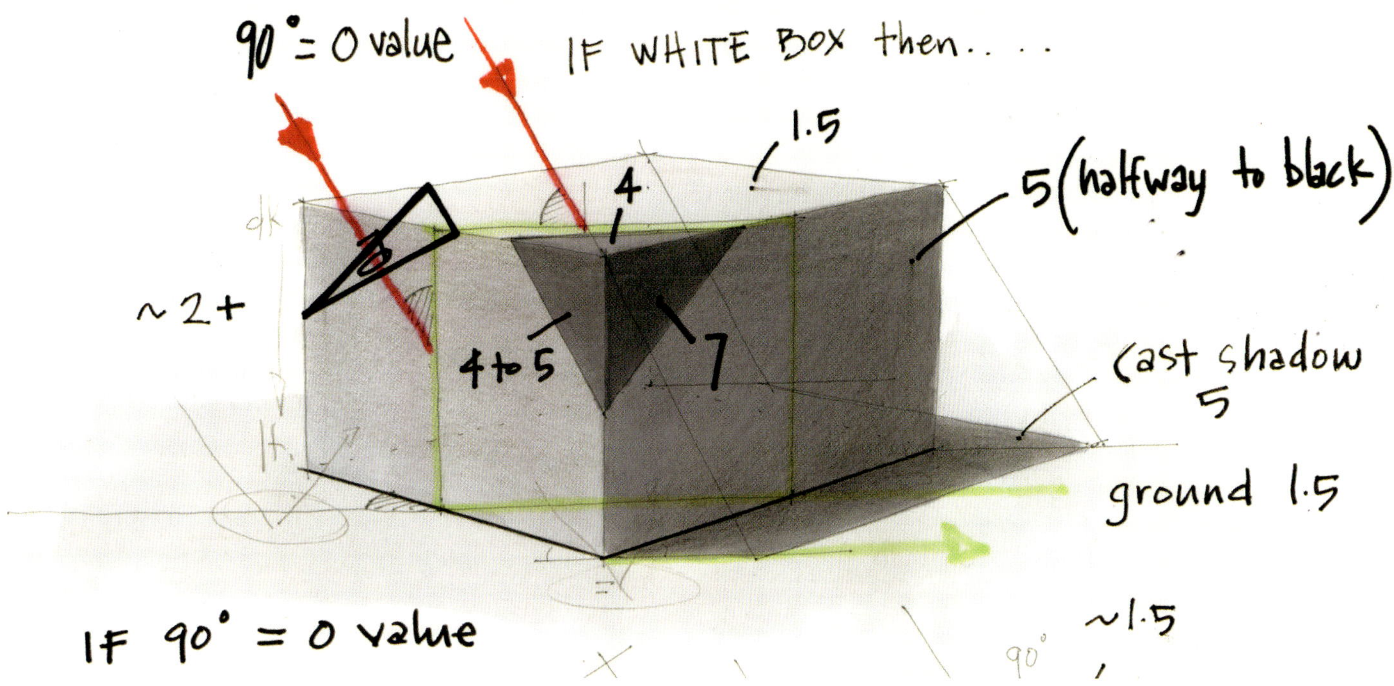

CHAPTER 03
RENDERING THE GEO FORMS

Being able to predict and draw correct cast shadows is the first step in creating realistic-looking renderings. Assigning the correct values to the surfaces of a volume is another essential skill to master and is the subject of this chapter.

The human brain interprets three-dimensional form in gray-scale values. Value, in art, is generally defined as the relative lightness or darkness of a color. However, color is actually not important for the brain to understand shape and form; we understand the three-dimensional volumes of the world around us through variants in light and darkness. So it is best to think of value as the *many shades of gray between black and white.* When drawing or rendering, especially from the imagination, it is crucial to be able to communicate your design using value. In this chapter we will assign values to three-dimensional (3D) geometric forms as a basis to create more complex volumes later. 3D Geo forms provide a good starting point to observe and study basic lighting scenarios so that more complex arrangements and form combinations begin to be understood.

When applying value, the goal is to make the form of an object clearly understandable to others. The tools for assigning values represent a jumping-off point; values can continue to be adjusted while a rendering is being created. Bend the rules, but don't break them. Breaking the rules will make your object unrealistic and difficult to understand for the viewer.

CREATING AN IDEAL LIGHTING SET-UP 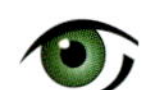

1-2-3 Read

The "1-2-3 read" of an object is a basic concept to be kept in mind at all times when choosing the best possible light direction to explain the form of an object with value. When rendering with value, no outlines are used to separate the surface changes of an object, all surfaces are defined by their differences in value. Having three clearly different values on an object will make it read well. In real life, objects are not surrounded by outlines, our brains discern form through light, medium and dark values. The strong shift in value between the three visible sides of an object is referred to as an object's 1-2-3 read.

Fig. 3.1

Number 1 surface is the lightest surface and is usually the top surface. A light source could be situated very low to create a Number 1 surface on the side, but this is less common as most environments we occupy have the light sources above our heads.

Number 2 surface is the mid-value surface that is still in direct light, but ideally dark enough that it clearly differentiates the edge/surface change compared to the Number 1 surface.

Number 3 surface also called the "shadow side," is the darkest surface. It should also clearly differentiate the edge/surface change between itself and the Number 2 surface.

A strong 1-2-3 read such as this ensures that the viewer will naturally read all surfaces clearly and the volume will appear three-dimensional.

The 1-2-3 read applies to organic shapes as well.

fig. 3.1

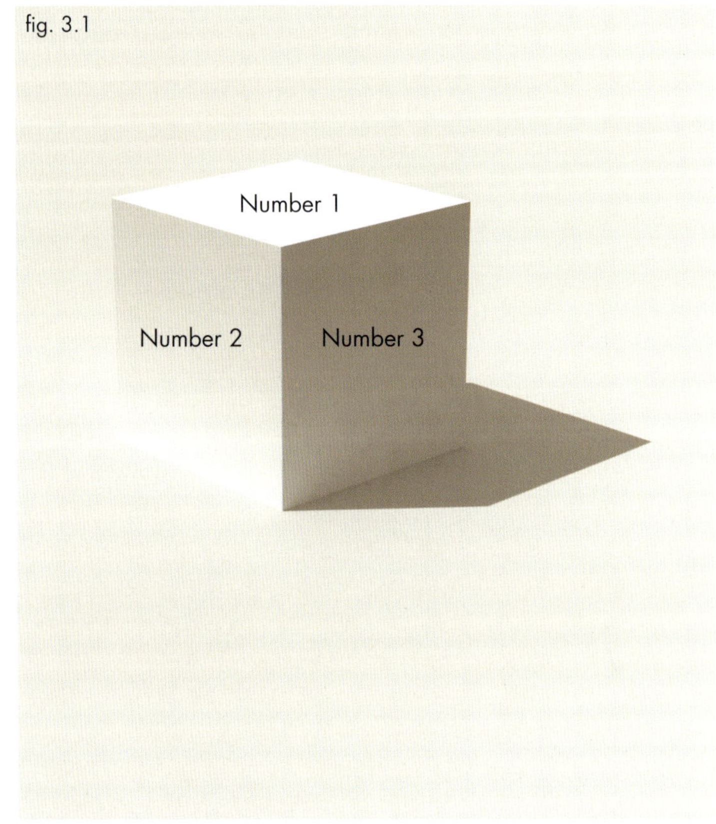

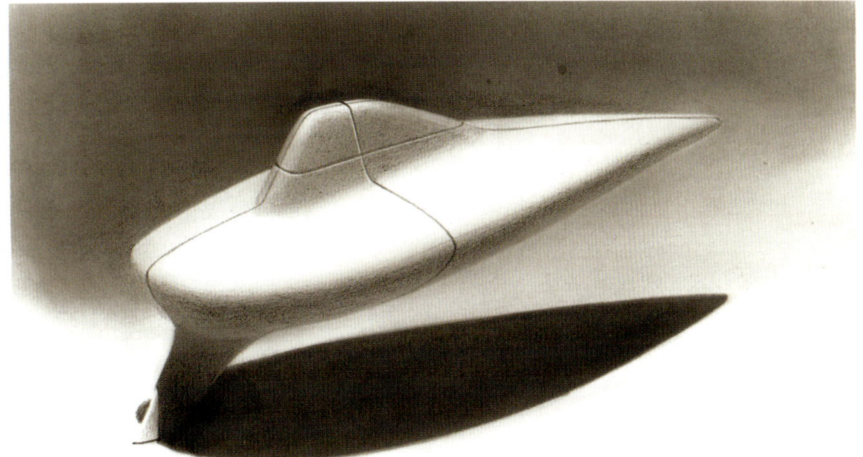

fig. 3.2

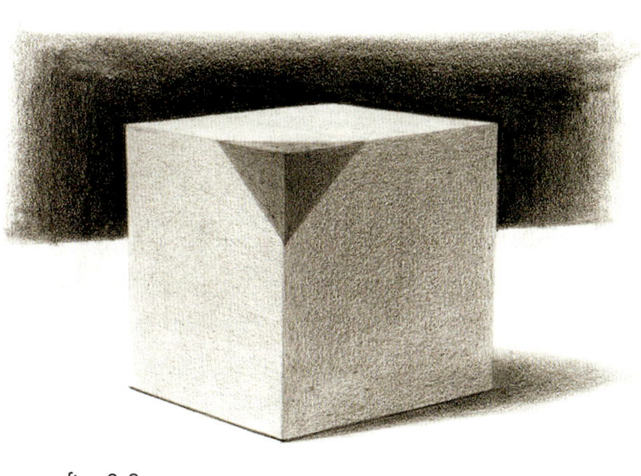

fig. 3.3

Figs. 3.2 and 3.3 by Joon Ahn
To see Joon's current work go to: http://cargocollective.com/joonahn

Having a lighting strategy that keeps the object's 1-2-3 read in mind is vital to rendering an understandable form. Of course there are exceptions that call for keeping a form less defined, such as having dramatic backlighting in which only the silhouette of the volume is seen. These situations have a good place in creating story but are less common. For now, the main goal is to render forms clearly and understandably.

Fig. 3.4: The backlit cube is missing the "Number 2 read" so only a "1-3-3 read" is achieved. The two "Number 3 read" surfaces have very little value separation.

Fig. 3.5: The top-lit cube has a similar problem and is also missing a Number 2 Read side, resulting in a weak 1-3-3 read.

Fig. 3.6: The front-lit cube has a weak 1-2-3 read. The light angle creates very evenly lit surfaces on all sides.

Fig. 3.7: Placing the light high and toward the left generates a nice cast shadow and separates the values of all three visible surfaces well. This creates a good 1-2-3 read.

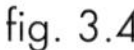

fig. 3.4

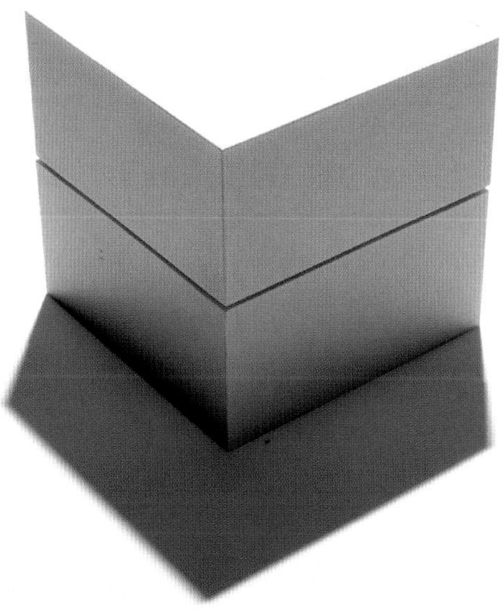

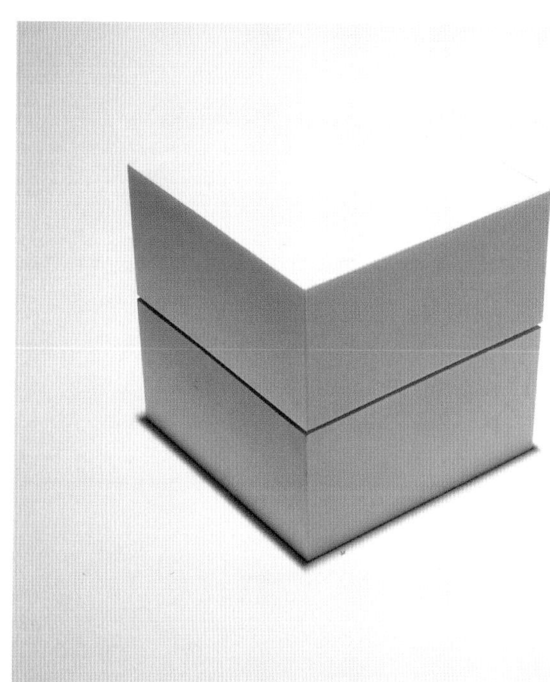

fig. 3.5

fig. 3.6

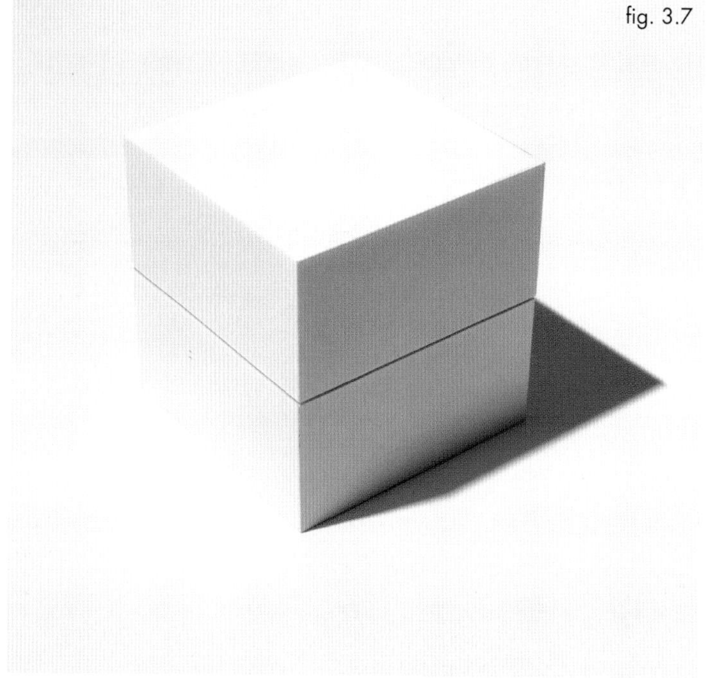

fig. 3.7

ASSIGNING VALUES FOR MATTE SURFACE OBJECTS 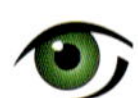

The basis for assigning values – the different shades of gray between white and black – is established by first considering sunlight around late morning or early afternoon. This is the best lighting condition for showing off an object's form. The sun is almost overhead, which creates not only short shadows but strong value changes for each side of an object. This scenario is the one in which we will begin to judge and assign the value changes of an object's surfaces.

Values are assigned on a scale from zero (white) to ten (black). Calculating values at ten percent increments is a precise enough way to start. Later, small adjustments to the values can be made depending on the surrounding environment. Don't get bogged down in numbers; use your eyes and your observational experience.

Shadow Value Observation: Halfway to Black

The "halfway to black" observation provides the method to determine the shadow value of matte objects. First, establish the "true value" of the object. This is the value of the physical color of the object, regardless of lighting conditions. To determine the value of the shadow side of an object, subtract its true value from ten (black) and divide it in half. This resulting value is the base value of the shadow side. Here are three examples:

Fig. 3.8: This cube is painted white, so its true value is zero. A value of 5 is the point halfway to black. Thus, the value of the shadow side should be around a value of 5, as is the cast shadow on the white ground.

Fig. 3.9: This cube's true value is 5, so now the value that is halfway to black is 7.5. Observe that the shadow on the ground still keeps its halfway-to-black point at value 5 since the true value of the ground is white (value 0).

Fig. 3.10: The true value of this cube is 8, making halfway-to-black value 9. Again, the shadow maintains its original halfway-to-black value from white and is a value 5.

"Value range" describes which values, from lightest to darkest, are involved with rendering an object. To render a white matte box, as in the first example, the value range spans from 0 to 5 (shown as an orange line). The darkest cube's value range spans only one step, 8 to 9, but the white cube's spans five steps. Conveying the form of darker objects requires the ability to render very small differences in value. Beginning with a lighter true value makes it easier to communicate the form of the object, since there are more values to work with.

fig. 3.8

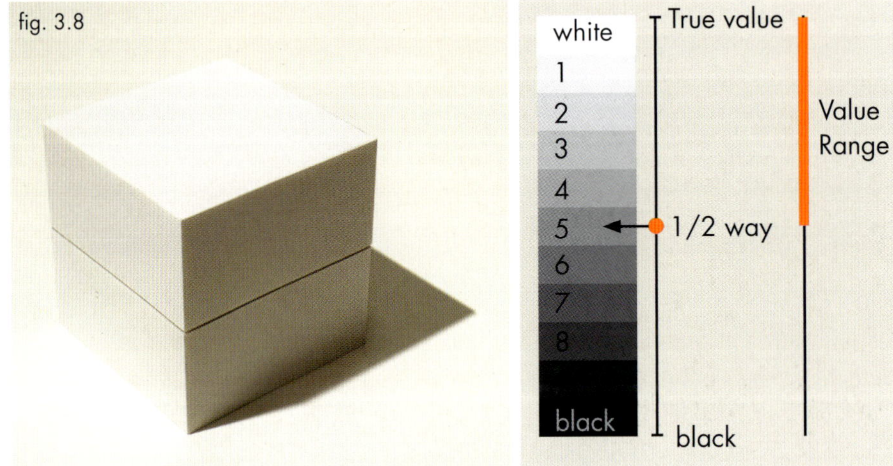

fig. 3.9

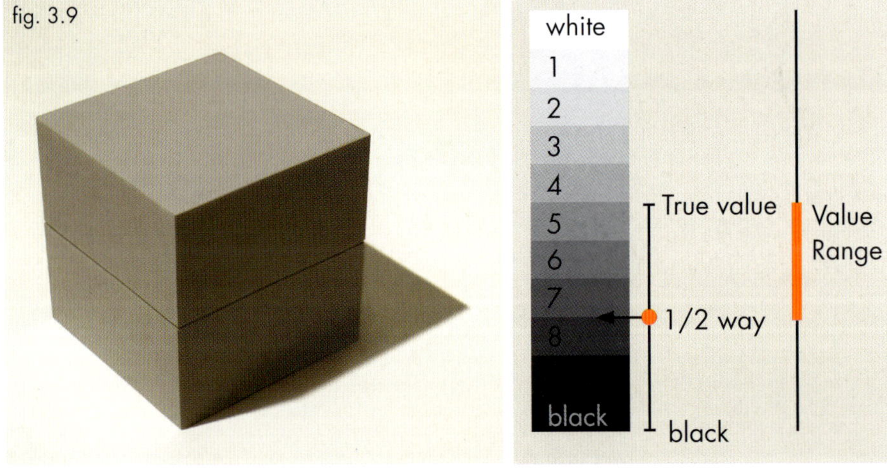

fig. 3.10

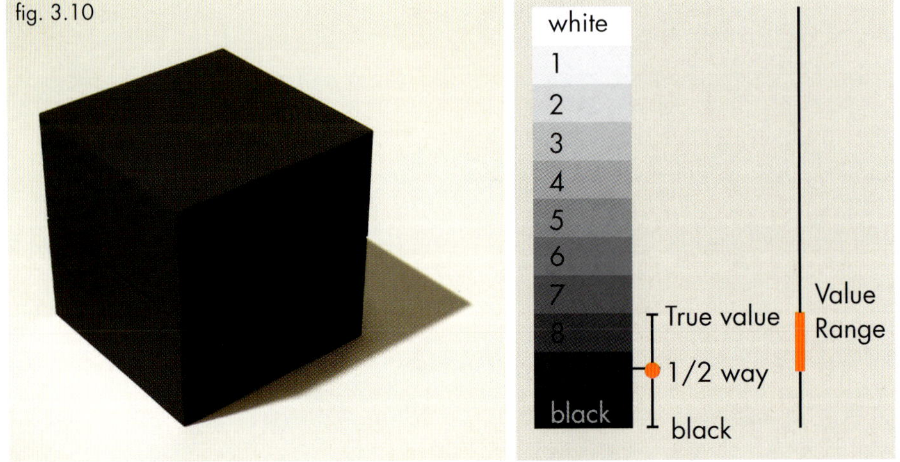

Object Value and Cast Shadow Value

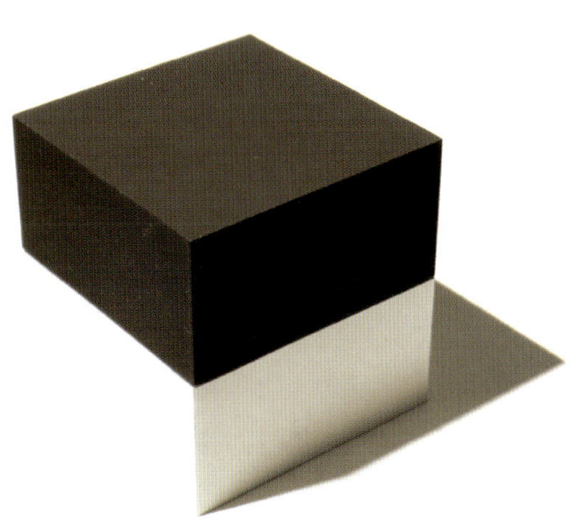
fig. 3.11

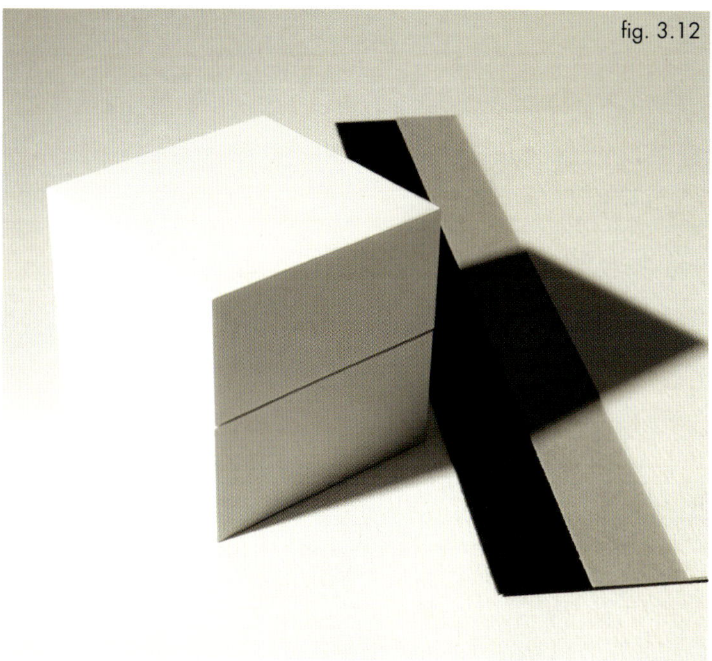
fig. 3.12

The value of a cast shadow follows the same halfway-to-black guideline. The true value to consider is the value of the surface upon which the cast shadow is landing.

Fig. 3.11: Even though the cube itself is black and white, the cast shadow on the ground appears no different than the examples from the facing page since the true value of the ground is uniform and white.

Fig. 3.12: Now the ground plane has three different values. Observe how the cast shadow values shift halfway to black on each of the stripes according to the stripe's individual true value.

Ambient Light and Shadow Value

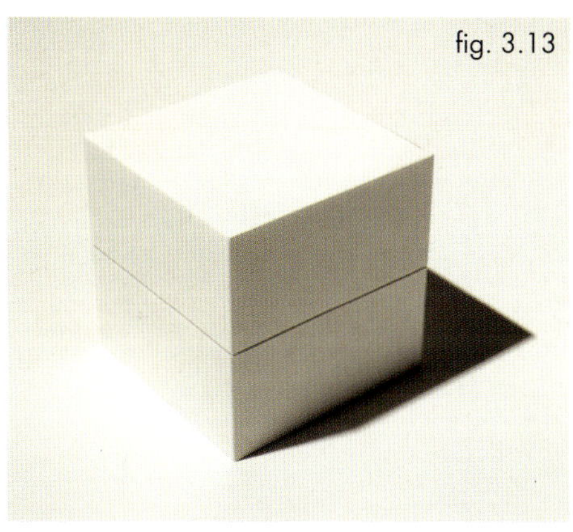
fig. 3.13

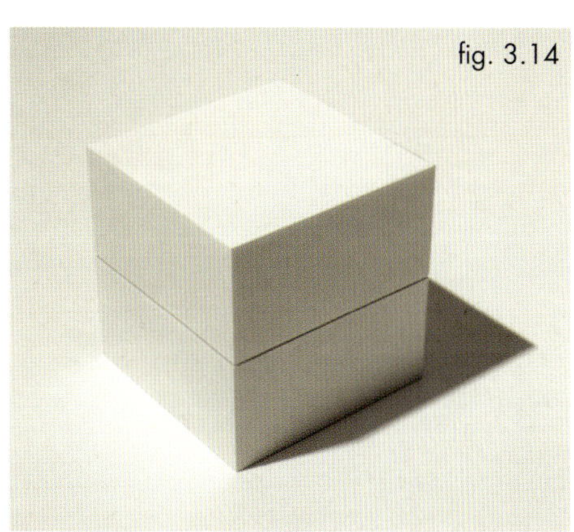
fig. 3.14

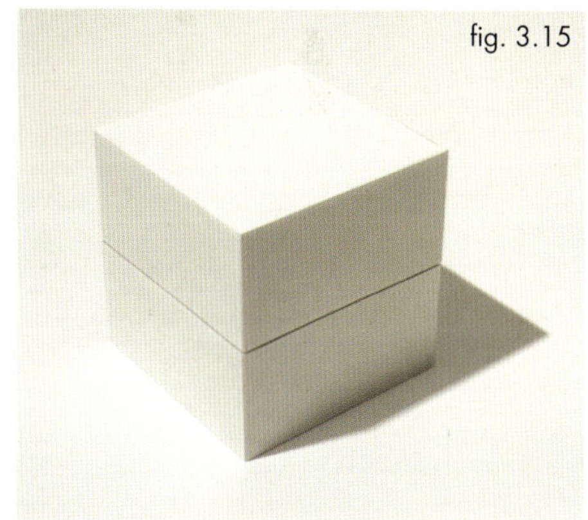
fig. 3.15

The lighting scenario for the halfway-to-black observation assumes midday sunlight and ambient light from a clear, blue sky. However, conditions in local light can be quite varied, since the strength of the ambient light can be controlled.

Fig. 3.13: Reducing the strength of the ambient light results in a darker shadow.

Fig. 3.14: Matching the strength of the ambient light to daylight makes the halfway-to-black observation occur again.

Fig. 3.15: Increasing the strength of the ambient light lightens the shadow well short of the halfway-to-black guide.

These changes need to be taken into consideration when working with reference images. These conditions are neither right nor wrong; they are just following different lighting scenarios. These changes to a lighting scenario need to be identified and consistently followed during the rendering process to create a coordinated, realistic image.

VALUE ASSIGNMENTS FOR OTHER SIDES 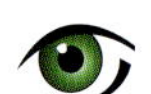

To recap, applying the halfway-to-black strategy establishes the value range for the shaded surfaces that are not in direct sunlight. A true value white cube has a shadow-side value of 5. The value-range for the other surfaces will fall between 0 and 5.

Determining the values for these other lit surfaces requires an assessment of the angle at which the light rays strike the surfaces. This angle of incidence dictates the value of the surfaces. A surface that is perpendicular to the direction of the light rays appears closer to its true value, while a surface with a more tangential orientation to the light rays appears closer to its shade value, halfway to black.

Observing and estimating the angles at which light rays strike a surface is the most efficient way of assigning values. It is too much effort to calculate all of the true angles of incidence in perspective. The best application of this knowledge is to use it as a conceptual tool when observing and rendering.

Top Surface, Number 1 Side

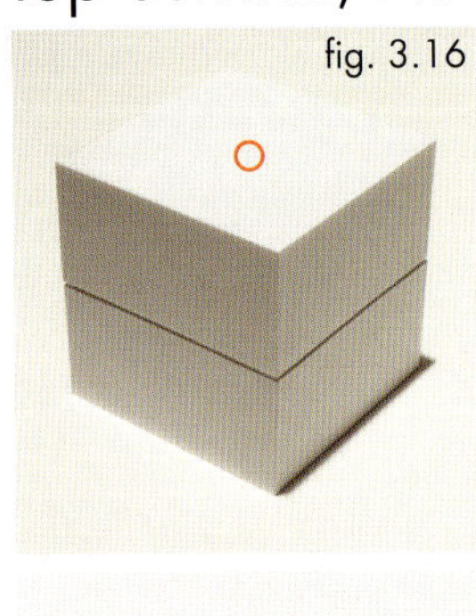
fig. 3.16

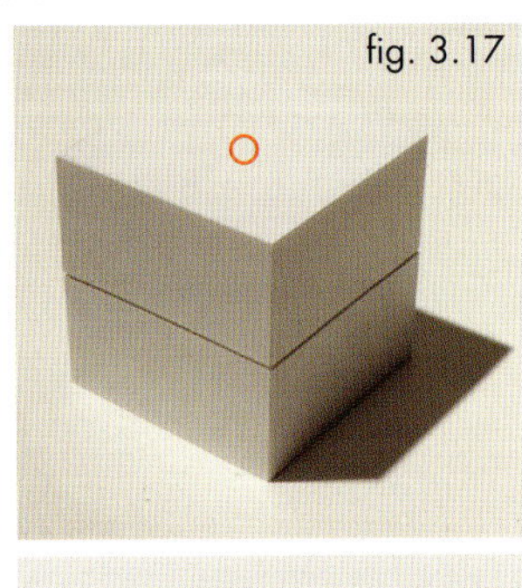
fig. 3.17

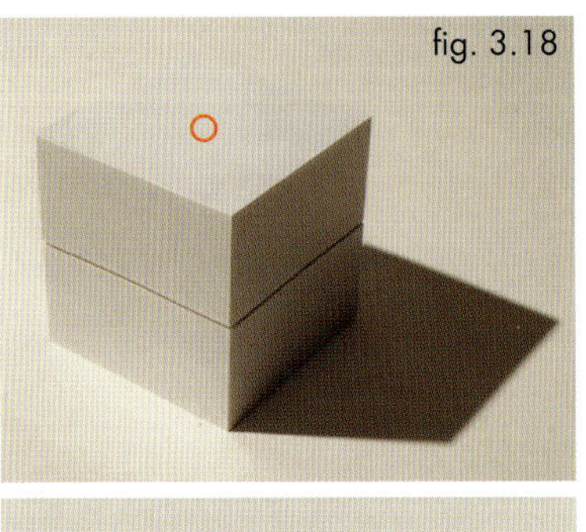
fig. 3.18

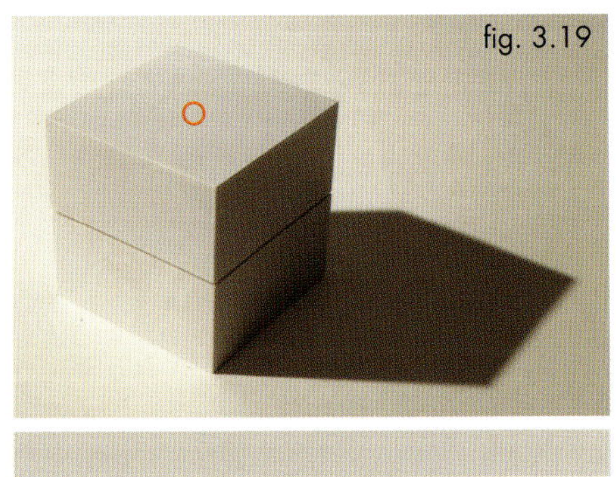
fig. 3.19

Focus on the top surface value in this set of cubes. The value swatch at the base of each image is a value sample taken from the area in the orange circle. The top surface value changes as the light moves lower, which is indicated by the lengthening cast shadow.

Top light creates the brightest top surface value (fig. 3.16), while a low sidelight position creates a darker value (fig. 3.19).

The value of the surface is dependent on the Light Ray Angle. When rendering a cube with sidelight, the top surface is never pure white but rather a light shade of gray.

Observe the ground plane. As the light moves, its value changes. A common error is to overlook the ground plane and leave the paper white.

Front Surface, Number 2 Side

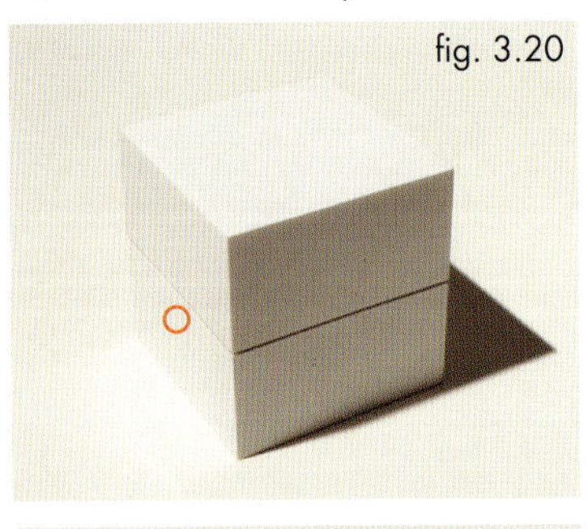
fig. 3.20

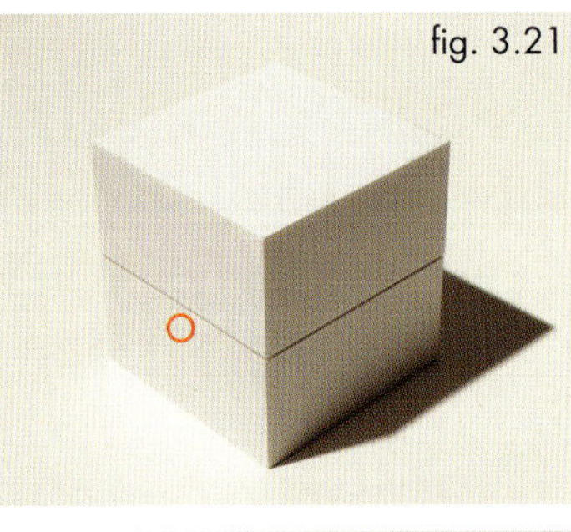
fig. 3.21

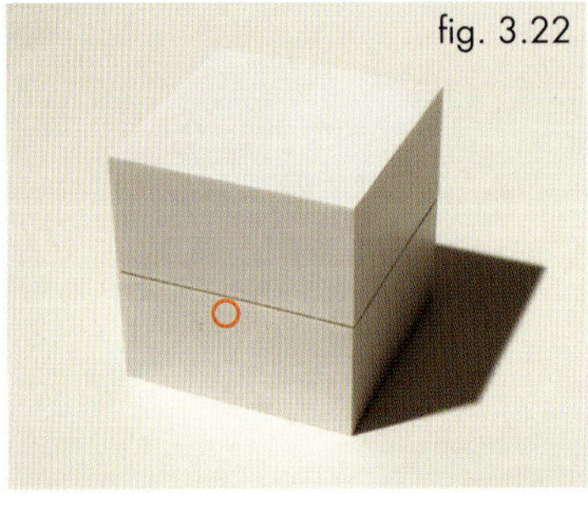
fig. 3.22

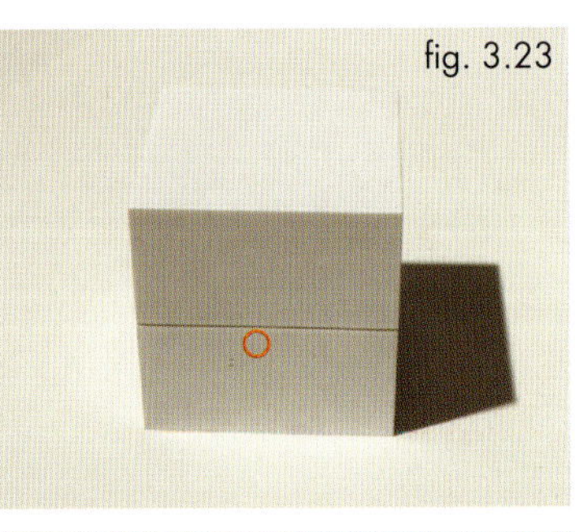
fig. 3.23

The Number 2 surface value in a 1-2-3 read of an object is also determined by the Light Ray Angle. In this example the light source remains in a static position as the cube turns. The orange circle indicates from where the value sample swatch below each image was picked. Keep in mind this Number 2 surface has a gradation on it from the reflected light and the value swatch represents an average for the side.

When the Number 2 surface turns away from the light source (less perpendicular) the value decreases as in figs. 3.22 and 3.23.

Observe that the Number 1 surface value stays static in every image, while the Number 2 surface value changes. Make sure to balance the value of the Number 2 surface to separate it well from the Number 1 and Number 3 surfaces to achieve a good 1-2-3 read.

Always invest the time to plan the value assignments before rendering. The planning can be done by estimating to the best of your drawing ability the Light Ray Angles striking the surfaces, assigning written values to surfaces and creating orthographic views to enhance the understanding of the lighting scenario.

The sketches on this page are study notes in preparation for a final rendering. With practice, these written notes will eventually be replaced by internal thought processes. The development of a more critical eye enhances the ability to judge finished renderings more accurately, whether done manually, or created by 3D computer-rendering programs.

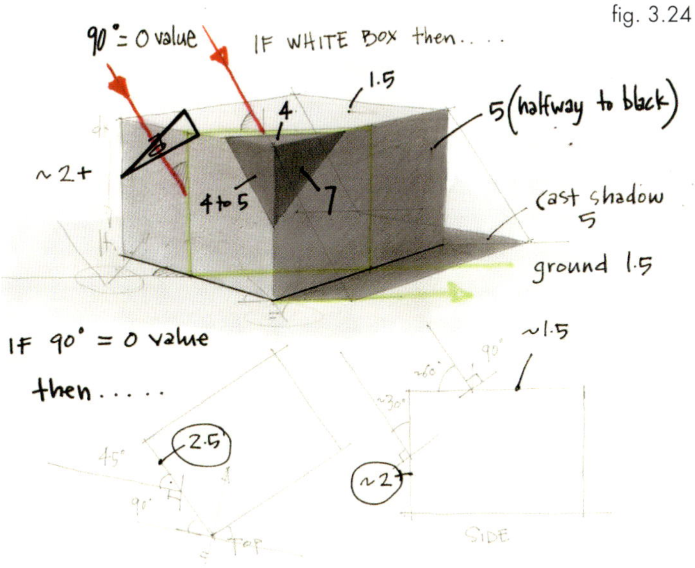

fig. 3.24

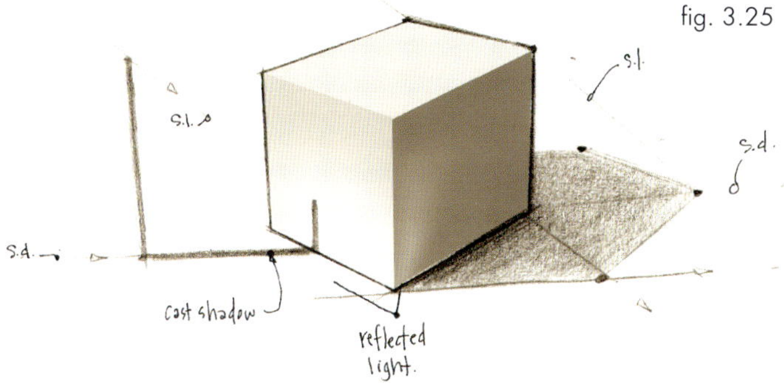

fig. 3.25

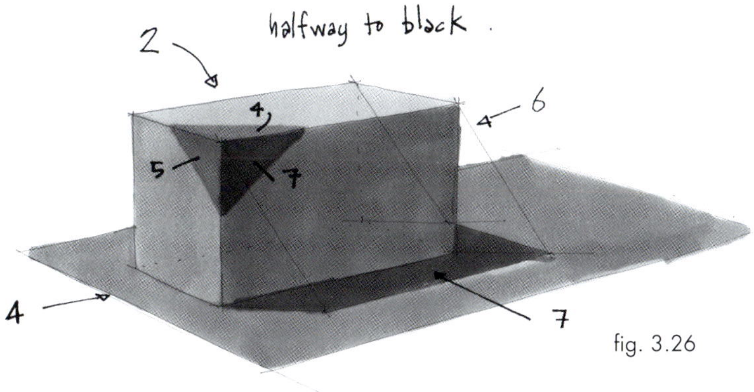

fig. 3.26

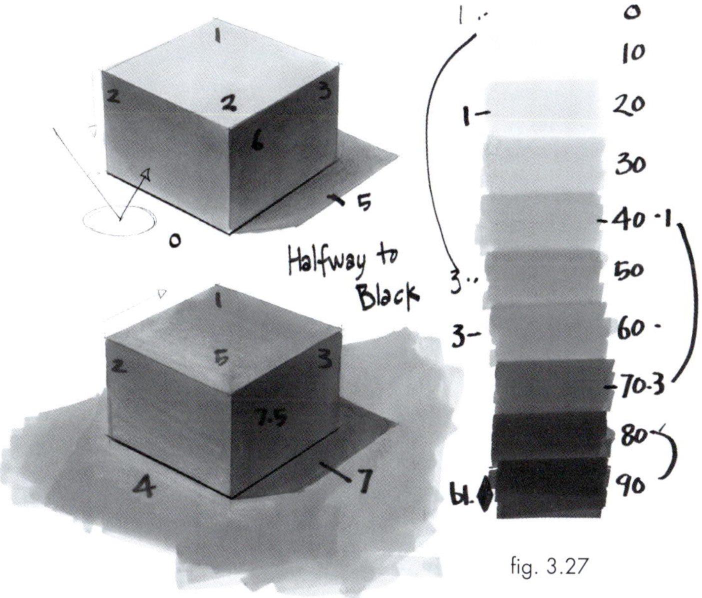

fig. 3.27

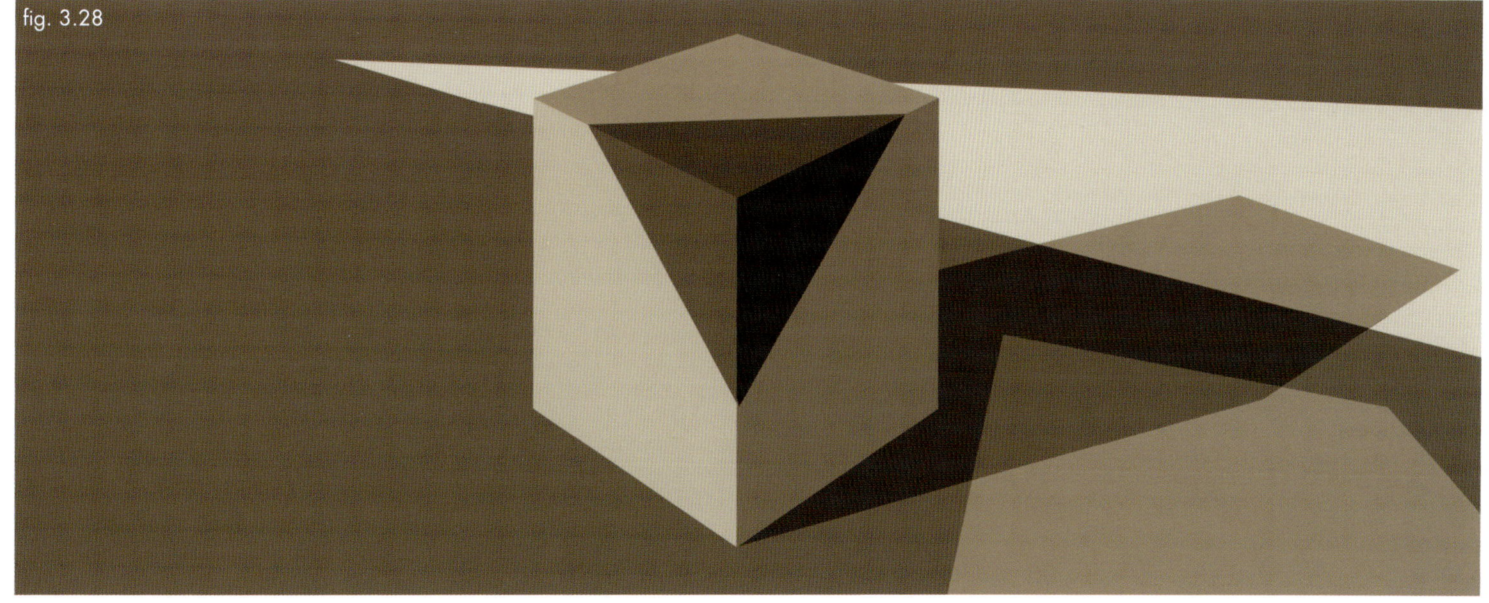
fig. 3.28

REFLECTED LIGHT

Reflected light helps bring renderings to life. Just rendering planar surfaces with flat values looks unnatural. The inclusion of reflected light remedies this by creating subtle gradations. Reflected light strikes one surface and then bounces to another area, lighting it more softly. Most materials reflect light in varying degrees. Reflected light is fundamental in creating realistic renderings because it ties objects together. It not only indicates the existence of other elements, but how they affect one another.

Reflected Light off the Ground

In these examples, as the value of the ground changes, observe how the reflected light from the ground to the Number 2 surface of the cube becomes more or less visible.

Figs. 3.30, 3.31 and 3.32 show how the Number 2 surface darkens when a darker surface is placed in front it. This is because the darker a surface, the less light reflects off of it. Light values reflect more light than dark values.

Fig. 3.32 uses the red paper to dramatically illustrate how the Number 2 surface of the cube is influenced by the reflected light, not only affecting its value with a gradation but also its color. The reflected light is red in this case because reflected light takes on the color of the surface it is bouncing off.

Figs. 3.33, 3.34, 3.35 and 3.36 show how the value of the ground plane influences the reflected light on the shadow side of the cube. While there is still some change in the cube's Number 2 surface, there is a much stronger influence on the shadow side. This is because the amount of reflected light on the Number 3 side, while less than on the Number 2, is comparatively stronger visually and effects the overall value of the Number 3 side more, due to the absence of direct light shining in that area.

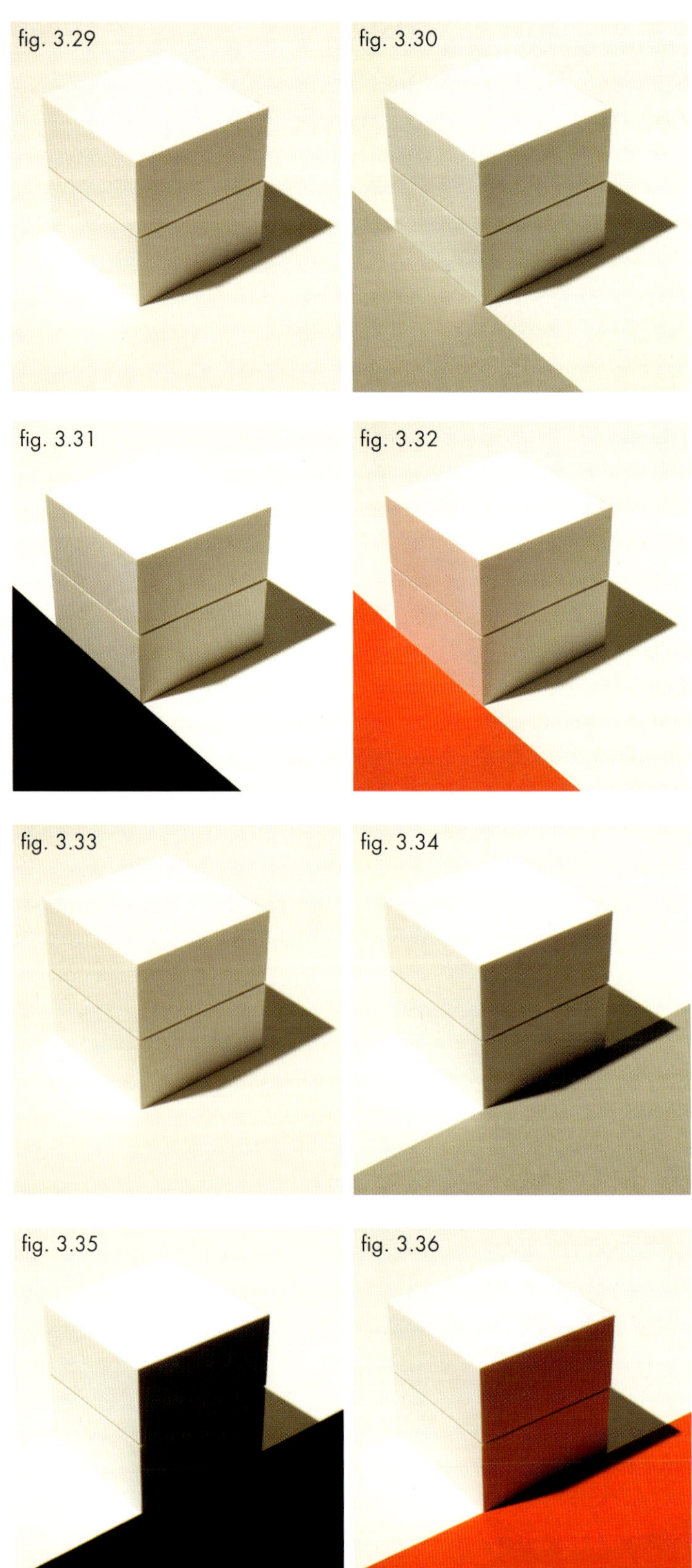

fig. 3.29 fig. 3.30 fig. 3.31 fig. 3.32 fig. 3.33 fig. 3.34 fig. 3.35 fig. 3.36

Reflected Light of Plane Reflectors

Plane reflectors, also called board reflectors or bounce cards, are surfaces used to manipulate the way the primary light source reflects into chosen areas of a scene. These cards are often outside the image frame and therefore not visible.

Figs. 3.37, 3.38, 3.39 and 3.40, when compared to one another, clearly show how the cast shadow and Number 3 side are influenced by the plane reflector. The cast shadow and Number 3 side are lightened or darkened depending on the value of the chosen card.

The red plane reflector in fig. 3.40 shows that both the cast shadow and the shadow side are affected. Think of the plane reflector as a stand-in for walls or other objects that you might render in the same image. Make sure that the fill light affects both the cast shadow and the cube, not just one or the other.

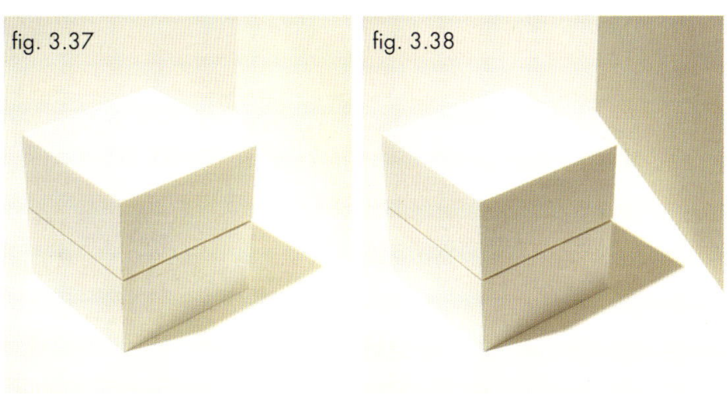
fig. 3.37 fig. 3.38

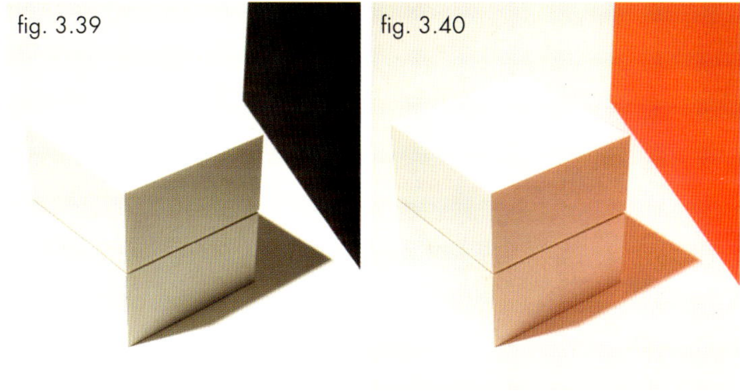
fig. 3.39 fig. 3.40

Reflected Light: Double-Bounce

Reflected light can bounce more than once, illuminating soft light back into the cast shadow as well. In fig. 3.43 the ground bounces light onto the underside of the cube and this underside of the cube returns the bounced light into the cast shadow creating a gradation on the bottom of the cube and in the cast shadow.

Fig. 3.42: The underside of the cube is not visible, but the cast shadow clearly shows the influence of the double-bounce light.

Figs. 3.43 and 3.44 show nice examples of occlusion shadows, where less and less of the neighboring surface receives bounced light as the bottom surface gets closer to touching the table.

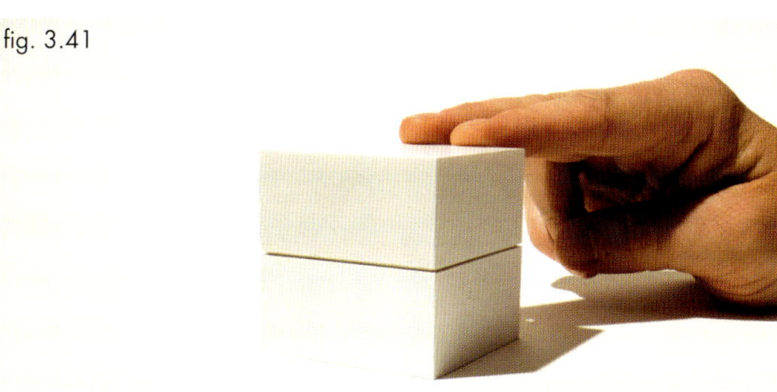
fig. 3.41

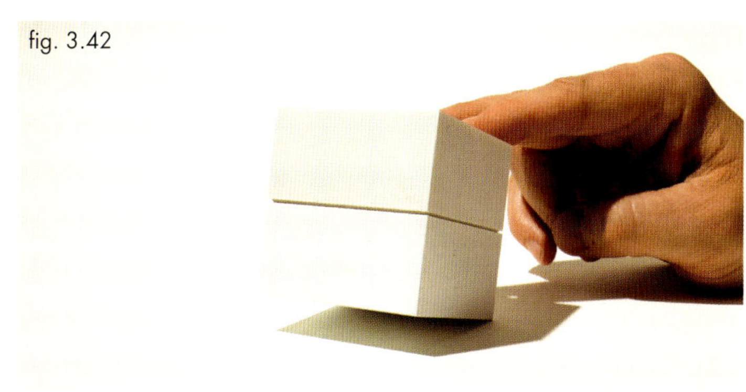
fig. 3.42

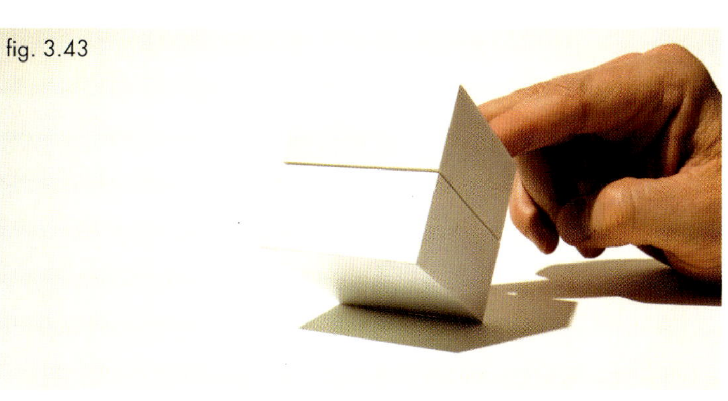
fig. 3.43

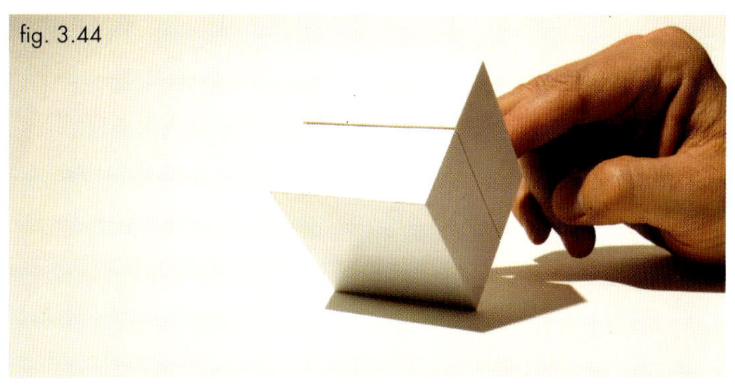
fig. 3.44

LOCAL LIGHT VALUE ASSIGNMENT

In a local light scenario, just as with sunlight, the light rays' angle of incidence is the key factor in assigning values. The main difference is that with local light, the light rays' angle of incidence keeps changing on a flat surface since the light rays are not parallel. If there is only one light source in an otherwise dark environment, reducing the amount of ambient light makes the cast shadows darker than would be observed with ambient light from a bright sky. Adjusting the strength of the ambient light determines the darkness of the cast shadows and the Number 3 side.

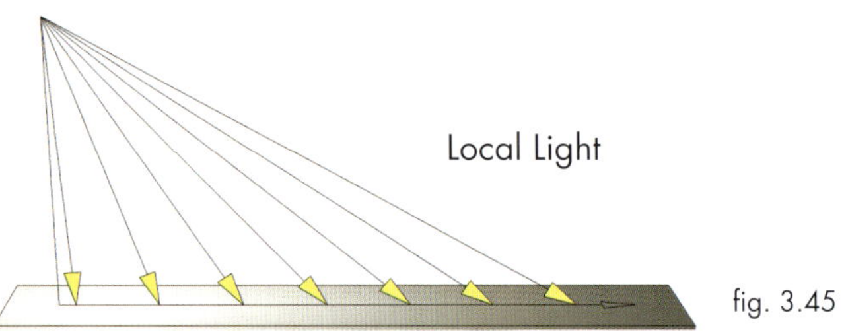

fig. 3.45

Fig. 3.45: With local light, because light rays emanate from a single point, their angles of incidence change along a flat surface. The lightest value is found directly under the light source, and cast shadows and shadow sides can become much darker than halfway to black.

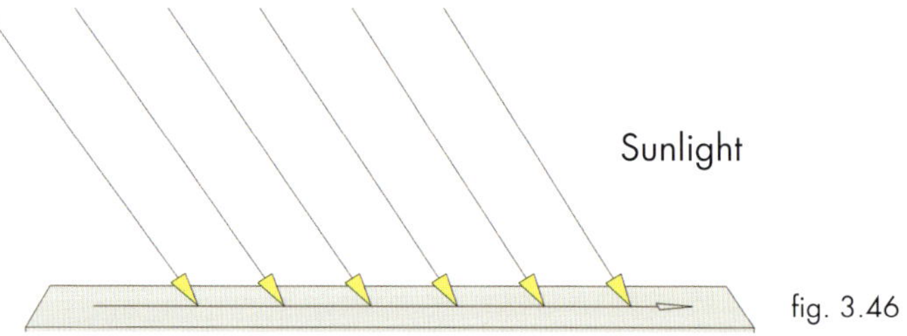

fig. 3.46

Additionally, the strength of the light source diminishes over distance creating a falloff effect, as discussed on page 22.

Fig. 3.46: By comparison, rays of sunlight, being parallel, always have a constant angle of incidence. This creates an even value across a flat surface.

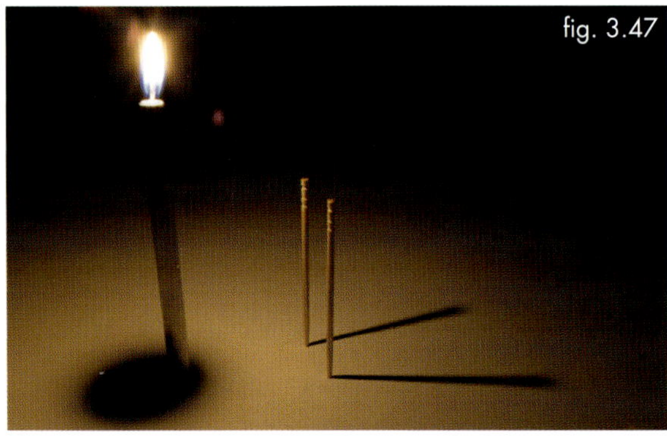

fig. 3.47

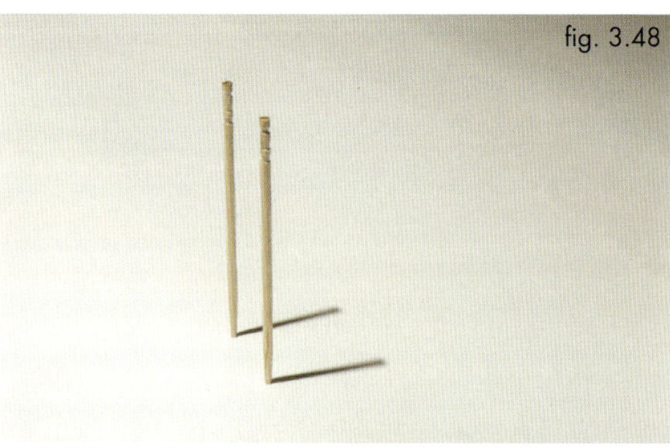

fig. 3.48

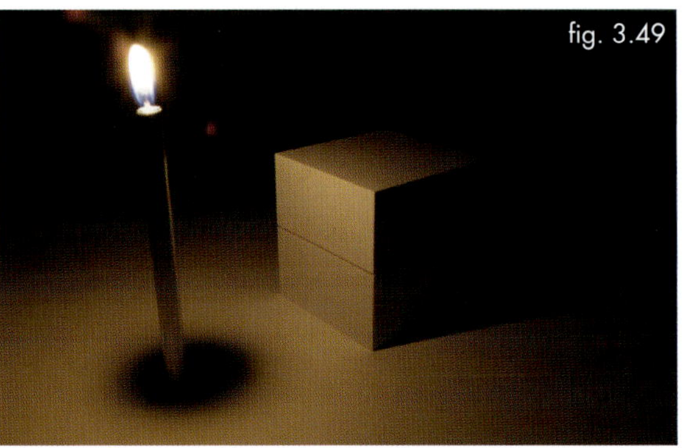

fig. 3.49

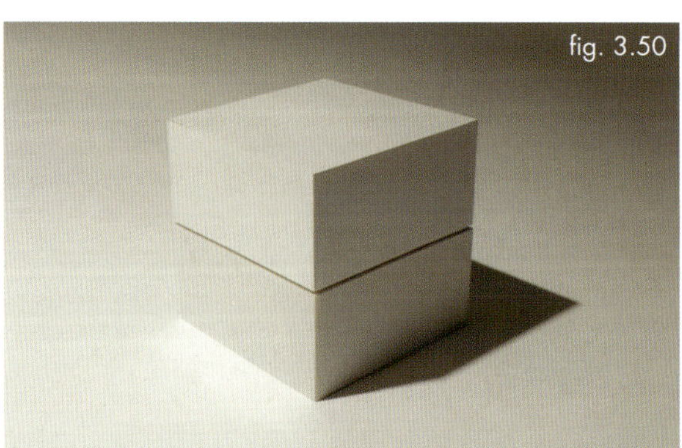

fig. 3.50

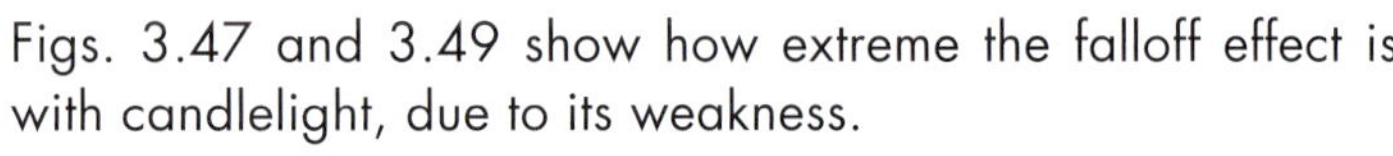

Figs. 3.47 and 3.49 show how extreme the falloff effect is with candlelight, due to its weakness.

Fig. 3.49 shows that falloff occurs not just on the ground, but on the top and side surfaces of the cube. The inclusion of these subtleties can make rendering with local light very attractive and dramatic.

For comparison, figs. 3.48 and 3.50 are sunlit scenes. In fig. 3.48, the ground shows very little to no gradation due to falloff. In fig. 3.50, the surfaces of the cube show gradations due only to the reflected light, not because of the falloff of the primary light source.

Observing real-life scenes such as these gives you the opportunity to make decisions on how best to render these situations. Using one strong light bulb has a different effect than stadium lighting. The main goal is to render the form well. Keeping a good 1-2-3 read will enable this in any lighting scenario.

Local Light Rendering Examples

Rendering local light becomes most convincing when the falloff is included on the surfaces of the objects, across the ground plane, and in the background. Reducing the amount of ambient light will darken the shadows and increase the 1-2-3 read.

Combining all of these elements makes a local-light rendering instantly understandable for the viewer.

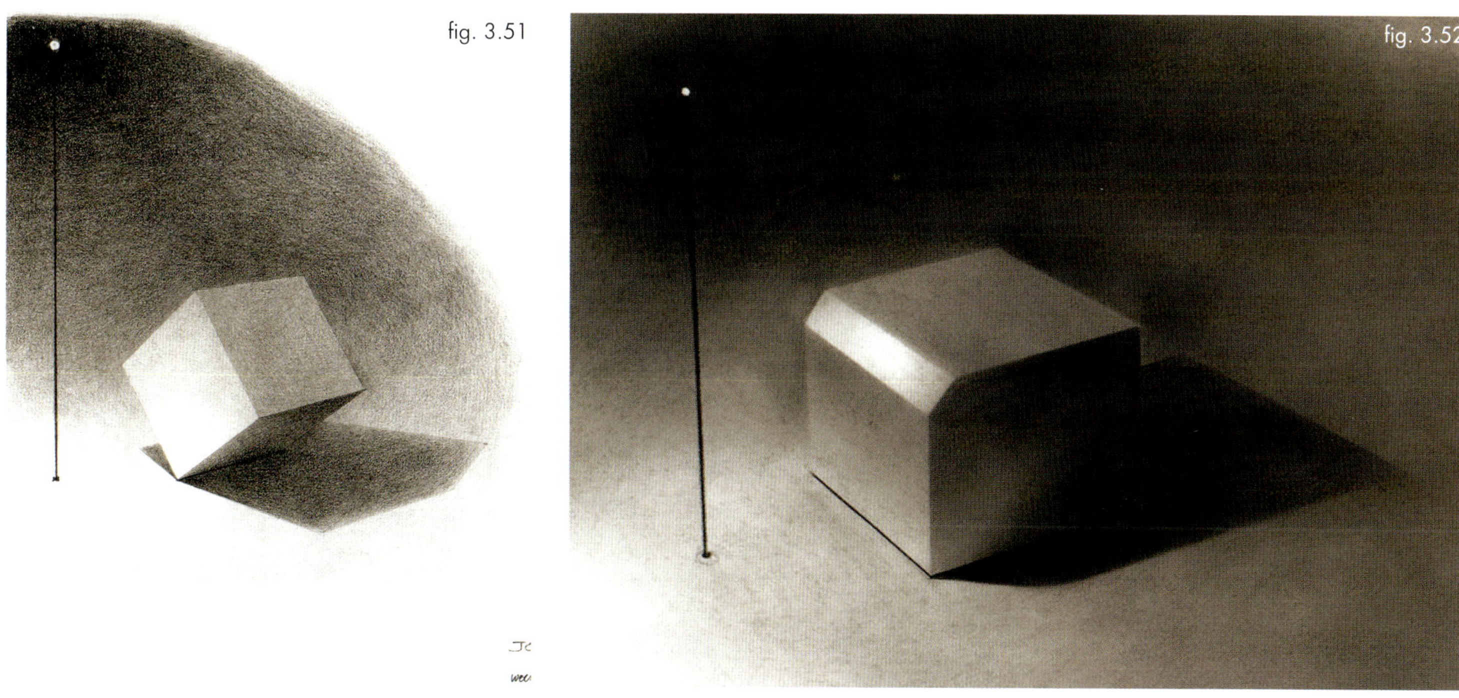

fig. 3.51

fig. 3.52

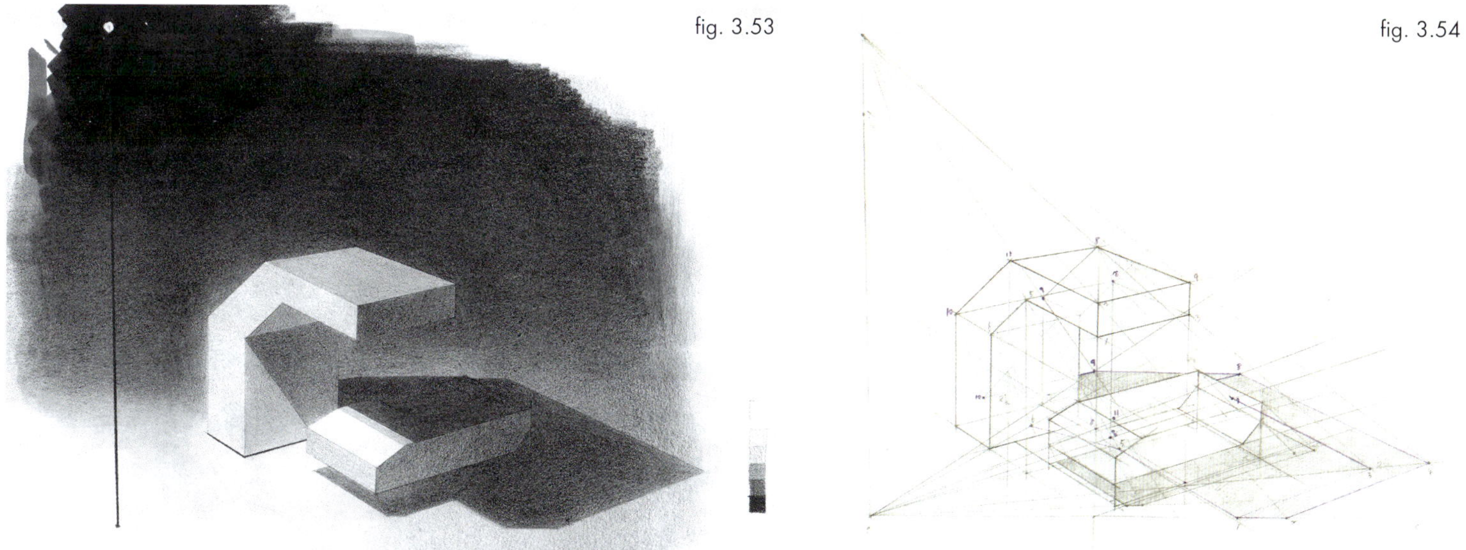

fig. 3.53

fig. 3.54

All drawings on this page by Joon Ahn

SLOPED SURFACE VALUE ASSIGNMENT

Taking the light ray's angle of incidence into account is even more important when planning the value assignments for sloped surfaces. After finding base values, making visual judgment calls is a good way to refine the values, since these surfaces can be very complex.

The goal is to differentiate surfaces against one another using changes in value. It is imperative to understand that a value change will read as a surface change, even when these changes are very small.

Planning and Approach to Value Assignments of Sloped Surfaces

STEP 1: Place the light where it creates the major value assignments for a strong 1-2-3 read on the surface that is most perpendicular to the light rays' direction (A) and on the vertical walls (C).

STEP 2: Plan and assign values to the sloped surfaces (B) based on the light rays' changing angles of incidence. Keep in mind that if the forms change, their values must change. Remember, our brains register a value change as a surface change!

STEP 3: Start to render by filling in the values. Don't forget the background and ground plane!

STEP 4: With all the values blocked in, adjust them with local contrast in mind. At this point, these are relatively small value gradations, so they require extra care while rendering.

fig. 3.55

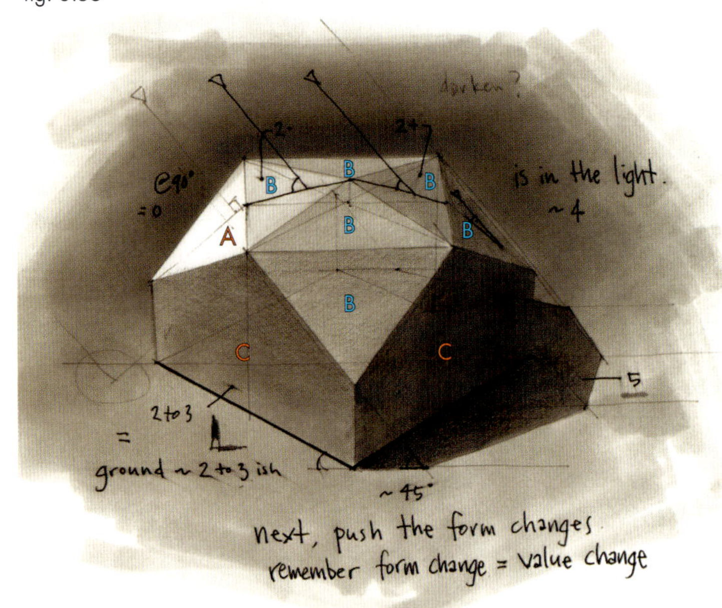

fig. 3.56

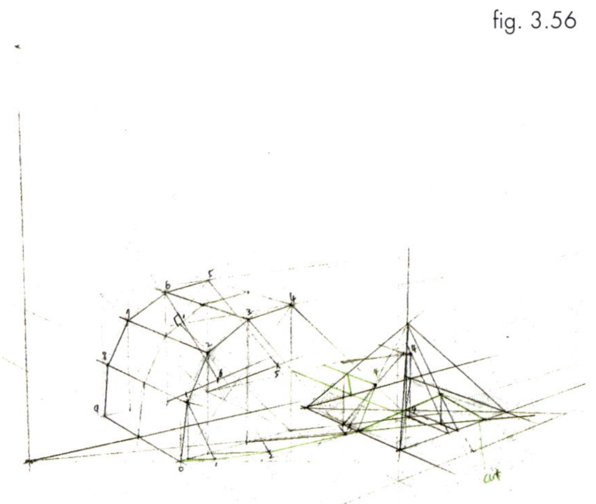

Joon Ahn

fig. 3.57

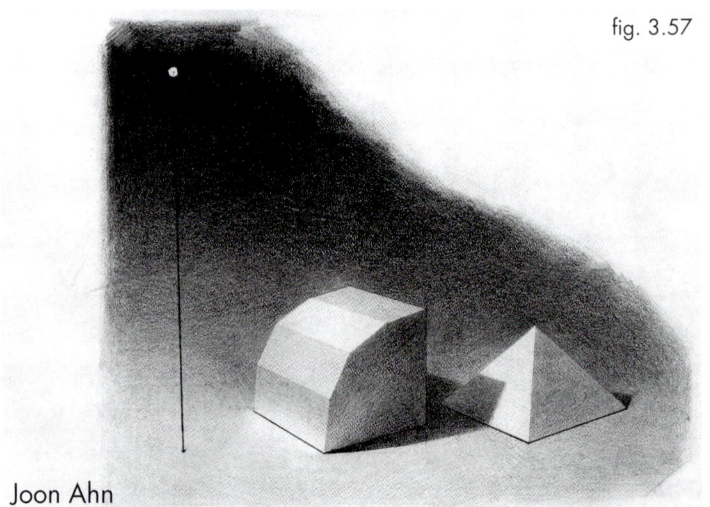

Joon Ahn

A combination of thoughtful planning and careful craftsmanship is the basis for great renderings, and there are many correct solutions and ways to achieve them. To take great renderings and make them exceptional though, a commitment to constant visual training is necessary. Skills can be continually enhanced by exercises such as taking photos and converting them to black-and-white images to see the values present in a color environment. Another strategy is to use the classic artist's trick of squinting your eyes to translate color changes into value shifts to understand a form.

Taking this real-life practice and adding it to the techniques explained will allow you to confidently develop your own design aesthetic as you render from your imagination.

Student Examples

Figs. 3.58 through 3.63 by Charles Liu. To see more of Charles' work, go to http://charlesliu-design.blogspot.com/

fig. 3.58

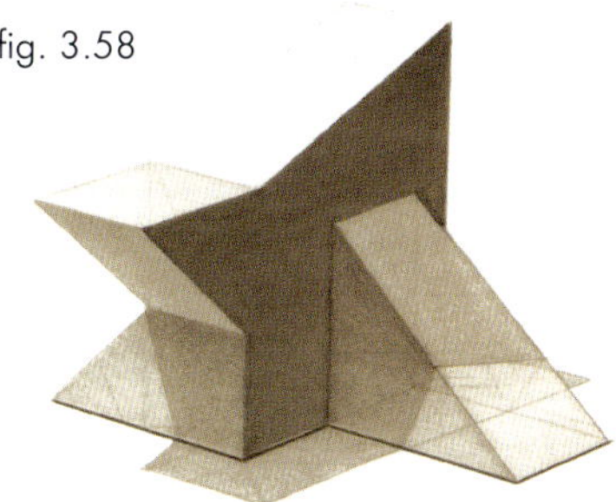

fig. 3.59

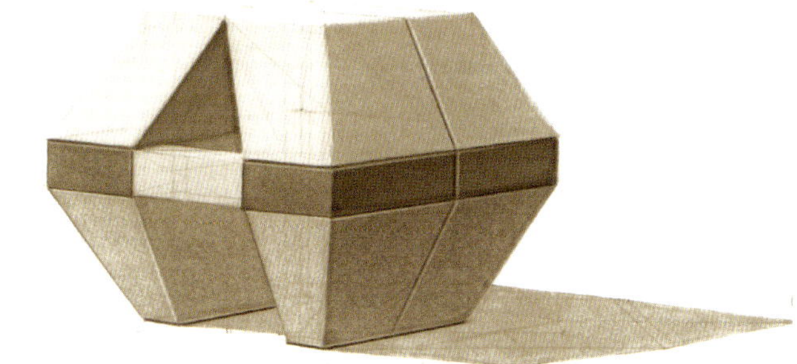

fig. 3.60

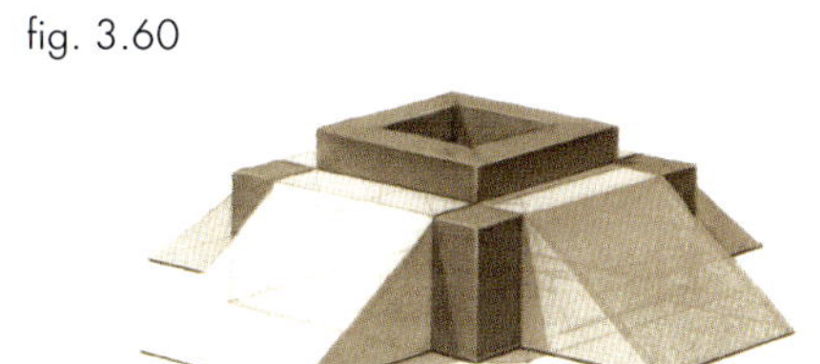

fig. 3.61

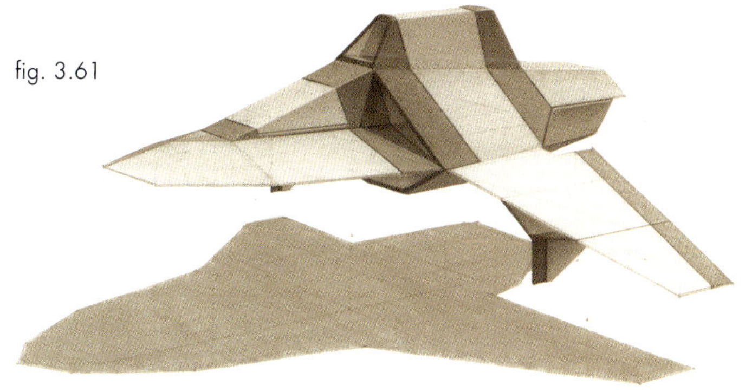

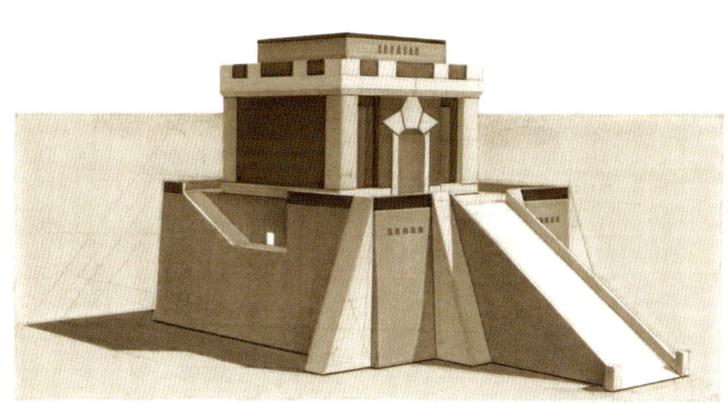

fig. 3.62

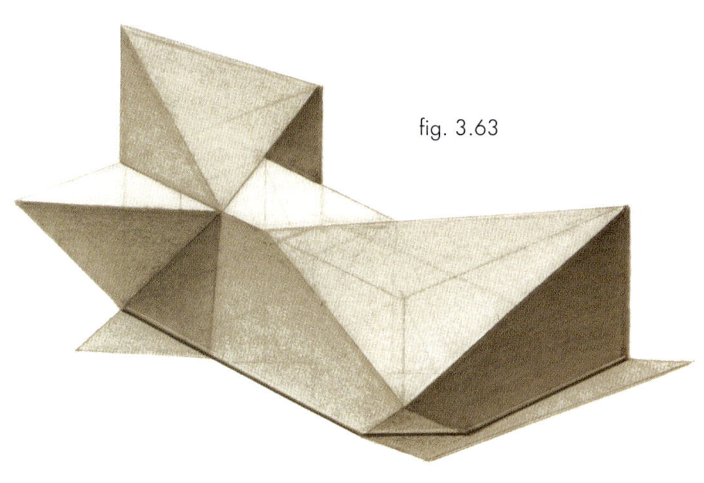

fig. 3.63

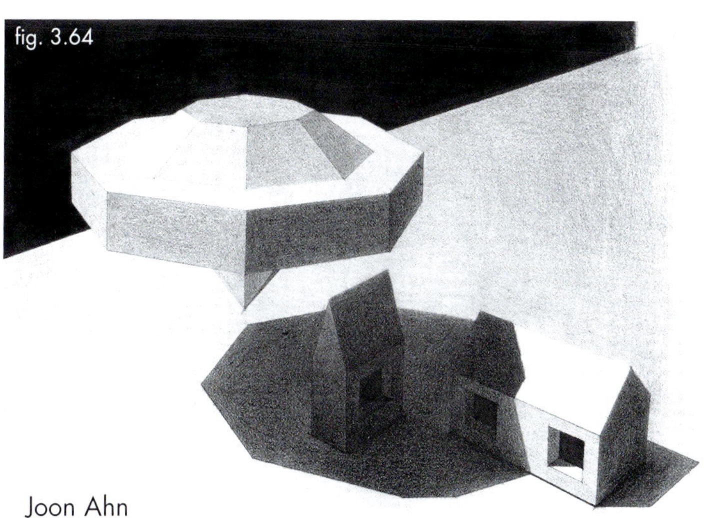

fig. 3.64

Joon Ahn

RENDERING CURVED SURFACES 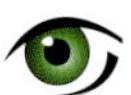

When rendering curved surfaces, it is essential to understand where the shadow side of an object starts. When working with flat surfaces and sharp corners, the shadow side is clearly defined by the edges. Curved shapes, however, usually do not have these edges to define the shadow side.

Terminology for Shading Curved Surfaces

A sphere includes all the elements that need to be considered when rendering a curved surface. It is important to be familiar with the correct terminology as shown below, and to notice how each element is affected by the others. Each term will be discussed in more detail on the following pages.

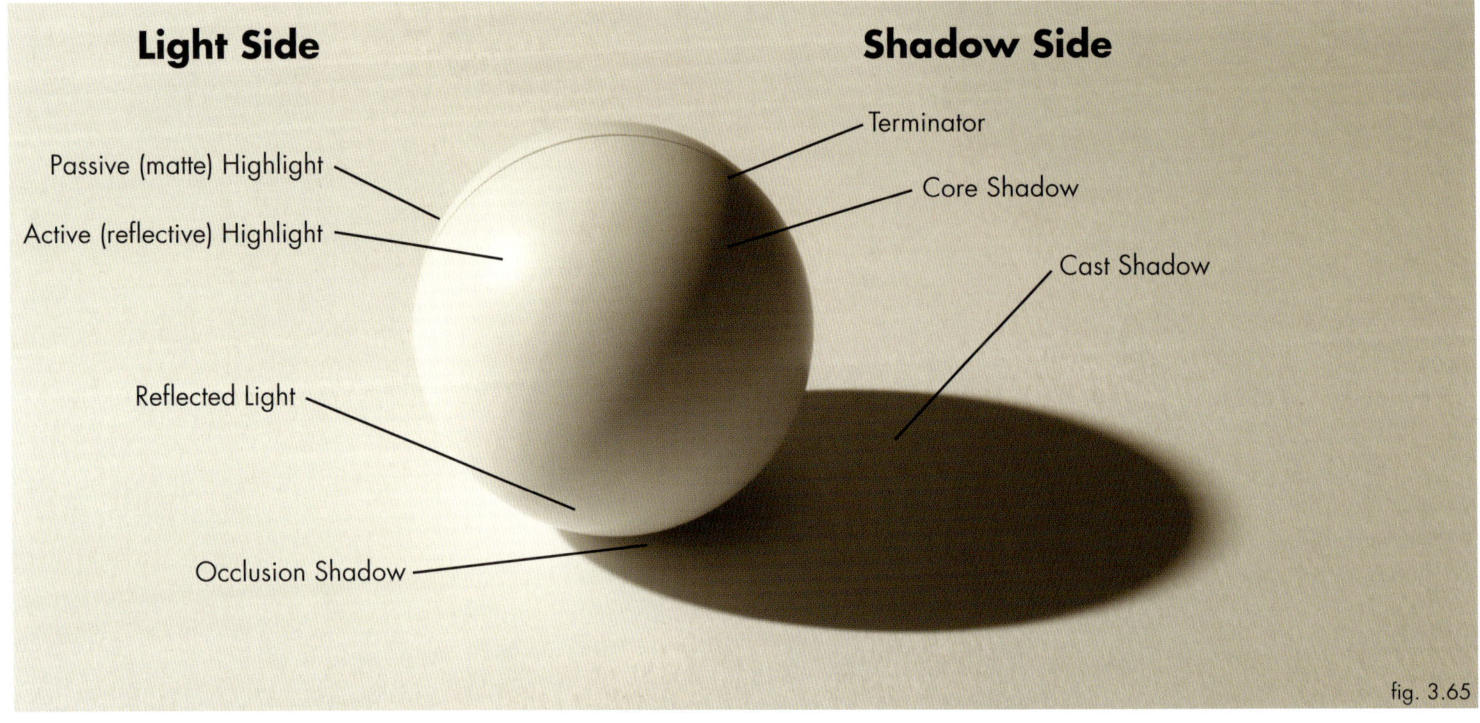

fig. 3.65

Light Side, Shadow Side and Cast Shadow

The light side, shadow side and cast shadow are the major graphic elements that the brain processes to understand form. Squint while looking at these images to see these major graphic elements more easily.

The **light side** is the area that is in direct light. This can be from either sunlight or a local light source. The **shadow side** is the area that is not exposed to direct light, but usually receives reflected or ambient light. This less-intense light creates value changes that make it possible to understand the forms of the shadow side of the volume. The **terminator** is at the transition between the light and shadow sides. The **cast shadow** shape is projected from the terminator to the ground plane.

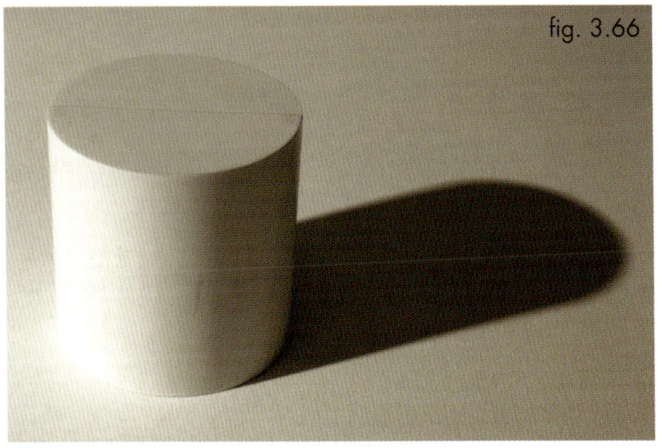
fig. 3.66

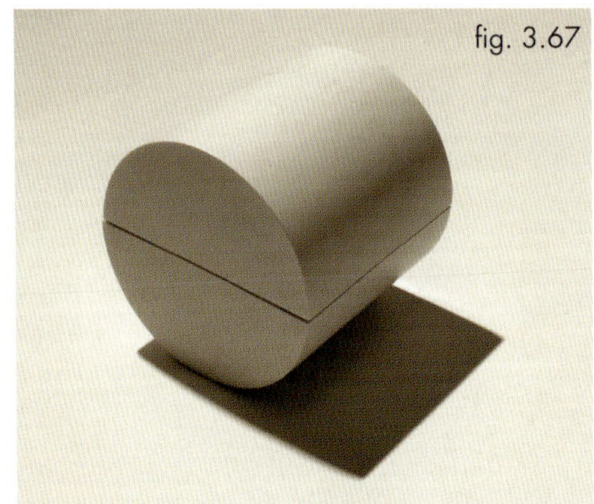
fig. 3.67

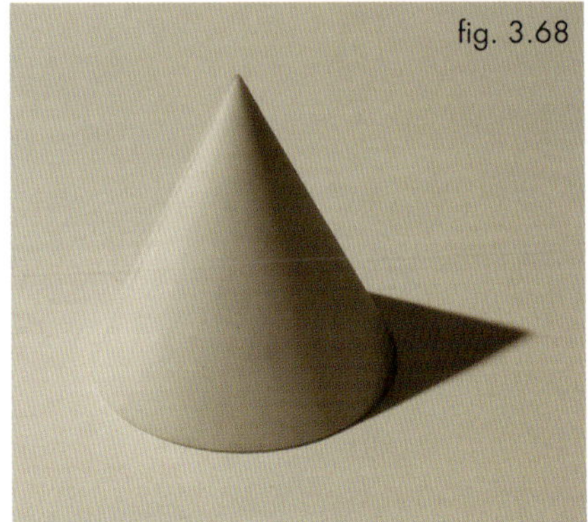
fig. 3.68

The terminator and the core shadow are closely related. The terminator is present where the light side transitions to the shadow side.

The core shadow becomes visible when there is ambient or reflected light affecting the shadow side of the object.

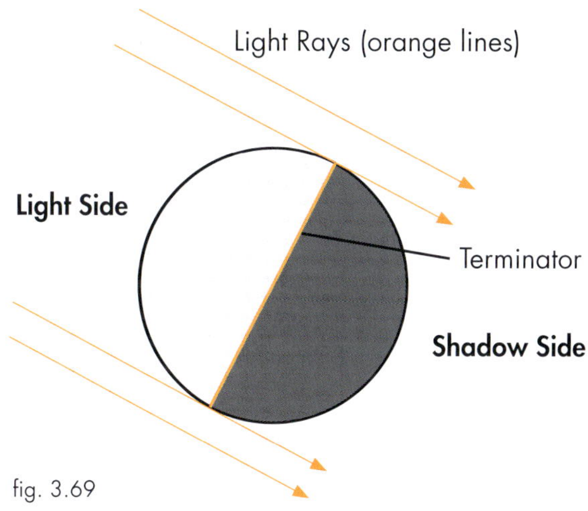

fig. 3.69

Terminator

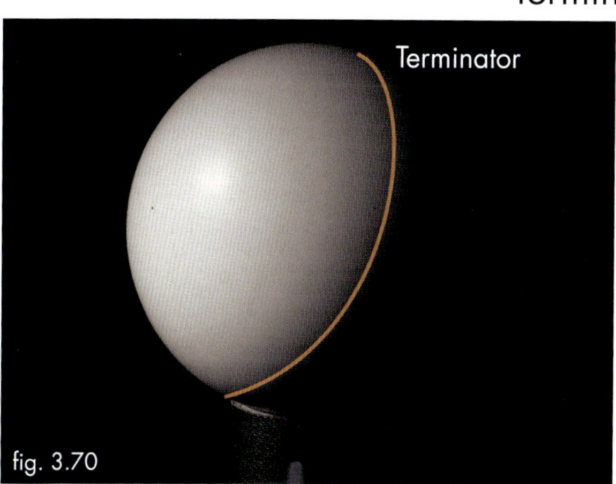

fig. 3.70

Fig. 3.69: The terminator (orange line) is where the light side transitions to the shadow side. It is located where the light rays are tangent to an object's surface.

Fig. 3.70: When all ambient and reflected light is removed, the terminator is very clearly observed. This is the case for objects in a completely dark environment, like outer space. In this scenario, the form of the light side is clearly perceived since there is enough of a value change across the surface, but the shadow side, with little or no value change, is barely defined. Any perceived form on the sphere's shadow side is simply the brain assuming symmetry of the volume.

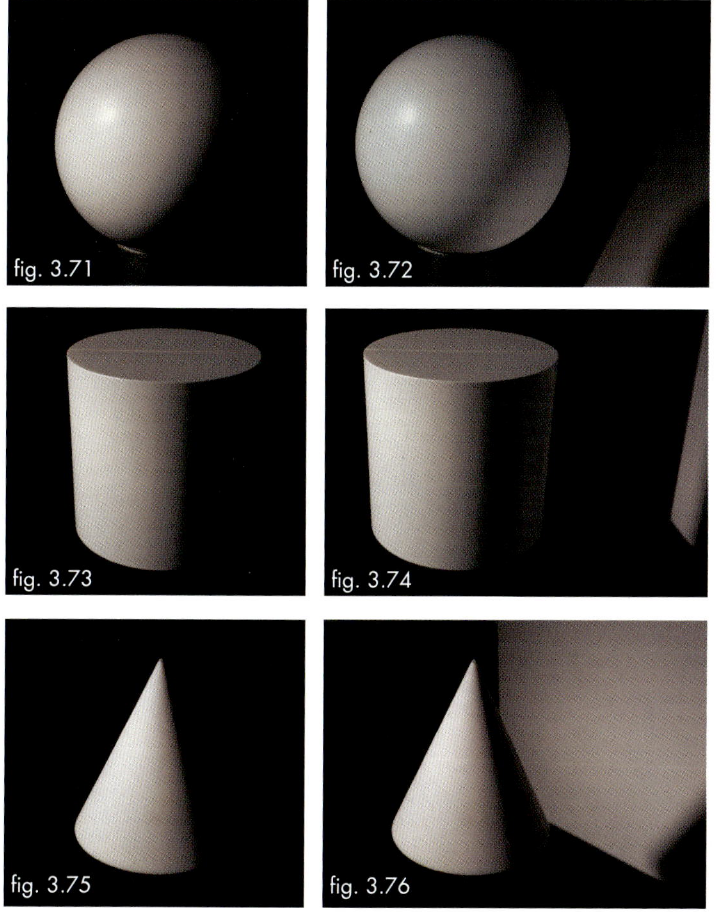

fig. 3.71 fig. 3.72 fig. 3.73 fig. 3.74 fig. 3.75 fig. 3.76

Core Shadow

To render the form changes on the shadow side of a volume reflected or ambient light must be introduced. This reflected light illuminates an object's shadow side, and when this happens a core shadow can appear. The core shadow is the dark band of value gradation that starts at the terminator and gets lighter onto the shadow side of the object as the reflected light increases in strength.

Compare the side-by-side images of geometric forms in figs. 3.71 through 3.76. On the right, the added reflected light creates a value gradation on the shadow side and enhances the perception of the dimensionality of the form. There is no reflected light from the ground, since the objects have either been suspended or elevated to remove the influence of this plane.

The addition of reflected light to create a core shadow is almost always done since it helps so much to describe the form of the shadow side and tie an object to a surrounding scene. Even if the shadow side might not be seen in a photograph, sometimes a rendering can be enhanced by taking some artistic license and adding reflected light to enhance the forms.

FORM HIGHLIGHTS 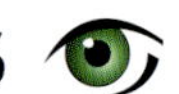

The brightest area of an object's surface is commonly called a highlight. This can cause confusion, since there are actually two highlights that can be observed on the surface of an object where the material finish has any level of reflectivity.

Passive Highlight

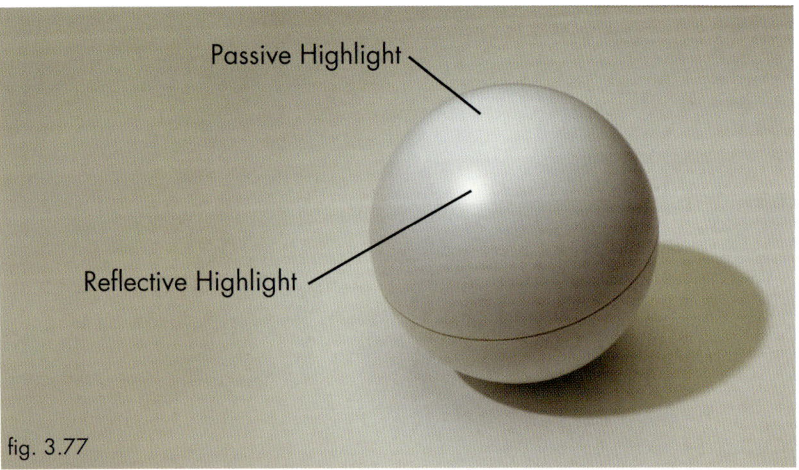

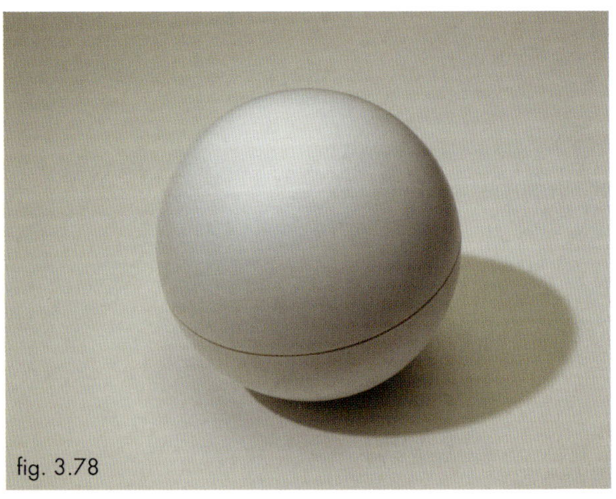

The **passive highlight** is the most brightly lit area of a matte surface, which occurs where the light rays are most perpendicular to it. Even when a camera or viewer moves around the object, the passive highlight stays exactly in the same position relative to the light source.

Fig. 3.77: The reflective and passive highlights are both visible. Observe that they do not match in position, but instead are located at different areas on the surface.

Fig 3.78: The reflective highlight has been removed; making the sphere look more like it has a matte surface.

Reflective Highlight

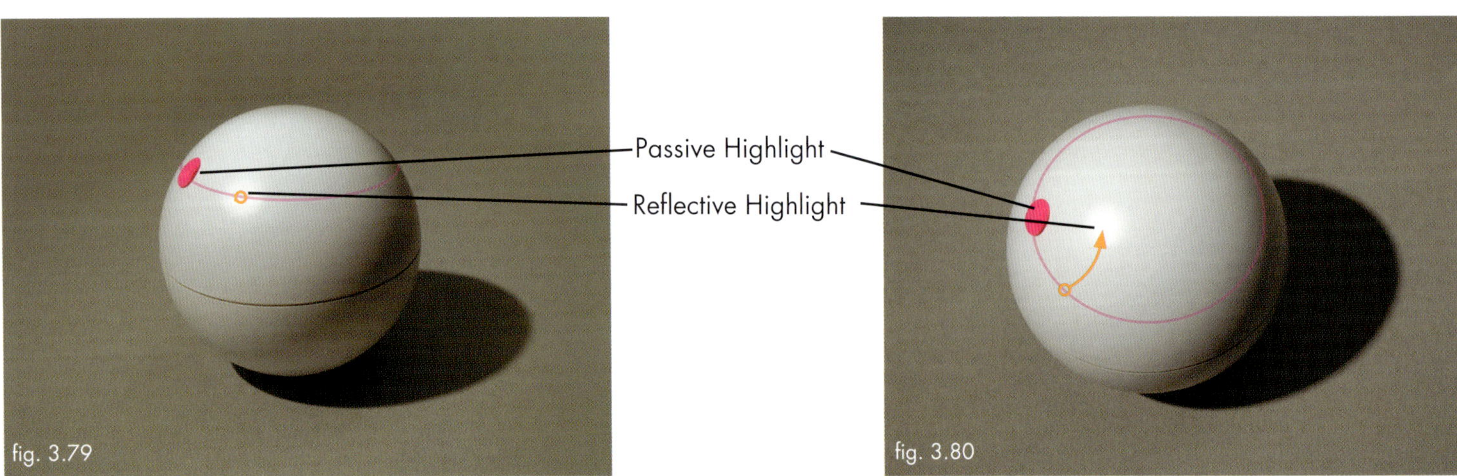

The **reflective highlight** is the reflection of the light source on the surface of the object. This reflective highlight moves whenever the camera or viewer changes position, since the line of sight's angle of incidence into the surface changes. (Much more about this later in Chapter 07).

Fig. 3.79: The location of the passive highlight is marked with a red dot. The reflective highlight's position is close to level with the passive highlight, relative to the red line.

Fig. 3.80: The camera has moved to a new position higher above the sphere, but the light source has not. Note how the passive highlight has stayed in the same original location on the surface of the sphere, since it only relates to the angle the light rays strike the surface. However, the reflective highlight has shifted up into a new location, since its position is relative to *both the observer's position (line of sight) and the light source.*

Reflected Light and Occlusion Shadows

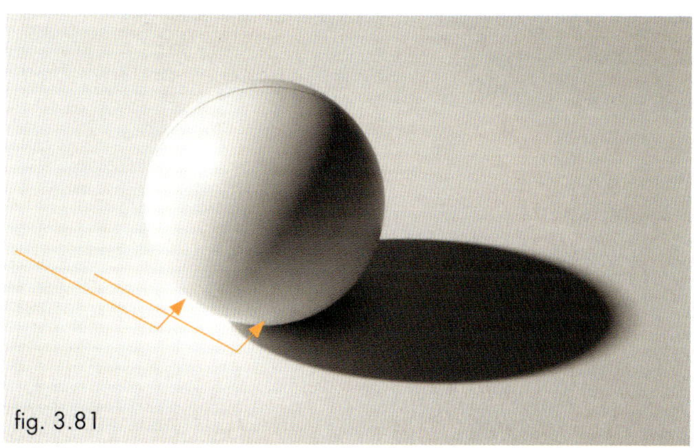

fig. 3.81

Reflected light and occlusion shadows bring a rendering of curved surfaces to life. These subtle interactions lend realism to a scene since the surfaces are now visually reacting to one another's presence.

Fig. 3.81: The **reflected light** interacts with the volume and the ground. Observe how the light reflects off the ground into the lower half of the sphere. The reflected light even lightens the value of the core shadow.

Fig. 3.82: Notice how the light reflects off the cylinder's surface onto the ground in front of the cylinder. Reflected light interacts equally with the surface of the object and the environment around it.

Fig. 3.83: The **occlusion shadow** is right next to where the cylinder touches the ground. It is the darkest part of the cast shadow due to a reduction of ambient or reflected light. Shadow areas that receive less ambient or reflected light are darker than the surrounding shadow areas that receive more. This is also the case for cavities and crevasses such as cut lines or intakes on a vehicle.

Reflected light and occlusion shadows can be observed in all three images. Now that you are more aware of them, start to study them through observation. Applying these principles when rendering forms is essential to achieving high-quality, believable images.

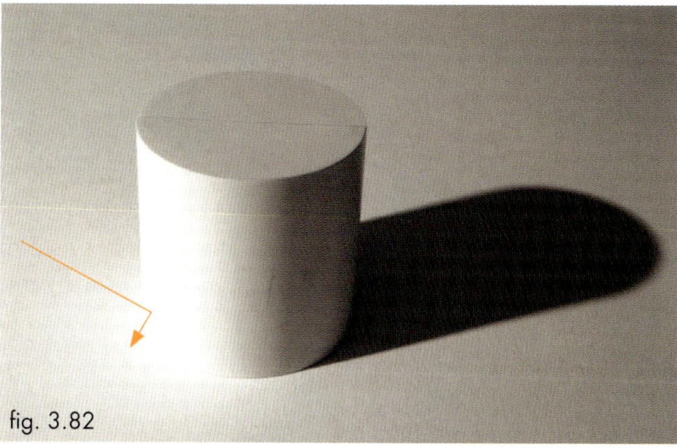

fig. 3.82

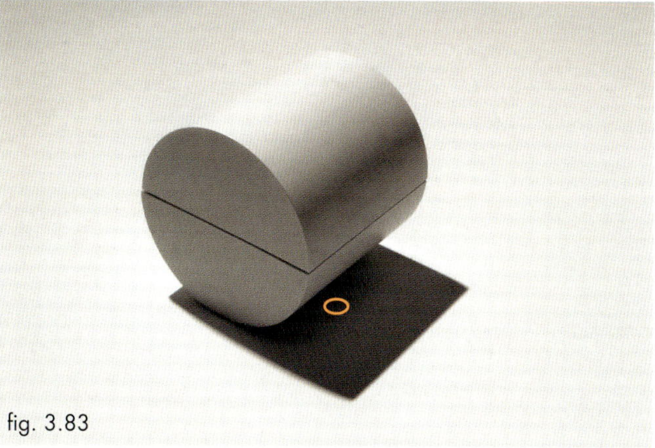

fig. 3.83

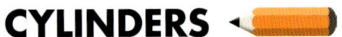

Shadow Construction of a Vertical Cylinder in Sunlight

Casting the shadow of a vertical cylinder requires finding the cast and core shadows of the cylinder. It is crucial to have the cylinder placed properly within an accurate perspective grid.

STEP 1, fig. 3.85: Set up the cylinder, the light and the shadow direction. Divide the cylinder into quarters, first with the light plane defined by the shadow direction and then with a perspective plane drawn 90° to the first one through the center of the top of the cylinder. An easy strategy to achieve this is to have the shadow direction align with the perspective grid.

STEP 2, fig. 3.86: Draw a line (A) tangent to the base of the cylinder in the shadow direction. This defines the terminator at the base (B).

Fig. 3.87 illustrates how the light rays of a light plane (C) extending upward from line "A" are tangent with the curved cylinder's side surface, defining the terminator.

STEP 3, figs. 3.88, 3.89: Cast the shadow of the top of the cylinder onto the ground by casting the shadow of each of the sticks at the quarter-marks, D, E and F. Place an ellipse defined by the quarter-marks on the ground with a vertical minor axis, lightly drawing through the cylinder. Should the shadow be short, the ellipse will be similar in size and degree to the one defining the base of this cylinder.

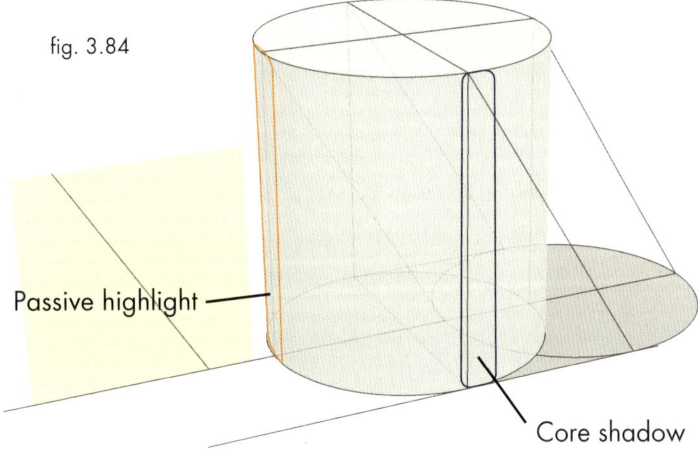

Fig. 3.84: The tangent terminator line on the sides defines the start of the core-shadow position. The core shadow is located a little behind the terminator line toward the shadow side. The width of the core shadow depends on the radius size of the curved surface: the larger the radius, the wider the core shadow.

The passive highlight of the vertical surface is found where the surface directly faces the light source, in top view. This is the brightest area of the vertical curved surface, but not of the entire cylinder. The top surface receives stronger light because it is more perpendicular to the light rays, making it the number 1 surface of a 1-2-3 read.

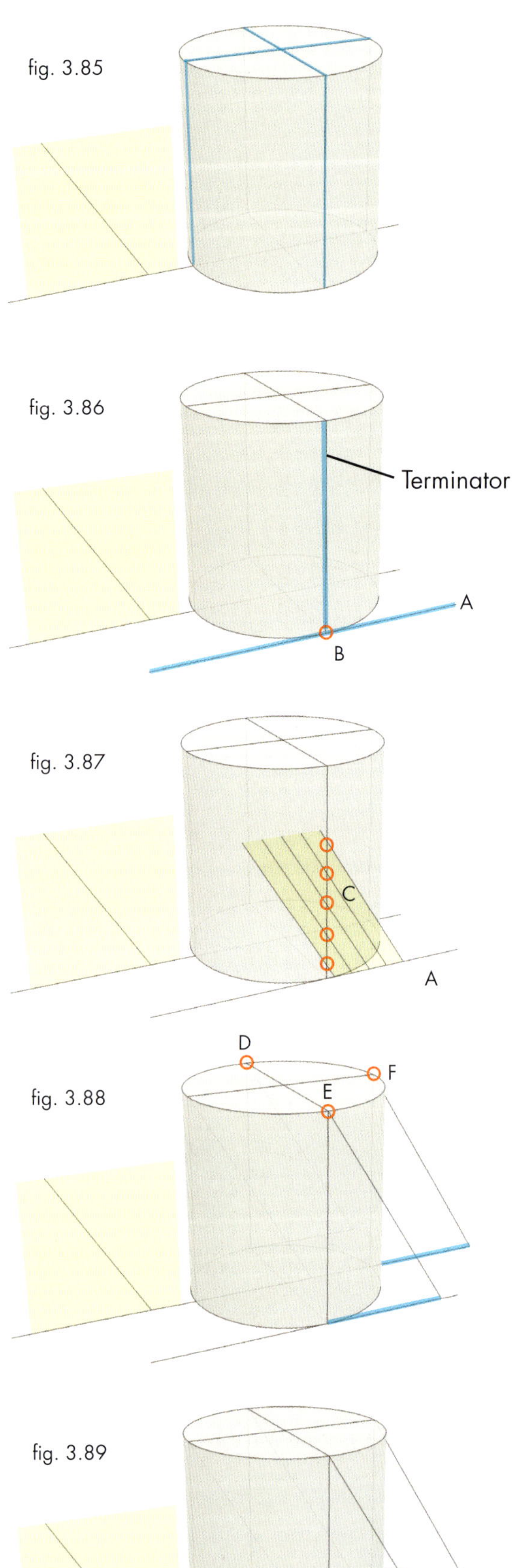

Shading of a Vertical Cylinder in Sunlight

Building on the skills already acquired for rendering flat-surfaced objects, here are the major points to remember when rendering a vertically placed cylindrical volume.

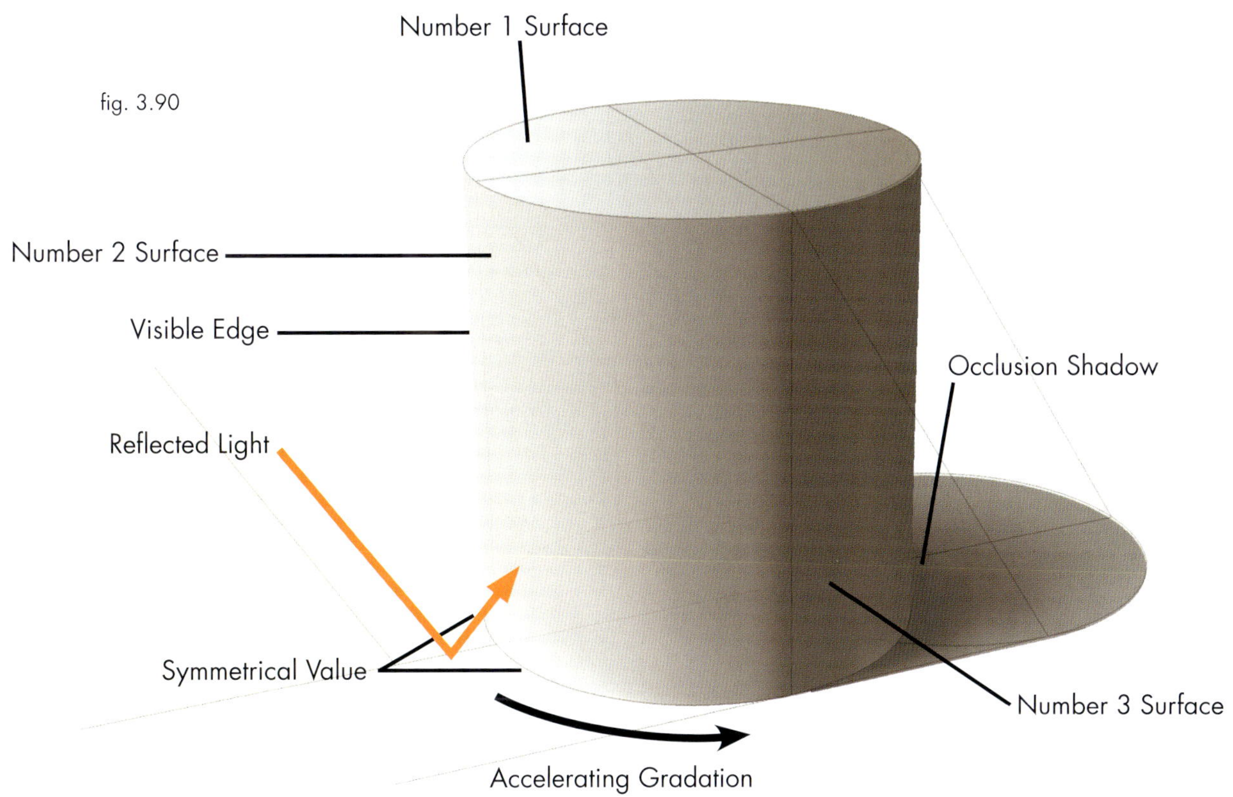

fig. 3.90

Major value assignment:
Make sure that a 1-2-3 read for objects with curved surfaces, such as this cylinder, is maintained just as well as for objects with only flat surfaces, like a box. Usually, the top surface is the brightest, and the vertical surface facing the light source is slightly darker as most light sources are located above scenes, as opposed to the side. The shadow side and the cast shadow still adhere to the halfway-to-black observation and practice. This would technically make them the same value, but the reflected and ambient light lighten the shadow side enough to provide differentiation between the cast shadow and the shadow side, adding variety to the rendering.

Vertical surface gradation:
An illuminated vertical surface has an accelerating gradation from the light side to the dark side. The values change more quickly as the surface gets closer to the core-shadow location. Avoid a linear gradation, as the form will read as less curved as a result.

Symmetry of value:
Imagine the cylinder split down the middle by the Shadow Direction plane. This split line is also the symmetry line for the values. This determines the value at the visible edge of the cylinder. Avoid adding a dark value to the cylinder's surface here to create contrast with the white paper, use background value instead if a stronger silhouette is desired.

Reflected light:
The reflected light that bounces off the ground brightens the vertical side of the cylinder most strongly where the side is most perpendicular to the reflected light source. This effect is greatest near the ground (the source of the reflected light) and then fades up the side. The same reduction of the reflected light effect can be observed as the surface turns away from the light.

Occlusion Shadow:
The cast shadow is the darkest closest to the area where the cylinder touches the ground. Remember, this occlusion gradation is not linear; it accelerates as it moves closer to the source of the occlusion, which is the juncture of the ground plane and sides of the cylinder in this case.

Shadow Construction of a Vertical Cylinder in Local Light

The construction of a cast shadow in local light is very similar to the construction in sunlight. The main differences are the shape of the cast shadow and the location of the terminator. In local light, the terminator does not divide the form in half as it did with sunlight. Also, the shadow side of the object is larger than the light side. By bringing a light source very close to the object, such as a candle or the light from a cell phone, one can observe this effect in the extreme.

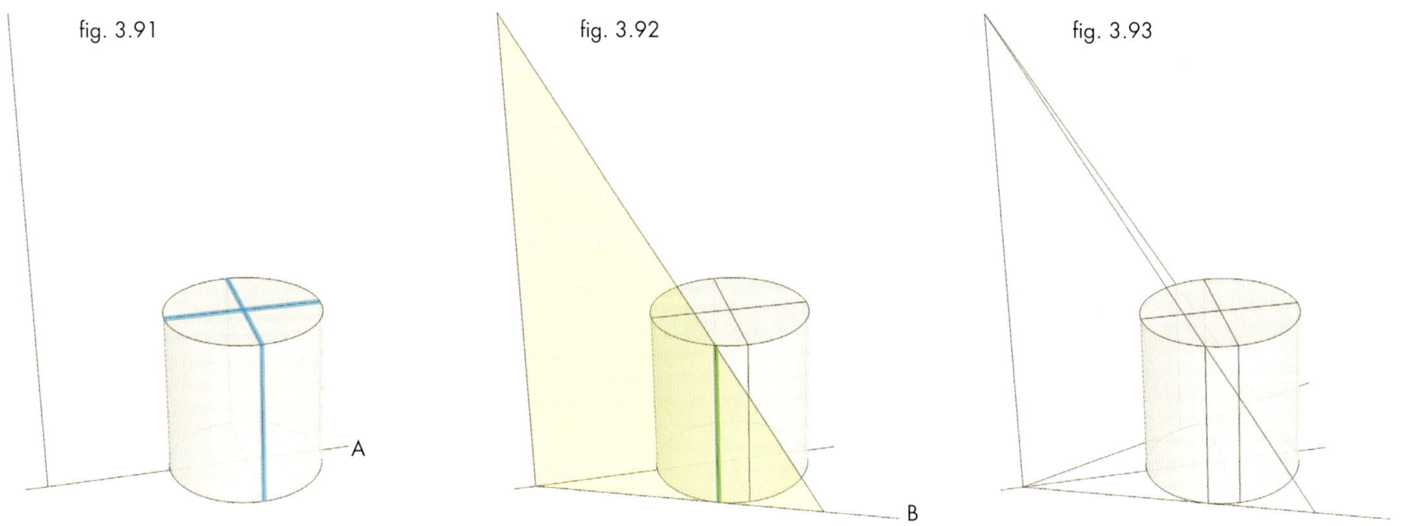

fig. 3.91 fig. 3.92 fig. 3.93

STEP 1, fig. 3.91: Set up the light source and the shadow origin. Draw a line (A) that runs from the shadow origin through the center of the base of the cylinder. Use this line as the basis to divide the cylinder into quarters in top view (blue lines).

STEP 2, fig. 3.92: To find the terminator (green line), draw a line from the shadow origin tangent to the base of the cylinder (B). Observe that the terminator location does not match the quarter line of the cylinder as it did in sunlight.

STEP 3, fig 3.93: Repeat this on the cylinder's other side.

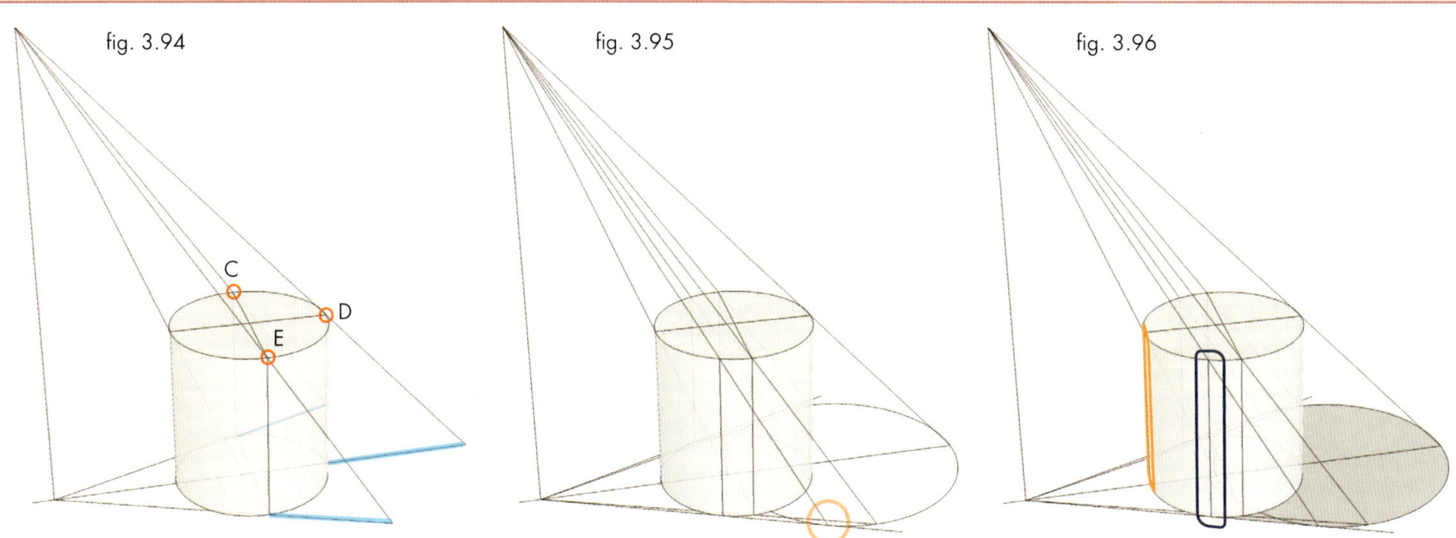

fig. 3.94 fig. 3.95 fig. 3.96

STEP 4, fig. 3.94: Cast/project the quarter points (C, D, E) of the top surface of the cylinder onto the ground.

STEP 5, fig. 3.95: Place the ellipse on the ground that defines the cast shadow of the top surface that touches tangentially to the cast-shadow line of the terminator.

STEP 6, fig. 3.96: Combine the cast shadow of the top surface with the cast shadow of the terminators to define the entire volume's cast shadow.

Observe that the constructed location of the core shadow does not align with the halfway point of the cylinder. Also notice the passive highlight area on the vertical side is perpendicular to the light rays when viewed in top view.

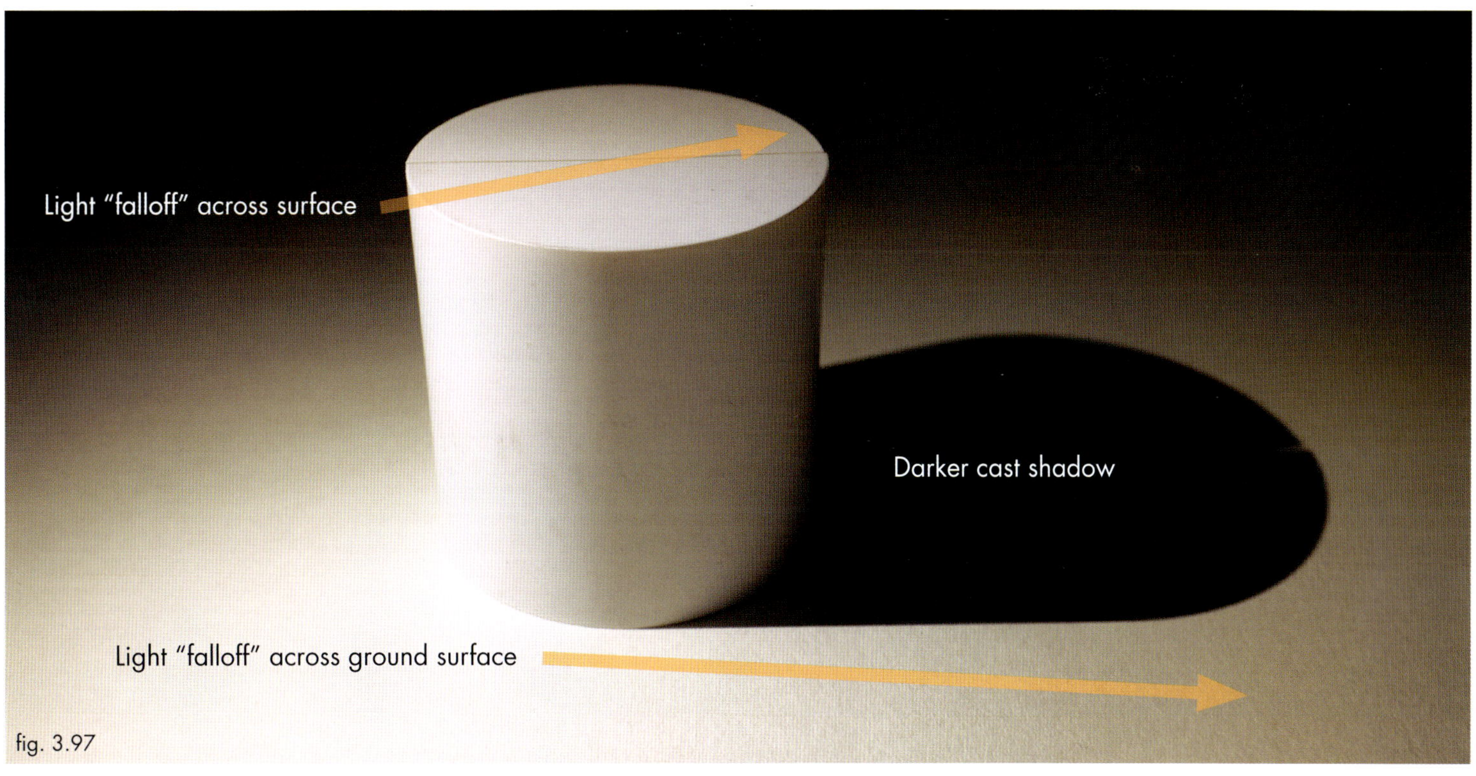

fig. 3.97

To render a vertical cylinder in local light, follow all the rules for rendering the cylinder in sunlight, while keeping in mind that the core shadow and the cast shadow will have different shapes and positions.

There are a few more other disparities. In Fig. 3.97, observe how the cast shadow can be much darker due to the lack of reflected or ambient light. The ground shows a gradation due to the falloff of the local light source. Note that this falloff affects the top surface of the form as well.

Observation of a real-life object is a smart way to approach a rendering, while comparing and contrasting local light conditions with sunlight.

Vertical Cylinder Rendering Examples

Tianxu (Tim) Guo
www.spikethehedgehogstudio.com

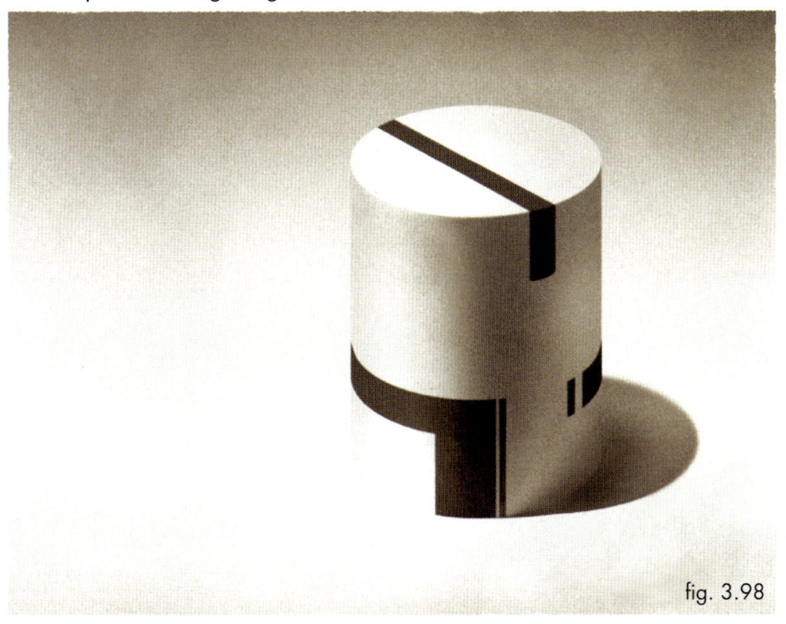

fig. 3.98

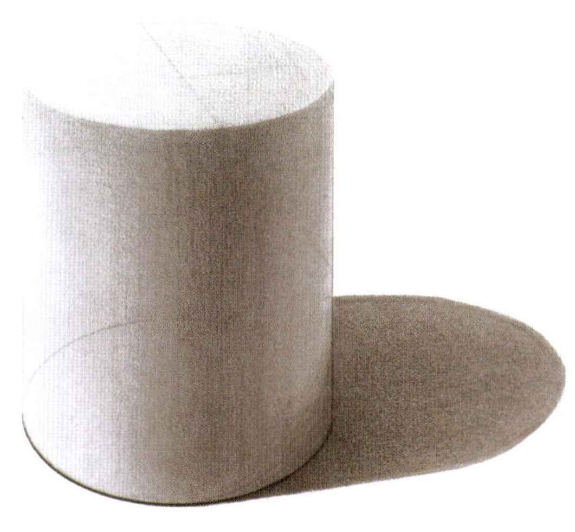

Charles Liu

fig. 3.99

Shadow Construction of a Horizontal Cylinder: Sunlight

To construct the shadow of a horizontal cylinder, it is necessary to know how to cut through the cylinder at an angle to create a section line (fig. 3.100), and how to cast the shadow of an upright standing disk (fig. 3.101).

In the following examples, construction lines were reduced in order to focus on the main rendering points.

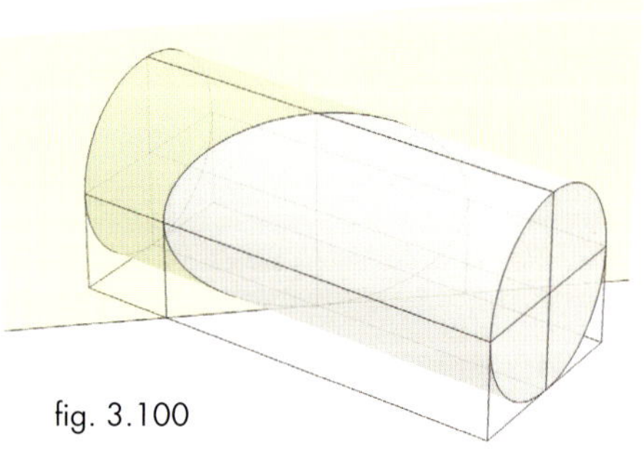

fig. 3.100

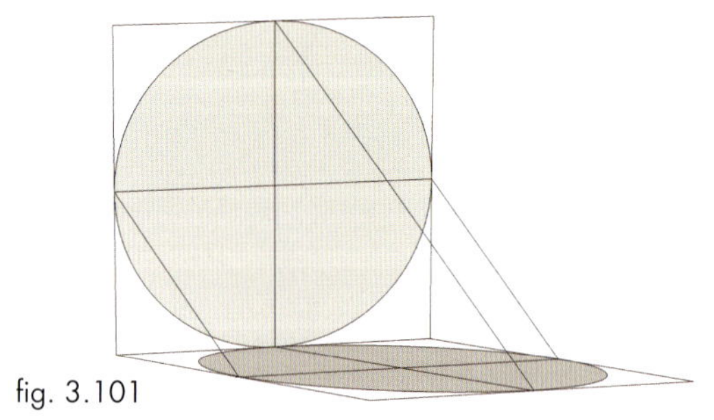

fig. 3.101

Terminator, Highlight and Cast-Shadow Edge of a Horizontal Cylinder

To construct the terminator and the highlights for a vertical cylinder, only the shadow direction was needed to find the necessary tangencies (figs. 3.102 and 3.103). For a horizontal cylinder, the location of the terminator is still defined by the tangency of the light rays to the curved surface, but because it has more possible orientations than a vertical cylinder, the shadow direction, light direction and light planes are all needed for the construction (figs. 3.104 and 3.105).

Fig. 3.104: The fact that the core shadows (A) and the passive highlight (B) run parallel to the axis of cylinder makes it possible to determine their locations by referencing just one section line on the surface of the cylinder defined by the intersection of the light plane with the cylinder's surface.

fig. 3.102

fig. 3.104

B

A

fig. 3.105

fig. 3.103

Top view

Light Plane section

With these principles in mind, let's look at different lighting scenarios that are commonly found when dealing with cylinders. These are as follows:

Sunlight from top: Light rays are perpendicular to the ground plane.

Sunlight at an angle: Light rays are not perpendicular to ground plane.

Local light: Light planes fan out from a local light source.

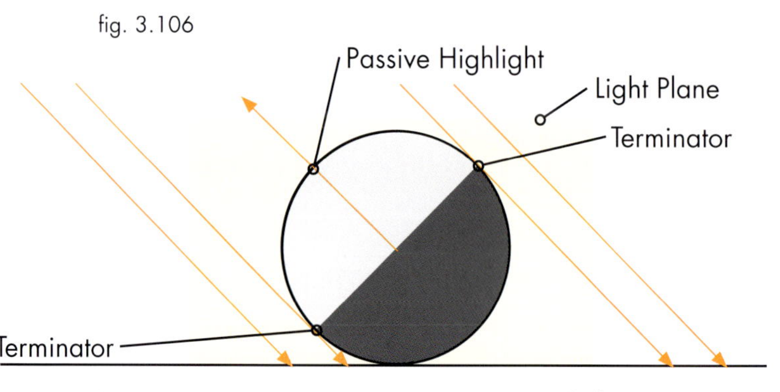

fig. 3.106

Side View

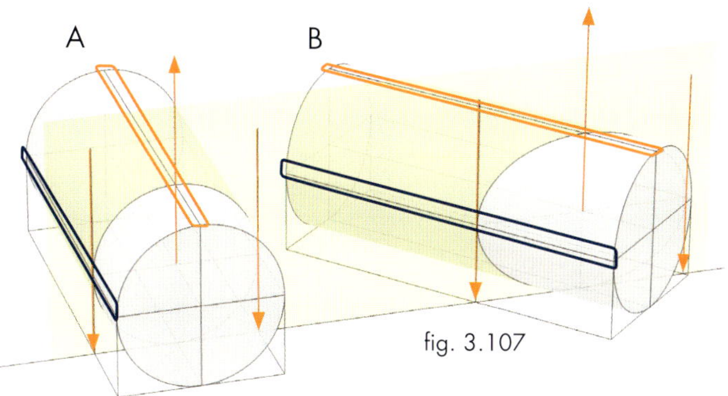

fig. 3.107

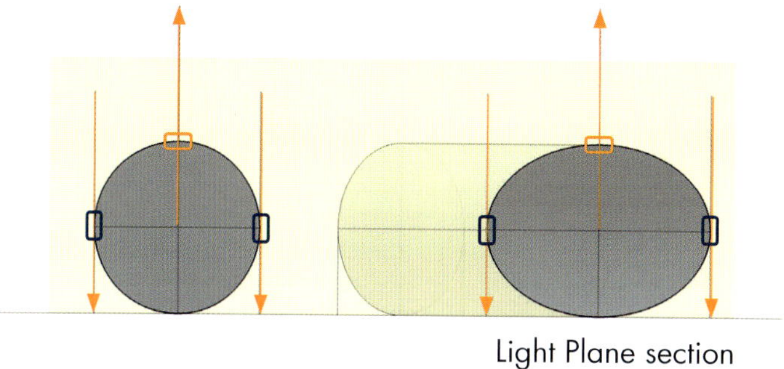
Light Plane section
fig. 3.108

Fig 3.107: In this scenario, since the light rays are perpendicular to the ground, there is no shadow direction. Choose any direction to place the light plane section for the construction. Find the terminators and passive highlight by intersecting light rays with this section line.

Looking at the two cylinders above, (fig. 3.107), the light plane is parallel to the end cap of cylinder A and intersects at a diagonal with cylinder B. In the cross sections (fig. 3.108) the locations of the terminators and the passive highlights are in exactly the same position relative to the ground plane, for both cylinders.

This makes it possible to use the section described by the end of a horizontal cylinder to construct the cast shadow, when the sunlight is coming from the top. This is the easiest way to create a cast shadow, since it will always align with the footprint of the cylinder, which is also the top view of the cylinder projected straight down to the ground plane.

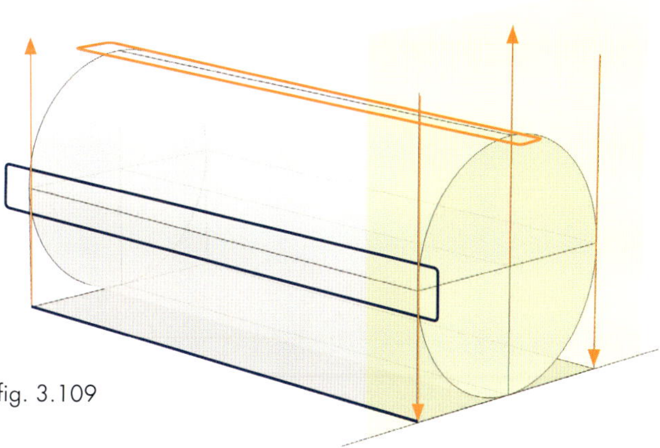
fig. 3.109

Construction:
Fig. 3.109: The light rays' tangencies define the terminator locations. The passive highlight is where the top surface is perpendicular to the light rays. This spot can be found for a cylinder by drawing a line from the center toward the light source on the light plane. The intersection of the light rays with the ground plane defines the size and the location of the cast shadow.

Value assignment:
Fig. 3.110: Determine the true value of the object, and then assign values based on the halfway-to-black practice.

Core shadow:
The core shadow is generated by reflected light bouncing off the ground plane and lightening up the shadow side of the cylinder.

Passive highlight:
The passive highlight is the true value of the object, since it is perpendicular to the light rays.

Occlusion shadow:
The occlusion shadow occurs where the cylinder makes contact with the ground. To ensure that the cylinder truly looks as if it is lying on the ground, it is essential to render this shadow.

End plane:
Since the light rays are parallel to the end plane, the cast shadow is a line. It can be rendered as either a Number 2 or Number 3 surface. A Number 3 surface maximizes the 1-2-3 read. Don't forget that the reflected light from the ground will light up the area on the end plane nearest to the ground (point A).

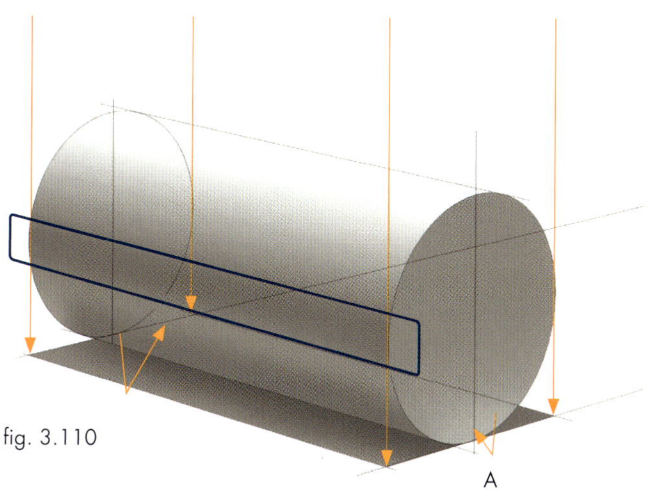

fig. 3.110

Rendering a Horizontal Cylinder in Sunlight: Side and Diagonal Light

fig. 3.111

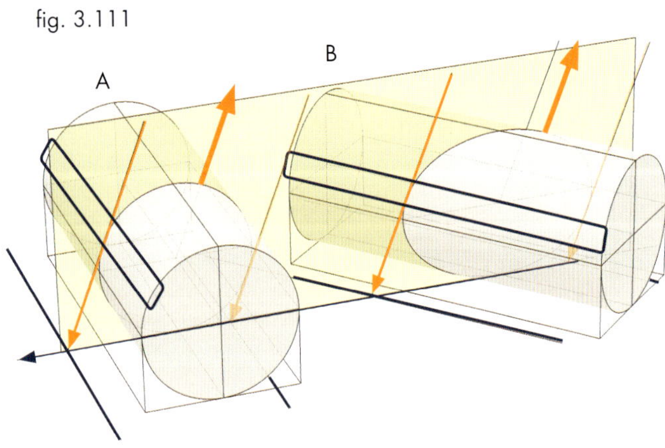

fig. 3.112

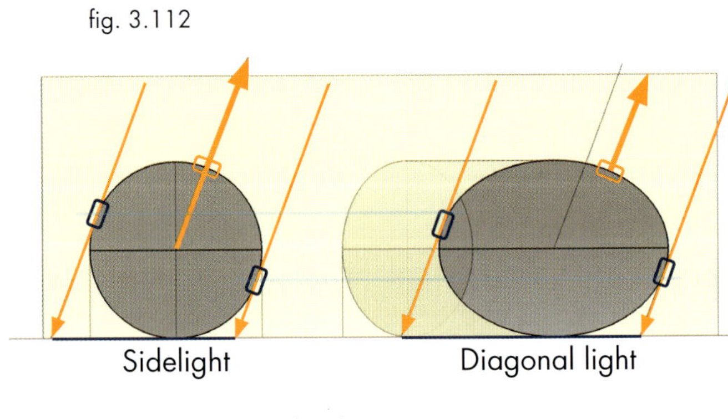

When light rays are not perpendicular to the ground plane, it becomes necessary to work with the shadow direction and light planes. The tangency of the light rays to the surface still determines the position of the terminators. With top sunlight, there is only one scenario since the terminators will always be in the same relative position; however, this is not the case when the light rays are at an angle.

Fig 3.111: Compare the sections of the two cylinders above. Cylinder A is perpendicular to the light plane; Cylinder B is at an angle to it.

Fig. 3.112: In the cross sections of the two cylinders, see that the terminator's positions change relative to the ground. In addition, observe that the cast shadows of the terminators on the ground are different in distance to the footprint of the cylinder. These differences vary in strength according to the light angle and rotation of the cylinders relative to one another.

Rendering a Horizontal Cylinder in Sunlight: SideLight

Construction:
Fig. 3.113: Any light-plane section can be chosen for this construction, so it is simplest to use the cylinder's end-plane section. Find the tangencies and draw toward the light source to construct the terminators, passive highlight and the cast shadow.

Rendering:
Fig. 3.114: To render the value gradations, core shadow, passive highlight and the end plane, follow the same rules as for the top-lit cylinder.

The main difference between the top-lighting scenario and this sidelighting scenario is that the positions of all the elements have moved. Due to this, the occlusion shadow can become darker and the reflected light on the shadow side of the cylinder is reduced in this view because the source of the reflected light (the lit area of the ground plane) is farther away. There is no reflected light bouncing out of the cast shadows.

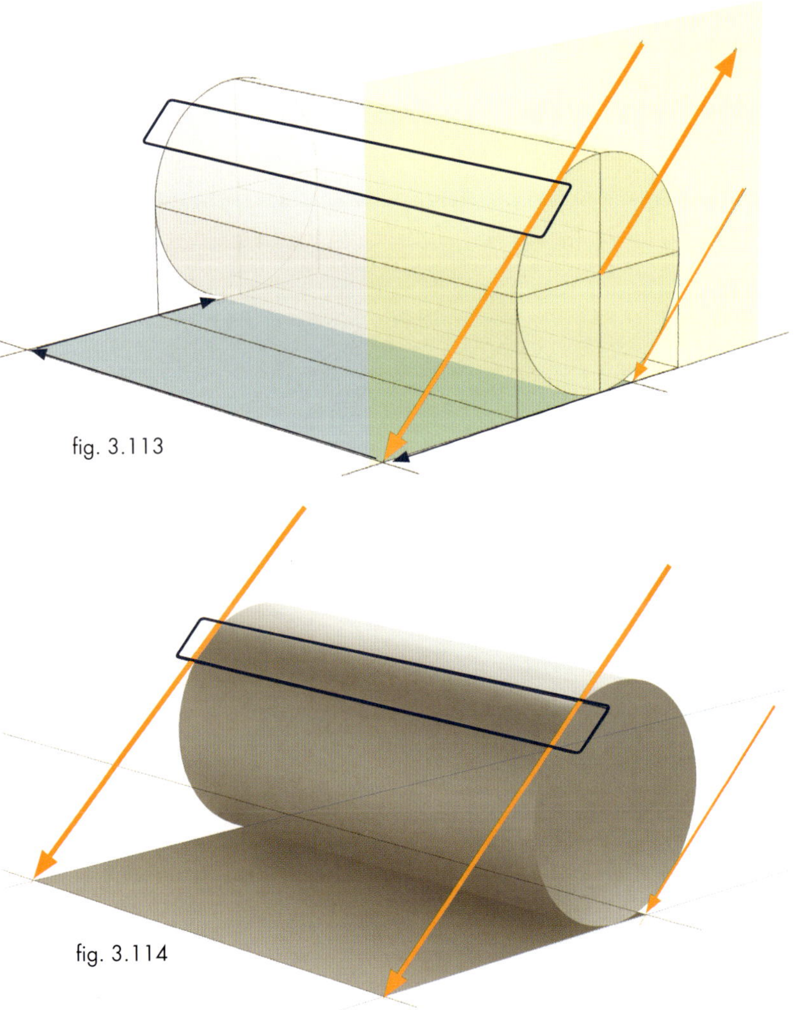

fig. 3.113

fig. 3.114

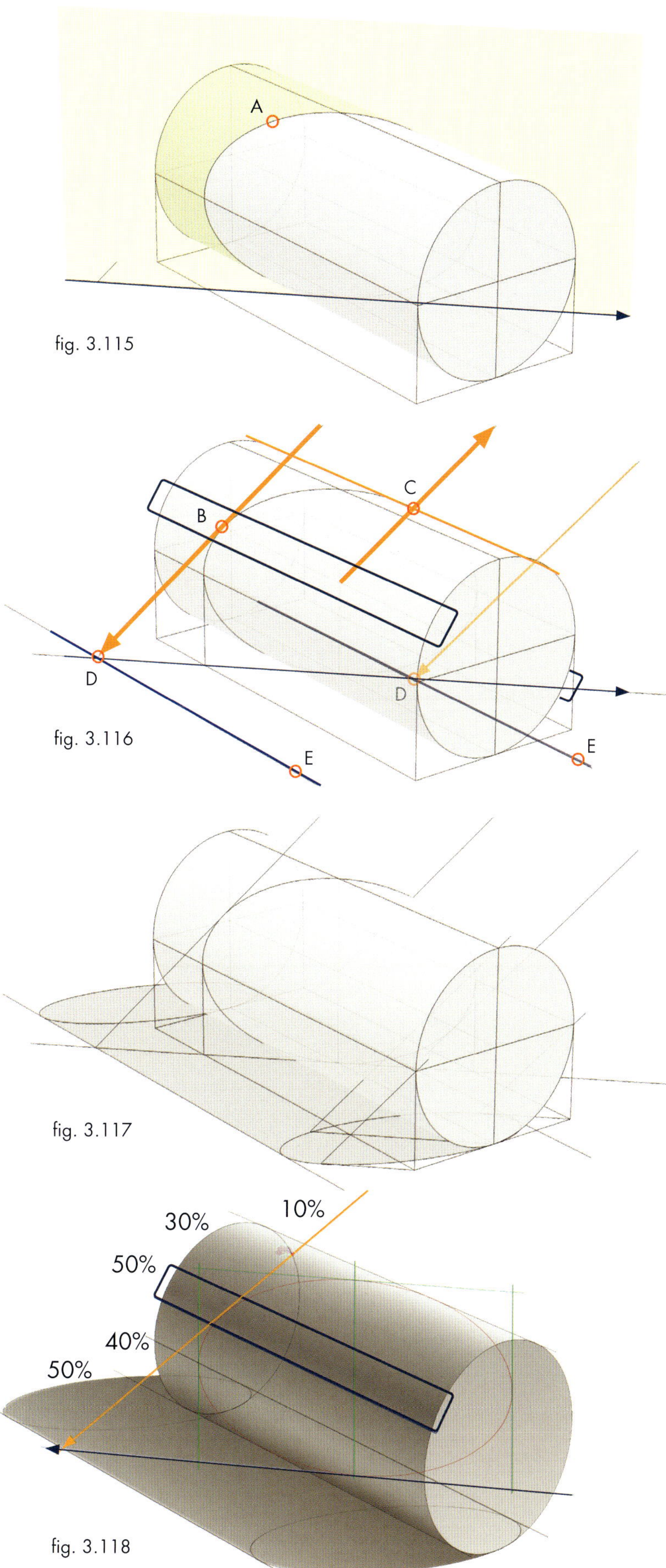

fig. 3.115

fig. 3.116

fig. 3.117

fig. 3.118

Construction:

Fig. 3.115: Choose a light plane that cuts through the cylinder, but avoid cutting through the end caps. Should the cylinder be too short, extend it for the construction. Mark the section line on the cylinder's surface (A).

Fig. 3.116: Find the terminators (B), passive highlight (C), and the width of the cast shadow (D) on the ground plane. Draw the width lines out far enough beyond the end planes, which will be necessary to cast their shadows (E).

Fig. 3.117: To cast the shadow of the vertical end planes, cast as many stick points to the ground plane as needed to make a good estimation. Since the ellipse will also be estimated, usually four points are enough. The cast shadow of the far end can be guessed based on the shape and location of the front end plane, since they both have the same shape. The only difference is that the far shadow might be slightly smaller due to perspective foreshortening.

Rendering:

Fig. 3.118: The value assignments are similar to the sidelit cylinder, except that the front end cap is now in light and therefore a true Number 2 side. The passive highlight is darker than the true value of the cylinder, due to the light rays hitting the surface at an angle that is not perpendicular.

It is helpful to assign numbers only to the biggest value steps, which are the passive highlight, core shadow, end planes and cast shadow. Avoid adding more detailed numbers; rather, experiment and observe to find the transitional gradations that communicate the volume most effectively.

Rendering a Horizontal Cylinder in Local Light

Fig. 3.119: In local light, the light planes fan out from the shadow origin. Each of the light planes cuts through the cylinder at a different angle. Choosing the plane that cuts perpendicularly through the cylinder (A) is the best choice for a construction, since this section is most similar to the end-plane sections. Should there not be a perpendicular section available because the cylinder is too short, consider extending it for the sake of the construction.

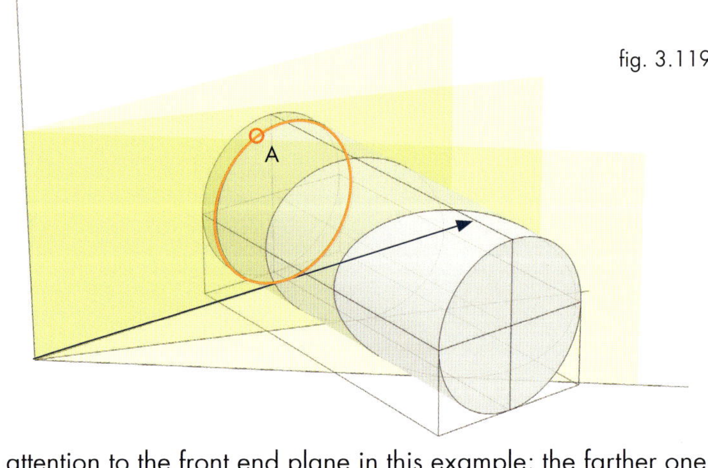

fig. 3.119

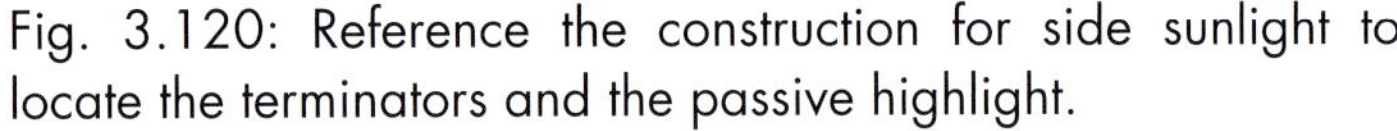

Fig. 3.120: Reference the construction for side sunlight to locate the terminators and the passive highlight.

Fig. 3.121: Use the terminators to determine the width of the cast shadow.

Fig. 3.122: Project as many points as needed to construct the cast shadow of the end planes of the cylinder. Pay closer attention to the front end plane in this example; the farther one can be created with an educated guess, since the shadow is thin and very little of it is visible.

Fig. 3.123: Rendering the cylinder will be similar to that of the sunlit cylinders; however, the falloff of the local light source must now be considered. The passive highlight and the ground plane show the light source's falloff most strongly.

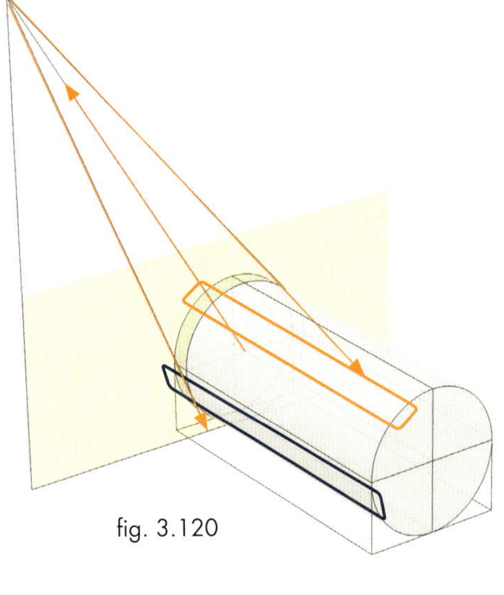
fig. 3.120

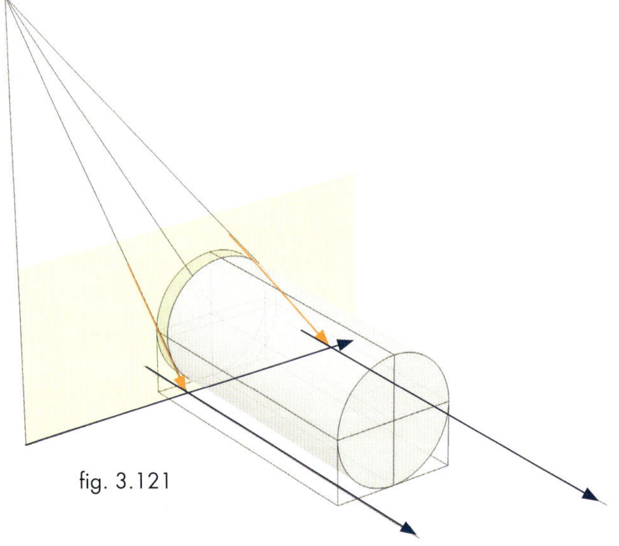
fig. 3.121

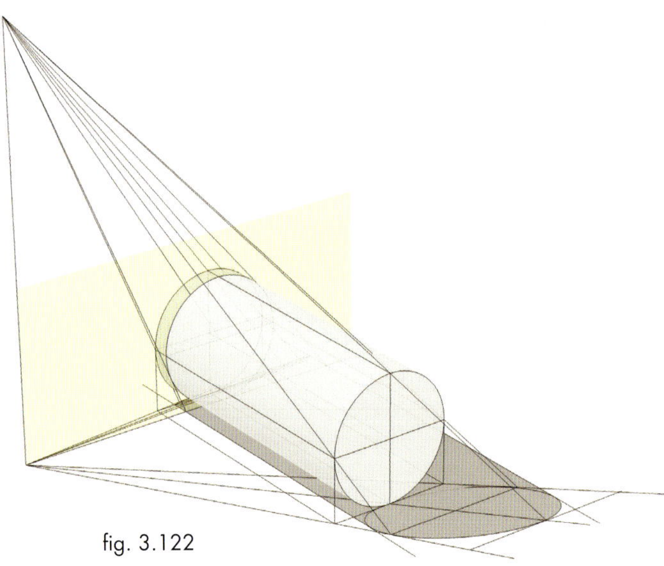
fig. 3.122

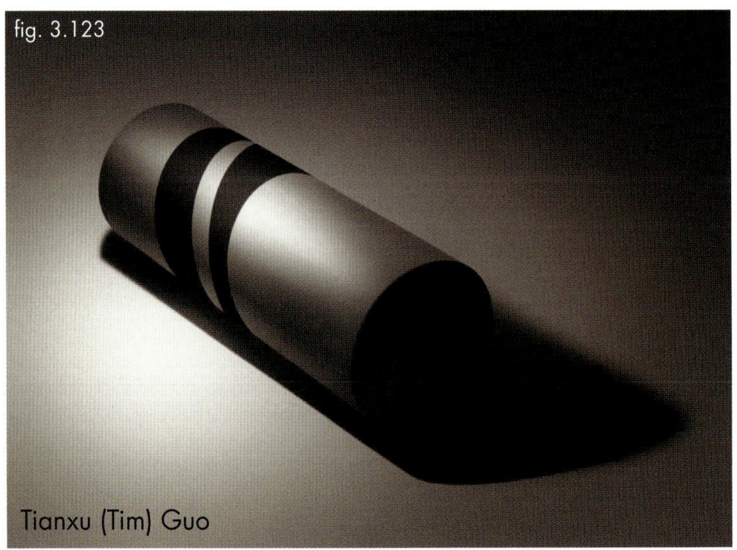
fig. 3.123

Tianxu (Tim) Guo

Cast Shadow Construction of a Tilted Cylinder

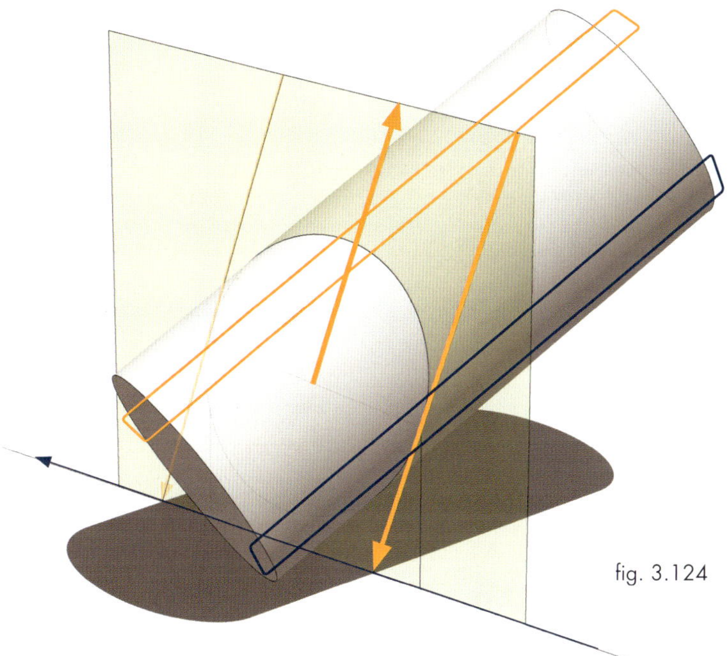

fig. 3.124

When a cylinder is tilted, its shadow construction follows the same rules for that of a horizontal cylinder. Fig. 3.124 shows a tilted cylinder in diagonal light, which encompasses all the principles for when not using top lighting.

To find the terminators, passive highlight and the cast shadow, a light-plane section is needed.

This construction can become very complex and requires knowing the exact position in three-dimensional space of the cylinder. Decide if it is worth creating a full construction or if making an educated guess is the better solution.

Sometimes, when working with more complex positions of geometric shapes, using simple materials to create a practical, observable set-up is the best way to preview the lighting with the least amount of effort. In this scenario, a cardboard roll, paper and a desk light were used to inform the rendering direction.

Creating Practical Set-Ups

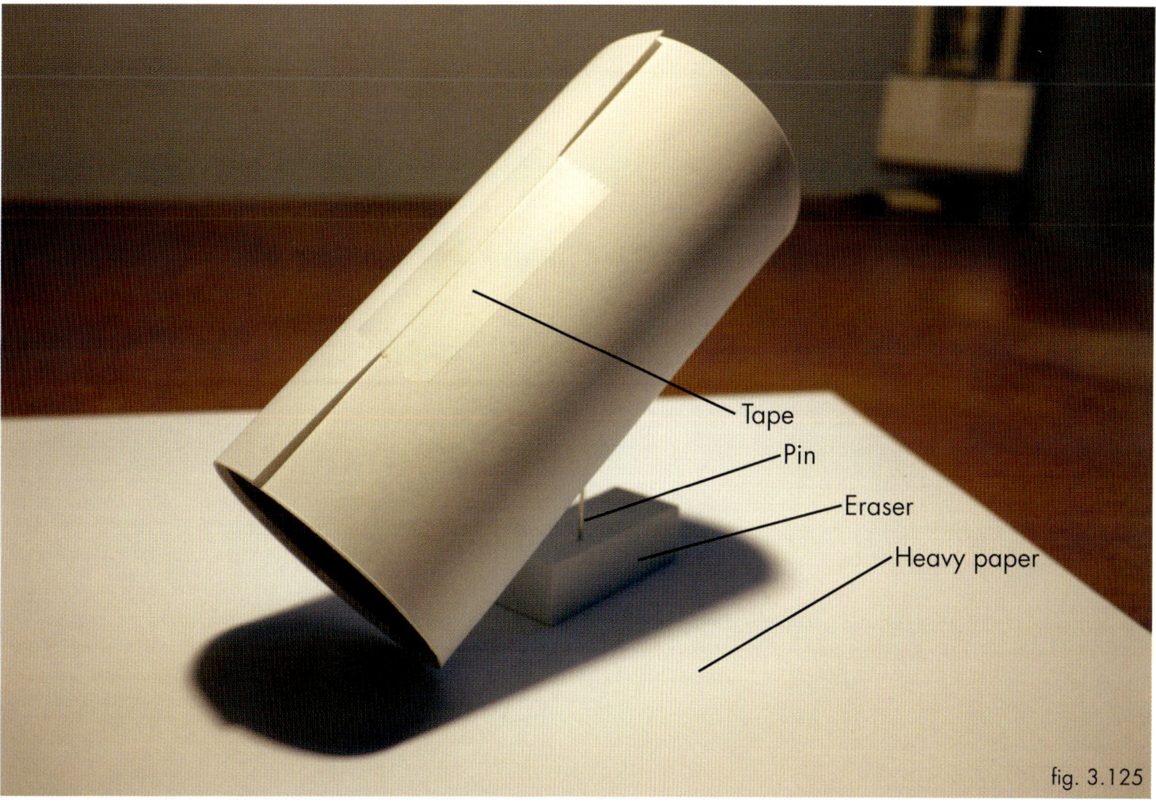

fig. 3.125

Create a physical set-up to gain needed information. Observing how a real-world object behaves in real-world light is often the most effective way to sharpen skills. Here, the core shadow and the cast shadow on the ground are the main areas of focus. Use white matte-finish paper to create surfaces that show highlights and shadows most clearly. Wrap a cardboard tube to create the cylinder and use available items to secure it. In this case, an eraser was used as the base and a pin worked for flexible positioning. Place it on heavy white paper to increase the contrast and the reflected light. The end planes were not as important, so the cylinder was open at the ends. The taped seam was placed in an area where it does not hinder any light and shadow observations.

CONES 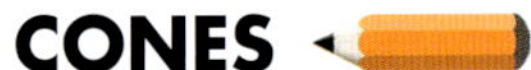

Rendering Cones in Sunlight or Local Light

After successfully constructing shadows of cylinders, the construction of the shadows of a cone will be simple...but there is one important difference to keep in mind! Drawing the shadow-direction tangents to the base of the cone does not properly locate the core shadows. A cone has a surface that tapers toward the top, so the core shadows shift in position when compared to a cylinder with sides that are parallel.

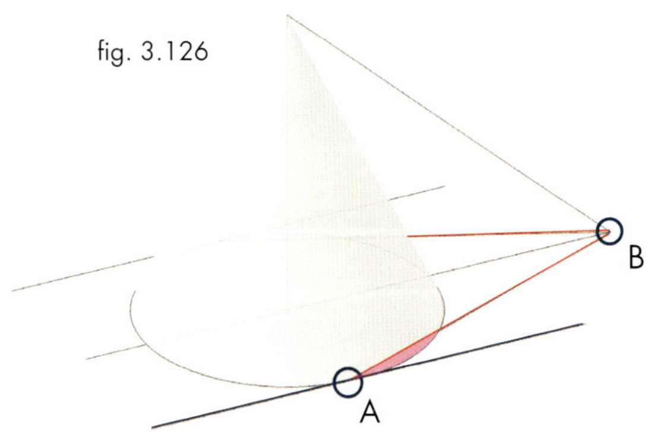

Fig. 3.126: If the points where the shadow direction intersects the base of the cone (A) were connected with the cast shadow of the tip of the cone (B), then the connecting lines run underneath the base of the cone. This is clearly incorrect.

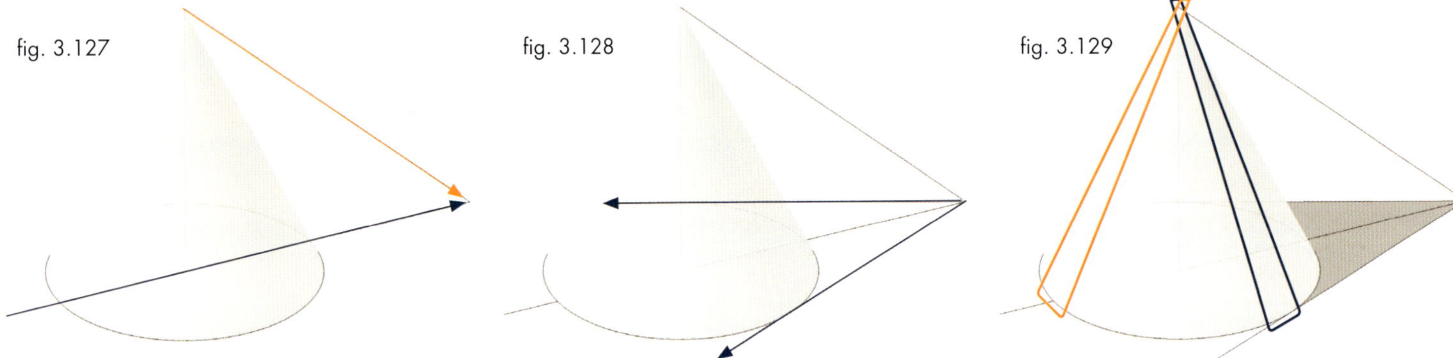

Fig. 3.127: So, to construct the shadow of a cone, first cast the shadow of the tip of the cone on the ground.

Fig. 3.128: Draw two lines from the tip's cast shadow, tangent to the base of the cone.

Fig. 3.129: The terminators and the cast shadow are now defined. The passive highlight is in the same location as on a vertical cylinder except the shape of it needs to taper. This technique works for both sunlight and local light scenarios.

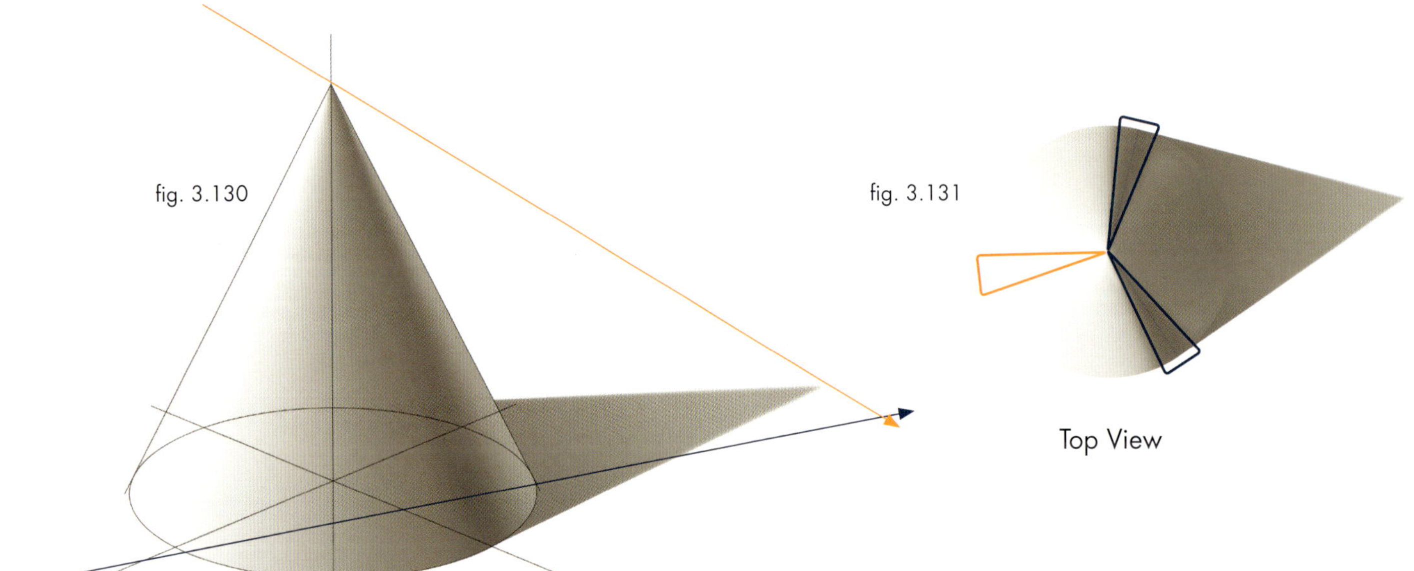

Rendering:

Fig. 3.130: Use the halfway-to-black technique to assign values to the core shadow and the cast shadow. The passive highlight's value will vary according to the light angle. In this rendering, the passive highlight area is close to perpendicular to the light source. Thus, this area is rendered with its true value.

Fig. 3.131: Observe how the core shadow and highlight taper toward the top of the cone. Keep in mind that with a cone, more of the surface area is in the light than is in shadow.

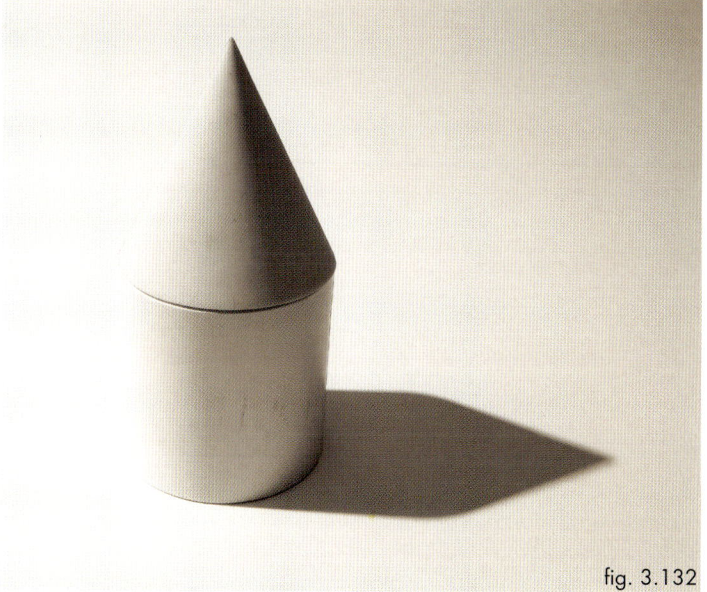
fig. 3.132

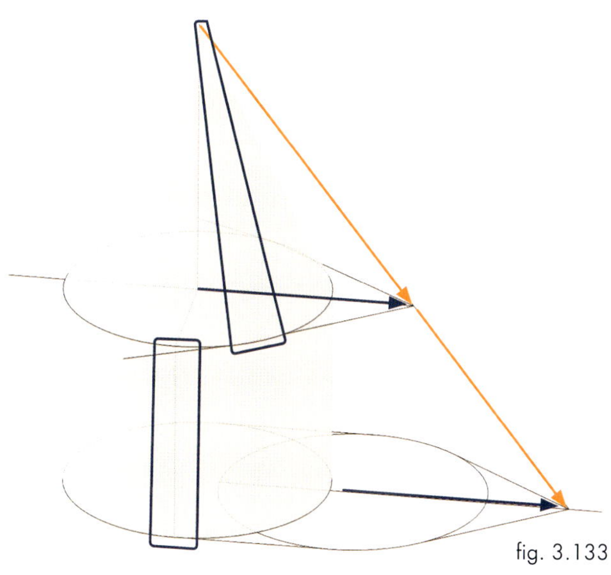
fig. 3.133

The difference in construction techniques for finding the terminators of a cone and a cylinder are made quite clear when these two geometric shapes are combined. The core shadows of the cylinder and the cone are offset, as the cylinder's shadow area is larger than that of the cone. In addition, the reflected light on the shadow side affects the cylinder more strongly than the cone, since the cone has a surface tilting away from the ground plane, which is the source of the reflected light. To combine both construction techniques, first cast the shadow of the cylinder. Then, cast the tip of the cone onto the ground to complete the full cast shadow. Finally, cast the shadow of the cone tip onto an extended construction plane the cone is sitting on, and use this cast point to determine the location of the cone's core shadow.

Casting the Shadow of a Horizontal Cone

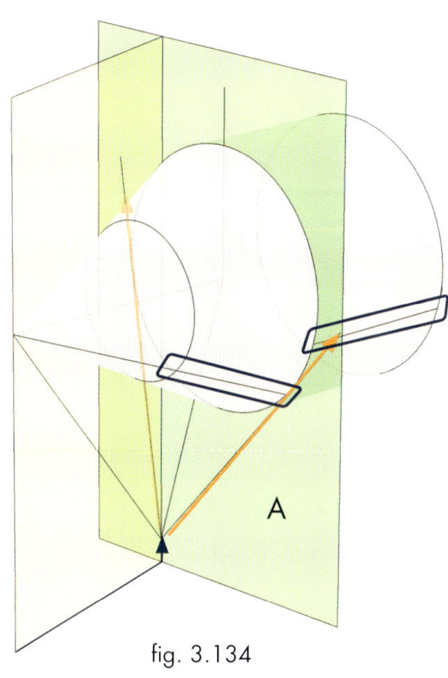

fig. 3.134

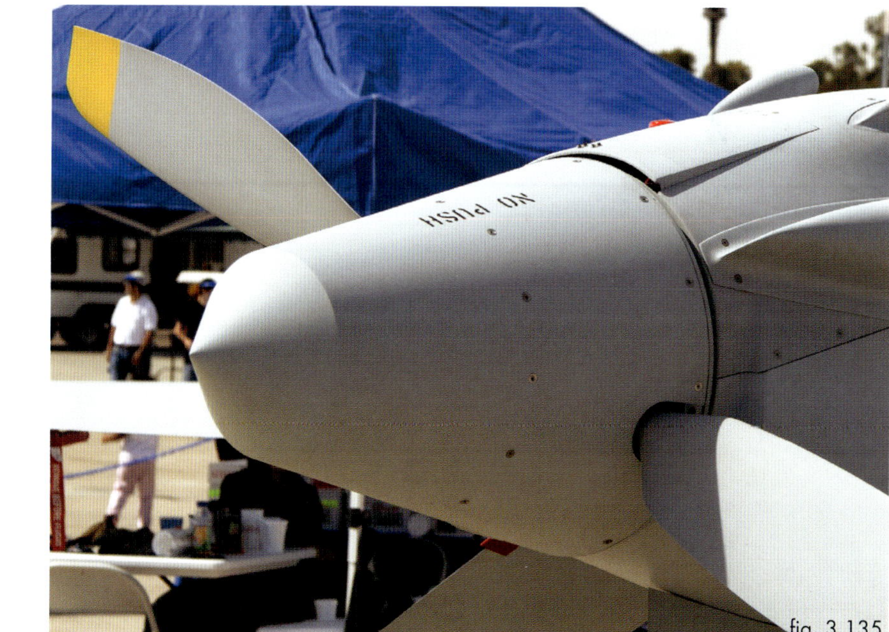

fig. 3.135

This technique is the same as that with a vertical cone, except that the "floor" is now a "wall." To cast the shadow of a horizontal cone, add a vertical plane (A) at the base of the cone. Cast the shadow of the tip of the cone onto this auxiliary construction plane. Should the tip be missing, extend the cone for the purposes of the construction. Finally, draw the terminator tangencies from the cast-shadow tip of the cone.

Be aware, as in the photo of the airplane above, that the core shadow "jumps" when the tapering angle of a cone changes.

SPHERES

When constructing and rendering a sphere, it is absolutely essential to assign the values consistently, since it has no surfaces that indicate what its orientation is. A cylinder has two flat end planes, but a sphere only has a position in space relative to the ground and the light direction, which makes the core-shadow/cast-shadow relationship a little more sensitive. The light direction determines how the cast shadow, the core shadow and the passive highlight are all aligned.

As with a tilted cylinder, these constructions may become quite complex since all variables must be under control. Based on the scenario and one's accumulated experience, making an educated guess may lead to faster results than performing a full construction. But before beginning to make these guesses, one must be confident doing a complete construction and accurately critiquing a rendering. Using and observing practical set-ups with real-world objects (such as, for a sphere, a ping-pong ball on a piece of paper), greatly helps.

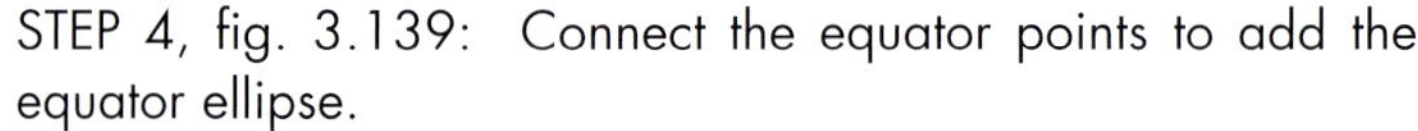

Constructing a Sphere

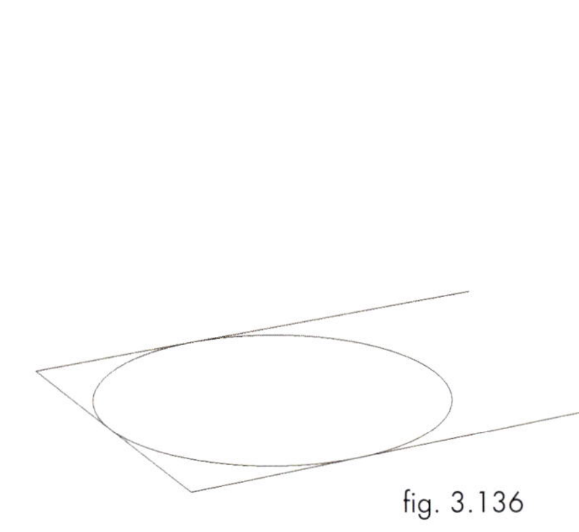

fig. 3.136

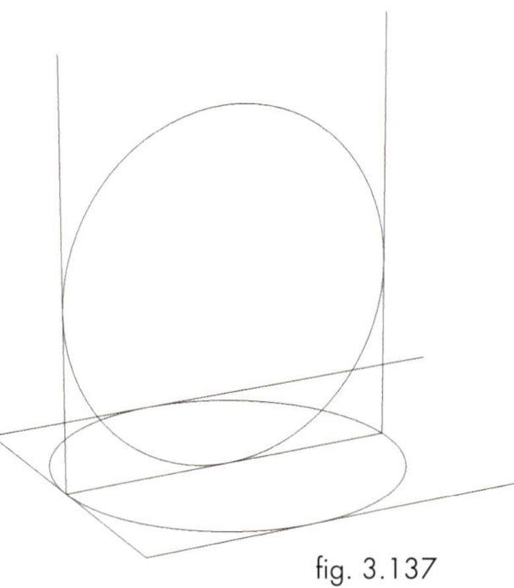

fig. 3.137

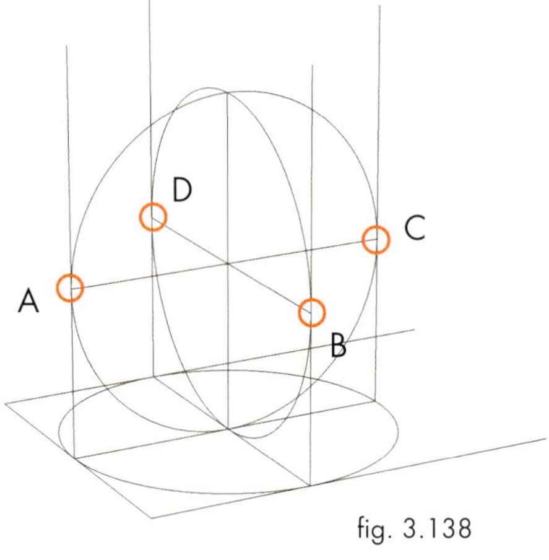

fig. 3.138

Before beginning, it is necessary to know how to draw a square in perspective by drawing an open-ended rectangle, and then placing an ellipse within it with the correct minor axis and degree. This technique is covered in detail in our book, *How to Draw*, in Chapter 05.

STEP 1, fig. 3.136: Define the footprint of the sphere on the ground by drawing the defining boundaries and ellipse (top view).

STEP 2, fig. 3.137: Transfer the width of the sphere onto a center plane, and then create a standing ellipse plane, in perspective, of the same size as the one on the ground.

STEP 3, fig. 3.138: Rotate the standing ellipse by 90° in top view. Draw lines that radiate horizontally from the center of the sphere to find the equator points (A, B, C and D).

STEP 4, fig. 3.139: Connect the equator points to add the equator ellipse.

STEP 5, fig. 3.140: Draw the outline of the sphere that is defined by these three elliptical section lines.

This is one way to approach the construction of a sphere. Alternatively, start with the outline and then add the elliptical section lines. Either way, these section lines are necessary for the sphere to be fully defined and to provide a good basis for the shadow construction.

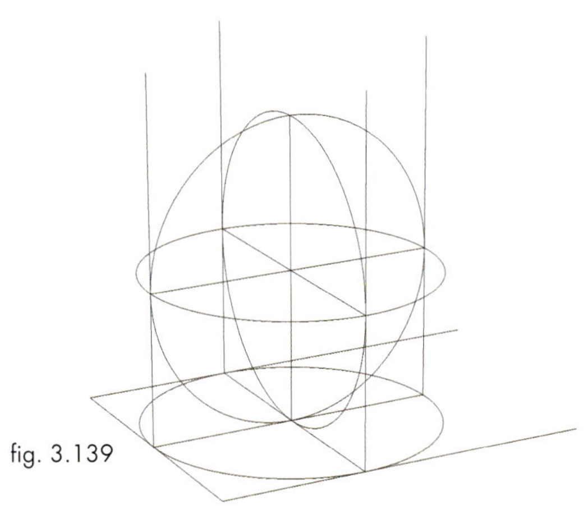

fig. 3.139

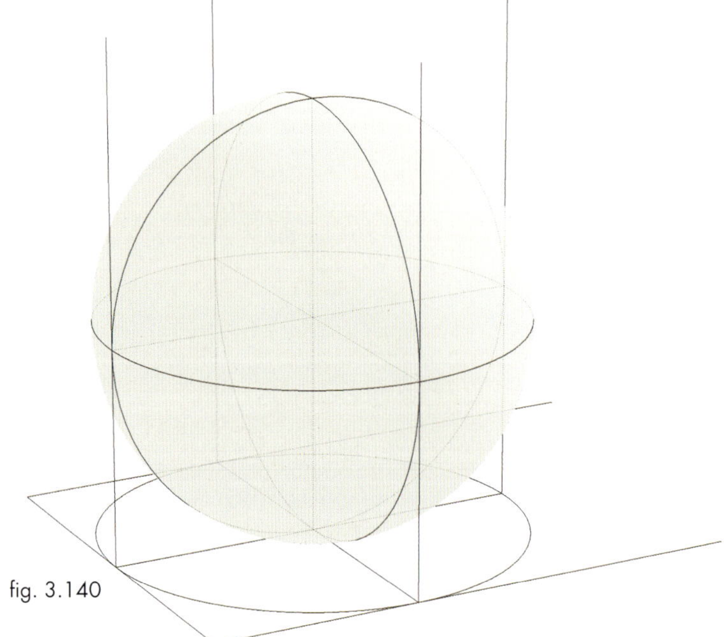

fig. 3.140

Rendering a Sphere in Sunlight: Top Light

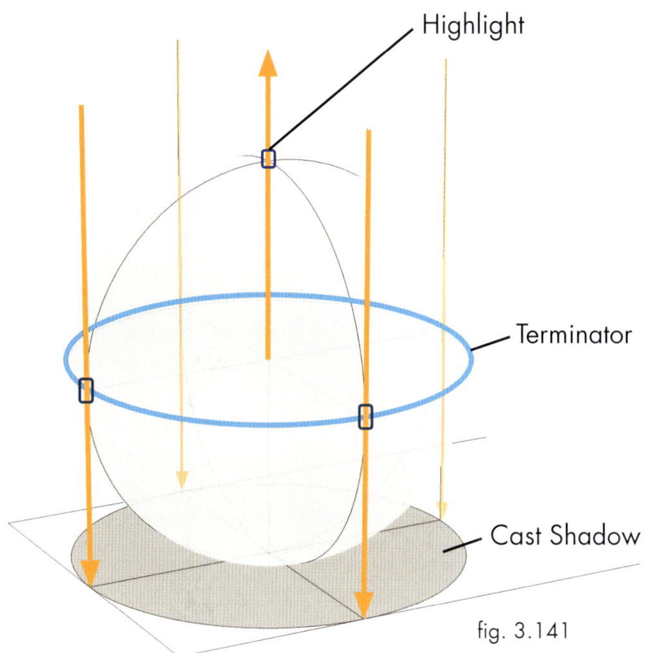

fig. 3.141

Construction:
Fig 3.141: If the section lines are accurate in the initial perspective sphere construction, then everything is already in place to cast the sphere's shadow in a top-lit sunlight scenario.

The terminator is found through the light rays' tangencies to the surface, which is the equator section line from the previous perspective construction. The terminator (blue line) is located at the equator of the sphere in this top-lit scenario.

The cast shadow is the footprint/top view of the sphere on the ground plane, which was the first ellipse in the perspective construction.

The highlight is found at the highest physical point of the sphere where the section lines meet and the surface is the most perpendicular to the light rays.

Rendering:
The value assignments for a sphere follow the halfway-to-black principle. The passive highlight is the true value of the sphere, and the core and cast shadows will be halfway to black. In fig. 3.142, observe the reflected light and the occlusion shadow where the sphere touches the ground.

Fig. 3.143: It is important to wrap the value shading around the minor axis that points from the center of the sphere toward the light source (fig. 3.143). Keep this in mind to ensure consistency of the gradation and proper shading of the form.

The reflected light off the ground can create an effective core shadow that wraps all the way around the sphere's terminator.

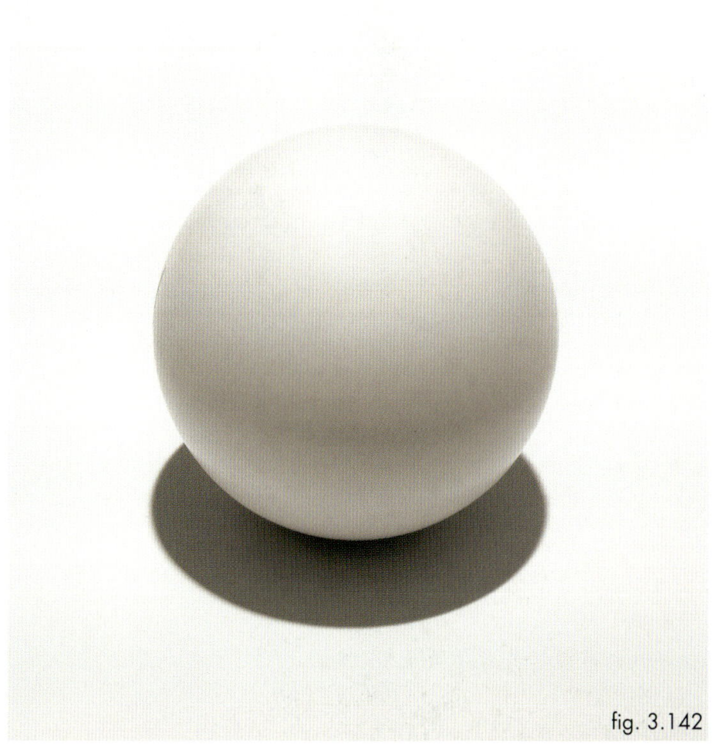
fig. 3.142

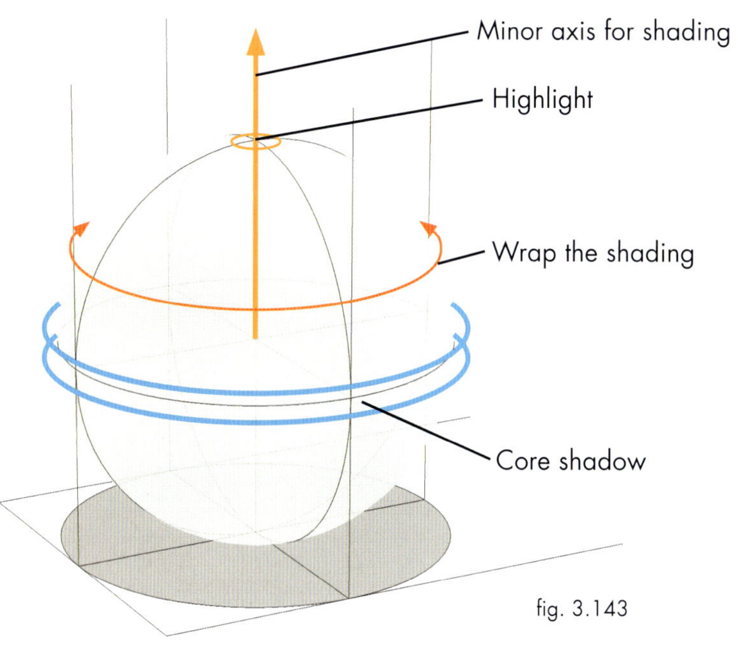

fig. 3.143

Rendering a Sphere in Sunlight: Sidelight

When sunlight comes more from the side, the core and cast shadows of a sphere are found almost the same way as for the top-lit sunlight scenario.

Construction:
STEP 1, fig. 3.144: Draw light rays tangent to the side centers that align with the shadow direction lines in top view (A and B).

STEP 2, fig. 3.145: Draw two more light-ray lines extending to the ground plane that are tangent to the vertical ellipse section shown.

STEP 3, fig. 3.146: Draw a line from the center of the sphere toward the light source to find the passive highlight position, which is also the minor axis of the terminator.

STEP 4, fig. 3.147: To draw the terminator ellipse, go through the four tangency points found in steps 1 and 2 using an ellipse with its minor axis aligned to the light direction.

STEPS 5-6, figs. 3.148, and 3.149: Project these four points onto the ground to help draw the shape of the cast shadow. Be aware that the shadow shape is not a circle in perspective (ellipse), but is still very elliptical. This means that if an ellipse guide is being used to draw a smooth curve, it might not match. Simply move the guide as needed to get a smooth and accurate curve.

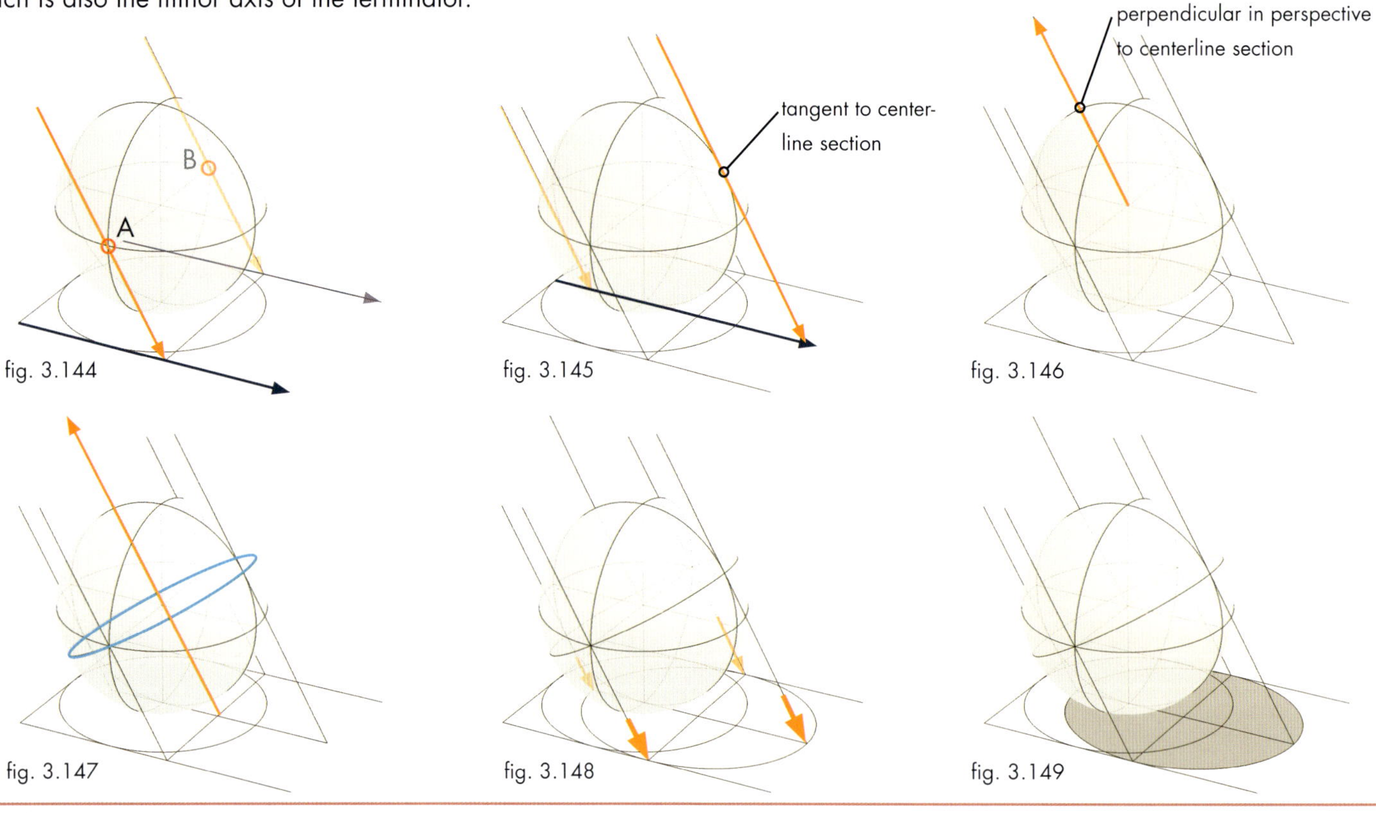

fig. 3.144 fig. 3.145 fig. 3.146 fig. 3.147 fig. 3.148 fig. 3.149

Rendering:
Fig. 3.150: In general, the sidelit sphere in sunlight is rendered in a similar way to the top-lit sphere in sunlight. There are, however, some subtle differences besides the changed position of the core shadow and the changed shape of the cast shadow. Notice that the core shadow is wider at point A on the shadow side of the sphere, but almost disappears at point B. This is because the reflected light off the ground is almost strong enough to remove the core shadow. It is possible to choose to make an educated guess about these values, instead of drawing a full construction. When this is the case, make sure that the minor axis for the core shadow retains the correct relationship to the passive highlight location, and that the cast shadow on the ground is symmetrical. If these three things align reasonably well, the viewer will accept that this is indeed a sphere.

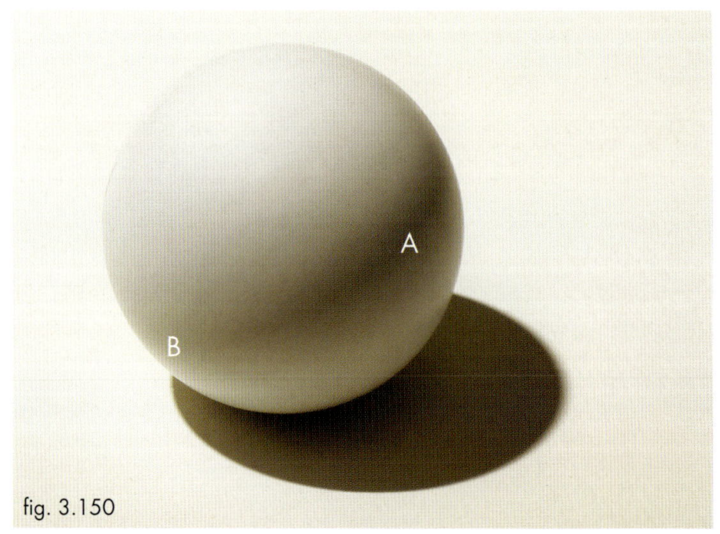

fig. 3.150

Rendering a Sphere in Local Light: Top Light

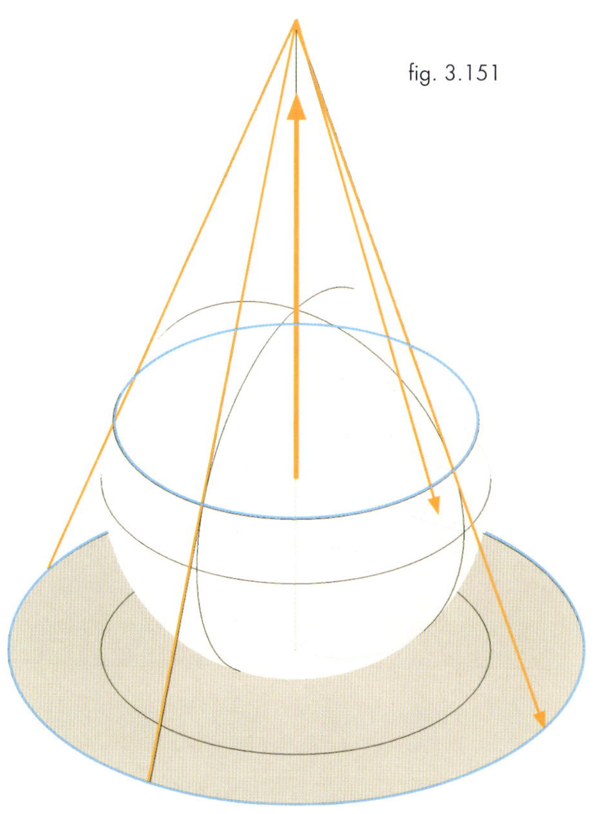
fig. 3.151

Construction:
Fig. 3.151: When top-lighting a sphere with local light the construction is very similar to that of a top-lit sphere in sunlight. Find the terminator, the passive highlight, and the cast shadow using the same techniques for the top-lit sunlight construction. Using the elliptical section lines, the tangencies to find the terminator can be constructed. The line from the center of the sphere to the light source shows the location of the passive highlight, and is also the minor axis for the terminator ellipse and cast shadow, all the same so far. The main difference that becomes apparent with local top-lighting is that in comparison to sunlight, the cast shadow has become larger and the terminator has moved away from the equator toward the light source. Now, just as with a cylinder in local light, there is more of the object in shadow than is in the light.

Rendering:
Fig. 3.152: Again, apply the same principles from rendering a sphere in top sunlight, but keep in mind the following differences: The core shadow can become wider as it is further from the reflected light source, the ground. The cast shadow receives less ambient (sky) light and is darker than the sunlight cast shadow.

We strongly recommend creating a practical set-up with real-life objects to observe and compare sunlight and local light scenarios. Being able to manipulate the object and light as needed while noticing how all the variables are affected is an invaluable way to take one's skills to the next level.

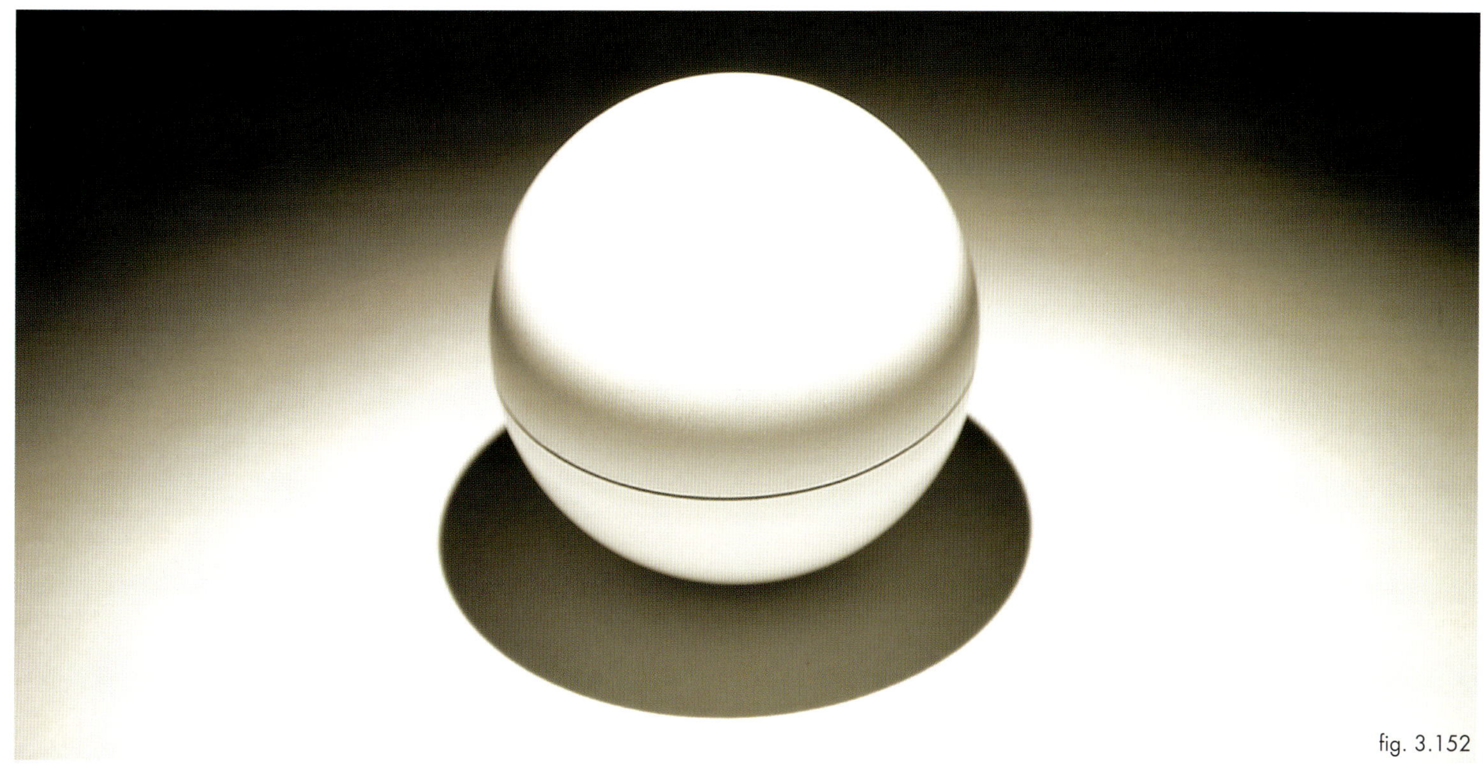
fig. 3.152

Rendering a Sphere in Local Light: Sidelight

The construction of a sphere's values in local sidelight can be quite time-consuming. In this case, it is often faster, and just as effective, to make an educated guess. The challenge in this construction is to position the core shadow correctly, and this requires familiarity and confidence with placing ellipses in perspective. This is covered in detail in Chapter 01 of our first book, *How To Draw*.

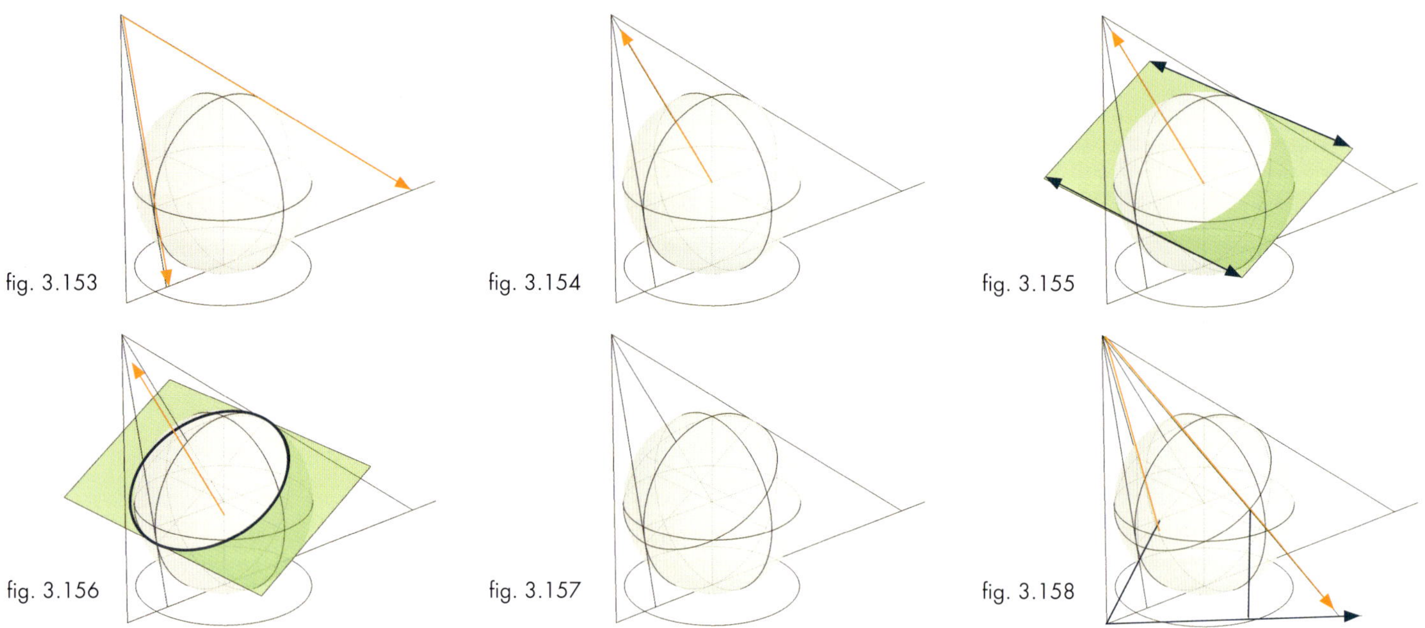

fig. 3.153 fig. 3.154 fig. 3.155

fig. 3.156 fig. 3.157 fig. 3.158

Construction:

STEP 1, fig. 3.153: Determine the light-source location and align the section lines of the sphere with the light source, so that the section lines divide the sphere in half. Draw the tangent light-ray lines down to the ground for the midsection of the sphere.

STEP 2, fig. 3.154: Draw a line from the center of the sphere toward the light source. This line determines the passive highlight location and is also the minor axis for the terminator.

STEP 3, fig. 3.155: To draw the ellipse for the terminator, the minor axis and two boundary lines that determine the degree of the ellipse are needed. For these boundary lines, draw two lines from the tangency points of the light rays perpendicular to the imaginary light plane, or center section of the sphere. This will result in the plane shaded green.

STEP 4, fig. 3.156: Using this construction plane, place the terminator ellipse.

STEP 5, fig. 3.157: There is only one solution for this ellipse in order to match the given perspective constraints. Observe that the terminator is closer to the light source, than in the previous local top-lit construction. Again, the shadow side covers more surface area than the lit side of the sphere.

STEP 6, fig. 3.158: Construct the cast shadow by projecting the terminator disk onto the ground using several vertical sticks.

STEP 7, fig. 3.159: Often, a good educated guess will take care of the construction. Having a rough idea where the terminator is located is usually enough, but also ensure that the minor axis in the construction is correct.

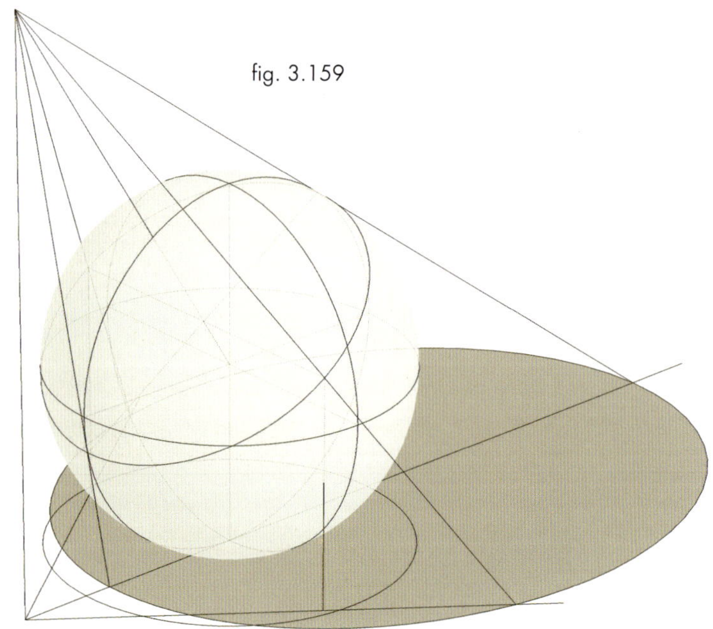

fig. 3.159

Rendering:

To render a sphere in local sidelight, combine the techniques for rendering a sphere in side sunlight, and a sphere in local top light. Be mindful of the effects of the falloff on both the ground plane and the spheres. Observe how the ground shifts to nearly black and the cast shadow has a value structure darker than halfway to black. The reflected light off the ground reaches a little past the equator cut line of the sphere. This causes the core shadow to disappear in the half facing the ground, while a dark wedge appears in the upper half on the shadow side of the sphere. Compare this to the sphere in side sunlight on page 82, where the ambient sky light instead brightens this part of the sphere.

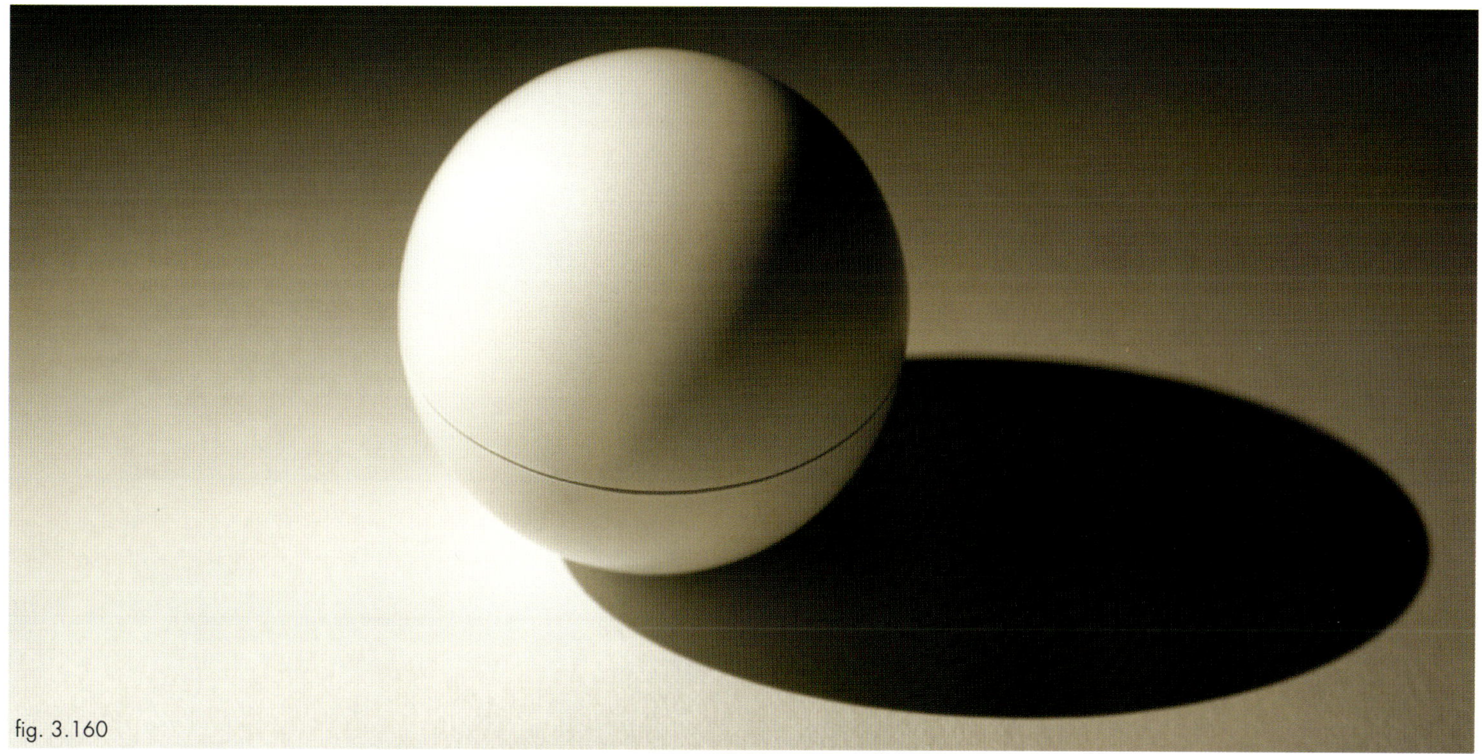

fig. 3.160

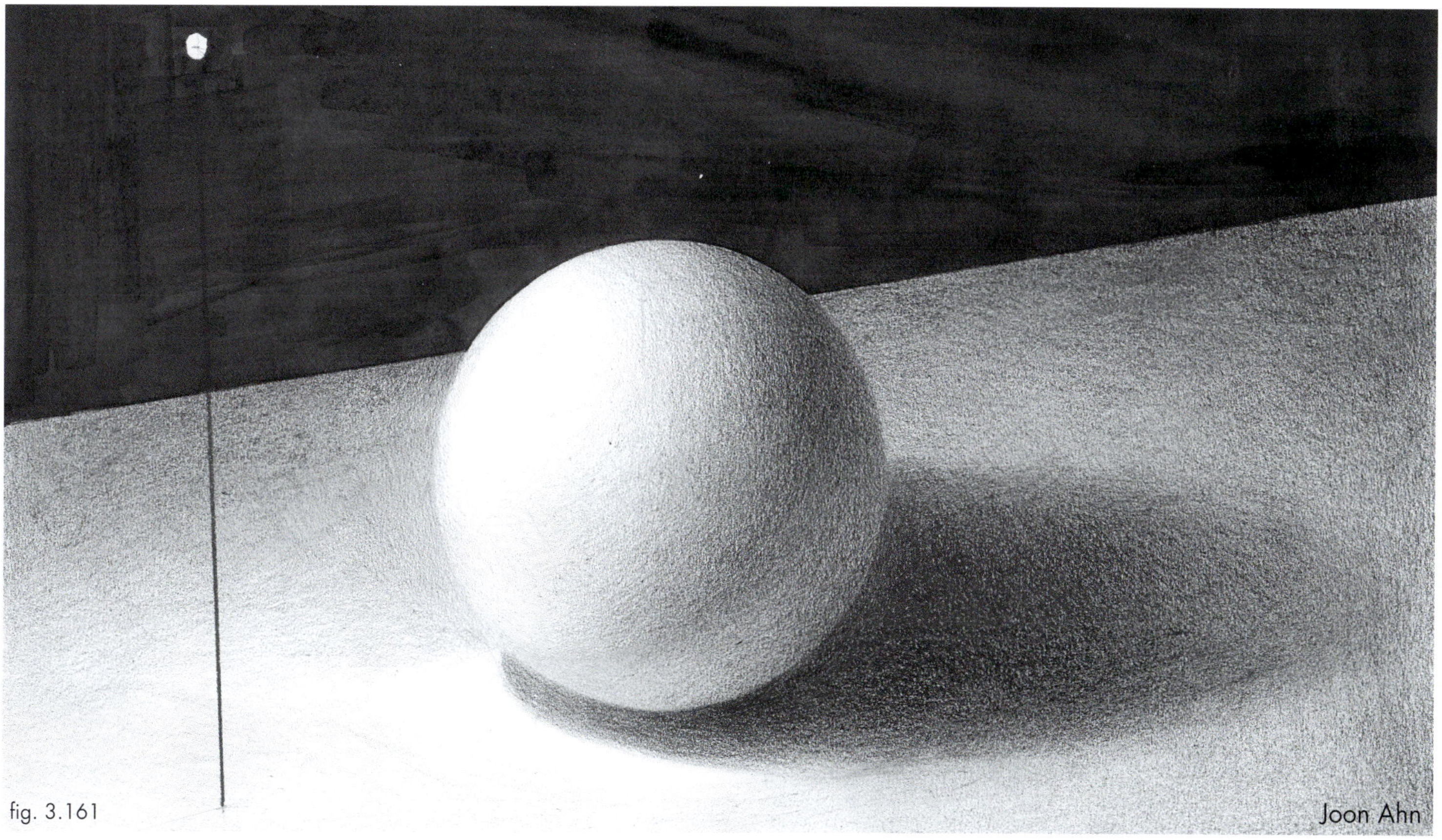

fig. 3.161

Joon Ahn

SHADOW EDGES

The sharpness of a shadow's edge indicates how far away the object is that casts the shadow. The farther away, the softer the cast shadow of that edge; the closer the distance, the sharper.

There are three important distances that can be observed in fig. 3.162:
Long – from the top of the light pole to the ground
Medium – from the top of the stool to the ground
Short – from the top of the cube to the top of the stool

Take a good look at the detail image (fig. 3.163) and spend a moment thinking about what can be observed about the cast shadows. First, the shadow edge of the light pole is the softest; the edge is blurred, and close to 4 inches or 10 centimeters wide. Second, look at the shadow the stool casts on the ground. Its shadow transforms from hard-edged, near the leg, to soft-edged at the seat. Third, observe the cube. The shadow edge is razor sharp all the way around.

Perspective foreshortening also sharpens a shadow's edge. In fig. 3.162 the pole's shadow looks crisp, but in fig. 3.163 the shadow is very soft.

Building the practice of observing surroundings with an artist's eye can really advance rendering skills. Look around the environment and notice the edges of shadows to see how they show scale. For example, find a high-rise building and follow its shadow away from it to see how it softens, while also observing that the windowsills cast short, crisp shadows all along the building's side.

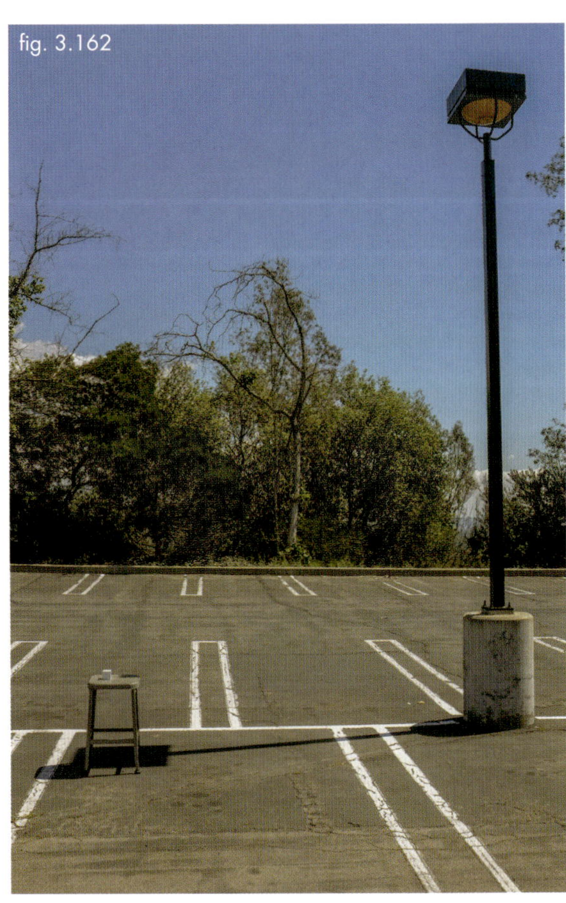
fig. 3.162

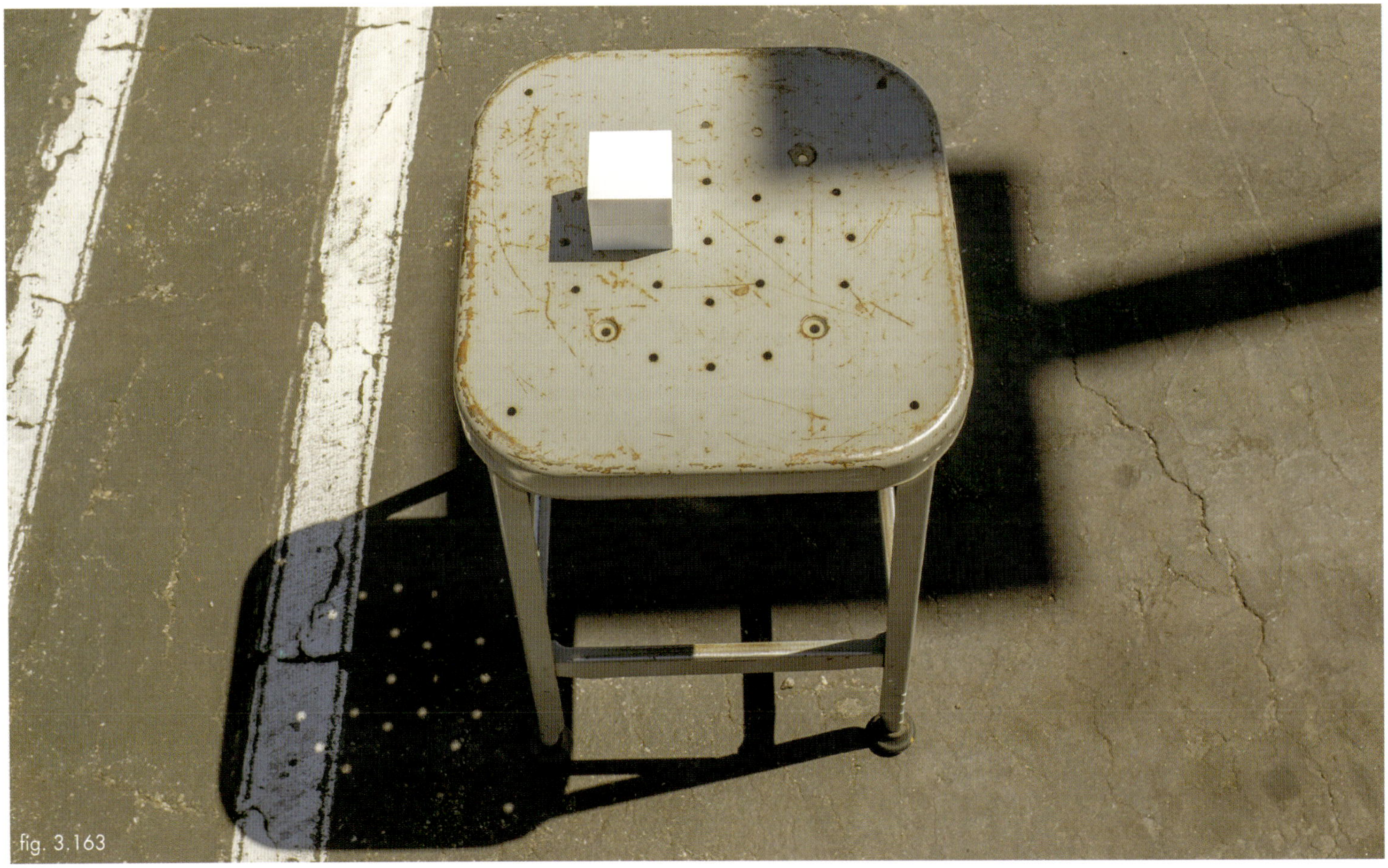

fig. 3.163

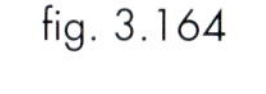
fig. 3.164

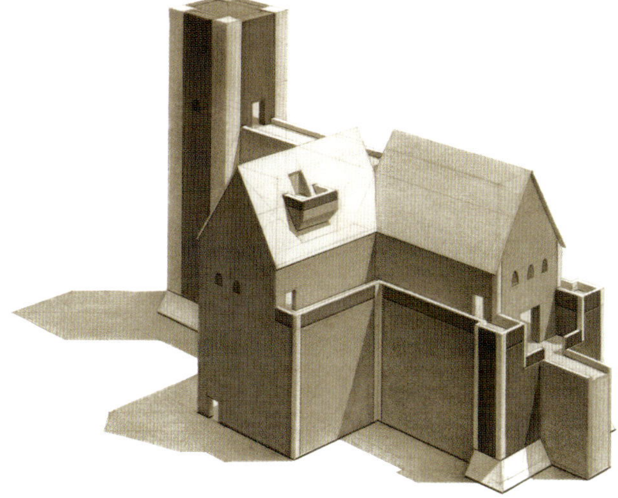

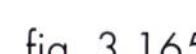
fig. 3.165

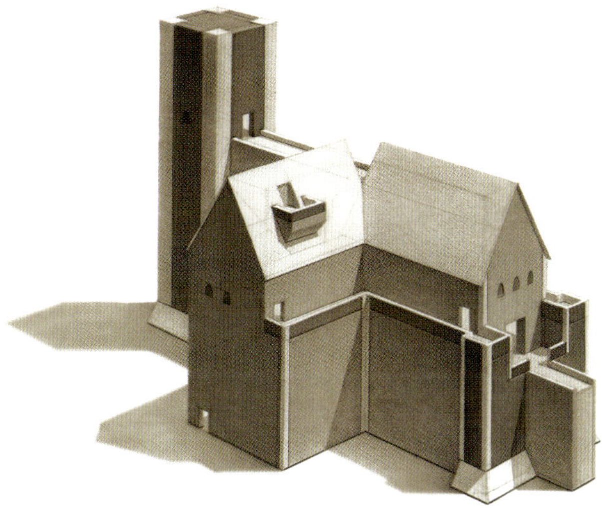

Charles Liu
www.cliuportfolio.blogspot.com

Tianxu (Tim) Guo
www.spikethehedgehogstudio.com

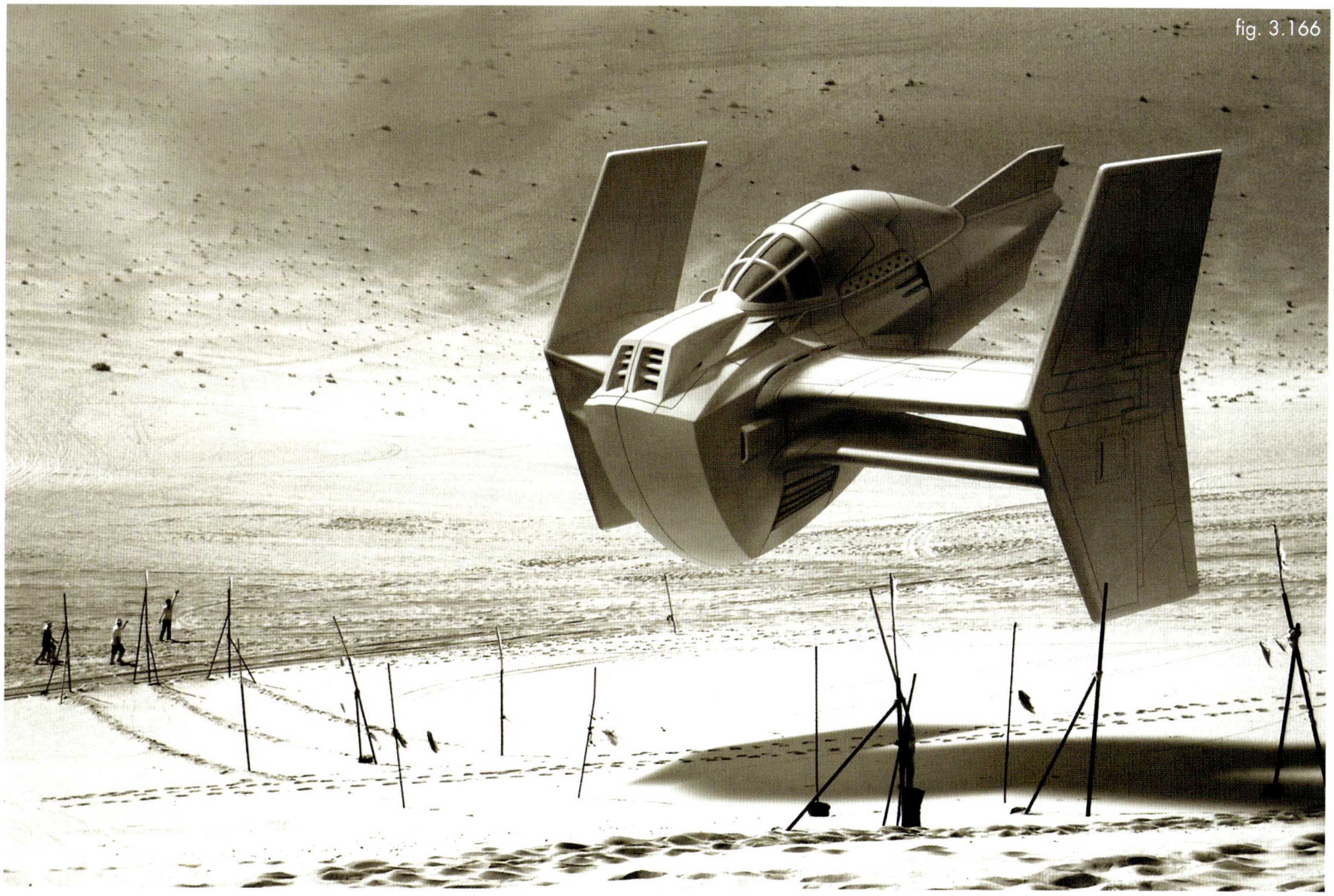

fig. 3.166

BACKGROUNDS 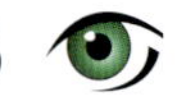

Backgrounds have a huge influence on how light or dark an object appears. Comparing figs. 3.167 and 3.168, the walker appears darker against the white background and brighter against the dark background. Adjusting the background is a way to control how strongly the silhouette reads. In fig. 3.169, the background itself has some value changes that serve to emphasize the silhouette of the upper walker forms and to minimize the strength of the silhouette of the legs.

Observe these relative darkening and brightening effects in photographs and in nature. In fig. 3.170, the rainbow stripes of the left balloon remain the same value as they stretch across the sky, but the stripes *appear* to darken as the background brightens. Adjusting the values of the background extends the illusion of the range of values of the objects in the foreground.

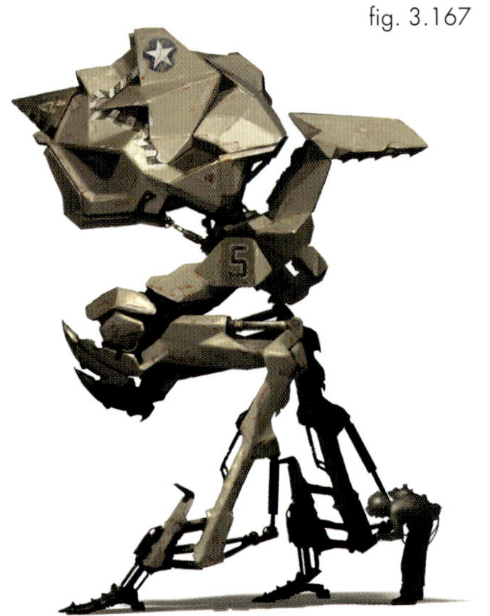

fig. 3.167

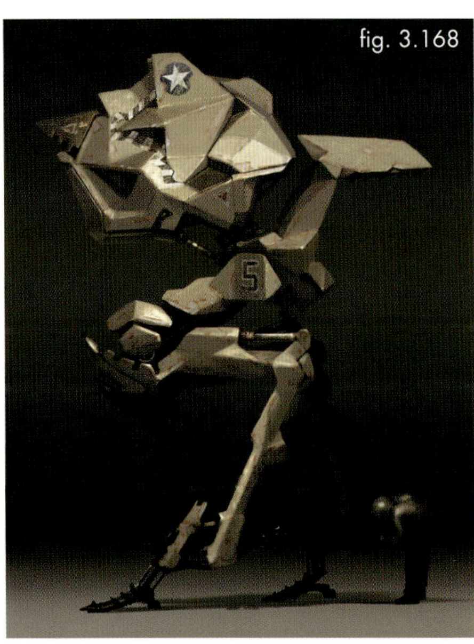

fig. 3.168

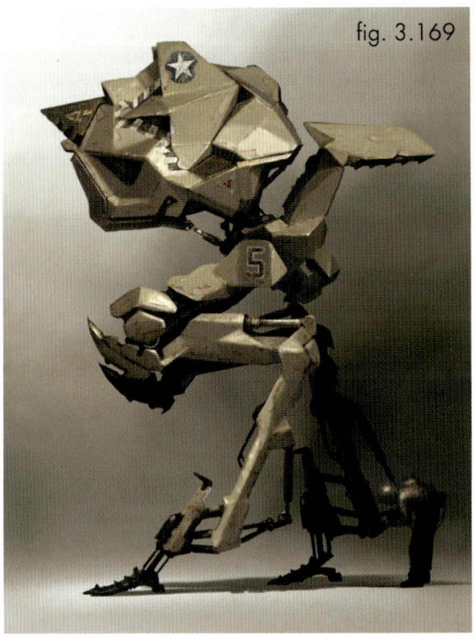

fig. 3.169

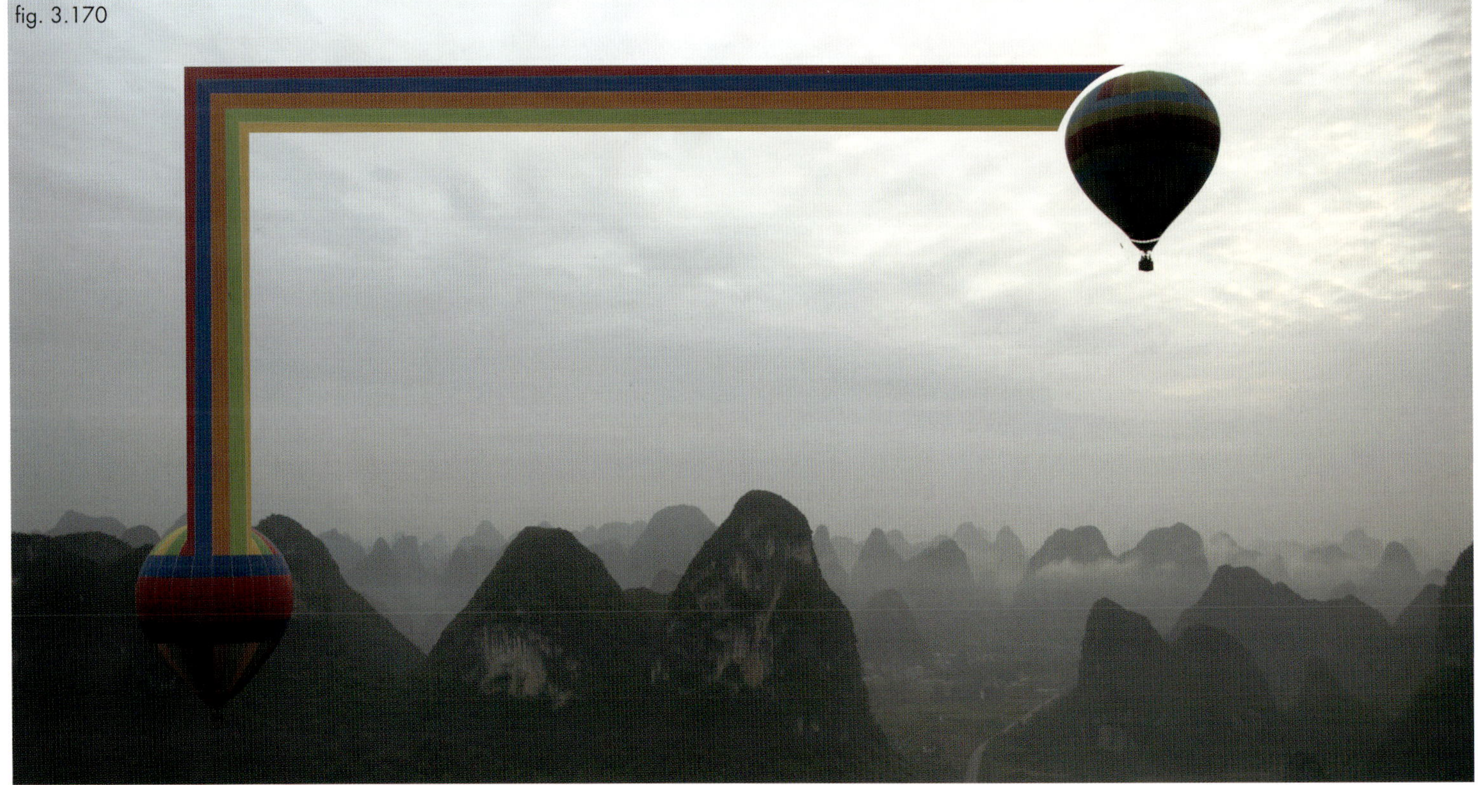

fig. 3.170

WHY THE BACKGROUND IS SO POWERFUL

This illusion, which makes a constant value look different depending on the value of the background, can be used to great effect. Fig. 3.171 shows an elliptical shape of a constant gray value against a background that has a dark-to-light gradation. The change in the background values creates the *illusion* of an inverse value gradation in the gray ellipse, from light at the top to dark at the bottom. The constant gray value is more easily apparent in the vertical band at the right. Notice how this gray appears to change as it moves from the white background onto the background with the gradation.

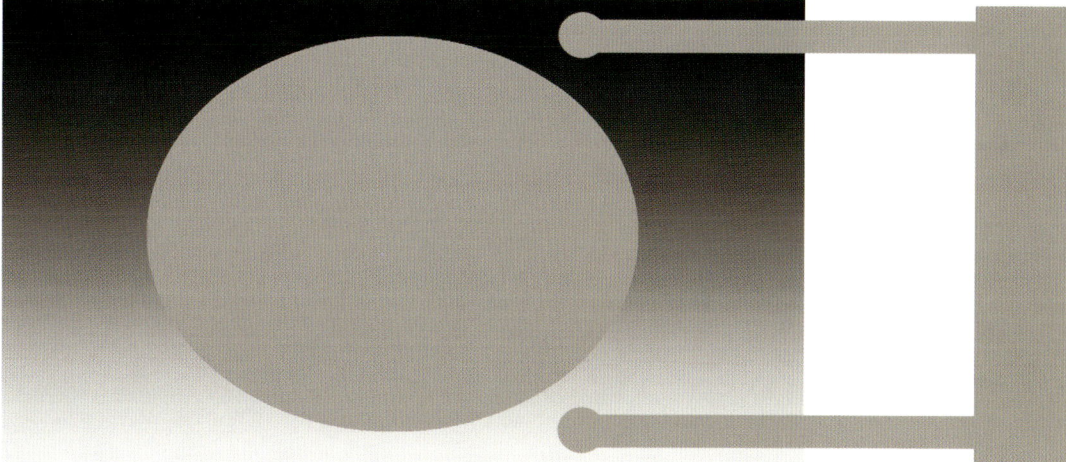

fig. 3.171

Figs. 3.172 and 3.173, below, use this effect to control which areas of the walker appear lighter or darker. They show the same rendering on two different backgrounds. In fig. 3.172, the top of the walker appears lighter than the one in fig. 3.173. Simply, to make an object look brighter, put it on a dark background; to make an object look darker, put it on a light background.

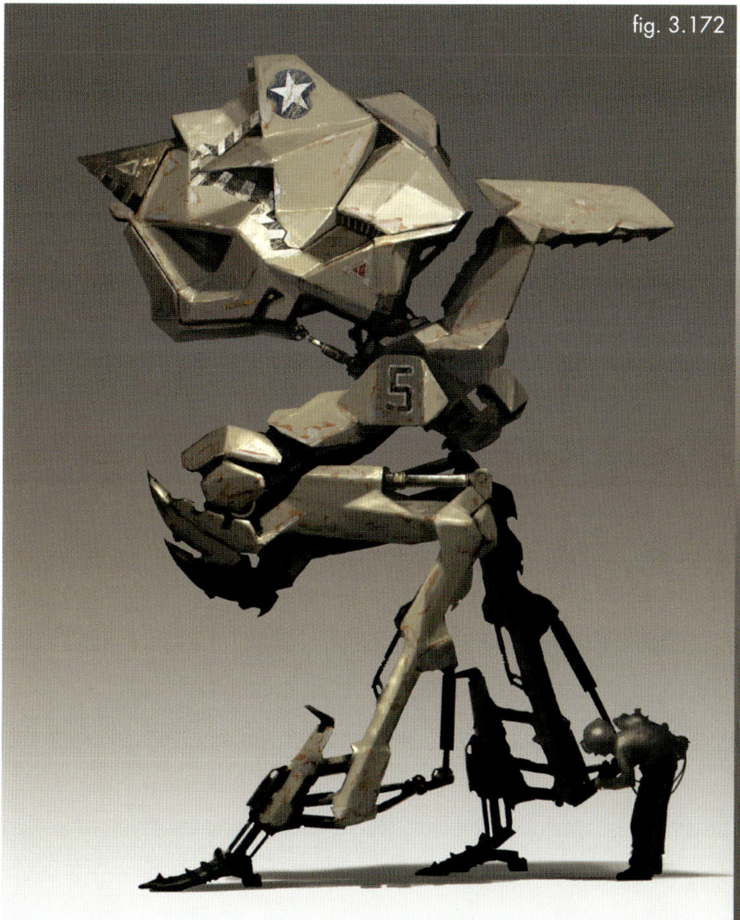

fig. 3.172

fig. 3.173

TRADITIONAL VERSUS DIGITAL RENDERING TOOLS

A powerful computer with an expensive software package cannot make a truly great rendering on its own. As a tool, the computer is only as good as the knowledge and skill of its user. It is the designer and/or artist who must patiently study the physics and practice the techniques until they are mastered. It is extremely important to start with traditional media. Not having the option to "undo" necessitates pre-visualizing an image and improves the decision-making process. Learn to mix these traditional mediums; they do not have to be used on their own. Also, as you become more proficient with both digital and traditional tools, experiment with combining them.

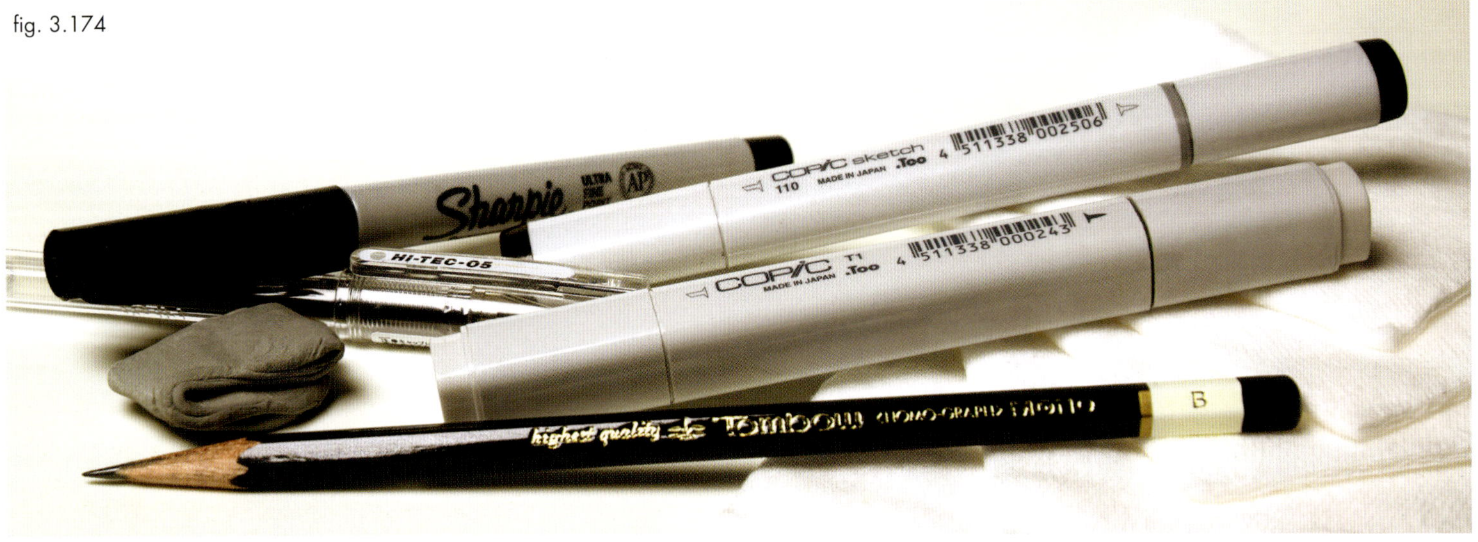

fig. 3.174

TRADITIONAL TOOLS OF THE TRADE

When choosing rendering tools, several things are important to consider. First, think about what tools you are already good at using. Most artists grew up using graphite pencils in school, so they are familiar and very accessible. These pencils provide a large amount of control when rendering smooth gradations, and erasing is easy, which is helpful while learning the basics. Eventually, erasing can be used as a rendering technique. Watch the video tutorial linked to page 92 for a demonstration.

Another thing to consider is how will the renderings be used? Do the renderings need to be archival? Will they possibly be sold in the future? If so, then choose rendering mediums that will not fade or blur over time. Work on "acid free" surfaces. Most sketchbooks and illustration boards indicate if the surfaces are acid-free. This prevents any mediums that are added to them from degrading.

I have a lot of old project renderings and class demos that were done on vellum with markers, wax (colored) pencils, gouache and NuPastel chalks. The markers have faded and blurred over time and the wax pencil has oxidized. Using a spray fixative, like those from Krylon, can help to slow this, but markers on vellum are not considered archival. If these mediums are known to fade and are not archival, why use them? It is because of the nice way the vellum accepts the chalk and pencil, along with a technique of rendering on both sides of the vellum to create translucent effects, which looks so good when it is new.

Unlike illustrations, which are finished products, design renderings are not meant to be sold as art, or appreciated long into the future. Renderings for most design jobs are a way to communicate concepts and to indicate how future objects will look. The fact that many of these renderings are skillfully executed pieces of art is secondary. For example, as movie studios change locations and need more space, old design renderings, living within large flat files, usually get the heave-ho into the trash bin. They have served their purpose to help communicate the concepts of a designer, and are no longer needed. The renderings were never the finished product, but they were a very integral part of the design process.

Pencil Handling

Once the ability to cast shadows and assign values has been learned, it is time to focus on controlling the medium. There is no absolutely right or wrong way to hold a pencil or marker, what matters most is that you are comfortable and can confidently shade the values as needed. That said, these are the two most common ways to hold a pencil. A "writing grip" (fig. 3.175) is useful when adding lines and details, and an "overhand grip" (fig. 3.176) is better for loosely laying out a concept, or for figure drawing. With the overhand grip, it is helpful to hold the pencil a long way back from the point, in order to sketch with a lighter touch and use more of the side of the graphite to add smoother gradations. Both of these techniques are explained in more detail in the video tutorial linked to page 92.

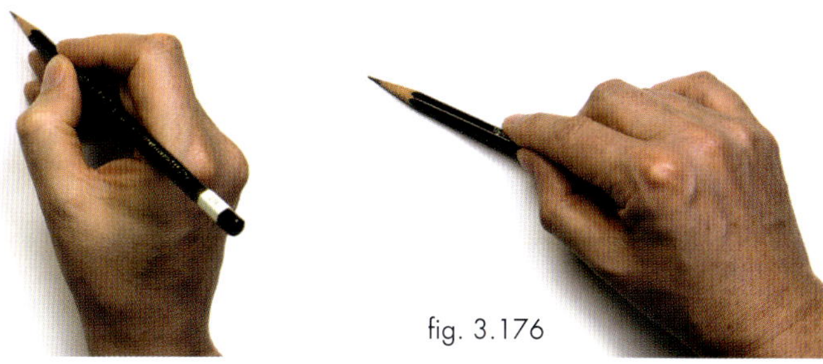

fig. 3.175

fig. 3.176

Graphite and Wax Pencils

There are two major types of pencils to consider for rendering. One is the ubiquitous graphite pencil, the other is a wax pencil. Wax-based pencils can be found in art supply stores. Purchase black ones for the strongest contrast shading. Buy several and keep them all sharpened beside your workspace so that the flow of drawing is not interrupted by sharpening pencils. Find an electric or hand-cranked sharpener that creates a steep cut (fig. 3.178, pencil A). Small, cheap, manual twist sharpeners do not create a steep enough tip (fig. 3.178, pencil B).

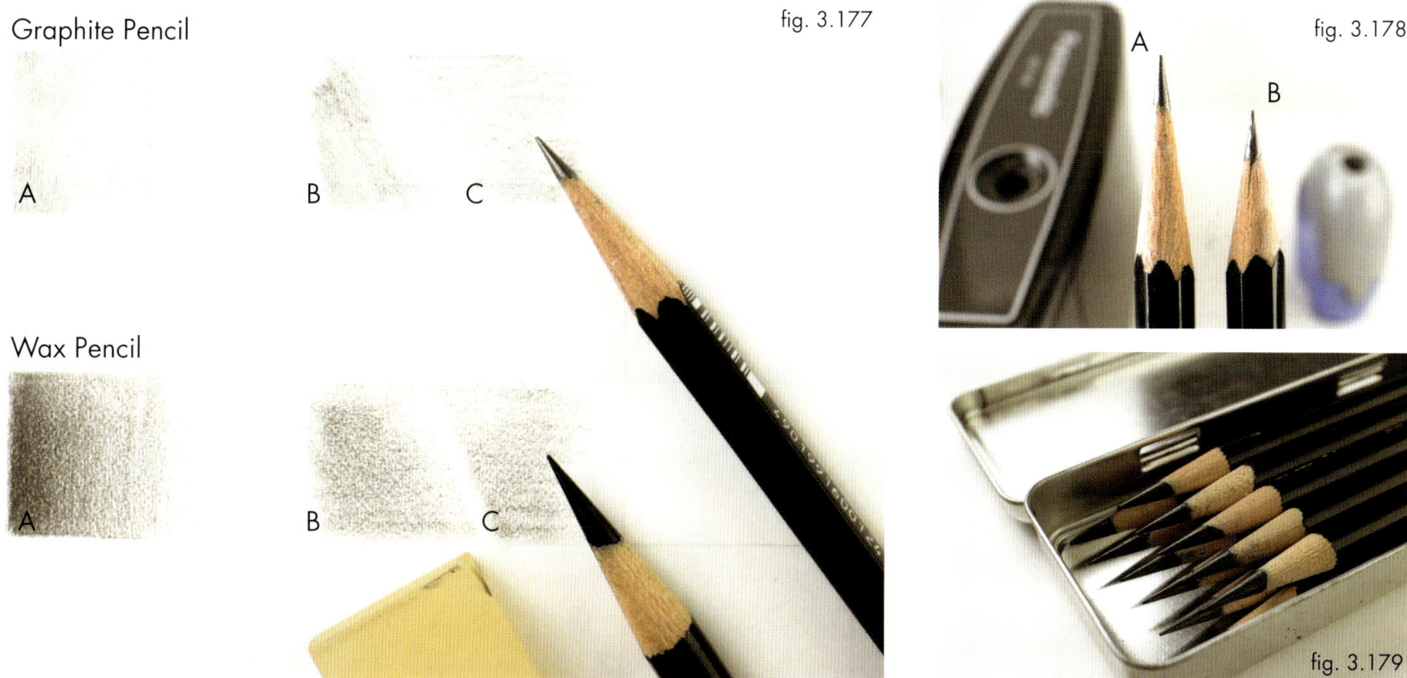

fig. 3.177

fig. 3.178

fig. 3.179

Whether using graphite or wax pencils, great results can be achieved. However there are some differences in handling and application, as illustrated in fig. 3.177.

A) Value Range: Wax has an advantage over graphite because it can render all the way to deep black. Eventually, a graphite pencil starts to polish the paper instead of getting darker.

B) Smearing: Graphite pencils smear more easily than wax pencils. This can be helpful when creating value gradations, but graphite pencils are also more likely to rub off on other artwork...or your hands.

C) Erasing: The graphite pencil erases well, while the wax pencil stays on the page. Building the value up slowly is important when using wax-based pencils, since erasing is not so easy. Use a drafting brush to remove eraser dust.

Wax Pencil Application

Wax-based pencils, also known as colored pencils, have long been a favorite of the industrial design community. They are much darker than graphite and harder to smear, so sketches stay cleaner. When using them in combination with markers, always render with the markers first, let the ink dry, then add the wax pencil. Otherwise, the alcohol in the marker will dissolve the wax, clogging the nib of the marker.

Graphite Pencil Application + Paper

Favorite brands of graphite pencil are listed on page 14. Look for a lead that is nice and smooth, and a hardness that feels comfortable to you. Graphite smears easily and erases well. This is both positive and negative. Smearing makes rendering lots of gradations possible, but it can be hard to keep the pages looking clean.

Choosing the right paper for the work is just as important as choosing the pencil. If the paper is too soft, it will not stand up well to erasing. If it is too hard, it might not accept the pencil very well and it will be hard to achieve the darker values. A paper that gets dark very quickly might be too rough and show more texture than desired, which is what happened with the pencil sketches below. Experiment to find your favorite pencil and paper. The choice is very personal and there is no one perfect combination for everyone.

fig. 3.180

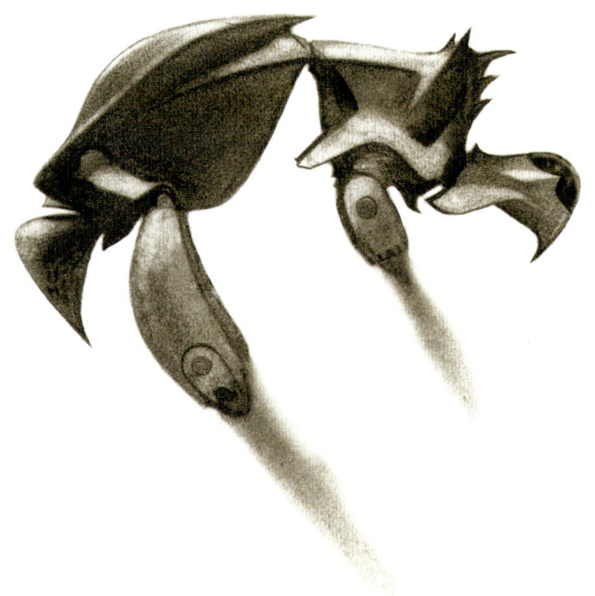

fig. 3.181

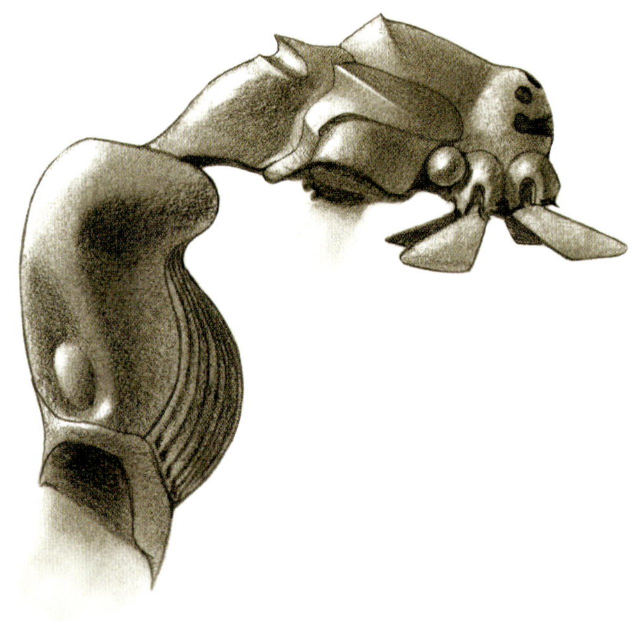

fig. 3.182

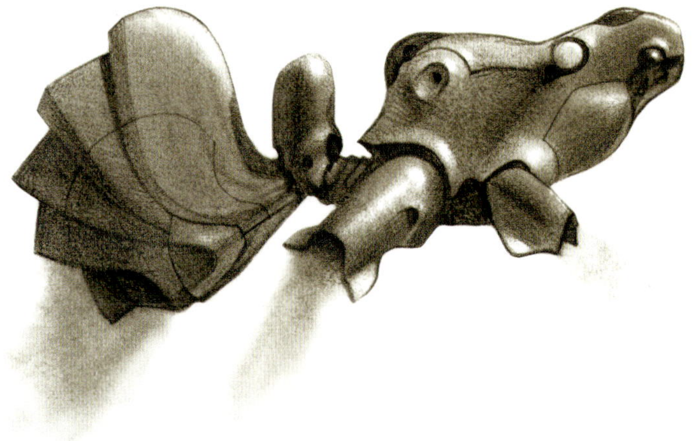

fig. 3.183

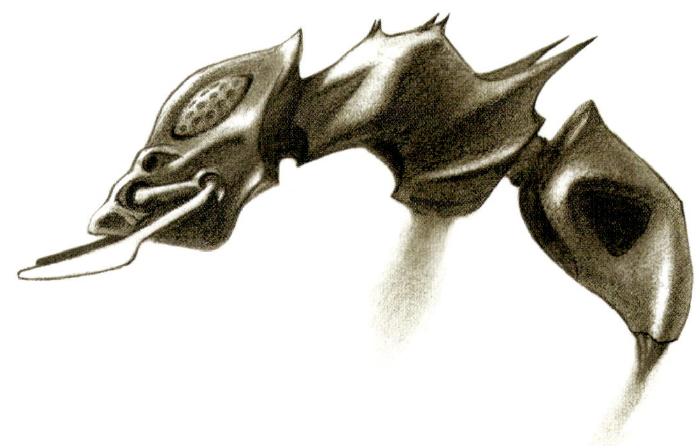

An excellent exercise to control the handling of pencils is to create a gradation with edges (fig. 3.184). Aim for the gradation to be smooth, without showing any pencil strokes. Stay within the edges, preferably *without any erasing*. Next, apply these skills to rendering geometric shapes (fig. 3.185). From there, the only limit is the imagination, as illustrated by Joon Ahn and Jason Kang.

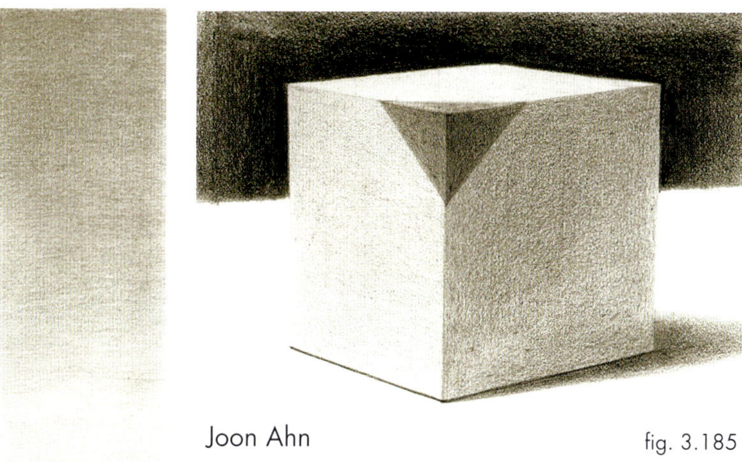

Joon Ahn

fig. 3.185

fig. 3.184

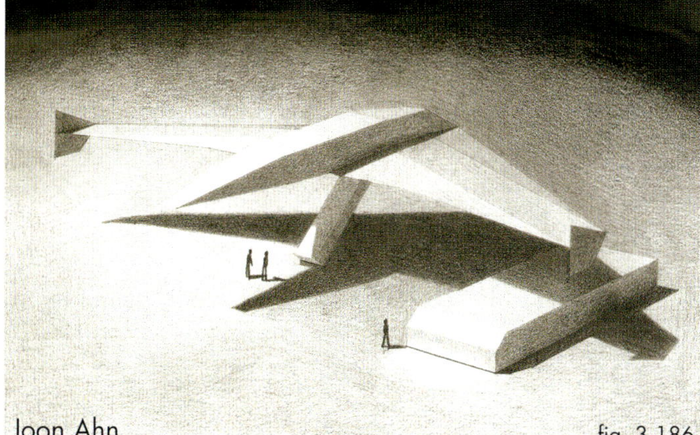

Joon Ahn

fig. 3.186

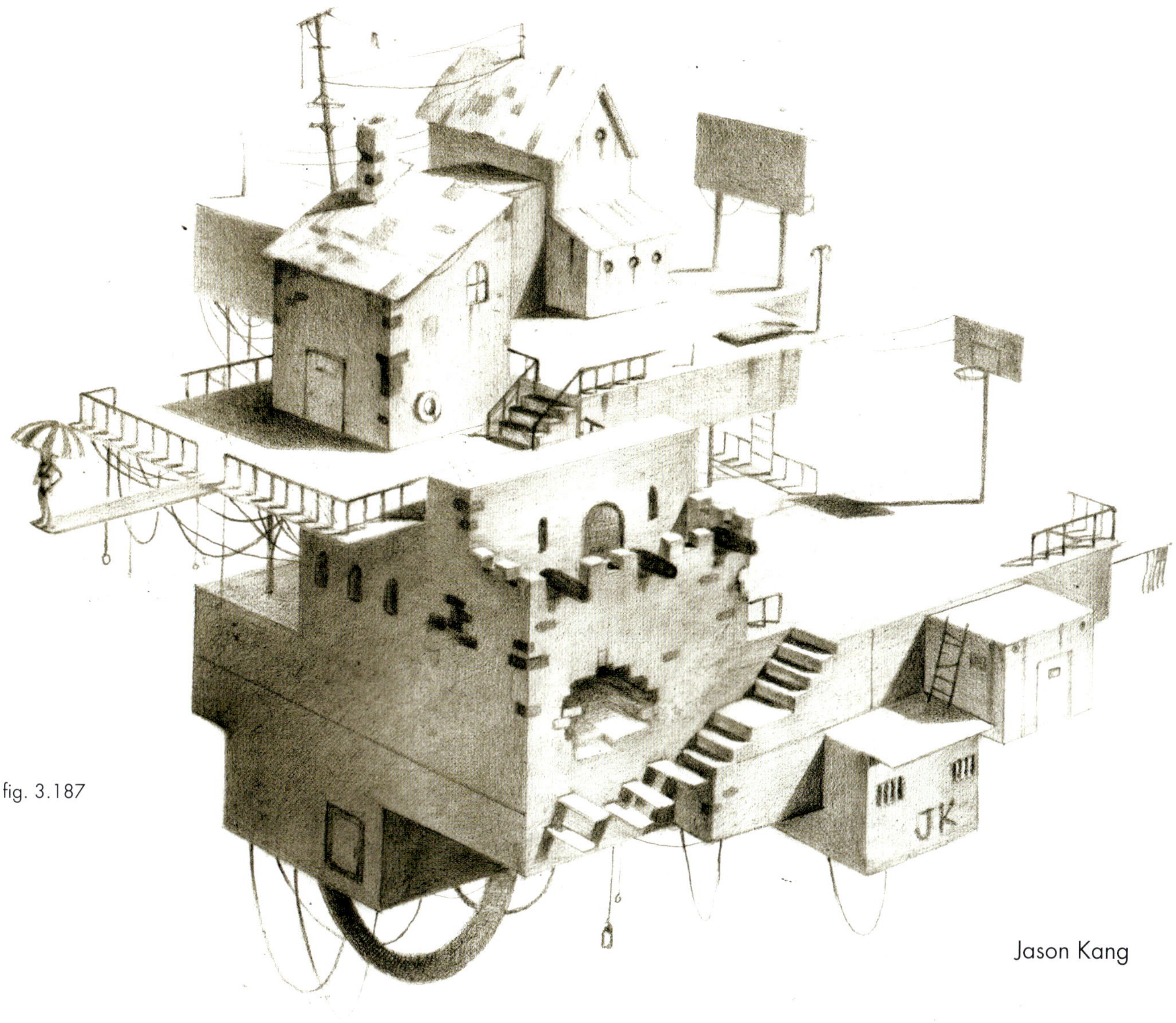

fig. 3.187

Jason Kang

MARKERS

Markers are used widely and were originally developed for the commercial-arts industry as a faster-drying alternative to watercolor. They are not very archival, but that is not the point of using them. Speed is their greatest advantage, as they dry very fast and permanently. This permanent nature of the inks makes them a bit of a challenge to use since, unlike pencils, they cannot be erased. Also, it is very hard to render nice gradations with makers. For this reason, NuPastel chalk or colored pencils are often used over the top of markers to add gradations. Watch the video tutorial for the sketch below (fig. 3.188) to see these techniques in action.

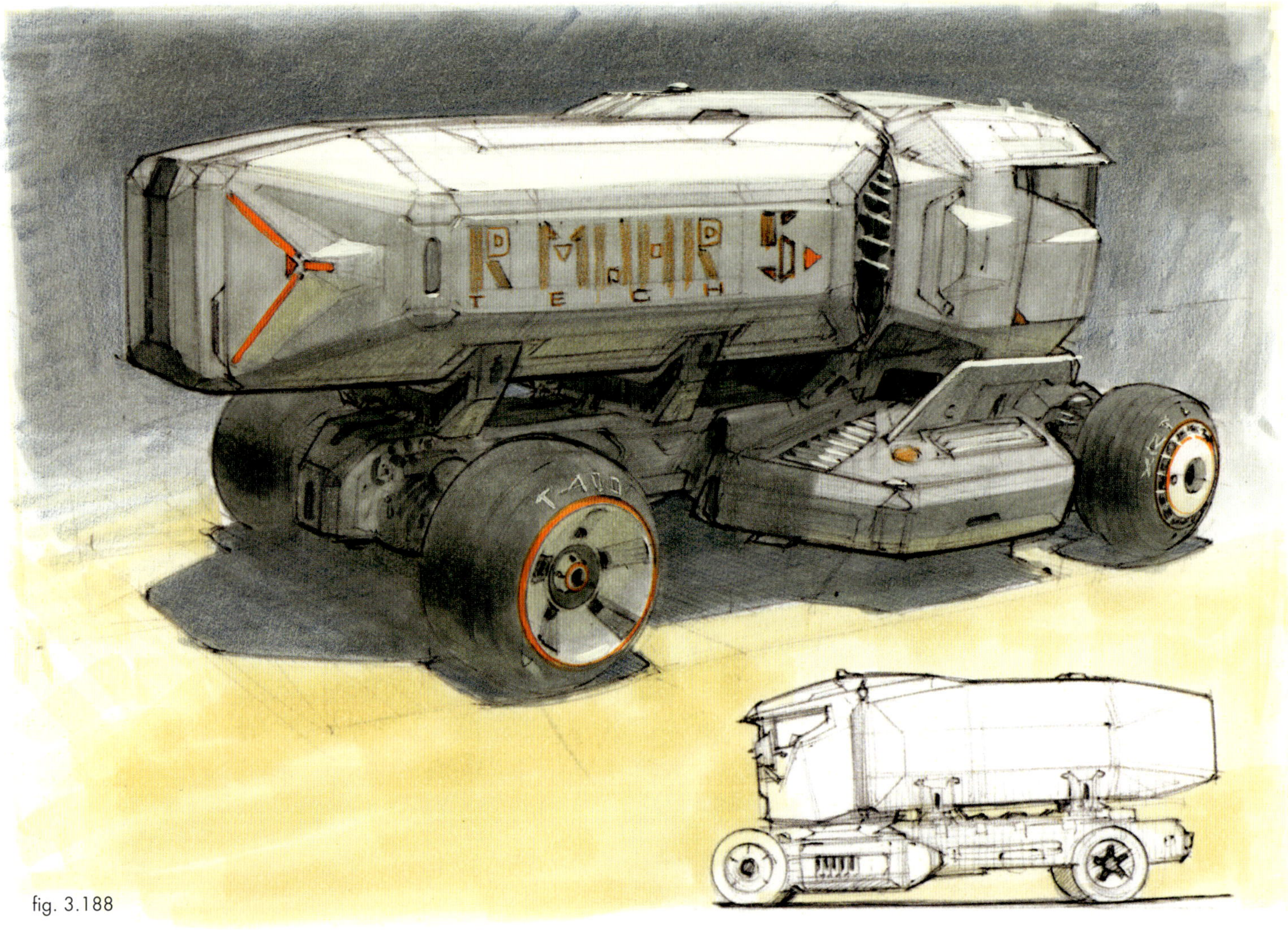

fig. 3.188

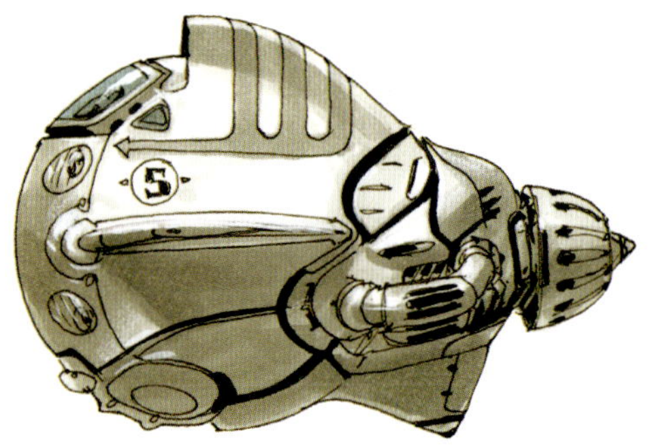

fig. 3.189

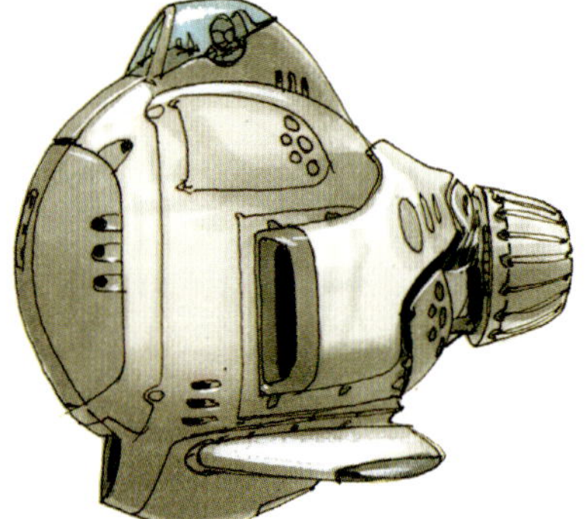

fig. 3.190

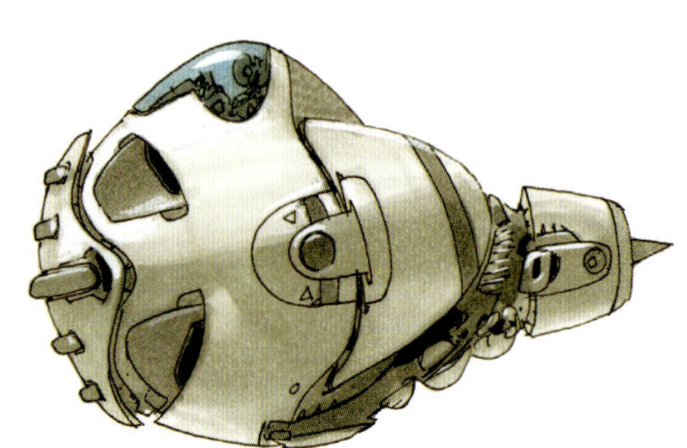

fig. 3.191

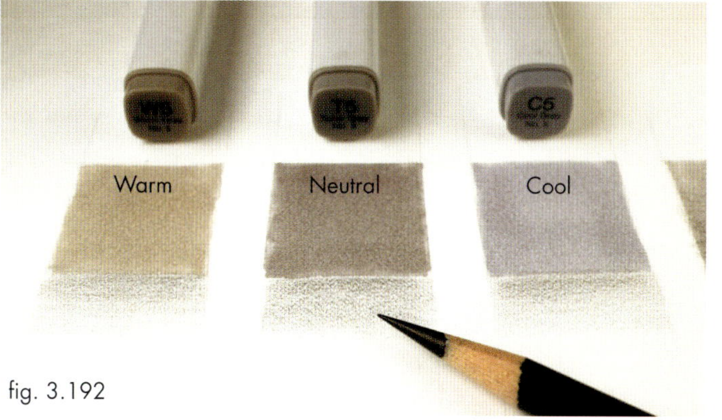

fig. 3.192

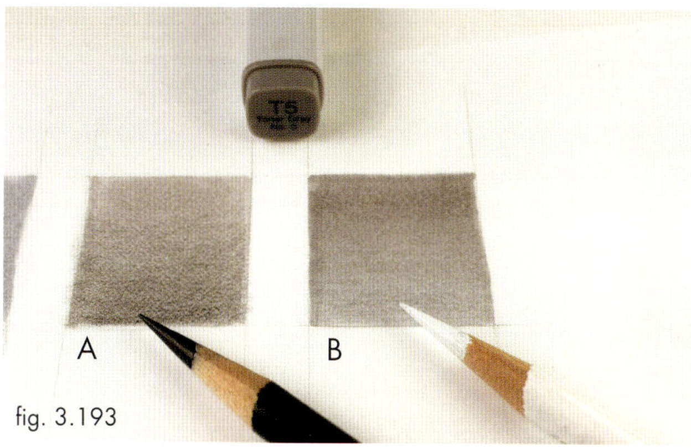

fig. 3.193

Gray markers come in warm, neutral and cool. Choose a set of gray markers that matches the hue of your pencils. Usually, neutral gray is the best match. Making gradations with markers is possible but it does require practice. Start by putting down a flat value and then adding a gradation with a pencil. Use a black pencil to scale progressively darker (fig. 3.193, A) or a white pencil to go lighter (fig. 3.193, B).

fig. 3.194

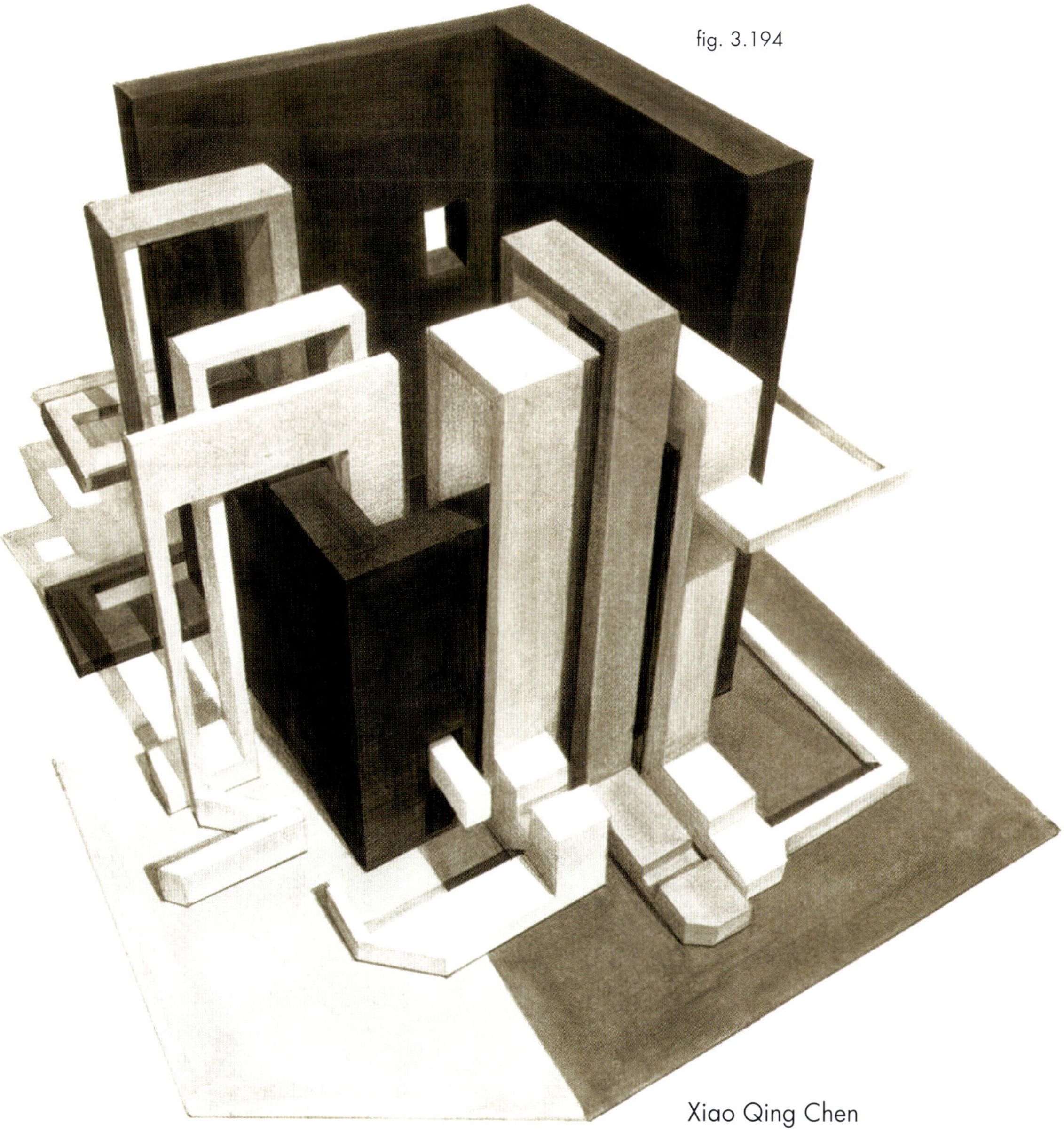

Xiao Qing Chen

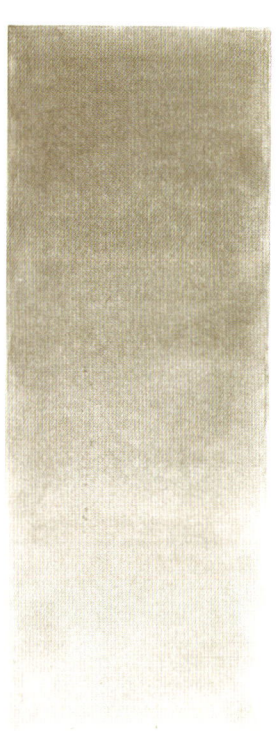

fig. 3.195

PASTELS

Pastels create the smoothest gradations and the deepest blacks. Like everything else, mastering them takes practice. Use a pastel that can be ground or scraped into a fine powder. NuPastel sticks from Prismacolor have this property. Scrape a knife or blade (A) over the pastel stick (B) to create the powder (C). For a smoother application, mix the pastel powder with talcum powder (D) using a flat, rigid tool, such as an old credit card (E). The resulting mixture (F) prevents the pastel from becoming too dark, and the talcum powder makes the pastel less grainy. Use a soft, lint-free pad as an applicator (G). Webril Handi-Pads are best and can be found at art or photography supply stores. To create sharp edges (H), use masks cut out of paper (J). Photocopy your line drawing and cut the masks from the copies. Finally, a kneaded eraser (K) is the best way to remove the pastel chalk. To prevent streaks, dab the eraser gently; do not rub it across the paper's surface.

Pastels can be a bit tricky and messy at first, but they are well worth the time and effort to learn, as they make gorgeously subtle and flowing value changes.

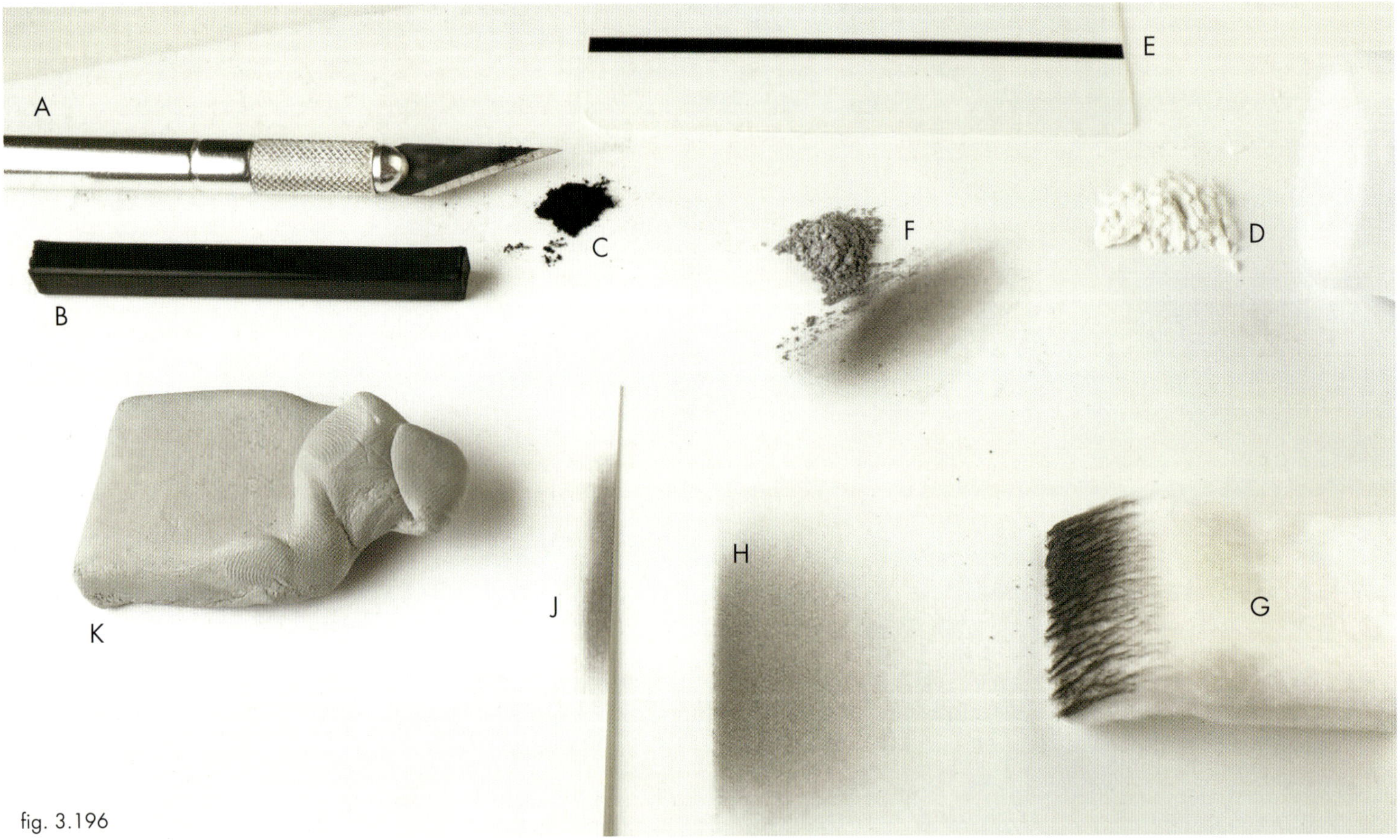

fig. 3.196

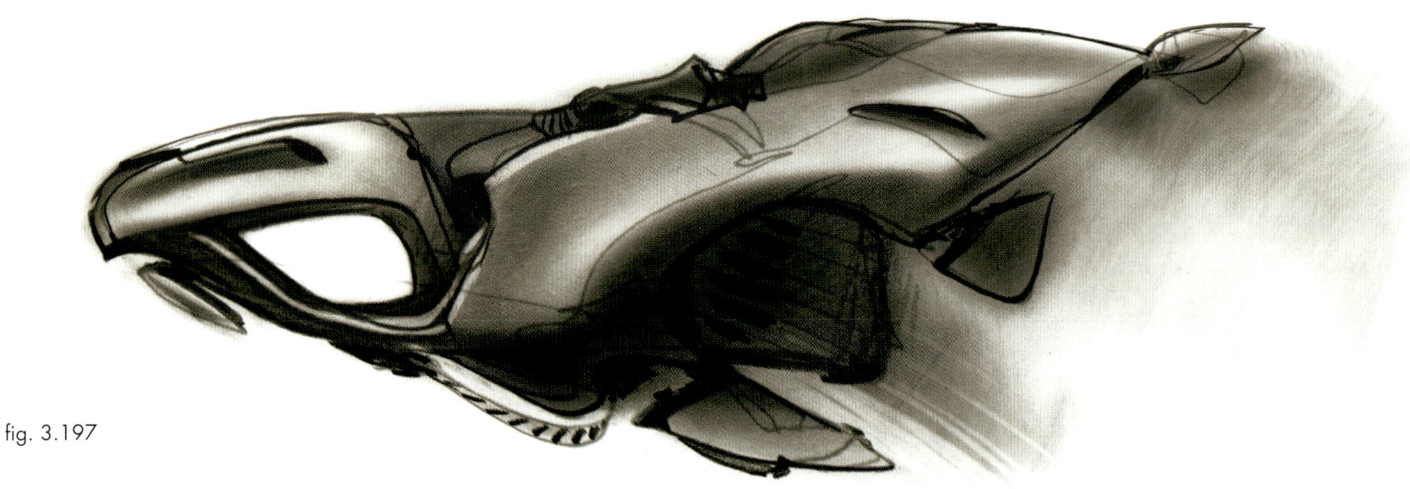

fig. 3.197

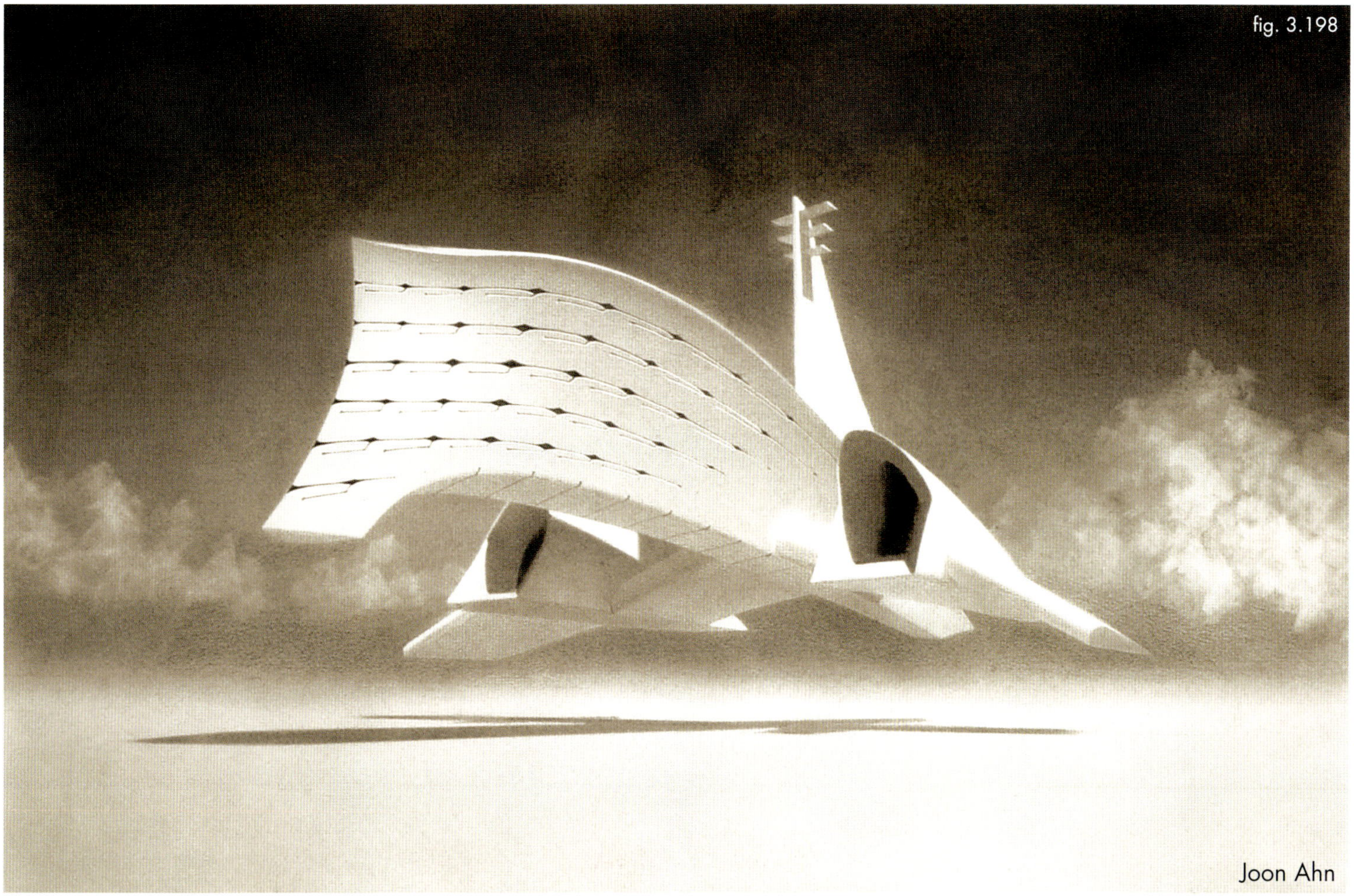

fig. 3.198

Joon Ahn

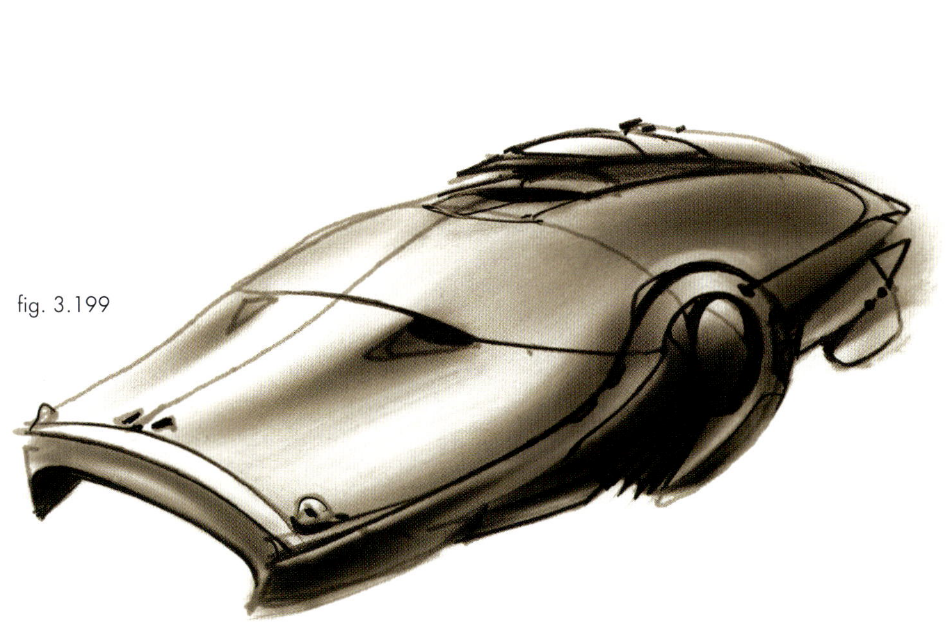

fig. 3.199

fig. 3.200

Xiao Qing Chen

WET MEDIA AND OTHER TRADITIONAL TECHNIQUES

Wet mediums such as gouache, watercolor, acrylic, and oil paints are considered archival, and are perfect for creating one-of-a-kind original works of art, but are not used much anymore for commercial art because they take too long to dry and, honestly, they are much harder to use. There are many books dedicated to mastering these wet mediums. All of the fundamentals from this book about light, shadow and reflectivity can be applied when using those more archival wet mediums. Neither Scott nor Thomas uses wet media beyond the occasional highlight or background. The limited way Scott uses wet media is explained on the next page.

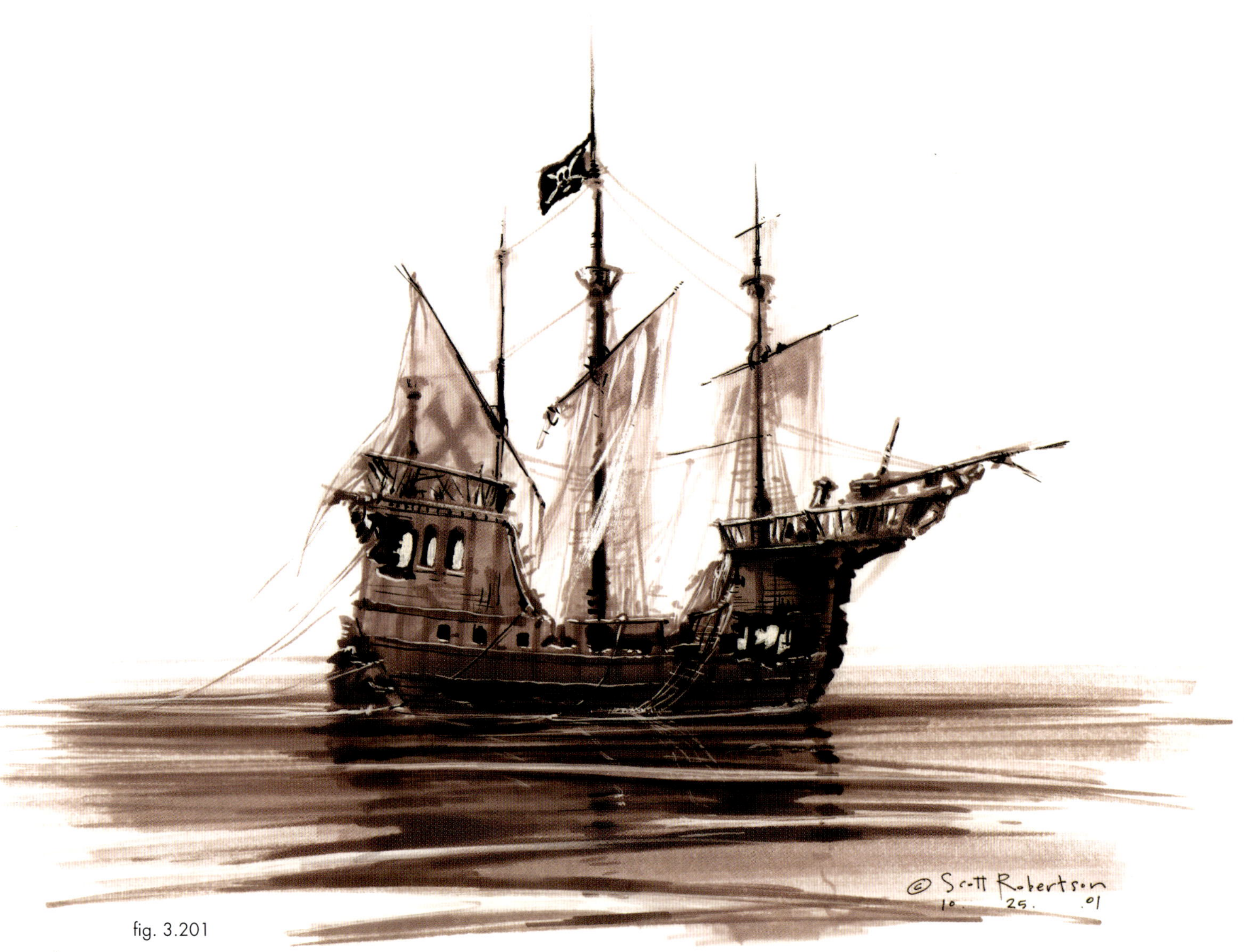

fig. 3.201

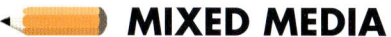

As should be apparent by now, every traditional medium has its pros and cons. The way to get the best out of all the mediums is to mix them. Use pencils for detailed work like cut lines, chalk or pencil for gradations, gouache for opacity and markers for speed and permanence. The three sketches on this page mix media, each on different paper with varying results.

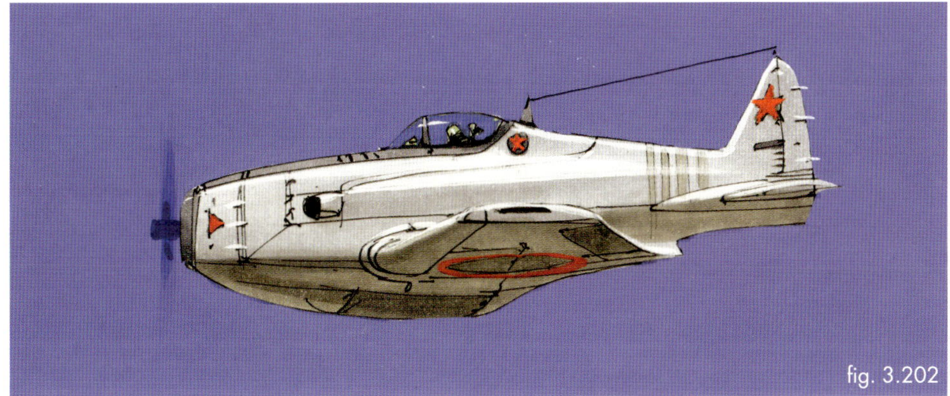
fig. 3.202

Fig. 3.202: This airplane was a loose marker sketch from the first book in this series, *How To Draw*. Scott created this rendering as a way to test if the paper from that book might work well for a sketchbook he is developing. The line drawing was done with a Pilot HI-TEC pen, the airplane's value work was done mostly with COPIC markers and the highlights and blue background were done with gouache. Check out the video tutorial for this rendering.

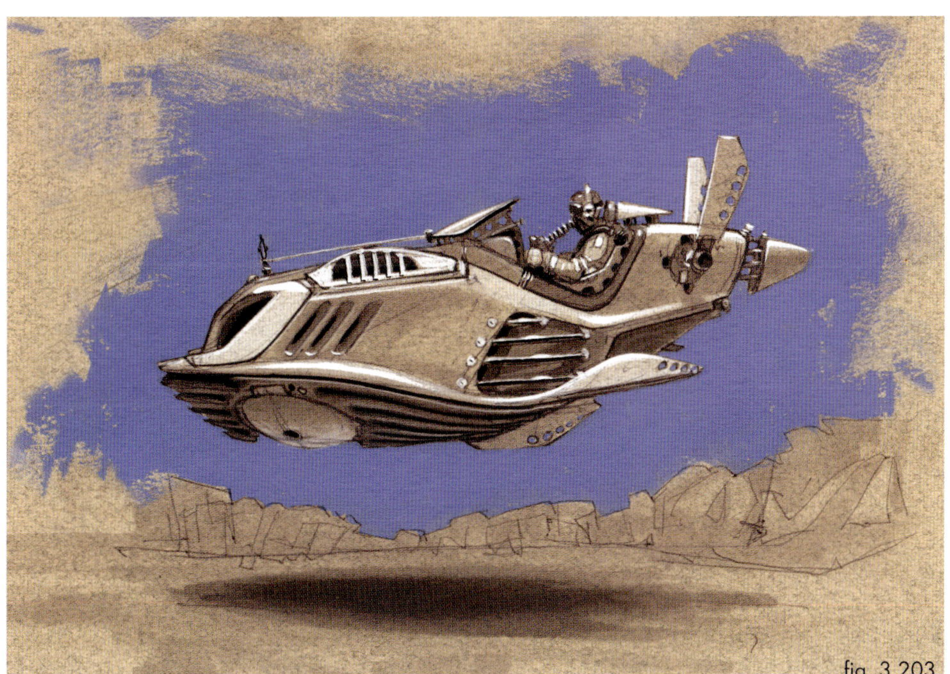
fig. 3.203

Fig. 3.203: Using the same basic techniques, this sketch varies in that the paper was a Canson toned stock and the pen was a ballpoint. When using ballpoint pens with COPIC markers, be careful because the markers will turn the ballpoint's ink purple. To avoid this issue, do most of the marker work first, then add the line work.

Fig. 3.204: This sketch of a powerboat was done on drafting vellum using mostly NuPastel black chalk, black and white colored pencils, black marker for the black stripe, and white gouache for the highlights and the stripes on the forward deck.

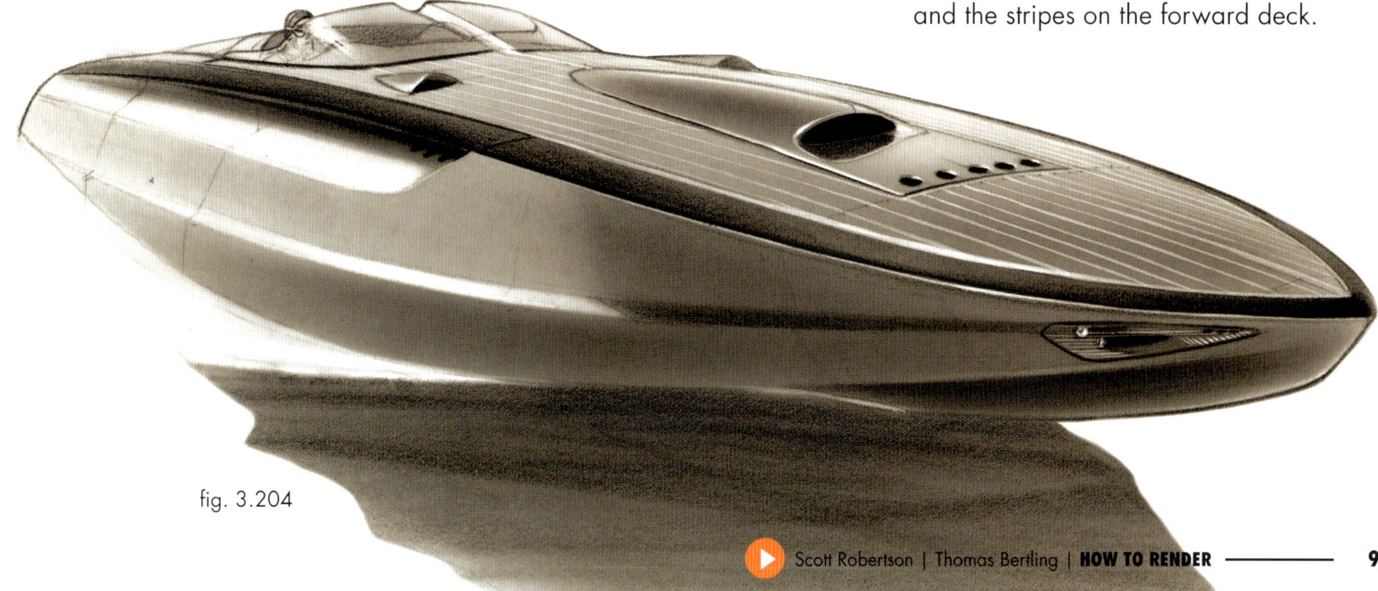
fig. 3.204

THE IMPORTANCE OF MASTERING THE FUNDAMENTALS

Practice, practice, practice. Slowing down and really devoting time to learning the fundamentals of rendering will make all of the various mediums easier to use. This includes digital rendering tools like MODO by The Foundry, SketchBook Pro by Autodesk and Photoshop by Adobe.

Sometimes students fear that a rigorous education, focused on improving their foundation skills of rendering light, shadow and reflectivity, will somehow limit their creativity. This confusion probably stems from a misconception that learning about how the brain visually interprets the world around us, and learning about rendering based more on physics and science than expressive illustrative or fine-art techniques, will somehow override or squash their creativity, making all the work look the same in the end. However, the exact opposite is true. Students who master the fundamentals become empowered by their ability to communicate their design ideas more clearly.

The physics are the physics, there is no way around that. What creative artists and designers choose to do with this knowledge is up to them. That is why there is such a term as "artistic license." This is when an artist or designer decides to bend or abstract the laws of physics for stylistic purposes. This happens all the time in Scott's work and in the works of others he admires. It does not mean they do not know how to render an object in a photo-real fashion, they do, but they have chosen to modify their renderings in pursuit of an illustrative style that works better for their own goals. Having a strong set of foundation skills, and then choosing to stylize, is a great position to be in, as opposed to always having to fake the proper construction of a cast shadow or reflection because the appropriate amount of time was never spent learning the often mundane, technical side of rendering. After mastering the fundamentals, weaving photography or 3D digital renderings together becomes much easier. Without this knowledge, the options for creative image-making become very limited.

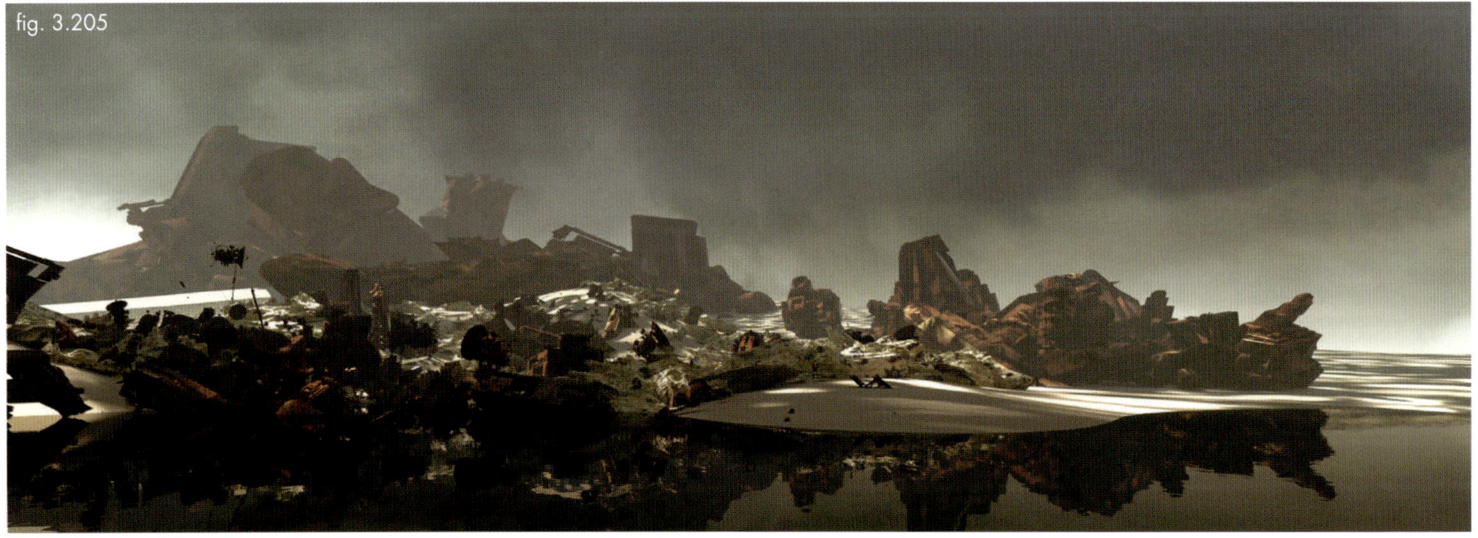
fig. 3.205

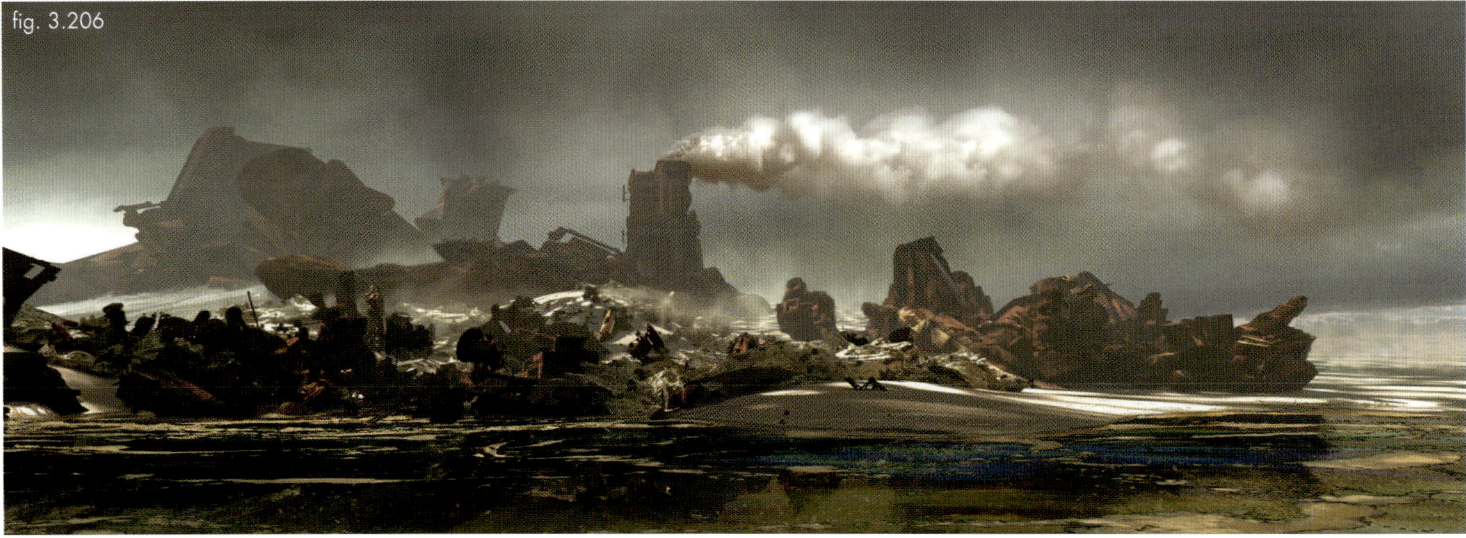
fig. 3.206

What about digital rendering programs? Haven't they made rendering with traditional mediums obsolete? Yes and no. For commercial purposes, like industrial design rendering, digital programs have indeed replaced colored markers and mixed media in much the same way that markers once replaced watercolors. Digital rendering tools are easier to use than traditional tools and revisions to the images are much less burdensome. However, in the world of fine art, the value of the single, original piece of art has not been surpassed by digital artwork. Owning an original piece of art is still much preferred over a printed, digital piece.

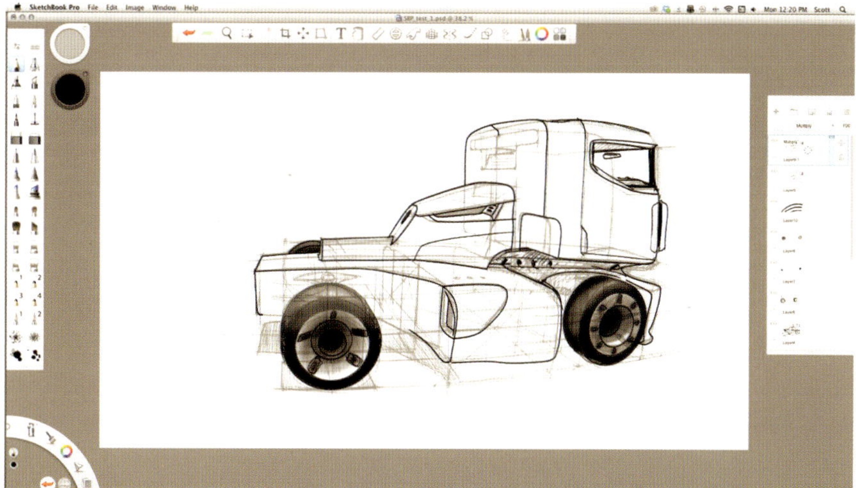

fig. 3.207

Whether working with digital or traditional media, the fundamentals of light, shadow and reflectivity remain the same. Since learning any of the digital programs in-depth could easily consume all of the pages of this book, how to use specific programs will not be taught here. What will be covered are the high-level concepts of working digitally. The video tutorials linked to this page provide an overview of working within two excellent programs: SketchBook Pro (fig. 3.207) and Photoshop (fig. 3.208).

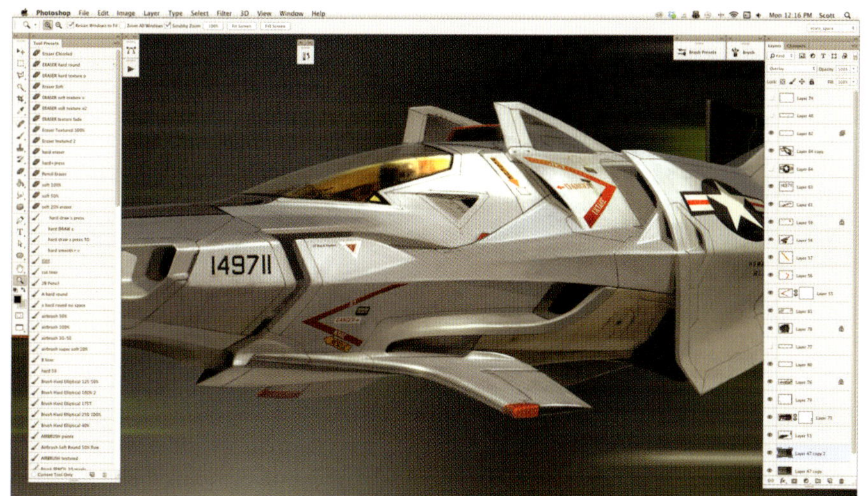

fig. 3.208

For more in-depth training on 2D or 3D programs, check out some tutorials from online schools such as the Gnomon Workshop or Schoolism. As your proficiency with digital tools increases, rendering will become easier. Be patient, mastering any new medium takes time.

MATCHING VALUES TO A PHOTOGRAPH

To merge a photograph and a traditional rendering, all skills and techniques can be combined to achieve a believable and satisfying result, as seen in the images on both pages by Thomas's student, Xiao Qing Chen.

Match Perspective

Fig. 3.209: Take a photograph with an object, in this case an electrical plug, embedded that serves as a placeholder for the perspective, light direction, and angle that needs to be matched.

Match Value

Fig. 3.210: When doing the marker rendering, it is essential to keep in mind what the values of the background will be. All of the items in this drawing were generated with flat surfaces, using the construction techniques that have already been covered.

Merge the Photograph with the Rendering

Fig. 3.211: To increase realism and create a believable union, apply the same lens blur of the photo to the marker rendering. In this case, the work was merged in Photoshop, but any software that offers layers and blurring will suffice.

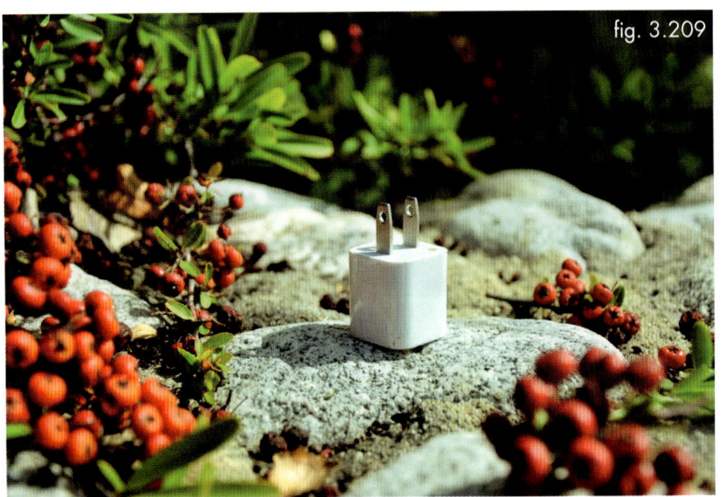

fig. 3.209

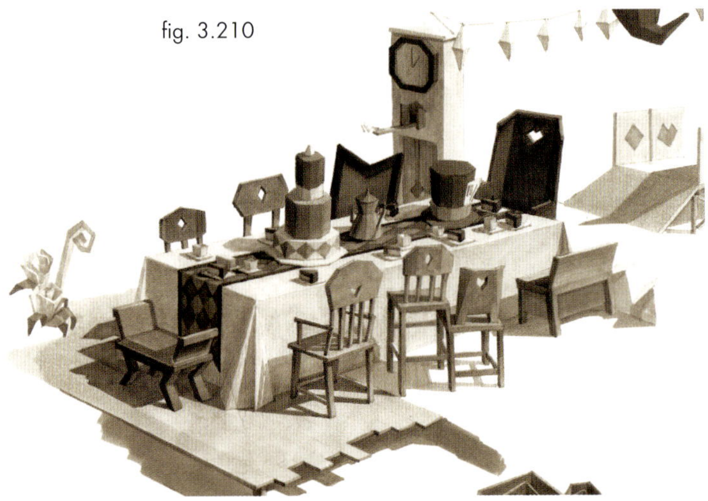

fig. 3.210

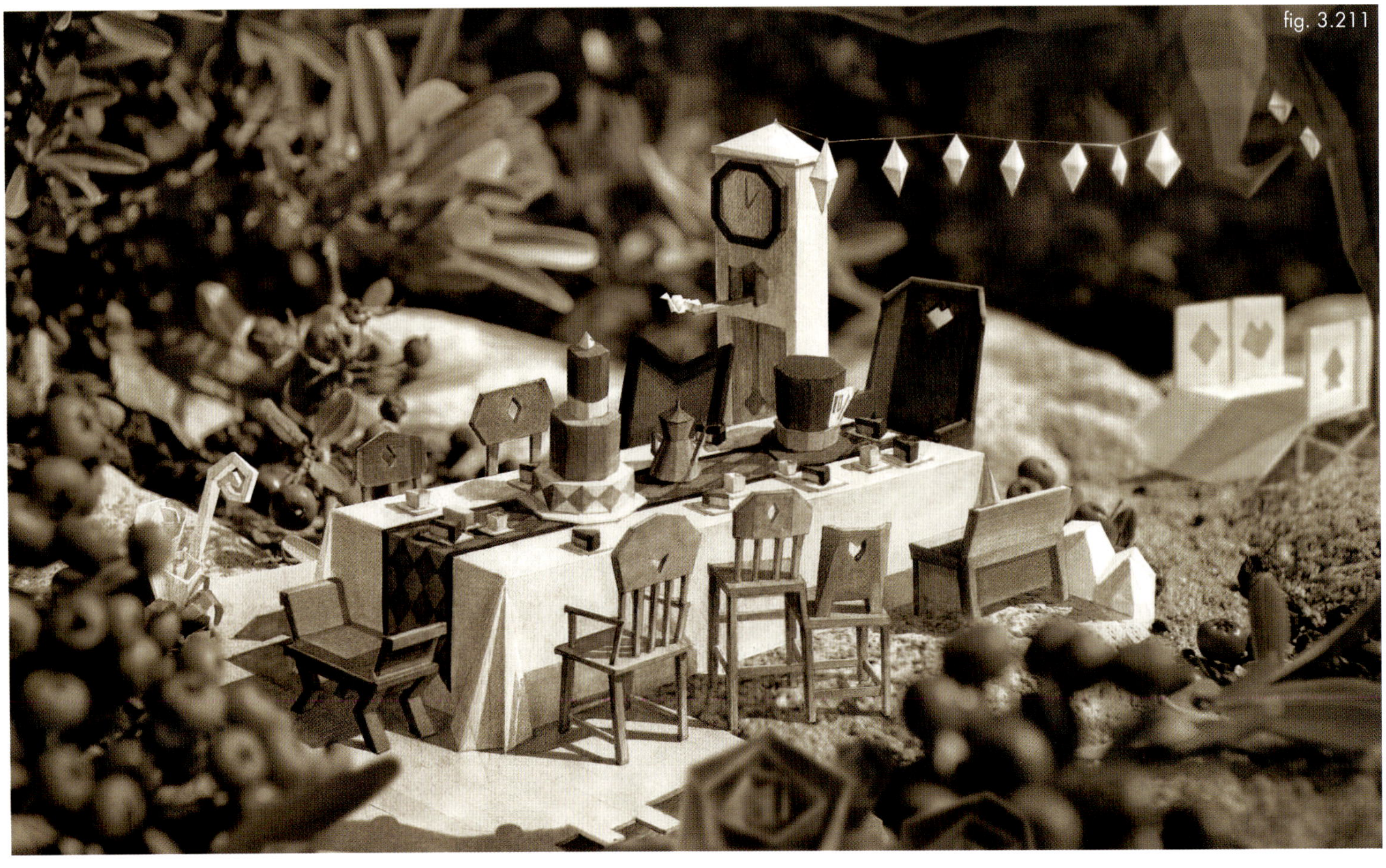

fig. 3.211

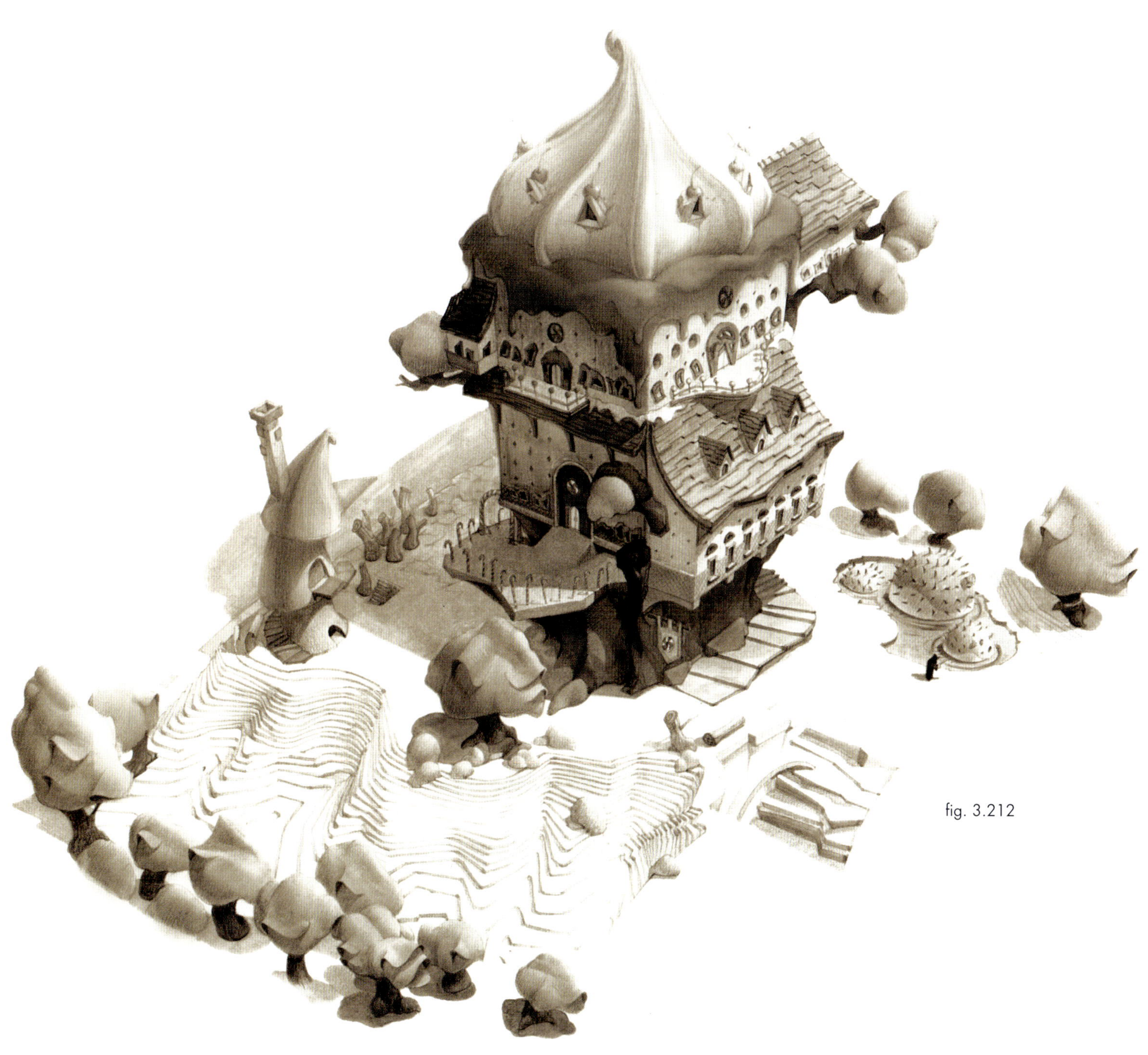

fig. 3.212

Chen Xiao Qing
www.chenxiaoqing.tumblr.com

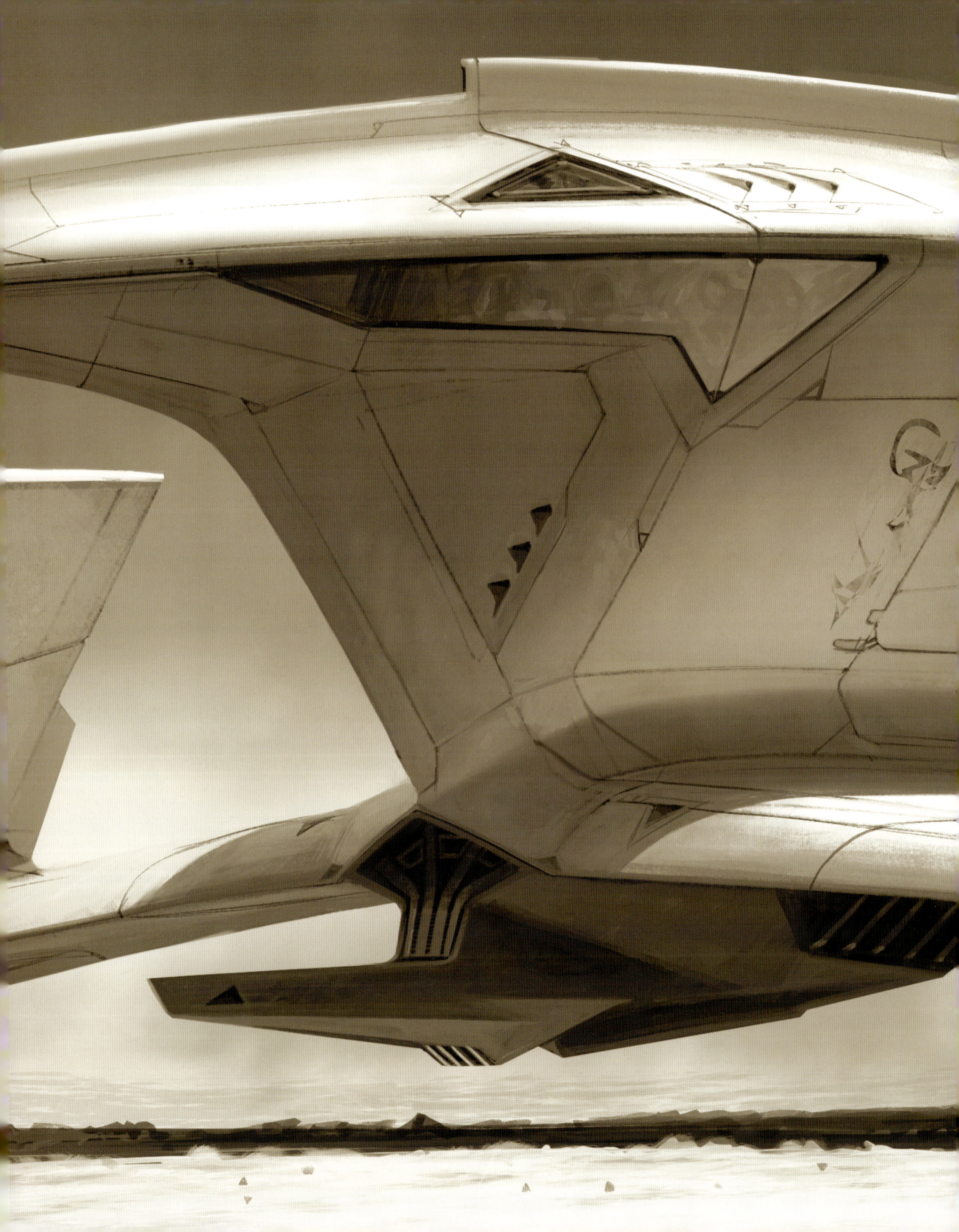

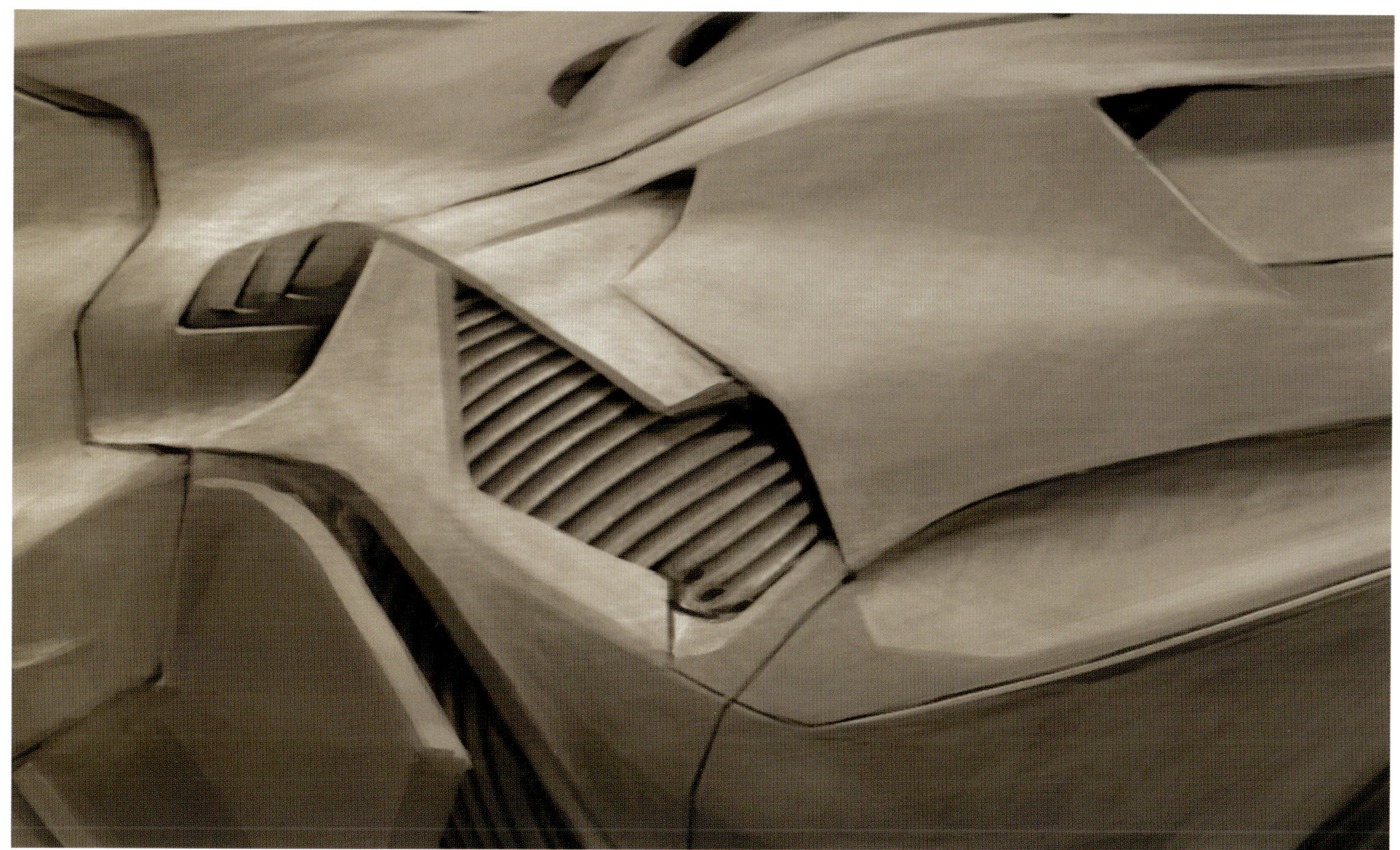

CHAPTER 04 RENDERING COMPLEX VOLUMES

This is where the real magic starts to happen. All of the lessons thus far will be applied to more complex forms than the geometric forms of the last chapter. When rendering these more complicated volumes, described by X, Y and Z sections, think of each section as part of a simple, curved panel. This will make it easier to figure out which values go where, and then the focus can be on blending them together.

Regardless of the chosen medium, the process becomes very fun when all of the values are working just right so that the surfaces start to come alive. Rendering them can almost feel like sculpting clay. By making simple and calculated adjustments to the values, the surfaces can be bent, twisted and bulged at will. Even though the volumes are still being rendered as gray-scale matte surfaces, what can be communicated about the forms is unlimited.

Although it can be tempting to jump too soon into rendering with color and reflectivity, it is extremely important to be skilled at rendering complex volumes in a monochromatic palette. Master the knowledge in this chapter, and you will have the ability to communicate clearly any form imaginable.

SHADING SIMPLE CURVED SURFACES

A simple curved surface bends only in one direction, much like sheet metal or a piece of paper that flexes in a single way. They are the building blocks of compound surfaces called X-Y-Z volumes or forms.

Fig. 4.1: The best way to find the key rendering elements of a simple curved surface is to think of it first like a section of a cylinder.

Fig. 4.2: Be aware that these surfaces do not change in a linear fashion, but rather decelerate and accelerate. As the surface flattens or bends, the gradation reacts accordingly by changing more slowly or more quickly.

Fig. 4.3: Create consistency within renderings and a method for communicating the form changes of a surface by assigning the same value to surfaces that share the same relative orientation to the direction of the light rays.

Fig. 4.4: To determine where the shape casts a shadow on itself, create a section in the direction of the light plane. Use this section to project the terminators and edges onto the light-plane section line to see if there are any intersections (A).

Always take into account the reflected light that bounces off the ground plane, as well as the light that reflects off of the form itself. Surfaces that are in shadow and face downward toward a reflected light source can actually be lighter than upward-facing surfaces within the same cast shadow.

Figures 4.2 – 4.4 are by Charles Liu.

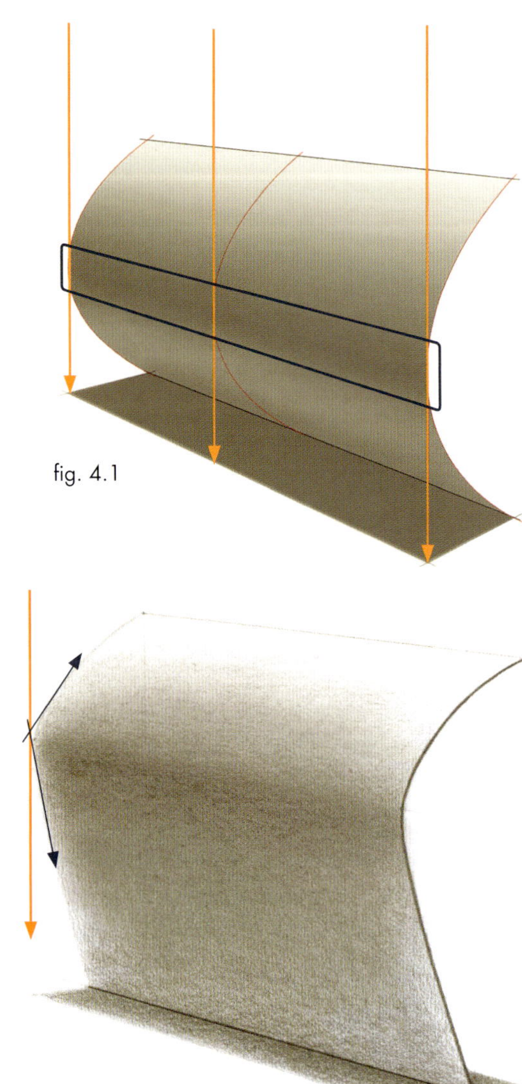

fig. 4.1

fig. 4.2

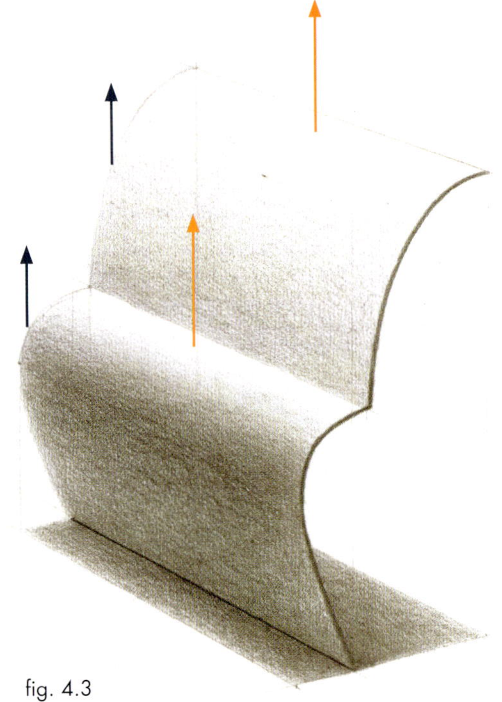

fig. 4.3

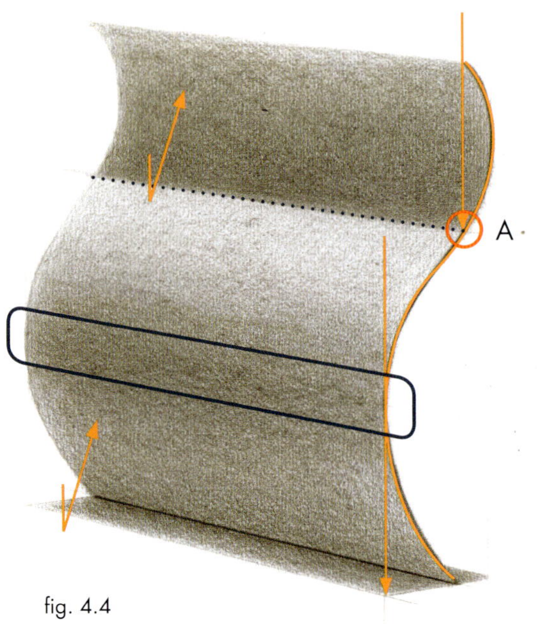

fig. 4.4

To calculate the values and cast the shadows of complex compound surfaces, it is crucial to be able to define a shape in perspective first. This requires a confident ability to draw the form/volume in a three-axis (or three-dimensional) coordinate system, in which each axis is named X, Y and Z. Forms constructed in this way are called X-Y-Z volumes or forms. The construction of these volumes is explained in detail in Chapter 06 of our first book, *How to Draw*.

Fig. 4.5: Working with an X-Y-Z form that is an extrusion of an arch uses the same rules as those for working with a horizontal cylinder in constructing the terminator, cast shadow and core shadow. Since the shape is a straight extrusion with two flat end planes, only one point is needed to find the terminator and the passive highlight. These features run parallel to the extrusion direction. (However, this is not the case for a compound X-Y-Z shape.)

Fig. 4.6: The passive highlight's position cannot be found by drawing a line from the center of the shape, since the volume does not have a perfect, circular cross section like a cylinder. To find the passive highlight, draw a perpendicular line, in perspective, to the light-ray direction that also touches tangent to the cross section created by the intersection of the light plane.

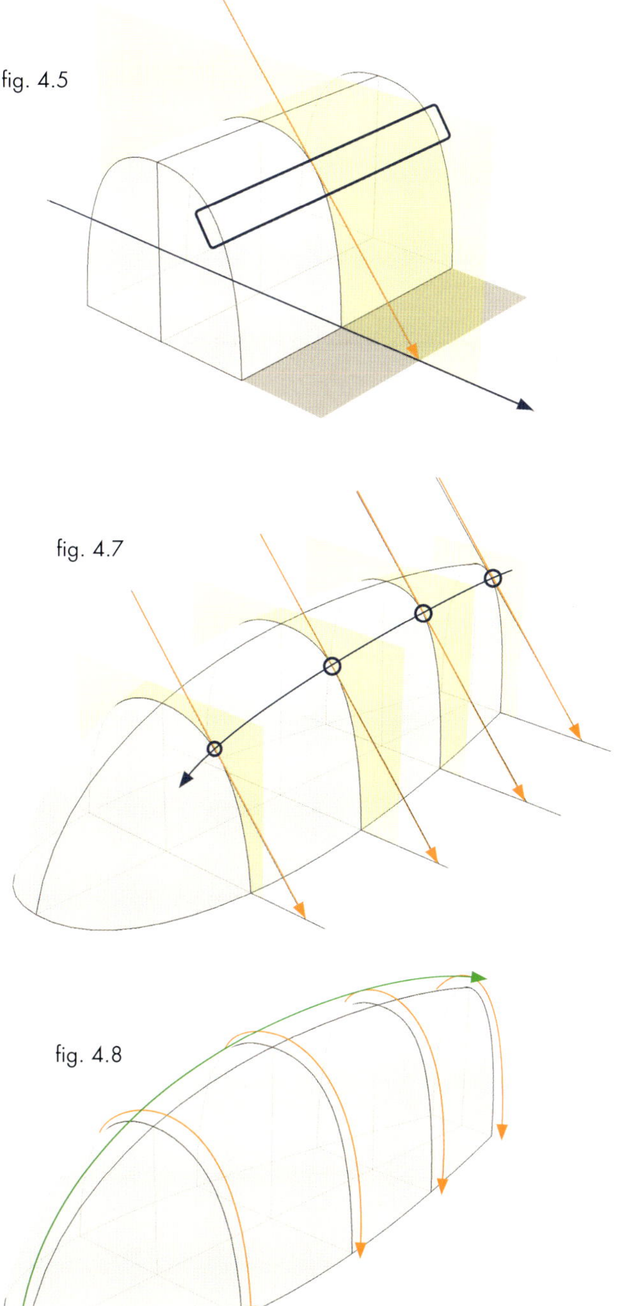

fig. 4.5

fig. 4.7

fig. 4.8

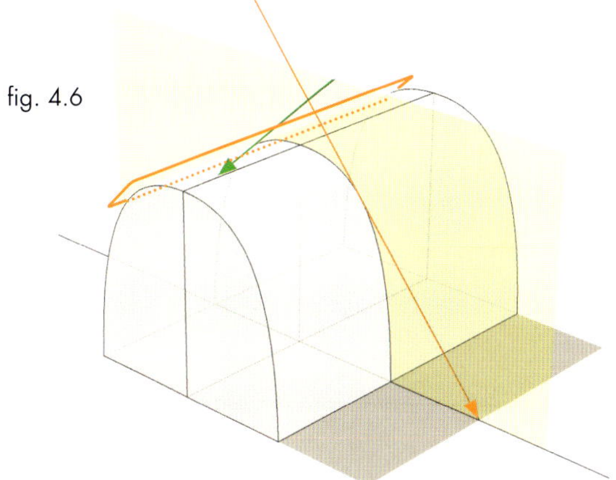

fig. 4.6

It is best to approach the construction of X-Y-Z forms in sections, since the shape as a whole can be too complex to tackle all at once.

Fig. 4.7: To construct the terminator of a compound X-Y-Z form, multiple sections are needed. Draw enough sections to project as many points onto the ground plane as necessary to be able to draw the cast shadow on the ground. Keep in mind that a smooth-surfaced volume like this one means the highlight and core shadow will also be smooth and flowing across the surface when rendered.

Fig. 4.8: When preparing the X-Y-Z shape for rendering, remember that the entire surface is changing form in all directions. Concentrate on the key points, since it is not practical to construct every point of the surface. Consider the value changes along the long, green centerline and along the shorter, orange section lines. Once these are rendered, simply connect the remaining surface areas with value transitions between these known areas. Examples to follow on the next pages.

X-Y-Z Form: Sunlight from Directly Above

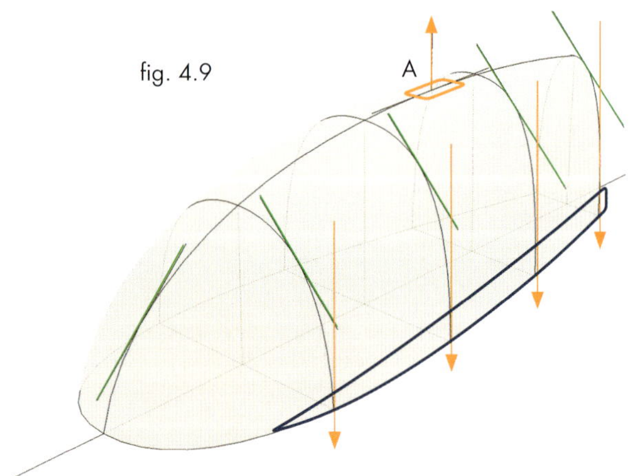

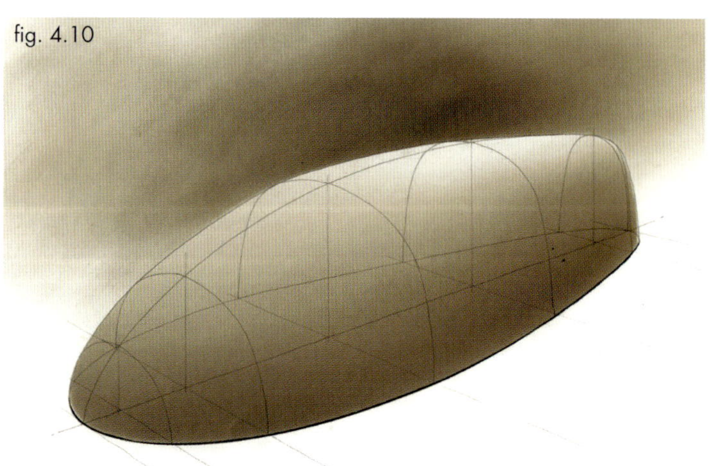

Fig. 4.9: When an X-Y-Z form is top-lit with sunlight, the core shadow can disappear into the ground, with only part or none of it showing. Imagine the continuation of the shape under the surface of the ground plane to create the correct tangencies. The highlight is where the top of the shape is parallel to the ground plane. This point can be found by drawing a perspective line tangent to the centerline that is also parallel to the ground (A). Draw tangent lines at 45°, or any constant angle to the section lines in perspective (green lines here), to help find common value points on the surface.

Fig. 4.10: The chalk rendering shows these construction concepts applied. Observe, however, that there are no visible light-ray lines. The more familiar the concept of top light becomes, the less necessary these lines are. This familiarity will enable the making of good educated guesses regarding value.

X-Y-Z Form: Sunlight from the Side

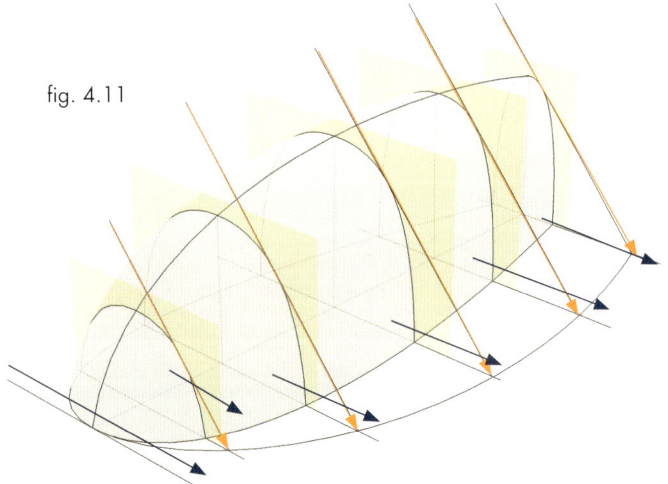

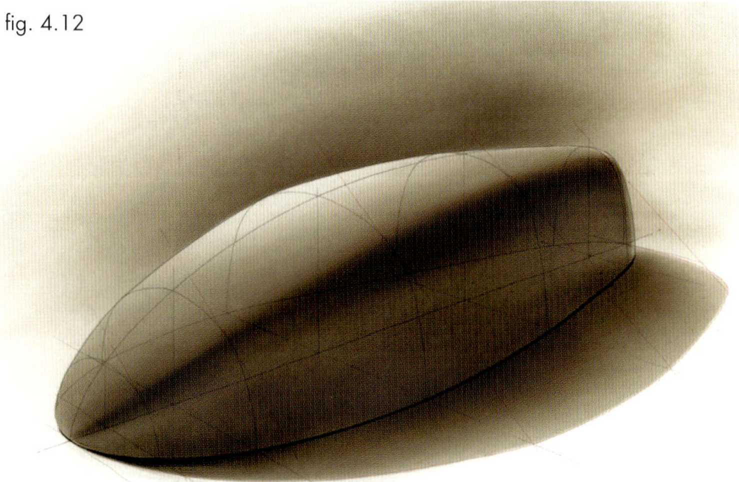

Fig. 4.11: With sunlight coming from the side, use the existing section lines to find the terminator and the cast shadow. The flat end plane does not need construction since its shadow is just a straight line. The cast shadow at the front starts where the shadow direction is tangent to the footprint, or intersection of the form, with the ground plane.

Fig. 4.12: Notice how the core shadow's width in the rendering changes according to the strength of the curvature of the section lines. The reflected light that influences the shadow side of the shape helps in the creation of a core shadow. Be aware that in this example, the passive highlight is not visible, but must still be considered in creating the gradations.

Sunlight from the side is a preferred lighting scenario because the construction of the core and cast shadows does not require additional section lines to be drawn, and it can provide a solid 1-2-3 read with a relatively minimal amount of effort. In addition, if the rendered object has openings and/or cut lines, side sunlight provides great opportunities to cast shadows from these overlapping surface features onto itself.

X-Y-Z Form: Diagonal Sunlight

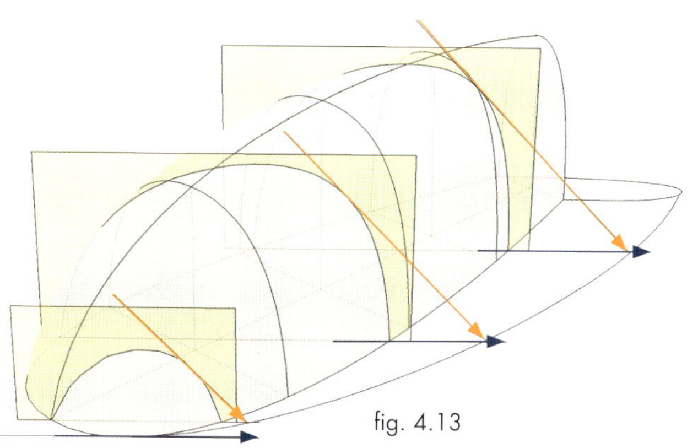

fig. 4.13

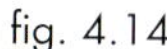

fig. 4.14

A

B

Fig. 4.13: Rendering an X-Y-Z form in diagonal sunlight requires the creation of additional cross sections using light planes, since the existing perspective construction sections do not already match the light/shadow direction as they did with the sidelight example. Use these new sections to determine the terminator and the cast shadow.

Fig 4.14: To render the shape, take into consideration that there might be no visible area that has the true value of the object, since none of the surfaces are perpendicular to the light rays. It is easiest to treat the shape as part of a "stretched" sphere and estimate the location of the highlight on that basis. To bring out the core shadow, increase the reflected light in the shadow area where the cast shadow is short (A) and it does not block the light hitting the ground right next to the form, unlike where the cast shadow is longer and prevents the bounced light from reaching the lower part of the shape (B).

X-Y-Z Form: Local Light

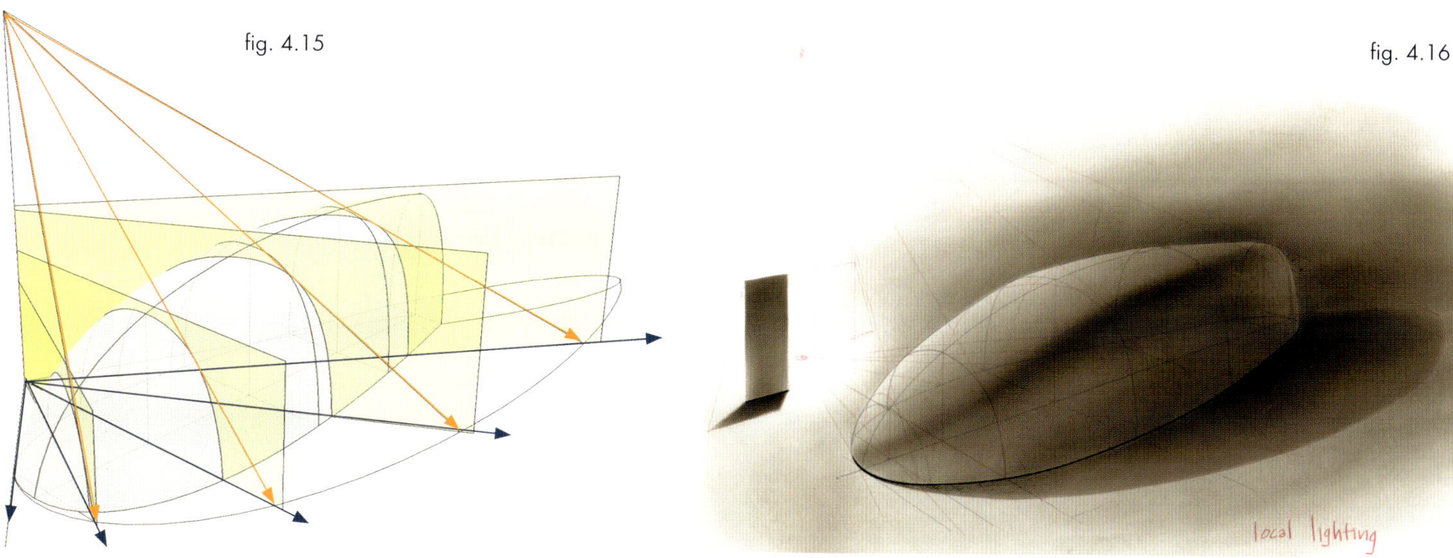

fig. 4.15

fig. 4.16

Fig. 4.15: Rendering an X-Y-Z form in local light requires multiple light-plane section cuts, just as diagonal sunlight does. Remember, with local light, the light planes fan out from the light source instead of being parallel. This makes the position of the core shadow and the terminator rather difficult to guess. Thus, for local light scenarios, a full construction and/or a practical set-up is usually the best way to work.

Fig. 4.16: When rendering the values of a locally lit X-Y-Z form, consider the orientation of the surface toward the light source. Observe that both the ground and the object are affected by the light's decay.

X-Y-Z FORMS: PLANNING A RENDERING

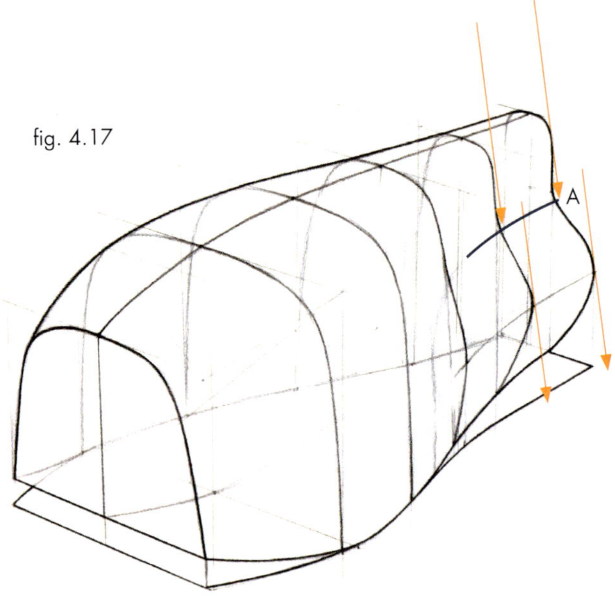

fig. 4.17

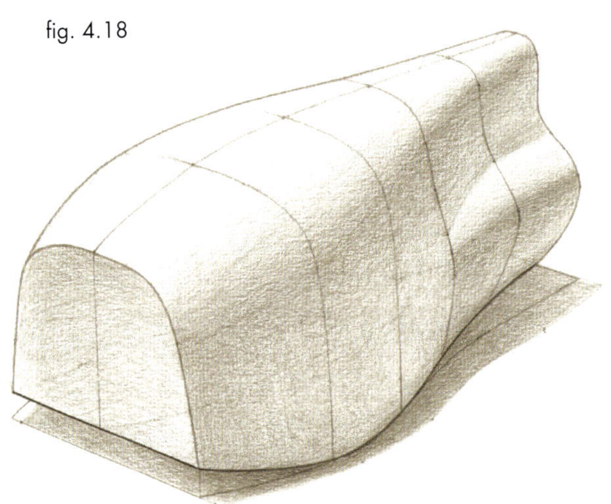

fig. 4.18

It is absolutely crucial to create a clear and correct line drawing of the X-Y-Z form before rendering. Making copies of this line drawing is very useful, to have as a solid basis to return to if things go wrong with the rendering, and also if it becomes desirable to create an alternative lighting scenario.

When the focus is on design, look for opportunities to reinforce the dimensionality of the rendered object. For example, in figs. 4.17 and 4.18, the light direction makes it possible to cast a shadow on the form change at the rear of the shape (A).

fig. 4.19

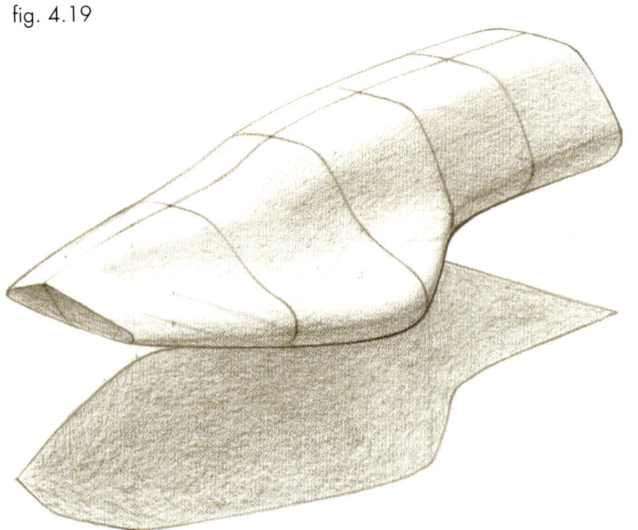

fig. 4.20

Fig. 4.19: Sunlight from the top is often a good default lighting solution, since it is easily understood by the viewer and requires the least amount of additional perspective construction. Strategies such as this can greatly enhance the speed with which the rendering can be executed.

Fig. 4.20: With an object that is mostly in shadow, remember to include some reflected light within the shadow side. This reflected light provides the value changes that make it possible to communicate the form of the shape's surfaces, even within the shadows.

Figs. 4.21 and 4.22: All of these basic lighting and rendering principles still apply when working on increasingly complex forms. While shading, remember to assign similar values to surfaces that have the same orientation to the light source.

Illustrations on both pages are by Charles Liu.

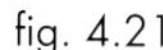

fig. 4.21

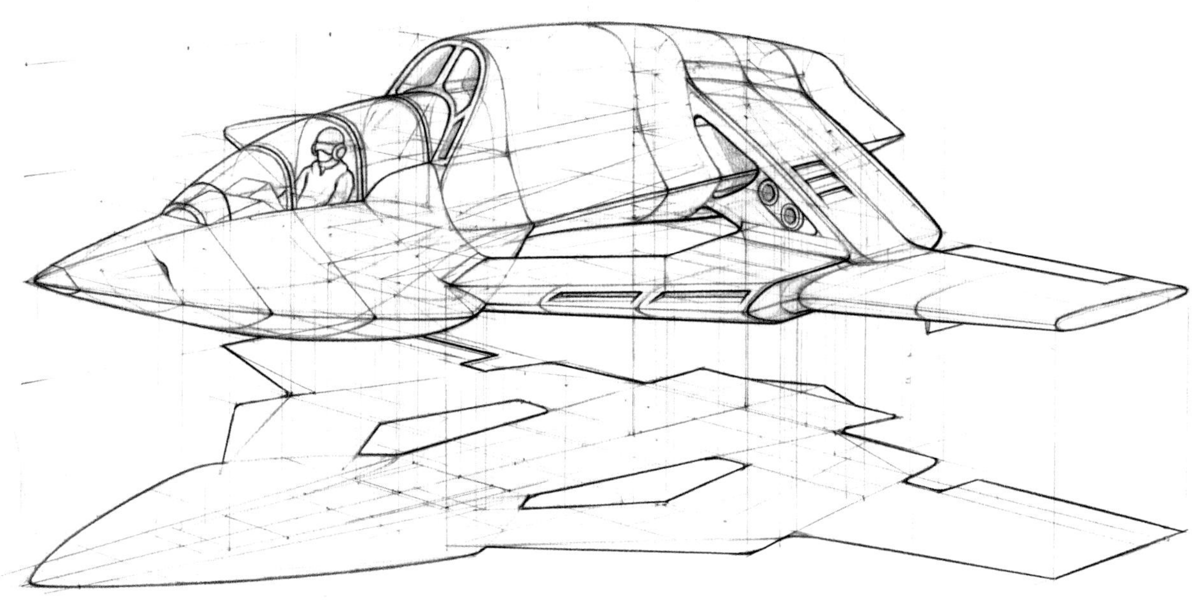

fig. 4.22

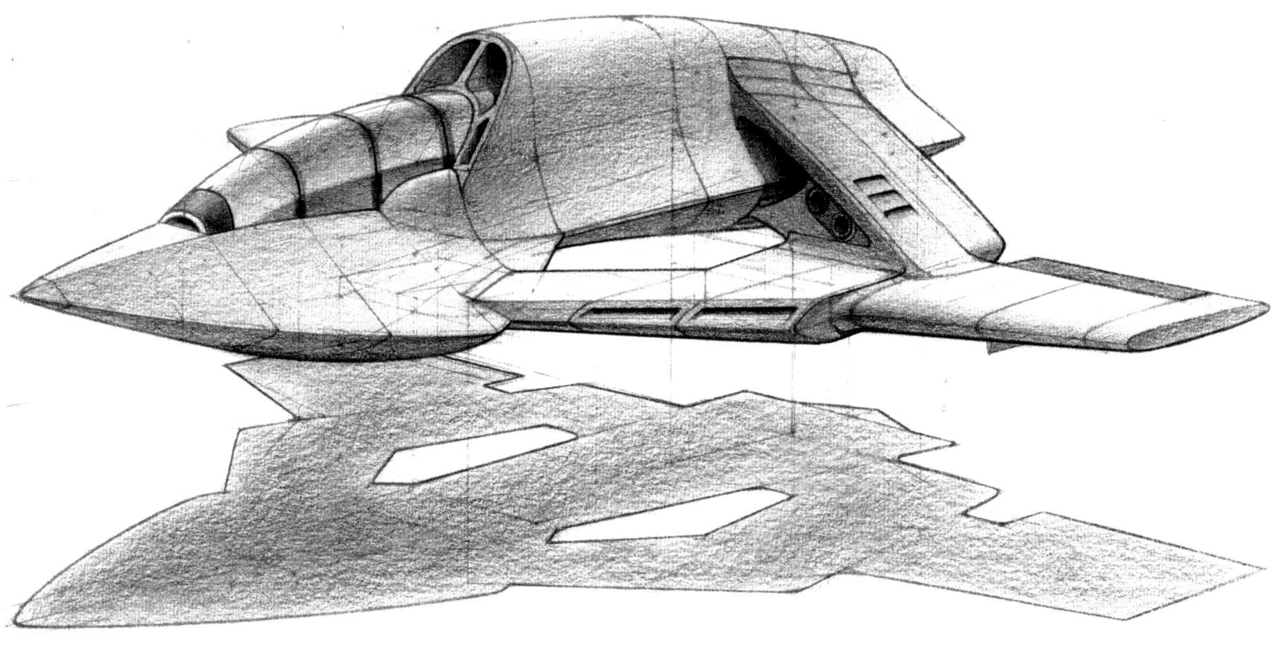

RENDERING WORKFLOW

Creating and understanding one's own workflow is a major asset. Constantly experiment and fine-tune what works best to keep the creative output flowing in a way that makes sense and is comfortable. This can be found through mastering a favorite rendering medium or discovering the right paper for a certain look.

For example, here is one workflow option for quickly rendering an X-Y-Z form.

Fig. 4.23: First, rough in the core shadow and cast shadow. Next, identify key areas and label the values to consider. This is a way to better understand and visualize how these areas affect one another, and how best to show the form with value changes when the line drawing is no longer visible.

Fig. 4.24: Render the primary shapes and basic value changes, while adding a background. Here, it is essential that the general X-Y-Z form be rendered correctly. Adding details at this point can distract from creating a strong 1-2-3 read.

Fig. 4.25: Finally, fill in the different gradations of value and other design details. Having already made a strong rendering of the primary shapes, the details now enhance the quality and reality of the rendering, since they are embedded into the well-resolved value structure.

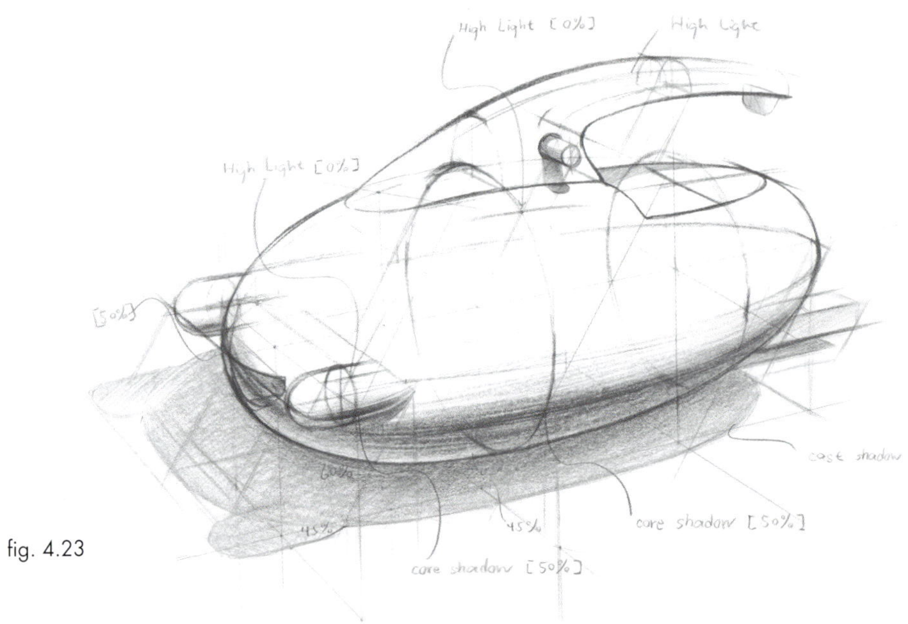

fig. 4.23

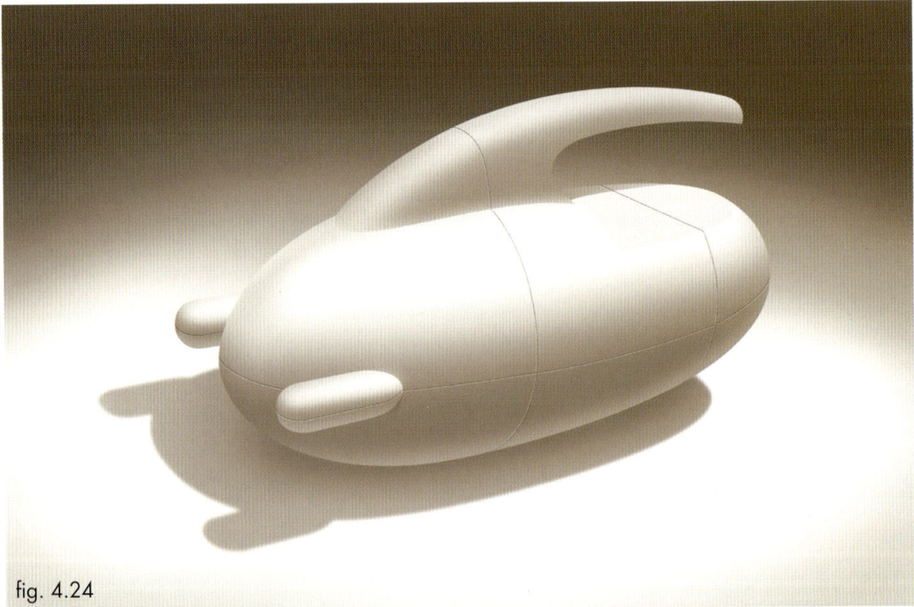
fig. 4.24

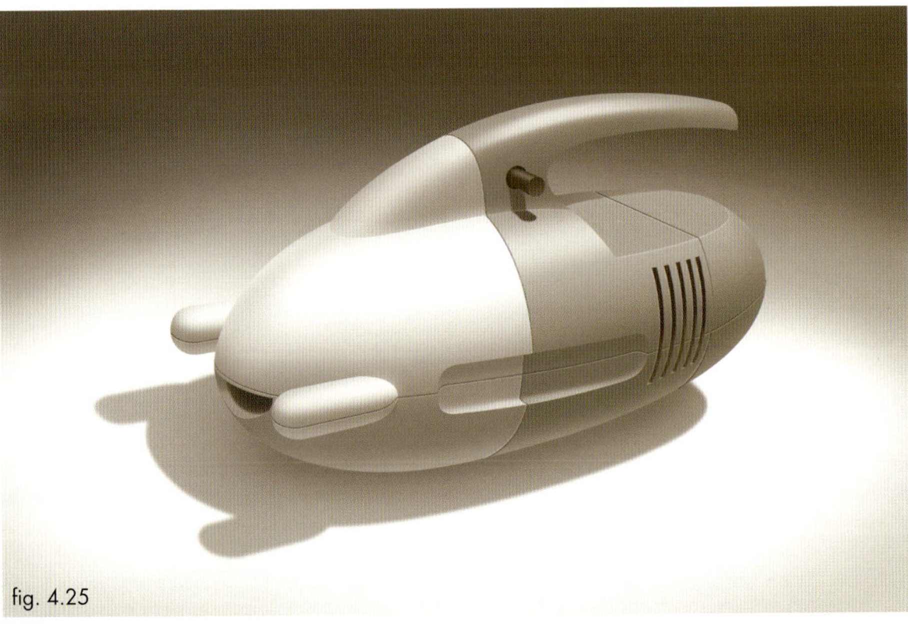
fig. 4.25

Illustrations on both pages by Baoqi Xiao

Student Examples

As your experience and confidence increases, so will the logical flow of the creation of a rendering from concept to completion. Observing and refining your own process is an important skill to think about and master.

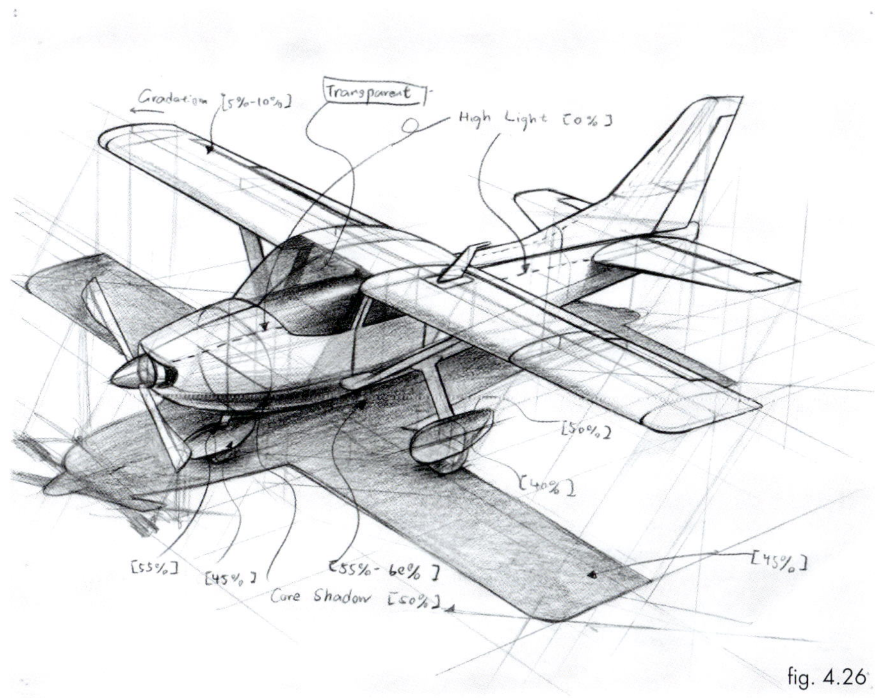

fig. 4.26

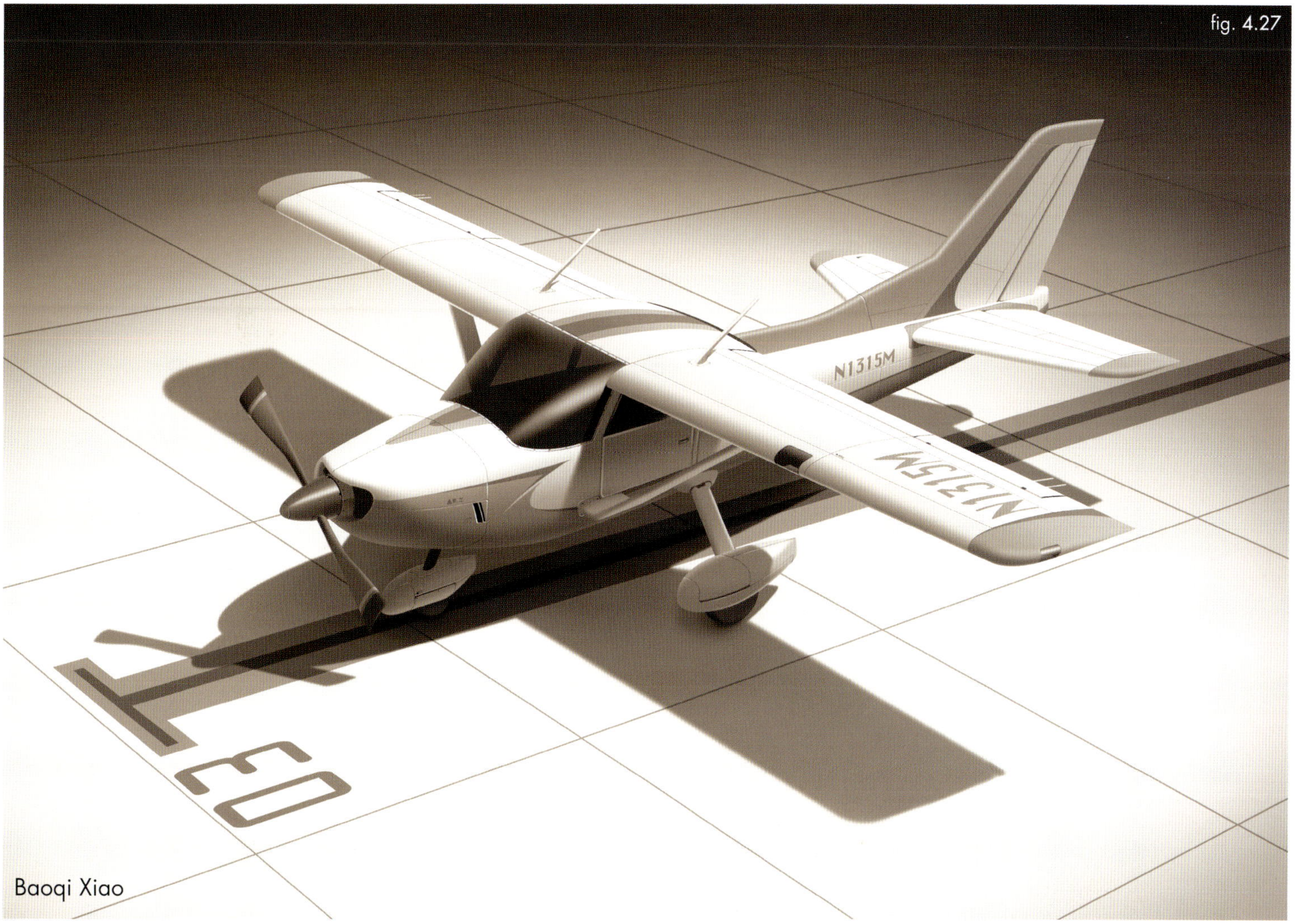

fig. 4.27

Baoqi Xiao

DIGITALLY RENDERING AN X-Y-Z FORM

Let's use Photoshop to render the forms of this volume from *How To Draw* (pages 094 – 095). When a volume has the section information defined in a line drawing, (fig. 4.28), it makes it both easier and harder to render. The easy part is knowing what shape the surfaces are supposed to be. The difficult part is assigning exactly the right values to the sections of the line drawing, making it harder to cheat. If the volume has no section-line information, try adding some value to define the surfaces, then add section lines.

fig. 4.28

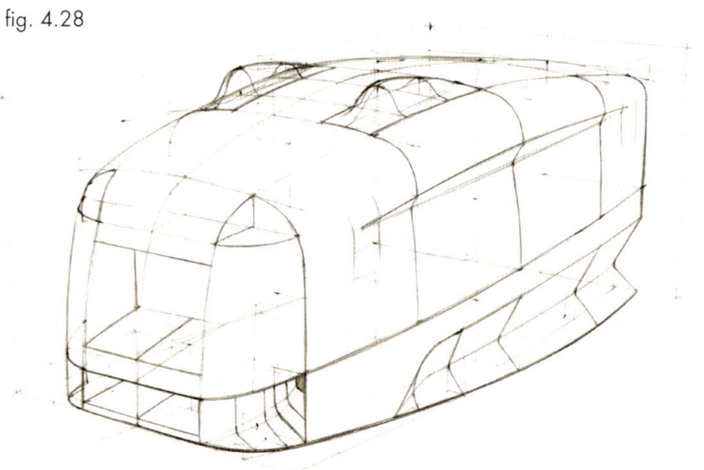

fig. 4.29

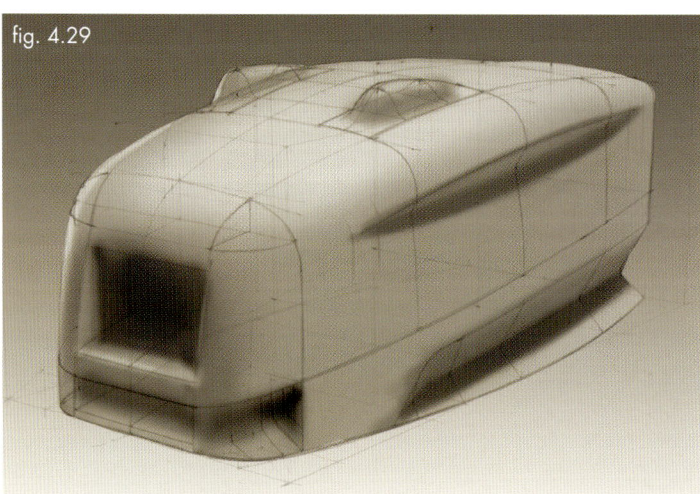

Most of the time, though, if the starting point is a line drawing, then the surfaces are pre-defined by section lines and quite a bit is known about the form to be rendered. Before getting into the details and rendering the smaller surfaces, do a couple of quick lighting studies to find the right light direction and rendering strategy. Fig. 4.29 shows a value estimation of a soft light positioned above the object, with a less intense secondary light placed just out of frame to the left to help edge light the silhouette.

fig. 4.30

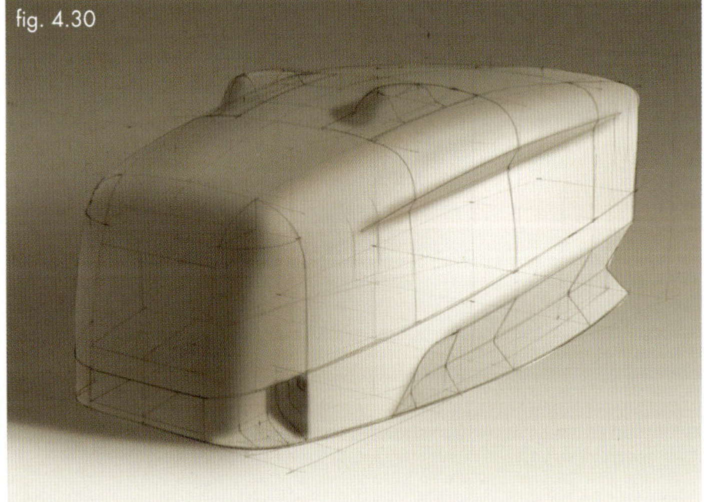

fig. 4.31

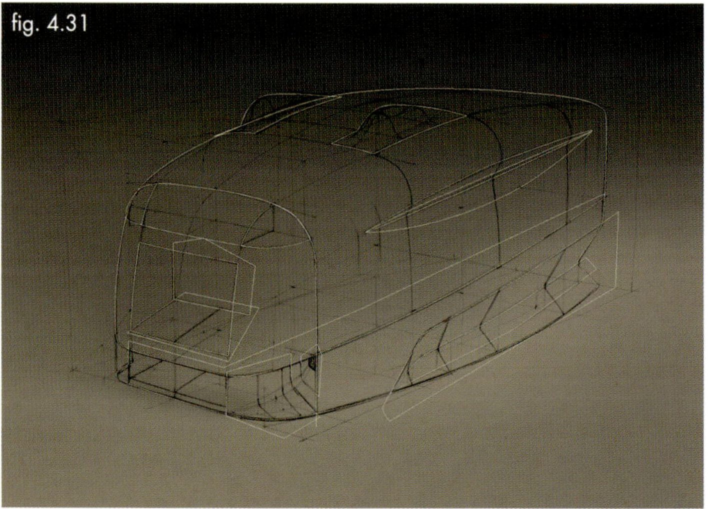

Before deciding whether or not this is working well, try another lighting set-up. Fig. 4.30 shows a value estimation of a soft light source placed at a lower angle on the nearside of the object and slightly behind it. This scenario is more dramatic with a stronger 1-2-3 read than in fig. 4.29, but a lot of the front details are lost due to the shadow, minimizing the form changes down the Number 1 side. Even though fig. 4.30 has the stronger 3D illusion across the front corner, this exercise will continue by rendering fig. 4.29 because of its strong form changes along the side and front.

When designing a lighting strategy for any surface, it is almost always about trade-offs. Staying true to what is most important to communicate will lead to the right choices. If the goal is to make the object look very photo-real, then get rid of the line drawing at some point. This can be done by making paths in Photoshop, creating very clean edges around each element of the design. In fig. 4.31, each path that describes a feature of the form was placed on a separate layer, preserving the edges of those shapes. The entire process for this rendering was recorded in a video tutorial, linked at the bottom of the next page, or via the URL on page 270.

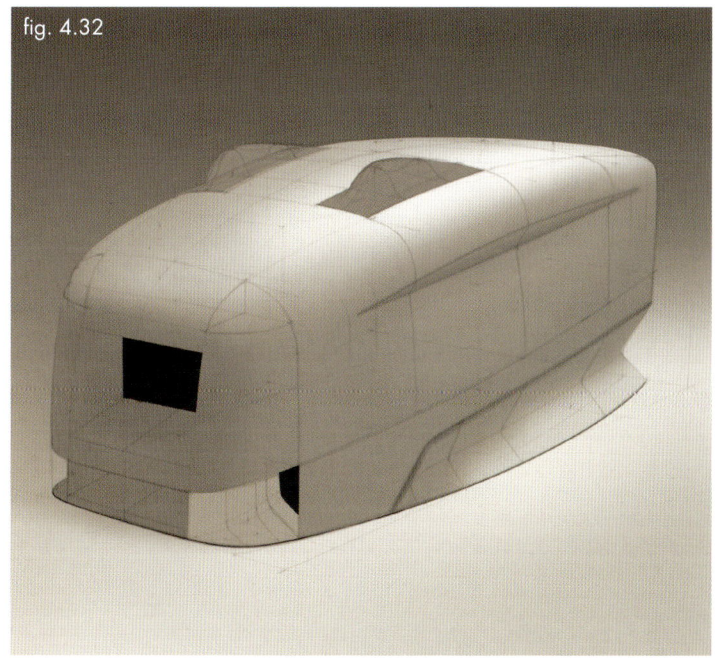
fig. 4.32

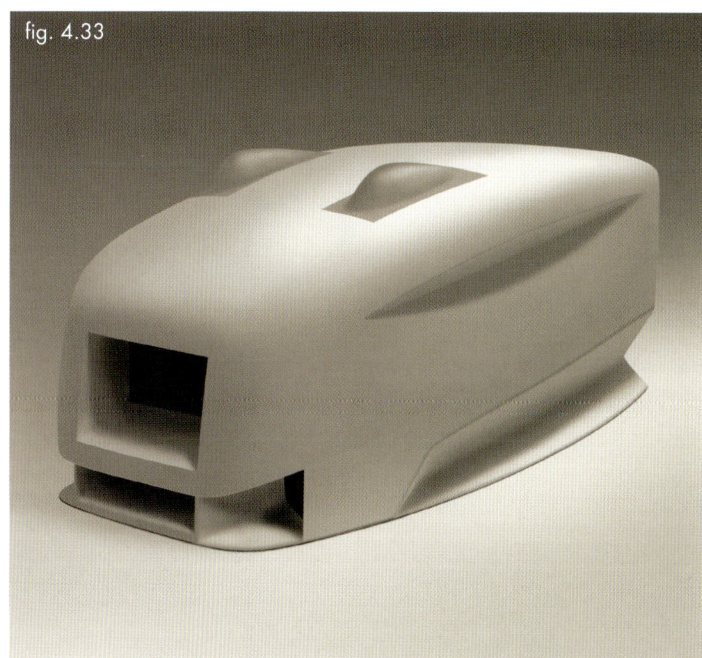
fig. 4.33

Once the layers have been made via the paths, it is time to render. Set the line drawing to "multiply" above the other layers (fig. 4.32) so the section lines can be referenced as a reminder that where the forms change, there must be a value change. Add some strong cast shadows (fig. 4.33) to make the sculpted sections of the volume more apparent. Then turn off the line drawing (fig. 4.34) and add some cut lines, along with a secondary light source to help define the silhouette. Remember, when rendering an object true-to-life, the trade-off is that the lighting may not be optimal. A little artistic license can be applied to enhance the value changes, at the potential expense of realism.

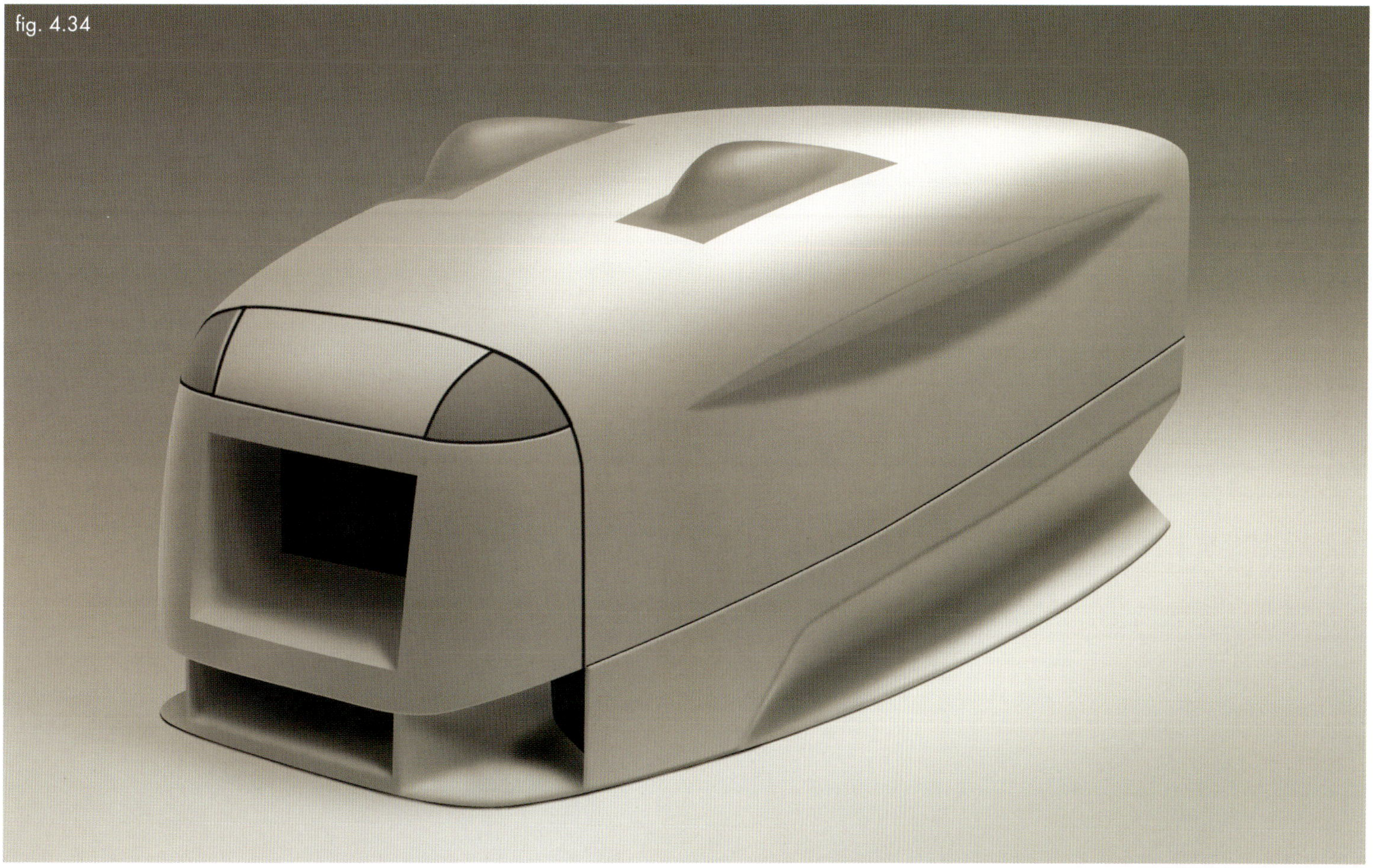

fig. 4.34

CUT LINES

In the previous rendering demo, the cut lines that were added near the end really make it easier for our brains to understand the forms, as they work just like the section lines of the original drawing. They also help to reinforce the direction of the light source and add realism to manufactured objects that have multiple panels. For this reason, cut lines are also referred to as "panel lines," "part lines" or "shut lines." Part lines can be thought of as two edges. The way the light catches these edges is important. Let's look at them in more detail in the photographs below.

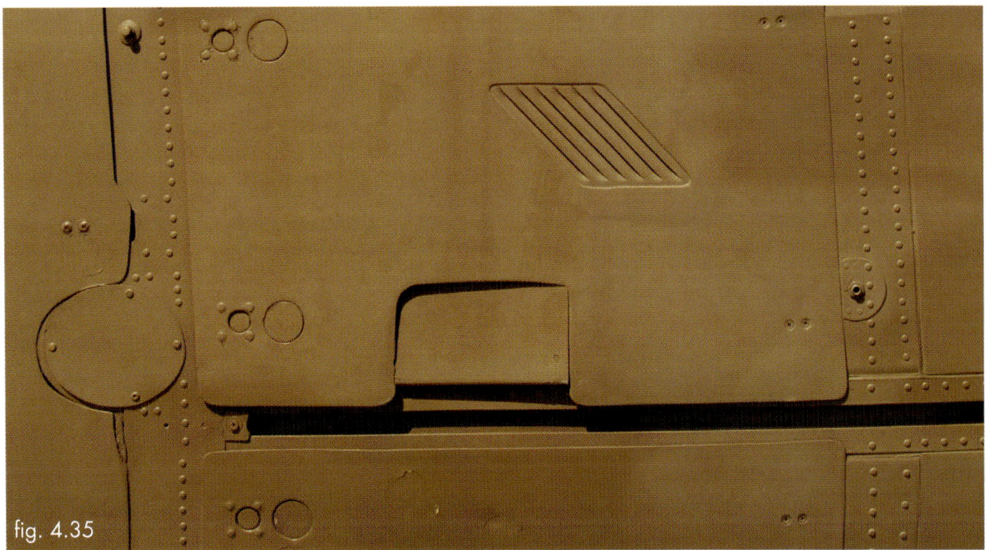

fig. 4.35

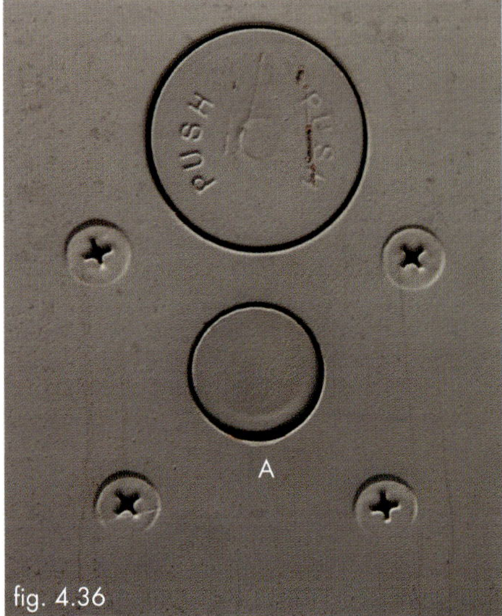

fig. 4.36

Fig. 4.35 is the side of a helicopter. The cast shadow indicates that the light source is from above and a little to the left. This is most visible when looking at the small bolts that stick out from the surface.

In fig. 4.36, notice how the circular part lines catch the light. The highlights fade based on the direction of the part line. Adding this fading highlight to renderings makes them look more real. The brightest spot of the highlight is where the light strikes at 90˚ (perpendicular) to the direction of the part line (A). If the light is from above, like it is in fig. 4.37, it illuminates the bottom edge of the part line and the top edge drops into shadow (B). It is very important to get this highlight on the proper edge and is a common mistake.

As a part line wraps from the light side of an object to the shadow side, the edge highlight fades to nothing. An exception is if there is a strong, reflected light source that makes another highlight fade in a different direction, but it should not be as brightly rendered as the one from the primary light.

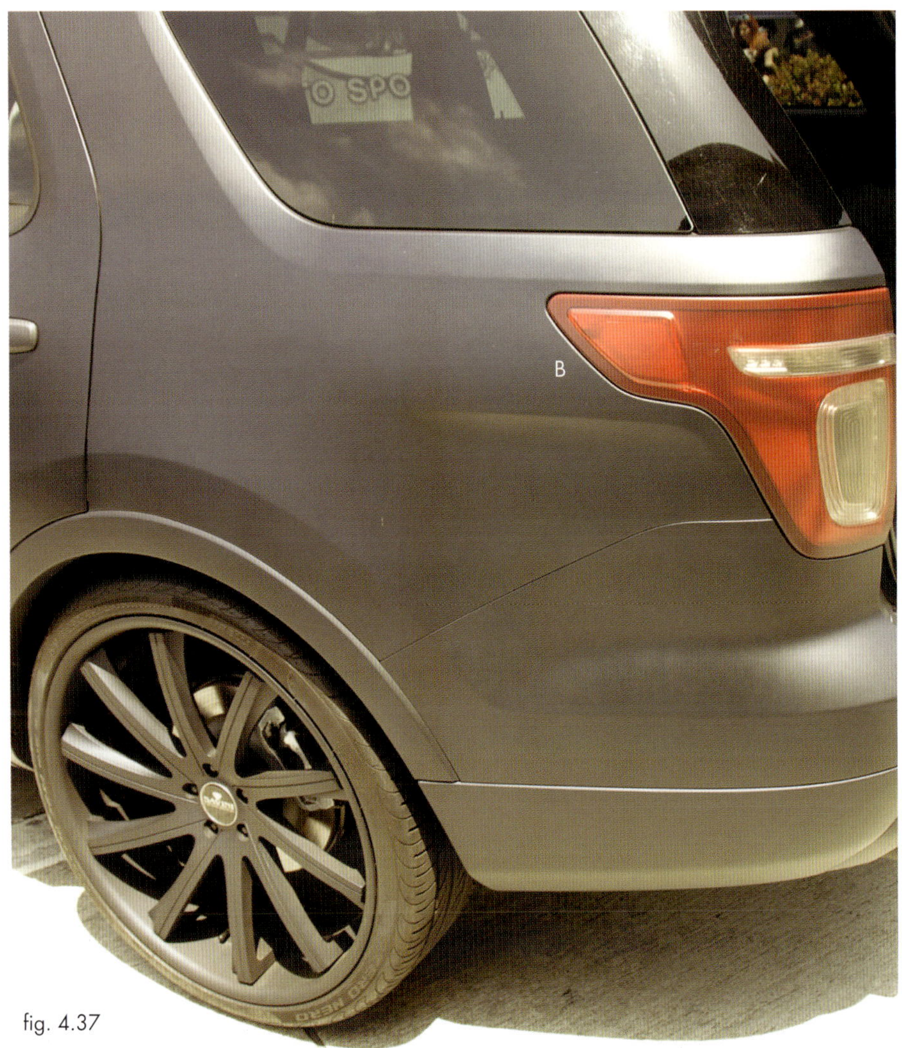

fig. 4.37

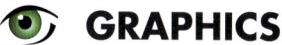

Graphics on an object's surface can present all sorts of challenges, and opportunities. Wrapping graphics like the yellow number 14 (fig. 4.38) first requires good perspective-drawing skills to wrap the graphic, and then the ability to render the proper value changes in the local values of the different colors. The stripe (fig. 4.39) actually works like two section lines and helps to show the form of the vehicle from end to end, whereas the graphic patterns on the pheasant (fig. 4.40) almost camouflage the form changes. When illustrating something like a butterfly (fig. 4.41) it is important to spend the time to get the graphic design of the pattern just right, because it is the element that the viewer will notice first.

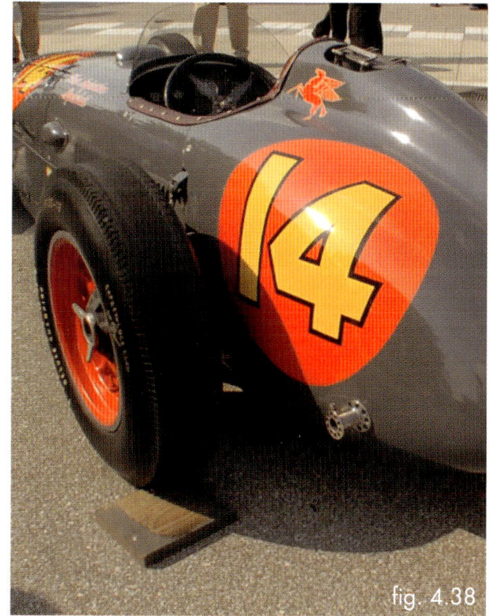

fig. 4.38

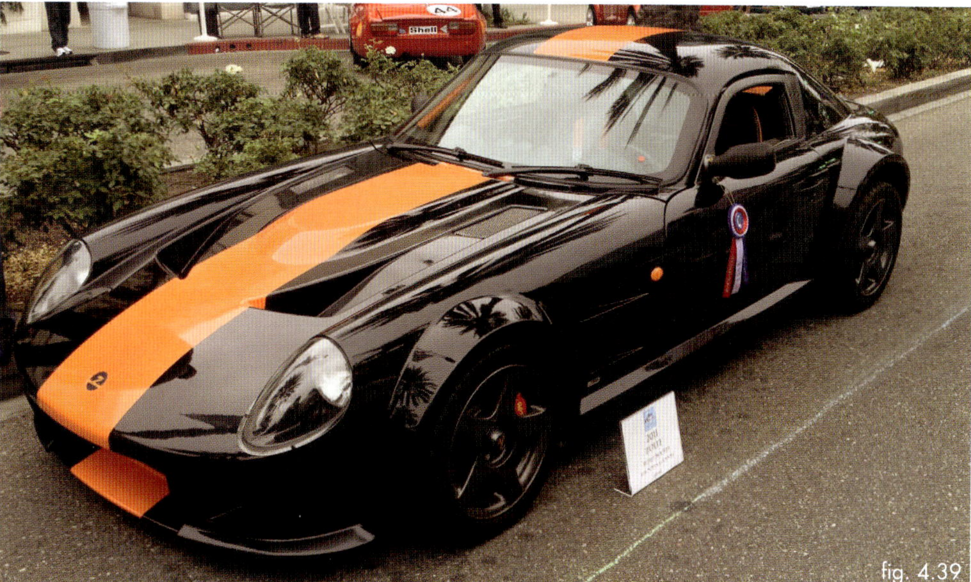

fig. 4.39

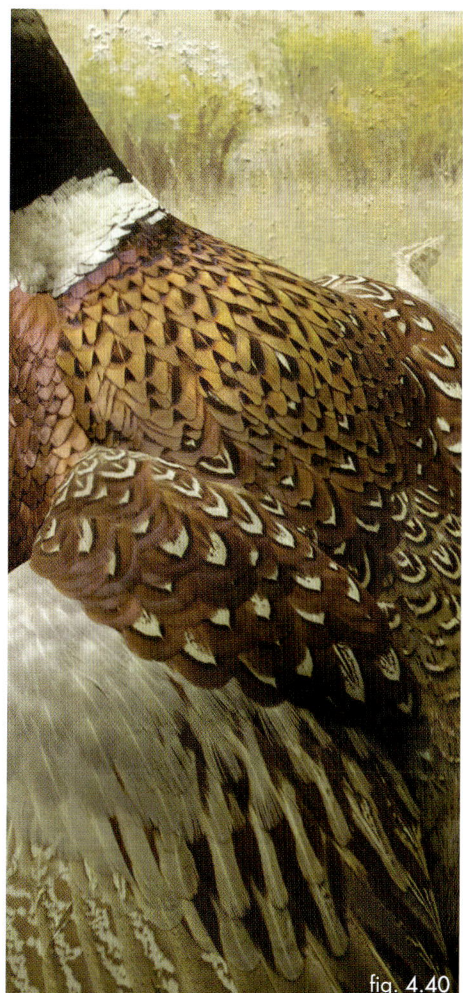
fig. 4.40

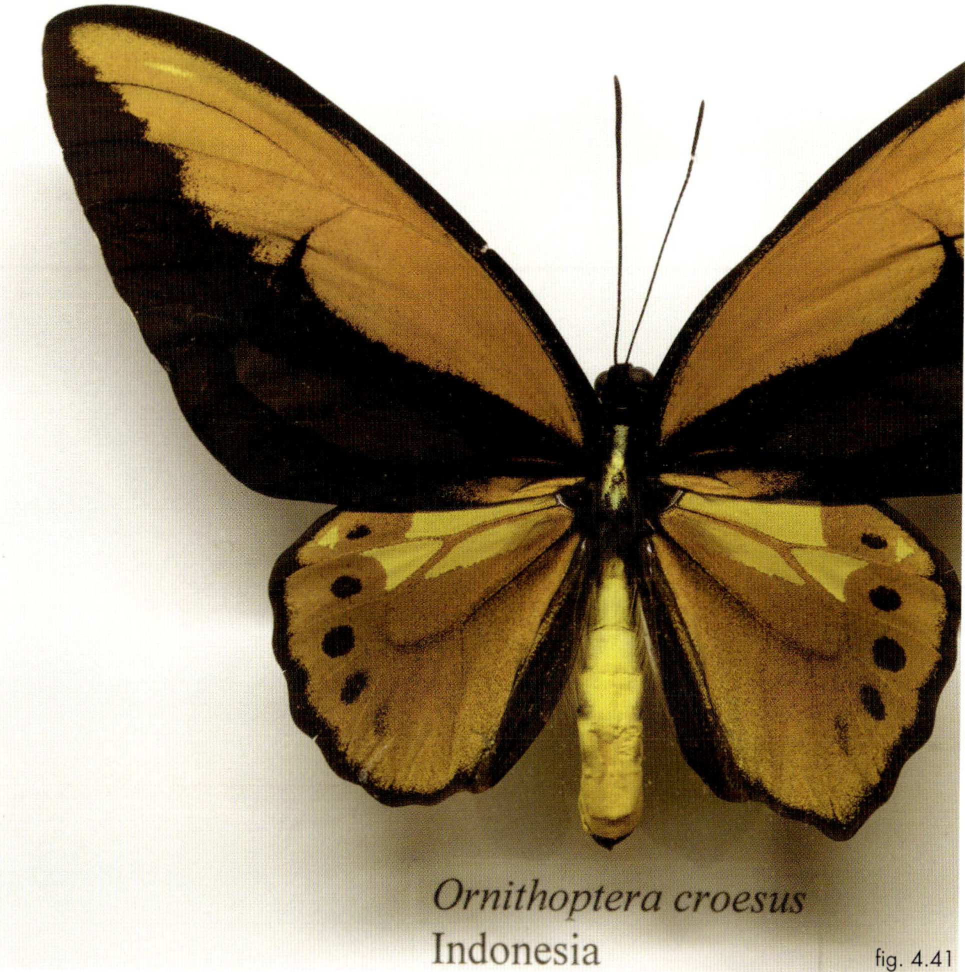

fig. 4.41

TEXTURE

Texture is the physical roughness of a surface. Texture is indicated by changes in value, and how noticeable it is depends upon how reflective the surface is. If a surface is reflective, like this detail of a steering wheel (fig. 4.42), then the texture becomes most visible where the light source reflects most strongly off its surface, increasing contrast. If a surface is very matte (fig. 4.43), then the texture is most visible on the Number 2 side, which on a round surface is just before the core shadow. In fig. 4.44 the textured skin of the lizard is most pronounced on the Number 2 sides, just above the core shadows, not on the shadow side or on the lightest side at the top of its head.

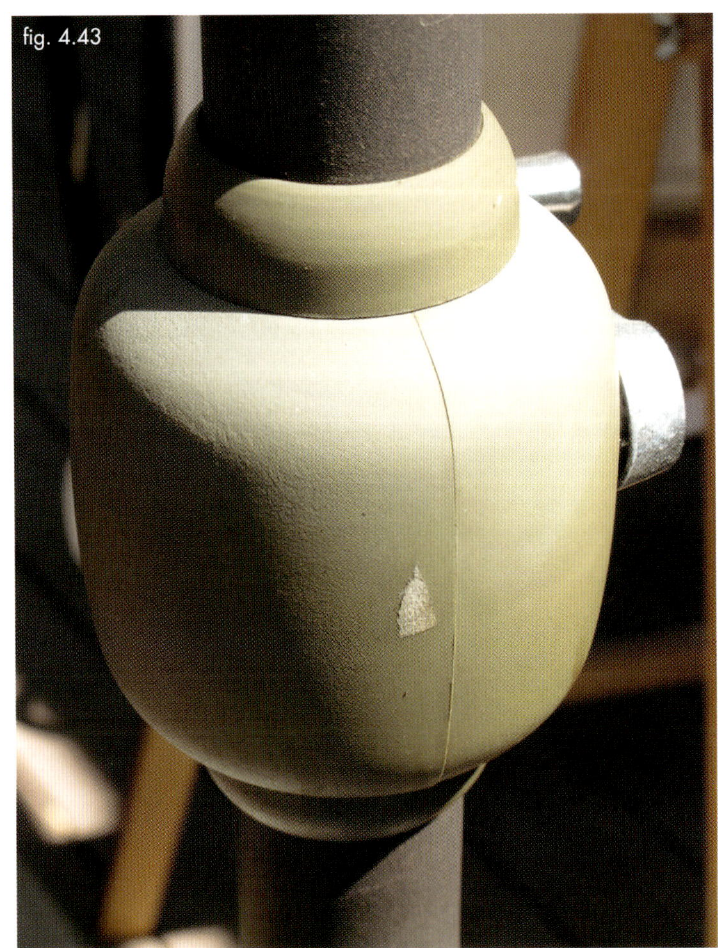

fig. 4.43

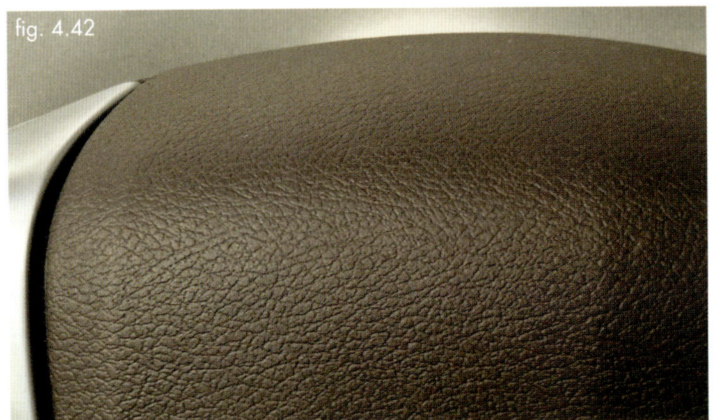

fig. 4.42

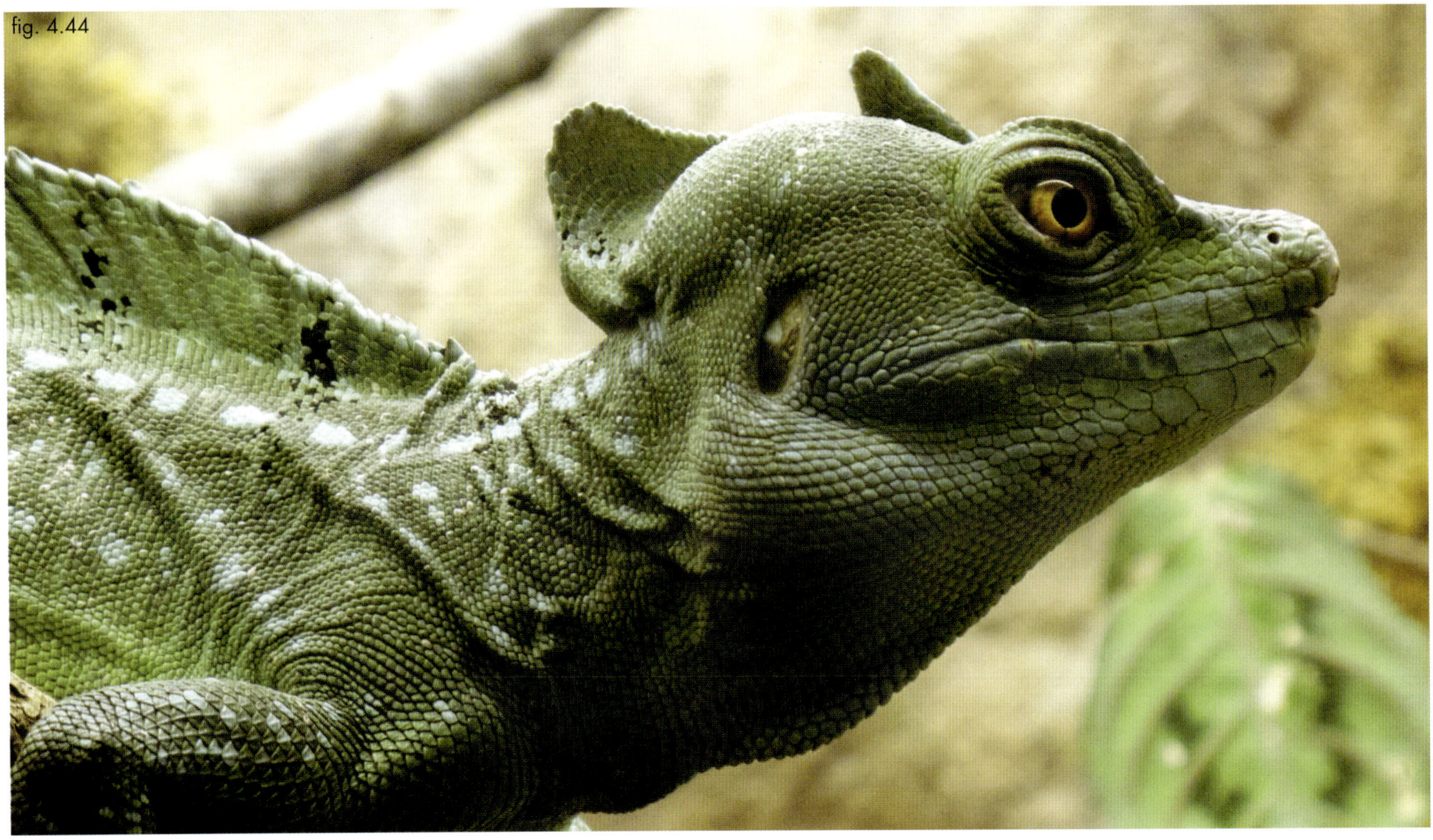

fig. 4.44

DETAIL AND TEXTURE GRADIENTS

fig. 4.45

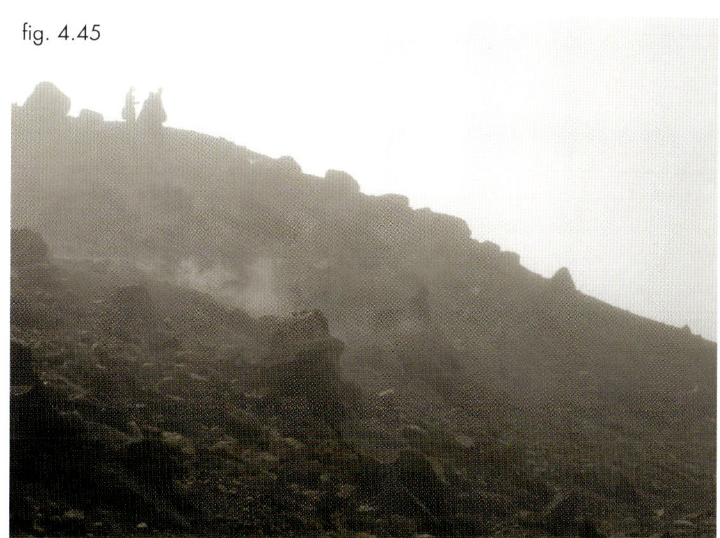

Texture gradient is the amount of detail visible on the surfaces of closer objects compared to objects in the distance. It gives a sense of depth perception. Fig. 4.45 shows that groups of objects appear flatter, the farther away they are. The foreground rocks have more apparent details and value range, whereas the rocks and people in the distance almost become one flat silhouette, with very little value range. When rendering, remember that foreground elements show the most detail. Linear and atmospheric perspectives make the textures in the distance appear more flat.

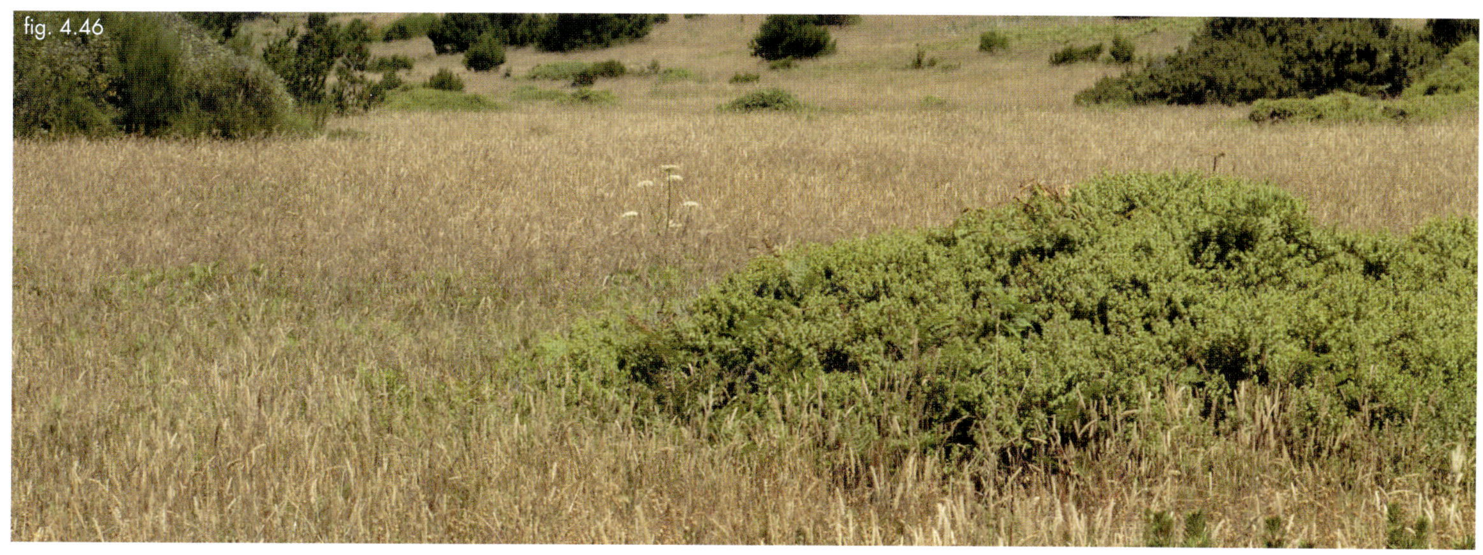

fig. 4.46

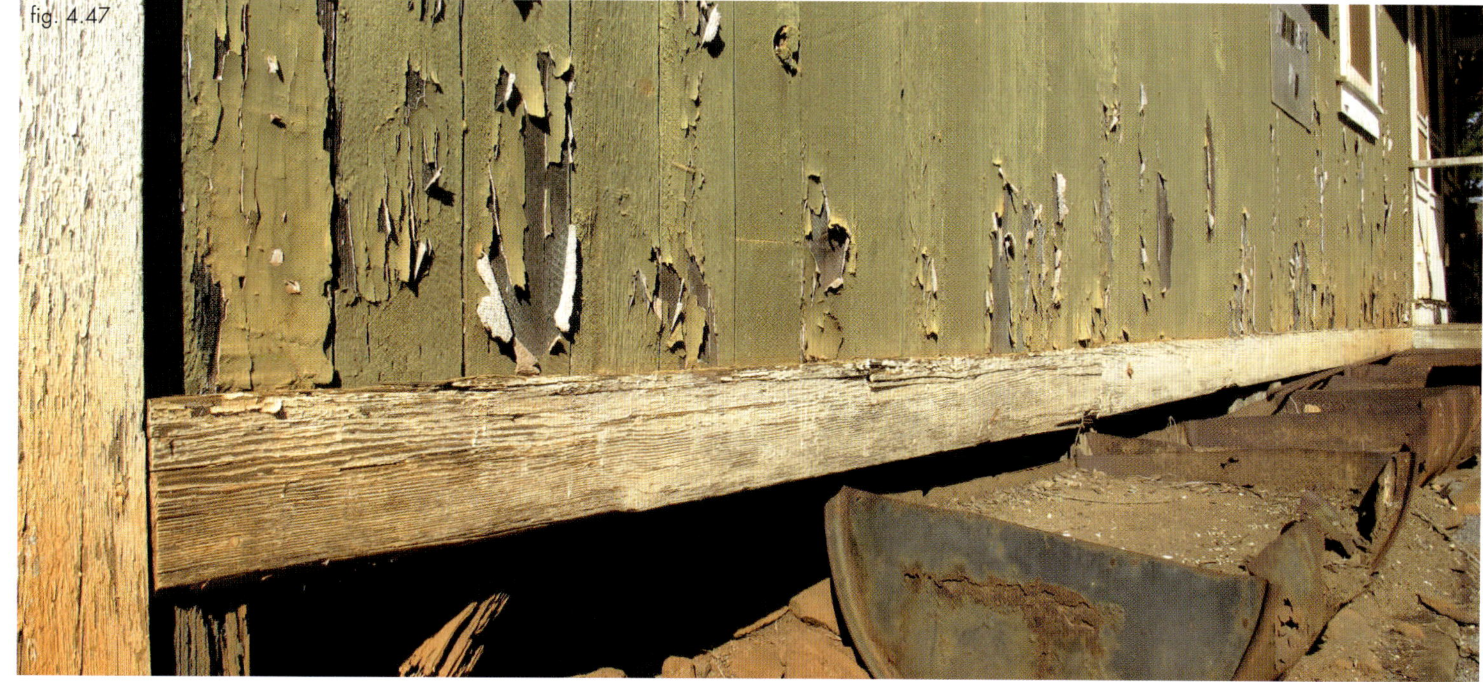

fig. 4.47

PRINT + PRACTICE

Here are a few line drawings of X-Y-Z forms that are posted on scottrobertsonworkshops.com. They can be downloaded via the URL link on page 270, then printed or opened digitally into a 2D rendering program. It is a good exercise to render someone else's sketches because, having not drawn them yourself, it forces you to think a little bit more about the sections. Experiment with the lighting and how to prioritize the focal point of one part of the form versus another. Add some graphics and part lines. Have fun; they are like little, lighting puzzles.

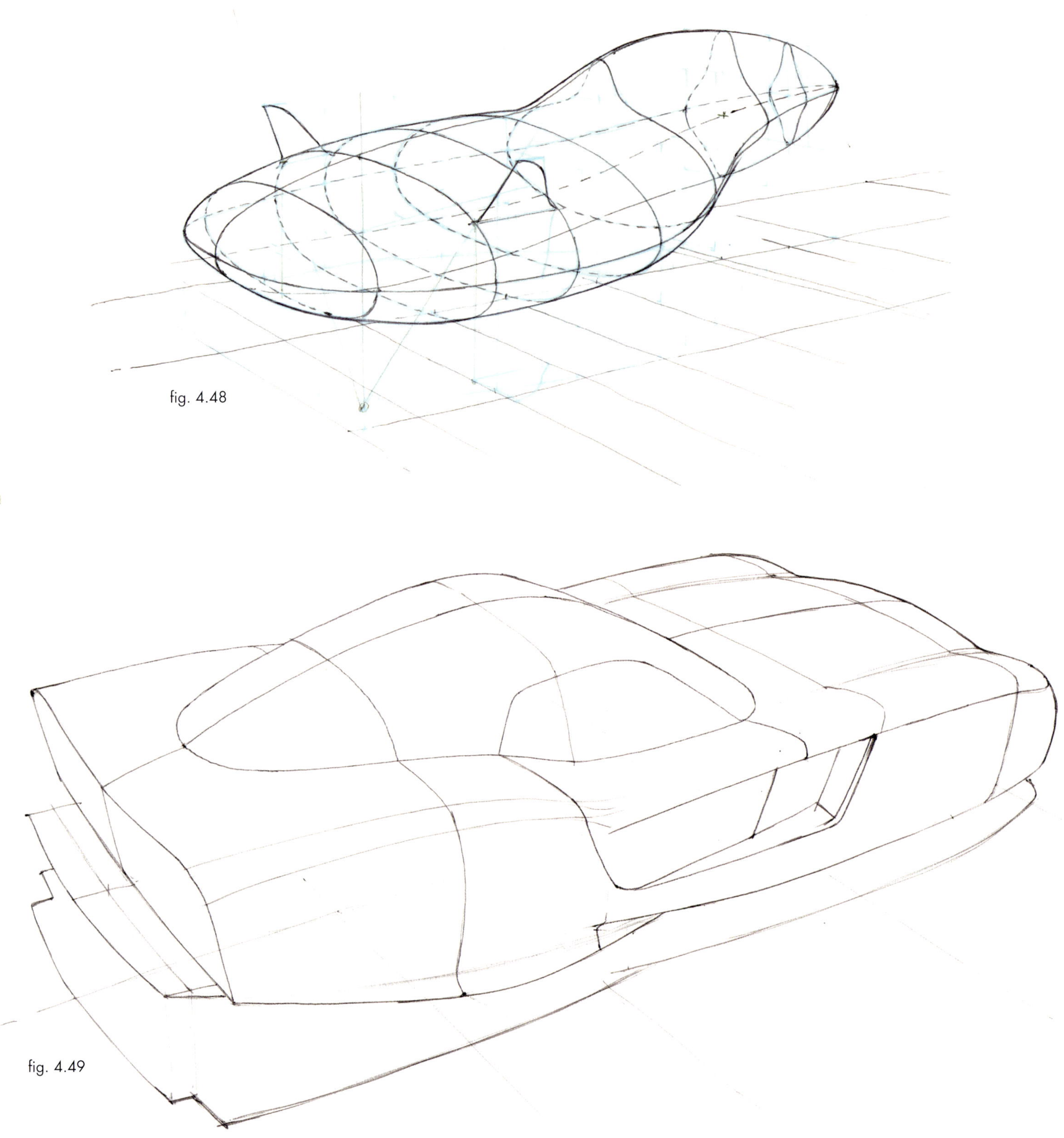

fig. 4.48

fig. 4.49

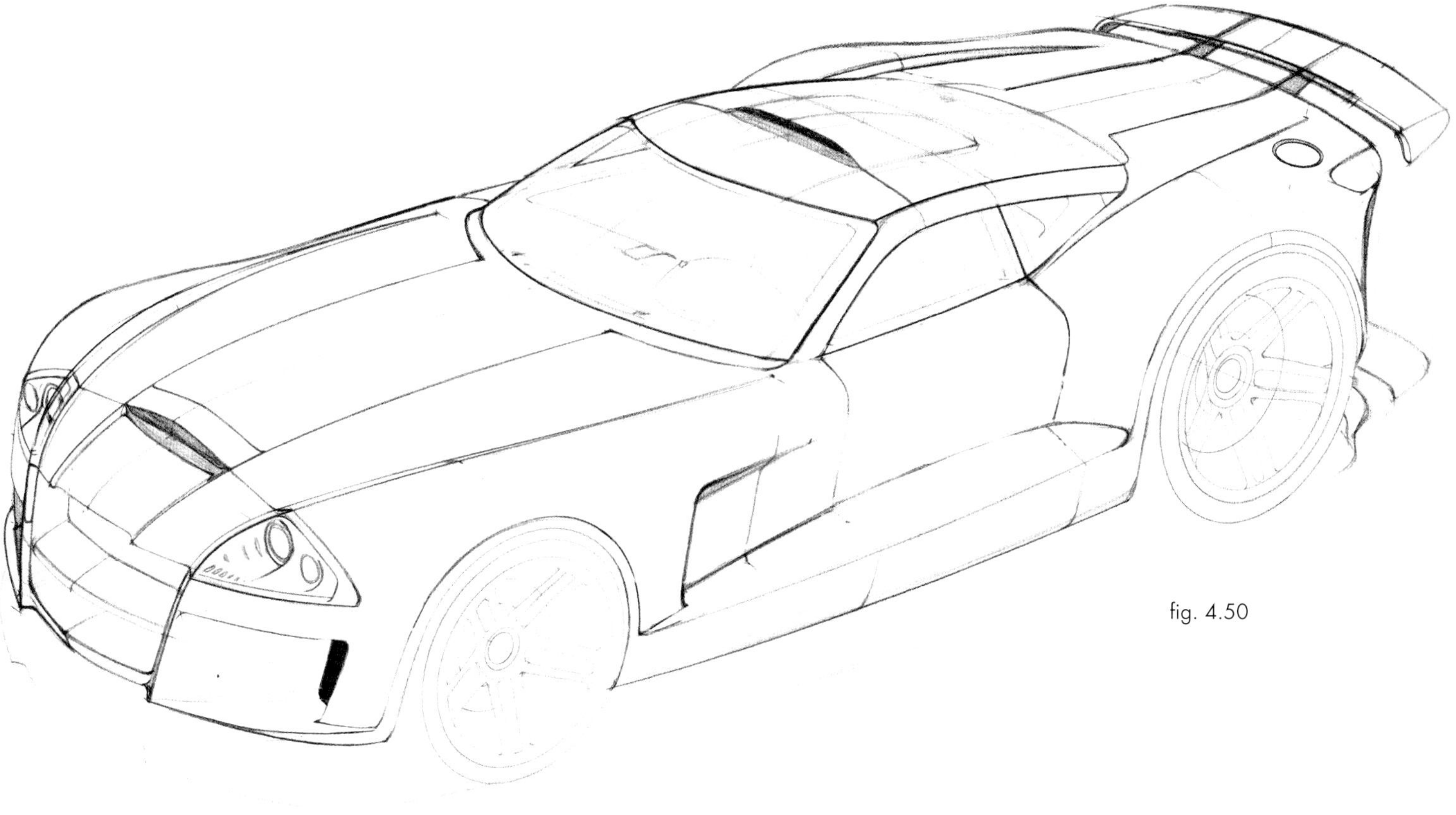

fig. 4.50

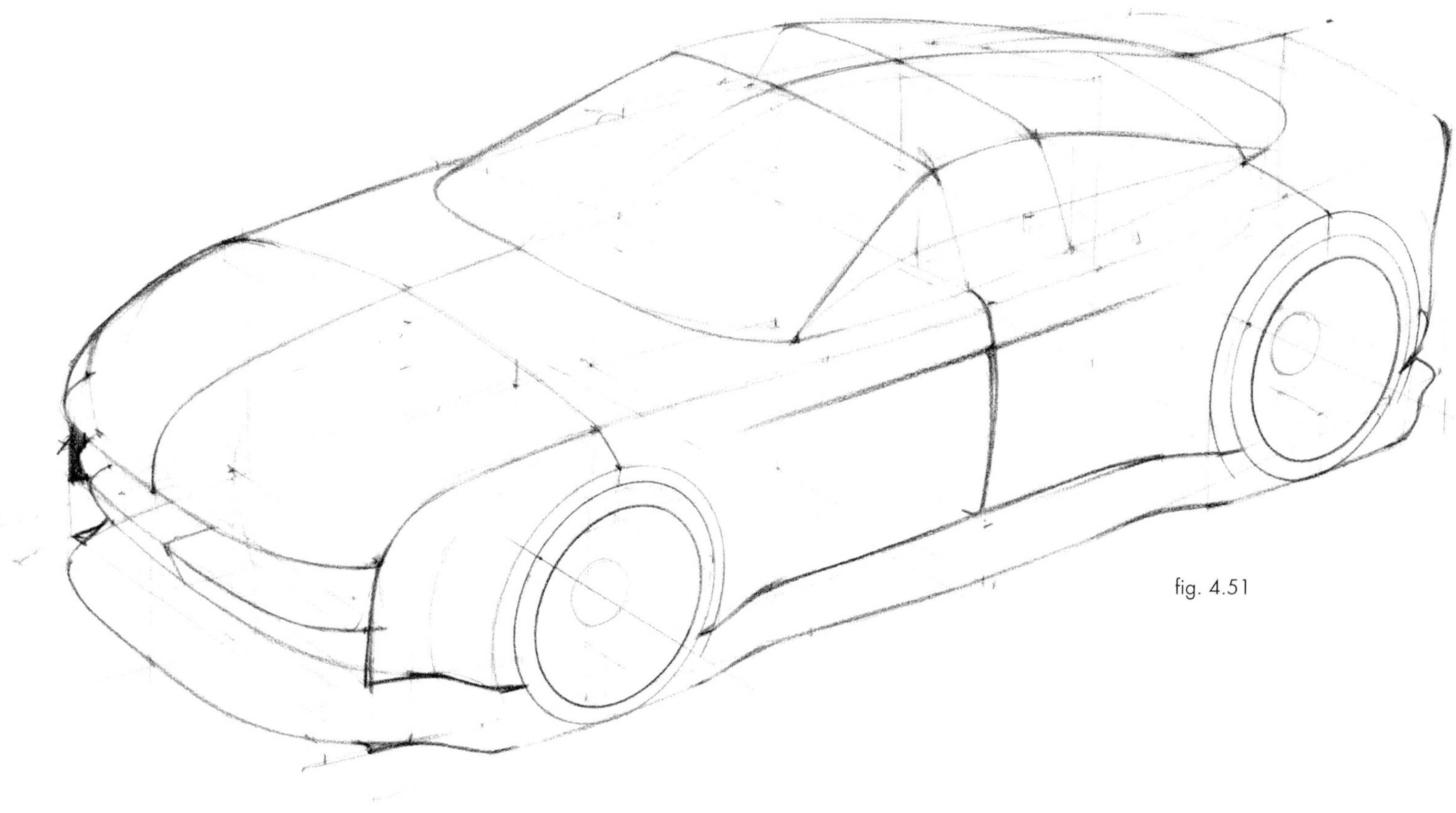

fig. 4.51

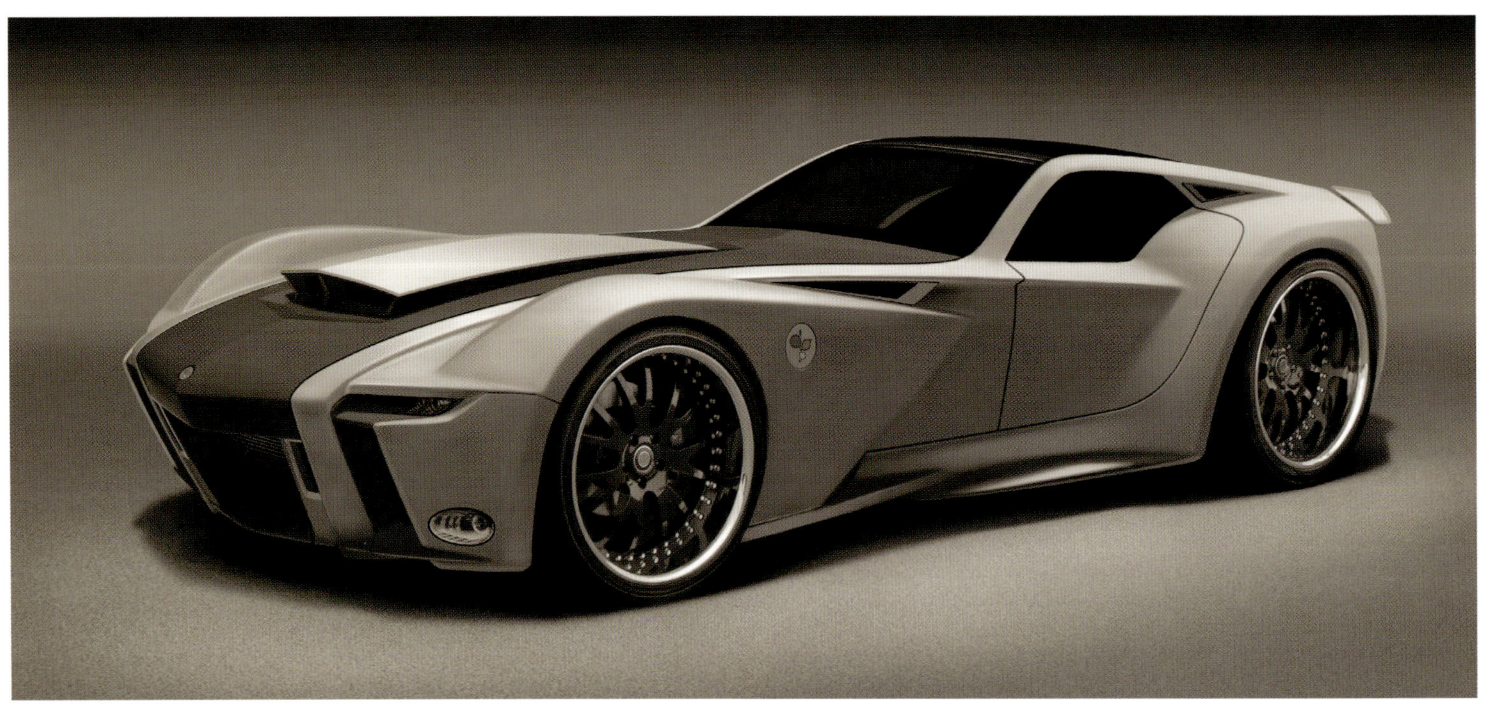

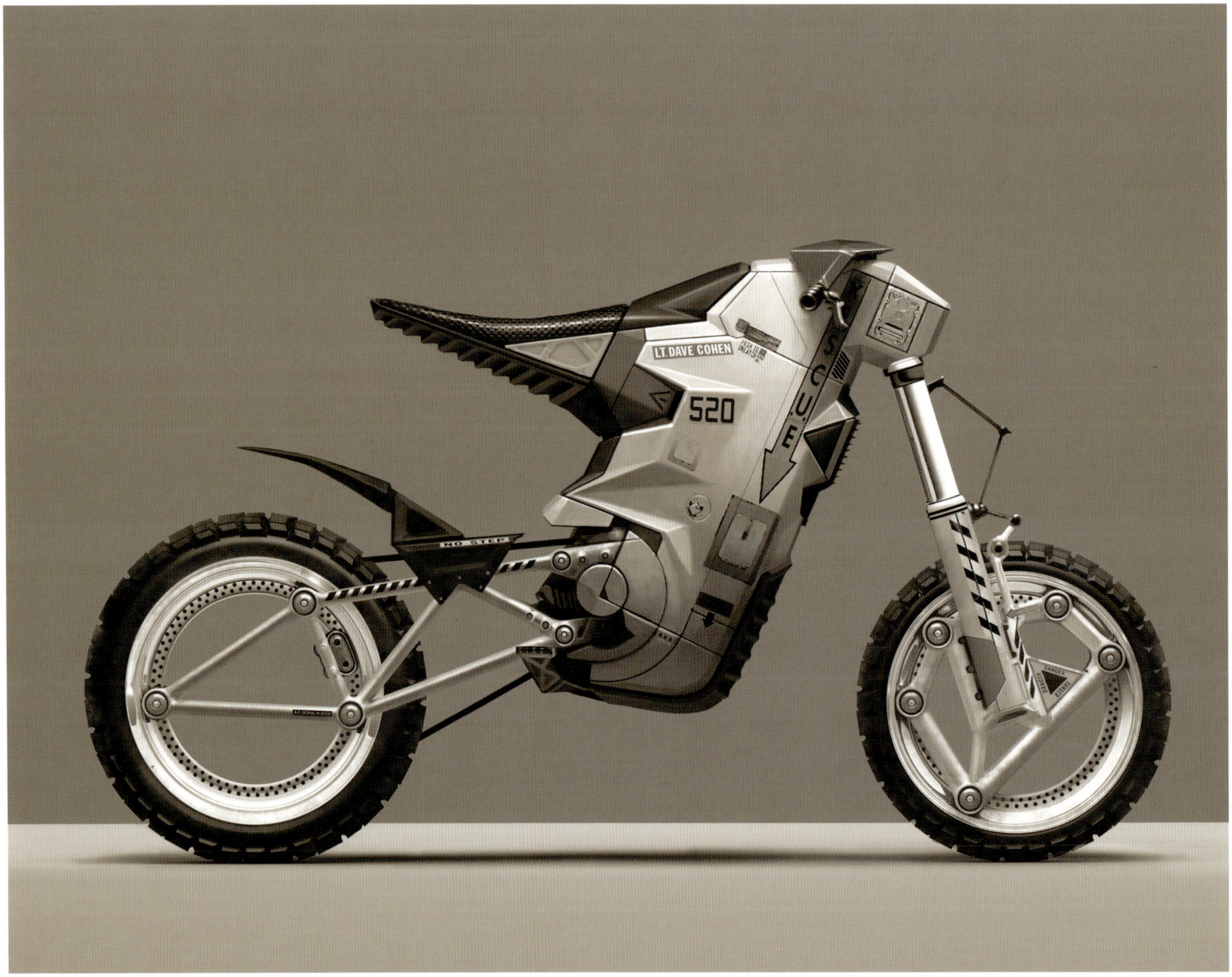

LT. DAVE COHEN
520
NO STEP

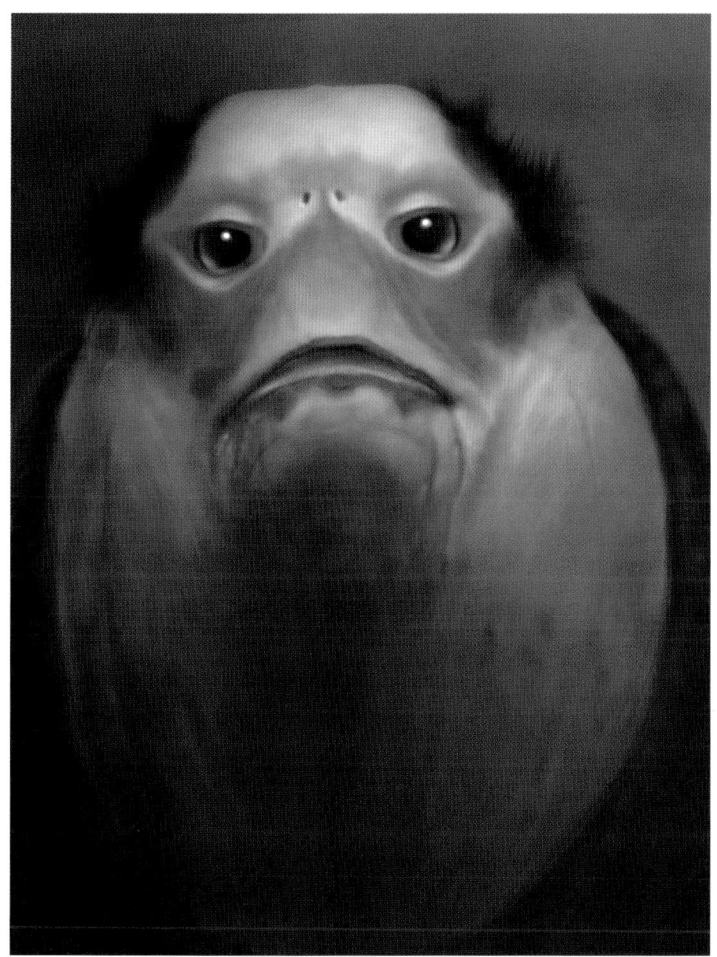

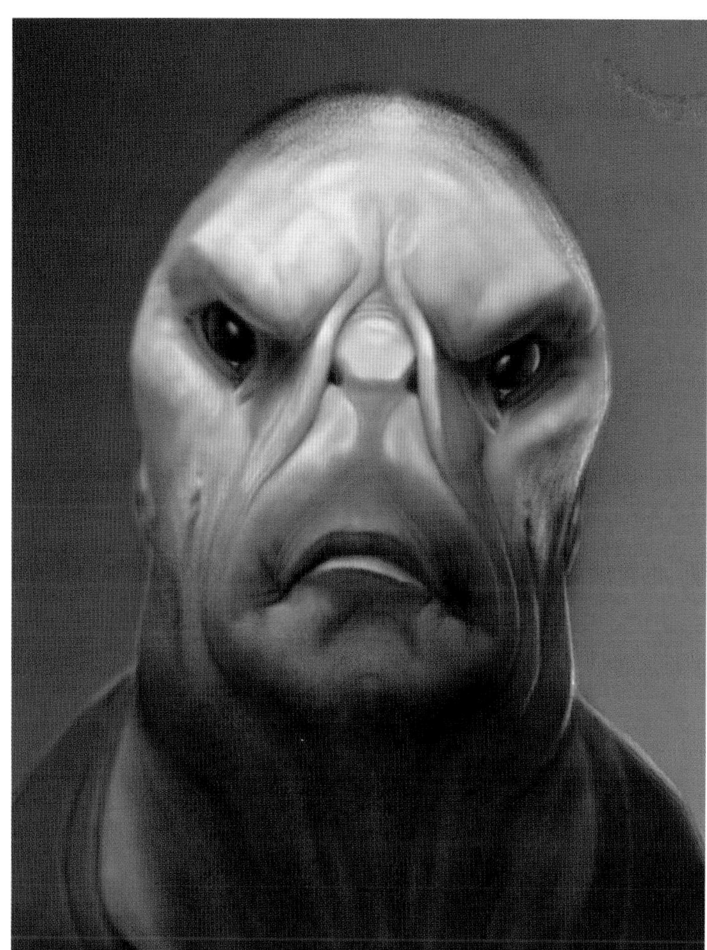

CHAPTER 05
RENDERING SPECIFIC OBJECTS

After a lot of explaining about how to render geo forms, simple curved surfaces and complex forms, let's put all of that knowledge to use in this chapter and render some specific objects. The most important thing to remember when rendering specific objects is that no matter what the subject, they are still just volumes, made up of surfaces with value changes, and the brain interprets those value changes as form changes. The previous chapters have thoroughly explained light, shadow and matte-surface value rendering. The rest is practice. Apply this knowledge to objects designed from your imagination before moving into color and reflectivity for the rest of the book. So now let's take a crack at applying the fundamentals from the first four chapters.

VEHICLE LIGHTING STRATEGIES

Like the demo on page 114, before investing a lot of time rendering a subject, it is a good idea to quickly experiment with lighting it in different ways. This helps you decide what to feature about the subject and how it will look best. Remember, the light can be controlled; it does not "just happen." It is beneficial to learn to think like a photographer who designs and stages the light. In fact, when doing a lot of renderings of a specific subject, like cars or people, buy a book on how to photograph that subject.

Figs. 5.1 and 5.2 show two possible ways to light the same vehicle. Fig. 5.2 is lit from above and fig. 5.1 is lit from the side. When doing these quick studies, use soft lighting so that plotting the shadows can be an educated guess. After picking a lighting strategy, refine the edges of the shadows for the final rendering.

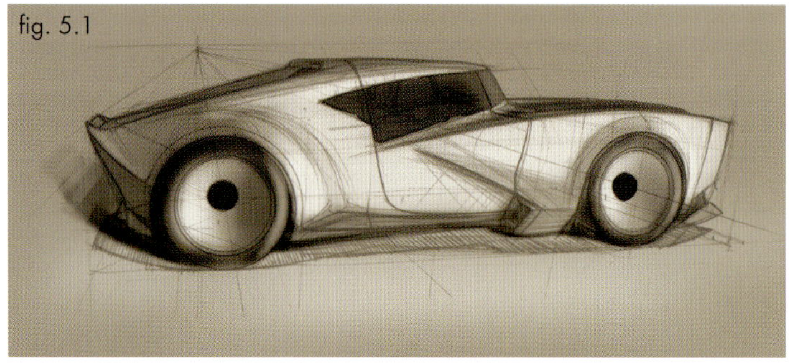
fig. 5.1

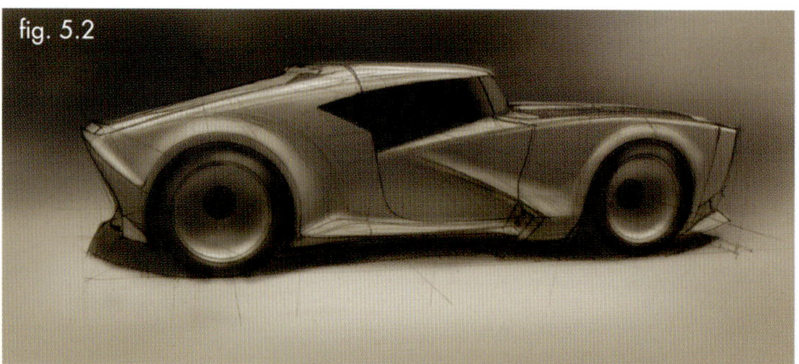
fig. 5.2

fig. 5.3

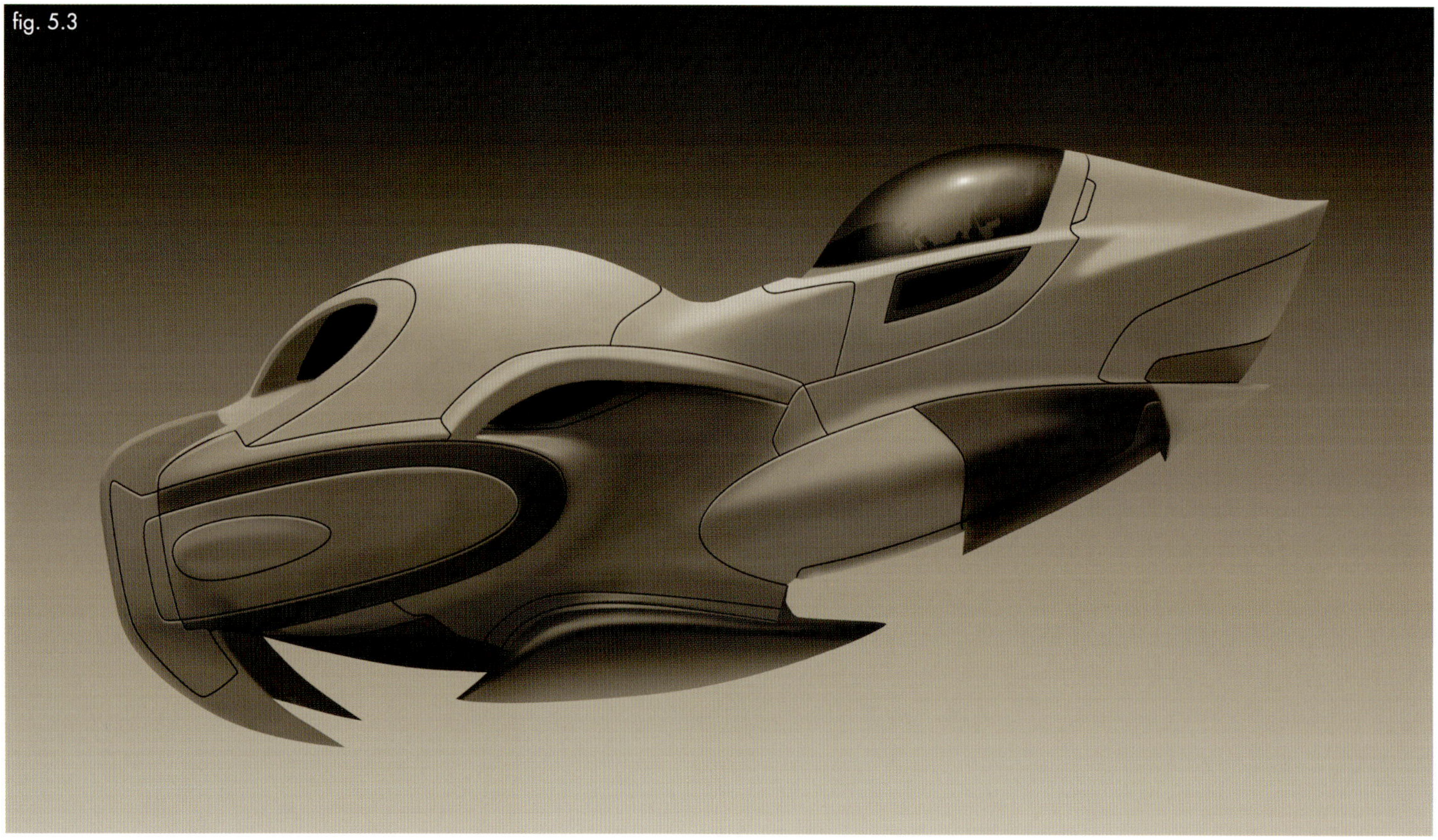

Fig. 5.3 is the subject of a video tutorial that goes into depth on how to set up layers in Photoshop, how to approach making clean edges and how to organize layers to have more control over the rendering process. It can take a little time to adjust to communicating a design without a line drawing holding it all together. Since real-world objects do not have lines around them, objects rendered without outlines immediately look more realistic.

GENERAL TO SPECIFIC FORM BUILDING

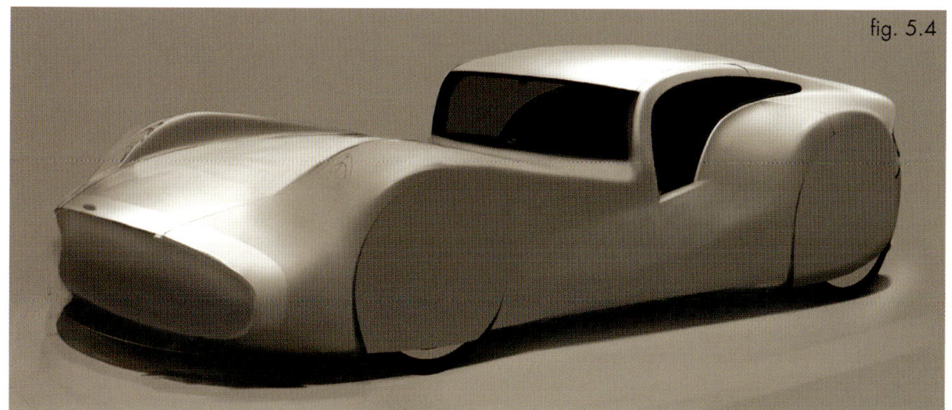

fig. 5.4

When building forms, start with the largest volumes. For this vehicle, the early stage of rendering provided a chance to adjust the overall proportions and the lighting.

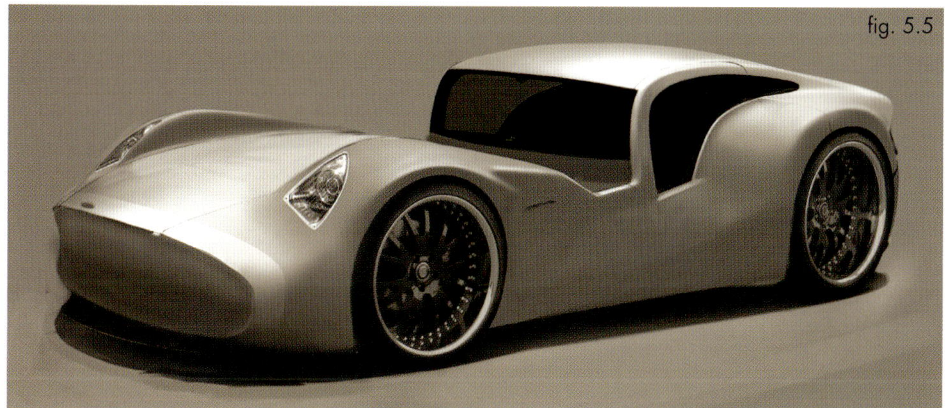

fig. 5.5

Next, the headlights, wheels and some smaller forms were defined just ahead of the base of the windshield. When doing a rendering like this, the surfaces of the car almost start to feel like clay, and value changes can be used to discover new forms. When working in a digital program, just add a new layer on top and modify values to see how new forms can be created.

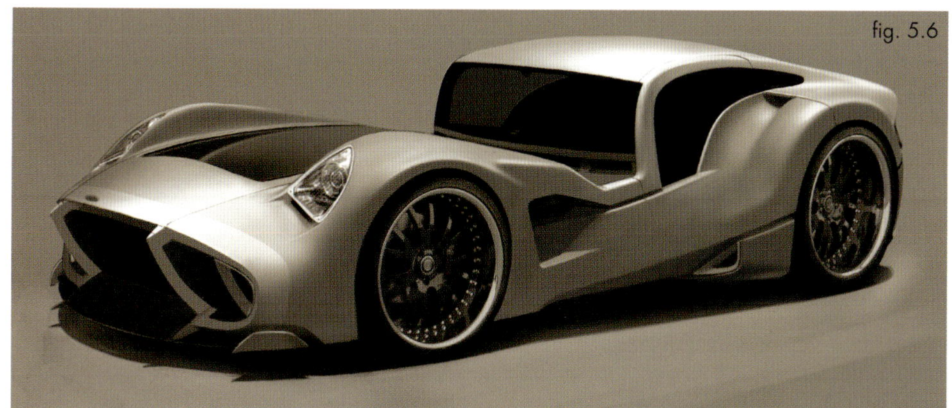

fig. 5.6

In this step, the remaining intakes and exhaust vents were added. Value levels were adjusted on a portion of the hood, which changed the local value of the painted surface to create the stripe. The bright spots on the top of the car and the front of the nose are starting to take on a metallic look. This material change will be explained in detail on page 210.

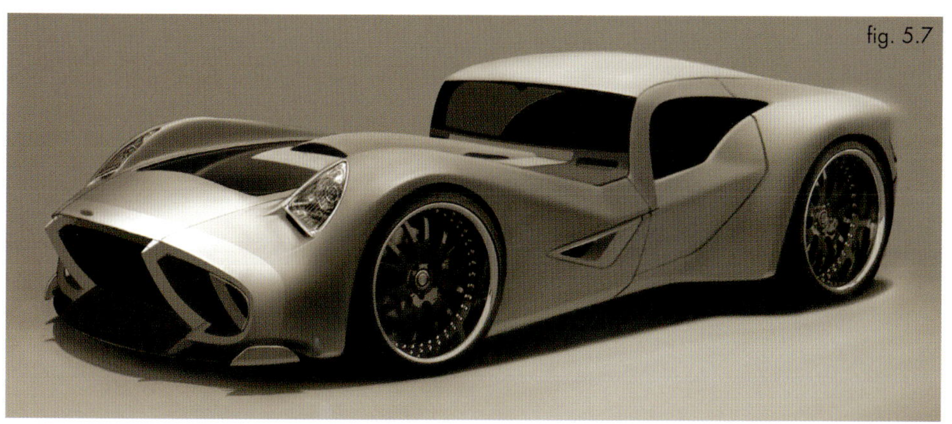

fig. 5.7

Here, a new layer was added and the car was rendered again in order to explore a new design direction for the side body and hood. This is a useful styling technique. Add a layer to do a study, and then another and another, until it works. This is all in gray scale, without color. Before introducing color, it is recommended to get the forms working by focusing only on changes in value.

VEHICLE RENDERING STEP-BY-STEP 

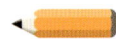

Let's walk through the steps of rendering a matte-surface car. This car is a drawing demo from *How To Draw*. Check out the linked video tutorial to observe how these discoveries about the forms were made.

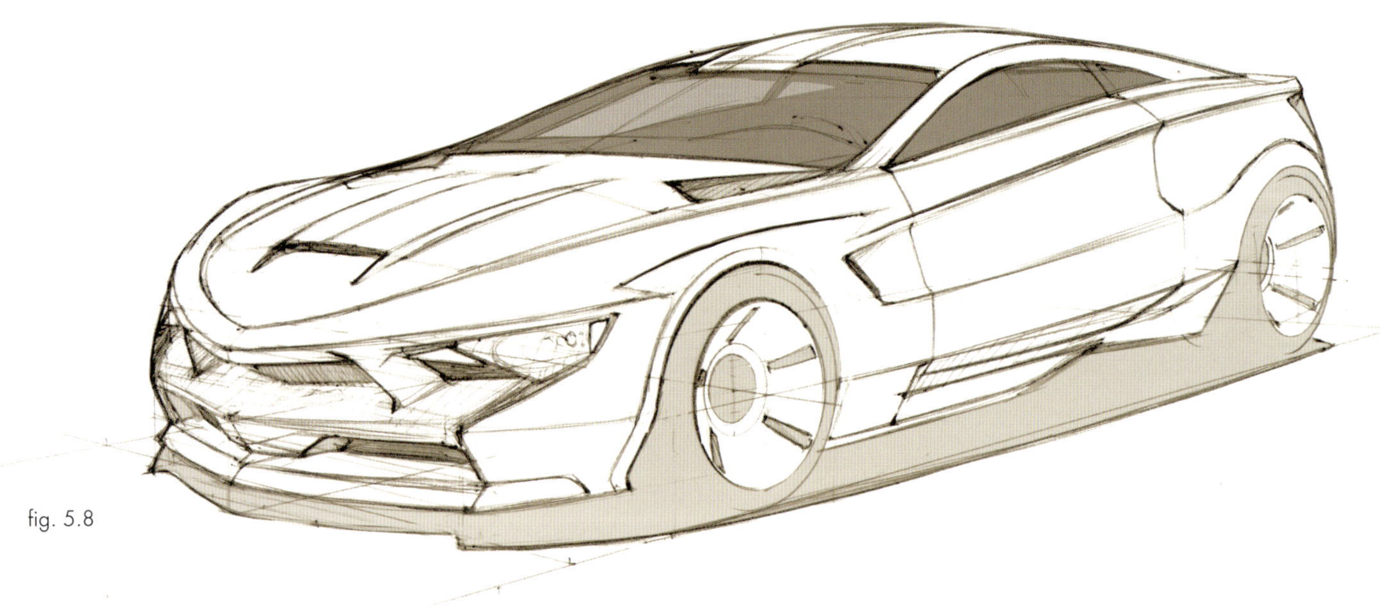

fig. 5.8

The goal was to do a tight rendering, with no line drawing visible by the end. After creating the line drawing (fig. 5.8), it was scanned into Photoshop and redrawn with paths (fig. 5.9). Redrawing paths is an opportunity to make adjustments to the design, if necessary. Put all of the paths on one layer and line them up to one another before making separate layers. Think about what surface will go on what layer, and then slightly overlap the edges of these layers in the layer stack. The layers should never meet edge to edge because then the background has a tendency to show through. A good, basic layering strategy is to make paths for each of the different materials: body, tires, wheels, glass, headlights, plus any small forms like a hood scoop.

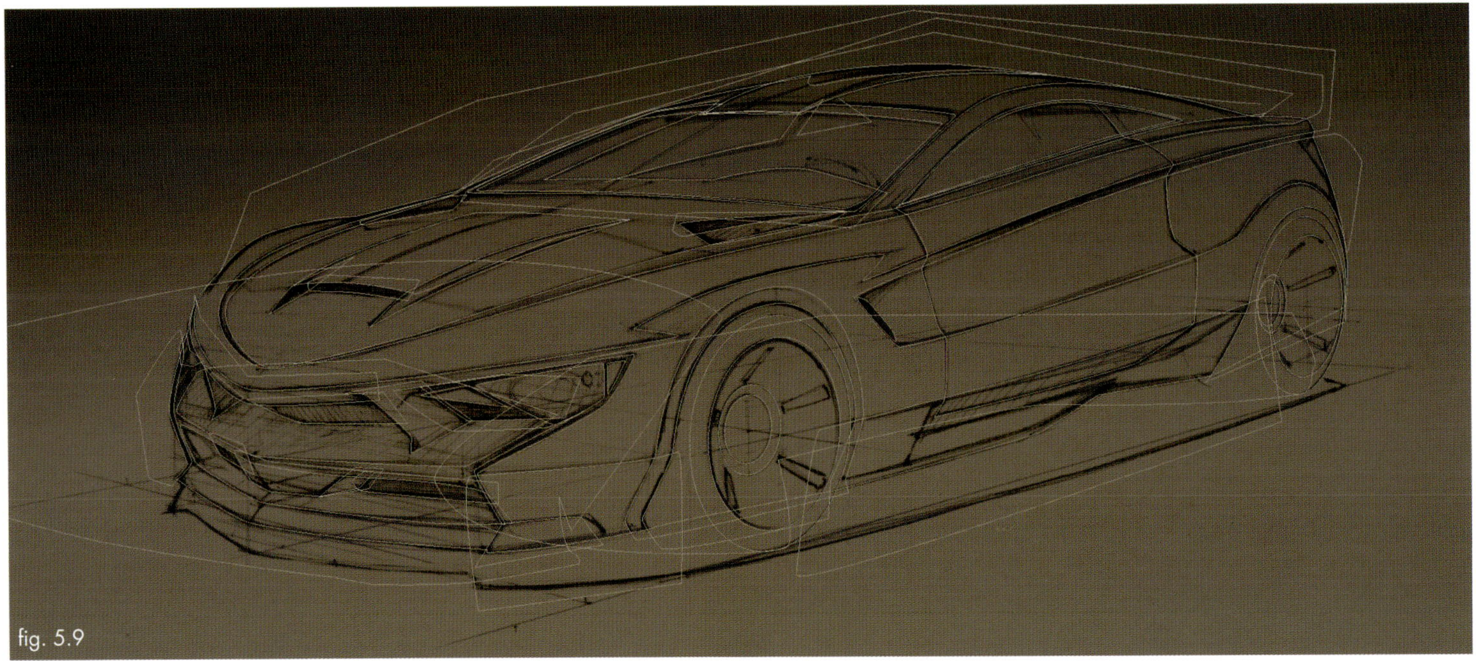

fig. 5.9

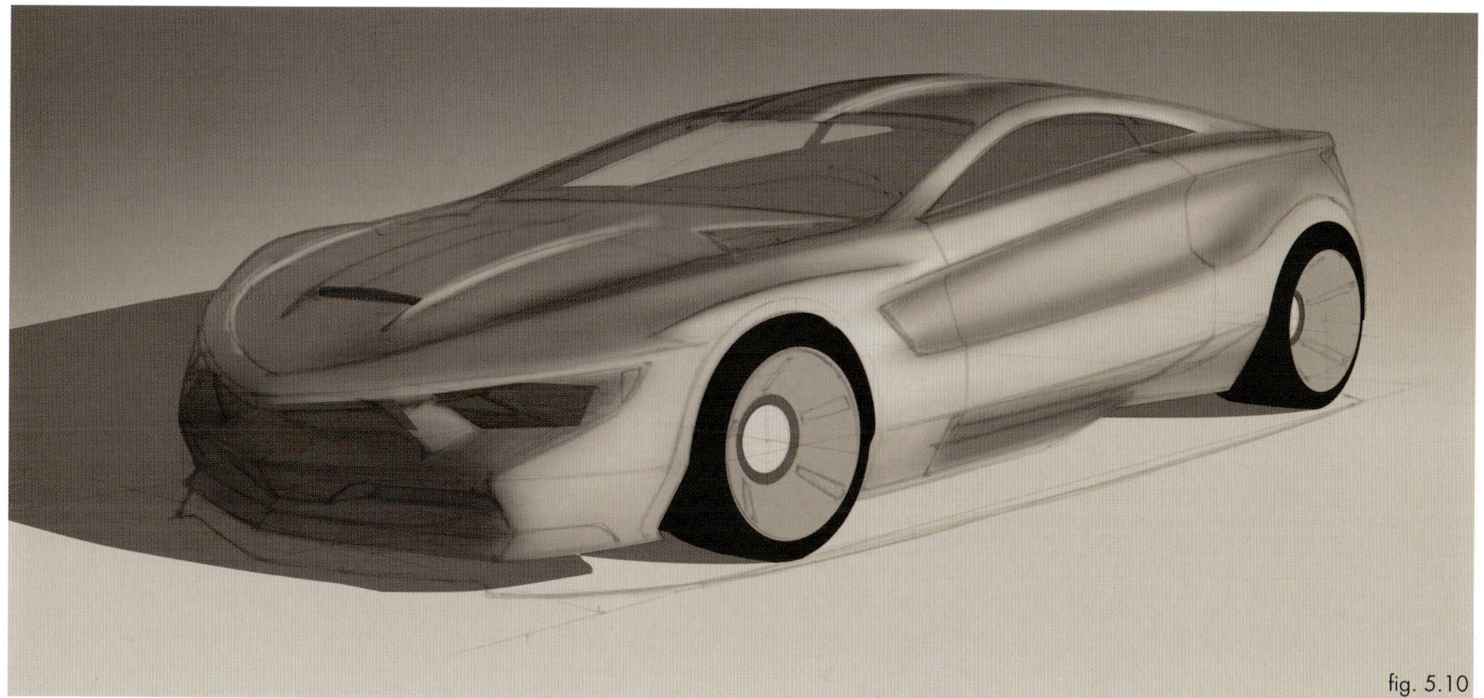

fig. 5.10

After all of the layers are made, and each is filled with a 100% opaque value to establish the silhouette of that layer, it is time to experiment again with lighting before committing to a direction. In fig. 5.10, sidelighting was tried and while it is dramatic, it makes the car's relationship to the ground a little difficult to understand because the low position of the light source throws the cast shadow under the car. On the positive side, it really highlights the headlight/air intake form notch at the front corner of the car. Again, it is all about trade-offs and compromises in designing lighting that supports why this perspective view was chosen in the first place.

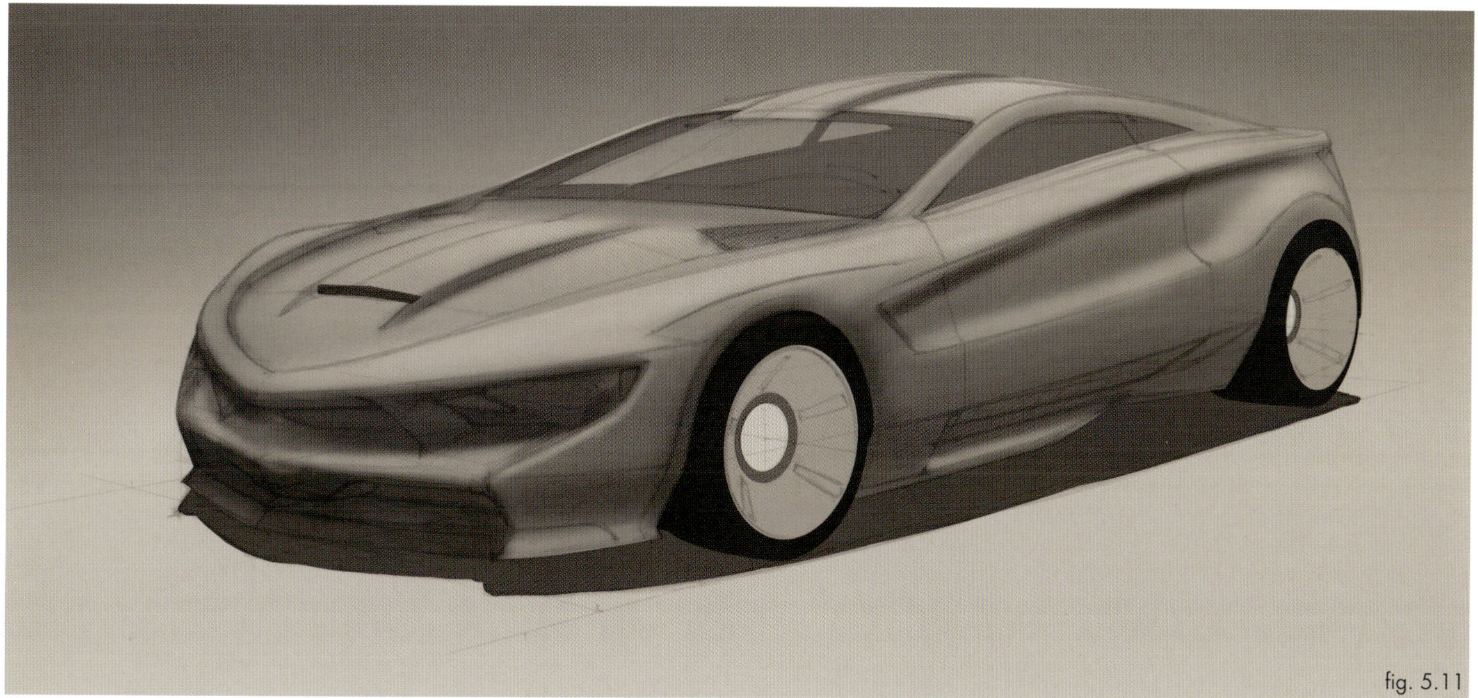

fig. 5.11

Fig. 5.11 shows a quick, rough test of positioning a soft light source at a high angle, knowing this would really play up the form changes from the top surface to the sides and make the sculpting down the side of the car more visible. If the light were directly over the car, the wheels would drop mostly into shadow. On the positive side, the cast shadow really anchors the car and makes the ground clearance and stance of the car more easily understood. When compared side by side, this was more important so this lighting scenario was chosen.

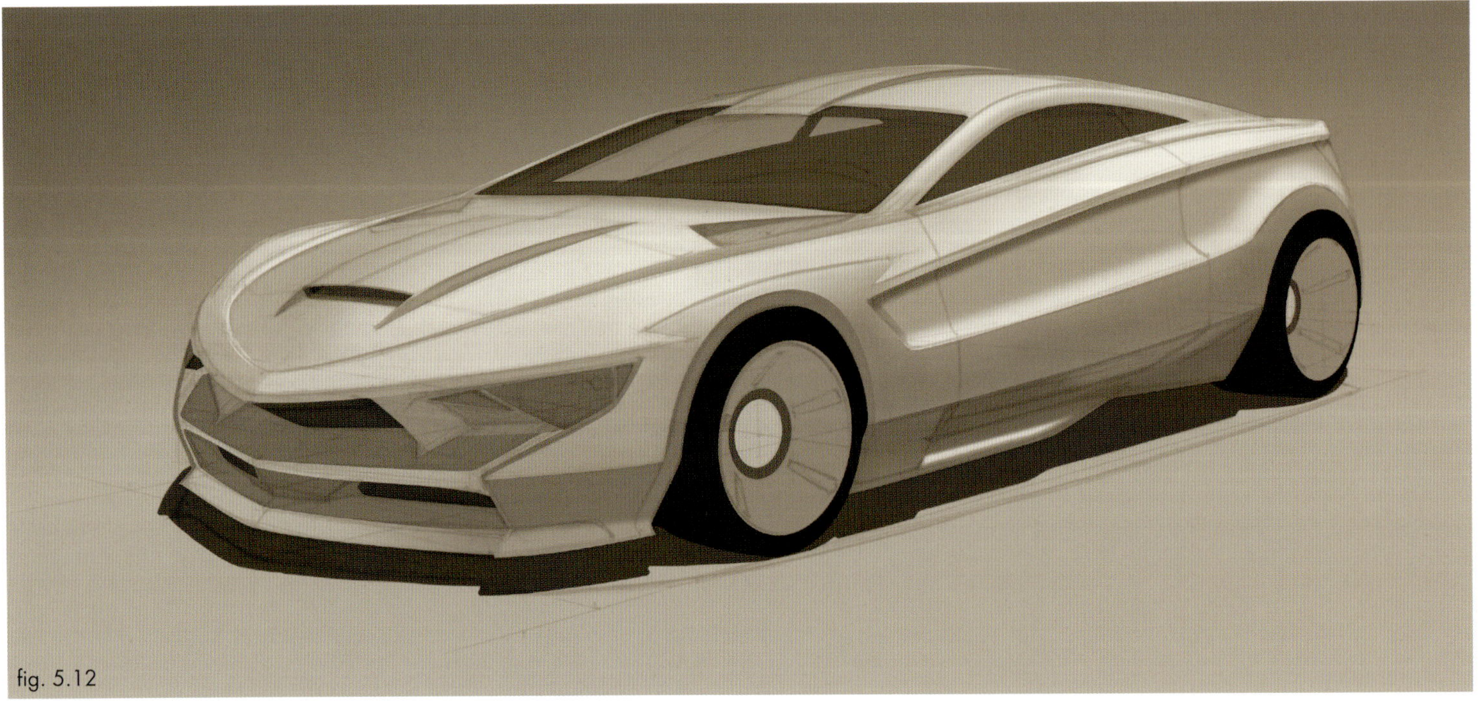
fig. 5.12

The next step is to start to let the sections define the values of the surfaces (fig 5.12). There is definitely some artistic license being applied here, and a fair amount of guessing, in the interest of working quickly. The original line drawing is floating over all the other layers as a multiply layer so the section information can be referenced, while trying to figure out what value to make each part of the surface. Relying on the shading techniques and value-assignment knowledge that was explained earlier in the book, best guesses were made. When in doubt, to create a form change, just change the values.

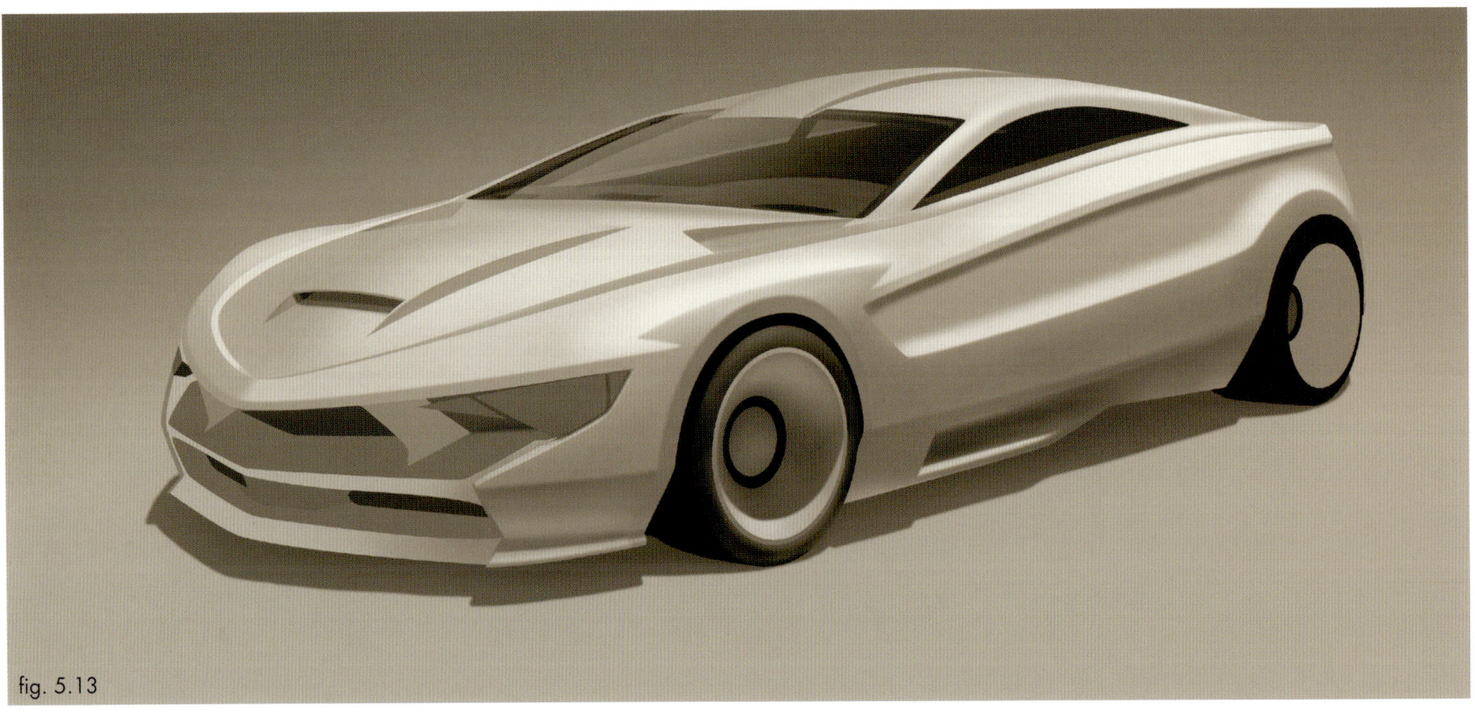
fig. 5.13

In fig. 5.13, some small adjustments were made to the lighting direction so that the light rays would illuminate the lower portion of the wheels. Also, since this vehicle is outside, the edge shadow is harder so that the light feels more like the sun, less like a softbox light that is commonly used for indoor photography. All of the sculpted features of the car body are still being rendered as hard-edged. The blending of one form into another can be accomplished by blurring any layer over the other. This is a great way to create small radii or fillets. For a refresher on radii and fillets, see *How To Draw*, Chapter 06.

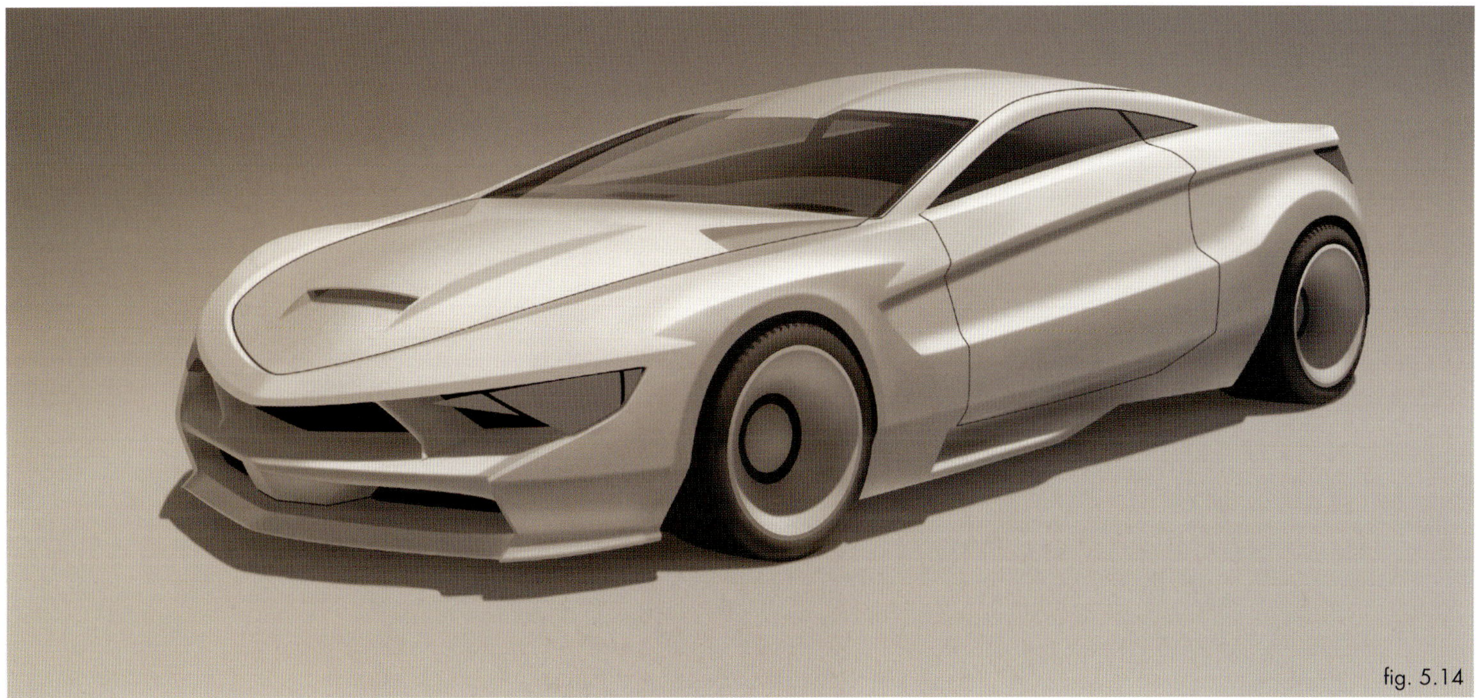

fig. 5.14

In fig. 5.14 the front end was addressed and the part lines were added. When working in the shadow areas, like within most of the front end of this car, use the reflected light from the ground to create value changes that communicate whatever forms are desired. By blurring the hood scoop, it now appears to have a small radius and fillet on it. This technique of rendering all of the add-on forms on separate layers with hard edges, and then blurring them to change the size of the radii and fillets, is a way to render very controlled surfaces.

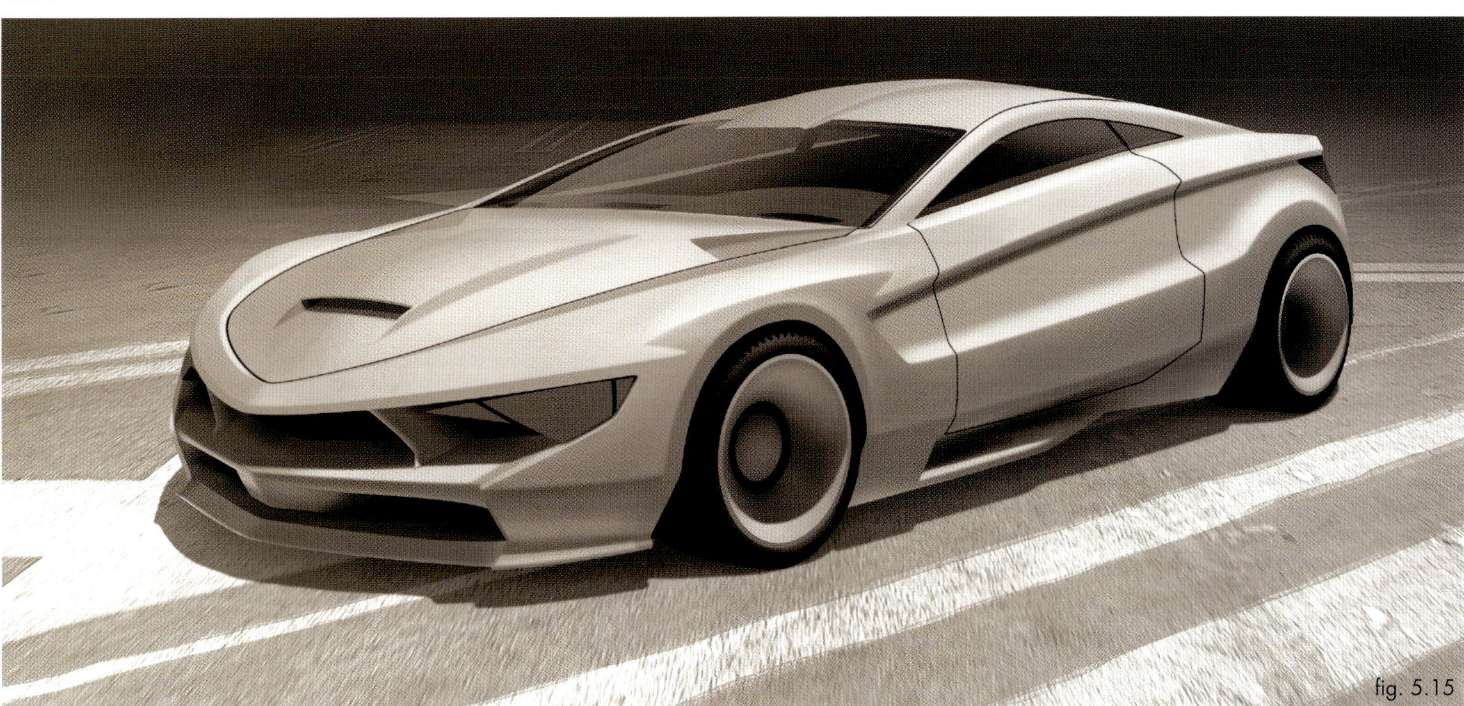

fig. 5.15

Lastly, a more photographic ground plane was added along with more contrast to the background. A few lines on the ground were roughed in, to better establish the perspective view of the car and to anchor it to the environment. The bright area of the ground, placed just under the car's right-front corner, provides a nice opportunity to increase the reflected light in the shadow area at the front of the car. Selectively locating elements like this is a natural way to generate more light in a scene. The cast-shadow layer is set to "multiply" so that the white stripes in the cast shadow are still visible.

SIMPLIFICATION FOR COMMUNICATION

Adding full-value backgrounds takes time. To communicate ideas more quickly, some level of simplification is required. By keeping the line drawing visible, it can be used to emphasize the silhouettes of the elements. If there is no time to plot the cast shadows properly by constructing a drawing, then use a soft light source so the cast shadows are very soft-edged and it is easier to guess where they belong. The two examples below demonstrate this simplified approach.

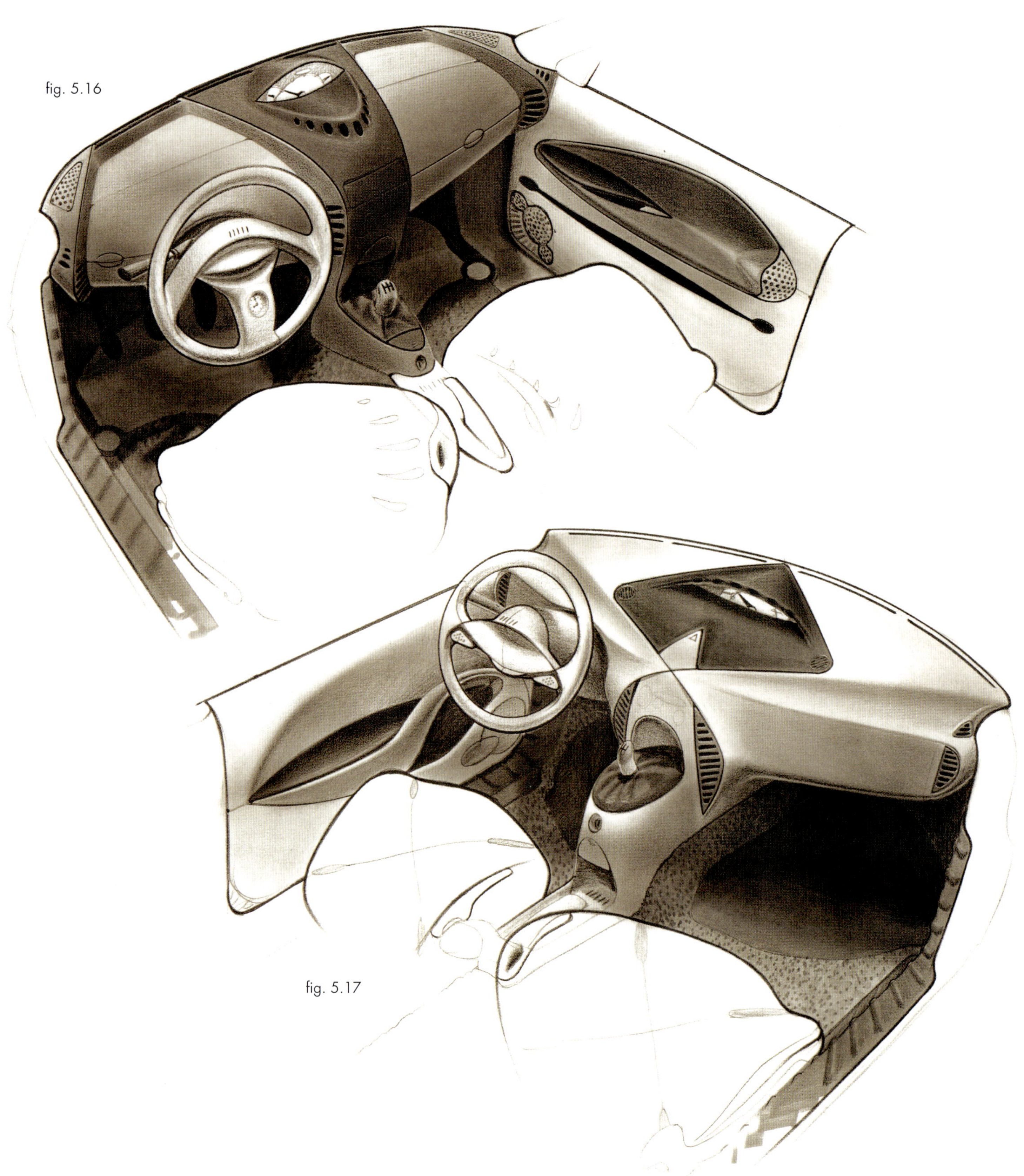

fig. 5.16

fig. 5.17

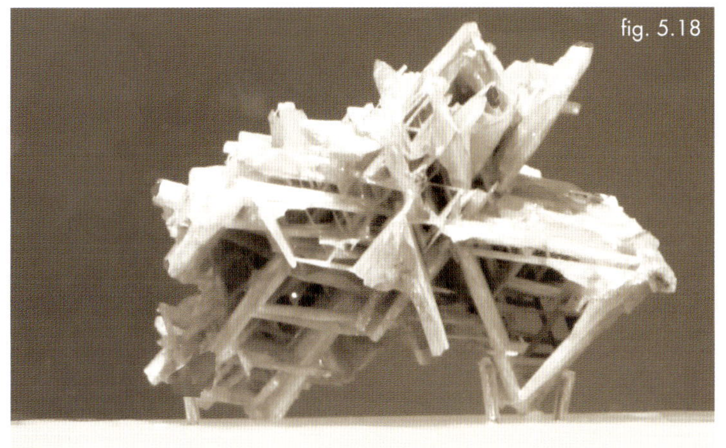
fig. 5.18

Fig. 5.18 is a photograph of the mineral cerussite. It was used as the basis for the concept sketch below (fig. 5.19). Simplification was achieved by rendering over the top of the photograph in Photoshop. A fun thing about working this way is that the value structure is already there from the photo, which can also provide inspiration for the forms themselves. By building upon the lighting that was already present in the source photo, the shapes of the walker were refined. Each time a fold or bend of the form was desired, all that was needed was to think back to the basic lesson from earlier in this book: Value changes equal perceived form changes. The cast shadows were best guesses based on the original photograph and rendered loosely to indicate the overlapping forms.

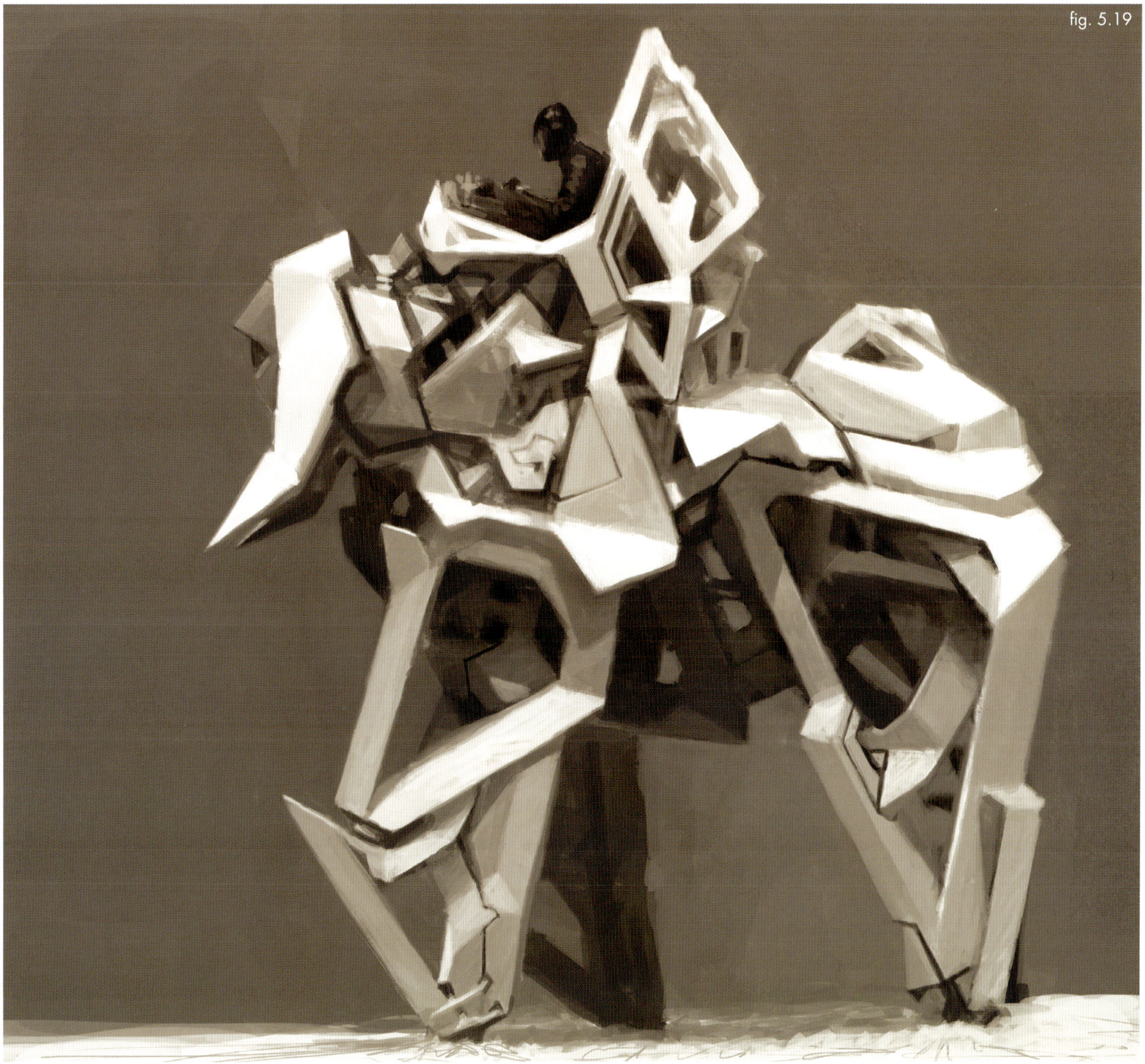

fig. 5.19

MECH STEP-BY-STEP

This rendering demo explains the application of value, without the use of a lot of paths or layers. The small sketch (fig. 5.20) was done with gray and black Pilot HI-TEC 0.5 pens. After scanning, it was selected off of the background in Photoshop using the "magic wand" tool, (fig. 5.21). With the sketch on a layer of its own, it was darkened and the background was added.

fig. 5.20

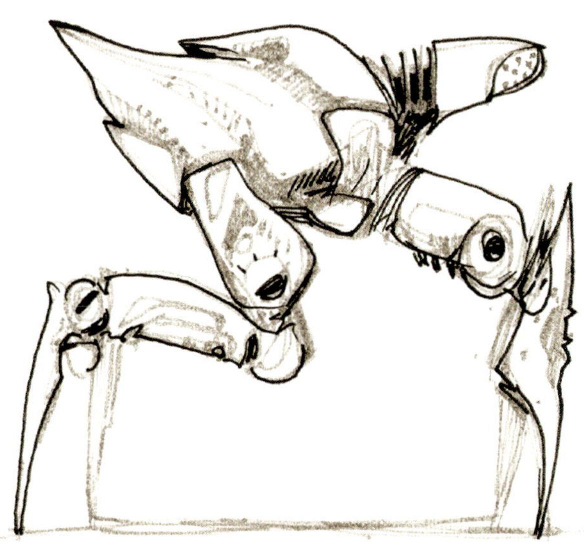

fig. 5.21

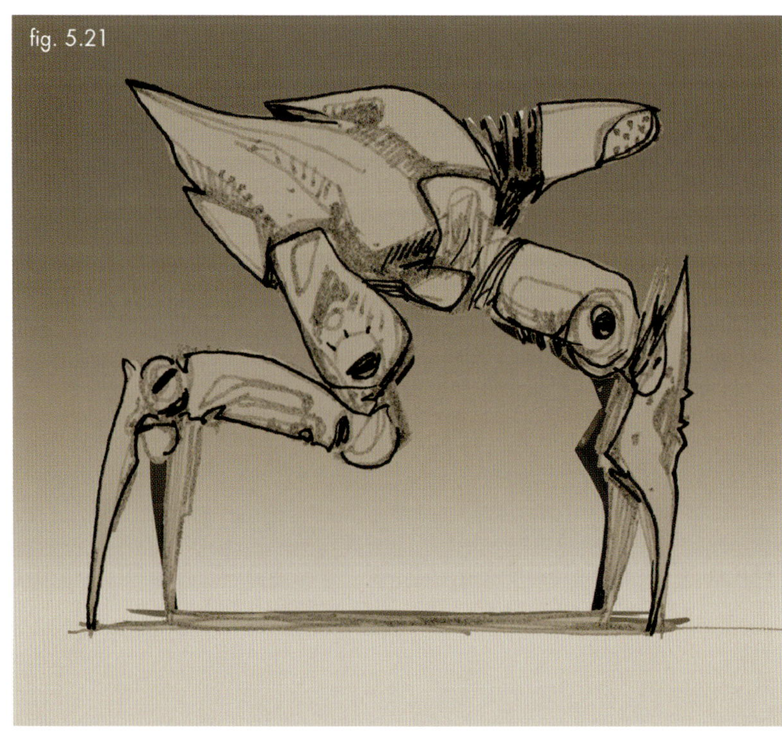

There are only two main layers when rendering an object in this style. In this case there is a layer for the far-side legs, (fig. 5.22), and an overlapping layer for the rest of the body (fig. 5.23). Setting up the far side and nearside on two layers makes it easier to preserve their edges by turning on the "preserve transparency" control button. By adding the background gradation right away (fig. 5.23), it starts to establish the top-lighting illusion explained earlier on page 88.

fig. 5.22

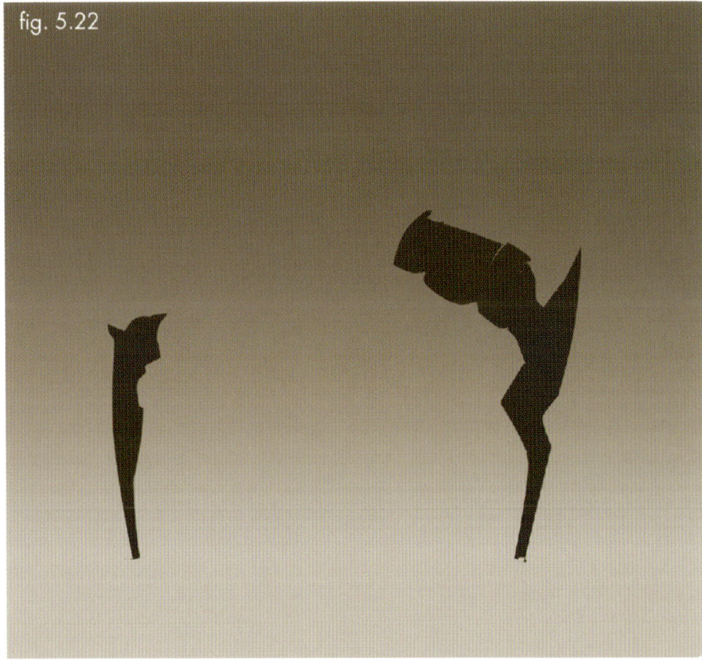

fig. 5.23

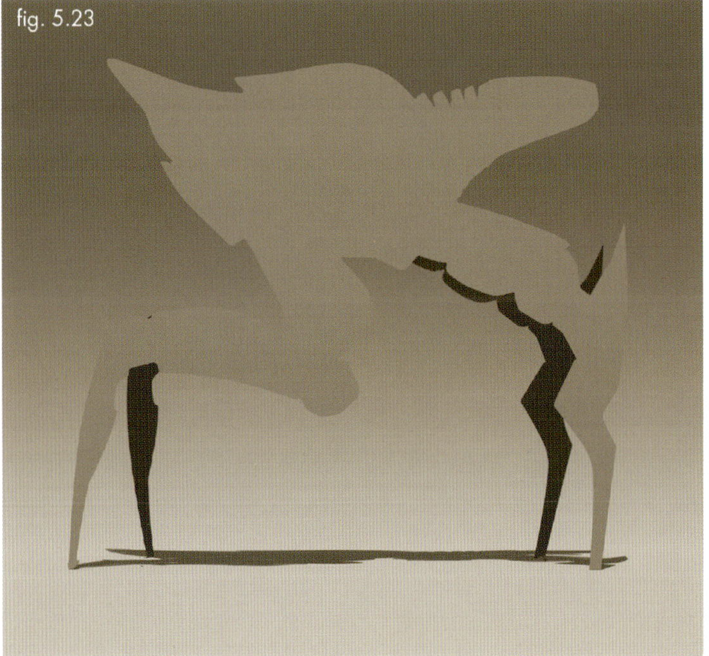

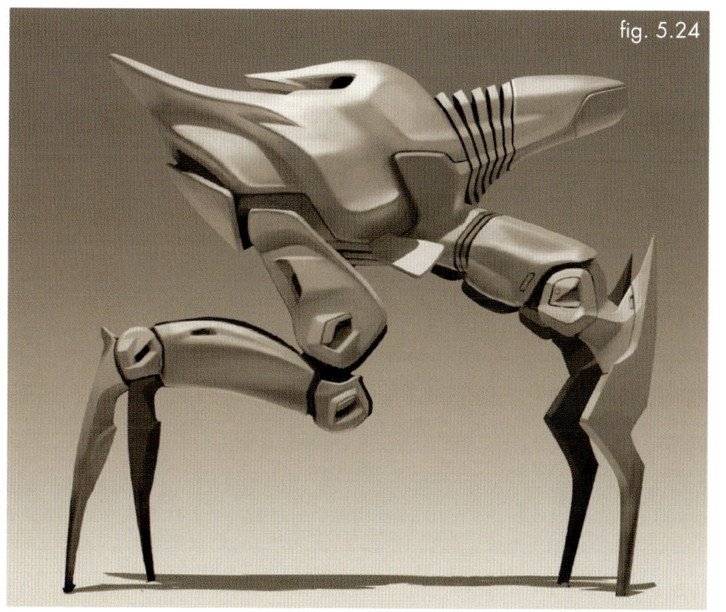
fig. 5.24

Once the silhouette of the nearside body layer is refined and the "preserve transparency" button is clicked on, it is time to start rendering the value changes. To see the entire process of rendering this mech, every single stroke of the airbrush, please watch the full video tutorial of this demo. Using a smartphone or tablet and the H2Re app, aim at the image below or go to the URL listed on page 270.

By now it should be evident that the better you understand the fundamentals of light and shadow, the easier it becomes to give objects the illusion of three-dimensional form. In fig. 5.24 the entire surface is just a series of value changes. On the Number 1 side are the surfaces that are most perpendicular to the rays of light. On the Number 2 side are the surfaces that are still in the light, but less perpendicular to it, and on the Number 3 side are the surfaces that are parallel to the light rays, where the core shadows reside. Lastly, the cast shadows and reflected light must be rendered as well.

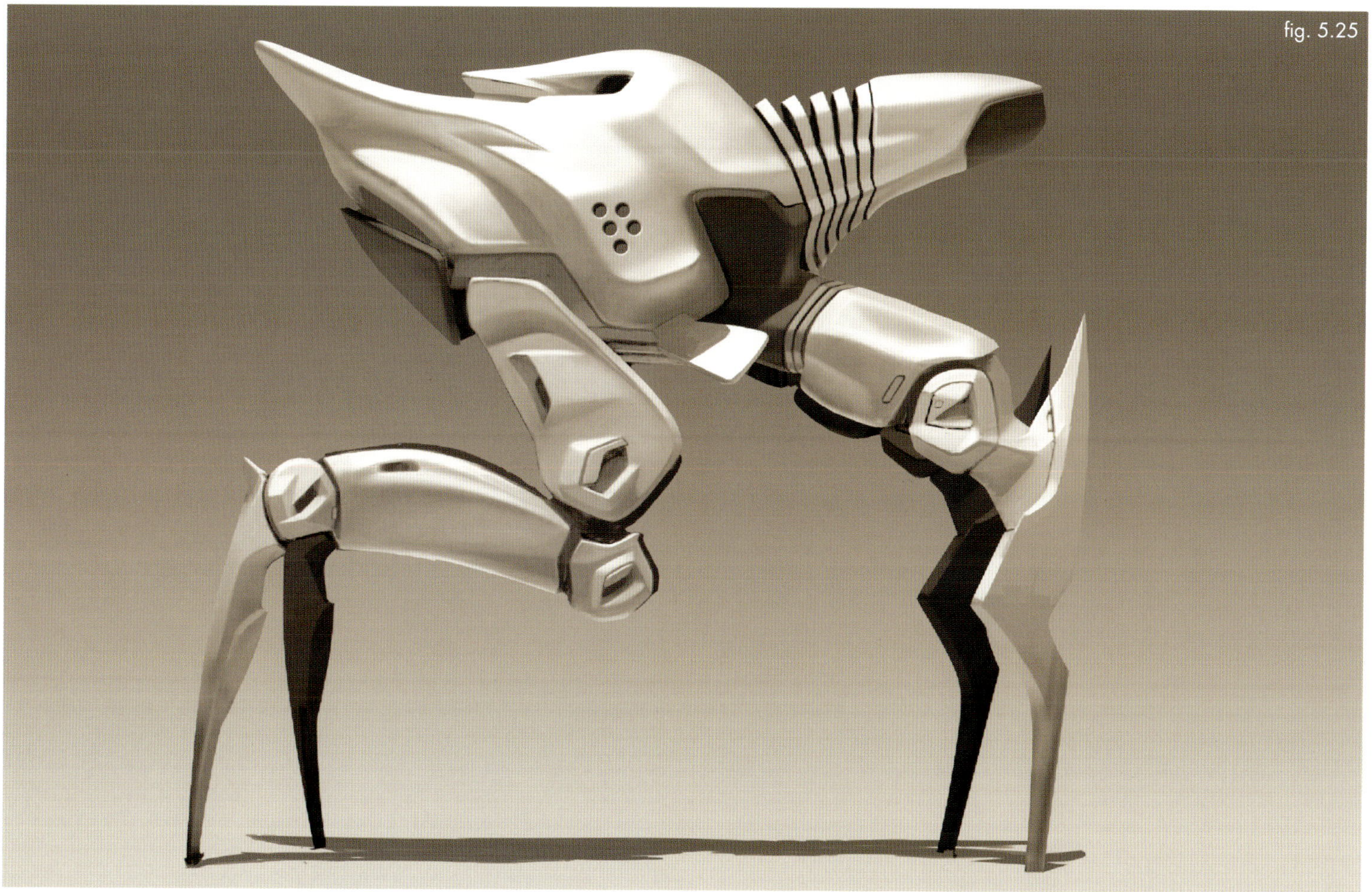

fig. 5.25

No accounting for the local value changes of any of the panels is considered until the end of the rendering process. By doing this, it makes it much easier to lighten or darker individual areas of the form as needed. In fig. 5.25, on the left side of the upper body, the somewhat triangular panel was made darker, as was a zone around the hip and the back, or right side, of the body. This was done simply by selecting these zones and pasting them to a new layer where the value levels could be adjusted. This method was also used to lighten the body overall.

CHARACTER STEP-BY-STEP

Next let's go through the steps to render this little rabbit sketch by Chris Ayers. Like most renderings, this one starts with an idea in sketch form. Finishing the sketch before starting the rendering process makes it a lot easier to focus on doing the rendering properly. It is much more difficult to try to come up with a design while at the same time rendering. Just like the walker on the previous pages, the sketch silhouette is first filled in on a layer with the sketch layer set to "multiply" above it, so the white paper of the scanned sketch disappears.

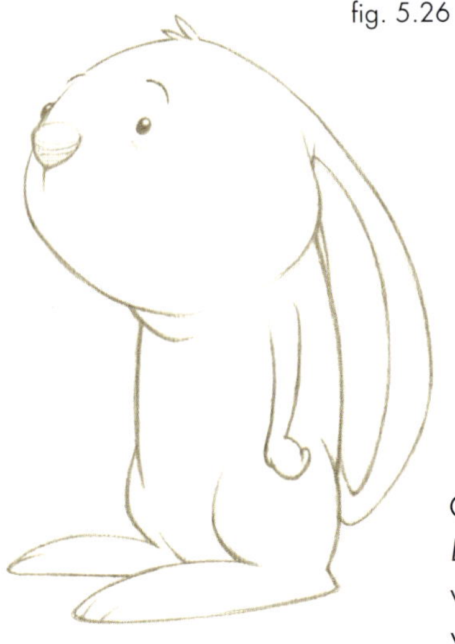

fig. 5.26

Chris Ayers, author of the *Daily Zoo* series.
www.chrisayersdesign.com
www.thedailyzoo.com

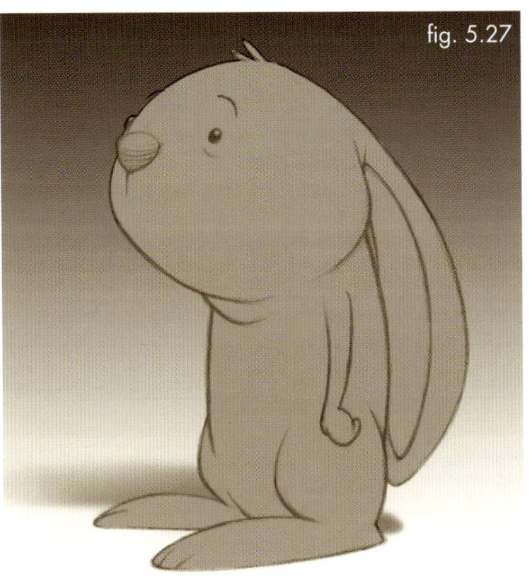

fig. 5.27

Now it is time to design a couple of potential lighting strategies. In fig. 5.28, the primary soft-light source is placed on the rabbit's left side. The shadow placement is an educated guess, based on years of doing the types of constructions at the beginning of the book.

In order to help hold the silhouette of the rabbit's nose and head, a secondary soft rim light is added that is weaker than the primary light source (fig. 5.29). Always remember, you control the light. Design the light in the way that works best to communicate the form you are rendering.

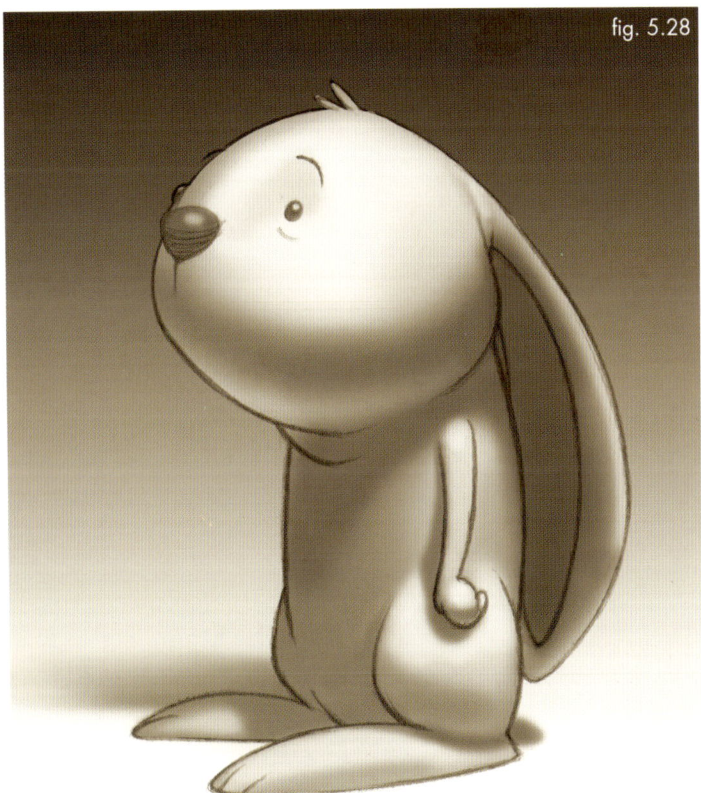

fig. 5.28

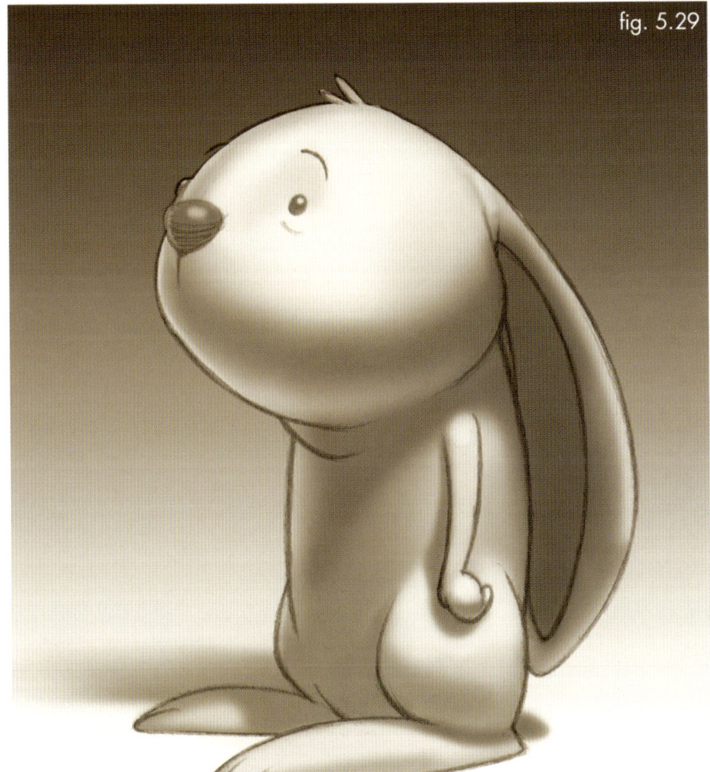

fig. 5.29

Fig. 5.30 illustrates a second option for the lighting. Now the primary light is coming from in front of the rabbit, and a weaker rim light comes from behind the rabbit. Each of these lighting studies was done rather quickly. They are just tight enough to get the idea.

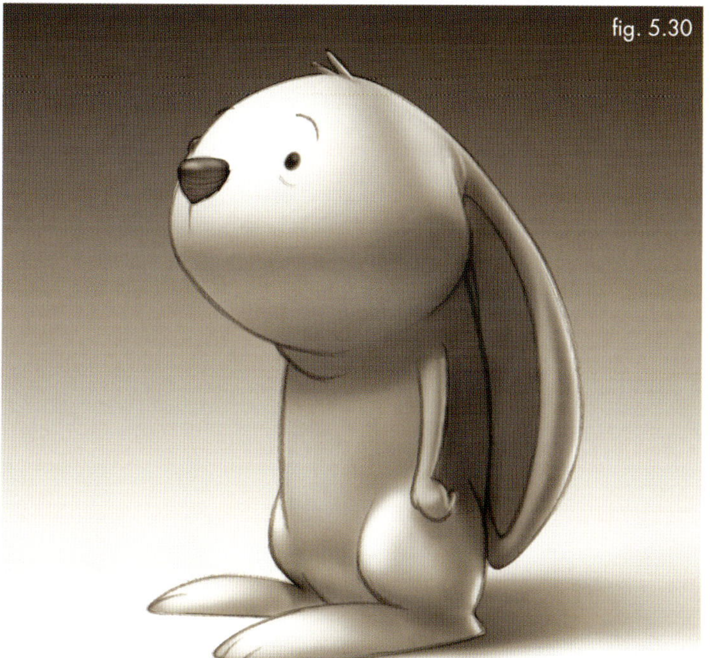

fig. 5.30

For a quick render with the line drawing still showing, this would be finished. But to take the rendering to a more photo-real presentation, paths are a good way to clean up the edges and to get rid of the line drawing. Fig. 5.31 shows the paths all drawn on one path layer.

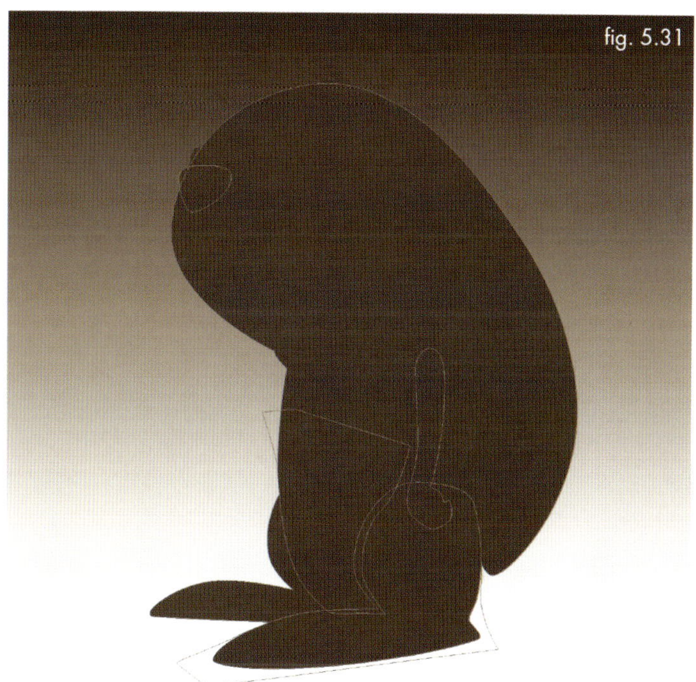

fig. 5.31

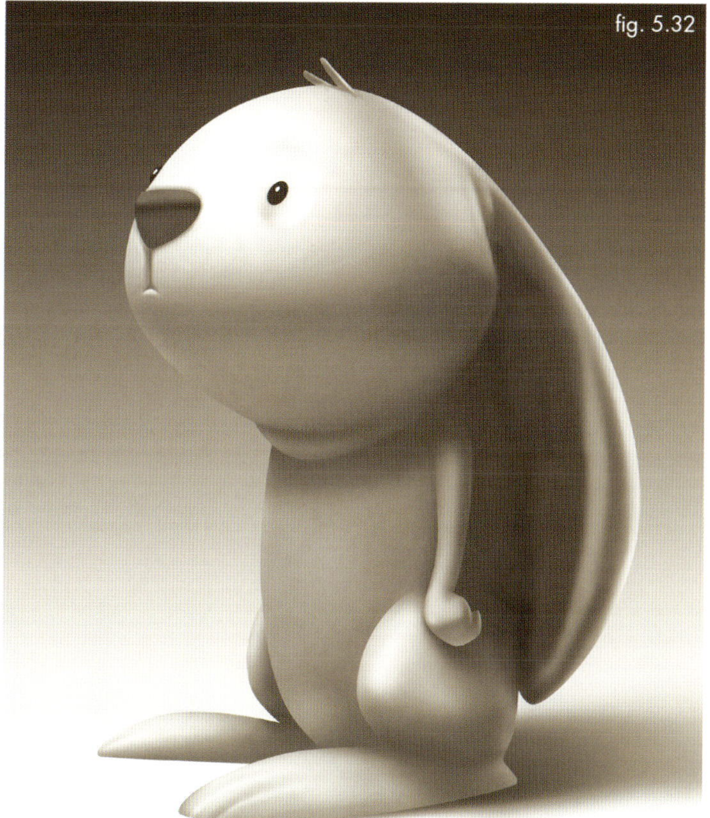

fig. 5.32

The second lighting strategy was chosen. The paths were used to make selections and new layers from the quickly rendered lighting study. Once these new layers were made with the refined silhouettes from the paths, the "preserve transparency" button was again activated so that only the areas already containing pixels would be affected when refining the forms of each layer shape. This terminology is used in a lot of 2D digital rendering programs. If you are working traditionally, you can ignore them. The ideas are the same, just the tools are different.

If this were a graphite or a colored-pencil rendering, to test the two lighting scenarios, start by making two copies of the original sketch. Once the better one is decided upon, there are a couple of options. One option is to keep refining the quick study into a more finished piece. Another is to do a very lightly drawn overlay of the original drawing and render that one as tightly as possible, so that the light line drawing would disappear and the rabbit would start to appear more photo-real. In fig. 5.32, the final rendering has no lines around it and all of the forms and silhouetted shapes are communicated only through changes in value. As renderings like this one get closer to being finished, trade-offs are made to show one edge against another, or let one edge disappear into the other. *Value rendering is a constant balancing act of relationships.* To make a volume or rendering more refined requires a critical eye and lots of very small adjustments to the values.

VALUE AND COLOR RELATIONSHIP 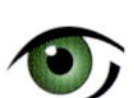

One of the most common challenges students face after mastering the ability to render matte-finish, gray-scale forms is colorizing them. There are a few, simple things to understand about the relationship between value and color. Typically what happens is a student tries to colorize the gray rabbit by using a 2D digital-rendering program like Photoshop, and fig. 5.34 is the result. It is just not possible to colorize the rabbit to a nice, saturated blue in the lightest areas when using this value structure. Many students will lament that colorizing a gray-scale image just does not work! All they ever seem to get are pastel colors lacking the saturation they are after. The problem is that the values are wrong for a saturated blue in the lightest areas of the surfaces. The value is too light.

As has been stated many times, value changes equal form changes. When working on a gray-scale rendering, the tendency is to extend the value range to help make the form changes more dramatic and obvious. This is done without any consideration of what color those values might eventually be.

The solution is easy. Choose the color that will appear on the Number 1 side. The value of that color needs to be known. Look at the photo of the blue car (fig. 5.41) and its gray-scale image (fig. 5.42). Note the value of the gray on the Number 1 side. It is not white like the Number 1 side of the rabbit (fig. 5.33). The lightest value of the rendering, also known as the "white point," needs to be darkened down to something closer

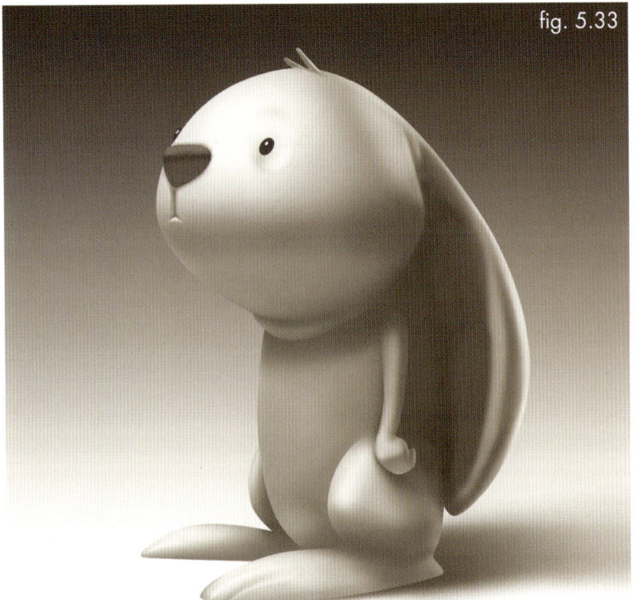

fig. 5.33

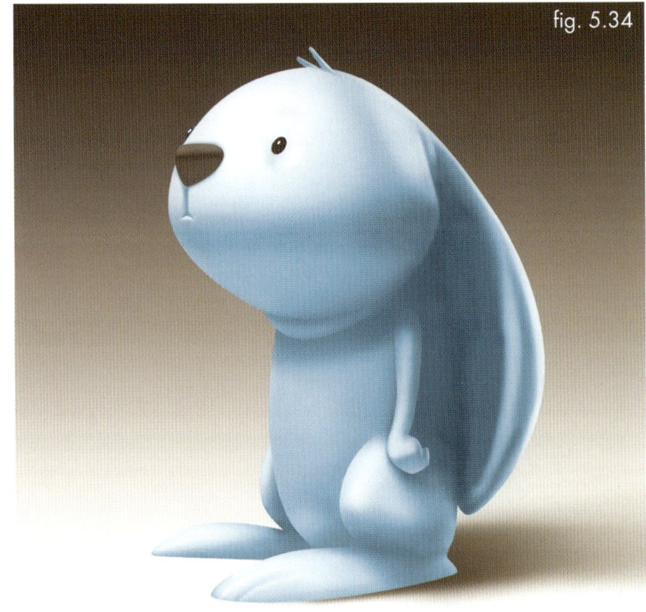

fig. 5.34

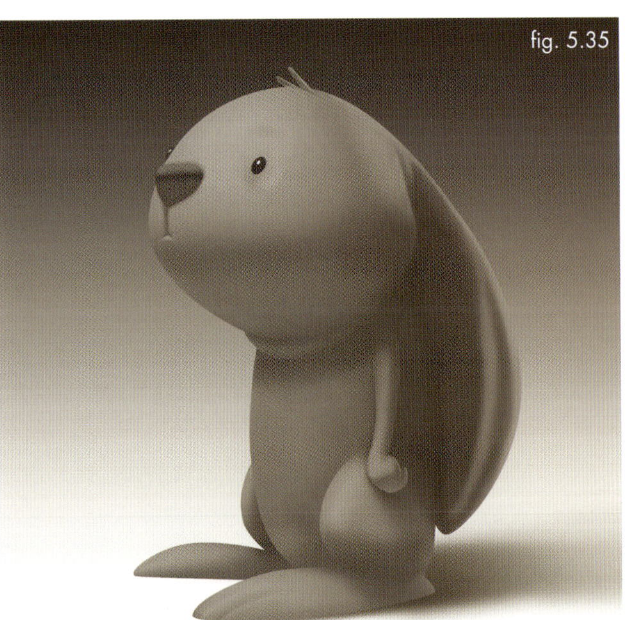

fig. 5.35

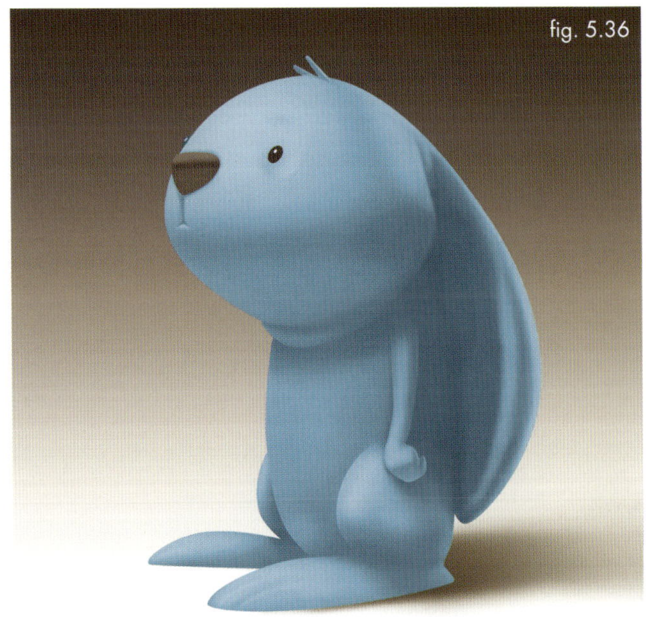

fig. 5.36

to what is observed in fig. 5.42. Once the values have been adjusted for the desired color, success will be achieved. The Number 1 side of the yellow ball is equivalent to a 20% gray (fig. 5.38), ignoring the reflection of the sun. The value of the red ball is even more shocking. Most people would not guess that the Number 1 side of a red object is equivalent to a 50% gray. In order to make the rabbit red, for instance, the brightest value on the Number 1 side can be no brighter than 50% gray. Without making this adjustment to darken the value, all that can ever be achieved is pink.

A good way to learn what the values of colors are is to scan an image of a favorite color, convert it to a gray-scale image, observe what the value range is and make note of it for future reference. This does not mean that the color of something has to be decided before rendering it in gray-scale, it just means that the values need to be adjusted before adding color. If a scanner is not available, make a black-and-white copy of a color image, or even take a digital picture and then make the image black and white. This is a very important lesson to learn before attempting full-color rendering in the upcoming chapters.

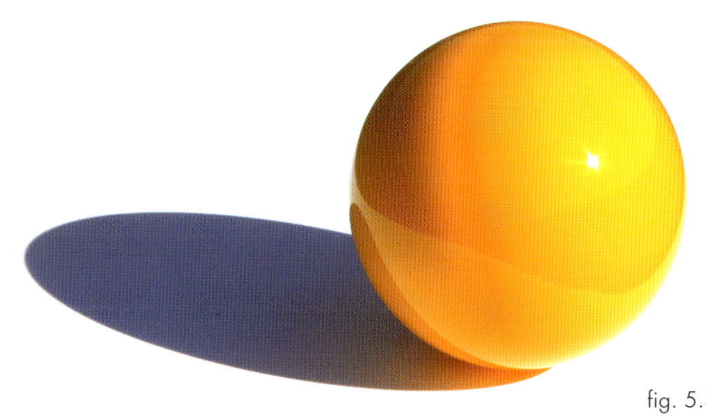

fig. 5.37

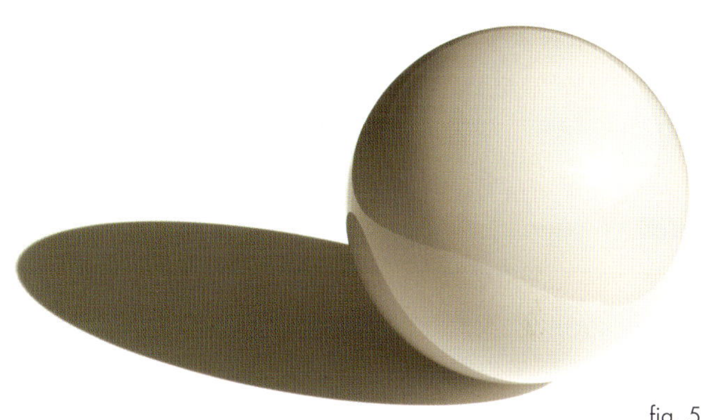

fig. 5.38

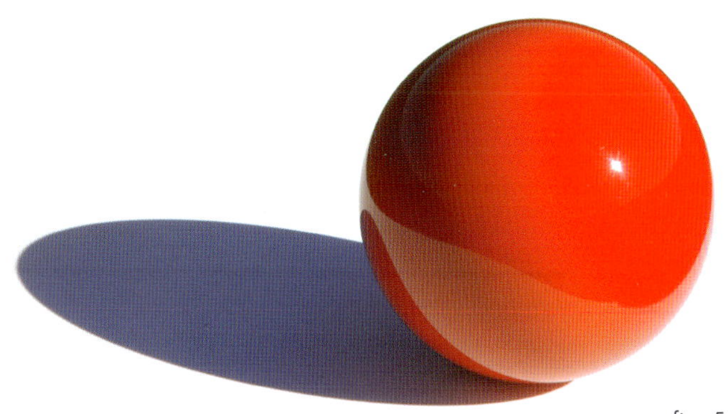

fig. 5.39

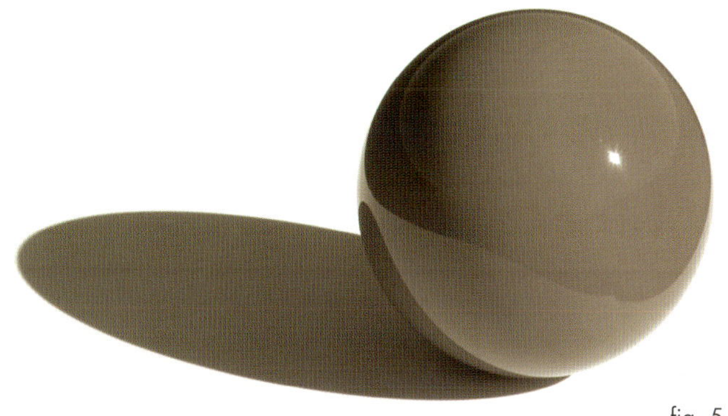

fig. 5.40

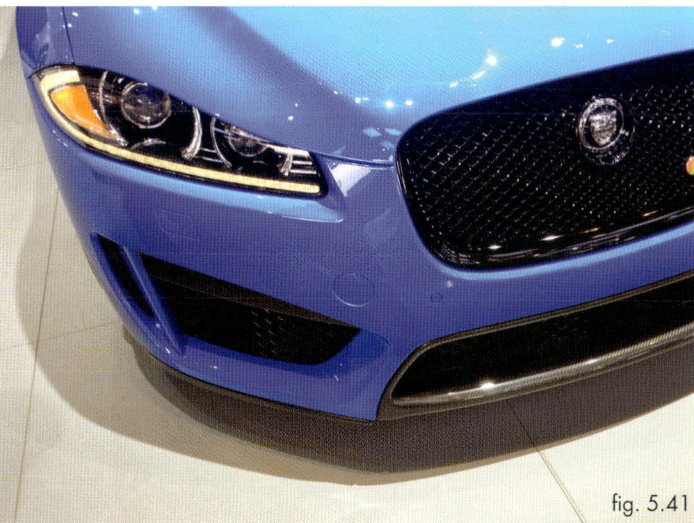

fig. 5.41

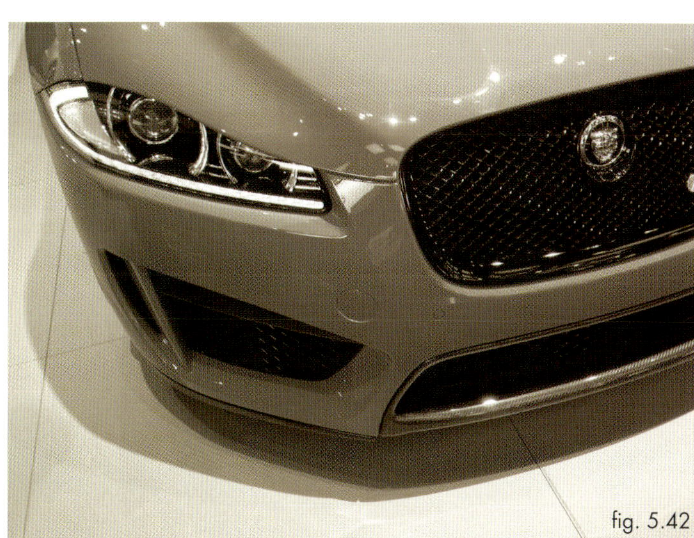

fig. 5.42

Creature and concept designer Neville Page shares his tips and tricks. See Neville's latest work at www.nevillepage.com.

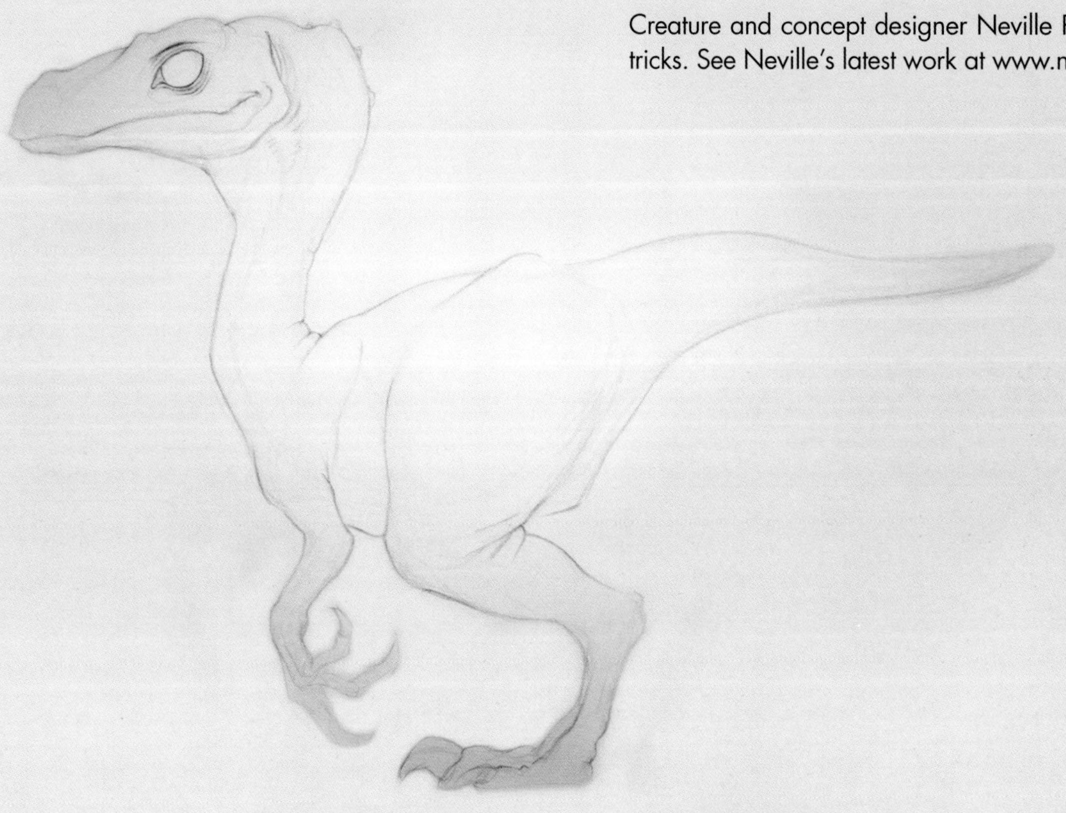

fig. 5.43

The point of this exercise is to understand how to render an organic character using the principles addressed so far in this book. I start by doing a simple drawing on Beinfang 360 Marker Layout Paper with black Faber Castell wax pencils.

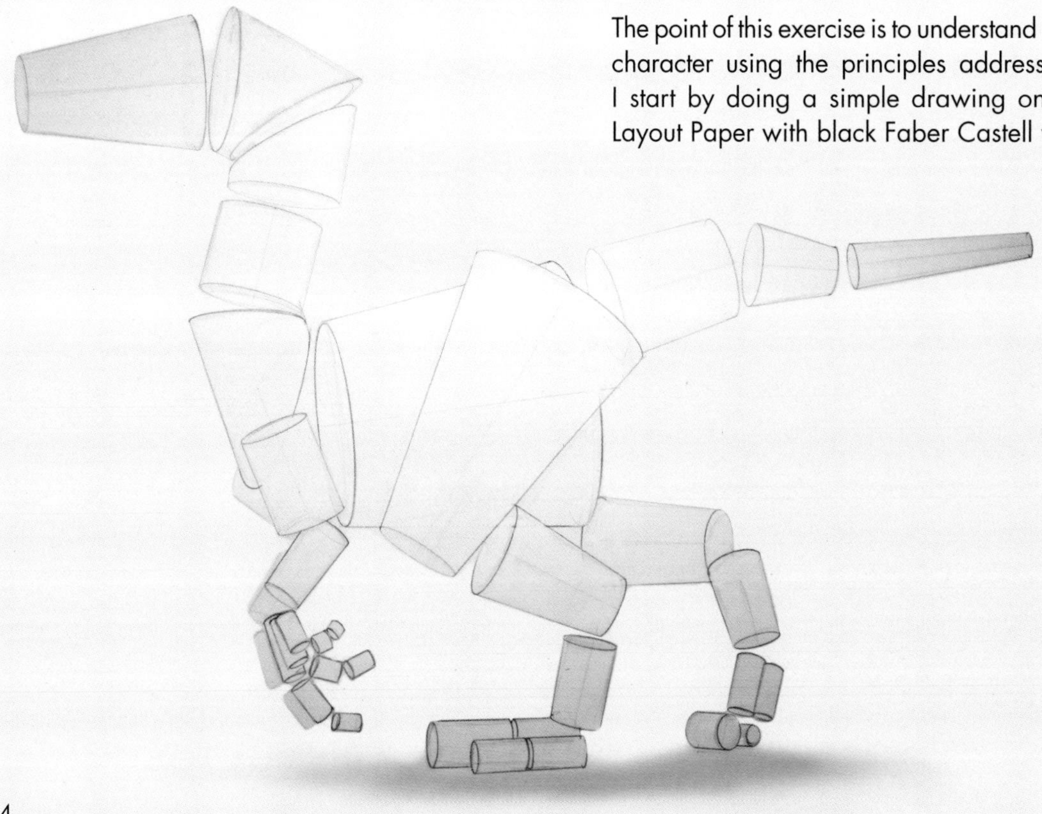

fig. 5.44

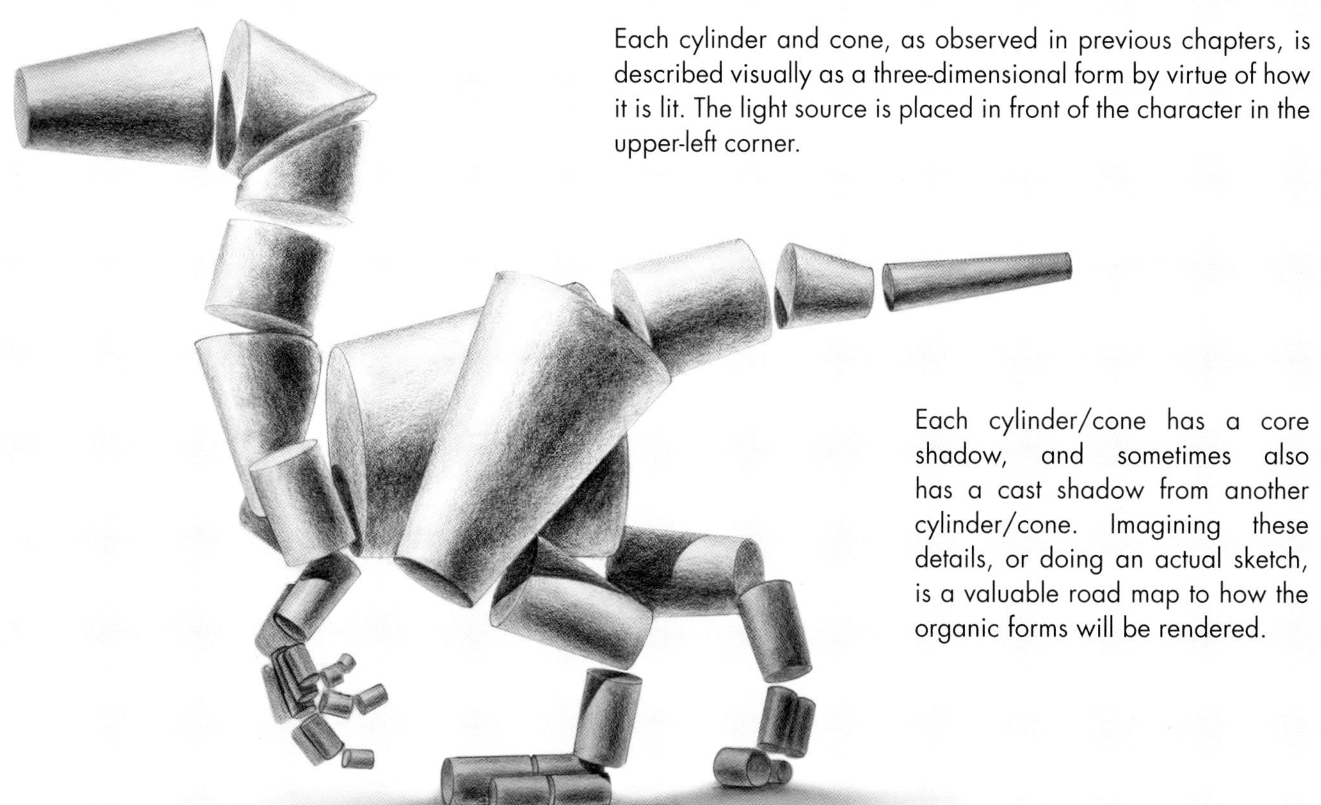

Each cylinder and cone, as observed in previous chapters, is described visually as a three-dimensional form by virtue of how it is lit. The light source is placed in front of the character in the upper-left corner.

Each cylinder/cone has a core shadow, and sometimes also has a cast shadow from another cylinder/cone. Imagining these details, or doing an actual sketch, is a valuable road map to how the organic forms will be rendered.

fig. 5.45

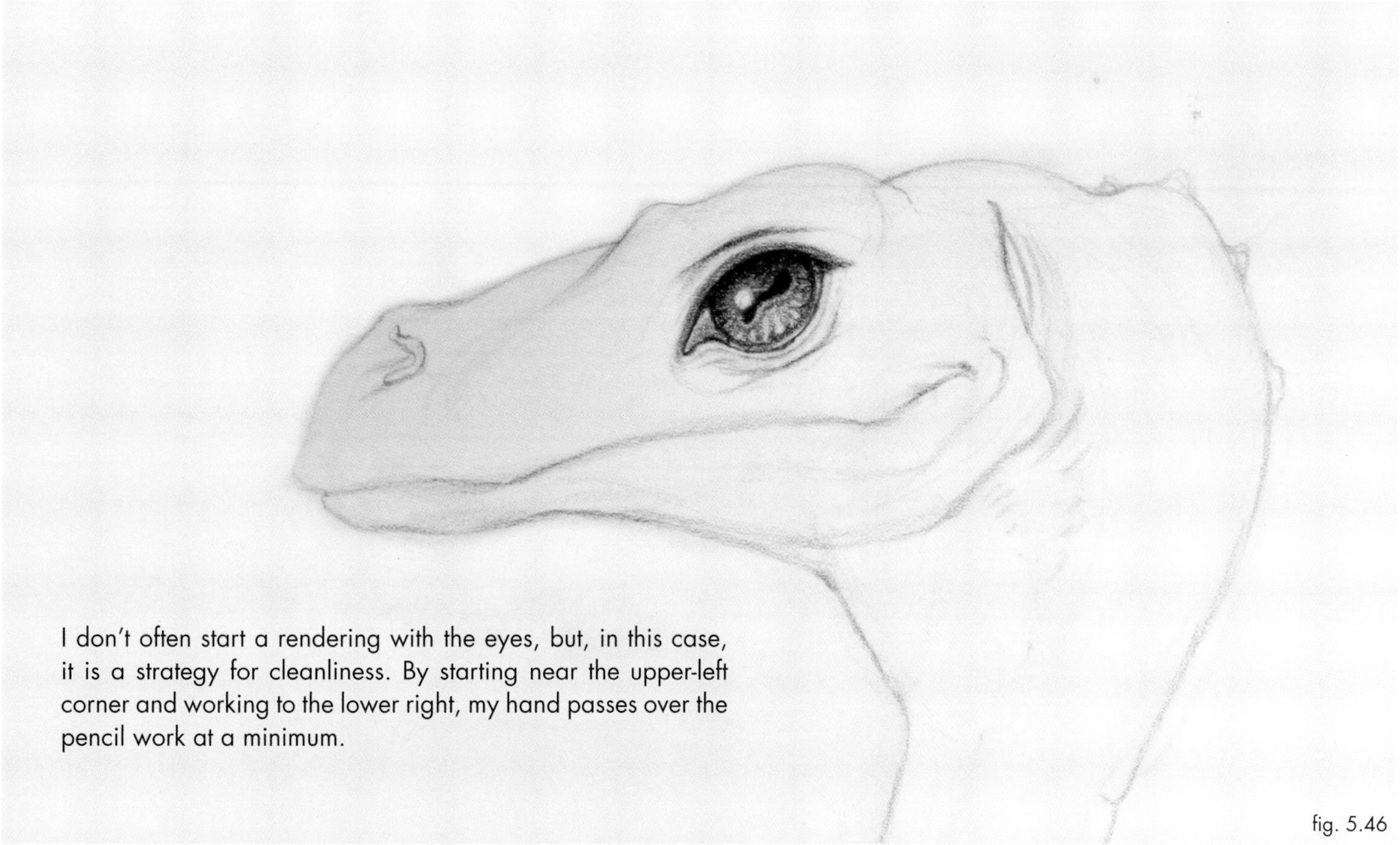

I don't often start a rendering with the eyes, but, in this case, it is a strategy for cleanliness. By starting near the upper-left corner and working to the lower right, my hand passes over the pencil work at a minimum.

fig. 5.46

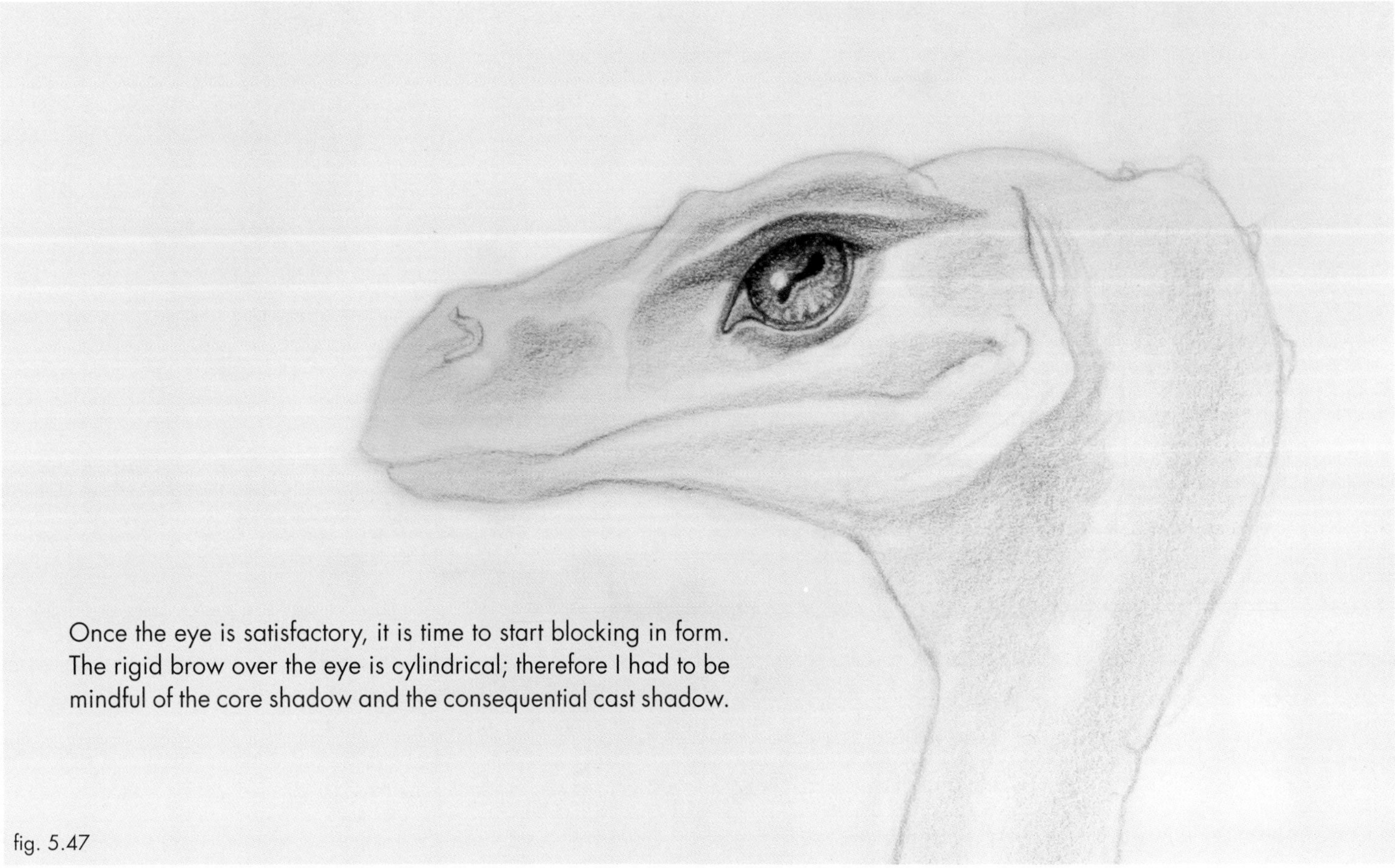

Once the eye is satisfactory, it is time to start blocking in form. The rigid brow over the eye is cylindrical; therefore I had to be mindful of the core shadow and the consequential cast shadow.

fig. 5.47

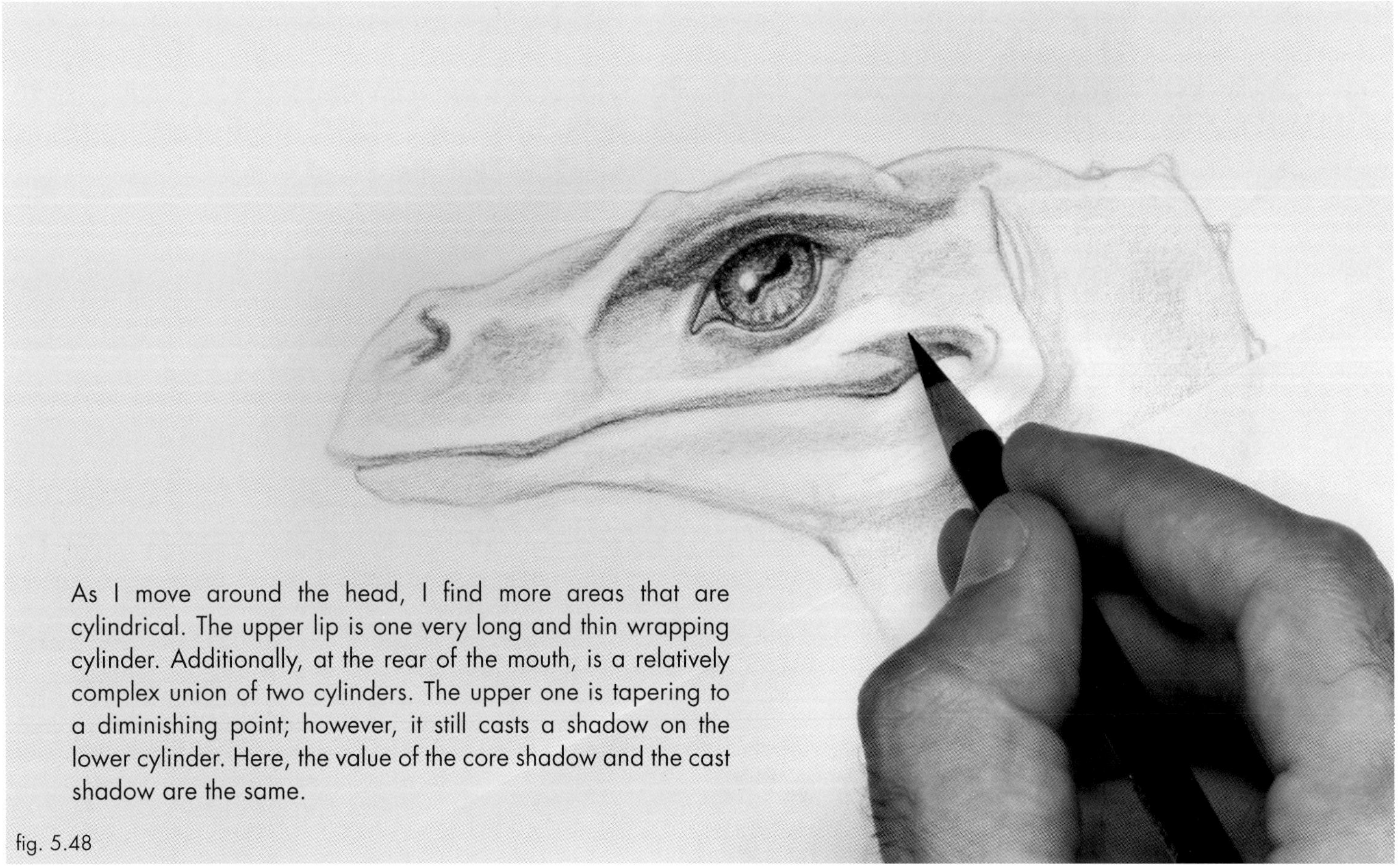

As I move around the head, I find more areas that are cylindrical. The upper lip is one very long and thin wrapping cylinder. Additionally, at the rear of the mouth, is a relatively complex union of two cylinders. The upper one is tapering to a diminishing point; however, it still casts a shadow on the lower cylinder. Here, the value of the core shadow and the cast shadow are the same.

fig. 5.48

Working down the body, the neck is mostly a vertical cylinder with a transition at the base of the head. This is one of the areas where the cylinder exercise comes in handy.

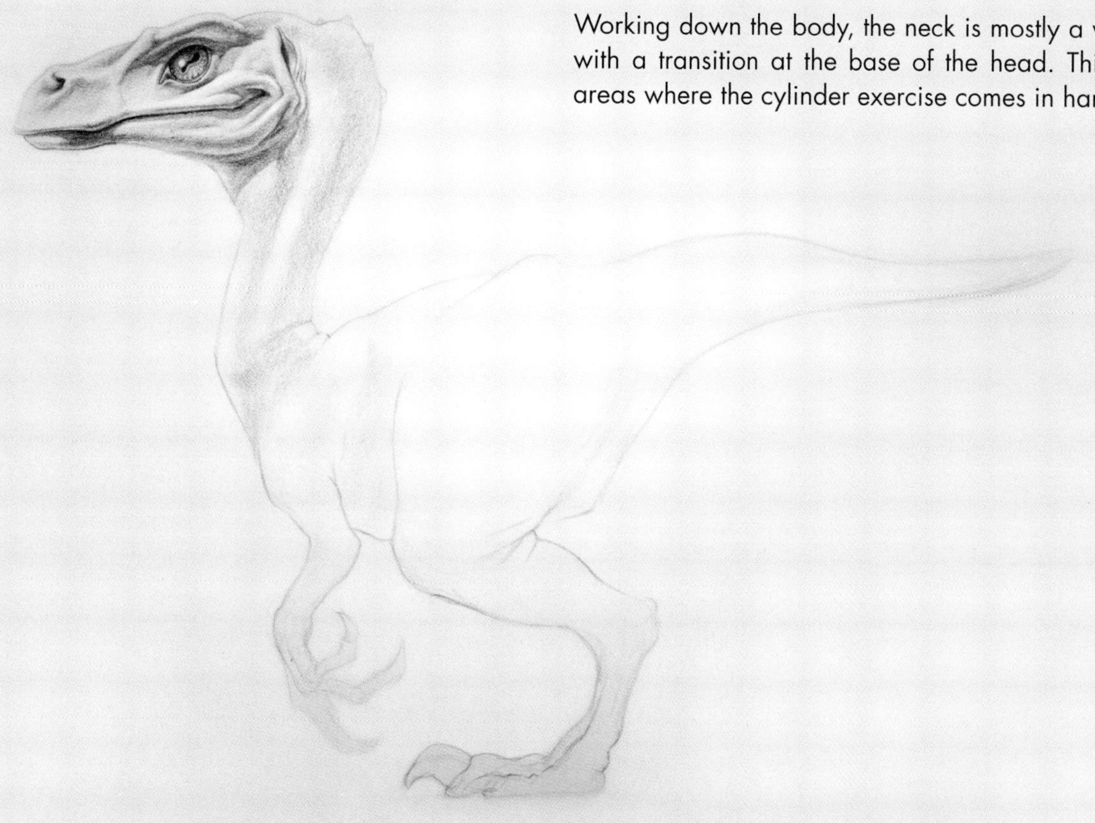

fig. 5.49

I continue to block in a bit of value on the entire body. It is just enough to begin to see where the blended core shadows will reside.

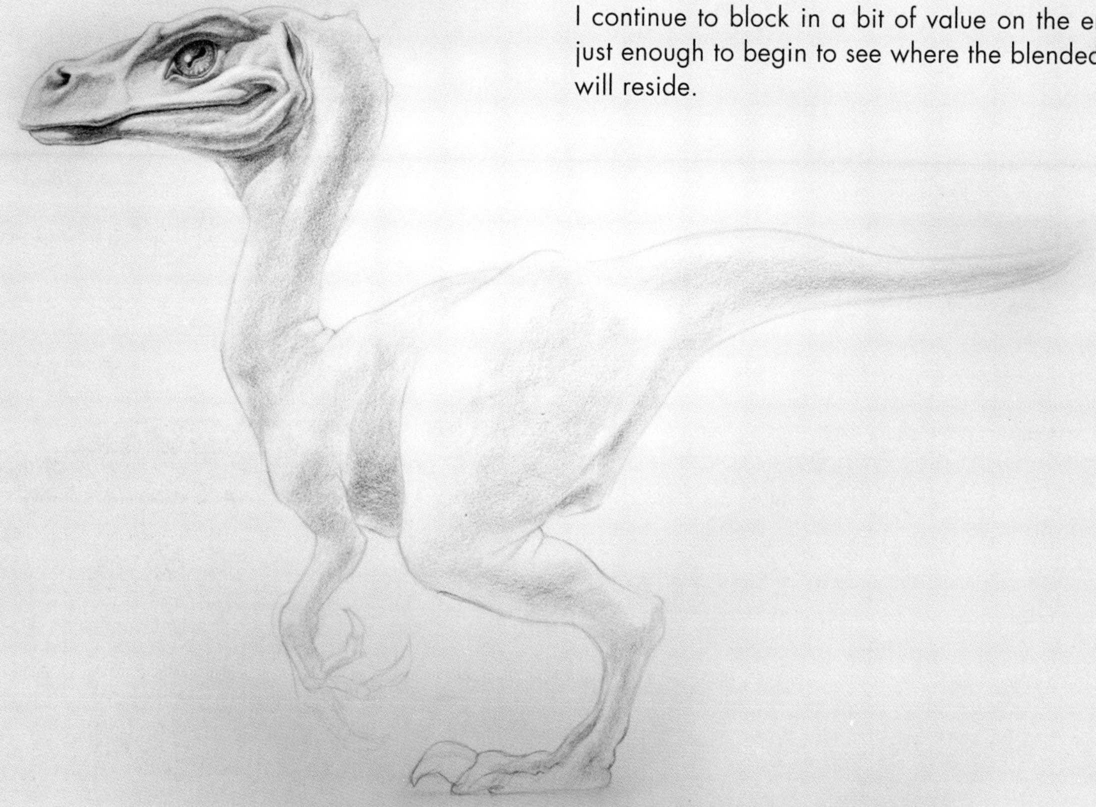

fig. 5.50

Again, the arms and hands are just a bunch of tapering cylinders at various angles. I am starting to include some blends and folds to imply a sense of skin. Interestingly, each new skin fold is an additional cylinder.

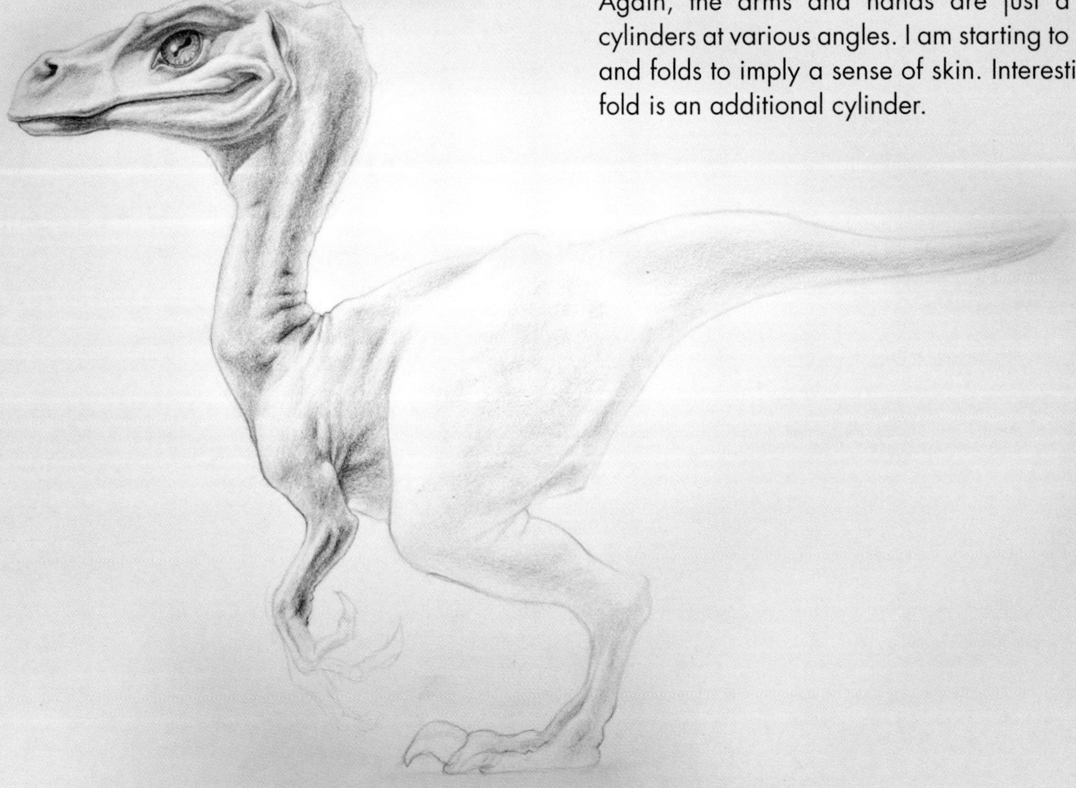

fig. 5.51

With a little more care I work my way toward the claws. I have to pay particular attention to how the shadows are cast. Now is a good time to refer back to the cylinder rendering. It is also helpful to grab a couple pencils, hold them in the desired orientation and shine a light on them. Nothing beats the real thing.

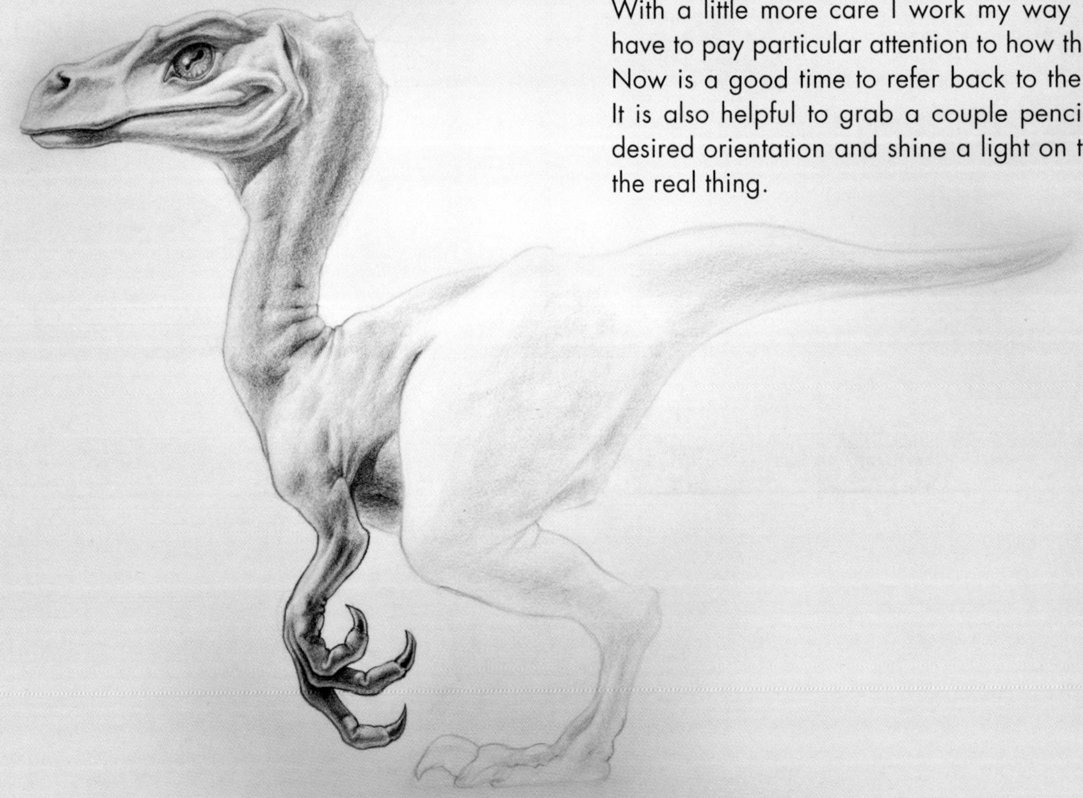

fig. 5.52

An outline has been added to the arm and hands, which is a fast and easy way to keep the surfaces separate from one another. It can be done with just value, but for a quick pencil rendering, this is an economical technique.

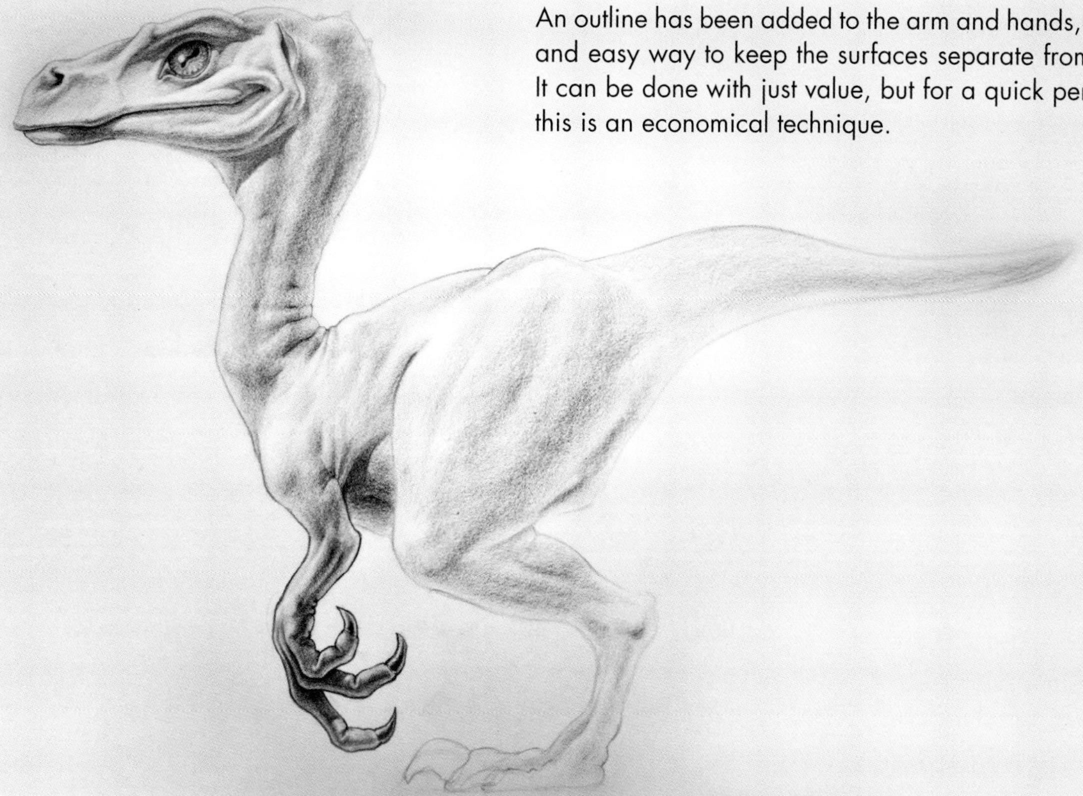

fig. 5.53

Time to work out the bigger forms. The thigh is a bit tricky. Although it is made up of a flattened cylinder, there are multiple "shallow" cylinders on the surface that describe muscles. Having a good sense of anatomy is key here.

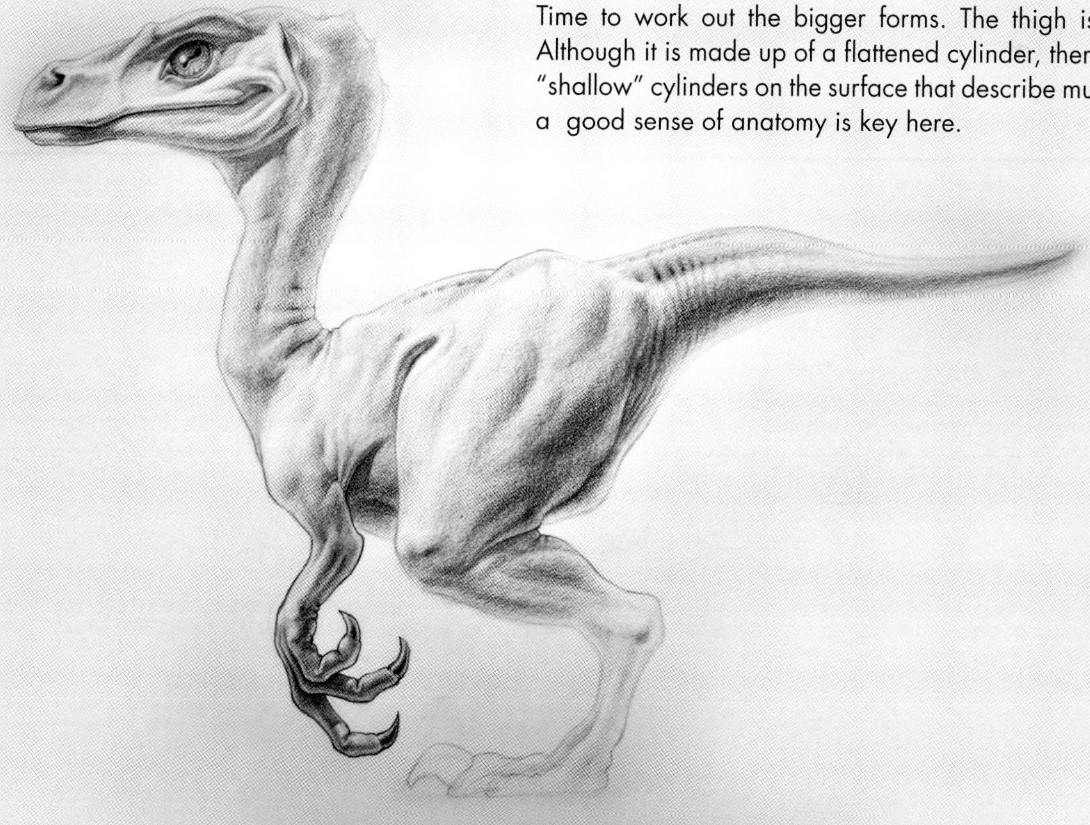

fig. 5.54

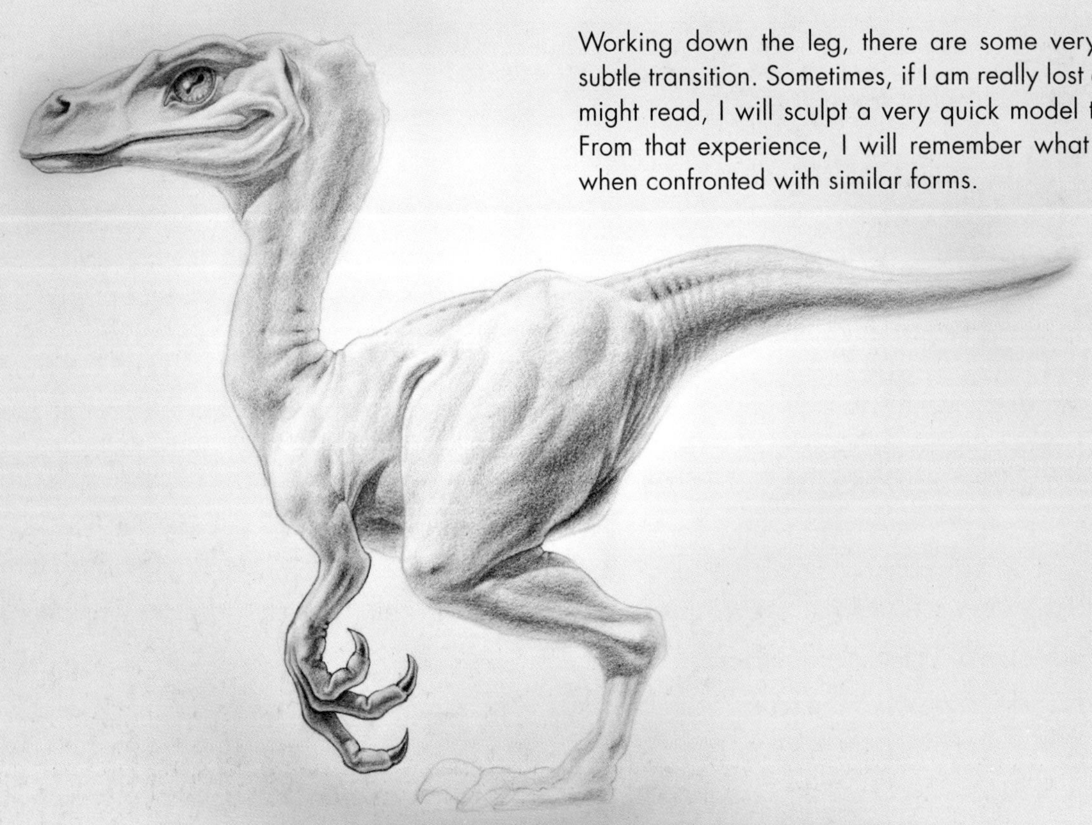

Working down the leg, there are some very tricky areas of subtle transition. Sometimes, if I am really lost as to how a form might read, I will sculpt a very quick model to see the forms. From that experience, I will remember what to do next time when confronted with similar forms.

fig. 5.55

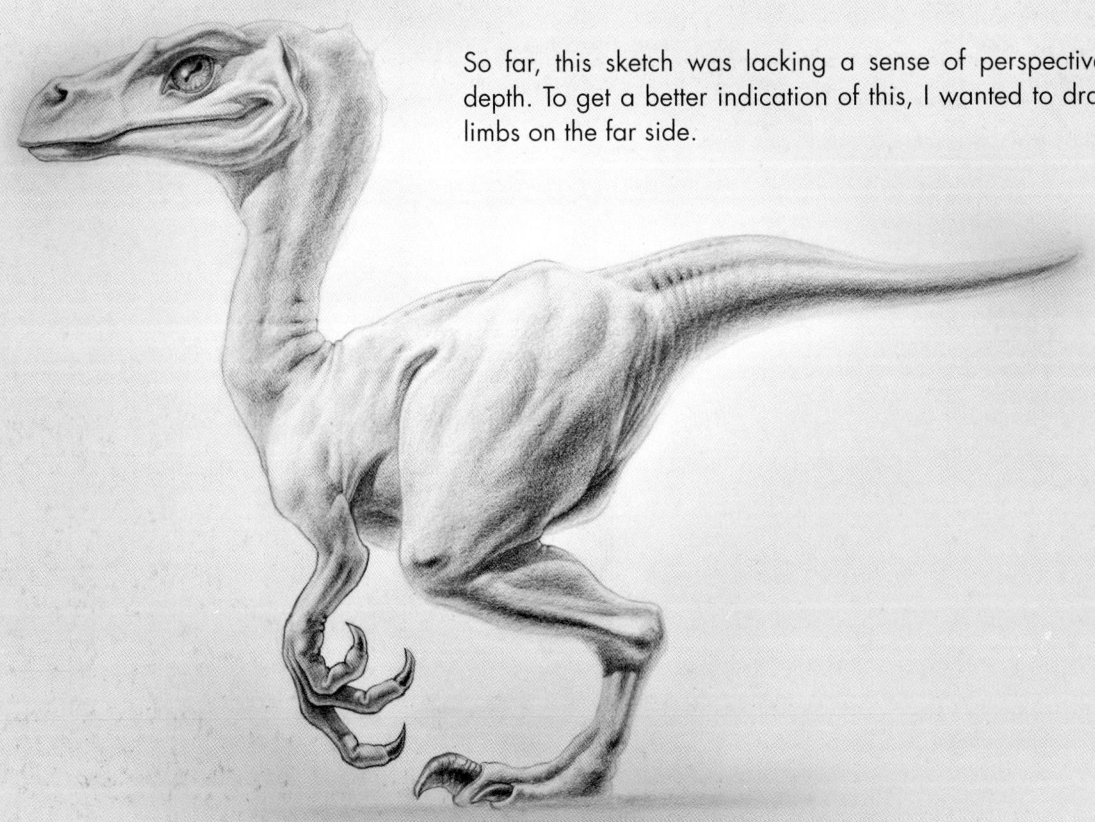

So far, this sketch was lacking a sense of perspective and depth. To get a better indication of this, I wanted to draw the limbs on the far side.

fig. 5.56

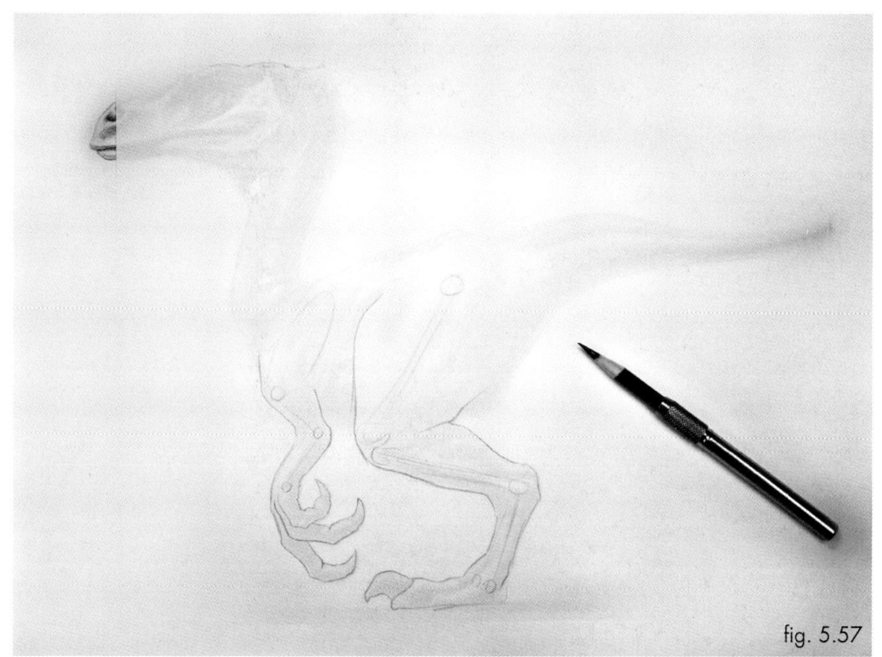
fig. 5.57

To figure out a relatively accurate placement, I layout the skeleton on a separate sheet of paper and then "articulated" the leg and arm to make them visible in a low eye-level perspective. It is all cheated but gives the character a better sense of foundation.

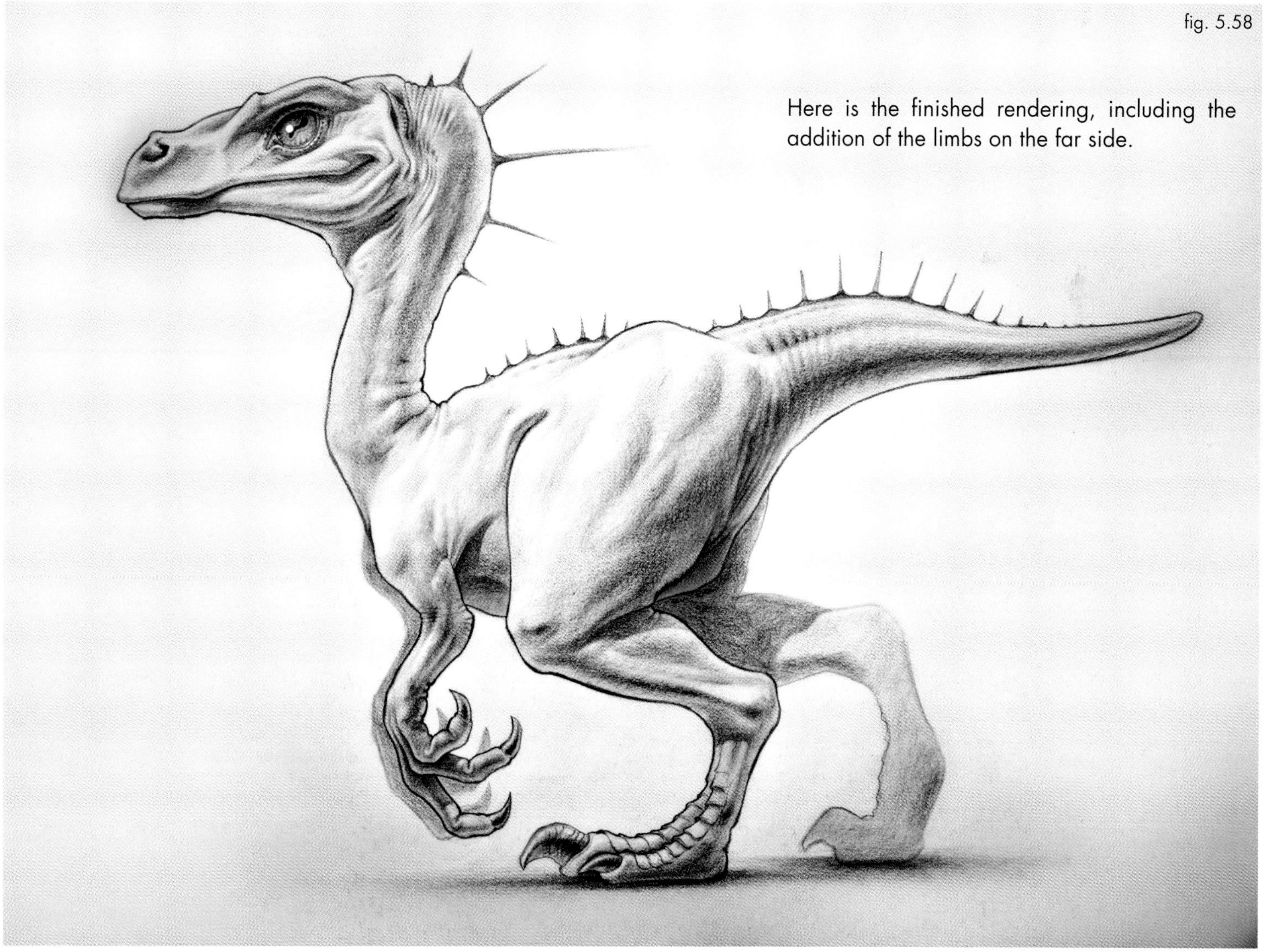

fig. 5.58

Here is the finished rendering, including the addition of the limbs on the far side.

ENVIRONMENT SKETCHING IN GRAPHITE WITH JOHN PARK

When John Park was working for me developing pitch art for a few of my own entertainment properties, we often worked up the concepts in graphite pencil before scanning the drawings and rendering them digitally in 2D. The following three pages were all drawn by John and nicely document the concept development process for an island, harbor and coliseum. See John's latest work at www.jparked.blogspot.com.

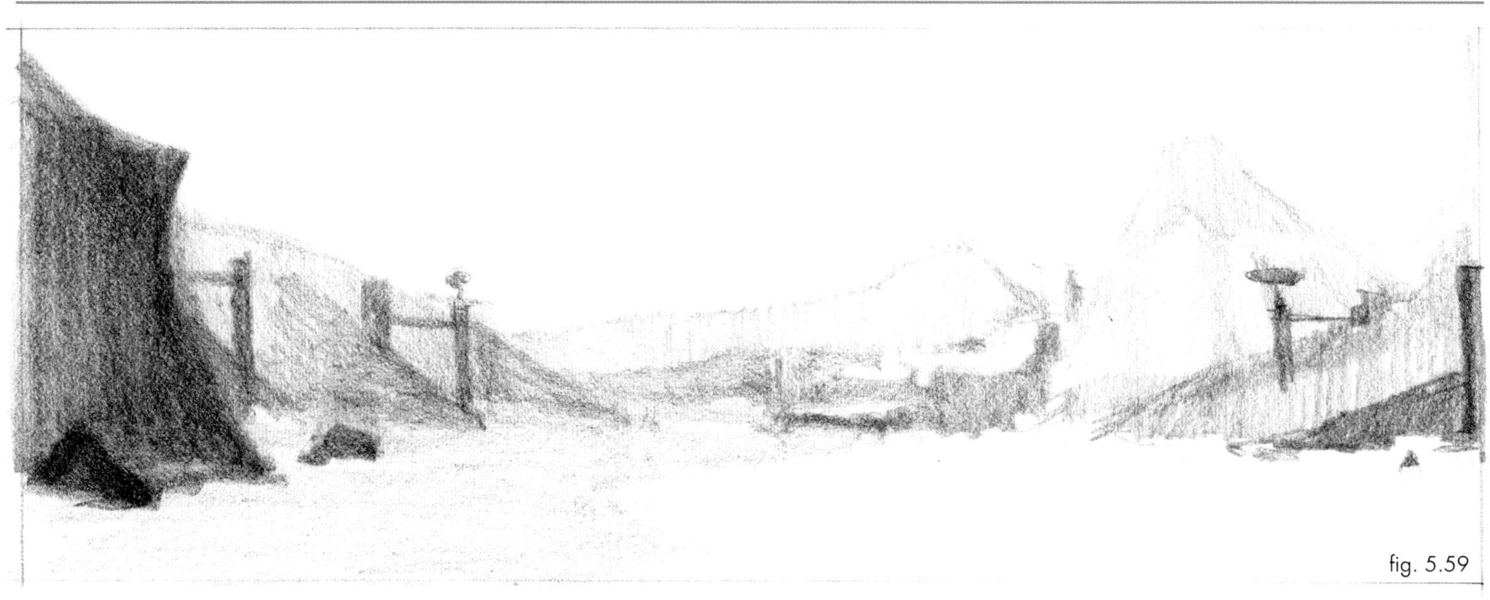

fig. 5.59

The basic brief for this environment was that it was an extinct volcano that became an island, so in top view it has a nice, circular form with an inner harbor. The harbor is reached by sailing under a large wall, and at the far end of the harbor, at the opposite side of the wall, is a coliseum where sporting events take place. In addition, there are several royal clans that all have elevated landing pads for their airships. The buildings are aesthetically inspired by ancient Greek architecture, with a bit of a fantasy spin. The top-view drawing can be seen on page 148, fig. 5.72.

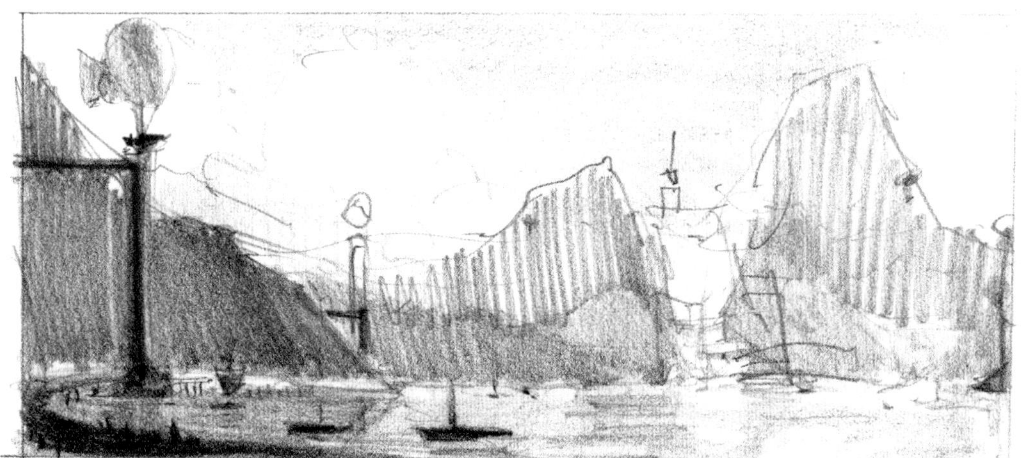

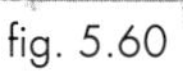

fig. 5.60

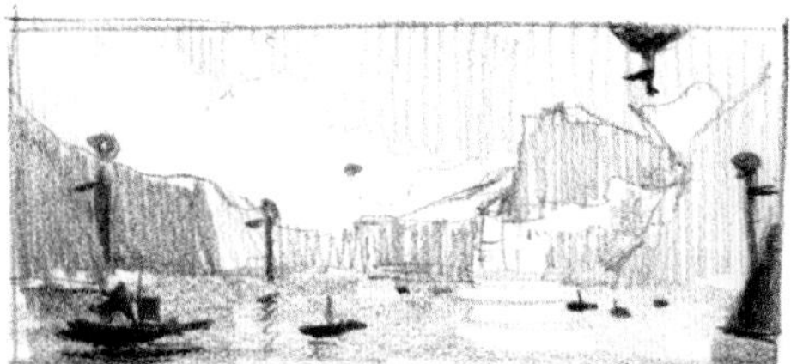

fig. 5.61

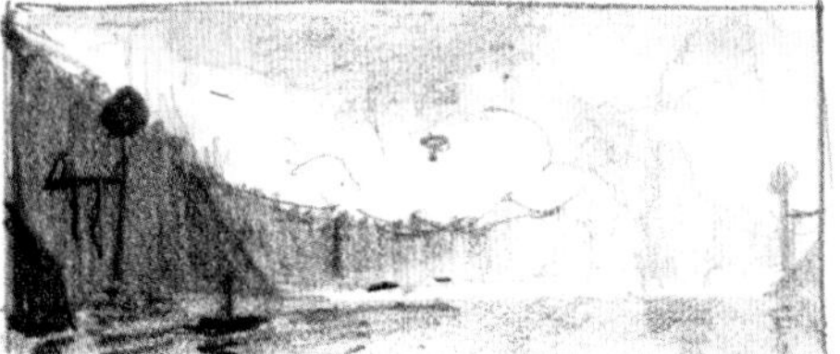

fig. 5.62

In the early stages, most of the effort goes into establishing the camera view and getting a feel for what view will work best to communicate the design of the environment. John is basically working with linear and atmospheric perspectives, and a very limited range of values. In each sketch he moves around the required elements and the camera position. Not much effort is put into modeling surfaces at this point.

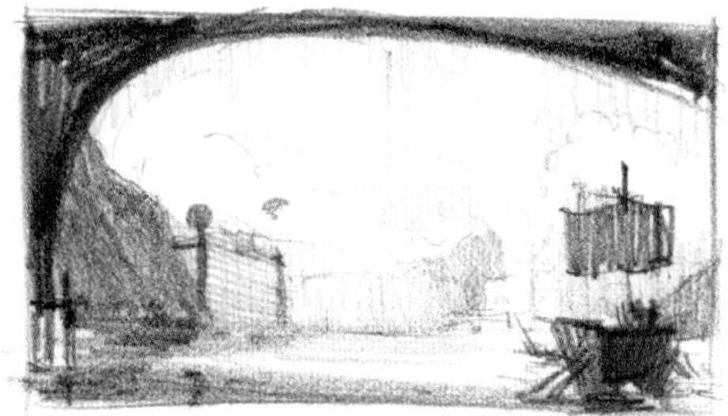

fig. 5.63

fig. 5.64

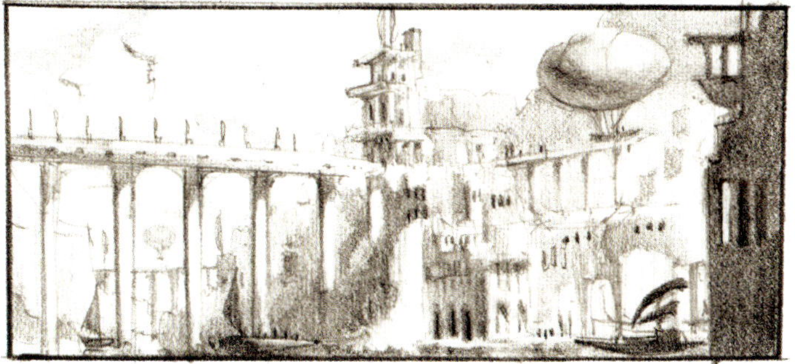

fig. 5.65

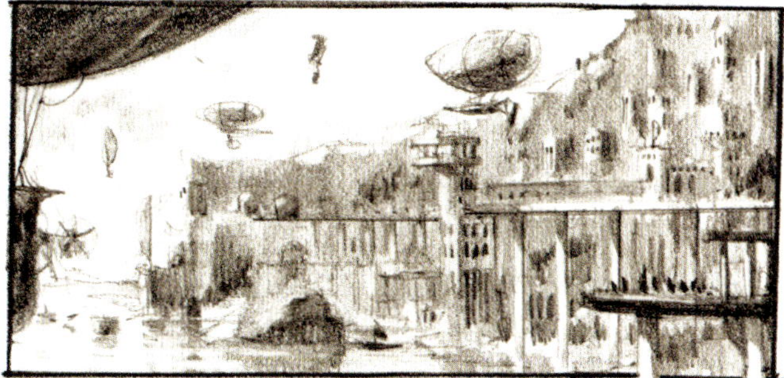

fig. 5.66

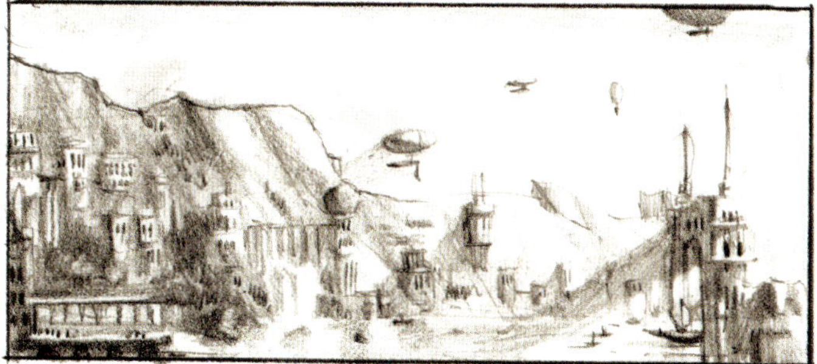

fig. 5.67

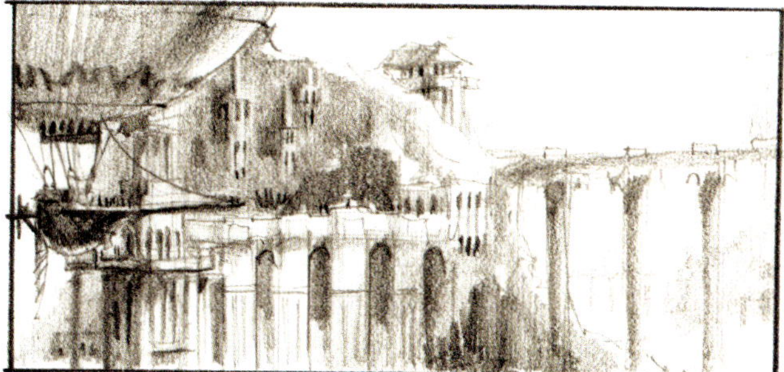

fig. 5.68

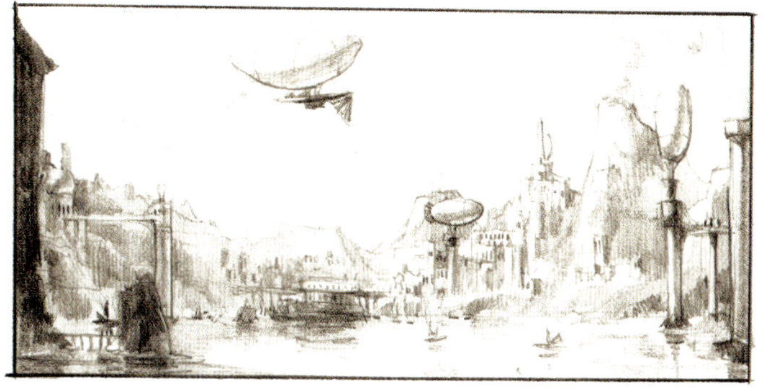

fig. 5.69

The components that make up the scene are treated, for the most part, as simple silhouetted shapes that overlap one another. By using atmospheric perspective as a tool to assign value, he can work quickly and still achieve a nice sense of depth. Above, the sketches start out looking like fig. 5.69, and then become more realistic by adding more values based on the presence and direction of the sun's rays, while keeping the overall atmospheric perspective working. Fig. 5.70 starts

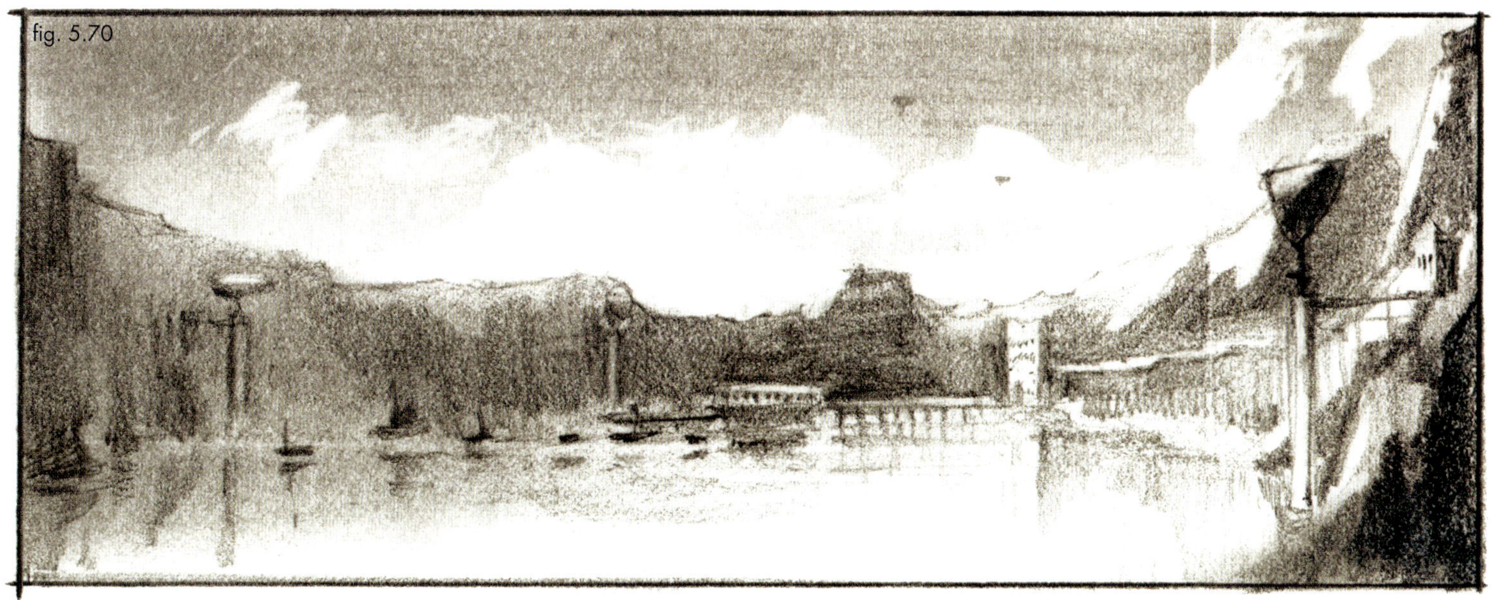
fig. 5.70

fig. 5.71

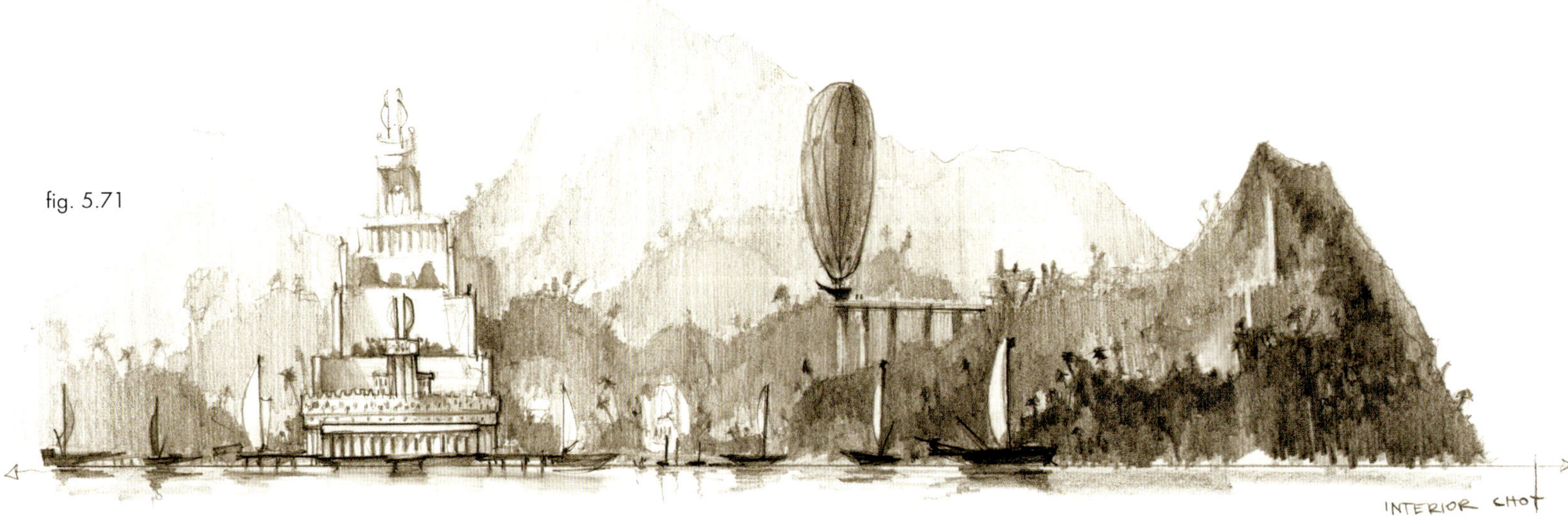

to incorporate the water's reflections and plays with light, illuminating only the right side of the image as the left side drops into shadow. This is a nice bit of lighting design to create a focal point on the airship and landing pad/bridge at the right side of the image.

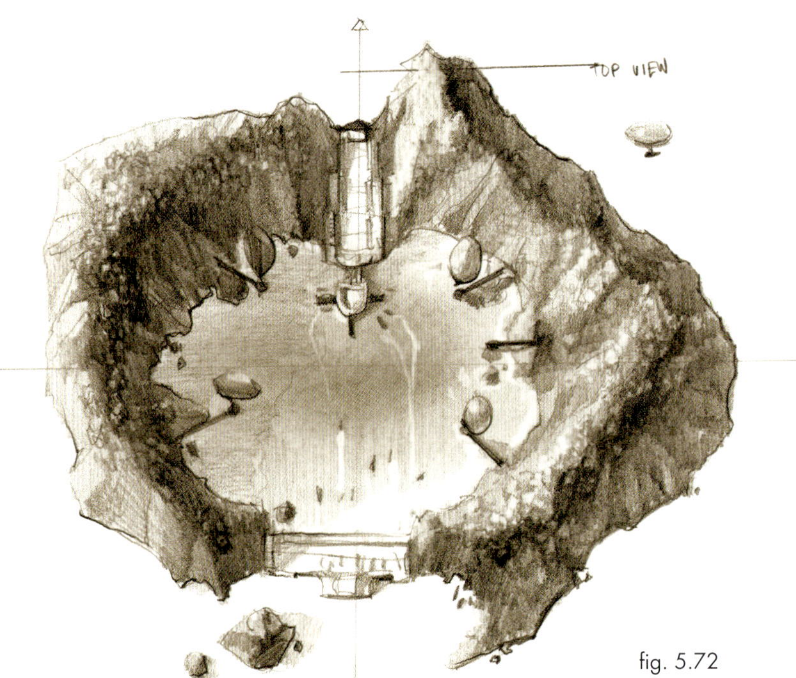

fig. 5.72

Fig. 5.71 is an even more refined sketch of the inside of the island and the first tight drawing of the coliseum set against the mountains. Remember, when trying to communicate shapes like mountain ridges, or any silhouette set against another, an overall value change is needed to define those edges. Take, for instance, the airship against the mountains: It is dark against light. The sails of the ships below it are light against dark and some rising mist is used on the right side behind the nearest mountain ridge. This sets up a value change to hold the silhouette of the dark trees. When applying values to mountain ranges receding into the distance, it is very common to add things like fog, or rising mist from crashing waves, that create a gradation from darker at the top of the mountain, to lighter down the hill. This value gradation makes renderings feel more natural because it is a common phenomenon observed in nature. Getting those value relationships correct, more than anything else, is what makes renderings feel realistic.

Even a draft view of the environment (fig. 5.72) can appear to have form. When the values are applied in a clever way, they establish a light side and a shadow side. This is a good example of applying the fundamentals explained earlier in the book.

With no linear or atmospheric perspective to rely on, using value in the form of cast shadows creates a strong 1-2-3 read, making forms appear to rise out of the flat piece of paper on which they are illustrated. Adding value to draft-view drawings is a lot of fun and it is amazing how well these values provide visual cues to the brain.

fig. 5.73

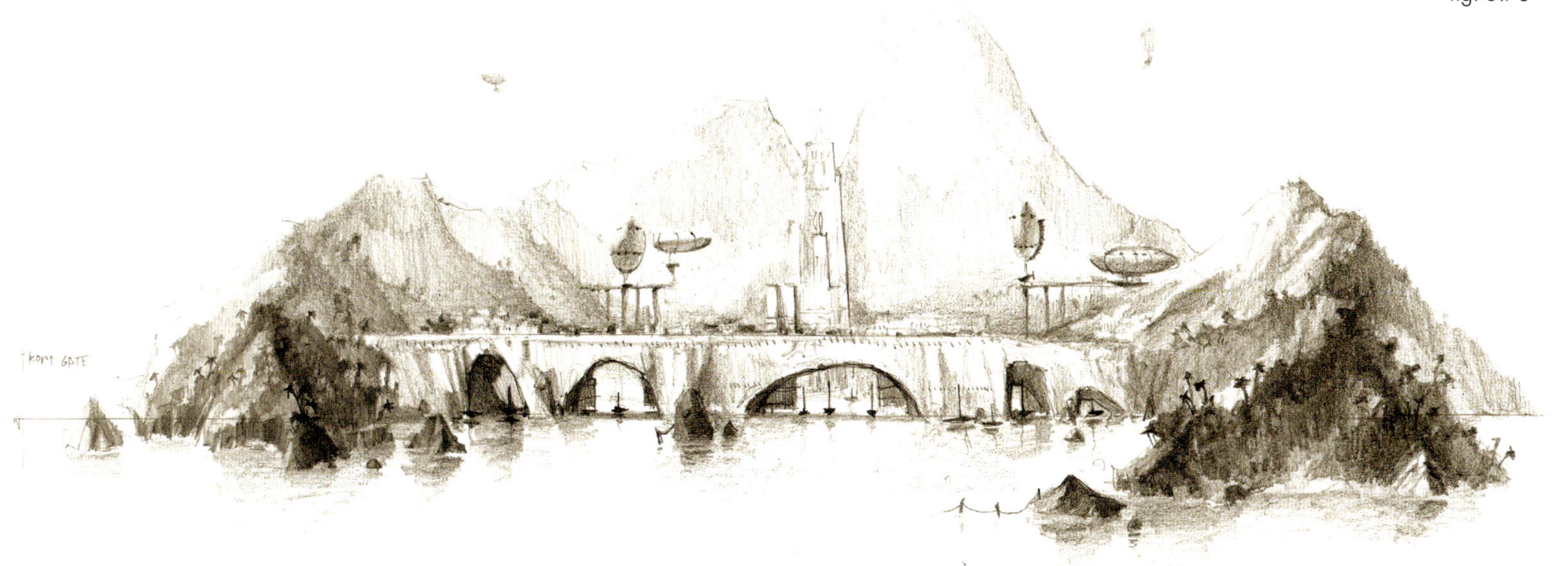

John's island sketches wrap up with the two images on this page. Fig. 5.73 is a nice overall approach shot of the island where the scope of the outer wall and glimpses of the interior harbor can be seen. The dark cast shadows in the arches of the wall and the value progression from dark to light within the mountains of the island itself, based on the distance from our viewing position, create nice depth. Notice how well the quick sketching of the 1-2-3 read of the airship volumes works and the water also starts to come to life by indicating a few simple reflections.

Here is the final presentation graphite drawing. It measures 10.5 inches wide. The discoveries made in the earlier sketches and John's ability to communicate form and depth with value, have all been put to good use. Occlusion shadows enhance the rafters of the wall arch across the top of the image. The airship in the middle ground is white so it pops out against the darker landscape. By designing the lighting so that the ship and sailors on the far right are in shadow, John was able to silhouette them against the background.

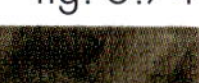

fig. 5.74

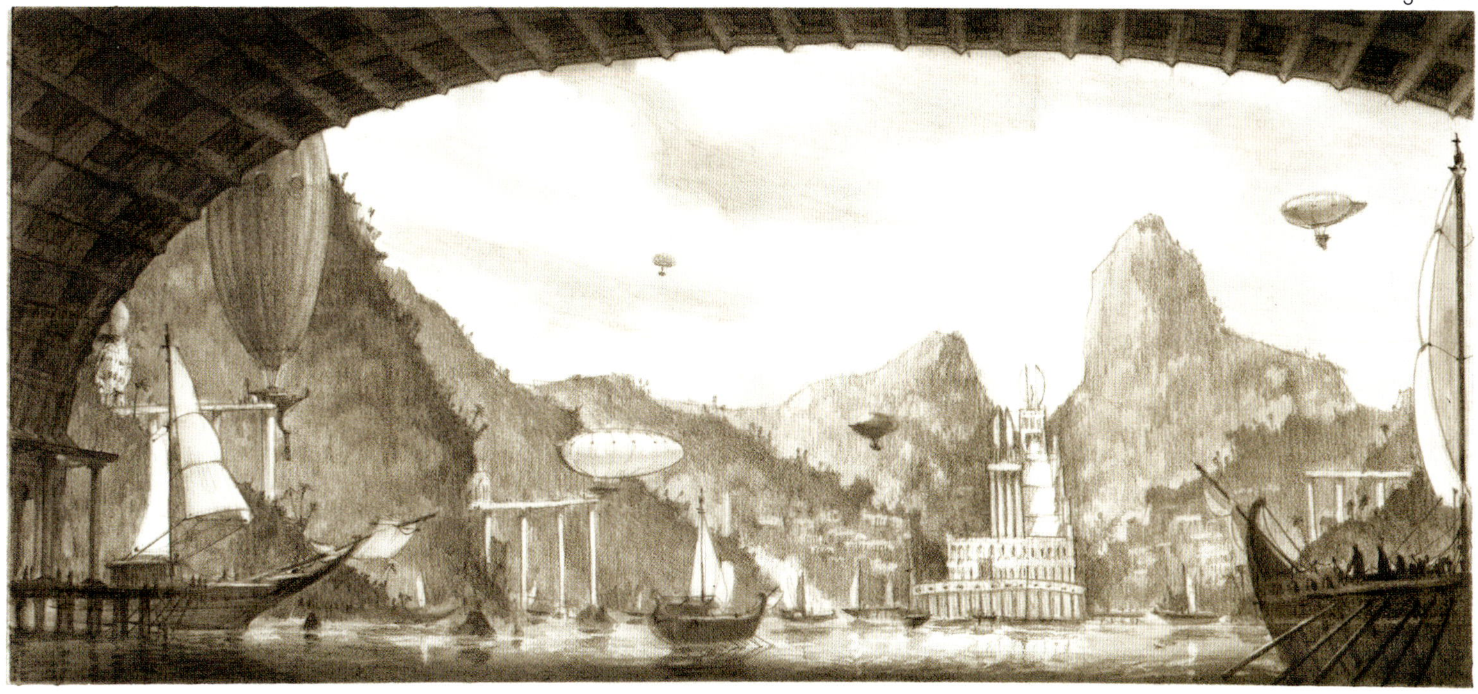

RENDERING ENVIRONMENTS BY ROBH RUPPEL

Art director and instructor Robh Ruppel shares his approach and techniques for rendering environments.

It is really important to start with a hand-drawn sketch. There is something about the natural rhythms the hand makes that cannot be duplicated. It is what makes art individual and unique. Even if a sketch is small, it is a good beginning.

fig. 5.75

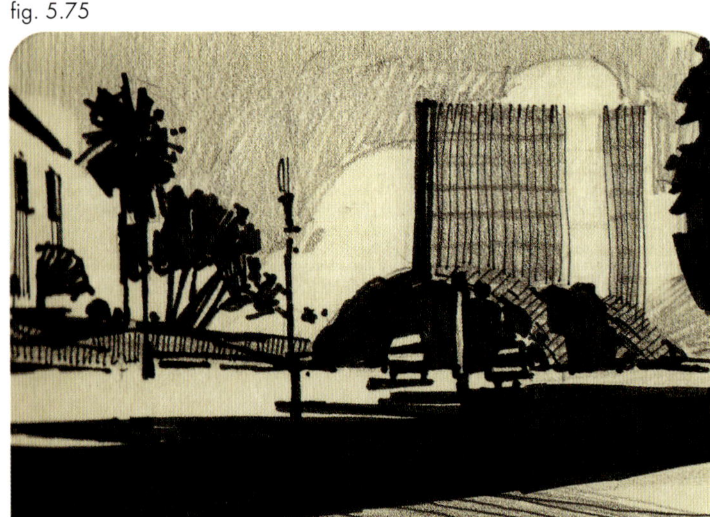

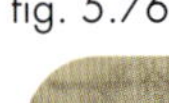

Sketching is the time to explore the shapes and rearrange the elements. We do not have to leave them as we find them. We are the artists. We should make it look the way it feels to us. We are NOT cameras.

fig. 5.76

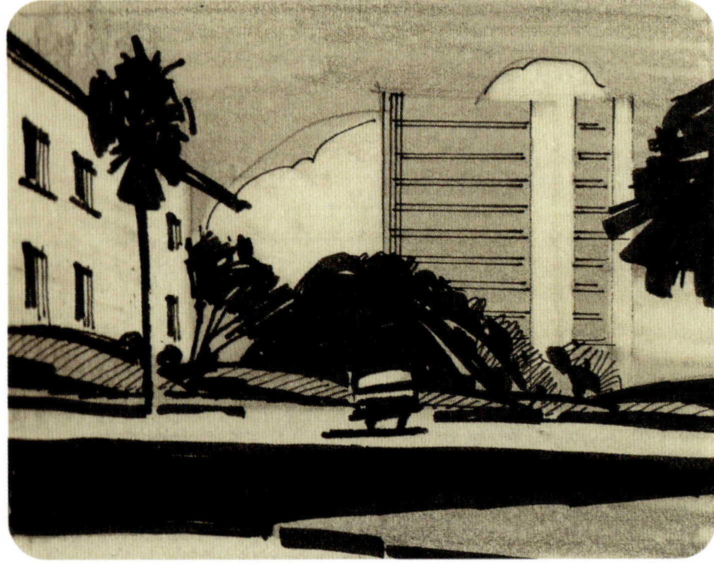

Robh Ruppel
www.robhruppel.com
graphicla.tumblr.com
broadviewgraphics.blogspot.com

fig. 5.77

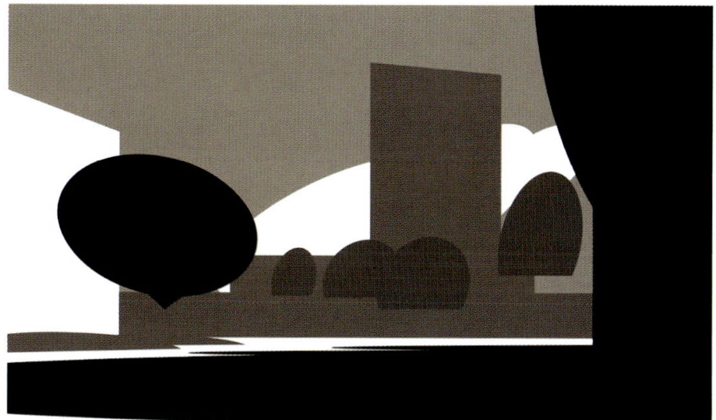

I start by making a simple design using as few values as possible. If it works at this stage, it will work all the way through. Test how abstract the shapes can be and still read as the objects. It can be pushed pretty far and still have the scene appear "real." That reality comes from the value relationships. If those are correctly relating to one another, that carries most of the weight of the "realism" of the scene. Getting and understanding those relationships comes from working outdoors, where these sketches were made.

fig. 5.78

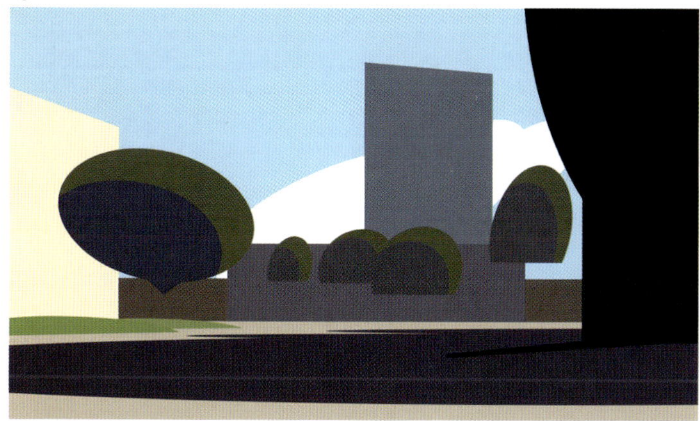

Next, assign colors to those values. Look for an average hue and work to maintain the relationships that were already established. Start to model the shapes and make choices about lighting and shadows, making sure the warm/cool bias works. Make thoughtful choices so that every step feels real.

fig. 5.79

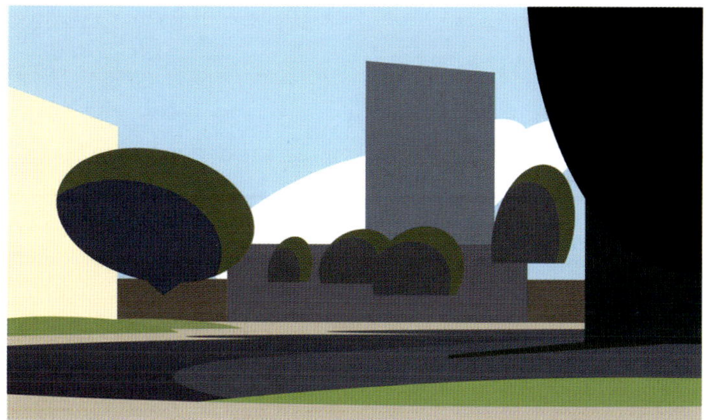

The large foreground shadow was divided into its individual elements: grass, sidewalk and pavement. They remain within the value range initially established. This is SO important, not only to maintaining design, but for learning how to use a narrow range of values to show changes in form.

fig. 5.80

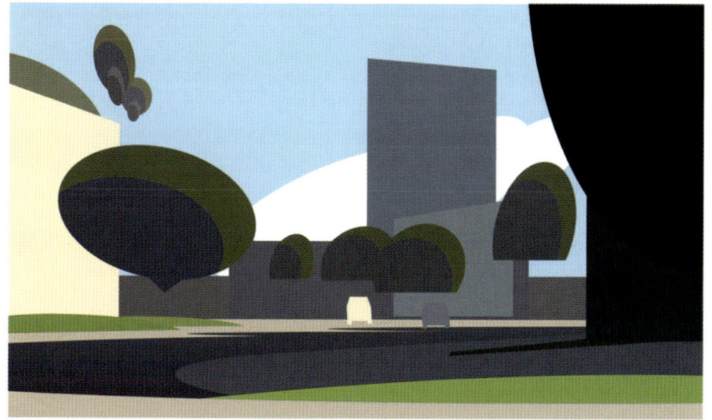

Now that the major elements are in, it is time to add a few details, still using flat values. These details must be "in tune" with the overall design and contribute to a harmonious whole.

It is time to start modeling the forms by adding half-tone colors and simple shadows. A few large, simple details were added on the architectural elements.

fig. 5.81

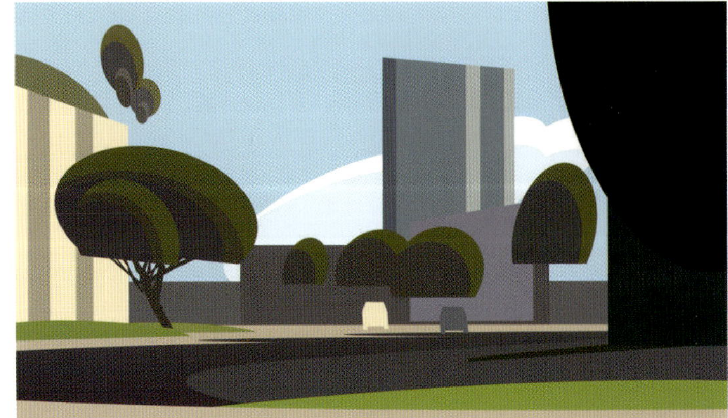

The fun begins by erasing into the tree shapes to make interesting patterns of negative spaces that mimic the structure of leaves. Tree trunks were added so that their rhythms flow with the overall scene. They need not be exactly what was observed in nature, but should be shapes that are true to their character and create an artistic statement.

fig. 5.82

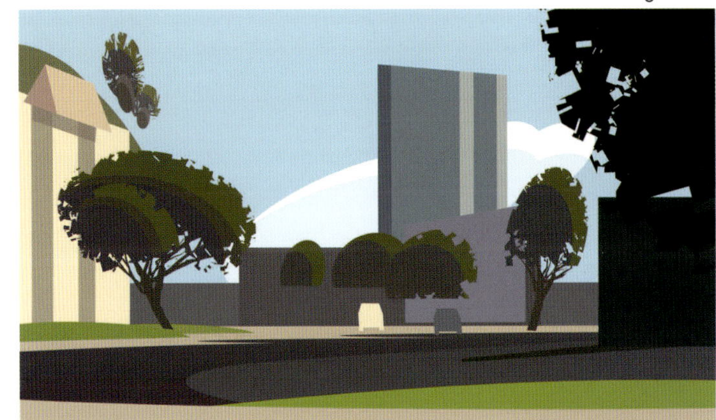

The forms of the cars are starting to be modeled and a few smaller elements were added in the shadow shape of the foreground.

fig. 5.83

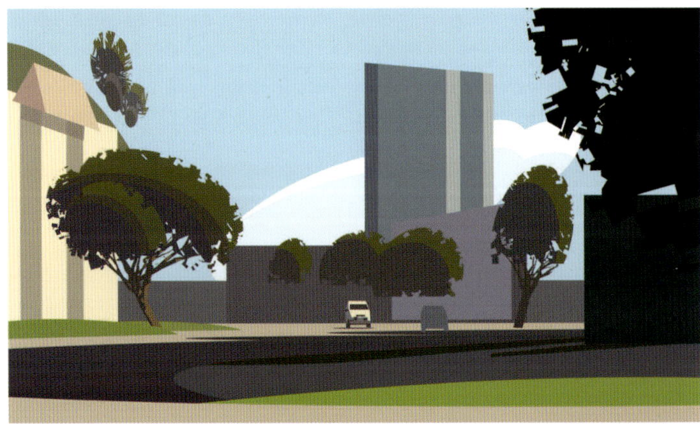

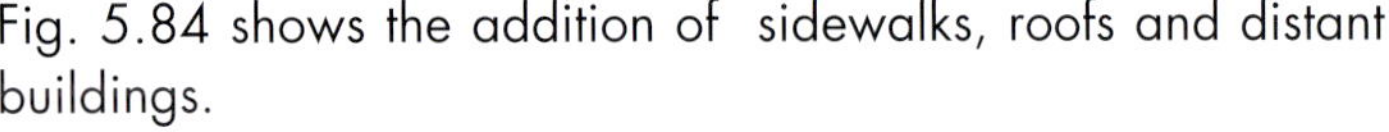

Fig. 5.84 shows the addition of sidewalks, roofs and distant buildings.

fig. 5.84

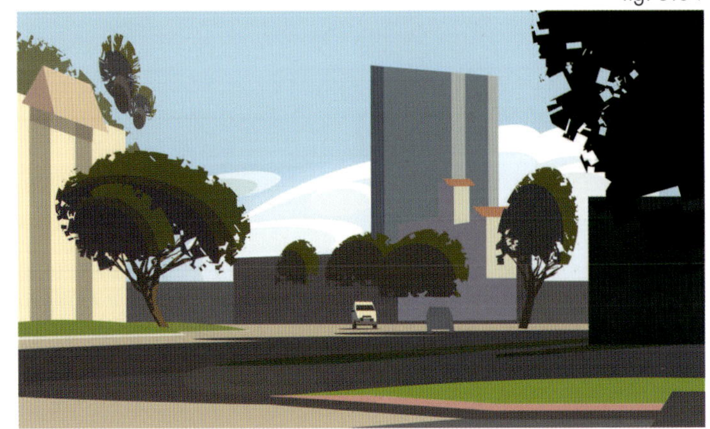

fig. 5.85

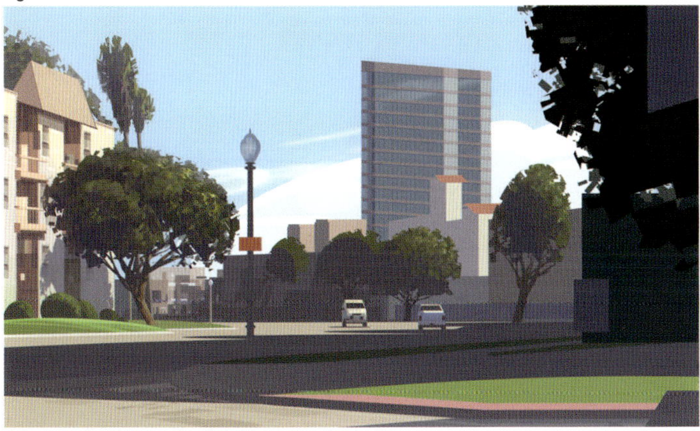

The details really get added in now: balconies, lampposts, separate floors on the distant building as well as a few reflections. These details are all still done in flat values, which mimics an older technique of architectural painting with gouache.

fig. 5.86

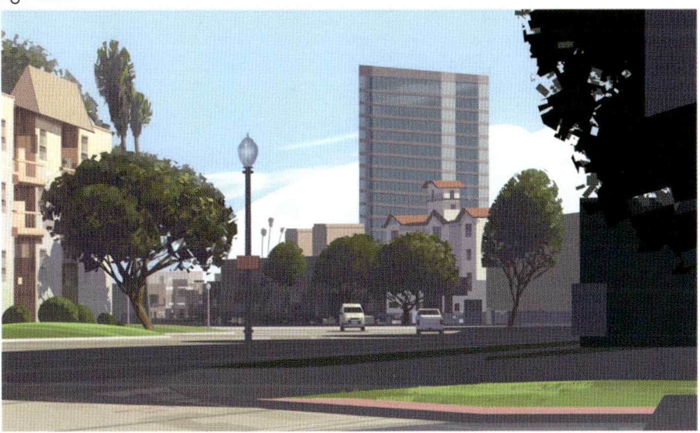

It is time to add in even more implied detail on the distant street-level structures. The details are "implied," because they represent the artist's impression rather than a photo-real depiction. No need to model every corner and post, but rather imply the details with random noise that is accurate to the visual information. Do this by squinting and noticing the relative values and colors. Reproduce that effect, rather than all the minute details and forms.

fig. 5.87

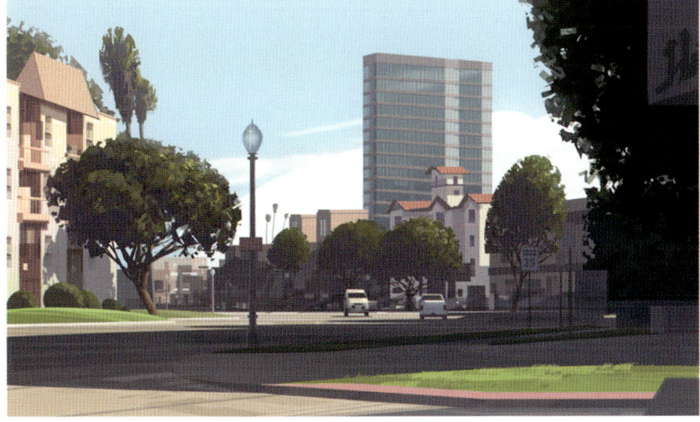

Atmospheric values were added around the upper-right tree. It is that subtle transition of tone that sits the object into the scene.

fig. 5.88

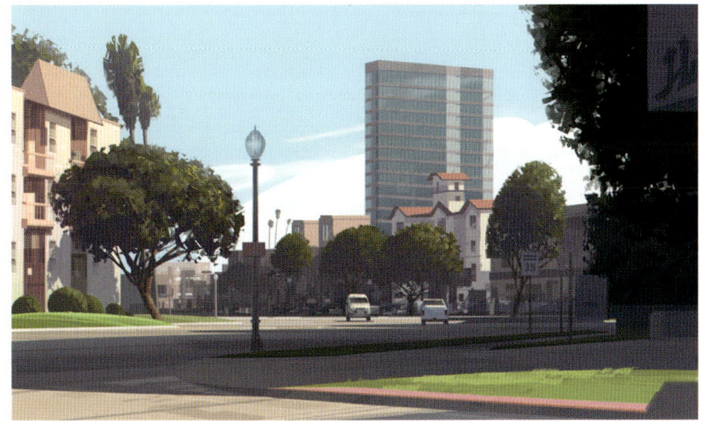

Subtle tonal adjustments tie the whole scene together. This succeeds only if the initial design and block-in were truthful to the relationships. If that was off, all the rest was for naught.

ENVIRONMENT RENDERING EXAMPLES BY ROBH RUPPEL

To learn more about Robh's process and see more of his work, please check out his book, *Graphic L.A.*

ISBN: 978-162465017-8

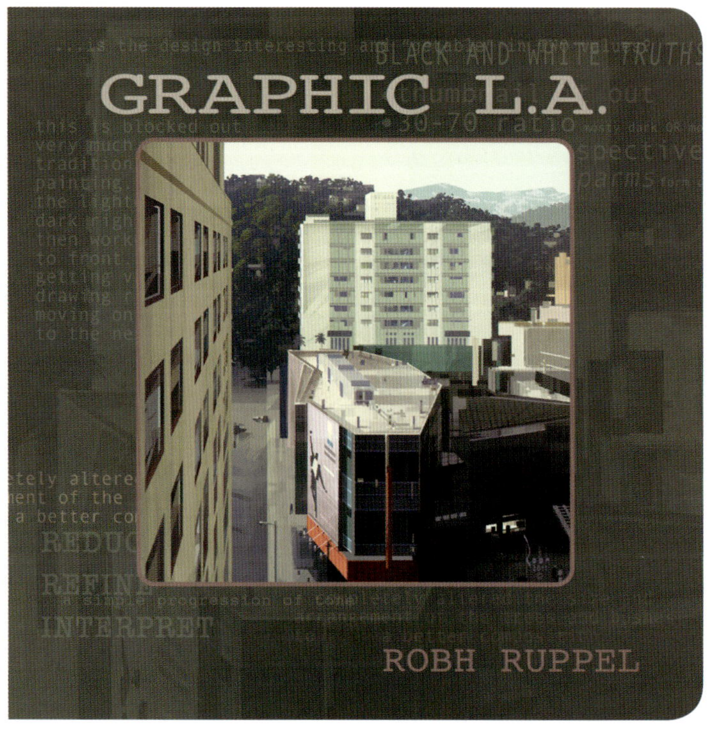

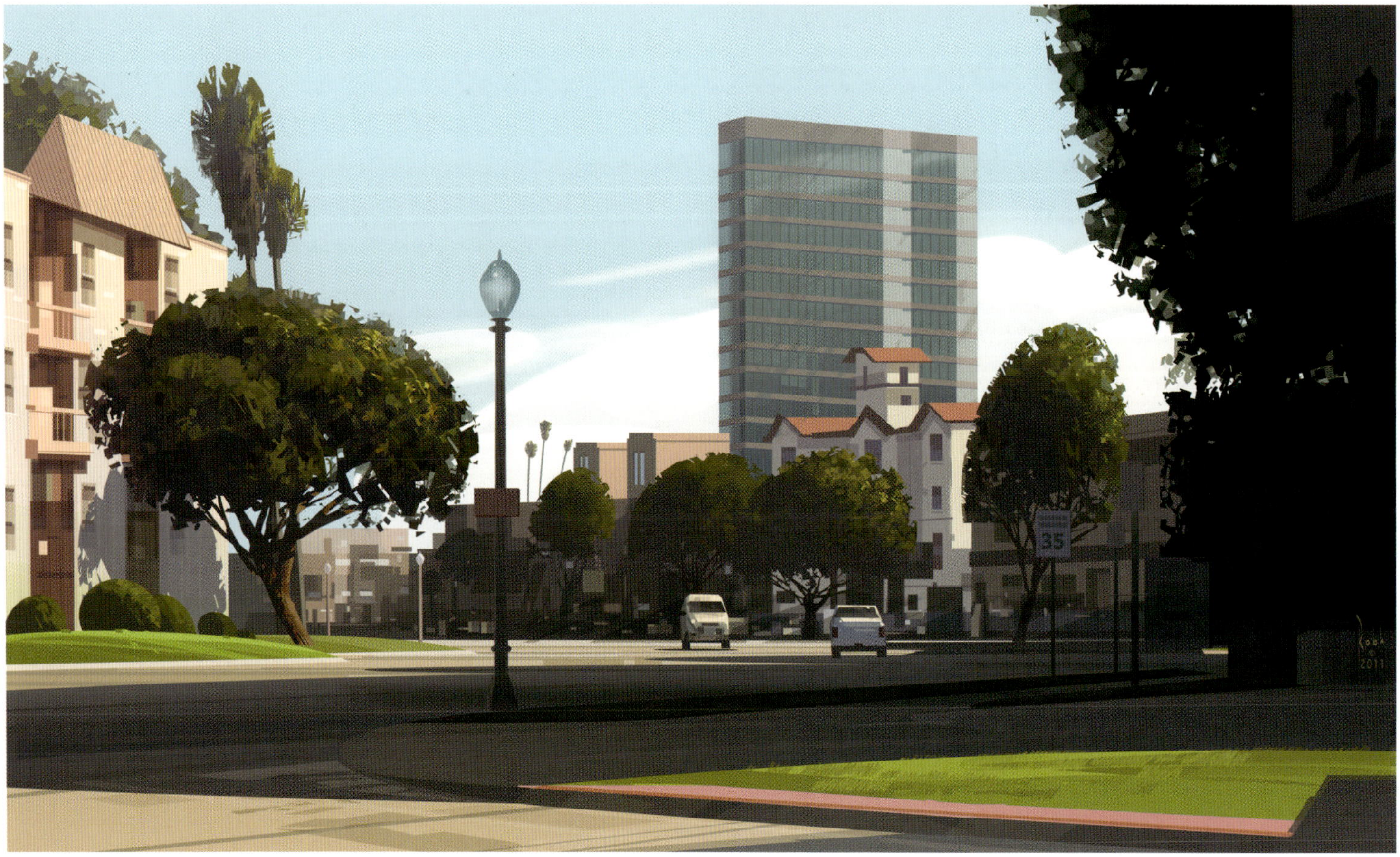

fig. 5.89

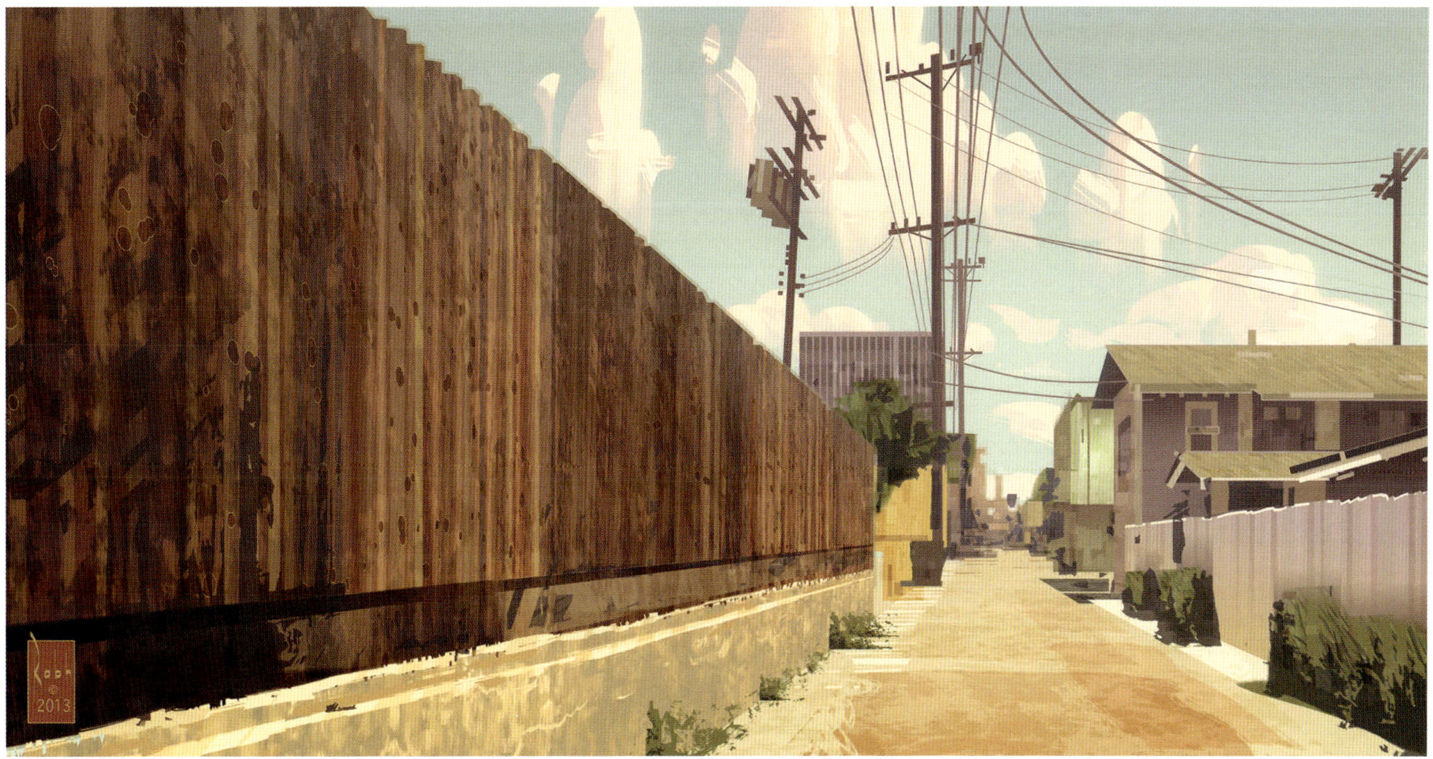

fig. 5.90

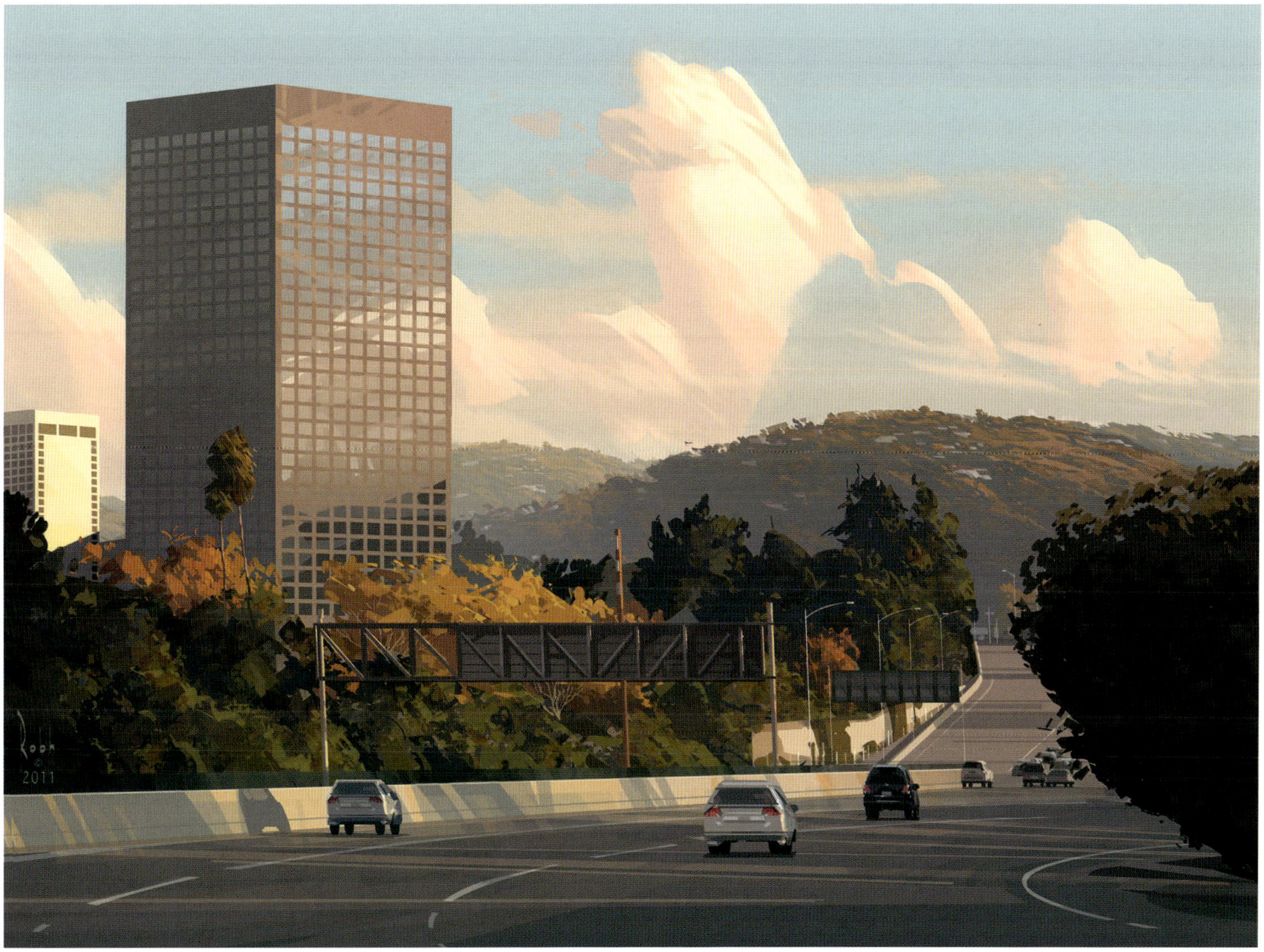

fig. 5.91

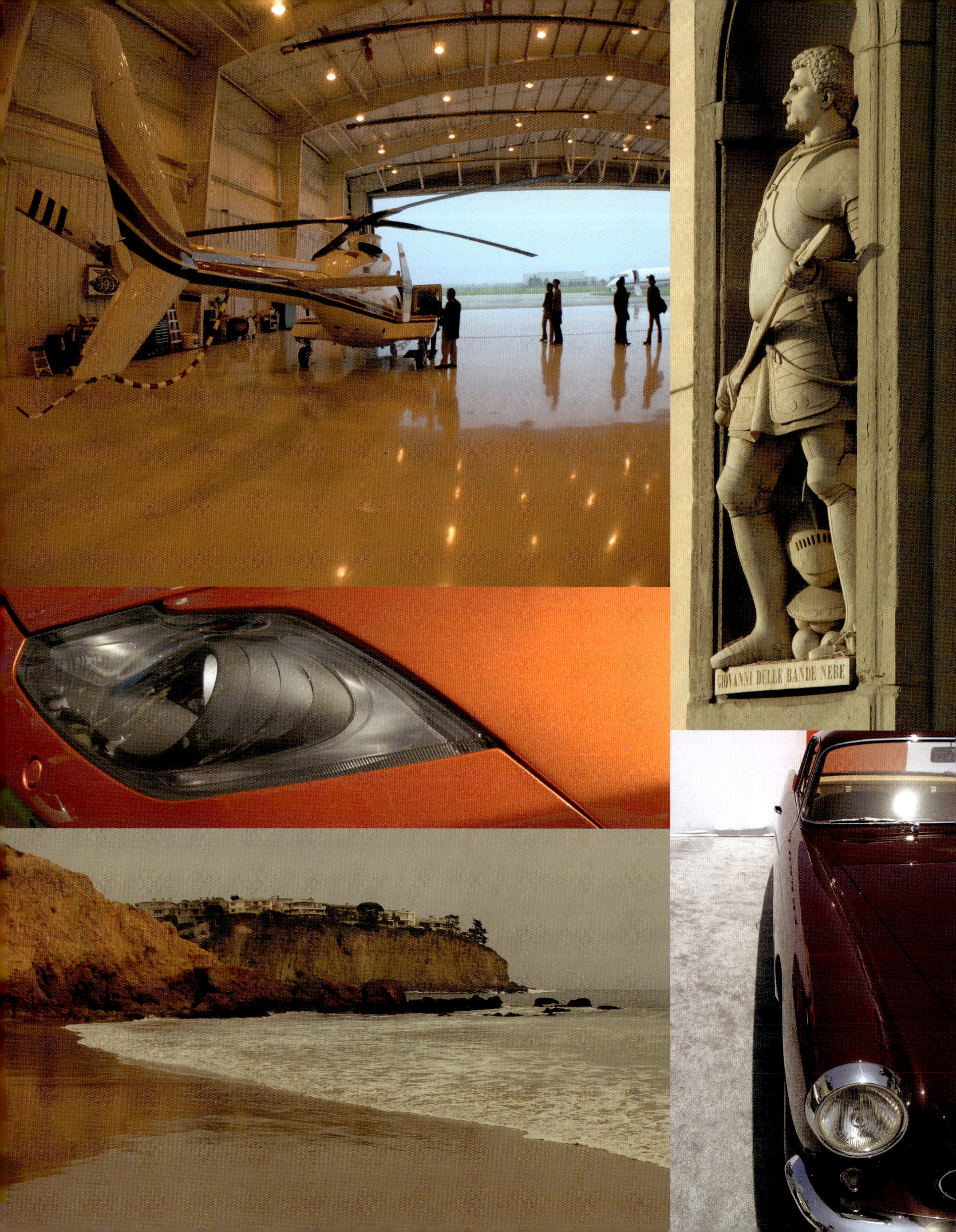
GIOVANNI DELLE BANDE NERE

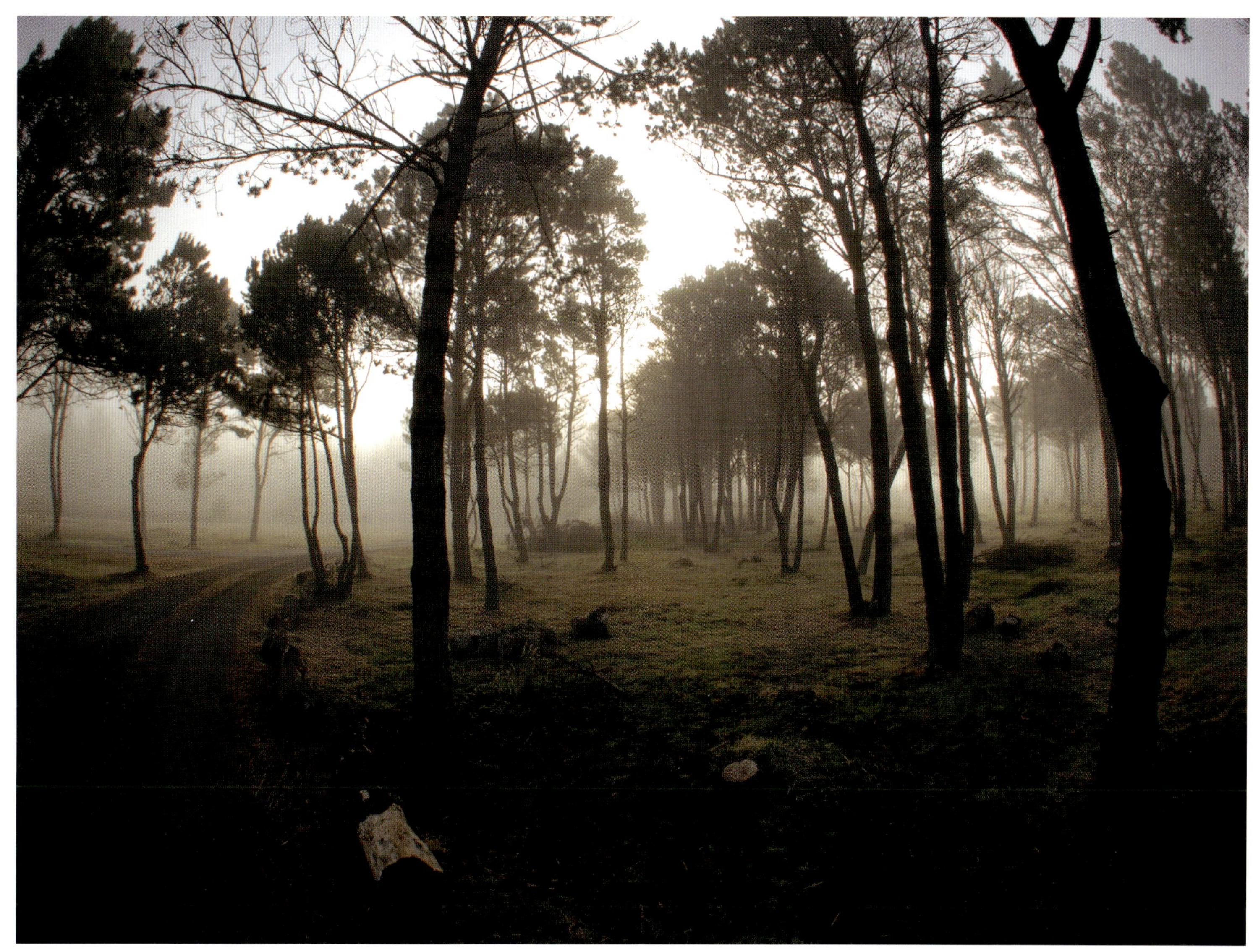

CHAPTER 06
PHOTO REFERENCE

When rendering people, places and things from the imagination, many designers fail to remember that light, shadow, color and materials do not also need to be reinvented. These imaginative designs, which may be of objects never before seen, should be shown in full color and made of proper materials that are already familiar from observing the world around us. Increasing observation skills is a very important step in improving rendering skills. Observing, combined with an increased understanding of the underlying physics, is the primary focus of this book. One way to start building this skill is to get out of the studio and to render from observation as Robh Ruppel described in his tutorial. Taking photographs while traveling and researching new materials and lighting strategies are great ways to help inform rendering work. Let's look at how best to use this reference material and organize it over time.

WHAT IS GOOD PHOTO REFERENCE?

Not all photographs are good examples of the research subject. References need to be qualified as they are categorized. Look at your photographs with a very critical eye and ask if the image is an ideal example of the subject? Weaker examples can be deleted, or thrown into a "catch all" folder. High quality images are more important than high quantity.

Say you need an example of a chrome-finish car. Just photographing one is not sufficient. Aim to build a reference library of only the very best examples. Be selective and constantly on the lookout for nice imagery. Inspiration for renderings is all around. The skill of seeing it and capturing it needs to be practiced.

HOW TO USE GOOD PHOTO REFERENCE

Photo reference was used to render the car in fig. 6.1. The real car (fig. 6.2) was photographed with an iPhone in the morning and then referenced that same afternoon while rendering the top car in Photoshop. The wheels were rendered in MODO and then added to the Photoshop image. In this case, the color and material finish are why the bottom car was photographed, as they were the desired color and finish for the Photoshop rendering. Wanting to invent new shapes, not new lighting, color or materials, the values and colors were sampled directly from the reference photo and applied to the new car's forms. A little artistic license was taken here and there, but referencing the image taken just hours earlier was a huge help at speeding up the rendering process. This is a very common practice in the commercial arts world, where speed and accuracy in conveying ideas are far more important than artistic, illustrative appeal.

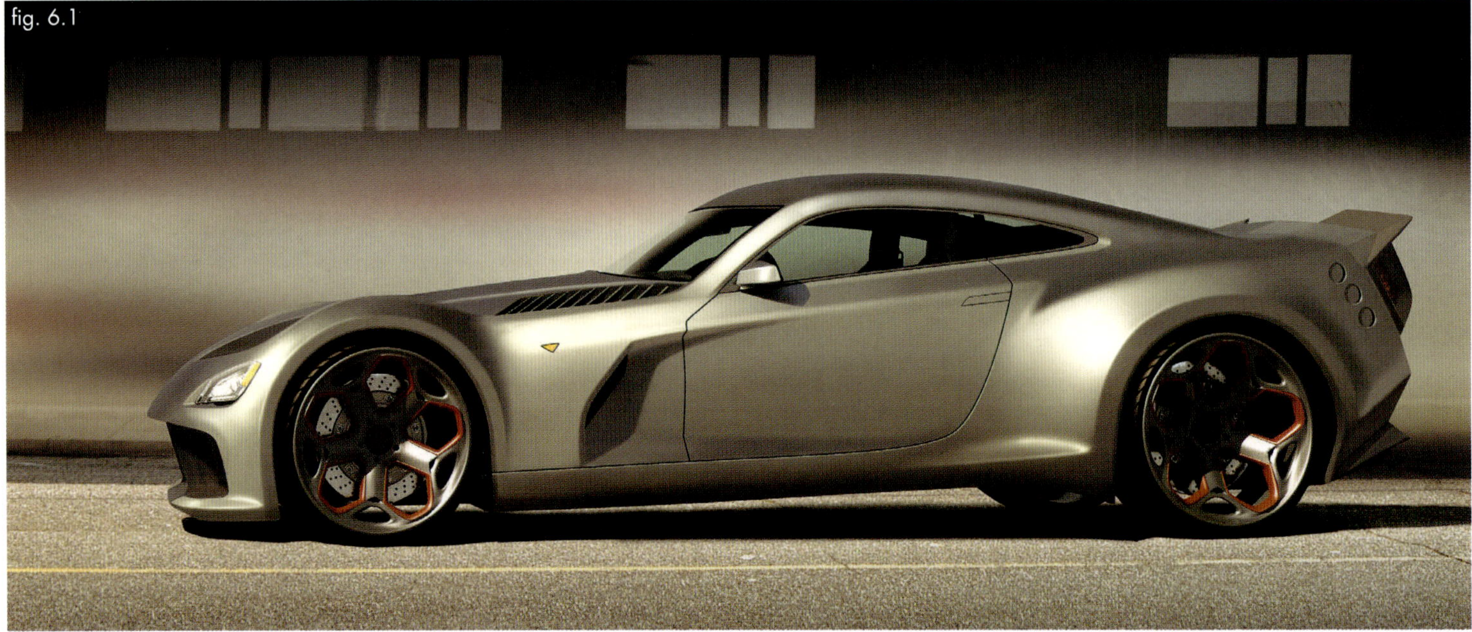
fig. 6.1

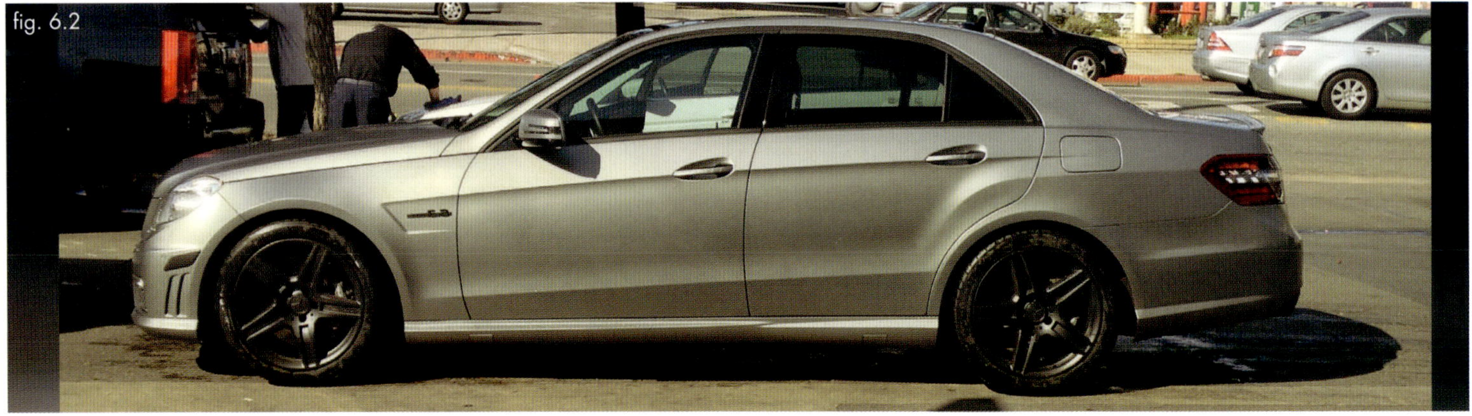
fig. 6.2

Below is a screen grab of the reference library created in Adobe's Bridge, when putting together this book. Approximately 15,000 photographs were considered to find the best examples to communicate the points for this book. The photos are of a wide variety of subjects, and each one does a nice job of communicating that all-important concept of form change equaling value change and vice versa. By reading the folder names, observe just how in-depth the research and preparation was for executing this book. All of the photos in this book were shot during the last 15 years or so. Never stop learning, always remain curious.

fig. 6.3

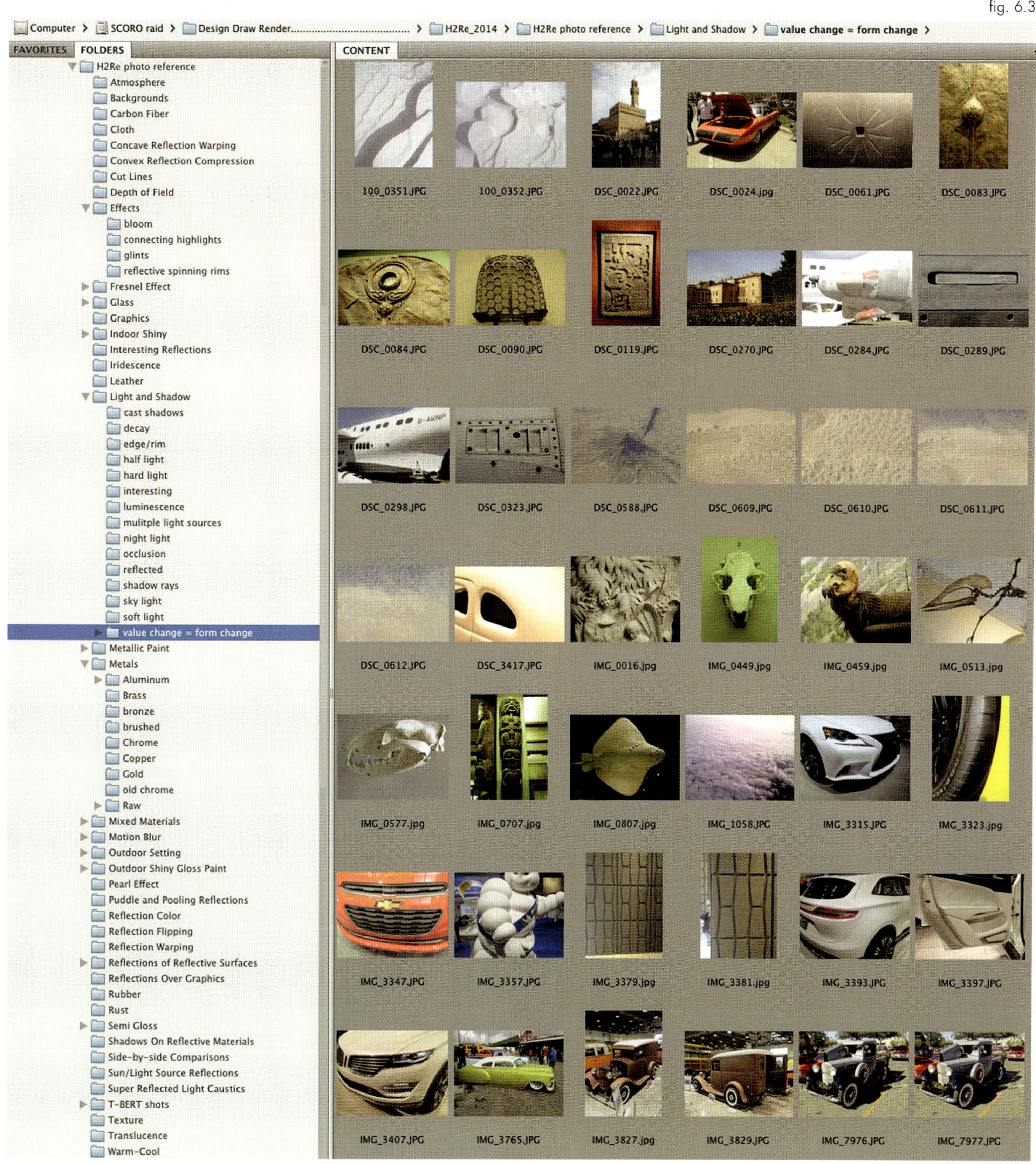

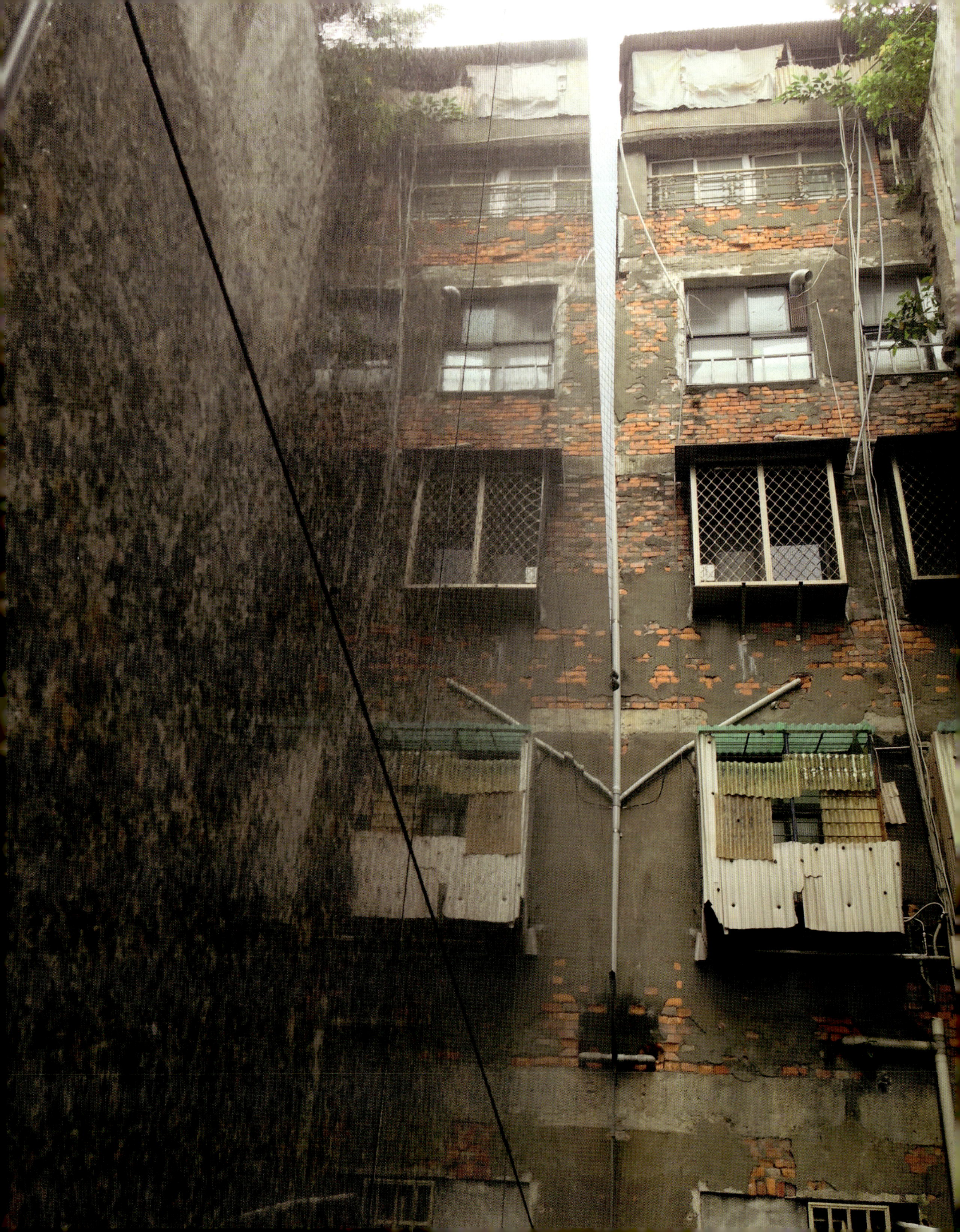

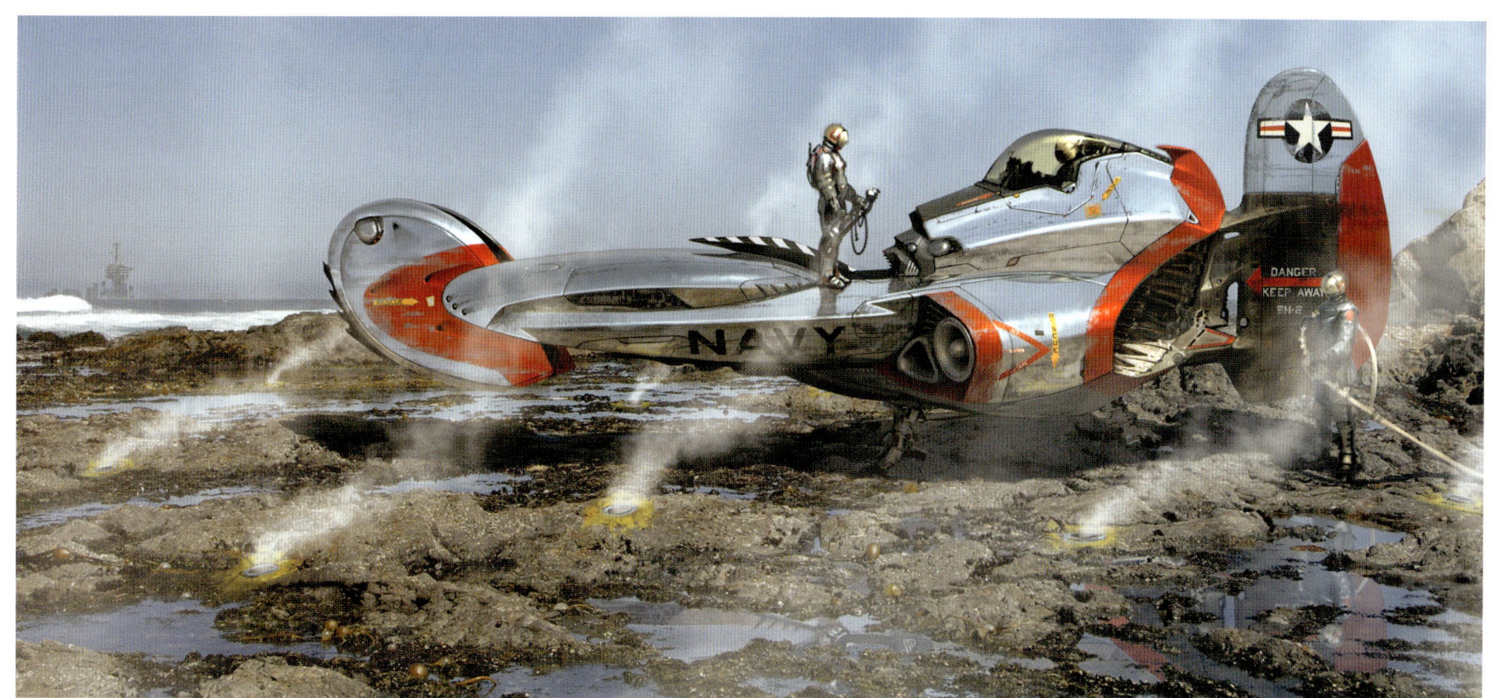

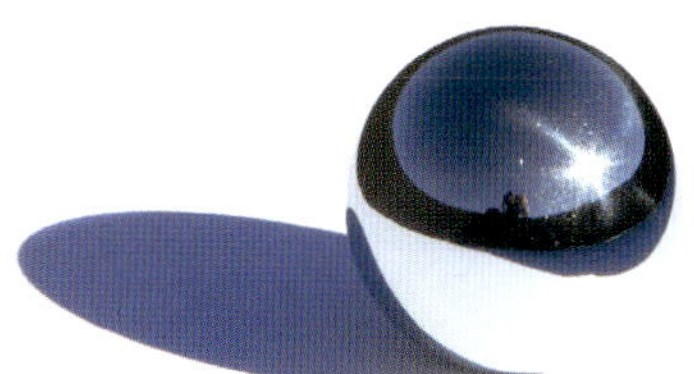

CHAPTER 07
REFLECTIVE SURFACES

Let's switch gears for the remainder of the book to study reflective surfaces. Rendering reflective surfaces requires learning different laws of physics that can be confusing because value assignments for reflective surfaces are very unlike those of matte surfaces. Learning how to observe and understand reflections, and then replicating them in new work, are very important skills to master because most surfaces have at least some reflectivity. All of the knowledge about matte surfaces is still needed, but now reflections will be added over those matte surfaces, merging the two material types together, across the same surface.

ANGLE OF INCIDENCE 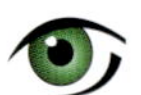

Rendering realistic-looking reflective surfaces can be a big challenge for any artist. The best way to improve is through a better understanding of the underlying physics that influence all reflective surfaces, from shiny cars to wet streets. Armed with this knowledge, the task of rendering shiny surfaces becomes much simpler. This chapter clarifies why reflections look the way they do and then demonstrates how to render imaginary shiny surfaces that look real, while at the same time communicating the form changes of the surface through the shape and strength of the reflections themselves.

The easiest way to start the discussion on reflections is to look at some reference photography.

The reflections, more importantly, the position of the reflections, depend on one thing and one thing only: The angle that the line of sight bounces off of the shiny surface, into the environment that surrounds that surface. (Yes, yes, I know sight lines technically work this way: If you're looking at an object, light reflects off it and enters the eye. But for the purposes of simplified instruction, let's imagine the opposite. To learn more about the science of the eye, feel free to Google it.)

In fig. 7.1, the environment reflects onto the shiny surfaces. This is the most basic thing to understand about reflections.

When the line of sight intersects a surface, the measurable angle created there is called the **angle of incidence.**

Simply put, when calculating where reflections go on a shiny surface, the line of sight's angle of incidence is always **"equal in–equal out."**

In other words, the angle at which a line of sight hits a surface is exactly the same as the angle at which it bounces away from the surface.

Fig. 7.2 shows a side-view diagram of a figure looking into a mirror. His line of sight is measured as equal in–equal out. This is absolutely the most useful information for plotting/locating reflections on shiny surfaces.

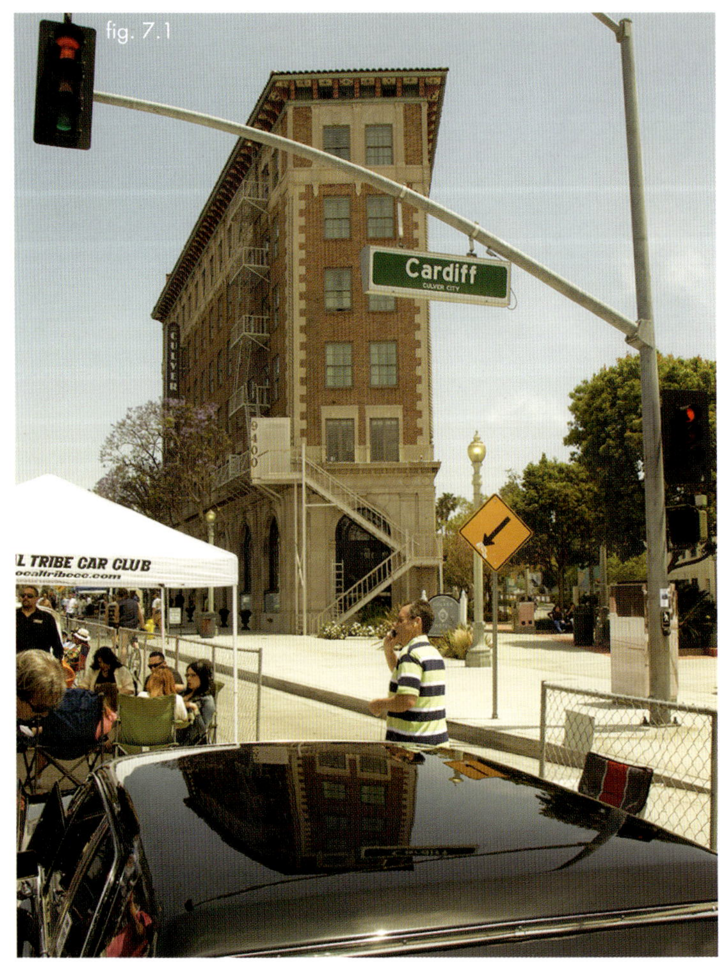

fig. 7.1

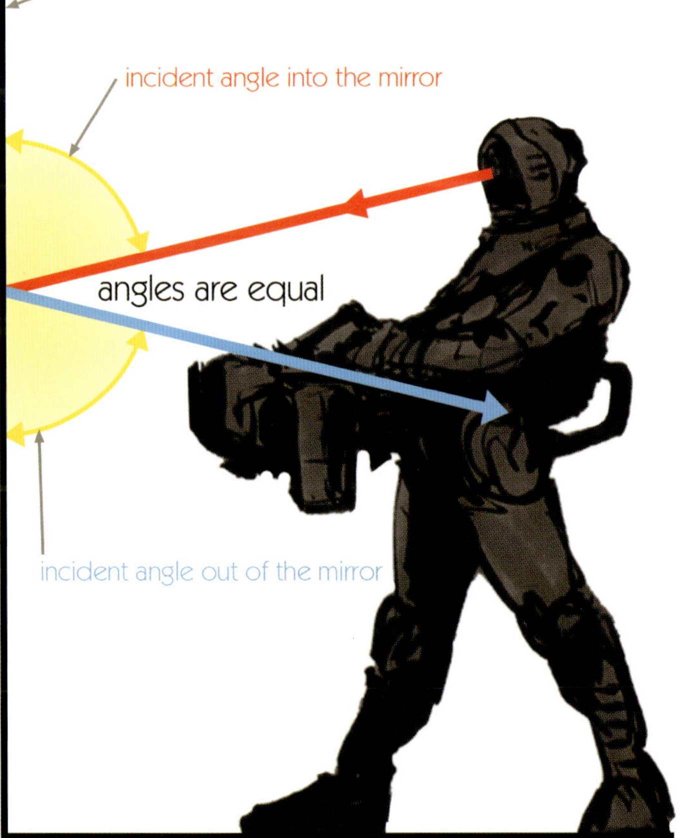

fig. 7.2

To calculate the position of an object's reflection in a mirror, use the same technique as the perspective drawing "mirroring" method explained in detail and demonstrated throughout *How To Draw*. Our previous book explains how to draw symmetrical forms by mirroring one side to the other. So if you have already studied that book, you already know how to plot reflections in perspective!

If the mirroring diagram (fig. 7.5) is unclear, please read *How To Draw* before going any further in this book. Seriously. Knowing how to draw in perspective and how to mirror is a prerequisite for understanding how to plot reflections. That book explains those skills in minute detail.

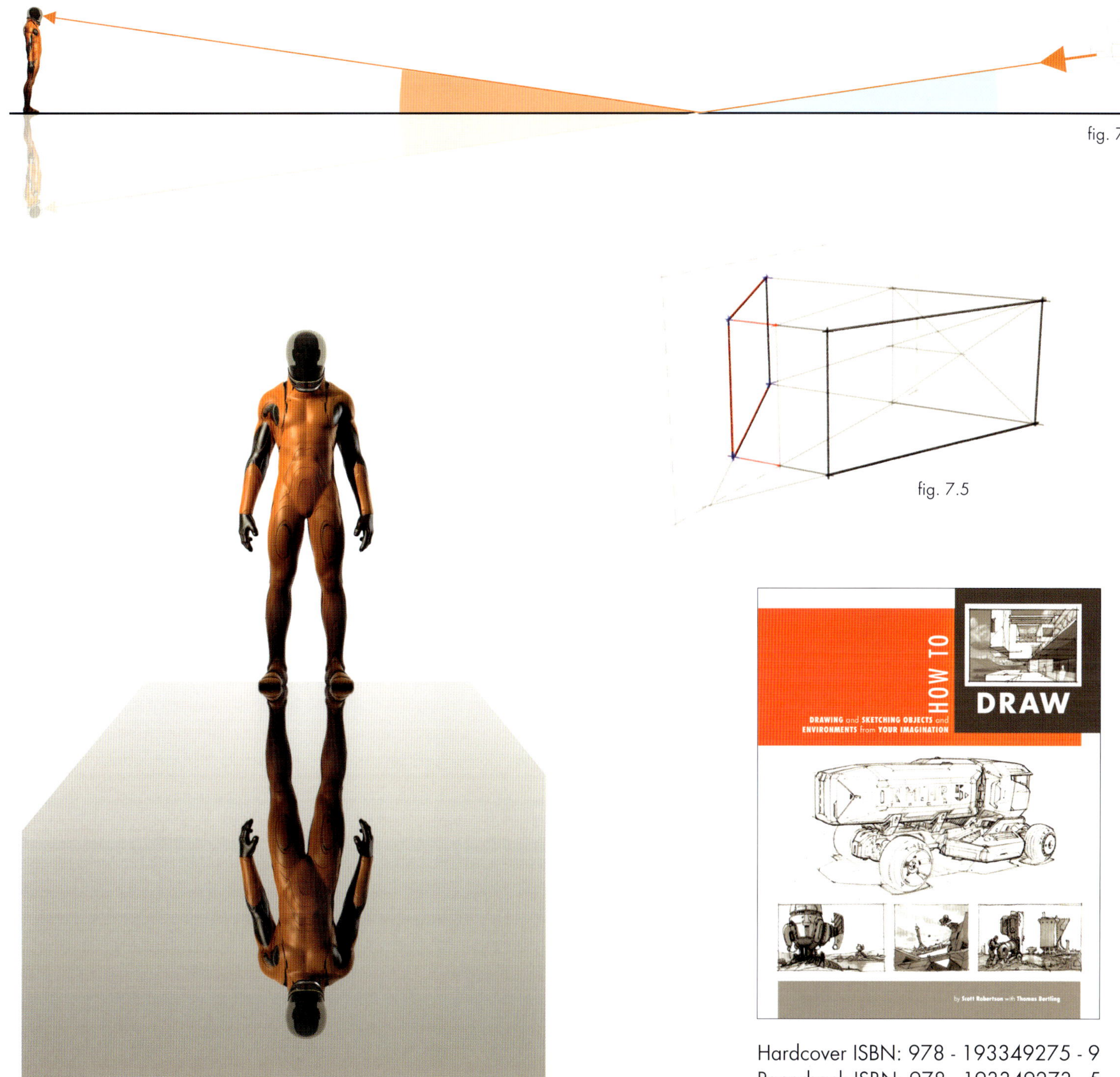

fig. 7.3

fig. 7.5

fig. 7.4

Hardcover ISBN: 978 - 193349275 - 9
Paperback ISBN: 978 - 193349273 - 5

PLOTTING REFLECTIONS

To accurately plot reflections of an environment across an object's surface, and then to plot that object's reflection onto a flat plane, requires some very advanced perspective drawing skills. In addition to knowing how to mirror objects and surfaces in perspective, knowing a lot about the X-Y-Z sections that make up the surface of a volume is necessary. Without this knowledge, the best that can be hoped for in plotting the positions of the reflections is straight-up guessing, as opposed to "educated guessing." It is important to plot reflections well enough that the average person, when shown the rendering, understands that the surface is shiny and also understands its transitional forms. Due to the time constraints of production rendering, it is often unrealistic to spend the time needed to plot all of the reflections properly. Instead, apply the physics of reflectivity and give it a best guess. Reading the rest of this book, doing the exercises and watching all of the related videos will prepare you to make some very impressive educated guesses about rendering reflective surfaces.

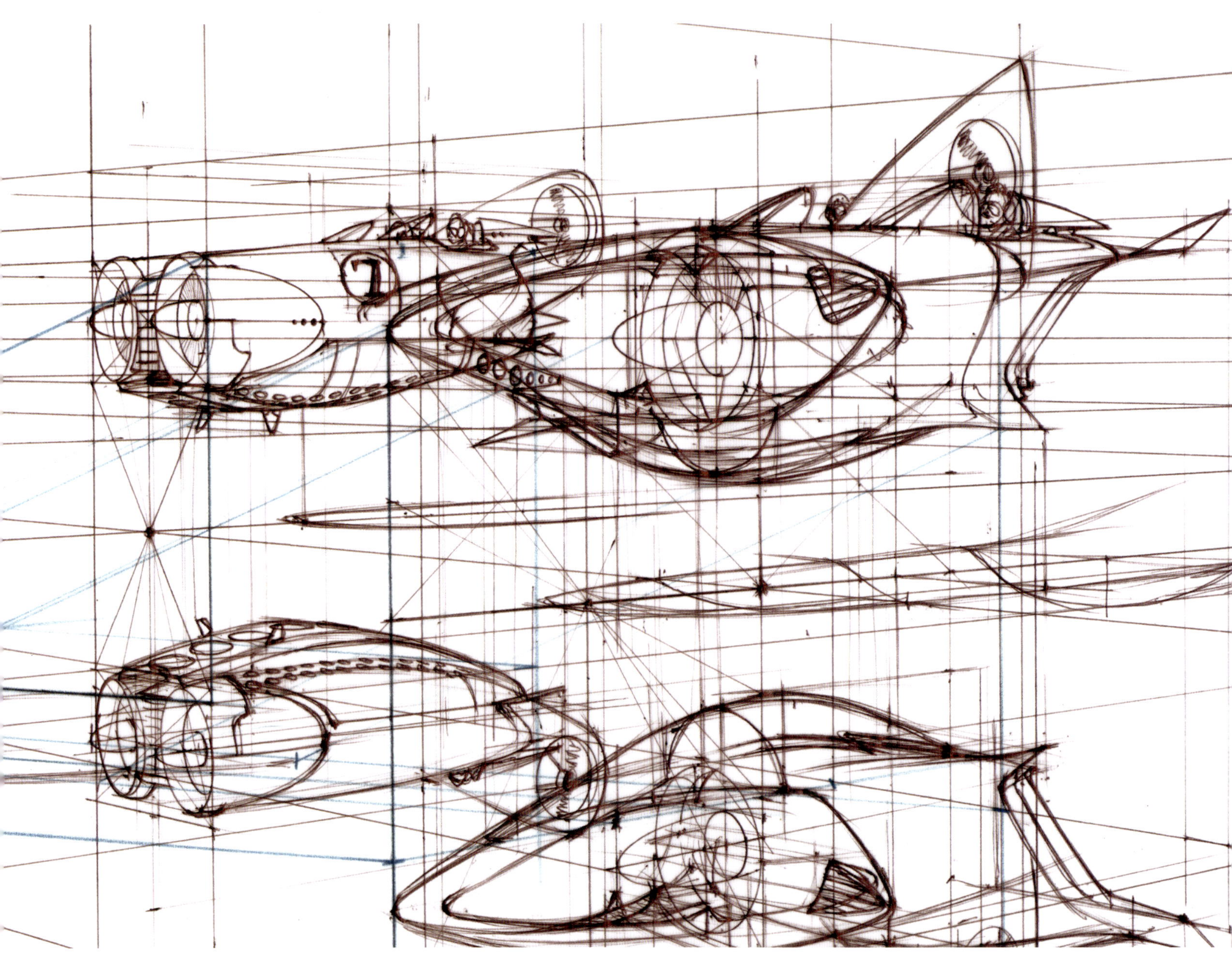

fig. 7.6

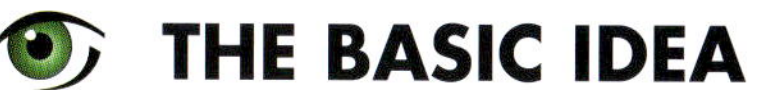

THE BASIC IDEA

Fig. 7.7 shows the basic idea of where the lines of sight will bounce/reflect off of the chrome form, into the environment around it. This is the most important thing to remember about rendering reflections, *the reflections are of what surrounds the reflective surfaces*. It sounds simple and obvious but many students start to render their reflective forms without first designing the scene around it. Even if the form is going to be rendered against a white background, the colors and values around the form still need to be known before the reflections can be rendered. Just like designing the lighting of the earlier scenes, now start to design the environments around the reflective surfaces to accomplish some very specific rendering effects. To summarize: Reflections are like a mirror, which is sometimes warped. Next up, rendering the environment surrounding a shiny object onto its surface. It is fun...really.

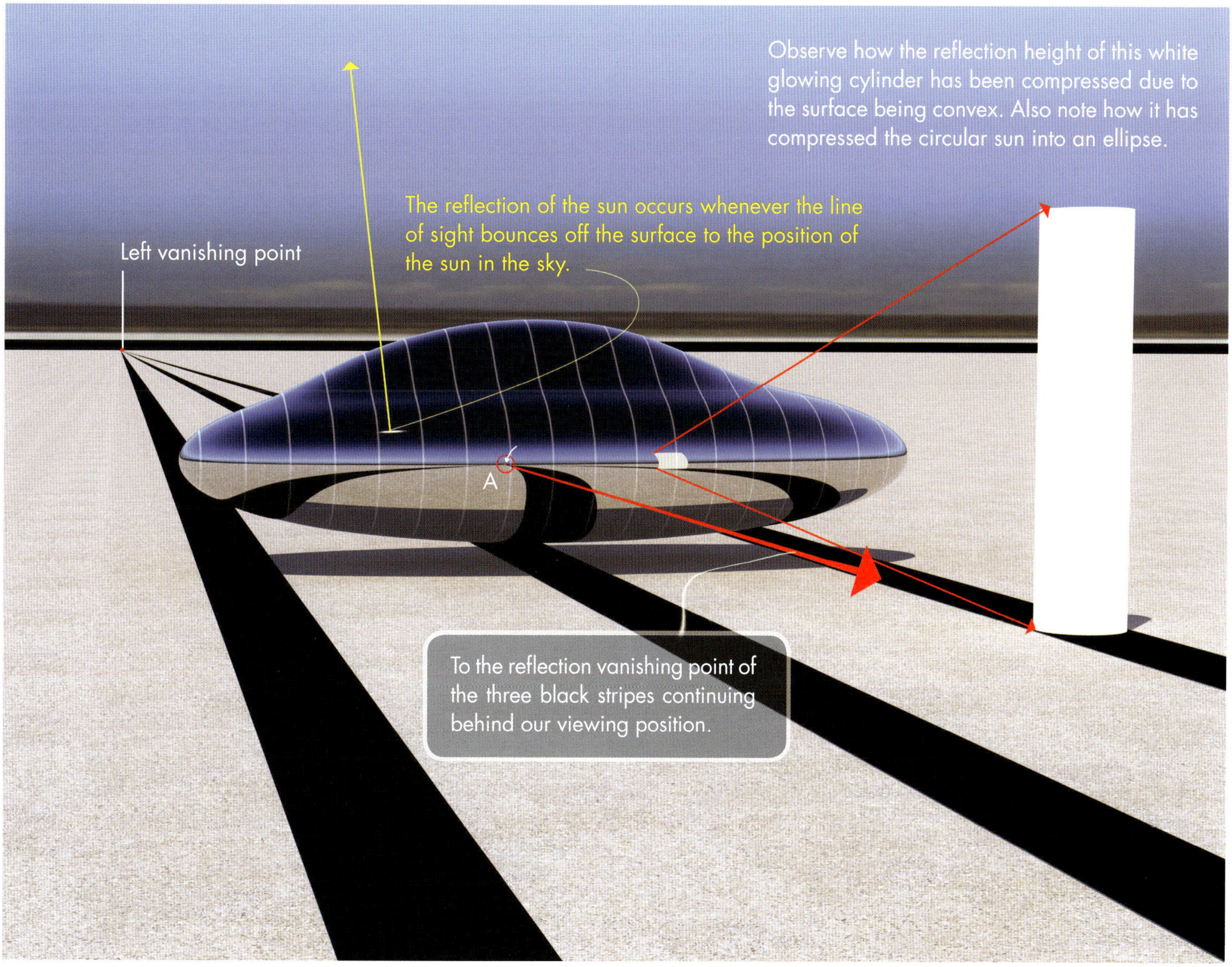

Observe how the reflection height of this white glowing cylinder has been compressed due to the surface being convex. Also note how it has compressed the circular sun into an ellipse.

The reflection of the sun occurs whenever the line of sight bounces off the surface to the position of the sun in the sky.

Left vanishing point

A

To the reflection vanishing point of the three black stripes continuing behind our viewing position.

fig. 7.7

Note: The horizon line is reflected in the surface where your line of sight bounces off of the surface parallel to the ground plane. The red line here is bouncing parallel to the black stripes on the ground plane. Parallel lines converge to a single vanishing point. In this case, this is a vanishing point behind where the camera is positioned in this scene, as you can start to understand by observing the entire reflection of the ground plane on the side of the chrome form (A).

PLOTTING REFLECTIONS: CONCAVE AND CONVEX SURFACES

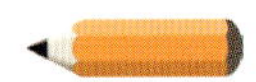

Anyone who has ever been in a "fun house" of mirrors knows that curved mirrors distort what they reflect. Here is the basic rule of thumb: A concave surface stretches the reflection, and a convex surface compresses the reflection.

In each of the images below, an astronaut stands in front of a curved mirror. The camera is in the same relative position in both of the views. Fig. 7.8 shows a concave curve; fig. 7.9 shows a convex curve. The line of sight bounces off of these two different curved panels to different parts of the environment.

In fig. 7.10, the line of sight bounces back to the astronaut across more of the surface, so his reflection gets stretched. In fig. 7.11, the line of sight bounces back to the astronaut over a narrower part of the surface, so his reflection gets compressed.

Next time you pass a shiny, curved surface observe how your reflection distorts and then look carefully at the curvature of that surface.

Remember: **concave stretches, convex compresses.**

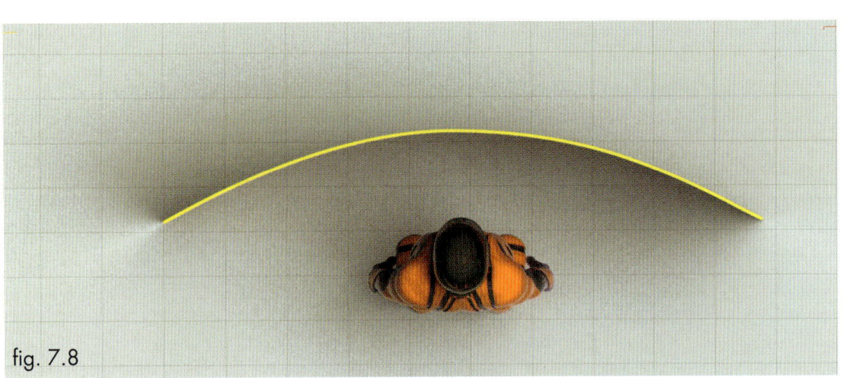

fig. 7.8

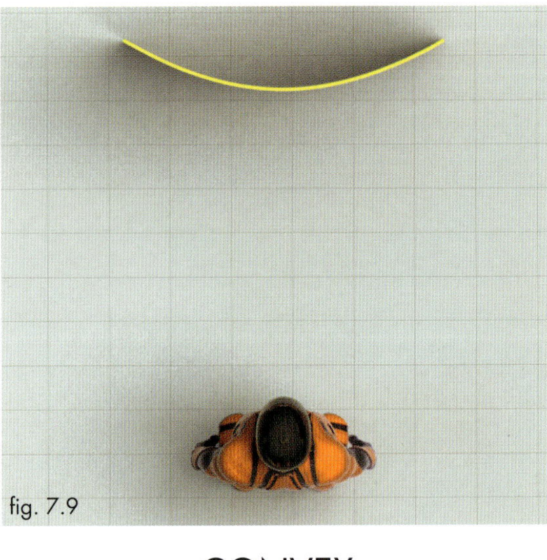

fig. 7.9

CONCAVE

CONVEX

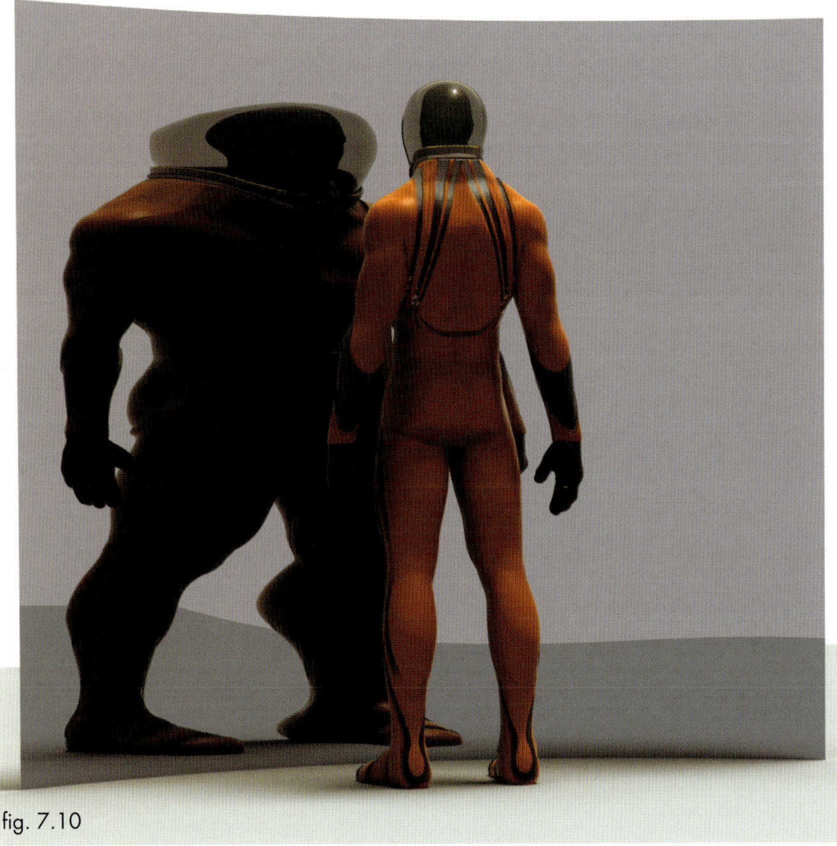

fig. 7.10

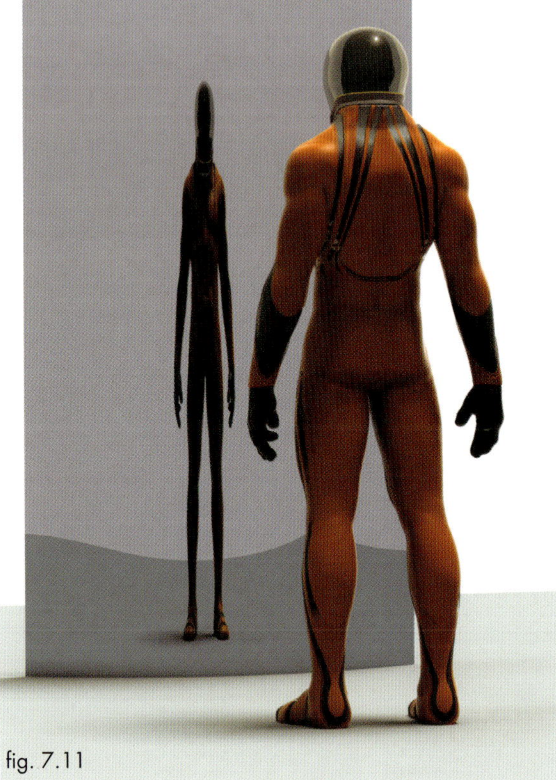

fig. 7.11

While sharpening some pencils, think about these reflective scenes of a shiny, horizontal cylinder and a vertical, convex mirror. Fig. 7.12 and 7.13 show draft views of the perspective view that the cameras see. By studying all of these draft views, it becomes possible to predict fairly accurately the reflection in the perspective view. Draw a bunch of sight lines from the camera to the shiny surfaces in the top and side views.

Next, estimate where the sight lines bounce back into the scenes. Trace back the position of the red ball and red cylinder on the reflective surfaces and then estimate their distorted shapes in the reflections. It takes some practice to think this way. The video that links to this page demonstrates this process, step-by-step.

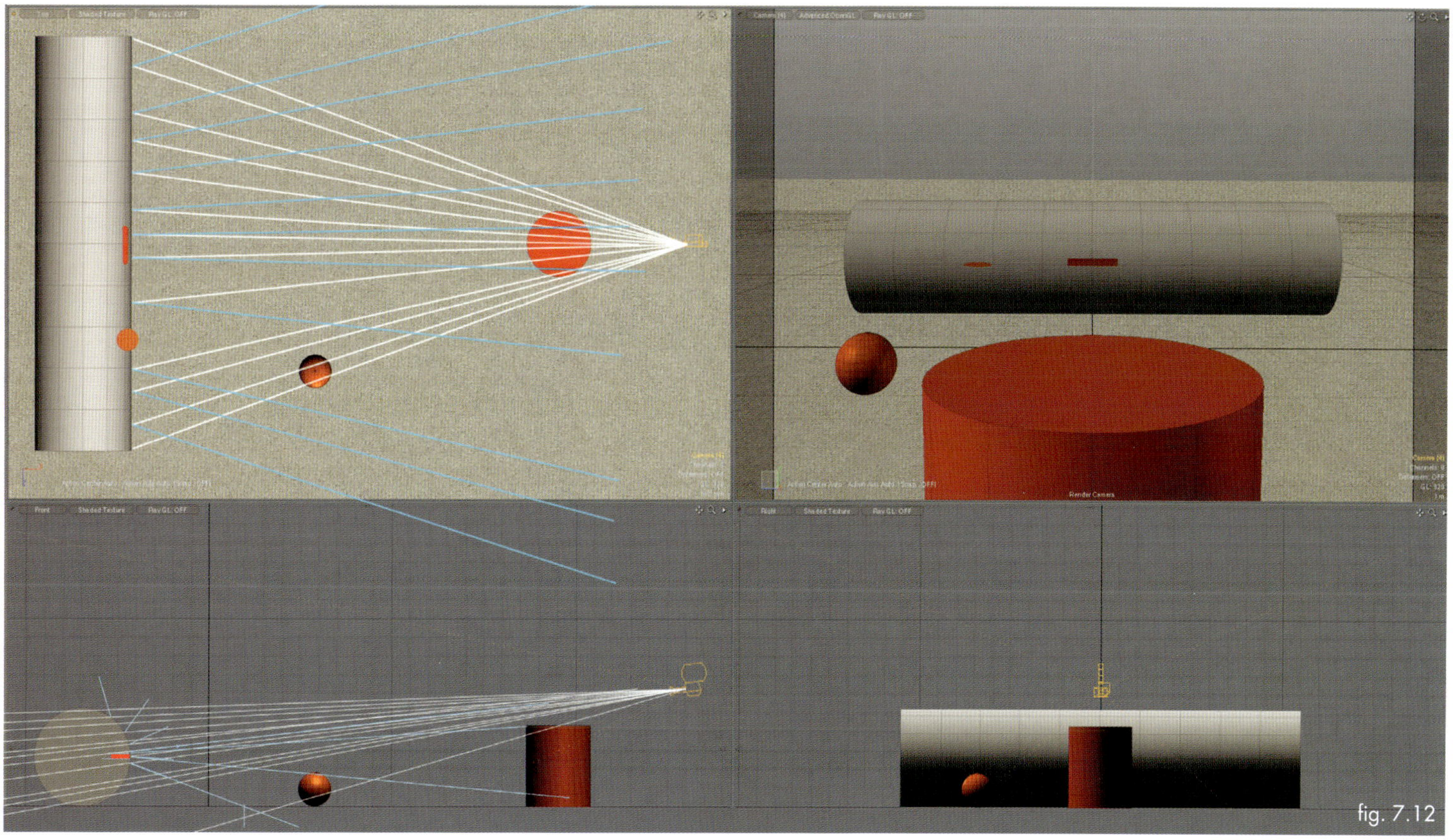

fig. 7.12

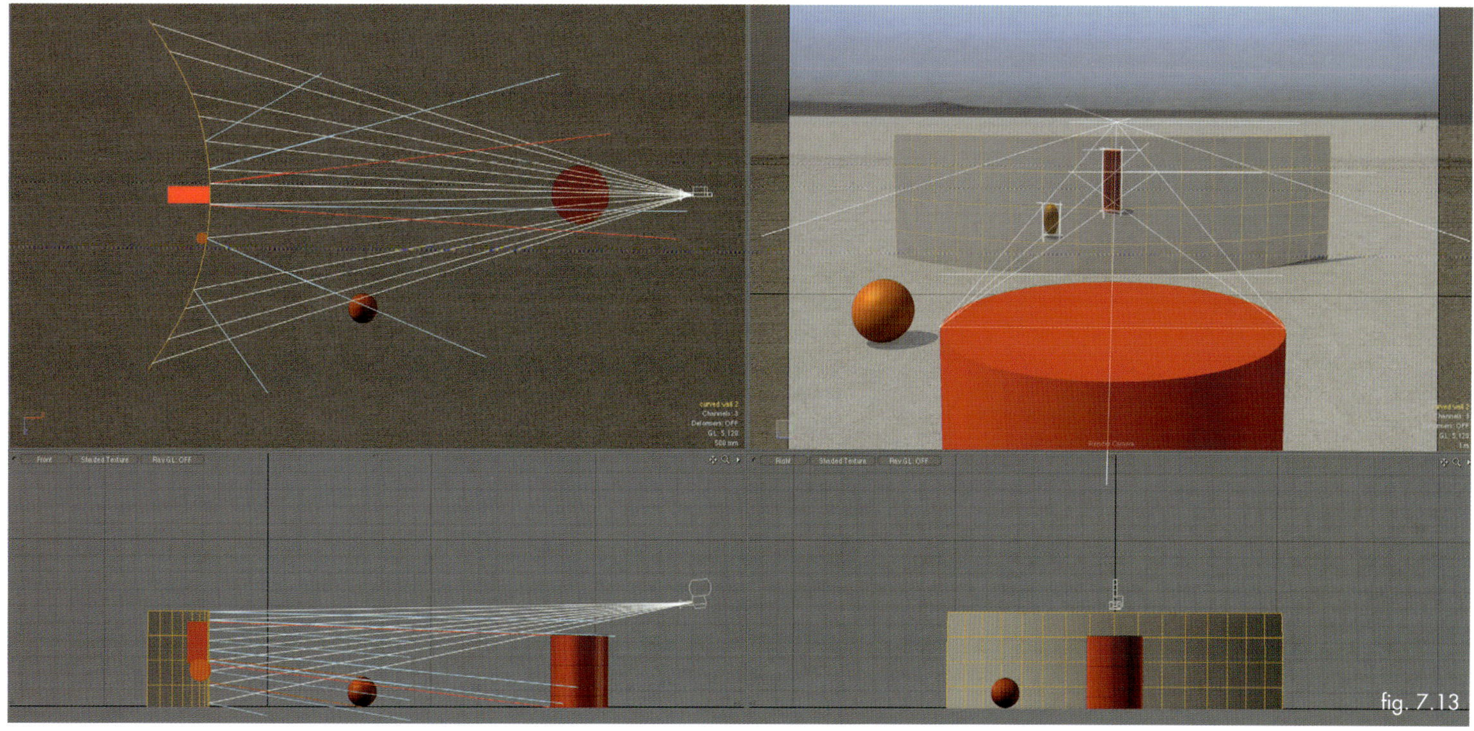

fig. 7.13

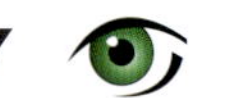

fig. 7.14

VALUE

0 5 10

% of matte surface visibility

100% MATTE

% of reflective surface visibility

100% MIRRORED

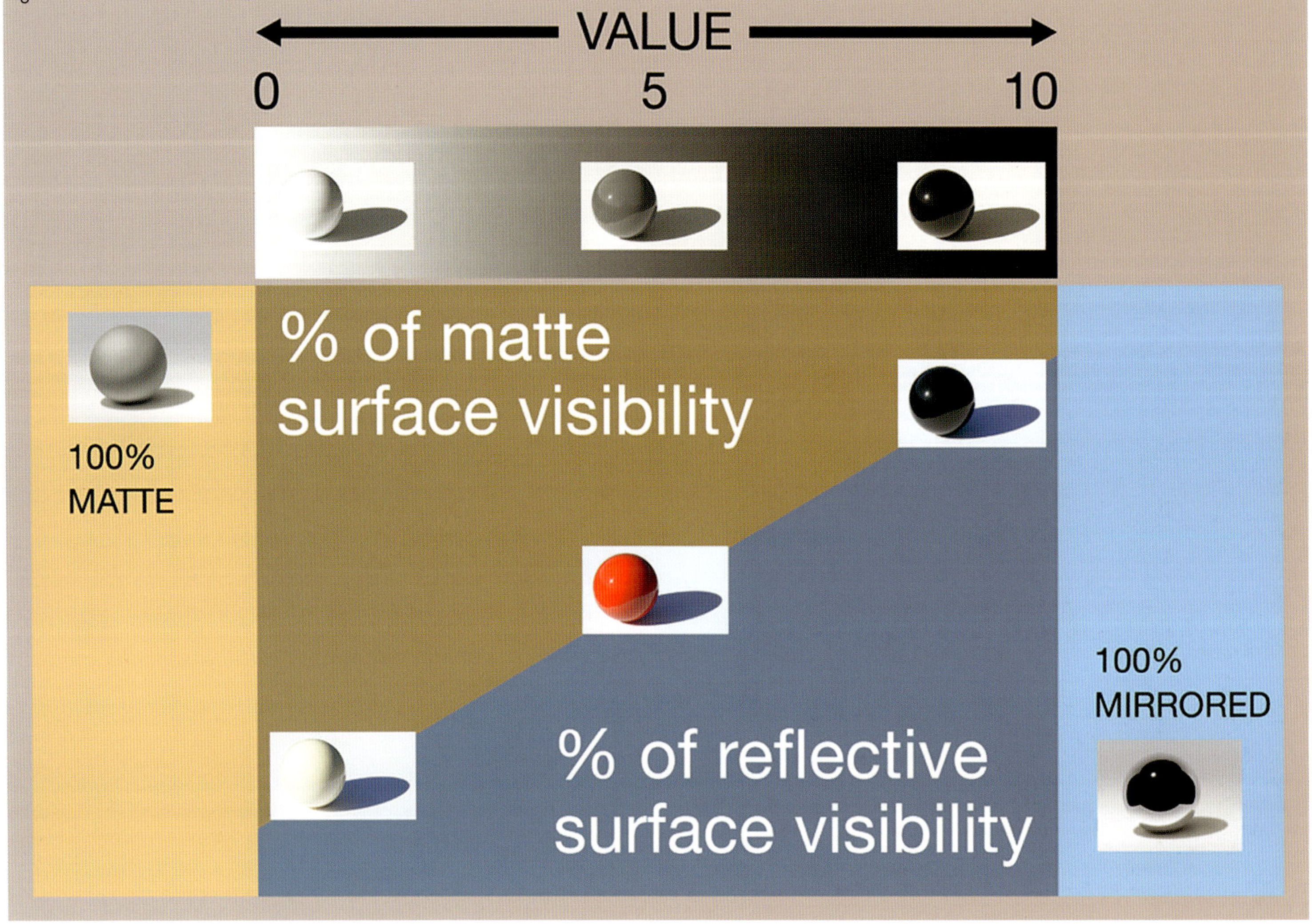

Let's look at these billiard balls of different colors and values. The chart above shows a 100% matte surface on the left and a 100% mirrored chrome surface on the right. Across the top is a sliding scale based on value, from white to black. When looking at shiny objects of different values, the brain perceives the amount of matte surface versus reflective surface differently. When looking at a white object, the brain perceives a mostly matte surface with a little bit of reflection. When looking at a middle-value color like the red ball, the brain perceives about equal amounts of matte surface and reflection. When looking at a black object, the brain perceives a very shiny surface with just a little matte surface information. In other words, almost no core shadow (or light side) is visible on the black ball, whereas the white ball seems to have a strong core shadow. This observation helps to build two different reflective rendering strategies, based on value. *The lighter the value, the less the perceived reflection*. The mind-bending part of this is that the balls all have the exact same reflective surface, and the environment around each ball is exactly the same. So what changed? Nothing is different about the physical surface of the white ball versus the black ball. What changes is the *perceived* visual strength of the reflection – not the true strength, but the perceived strength. Why is that? Mostly because when things reflect off of objects, the lighter values appear more opaque, like the sun's glare off of a car's windshield. When reflecting something dark, like a black building or a dark tree, the eye mostly sees through the reflection.

The light values shining off the black ball appear as white on black, very high contrast. The light values shining off the white ball appear as white on white, very low contrast. Both balls are reflecting the surrounding light values; it is just more noticeable on the black ball.

A good strategy is to prioritize matte-surface rendering when the surface is a light value, and reflective rendering when it is a dark value. Another consideration is that the matte surface underneath the reflections shows more value change on a lighter-value surface than on a darker one, which was explained in Chapter 03. Looking at the chart, which ball is the easiest to render?

Black is the easiest color to render.
What about white? Harder.

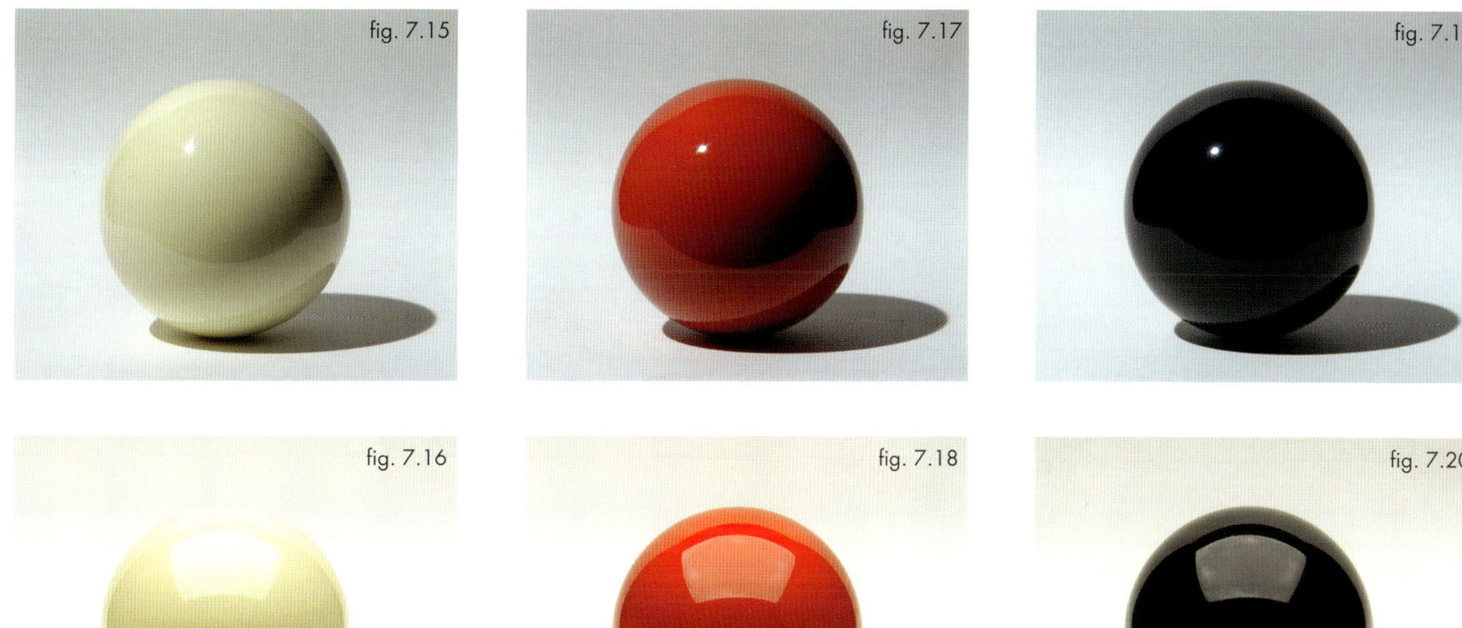
fig. 7.15 fig. 7.17 fig. 7.19 fig. 7.16 fig. 7.18 fig. 7.20

Red? The hardest of all.

Red is the most difficult because equal amounts of reflectivity and matte-surface qualities have to be rendered and the value changes of each can cancel out the other. Forms painted red, really shiny red, can appear very flat. Right where there might be a shadow, there might also be a lighter reflection, canceling it out.

Figs. 7.15 – 7.20 are a series of photographs that were taken to demonstrate exactly this phenomenon. When studying reflections, it is crucial to understand the environment and the lighting. Fig. 7.21 represents the environment for these examples: an indoor studio where it is black all the way around the sphere, with a white piece of paper going up the wall behind the sphere. There is a single spotlight placed to the left, and slightly in front, of the balls in the top row. The balls in the bottom row were lit with a softbox.

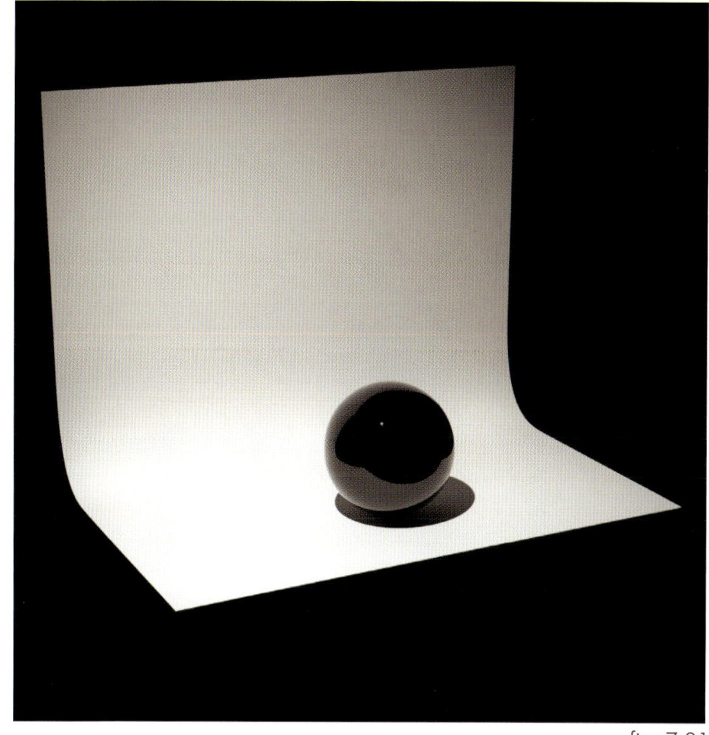
fig. 7.21

The light is brighter right around the object, creating a gradation on the paper background, thus there is a gradation of the environment's reflection of the paper onto the ball. Notice how the core shadow and the gradations are much more apparent across the lighter-value balls. On the black ball, the core shadow is invisible! For this reason, rendering a shiny black sphere is pretty easy. Doing the matte-surface part of the rendering of a shiny black ball only requires filling a circle shape with black. Super easy!

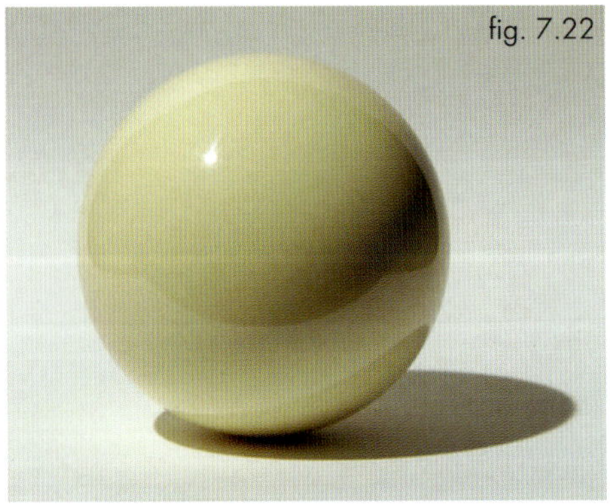
fig. 7.22

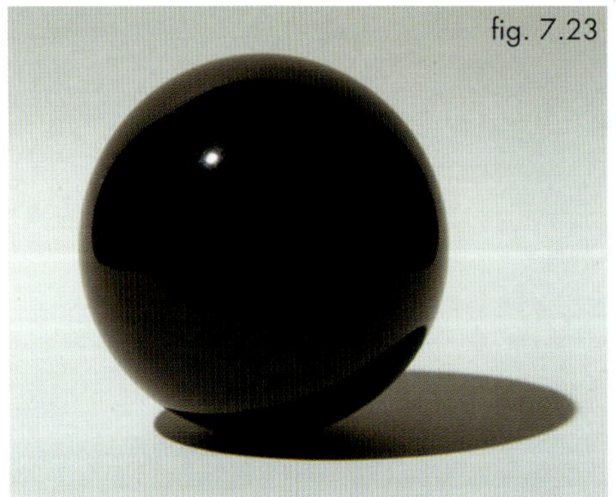
fig. 7.23

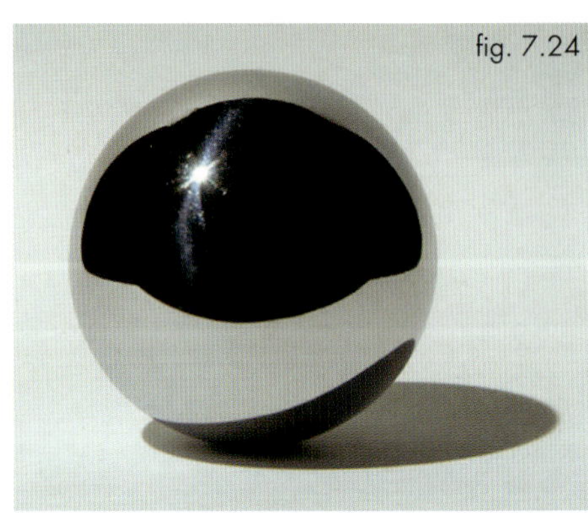
fig. 7.24

Chrome also has no matte-surface qualities to worry about, no core shadow, no light side, and no shadow side. If a mirror is very clean and you try to cast a shadow on it, the cast shadow will not be seen. Cast shadows do not appear on a mirror because all the light striking it is reflected away. If it is dusty, yes, a light shadow starts to appear. Generally speaking, though, perfectly clean chrome surfaces have no cast shadow information.

What happens in nature, on all surfaces other than chrome, is that the strength of the reflection changes based on the line of sight's angle of incidence into that surface. This is the most important thing to understand fundamentally with reflections on any shiny surface.

The strength of the reflection of the environment changes and grows stronger as the surface rolls away from the line of sight. Therefore, a rendering strategy for how to show form with the strength of the reflection emerges. Three-dimensional computer-rendering programs commonly refer to this phenomenon as the "Fresnel Effect" (pronounced freh-nel´, the "s" is silent).

When rendering chrome, take whatever colors and values are in the environment around the object, color-sample from that area, and just drop them right onto the surface of the object. Here is a basic rule to remember with reflections: **Where the line of sight is perpendicular to a surface,** which would be the center of the sphere in this case, **the reflection is weakest. Where the line of sight passes tangent to a surface, the reflection is strongest.**

Chrome is the exception to the rule. It is 100% shiny from any angle.

Before rendering anything shiny, the environment around the object must be known. Otherwise, the colors and textures from around the object cannot be mapped into the reflective layer. Any time anything shiny is starting to be rendered, first and foremost, design the environment.

As seen on the previous page, the ground plane is a white piece of paper that bends up the wall, with a single light source in the foreground. For these examples the reflections were controlled and simplified.

Generally speaking a simplified environment helps the viewer. When presenting a rendering to a client, the most important thing is the design, of course. But also try to show what color and material will make up the final product.

It is important to communicate if an object is shiny, but not so shiny that it reflects an oak tree next to an old barn, a scarecrow, a flock of birds and jet trails in the sky. That might look more photographic, but it could make it very confusing for the viewer. As a bonus, simplifying and controlling the environment can help work get done more quickly.

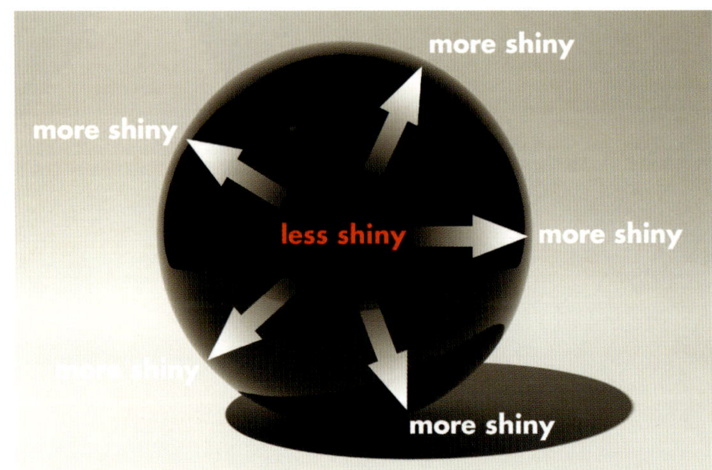

fig. 7.25

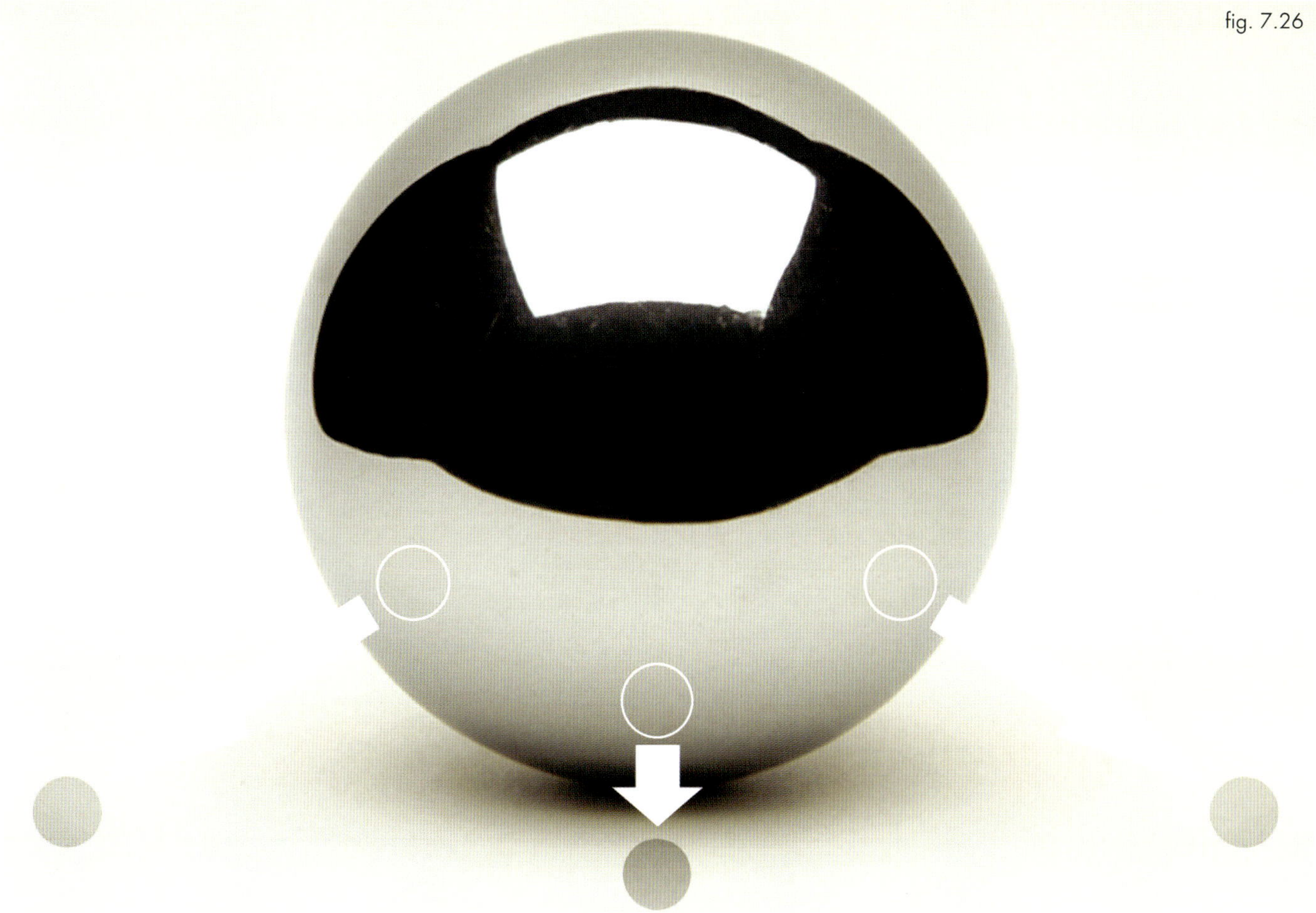

fig. 7.26

Chrome is usually one-half to one full value step darker than what it is reflecting. Look at the sampled swatches in fig. 7.26. They are a half step darker in value, maybe even one full step darker in value than the surrounding environment reflection. So when rendering chrome, literally everything can be taken from around the object, dropped onto the surface and then that whole layer can be darkened to provide a decent chrome surface. This simple rendering technique is easy to remember and also easy to do in Photoshop.

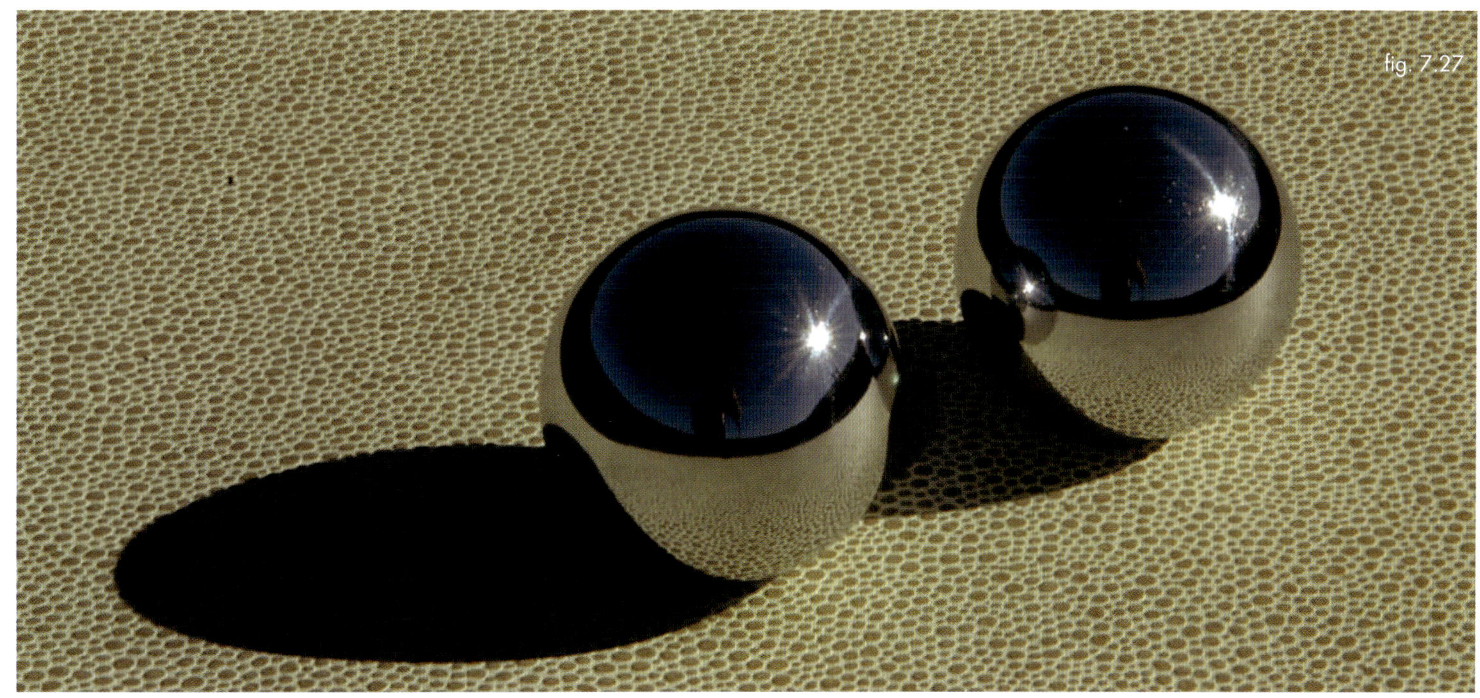

fig. 7.27

Fig. 7.30 is a photograph of a real black billiard ball. Fig. 7.31 is a rendered black ball created in Photoshop by layering the chrome ball (fig. 7.28) over the flat, black disc (fig. 7.29) and then carefully erasing it in places to vary the strength of the reflections due to the angle of incidence, or the Fresnel Effect. As the black sphere is revealed underneath, the way they mix together is very close to what happens in nature.

Just turning down the opacity of the chrome layer to start to make it look black, without erasing or masking the center of it, does not quite work. There will not be a big gradation across the sphere's surface, therefore a form change will not be observed. What should be done is to take the reflection layer and, with a soft-edged eraser, start erasing the chrome layer wherever the line of sight is perpendicular to the form. Where the line of sight is tangent to the form, leave the reflection layer more opaque. Changing the opacity of this reflection by erasing it reveals the black sphere underneath, but only where the line of sight is most perpendicular to the surface. This is exactly where the reflection would be least shiny. As a by-product of erasing it this way, it is more opaque at the edges and more transparent in the middle. A gradation has been introduced. What happens when there is a value gradation/change across a surface? We see a form change.

Always render the reflection of the light source on a separate layer. It is too strong to put on the same layer as the rest of the environment's reflection. The light source in this case is just a white light bulb with little scratches and imperfections showing up in the reflection. Look at how close it comes to reality without having to do much of anything, (fig. 7.31).

On the opposite page (fig. 7.33) a scene was staged in MODO to show the Fresnel Effect in action. Observe what happens to the same shiny plane when introducing small ripples or waves across it, like water (fig. 7.34). In the side view of the scene (fig. 7.32), the ripples are quite small, but with shiny surfaces, small changes to the angle of incidence of the viewer's line of sight can have large effects. The front of each ripple facing the viewer is less shiny because it is more perpendicular to the viewer than the back of each ripple. This causes variation in the strength of the reflection, which, in turn, results in a value change, which makes the brain perceive a form change.
Cool, huh?

In fig. 7.33, look at the A and B swatches pulled off of the reflective plane. The background behind the plane is a constant color and value, but the color and value of the swatches is very different. Once again, this is the Fresnel Effect in action.

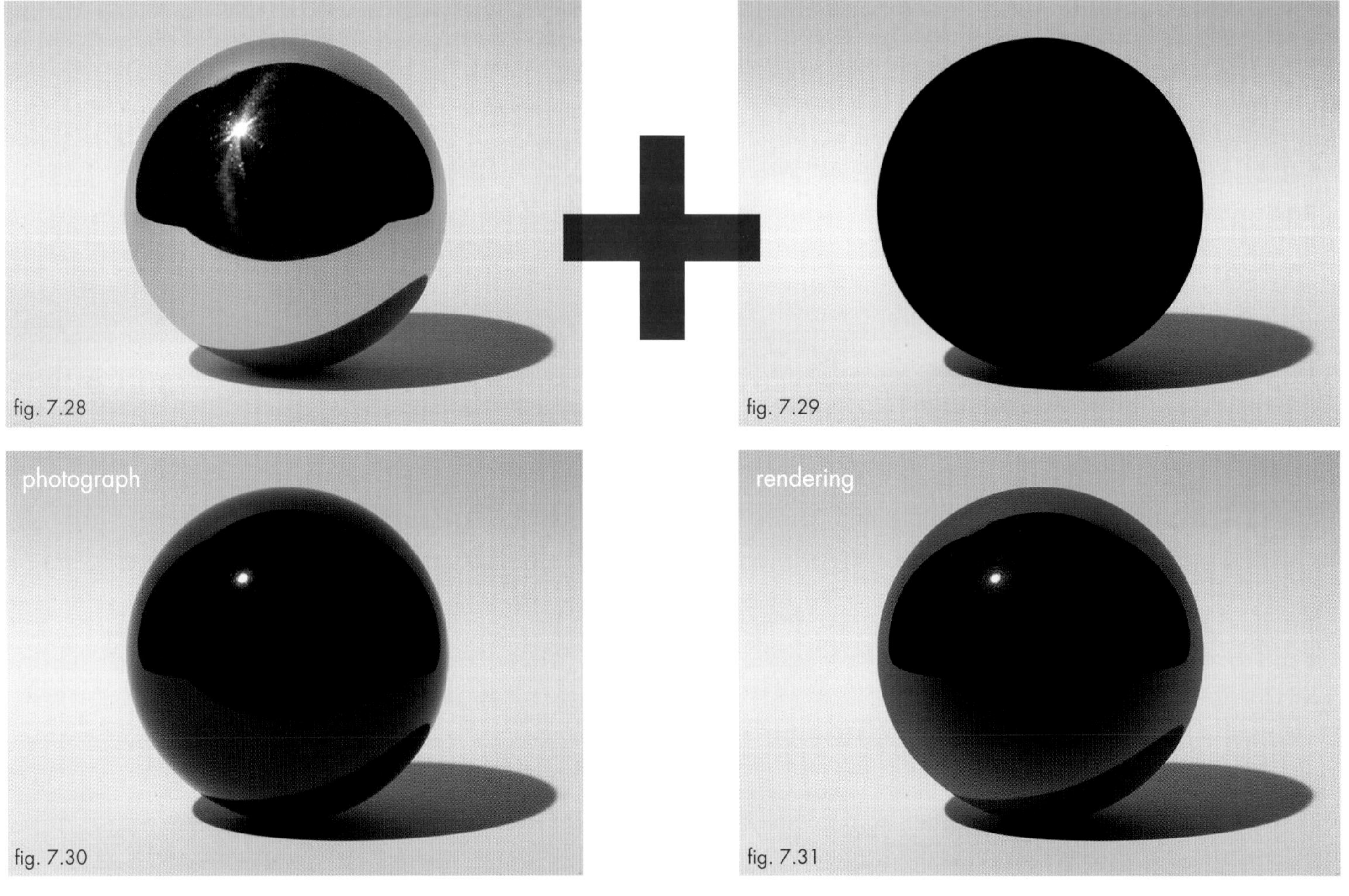

fig. 7.28

fig. 7.29

fig. 7.30

fig. 7.31

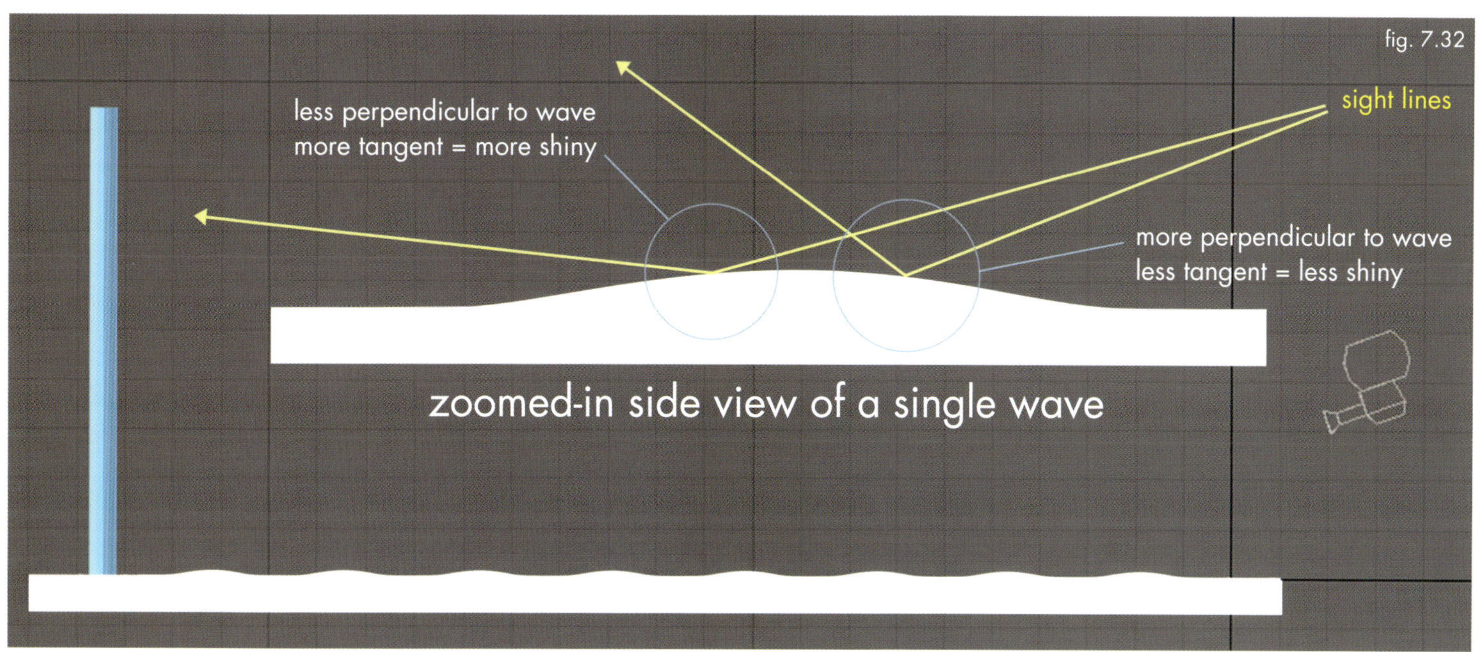
fig. 7.32
sight lines
less perpendicular to wave
more tangent = more shiny
more perpendicular to wave
less tangent = less shiny
zoomed-in side view of a single wave

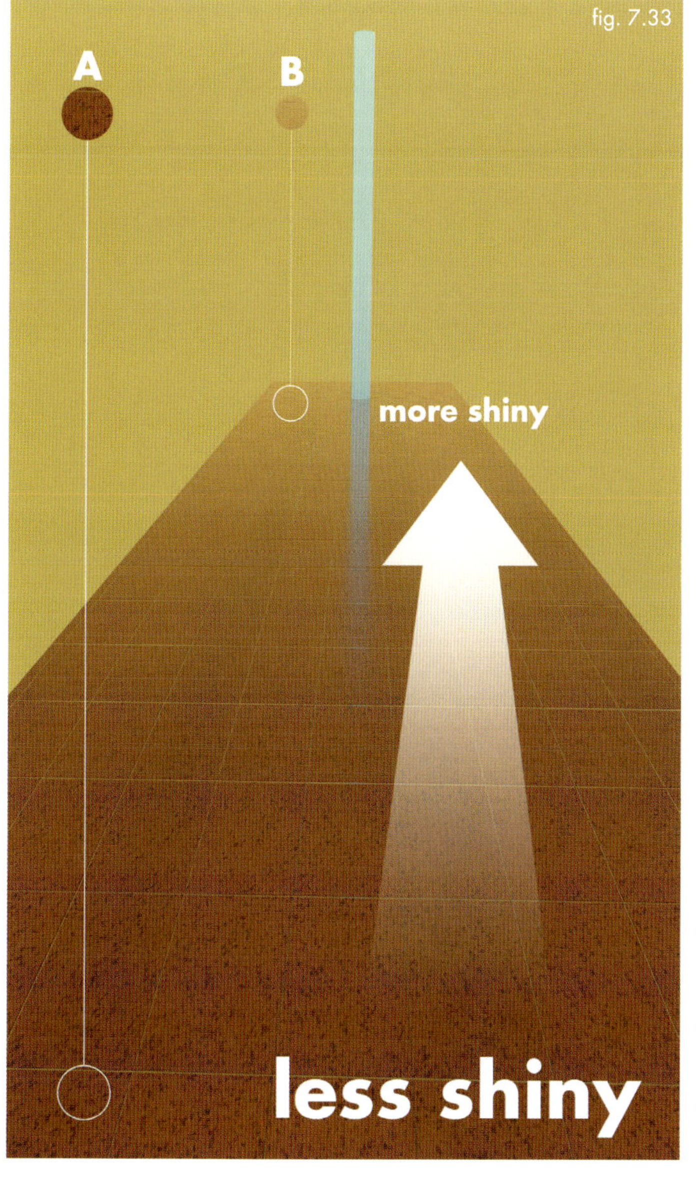
fig. 7.33
A
B
more shiny
less shiny

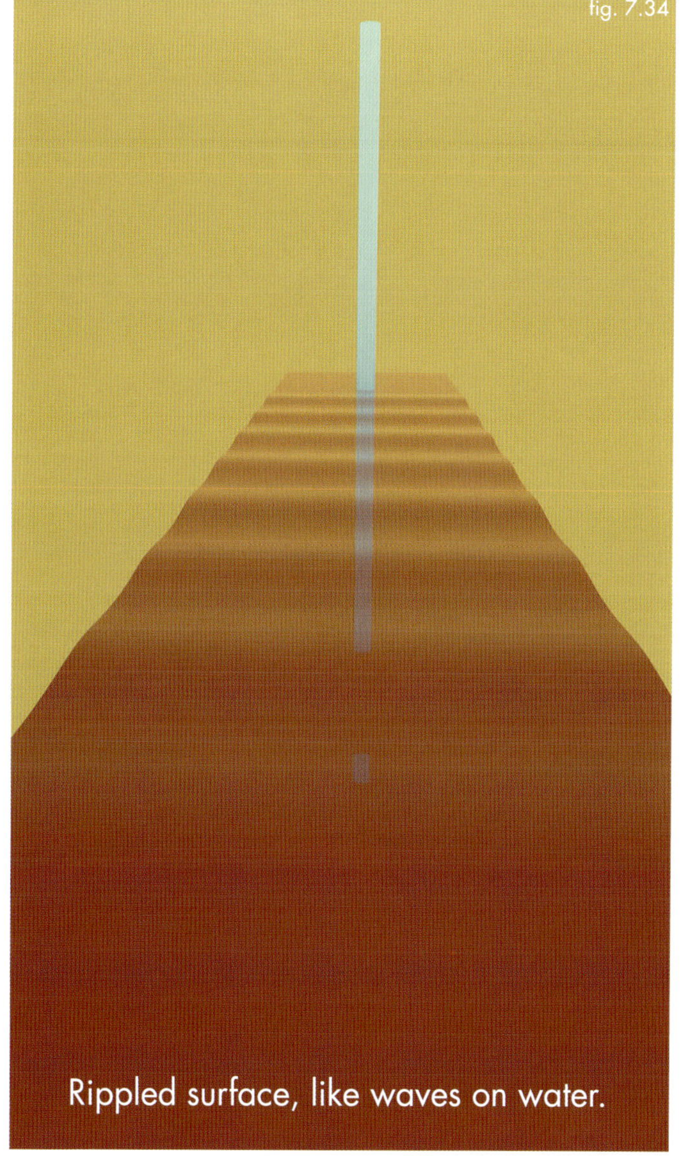
fig. 7.34
Rippled surface, like waves on water.

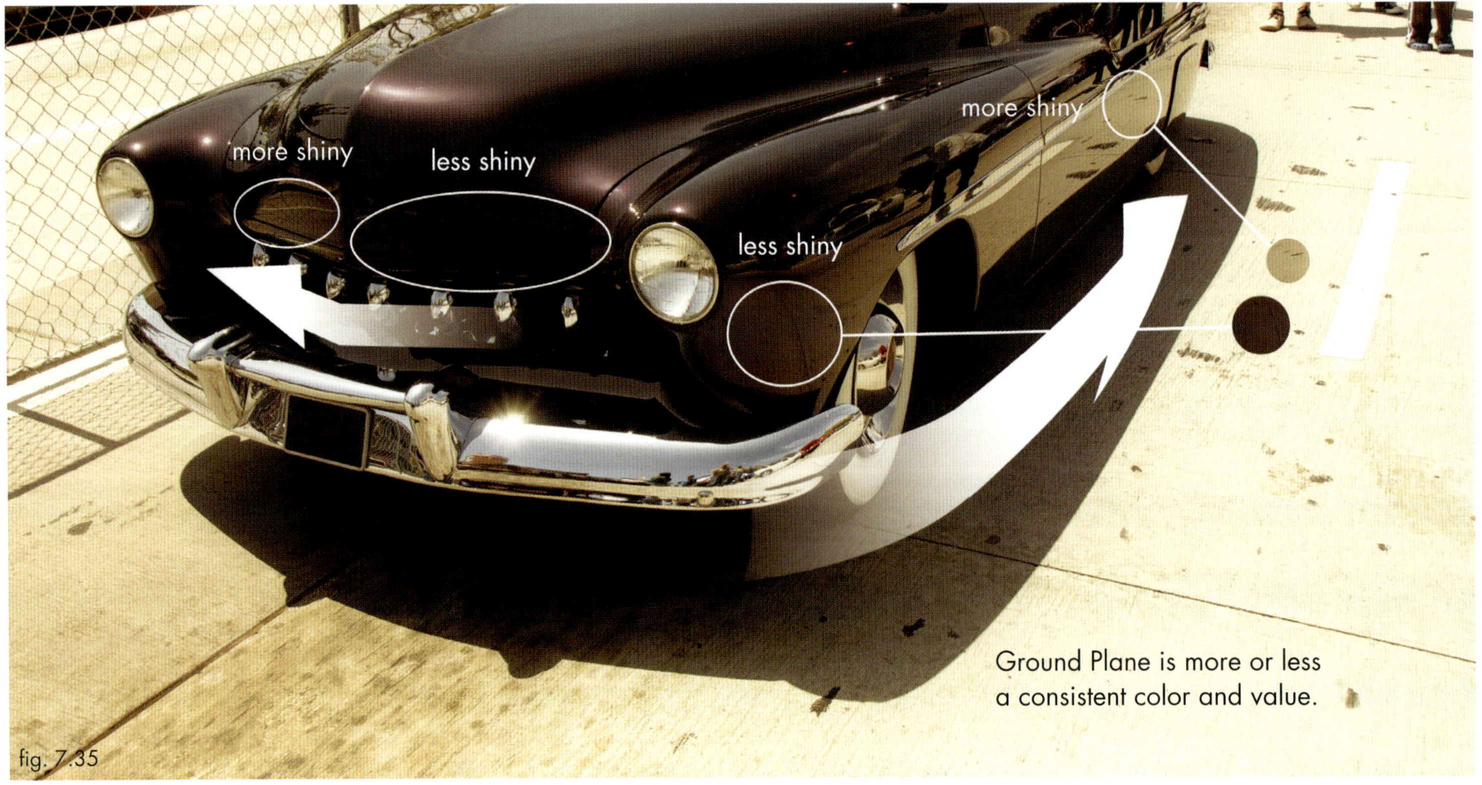

fig. 7.35

The Fresnel Effect can be observed quite easily. Next time traffic is at a standstill, look at the car just ahead. A clean black car is perfect. The back of the car will appear less shiny than the sides of the car, exactly like the billiard balls, because from this vantage point the back of the car is perpendicular to your line of sight. The back will look like a matte surface, like base-coat paint, almost all black. As you look at the side of the car, it appears very shiny, almost like chrome, because the angle of incidence of your line of sight is tangent to the car's surface. A light road reflecting onto a dark car provides the most contrast. As the strength of the reflection changes, value appears to change. The end result is that the brain, once again, is tricked into perceiving a form change.

Why does reflection strength change like that? At the microscopic level, even a polished, smooth surface has very small hills and valleys. When looking straight into them, perpendicular to the surface, the line of sight is bouncing in all different directions off of those irregularities. When looking tangent to it, the irregularities align and many more of the sight lines bounce off the surface in the same direction, which causes the Fresnel Effect.

Figs. 7.35, 7.36 and 7.37 clearly show the Fresnel Effect. In fig. 7.40 on the opposite page, this effect was rendered on the materials to make them look more reflective, simply by sampling the colors and values from the background behind the object and then rendering them over the base-paint layers in Photoshop. It is a simplified look, but the forms still read and the materials look shiny. Job done.

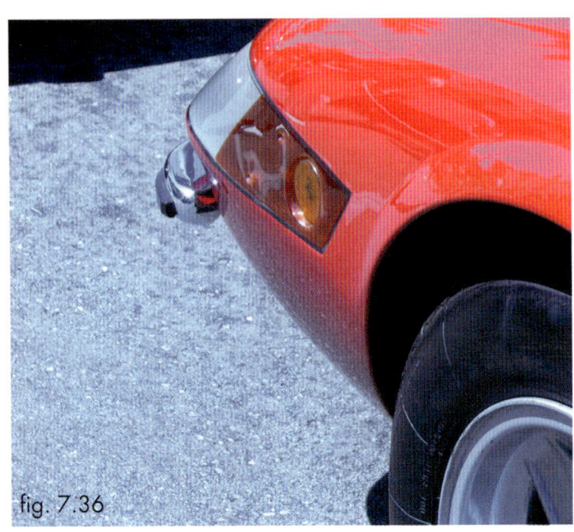
fig. 7.36

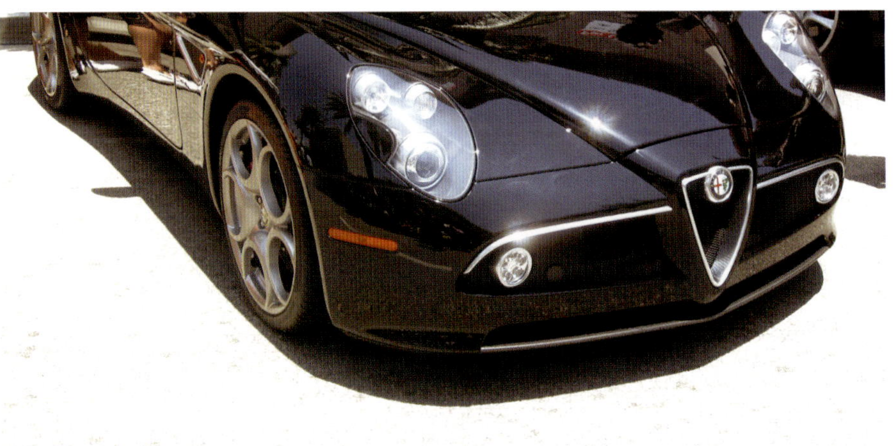
fig. 7.37

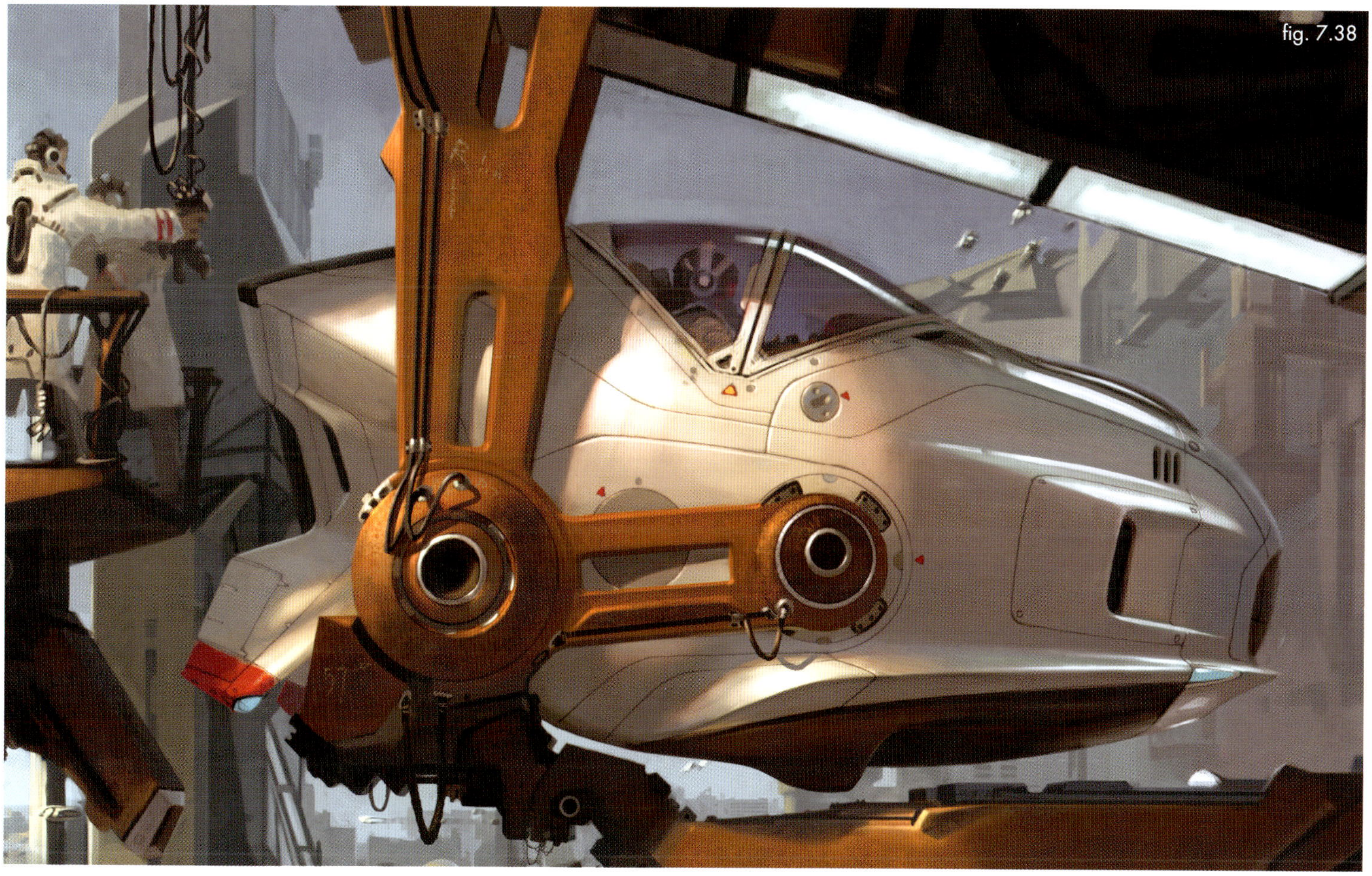

fig. 7.38

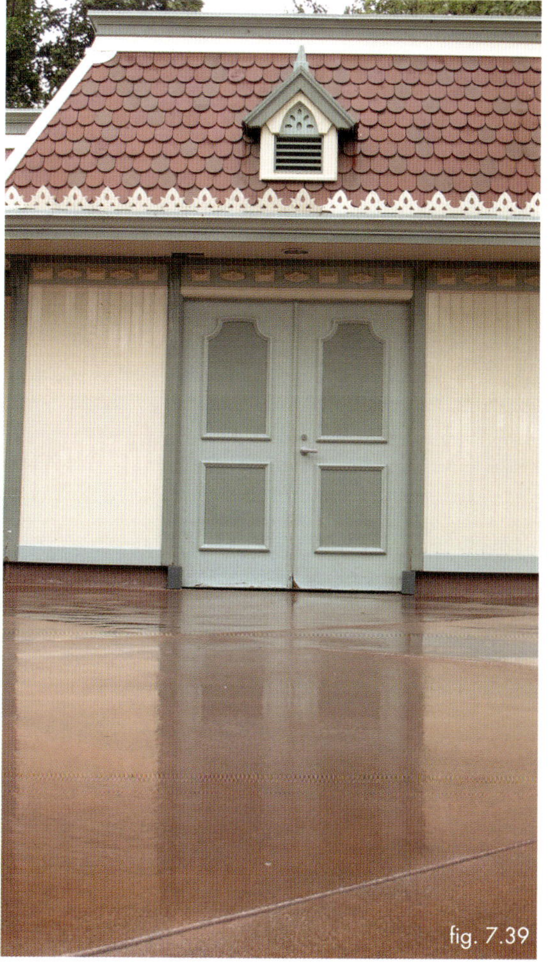

fig. 7.39

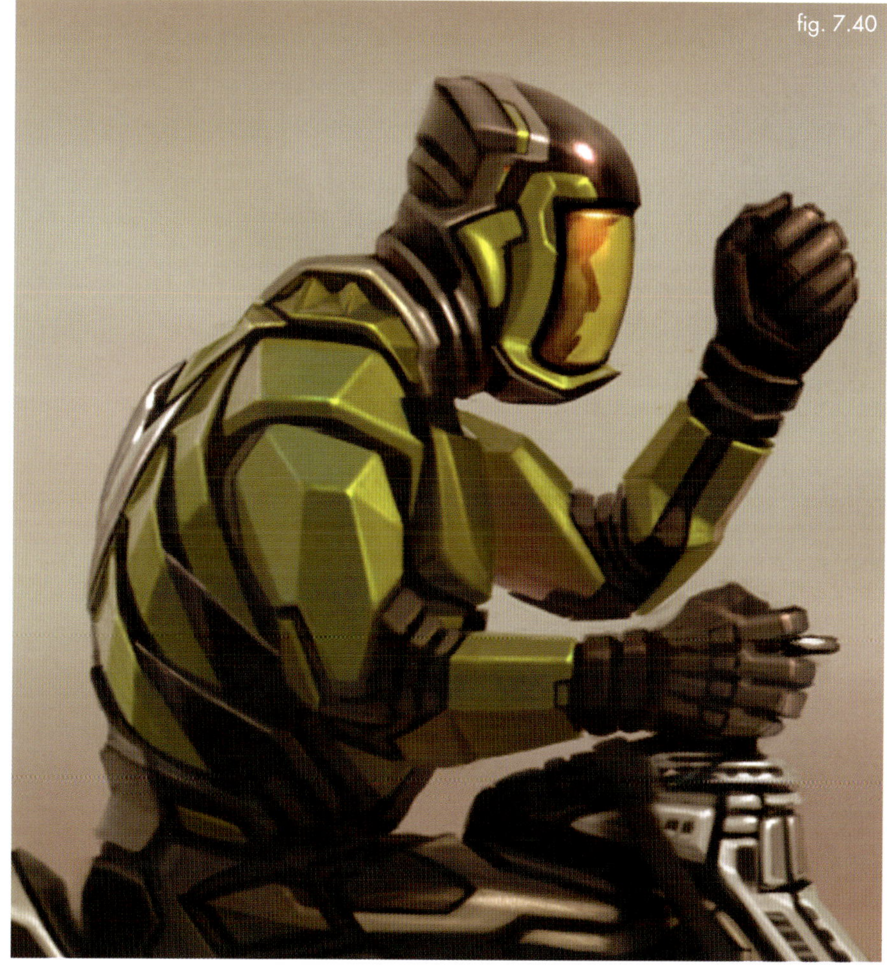

fig. 7.40

With water, the same thing occurs as with a clear coat over matte paint. Looking vertically down into water, it usually looks like its true color. When looking out across the water, it becomes brighter and brighter, more the light color of the sky (fig. 7.42). The same effect also happens over glass (fig. 7.43). All reflections behave the same and can be rendered this way on a separate layer, or layers, above the base matte surface in any 2D digital rendering program. If the rendering is not being done digitally with layers, go ahead and pre-visualize the end result and render the base matte surface and the reflective clear coat at the same time. This is how it was done before digital rendering came along!

Earlier in the book, the physics of light illuminating on matte surfaces have been pretty straightforward and hopefully easy enough to understand. What if the surface of the object is not matte, but reflective? The same physics unfortunately do not apply.

The form of a shiny object is communicated by varying the strength of the reflections. Look at the dimple and the dome on the black sphere (fig. 7.44). The value changes are subtle,

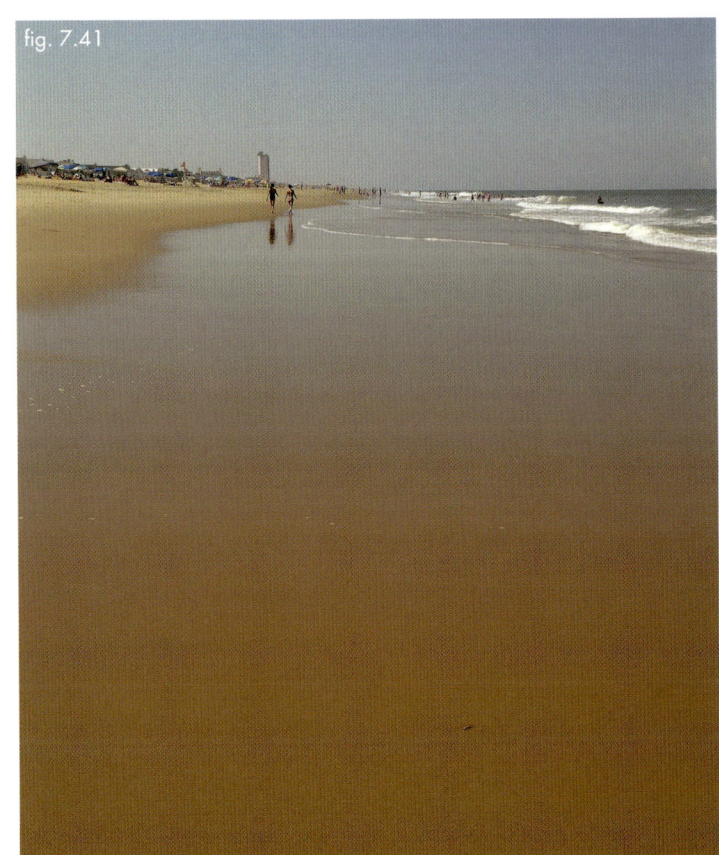

fig. 7.41

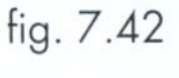

fig. 7.42

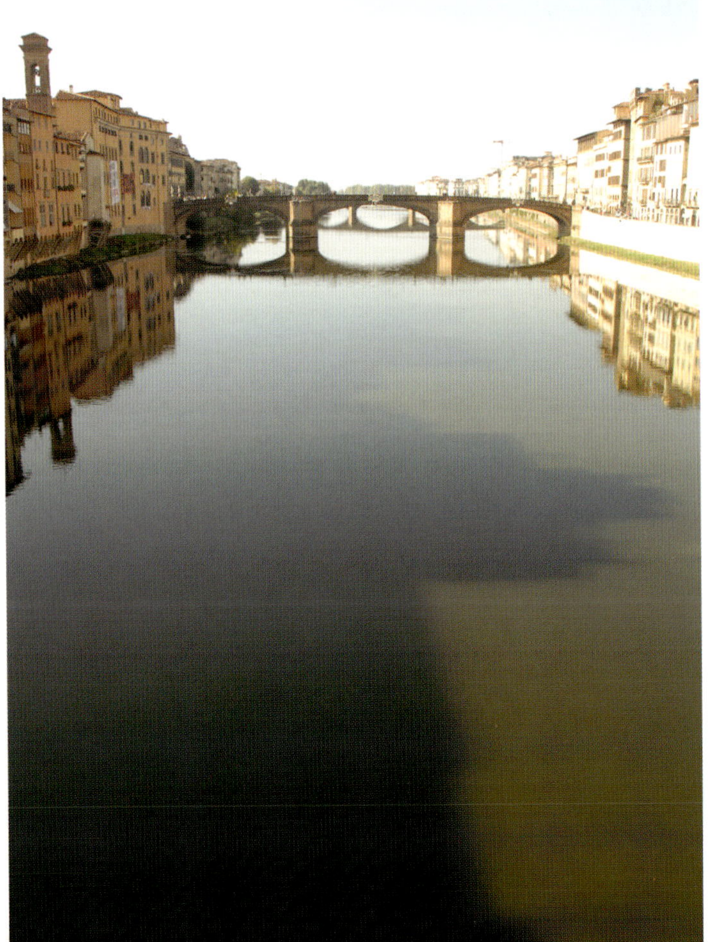

fig. 7.43

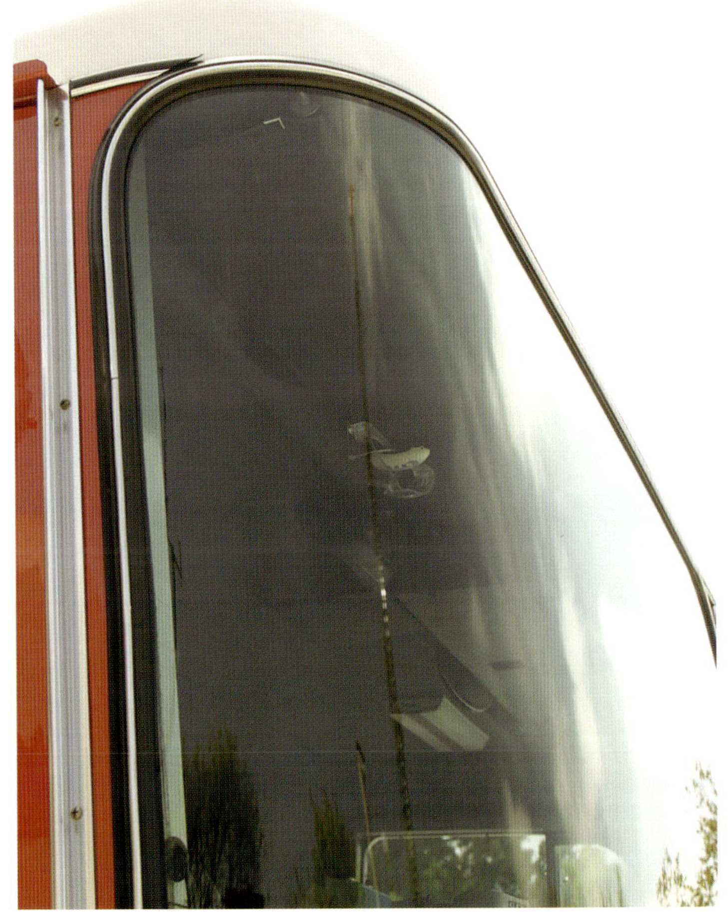

but the brain still understands the forms. Hopefully by now it is obvious that there is a strategy to showing form on shiny objects based on how the line of sight intersects that surface. It does not have anything to do with where the light source is coming from. It only has to do with the varying strength of the reflection of the surrounding environment. It is a very different rendering strategy and again the end result is a gradation, which tricks the brain into seeing the form. To make that surface "go bump" with a dimple or a dome, it has to have a gradation on it. If the value changes, the brain sees the form.

Here comes the tough part. What if the surface is chrome?

The form changes are not seen as easily. If the environment has gradations in it, like the sky, there will still be gradations in the reflection. Referring back to the photograph of the chrome sphere (p. 171, fig. 7.26), the piece of paper that surrounds the ball has a gradation on it, so this environmental gradation is visible in the reflection on the chrome sphere. That gradation shows up on the surface, but not as a result of the strength of the reflection changing.

Observe how the reflections behave in the photos of the cars below, (figs. 7.45 and 7.46). Think about how they can be rendered, perhaps by setting up a couple of layers. If the cars had truly matte surfaces, they would be lighter toward the top where the surfaces are more perpendicular to the light rays, and darker on the sides where they are more tangent. But these gradations are upside down. They are the reverse of what happens on matte surfaces. Remember, there is not much, if any, matte-surface information to worry about with these darker colors. The fender, in fig. 7.46, has a curving surface that rolls away from the line of sight so it gets shinier near the silhouette of the form. Observe how the same physics happen independently of scale. The same logic applies for the lip running around the wheel-well opening and the bottom of the car (fig. 7.46). Observe how less shiny areas show more body color, while in the shinier areas, there is more of a "chrome" look.

fig. 7.45

So when trying to render reflective surfaces, it is a really good idea to get a bunch of props and stage some photographic examples, like the ones shown throughout this book.

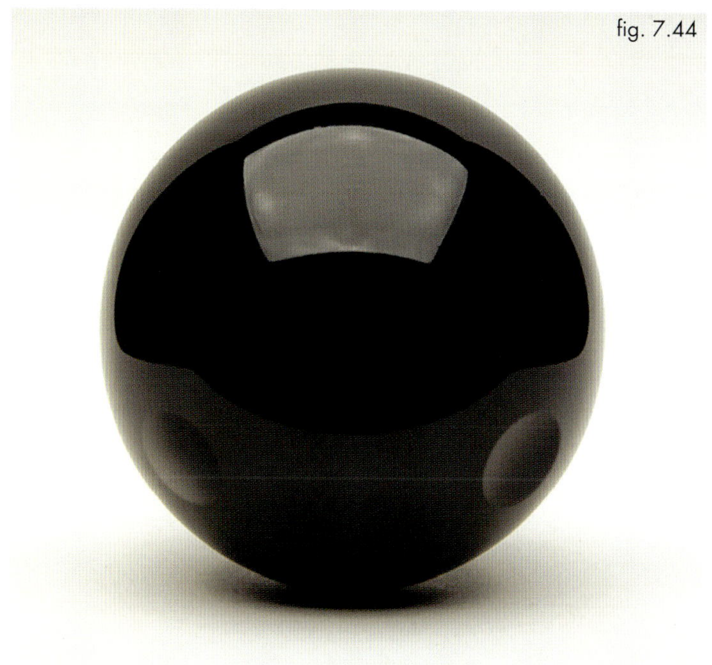

fig. 7.44

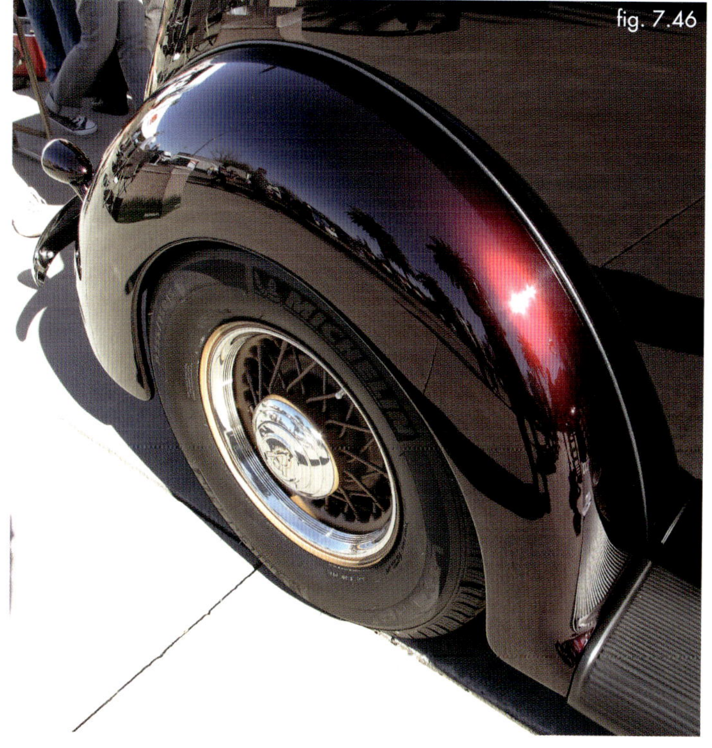

fig. 7.46

When shiny shapes touch or overlap what they are reflecting, they become the same color and value of that surface. Shiny objects become a mirror of their environment. Working hard to preserve the silhouette of an object will not look natural or real. When rendering shiny shapes, go ahead and lose their silhouettes, merging them with the background. That is what we observe in nature, and that is what will happen if the shiny object being designed is ever built. Outlining shiny objects with strong silhouettes kills the illusion of realism, as it will almost never look like that in reality. Let the edges disappear, like the chrome truck bumper in fig. 7.47. Artists have a tendency to outline everything like a line drawing, but that does not look natural. If retaining the silhouette of the form is more important than rendering it true to life, do not give the object a reflective surface finish. The surface needs to be dulled down to justify showing the silhouette.

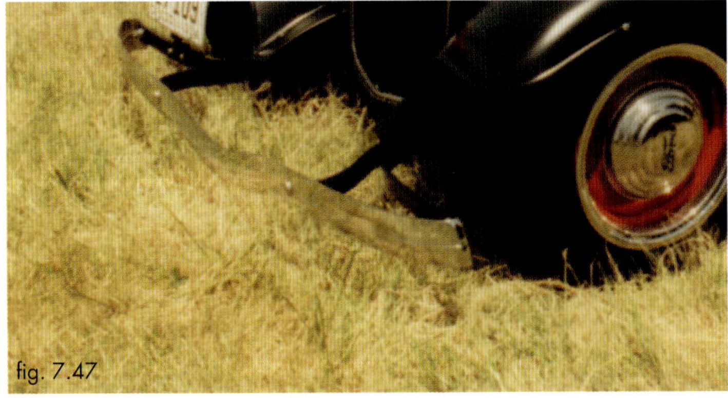
fig. 7.47

The carbon-fiber helmet shell (fig. 7.48) has a little bit of something else happening in the base layer. It is not so much about matte-surface shading anymore because there is a carbon weave under the clear coat. Literally, the clear coat is directly on top of the carbon weave and the reflection of the environment is on this clear-coat layer. Now look downward at the silhouette of the shell. It takes on the color and value of its environment. The middle of the helmet form looks much more black than the surface near the silhouette, which appears much more orange. See it roll away, just like a sphere that has been squashed a little bit.

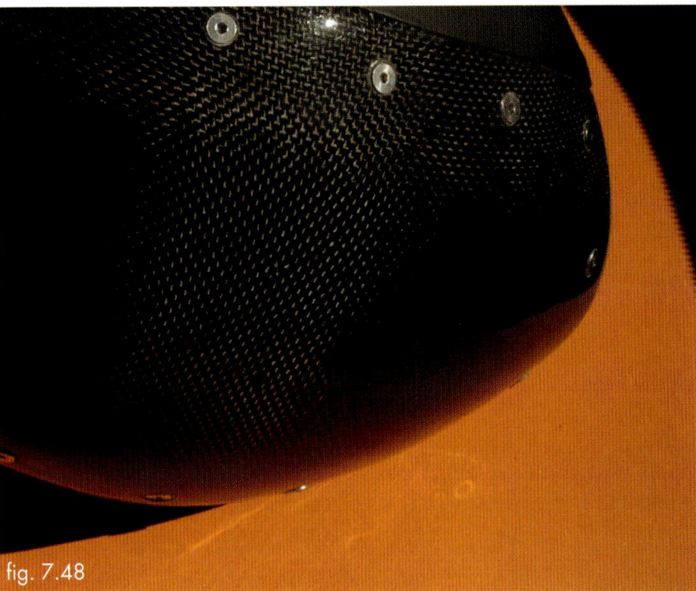
fig. 7.48

The 2D digital rendering strategy would be to put a new reflective layer over the top of the carbon-fiber layer. In this case, use the "cloning tool" to put the ground right over the base layer and then erase/mask the ground-reflection layer based on how the line of sight's angle of incidence changes. This is a rendering strategy to achieve the same type of reflection as seen in the photo. It is a little difficult to understand at first, but eventually you will notice it everywhere on a wide variety of surfaces, and then it will hopefully really click.

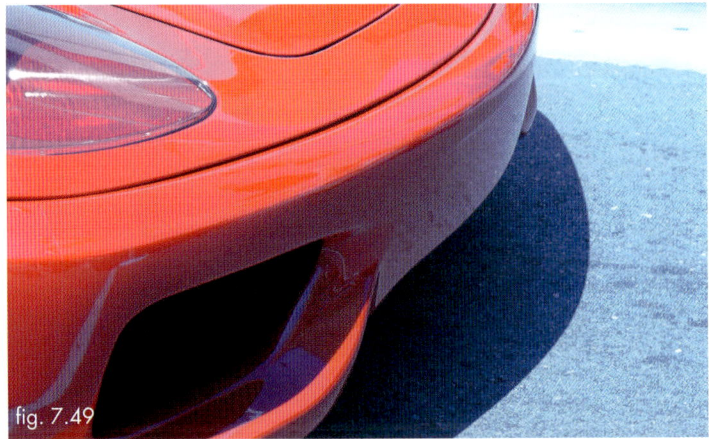
fig. 7.49

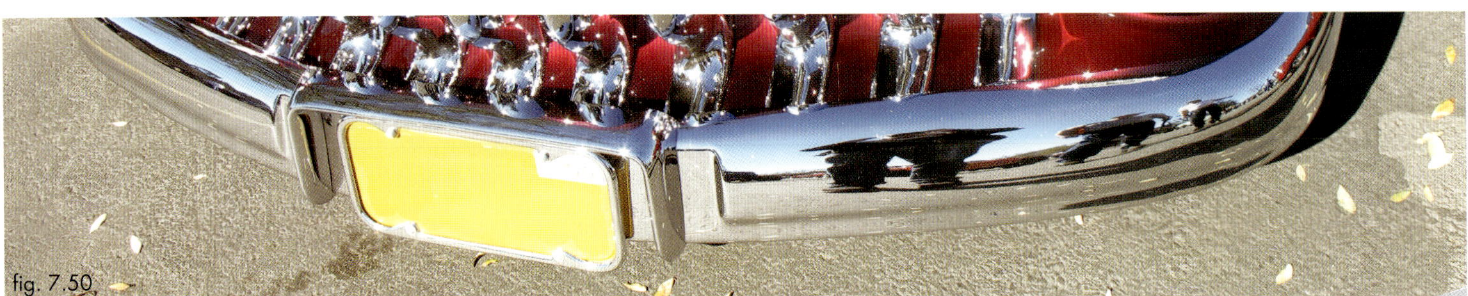
fig. 7.50

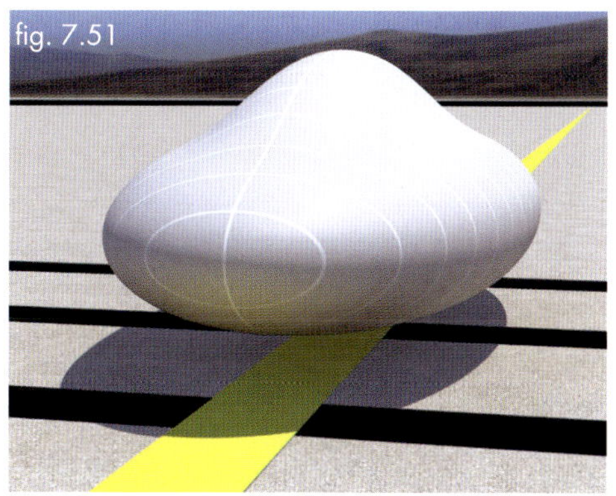
fig. 7.51

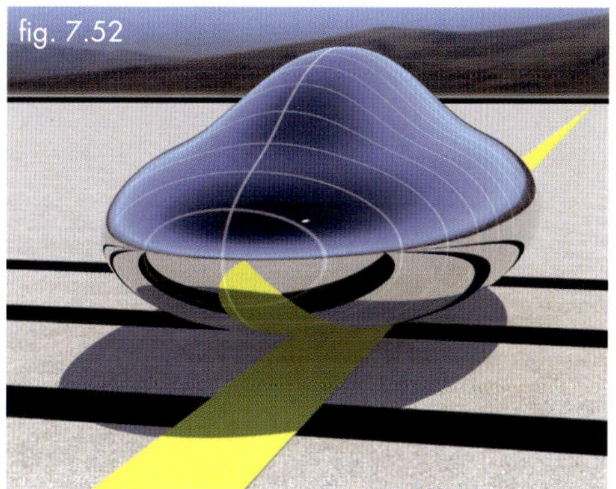
fig. 7.52

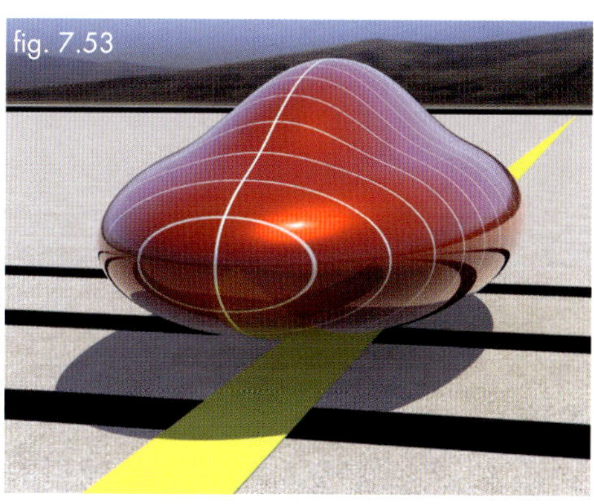
fig. 7.53

Building a digital rendering with layers is so much easier than having to render all the layers at once, like when working with traditional mediums. Above is a matte surface (fig. 7.51) and a chrome surface (fig. 7.52). Notice how much more visually complex the combined layers are on the red surface (fig.7.53).

The red surface also has some metal flake in the base layer, which will be explained in detail on page 210. It is important to observe how the matte base layer is visible at differing degrees underneath the clear-coat reflections, while also appearing to be red.

Below are the digital layers from the rendering demo on page 201. All of these layers look as they do in the render, except for the reflections, which were put on a black background so they would be more visible here. Where the image is black, the real reflection layer is transparent.

fig. 7.54

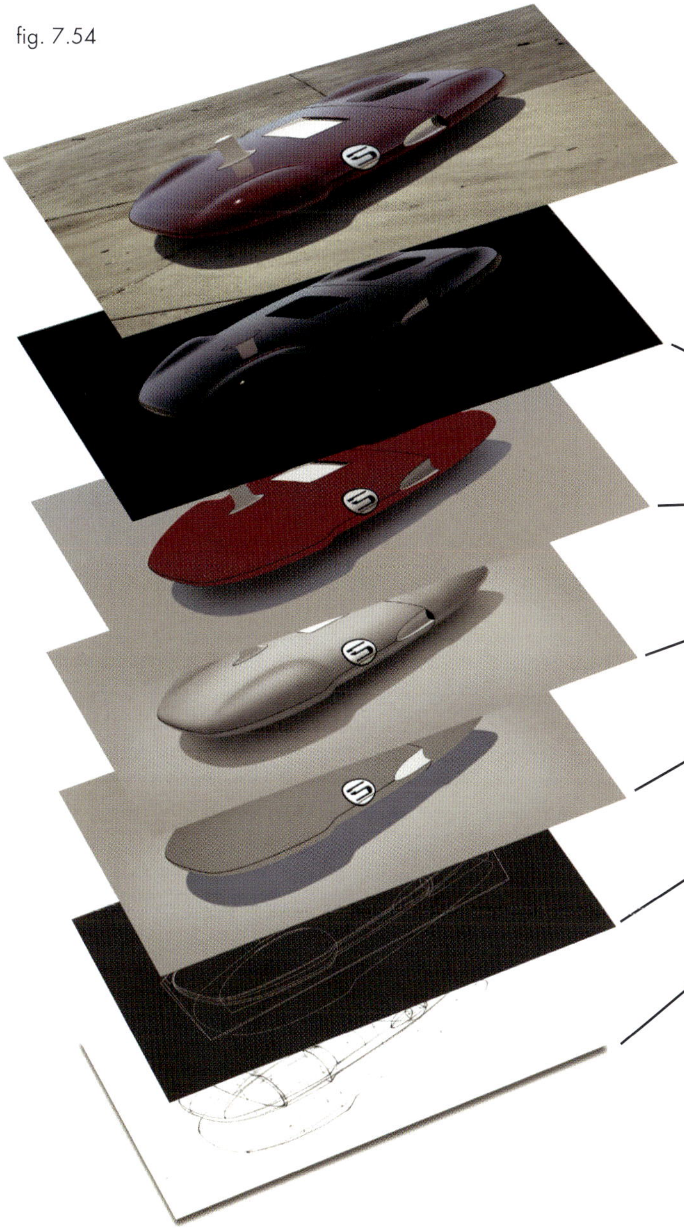

final composite image

reflections

colorized layers

value to communicate form

layer building

paths

line drawing

REFLECTION FLIPPING 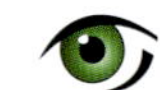

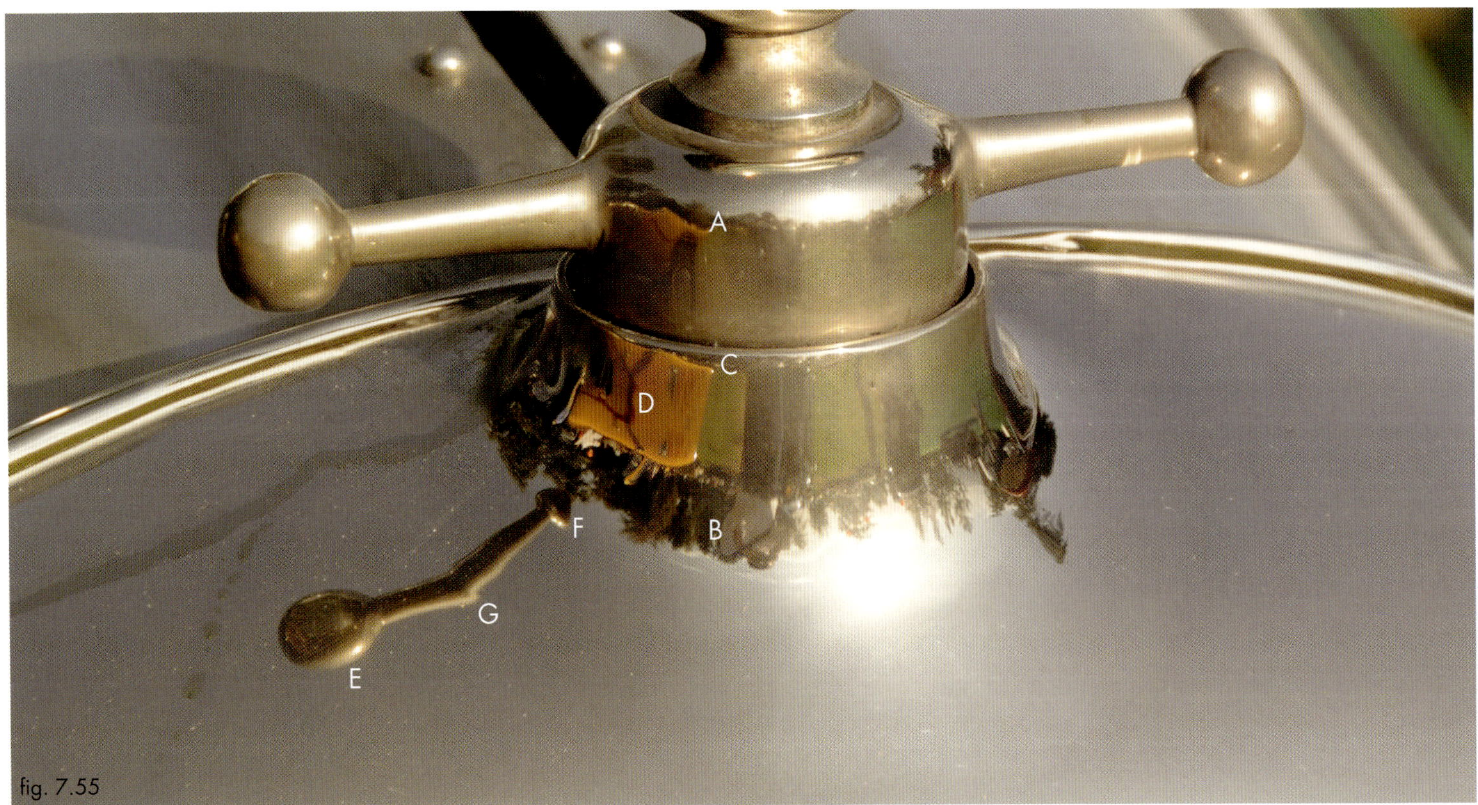

fig. 7.55

Reflection flipping is a very common occurrence when the line of sight bounces off a shiny concave surface. The reflecting sight lines make the reflection of the environment appear upside down. In fig. 7.55, there are two obvious reflections of the people standing around this part of the car. In the top radius, at point A, the people appear right-side up, just compressed, as expected because the surface is convex. At point B, the same people appear stretched because the surface here is concave. Whether convex or concave, the sight lines are bouncing back to the same surrounding environment so that the reflections will be the same, just inverted. Look very closely at point C, it shows the same reflection as the one at point A, but it is even more vertically compressed because the radius is even smaller. The inverted reflections between points B and C must, at some point, melt together into one reflection, as the surface goes from concave to convex. This can be seen in the reflection of the yellow car at point D. The same thing is happening in the reflection of the ball end back into the lower surface.

Fig. 7.56 loosely maps out where the sight lines are bouncing in a simple cross-section view schematic. *Where the reflection of the ball end is near points E and F, (fig. 7.55) it is supported by the bouncing sight lines off of the surface at points E and F to the right where both sight lines bounce back to the same ball end.* At point G is where the sight lines bounce back to the same point on the arm extending out to the ball end. This is the point where the reflection flips.

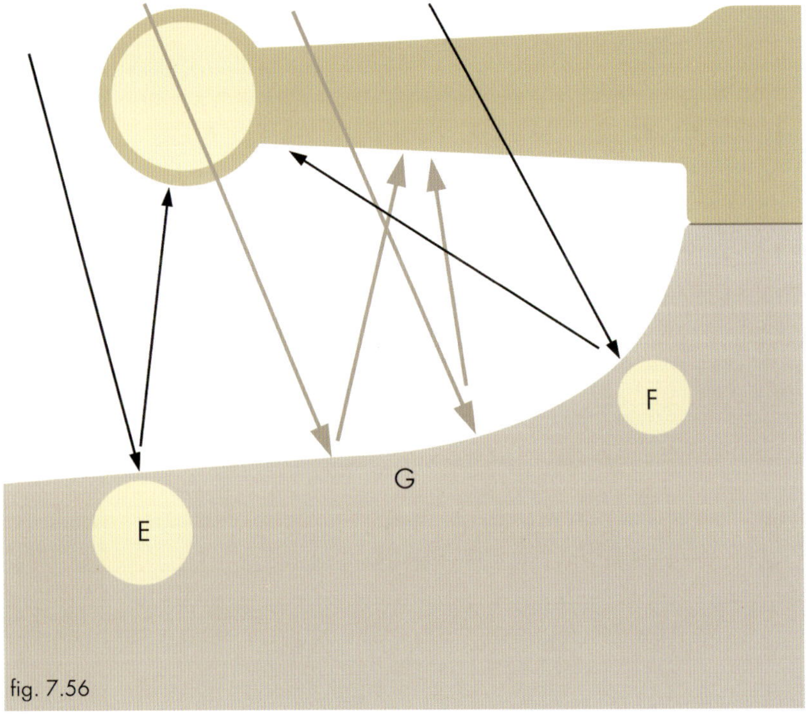

fig. 7.56

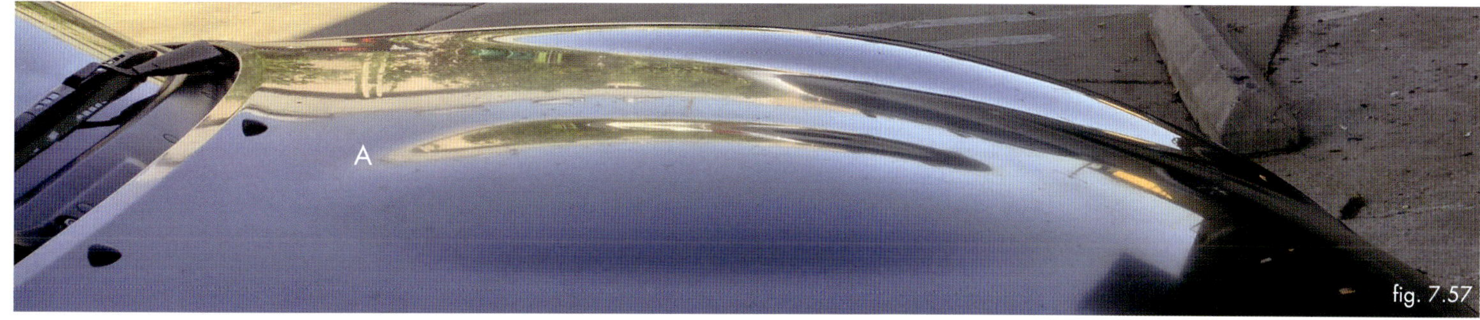

fig. 7.57

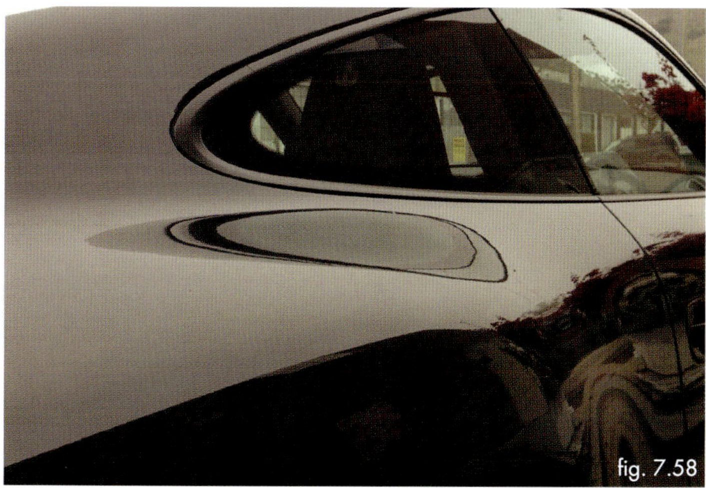
fig. 7.58

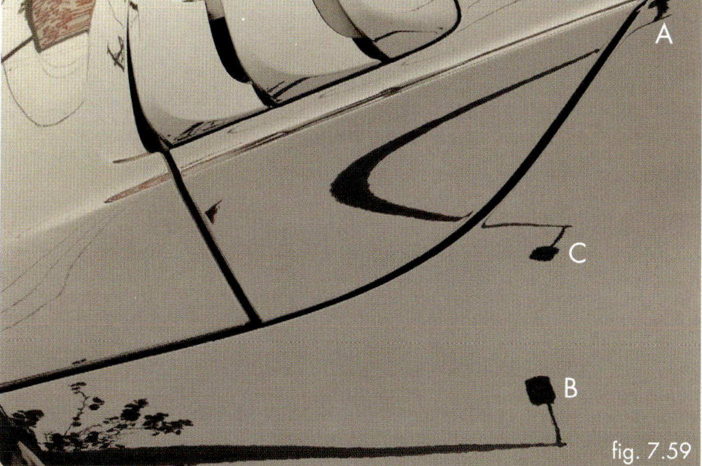

fig. 7.59

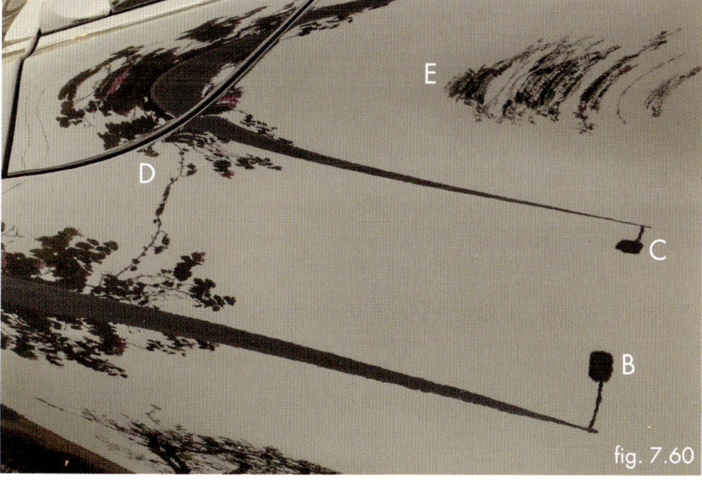

fig. 7.60

Pools and puddles are a more automotive-specific terminology. They refer to reflections of the environment or other parts of the car itself that tend to be like floating islands, pools or puddles on the reflective surface, due to reflection flipping. Even reflections of the sky can technically be thought of this way, like on the far-side fender bulge in fig. 7.57. More commonly though, it is thought of as the reflections floating within the reflection of the sky, like the one floating just on the far side of the hood bulge (A) also in fig. 7.57.

In fig. 7.58 a reflection of the car's side window and a little bit of the car body behind it can be observed reflecting as a puddle or pool on the fender. Why do they look like this? It's all about the sections of the surface and where the sight lines bounce. In the case of the pools and puddles it is just like the reflection flipping on the facing page. At the top and bottom part of the pooling reflection, the sight lines bounce to the same part of the car, the window, defining the upper and lower parts of the reflection, just inverted. Through the middle of the reflection the sight lines bounce to exactly the same spot of the window making the reflection appear symmetrical and melting the lines around the window into one continuous line around the perimeter of the reflection pool.

Look at figs. 7.59 and 7.60 and it might be more obvious as the light pole that is reflecting is just like the ball end on the previous page. The same pole is being reflected on two convex forms at points A and B and then again in the concave form of C. The reflection in the concave section of C is upside down when compared to the ones on the A and B convex forms. In fig. 7.60 the mirrored reflections and the melting of the two into one can be seen at points D and E.

Try to find more of these types of reflections occurring around you to increase your awareness of them.

REFLECTIONS OF THE LIGHT 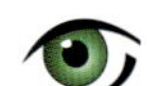

When a light source shines on a reflective form, it is almost always the most obvious part of the reflection layer. Thus, it deserves some special attention. Fig. 7.61 demonstrates very clearly that the position of the reflection of the light source can occur in a very different location than the lightest area of the matte-surface shading (passive highlight). Observe how close the reflection of the sun is to the core shadow of the stone sphere. Remember that the sun's reflection position is calculated by where the sight lines bounce off of the shiny form into the sky to where the sun is, not where the light rays strike perpendicular to the surface. That would be the way to calculate the matte-surface value of the lightest part of the sphere, which is just behind it and not visible from this angle. To test if something is a reflection or part of a matte surface, just move your head. If the values move around then a reflection is being observed. For example, if you were to walk around the spheres in the photograph, the core shadows and cast shadows would stay where they are but the reflections of the sun would move a lot. There is a big difference when thinking about reflective surfaces versus matte surfaces, even though most often they coexist on the same surface.

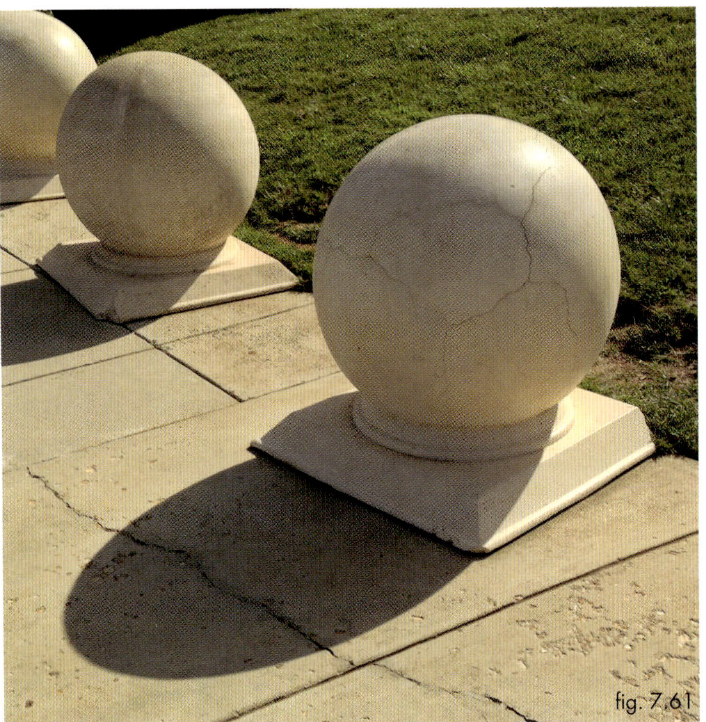

fig. 7.61

Fig. 7.62 shows how sensitive small scratches are on a shiny surface, and these are most obvious in the reflection of the light source, due to the high contrast of the reflection.

Fig. 7.63 is a nice example of the light-source reflection having a different shape because it is reflecting a series of long, narrow fluorescent light bulbs on the ceiling of an interior space. What is important to observe here is that the reflecting light lines help to communicate the form. They work like section lines in a line drawing, and the way they wrap and warp across the car's surface is an integral visual cue to the car's form. Designing the shape of the reflections, especially those of the light, cannot be emphasized enough.

fig. 7.62

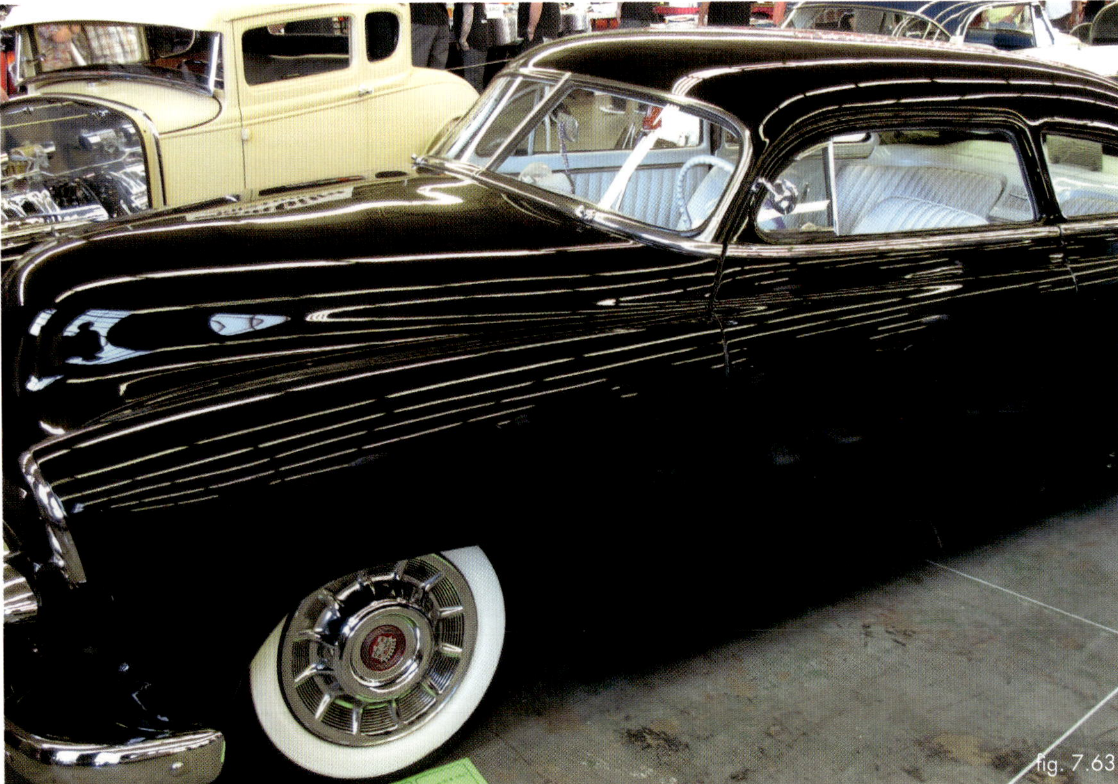

fig. 7.63

Rendering reflections of reflective surfaces is challenging because the correct true-to-life color and value of these types of reflections can be surprising. But with a little more knowledge they are much simpler to understand. Fig. 7.64 shows chrome velocity stacks (A), with a very green reflection on their sides. However, when these same chrome stacks reflect into the shiny black paint of the hood, they look blue (B). What is going on? This is a double-bounce reflection.

The simple section diagram (fig. 7.67) shows how the sight lines bounce off the sides of the nearly vertical chrome surfaces down to the green grass. Other sight lines bounce off the black paint and up into the chrome, and then bounce again up to the sky, which is blue. This is where the colors for those reflections come from, different parts of the same environment.

Figs. 7.65 and 7.66 show the same double bounces occurring. The reflections of both of the chrome parts show something yellow (C and D), but the reflections of the these chrome parts in the green paint are both blue (E and F), reflecting the sky in a double bounce. The easiest way to think about these types of reflections is done by sketching a simple section view, like fig. 7.67, to calculate the bounces. Another way is to imagine standing at the point of the first reflection bounce on the car paint and looking up to the chrome surfaces. Remembering the rule of "equal in–equal out," it is easy to imagine how that line of sight will bounce up into the sky, and not downward.

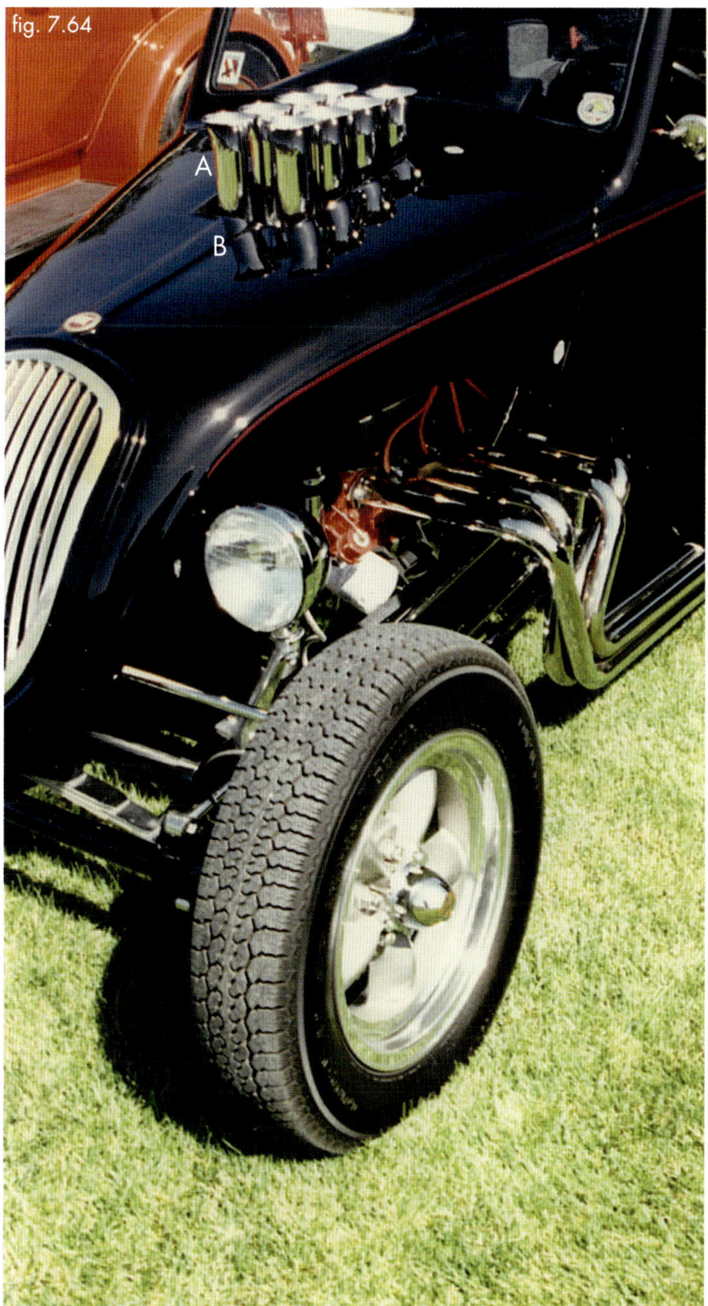

fig. 7.64

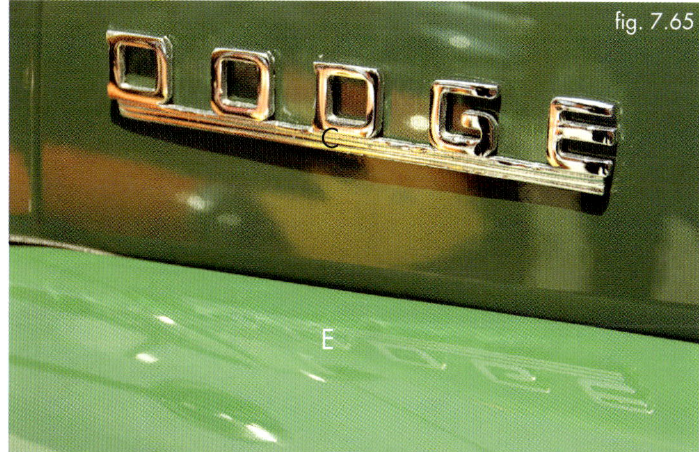

fig. 7.65

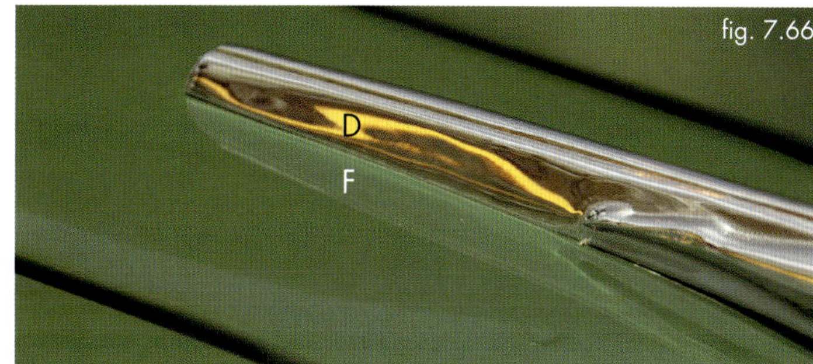

fig. 7.66

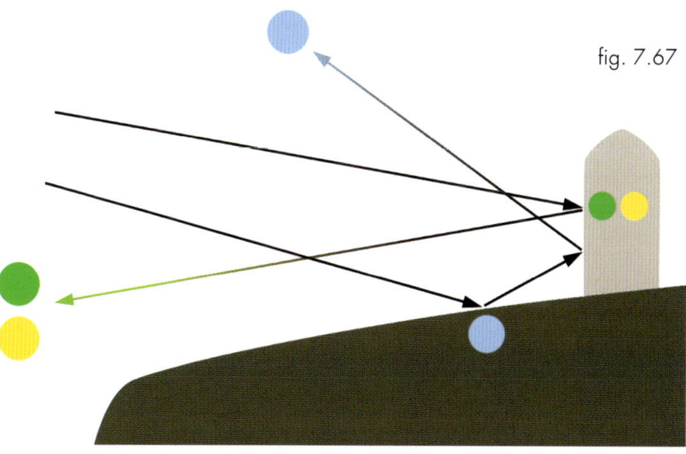
fig. 7.67

REFLECTIONS OVER GRAPHICS 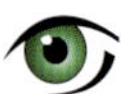

When rendering reflections, it is important to obscure any graphics that might be underneath them. In fig. 7.68 the flames extend along the side of the car but disappear as the strength of the reflection increases. In figs. 7.69 and 7.70 the reflections of the overhead indoor lights obscure the graphics in the base-paint layers. Also, note that the glare of the lights is so strong that even though the line of sight is very perpendicular to the tops of these cars, the reflections of the light sources are still very visible. Again, this is why it is a good idea to render the reflection of the light sources on a separate layer than those of the rest of the environment.

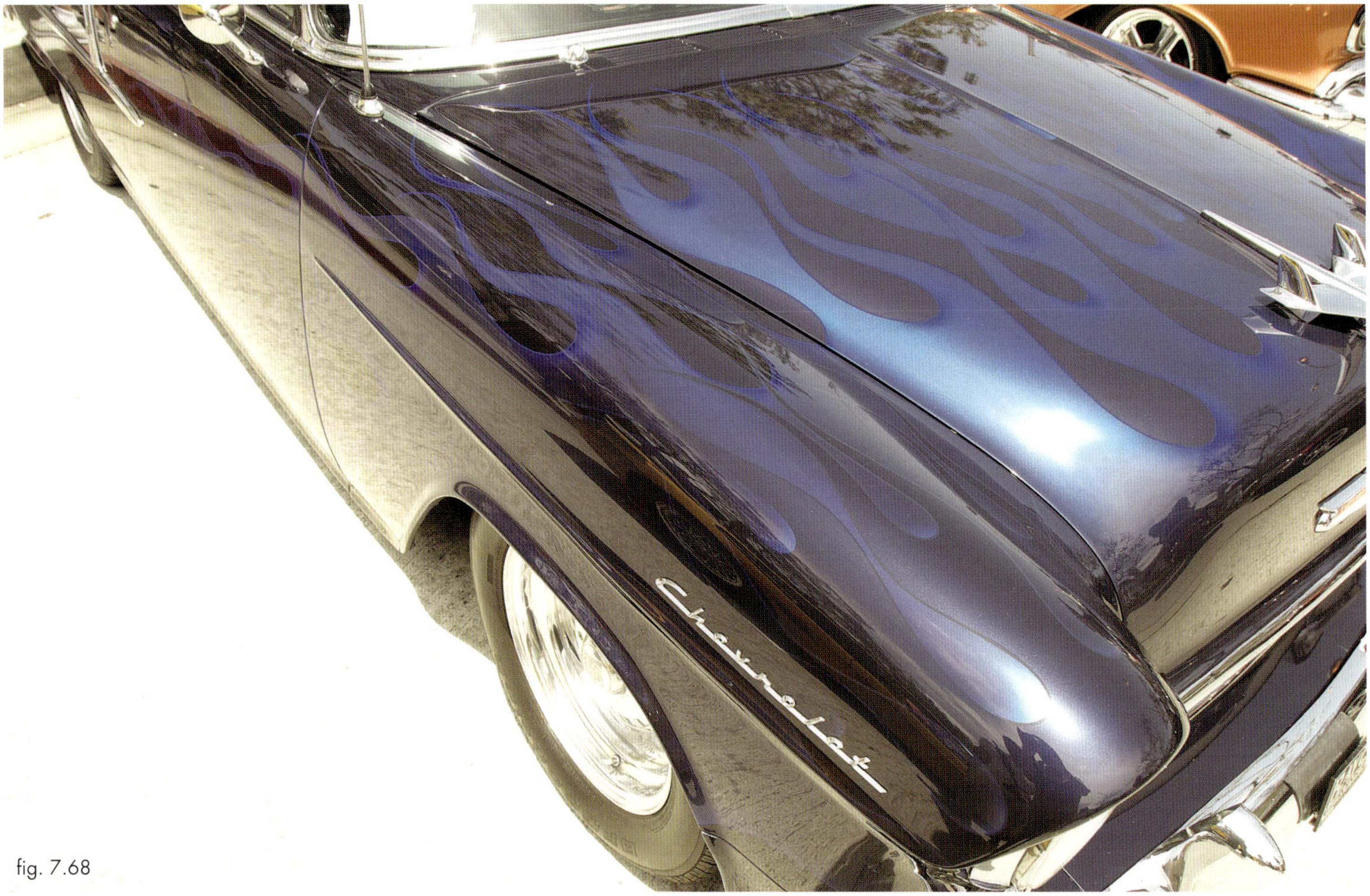

fig. 7.68

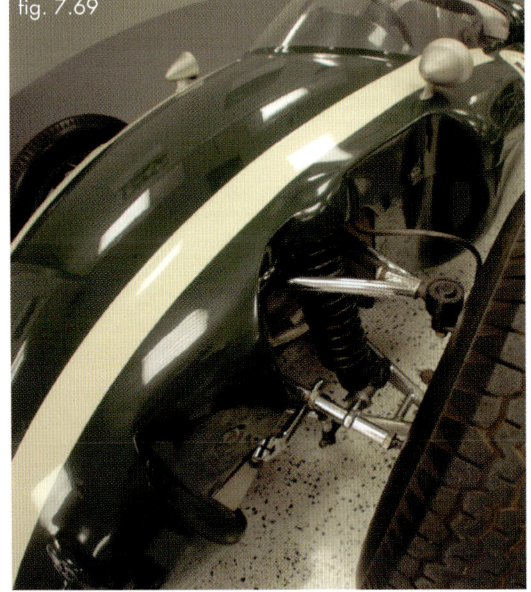
fig. 7.69

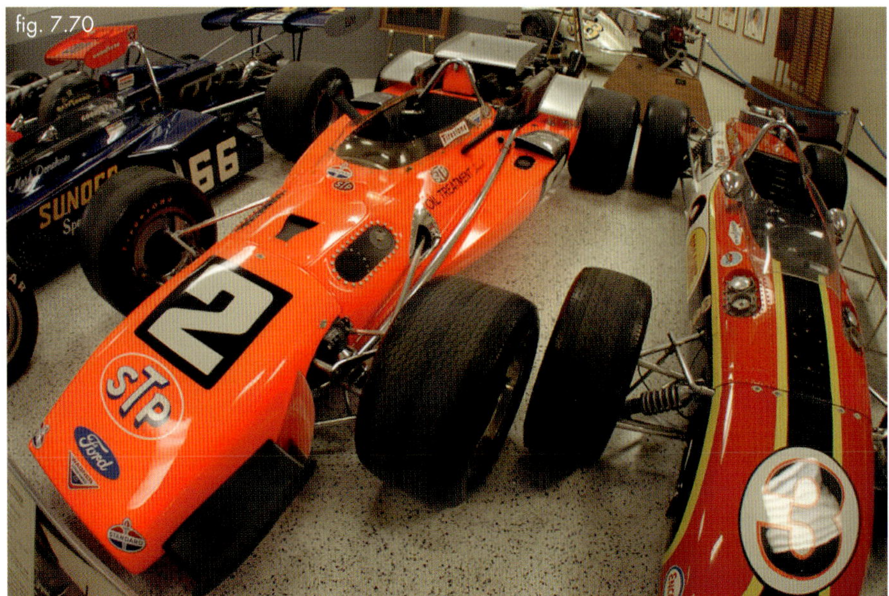

fig. 7.70

Rendering the cast shadows on a reflective surface is something that is often overlooked, especially when using traditional media. When doing renderings in a digital program and using layers, the odds are that they will be there as part of the matte-surface layer, which is a good start. Be aware of how the reflections change inside these cast shadows. The images on this page are all of metallic painted surfaces that really demonstrate how the color of the reflection in the cast shadows is very different than that in the sunny areas. Look how blue they are where they reflect the sky. To have more control over rendering these areas it is a good idea to isolate the cast shadow shapes and put them on a separate reflection layer. In fig. 7.71 the cast shadow behaves more like a black painted surface. With very few matte-surface value changes it acts almost like chrome. The Fresnel Effect is the same as usual, but since no sunlight is being reflected in these cast shadows, they become very blue.

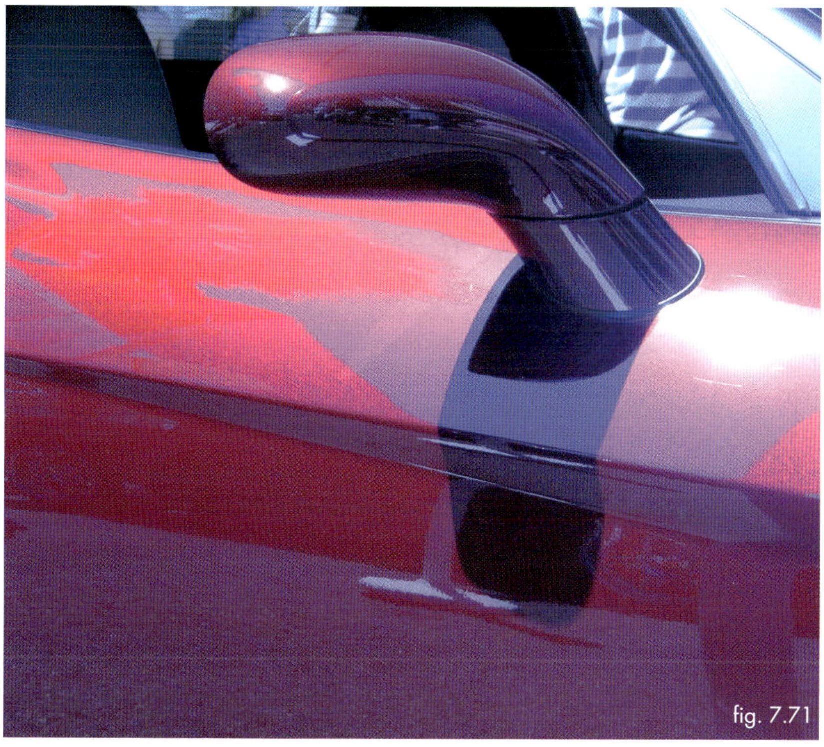
fig. 7.71

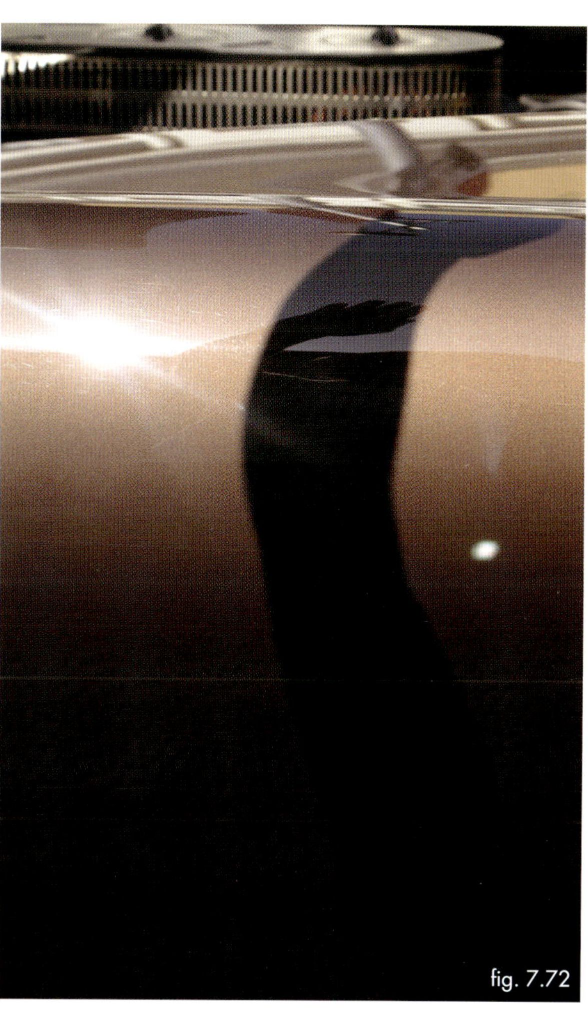
fig. 7.72

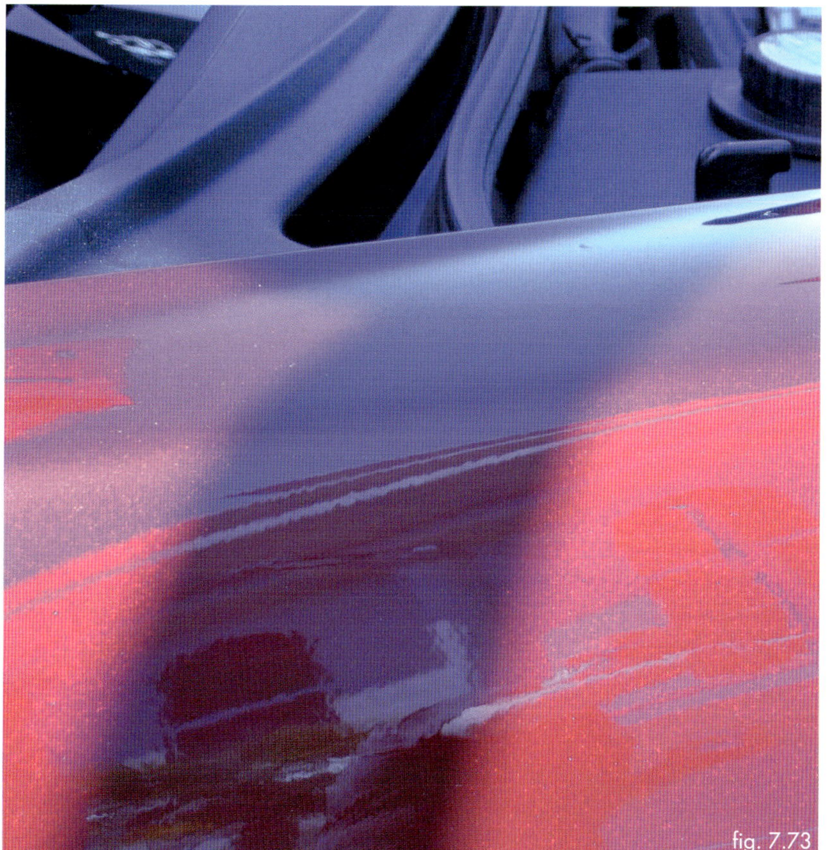
fig. 7.73

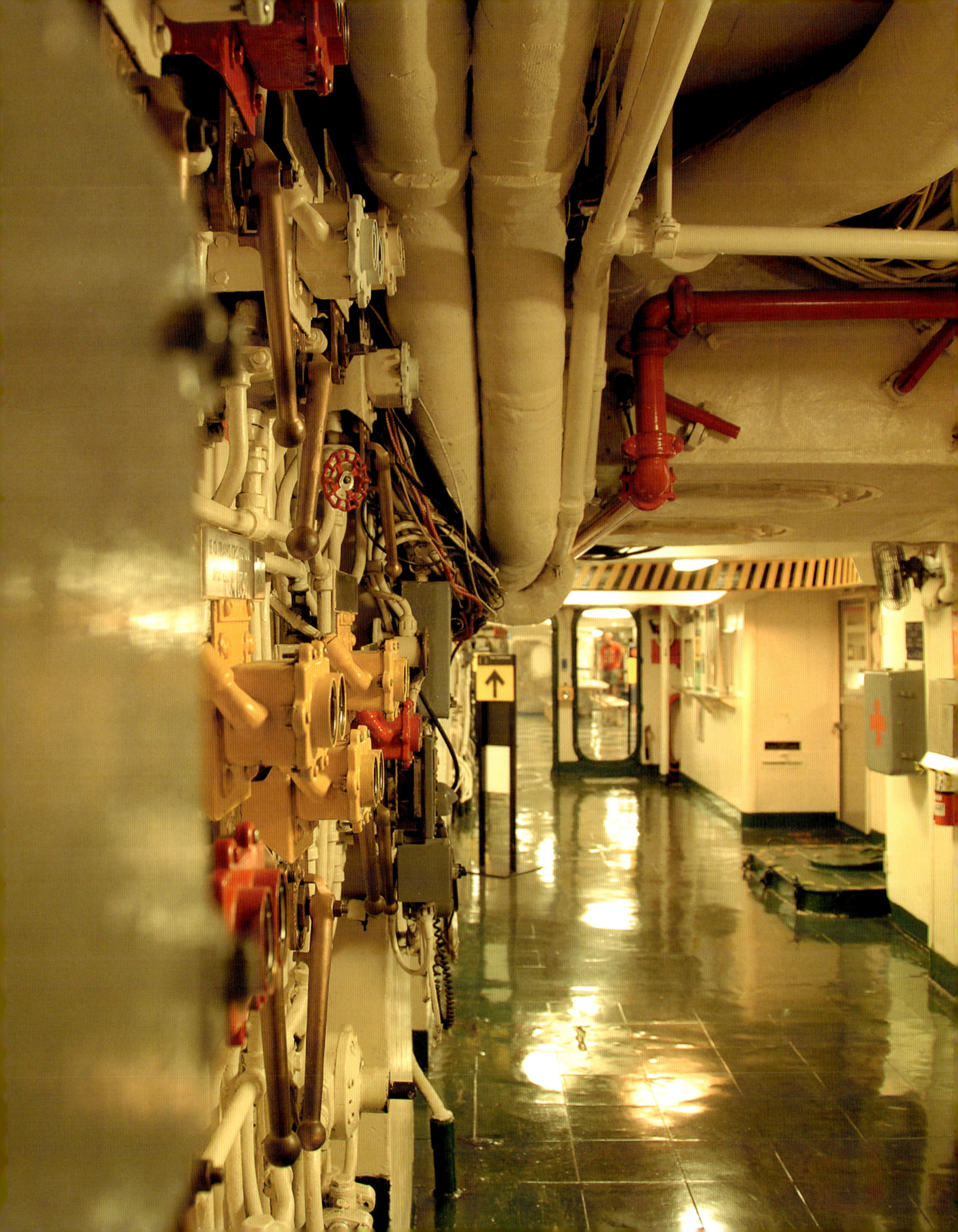

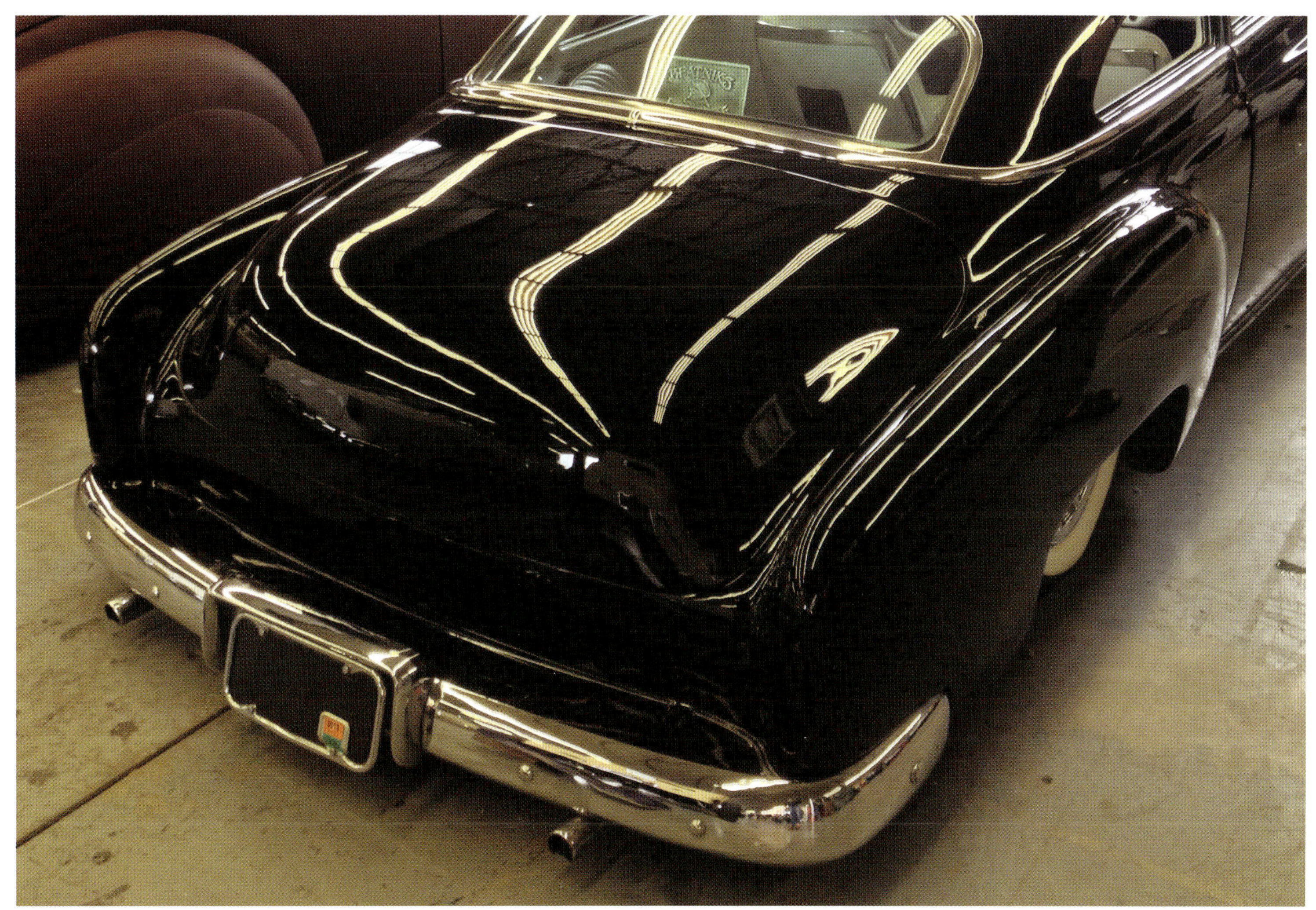

CHAPTER 08
REFLECTIONS: INDOOR SCENES

This chapter explains how to render a shiny surface in an indoor environment. Because the environment around an object is what drives the reflections rendered on the clear coat, the key to the success of these renderings is to design indoor environments that enhance the design of the objects.

UNDERSTANDING THE SCENE 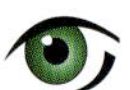

Knowing the surroundings and the shape and type of light sources in it makes rendering the reflections much, much easier. Fig. 8.1 shows a simple indoor studio scene in the 3D modeling and rendering program, MODO. It consists of a red egg-shaped volume floating above a ground plane that extends behind it, and then bends and goes up the wall. The volume is not perfectly egg-shaped; it has a ridge on one side. The studio surrounding the scene is black and there is one rectangular, overhead soft light source, shining downward onto the form.

Again, the scene can be better understood with a couple of diagrams to illustrate where the sight lines bounce into the environment, (figs. 8.2 and 8.4). This is a very simple set-up, which usually works best.

Fig. 8.3 is the final MODO rendering, showing the placement of the light-box reflections. The curvature of the large light reflection indicates that the surface is curved. The smaller reflection on the interior form indicates the part of the volume that is ridged.

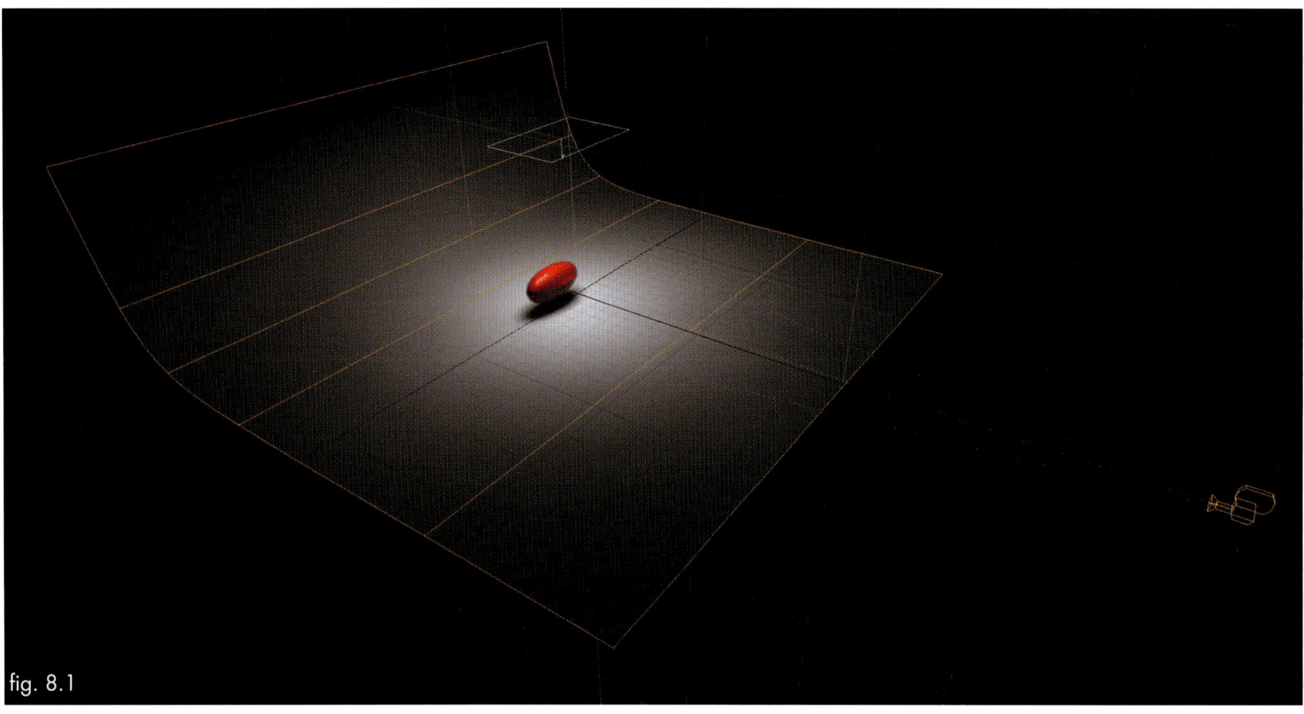

fig. 8.1

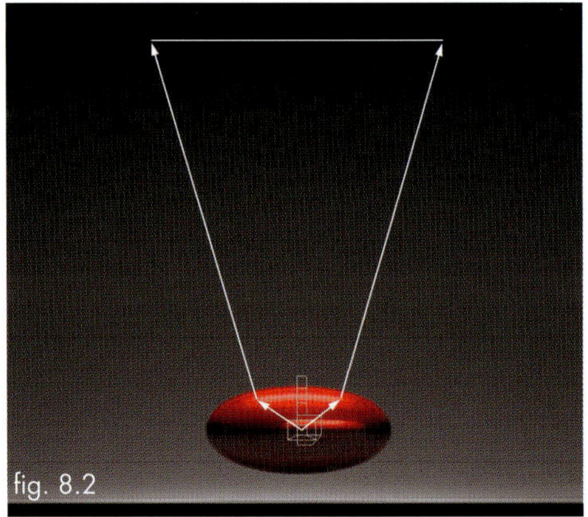

fig. 8.2

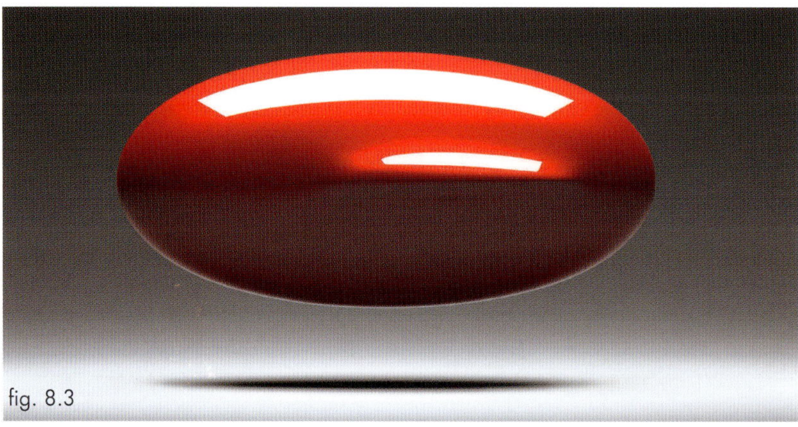

fig. 8.3

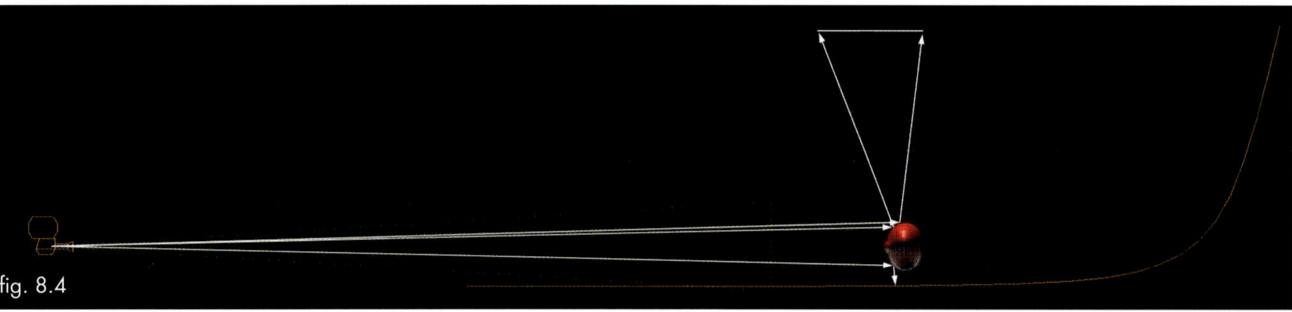

fig. 8.4

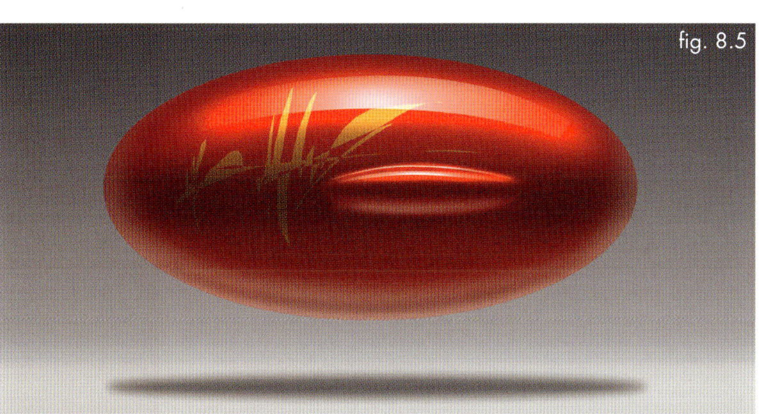
fig. 8.5

In fig. 8.5 the same shape was rendered in Photoshop, imagining the same basic environment as the one on the previous page. This rendering exercise is very much like the shiny billiard balls in Chapter 07. It is still a very simple layering strategy in Photoshop. First, create a base matte-surface layer, then make two reflection layers on top of it, for the reflection of the background and the reflection of the light source. There is a video tutorial for this page. Please follow along and try this exercise.

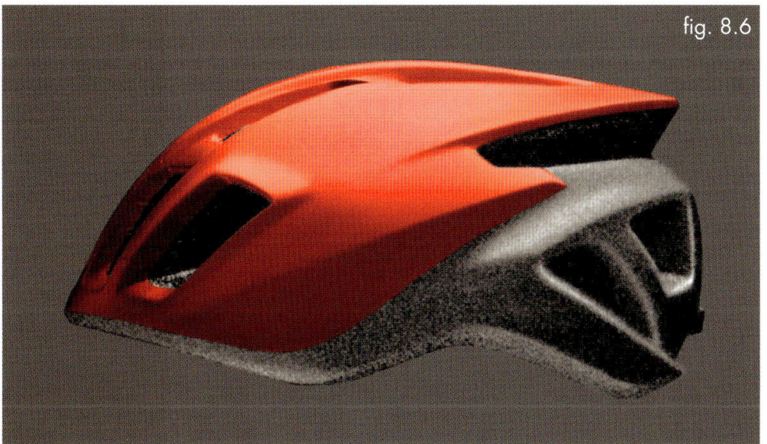
fig. 8.6

This helmet illustrates how easy it can be to make something look reflective. Everything about reflections comes down to two basic layers in Photoshop: one base layer that has all the form and matte-surface shading discussed in earlier chapters, and one reflection layer, in this case of a single white light box.

The base layer is rendered first, showing the local, or true, color of the object and its form (fig. 8.6). In this helmet example, the contrast is pushed pretty strong, using more than a halfway-to-black value range. Because there are no graphics or cut lines to worry about in the shadow areas, it is okay to push the contrast a little. To make it look shiny, just add a clear-coat layer (fig. 8.7) on top of the base layer. Most of the red layer is still visible, which helps to show the form of the object. With the clear-coat layer turned "on," it looks wet or shiny (fig. 8.8). It is just like taking a matte, red object and spraying a clear coat on it, or spraying it with water. All of the reflections of the environment around the helmet are on that clear coat.

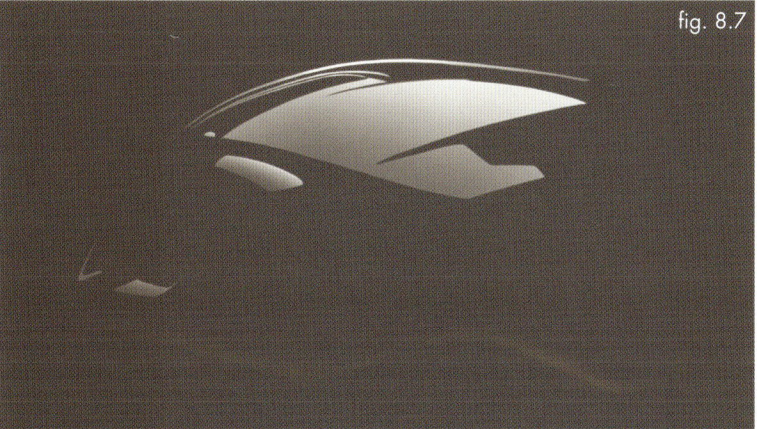
fig. 8.7

The helmet is in a dark, indoor studio environment, just like the one on the opposite page, with an overhead light box or skylight above it. The reflective layer is white only. There is just one light box reflecting into the helmet and that is all it takes to trick the brain into thinking it is shiny.

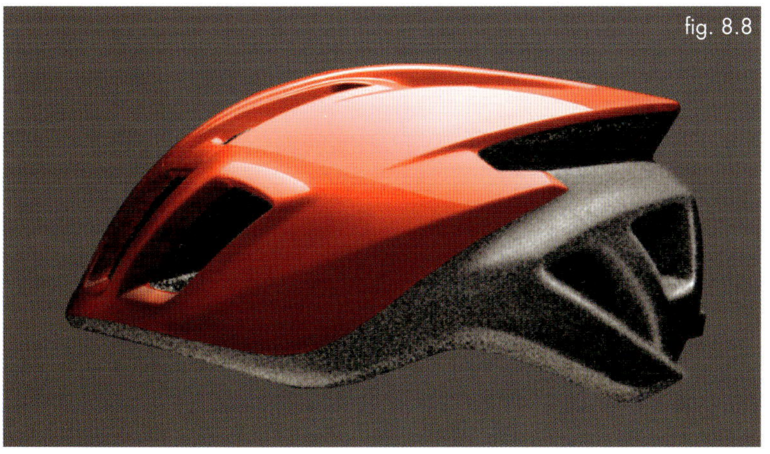
fig. 8.8

A big advantage of using Photoshop in this manner, with the reflections on their own layer, is that if the client does not like red, it can easily be changed to any other color and the rendering still works. Changing the color takes very little time, and many variations can be viewed quickly. In the days before digital rendering, varying the color meant that the whole thing had to be rerendered from scratch. Making the helmet blue meant starting all over again.

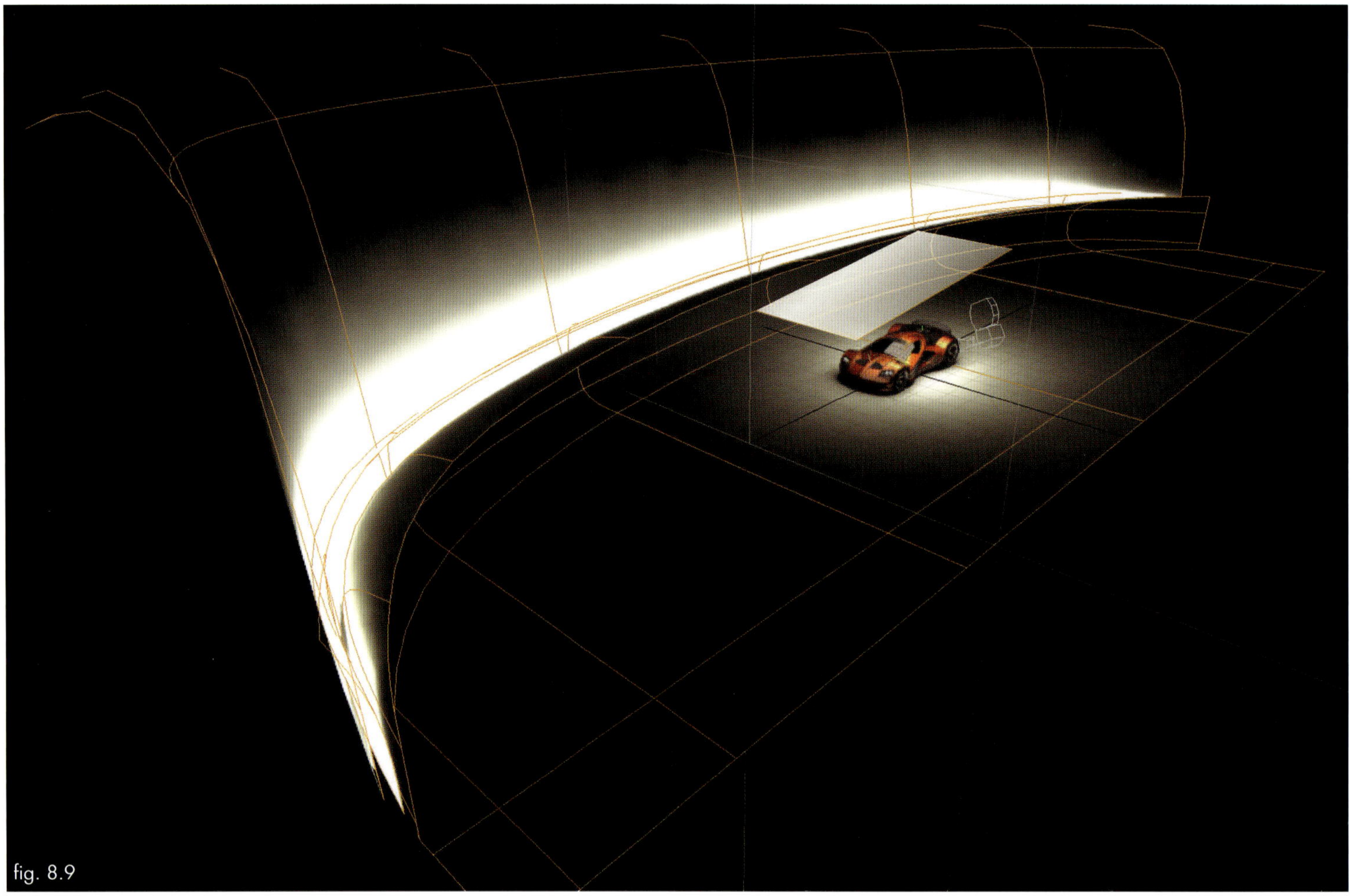

fig. 8.9

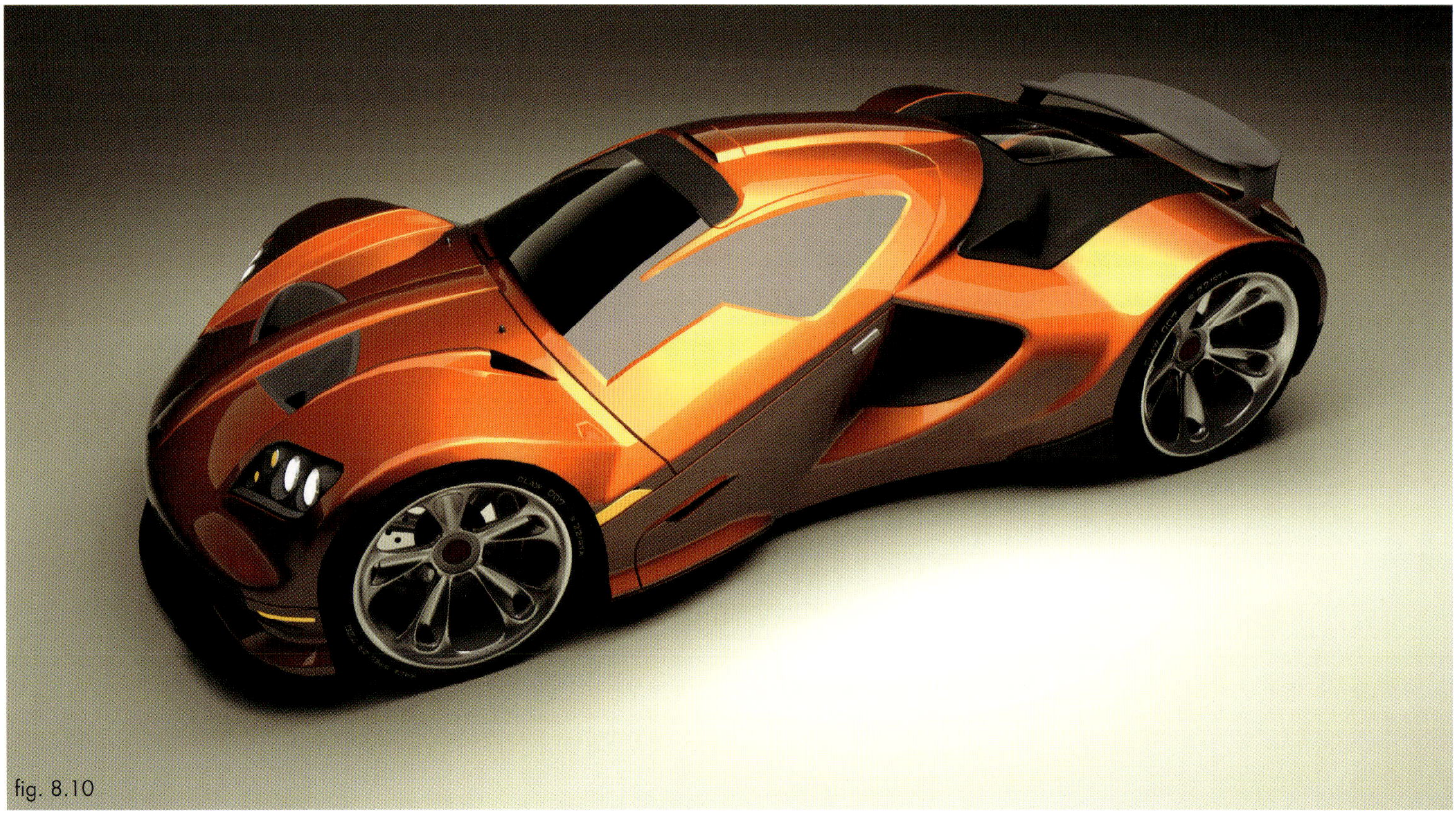

fig. 8.10

In this demo the same basic environment carries over from the previous pages with one modification, there are now two walls behind a car model with a light shining up the back wall, as seen in fig. 8.9. This environment was created based on the Photoshop rendering at the bottom of page 193 (fig. 8.16). The Photoshop rendering came first and this was the studio environment that was in mind while it was being rendered. With the knowledge in this book and *How To Draw*, an artist can learn to think and render in 2D, just like a 3D computer program such as MODO does in 3D. The 3D set up was moved around and staged to mimic the color, lighting and reflection patterns of the 2D rendering, with a different car model.

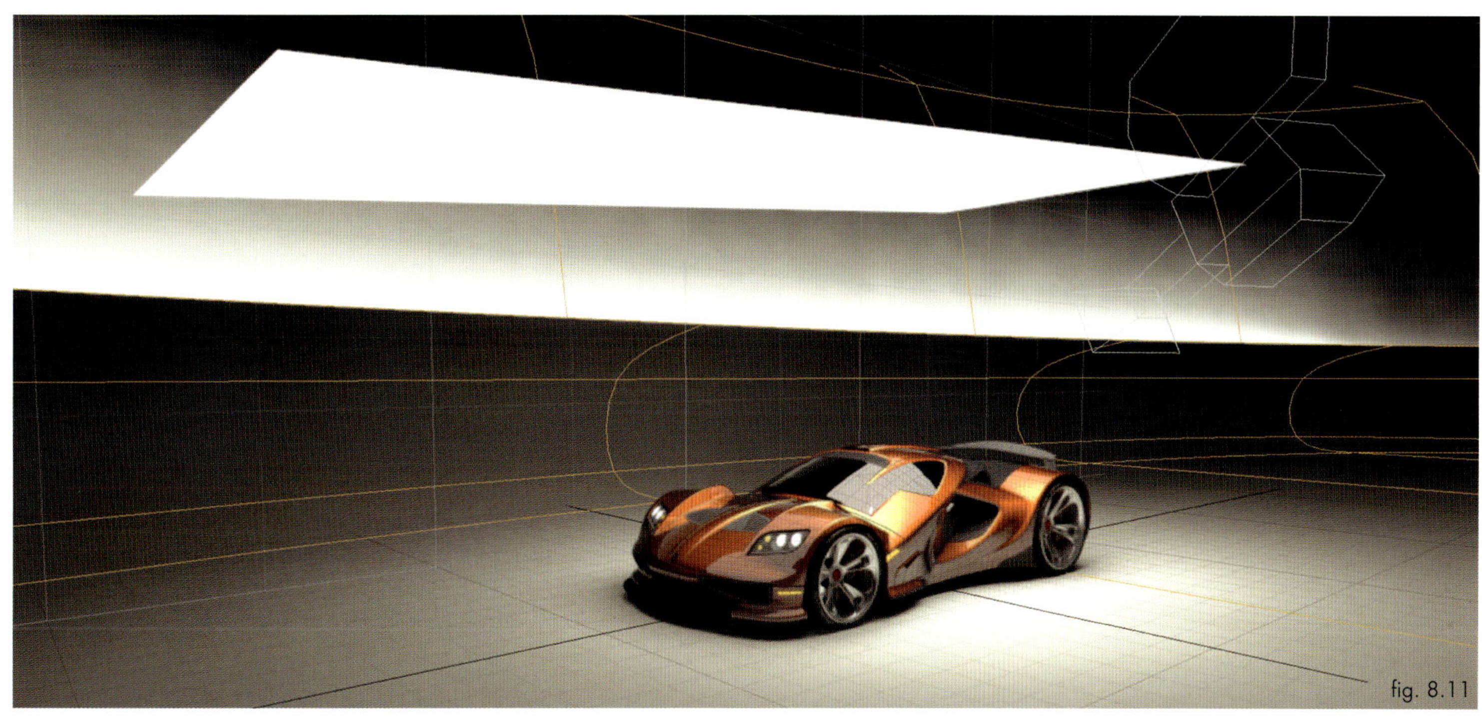

fig. 8.11

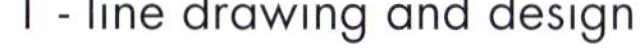

1 - line drawing and design
2 - know your surroundings
3 - staging and planning
4 - designing and placing the light sources
5 - matte-value rendering
6 - colorizing and adding details
7 - plotting and blocking reflections
8 - Fresnel Effect with layer mask
9 - background refinement
10 - final adjustment layers

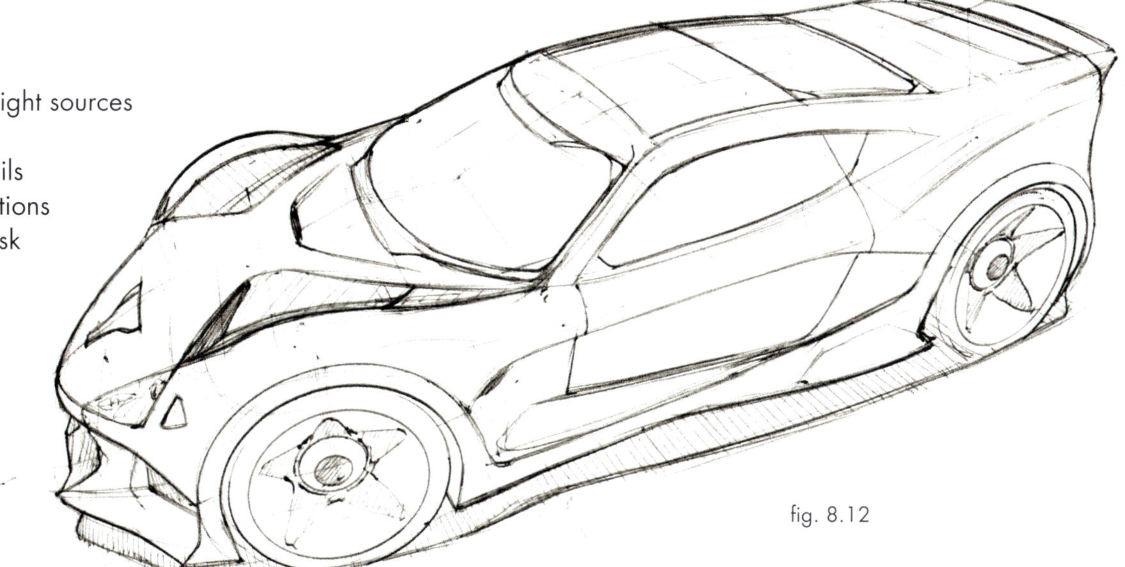

fig. 8.12

Above are the steps to follow when doing a reflective rendering of this type. Line drawing is the first step and it does not need to be any tighter than this. Scan a design drawing at a resolution of 300 to 400 dpi if the image is about 8 inches wide. Generally speaking, higher resolution is better to a point. That point is reached when the computer starts to slow down due to the larger file sizes of higher-resolution images. Each computer is different. In order to print the image later without a lot of bit mapping or ragged edges, the resolution should be around 300 dpi at print size. For instance, a 10" wide image at 300 dpi will be 3,000 pixels wide.

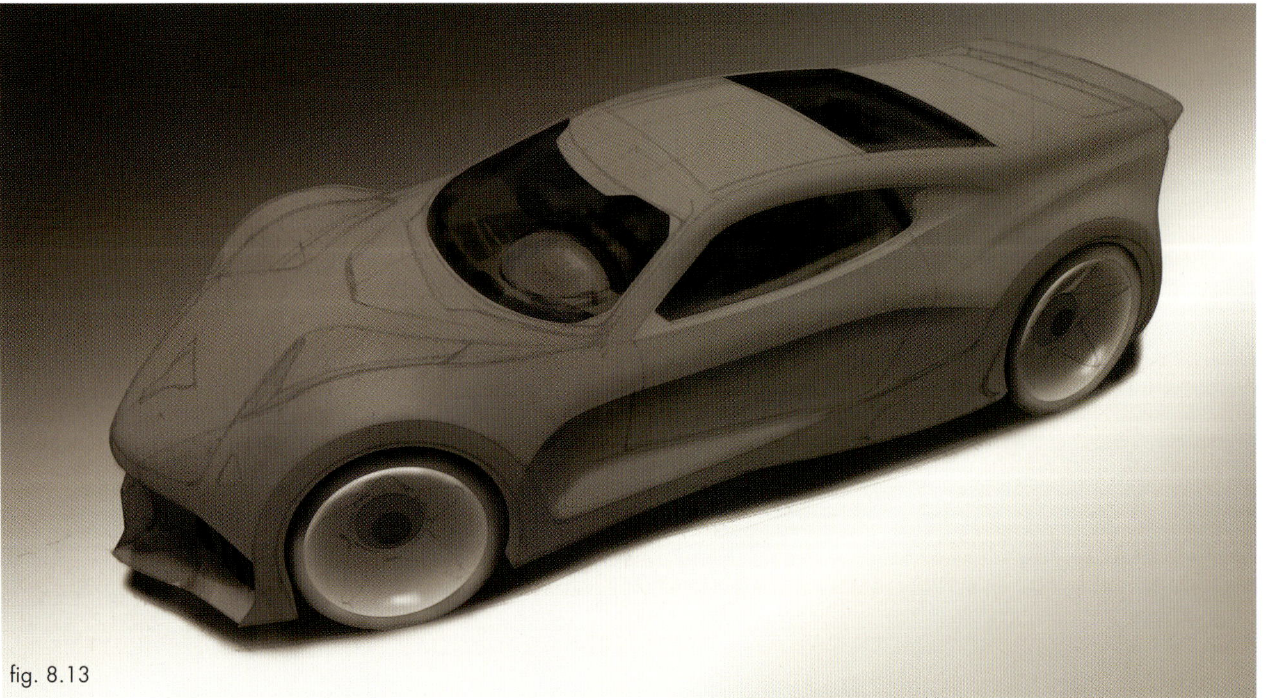

fig. 8.13

Fig. 8.13, above, shows steps 2–5. The surroundings and the lighting were described on the previous two pages. Here the matte-surface shading of a base layer under the line drawing was executed. Paths can be used to render clean edges, or it can be done all freehand. Render the base-paint layer as a matte surface, applying the knowledge from the first half of the book. Render the interior as well with soft light illuminating it from above, without any glass. A very dark saturated orange was always the intended color for this car, so therefore the value of the light side of the car is quite dark.

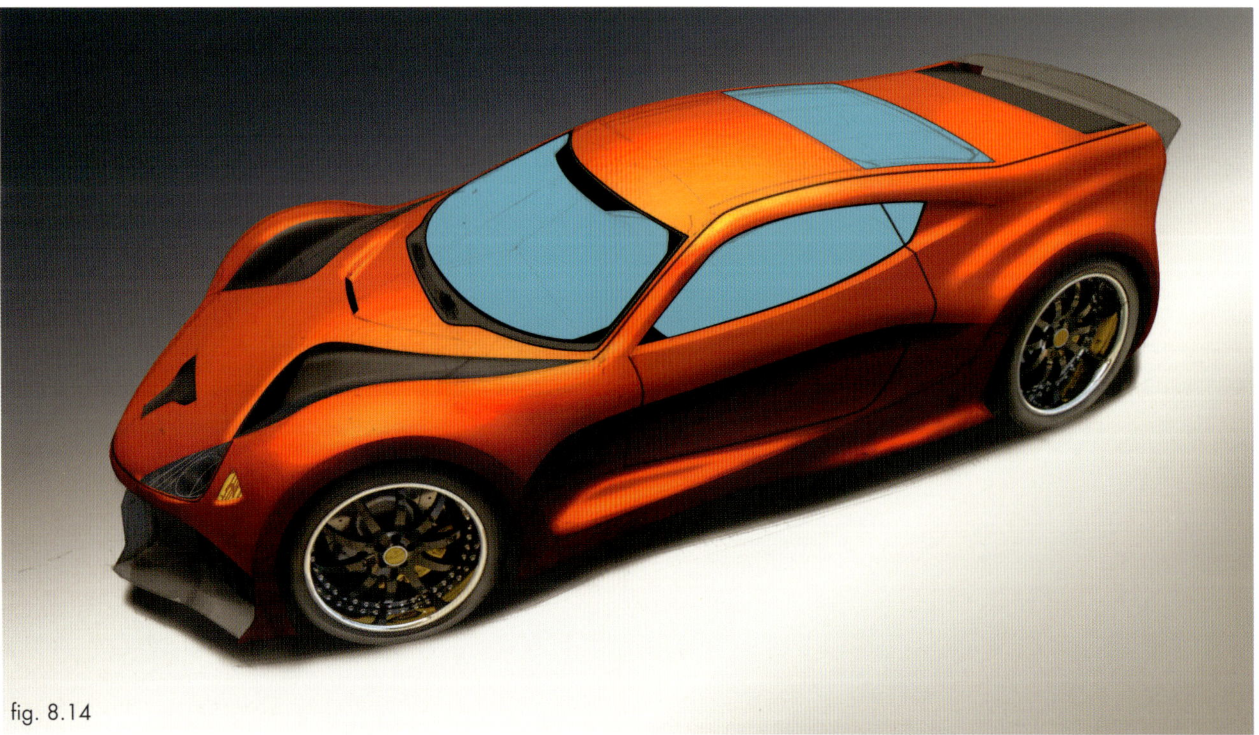

fig. 8.14

Fig. 8.14 shows step 6. Colorize the gray-scale matte-surface layer and add details. A set of photographed wheels was added as well, to increase the illusion of realism. Right now, the window glass is an opaque layer that later will be switched to the mode of "overlay" to achieve a convincing colored-glass feel. If the glass is to be clear and only tinted, use a dark gray for this layer and set it to "multiply." Both the color overlay and multiply layers can be used together. Then a "color dodge" layer was added to give the base paint more of a metallic finish. See page 214 to learn more about rendering a metallic finish.

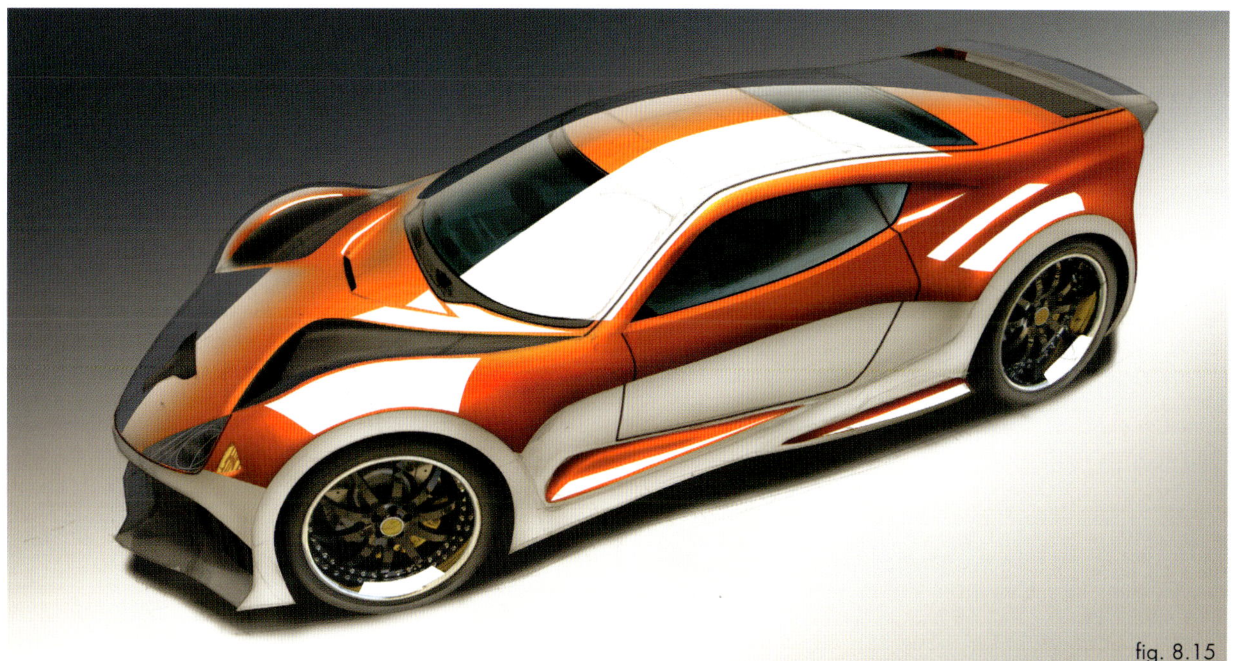

fig. 8.15

Fig. 8.15 shows step 7, where the reflection layers are added. Each reflection is on its own layer, with the light on one layer, the nearside ground on another, the back wall on another and the light gradation up that back wall on yet another. Never overlap the reflection layers. They should always butt up to each other, edge to edge. They can be controlled more easily when they exist on separate layers, although they all combine to create a single reflection of the surrounding environment.

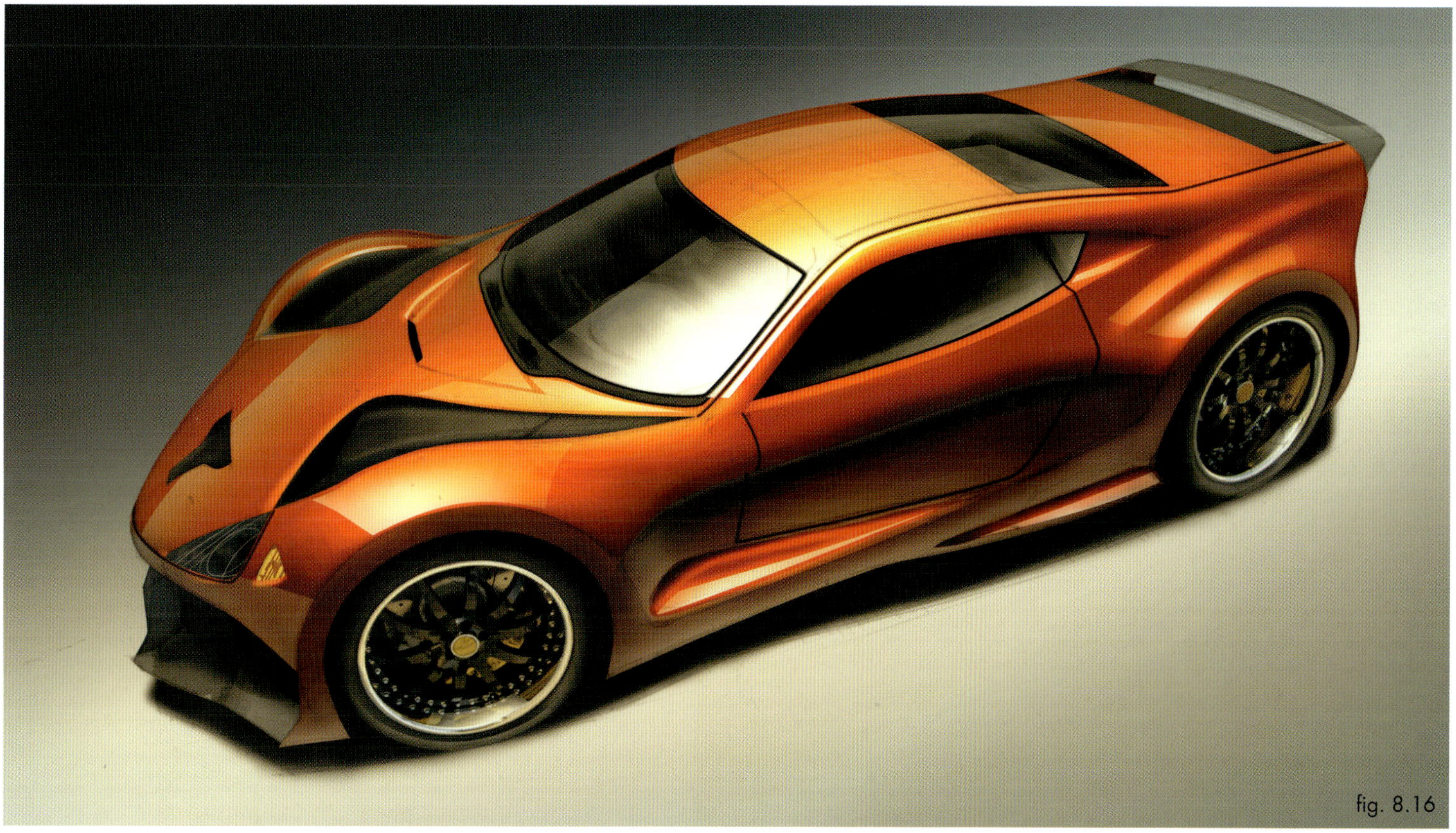

fig. 8.16

Fig. 8.16 shows steps 8–10. Each reflection layer has had a layer mask applied to it and the masks have been airbrushed to hide parts of each layer, creating the Fresnel Effect and the illusion of realism. Final color adjustments to the entire image and any tweaks to the background were made. In this case, the background was slightly lightened just behind the rear of the car to better indicate the silhouette.

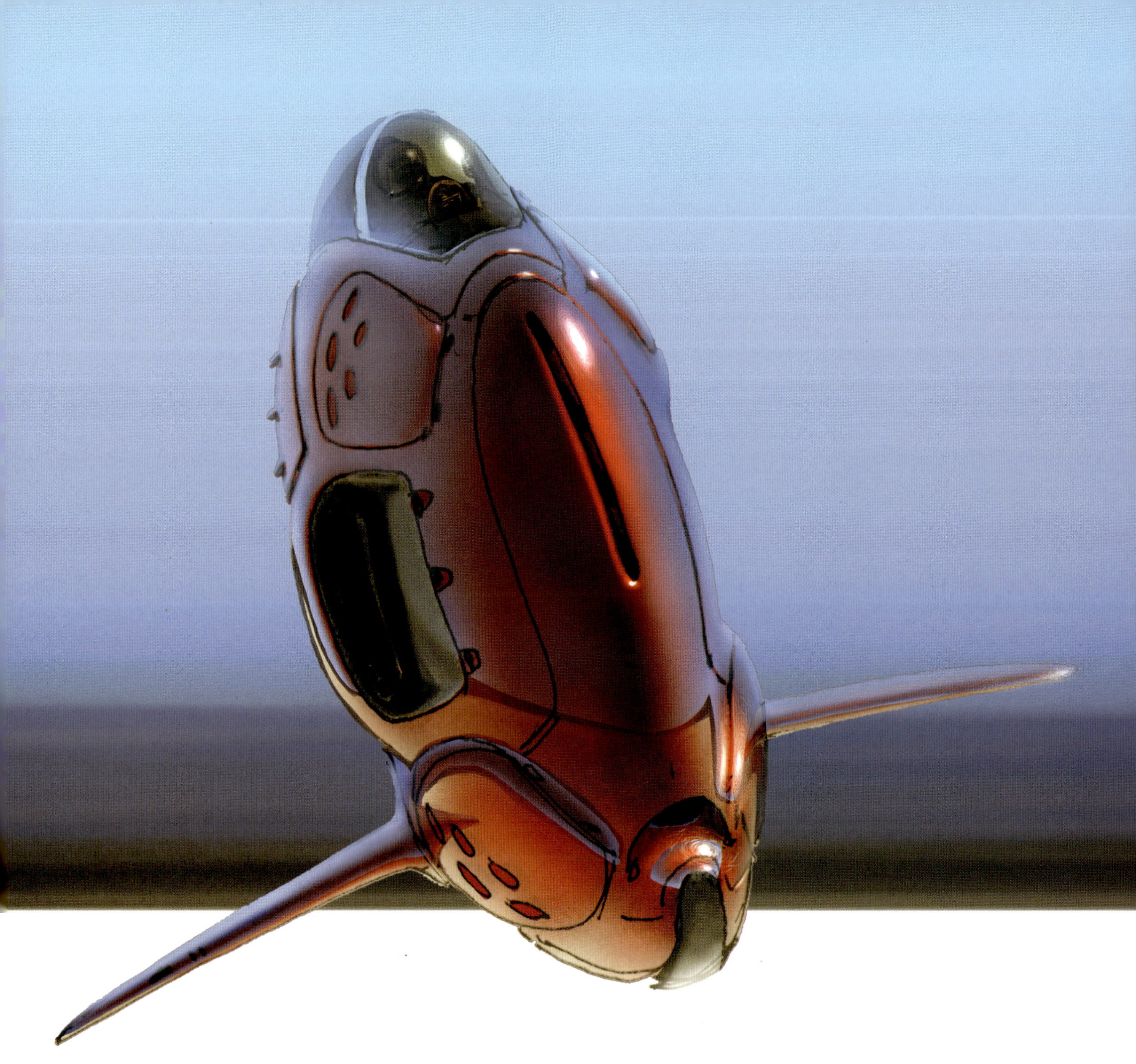

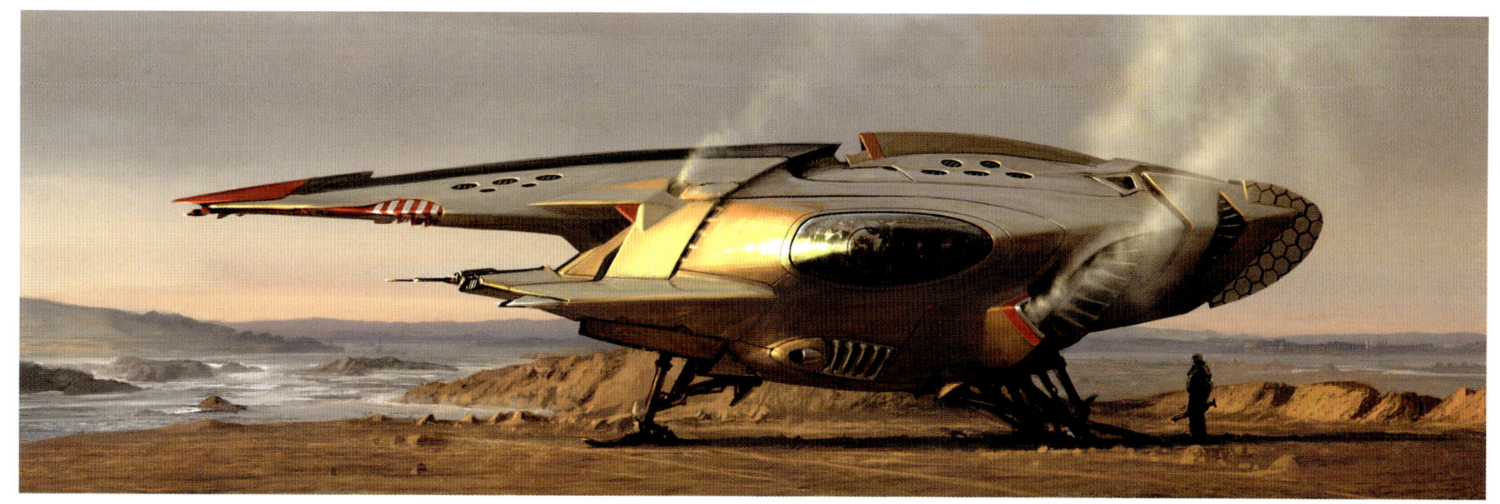

CHAPTER 09
REFLECTIONS: OUTDOOR SCENES

Rendering reflective forms in outdoor environments requires the same basic rendering steps as indoor environments, but with two variations. First, the light source is usually the sun, which is obviously a different shape than an overhead light box or skylight. Second, the sky is a big variable. Rendering reflections of the sky presents several twists because it changes color depending on the time of day, and it has a gradation in it. It is generally recommended to stick with a cloudless, blue sky when rendering. Let's get started learning about the adjustments that need to be considered to render shiny surfaces in an outdoor environment.

UNDERSTANDING THE SKY'S REFLECTION 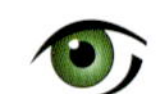

Figs. 9.1 and 9.2 are chrome surfaces. They literally reflect the exact environment around them, just a bit darker. Notice the gradation present in the reflection of the sky. Because this is chrome, the gradations are not due to the strength of the reflection changing across the form, like on a painted surface. The reflections have gradation because the sky itself has a gradation in it. It is darkest straight up and it is lightest at the horizon, where the concentration of dust, smog and moisture is highest (fig. 9.3). Because the gradation seen in the chrome is not caused by the Fresnel Effect, but actually is in the sky, this makes it a bit trickier to render.

To render the ground reflection, literally clone a piece of the ground onto the reflection layer. Remember to darken down the reflection at least one-half step in value to make it look like real chrome.

Figs. 9.5 and 9.6, on the facing page, show a large section of clear, blue sky. Notice that the sky just above the horizon is lighter than the sky directly overhead. This gradation exists no matter where the sun is in the sky. If the sky is very hazy then the sun will have a stronger glow around it and affect a larger area of sky than on a clear day. For rendering purposes, a clear, blue sky is much easier to render than one with clouds in it.

With this in mind, observe the dark maroon car (fig. 9.7). What appears to be a very strong Fresnel Effect on the clear coat is actually the Fresnel Effect plus the gradation of the sky. The line of sight bounces more straight up in the near part of the roof, to the darker part of the sky, and then proceeds to bounce to the lighter part of the sky as the roof surface rolls away from the line of sight. This reflection of a sky gradation adds a kind of doubling feel to the Fresnel Effect.

When starting to render reflective forms in an outdoor environment, the sky is the most difficult part. It takes practice to get good at estimating where the line of sight bounces off the form into the sky and also to know what value and color the sky is in that spot.

In the chrome sphere (fig. 9.1), the reflection of the sun is blocked with a piece of foam core so it is easier to see the reflection of the sky. Compare this area to that of the reflection of the ground and the back wall surrounding the sphere in the scene on page 171, fig. 7.26, and it becomes clear that the sky requires more effort to render.

Welcome to rendering reflective forms in an outdoor environment!

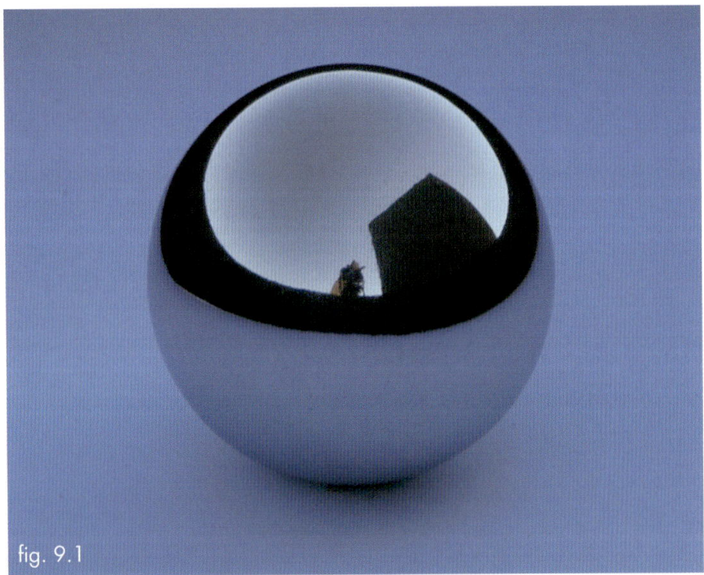

fig. 9.1

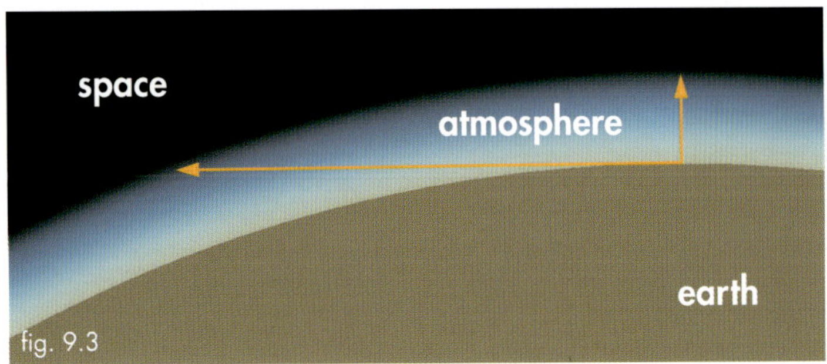

fig. 9.3

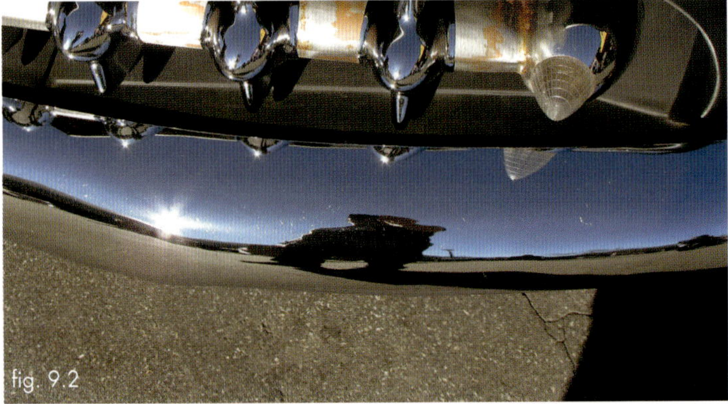

fig. 9.2

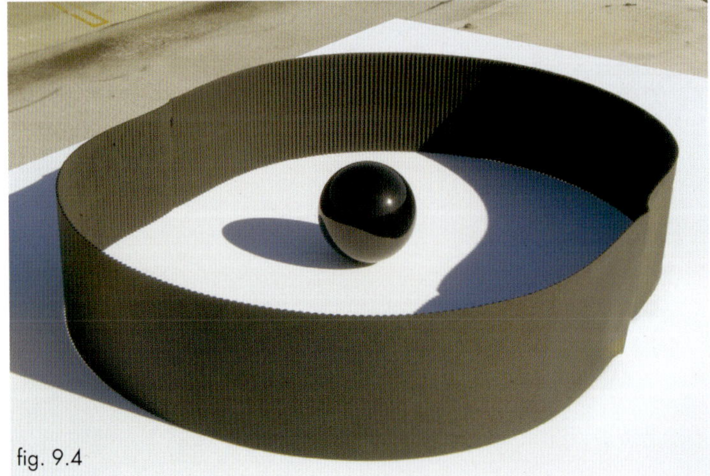

fig. 9.4

fig. 9.5

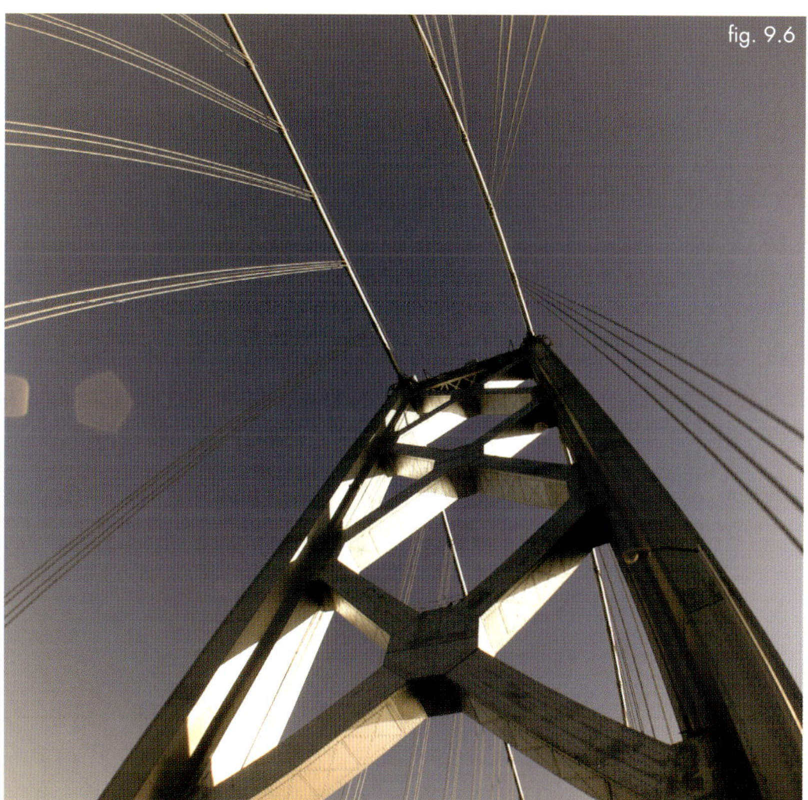
fig. 9.6

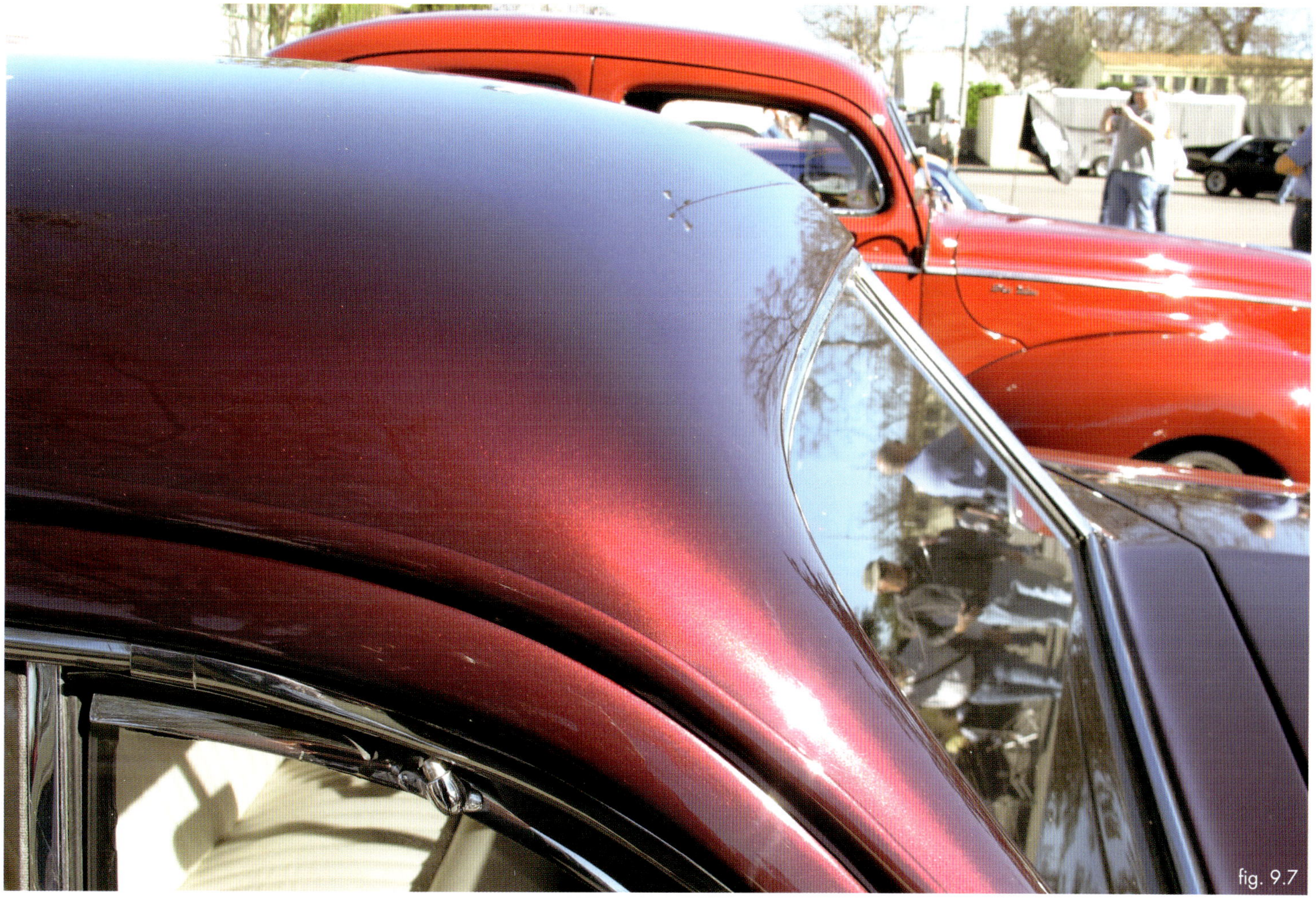

fig. 9.7

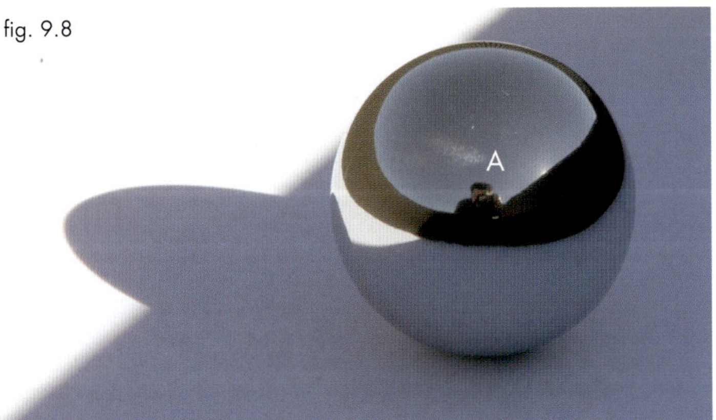

fig. 9.8

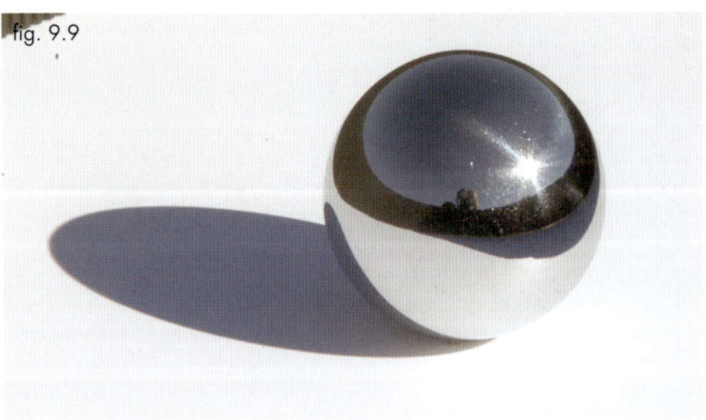

fig. 9.9

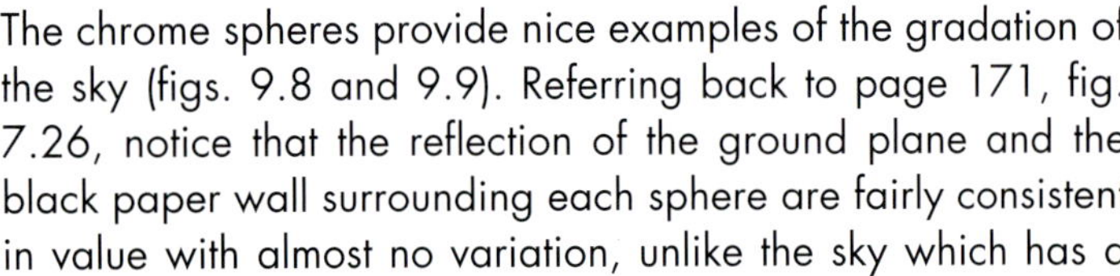

The chrome spheres provide nice examples of the gradation of the sky (figs. 9.8 and 9.9). Referring back to page 171, fig. 7.26, notice that the reflection of the ground plane and the black paper wall surrounding each sphere are fairly consistent in value with almost no variation, unlike the sky which has a large value change. Even when a cast shadow blocks out the sun, the reflection of the sky does not change (fig. 9.8). In this image look for the cast shadow on the chrome surface. It is really only visible where it blocks the reflecting sun, near point A.

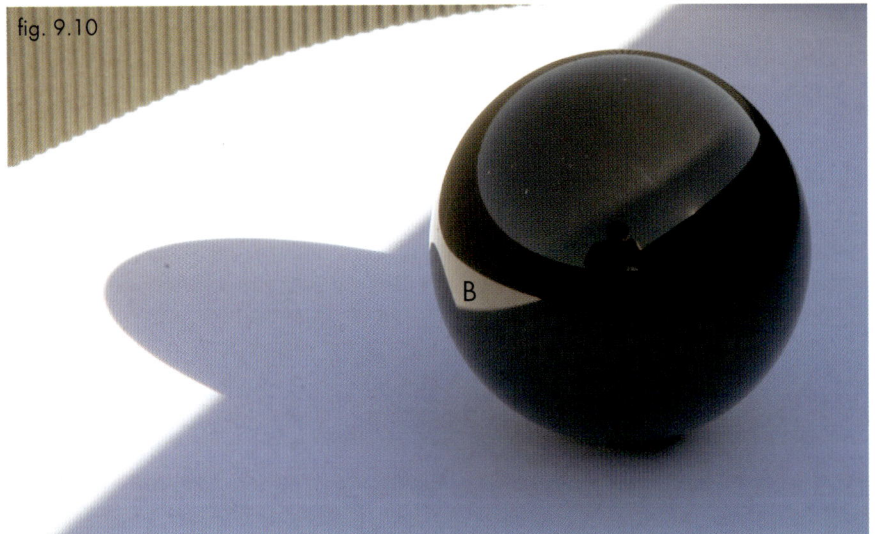

fig. 9.10

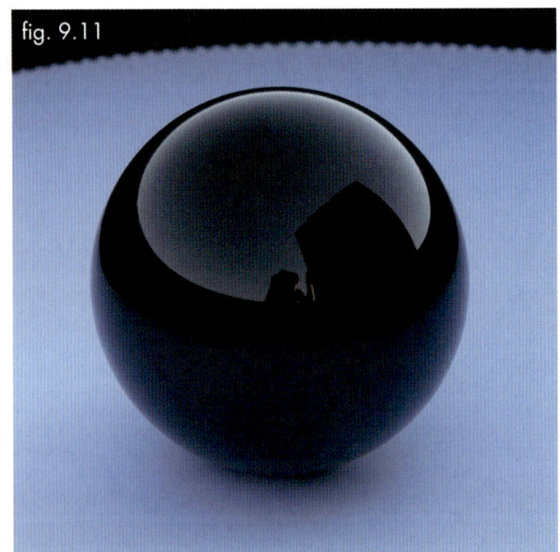

fig. 9.11

Just like the spheres in the indoor studio environment, it makes no difference what color or value the base paint is, the reflection of the environment remains the same. Figs. 9.10, 9.11 and 9.12 show the difference in the perceived strength of the sky reflection when the sun is shining on the sphere, or not. The sky appears to be strongest in fig. 9.11, where the entire scene is in shadow. This is due to the increase in contrast between the darker values of the base black and the lighter values of the sky. A similar effect can be seen in fig. 9.10 where the left side of the sphere has a very bright reflection of the sunny side of the ground plane, near point B. To increase the perceived reflectivity of a surface, increase the contrast of the reflection compared to the base-paint/local value.

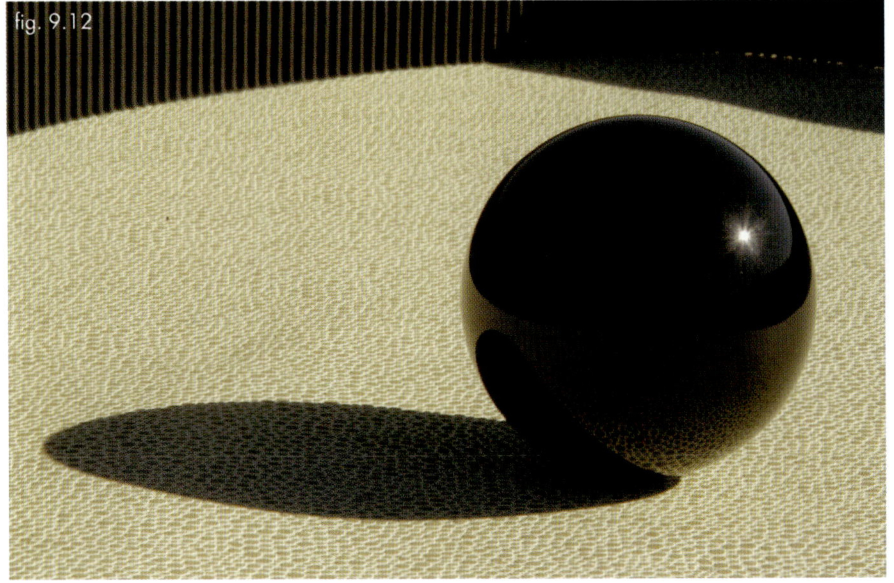

fig. 9.12

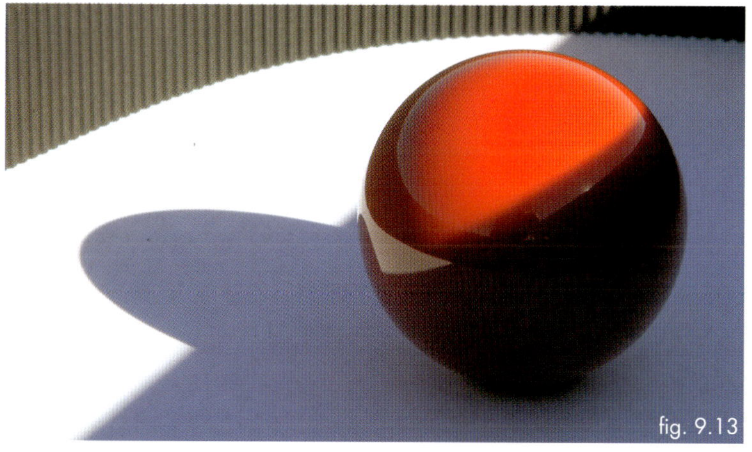
fig. 9.13

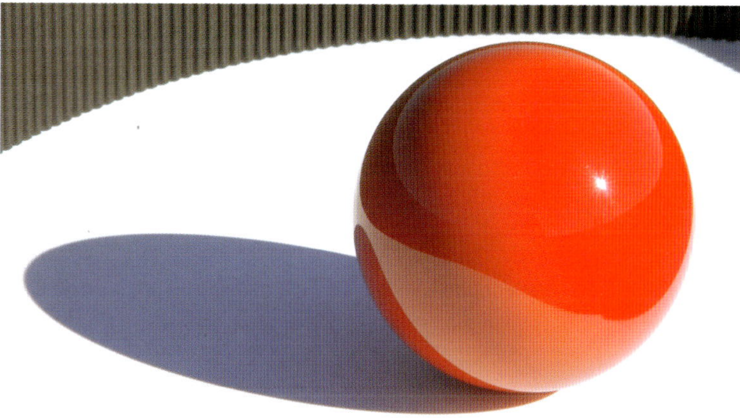
fig. 9.14

fig. 9.15

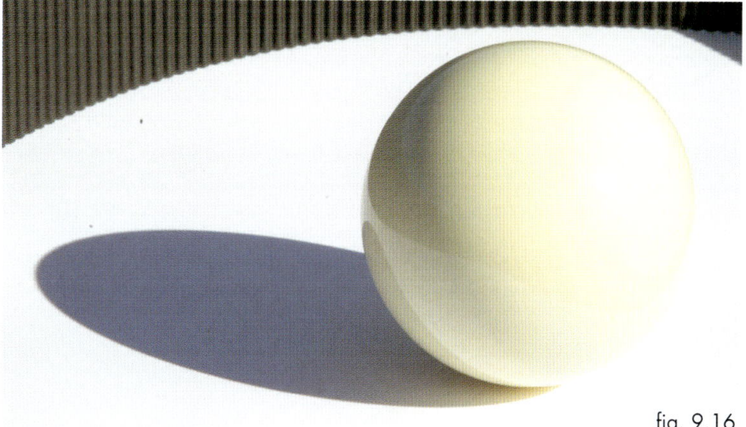
fig. 9.16

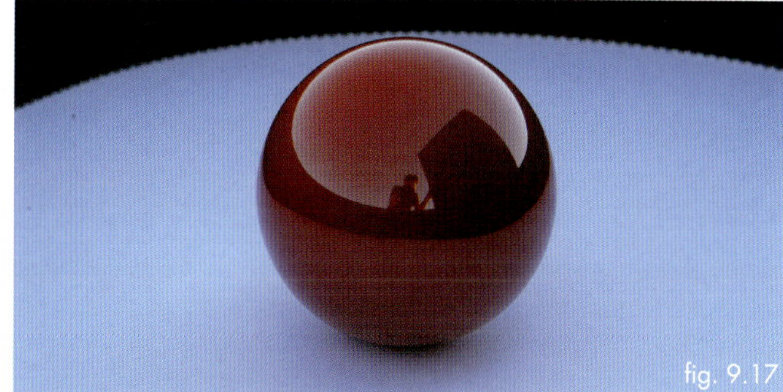
fig. 9.17

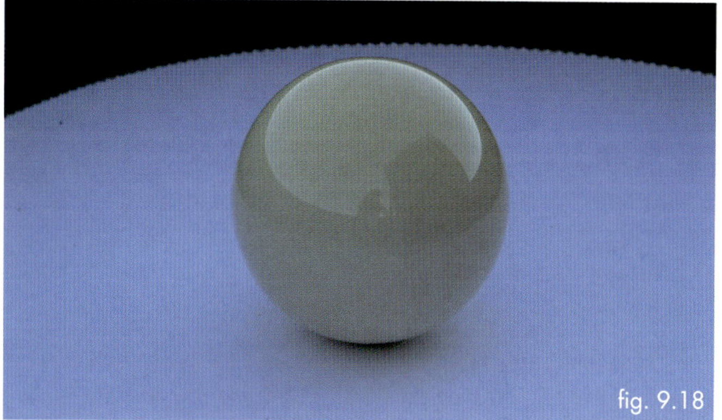
fig. 9.18

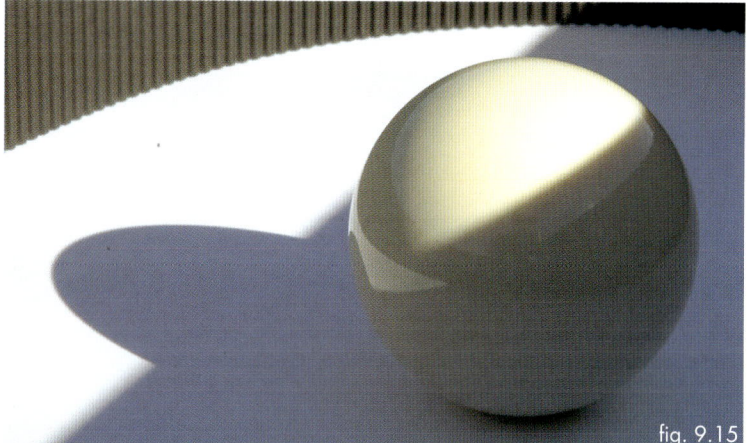

It is this relative difference between the value of the underlying color and the reflection that makes it appear like the surfaces of the spheres (figs. 9.13 – 9.16) are actually shinier on their shadow sides than on their sunlit sides.

Figs. 9.17 and 9.18, above, show that this assumption is false. The spheres are consistently reflective over their entire surfaces. When rendering reflective forms digitally, just layer the reflection over the matte surface as shown in Chapter 08. This illusion of the cast-shadow areas being more reflective happens automatically. With traditional media this can be a little more challenging and more planning and pre-visualization are necessary to achieve this same illusion.

LAYERING STRATEGY

Rendering these types of shiny surfaces as two layers, one as a matte surface with form and color and the other one with a chrome-like reflection, creates results so close to reality that there is no need for photographic reference. These skills come in handy when trying to invent something no one has ever seen before and maybe even putting it in a place that does not really exist.

Figs. 9.19 and 9.20 show two different layers in Photoshop. The sketch, which is a pretty loose one, can still be seen floating over the rendering. This, of course, takes away from a bit of the realism but it also means not having to worry about cleaning up the edges with paths/selections. The result (fig. 9.21) is still pretty convincing as a shiny, metallic-red car.

For those just starting out, it is a good idea to render a separate layer for the base paint. Then render the full chrome surface, as in fig. 9.20, to help see the surfaces as 100% reflective.

The chrome layer can then be erased or masked, revealing the base coat below. Using a masking layer in Photoshop is more efficient than erasing because masking allows for going back and making small refinements, increasing the opacity of the chrome layer later if needed. By erasing only, this opportunity is gone.

After masking the chrome layer, the finished result is convincing enough for a quick color sketch. There is a lot of nice color mixing happening between the base color and the reflection colors that is difficult to achieve if trying to render everything on one layer.

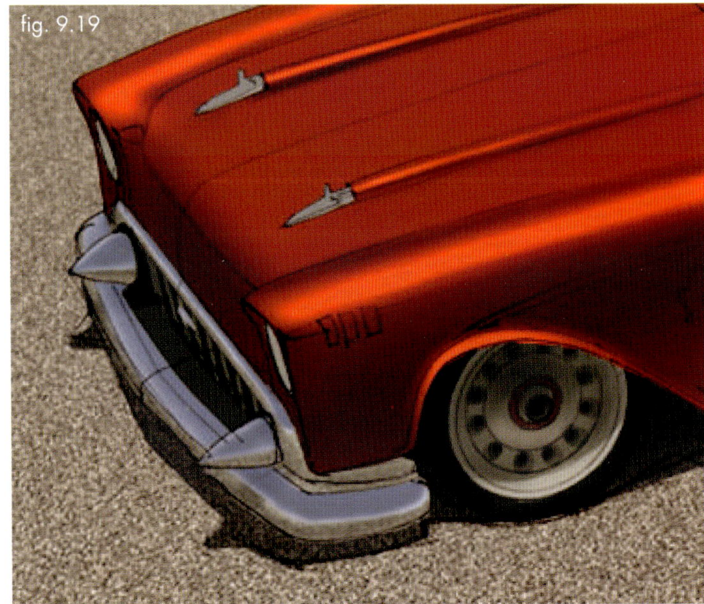
fig. 9.19

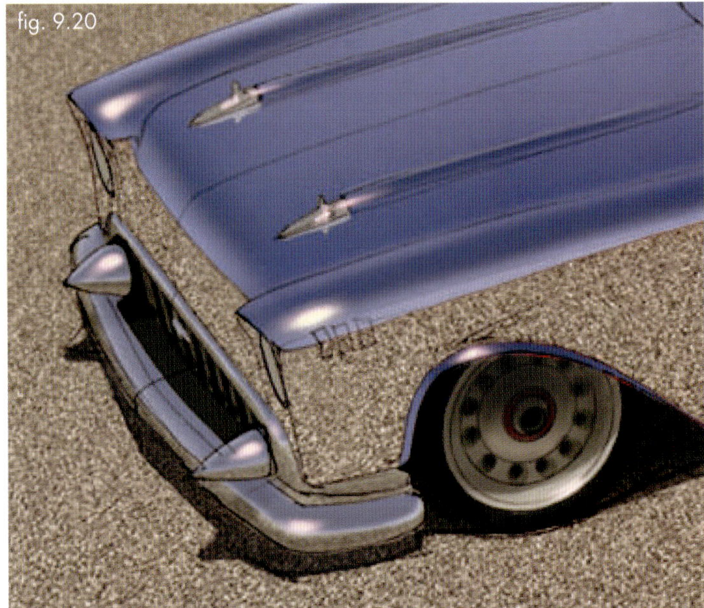
fig. 9.20

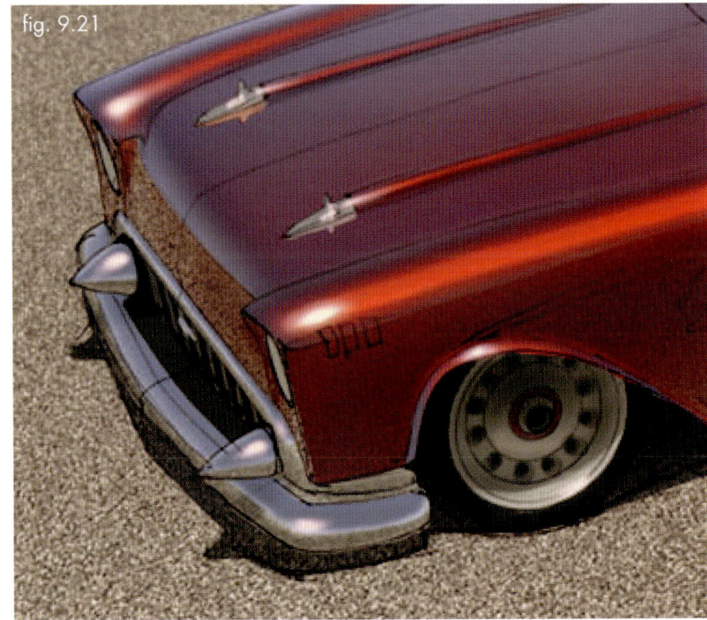
fig. 9.21

OUTDOOR SCENE RENDERING STEP-BY-STEP

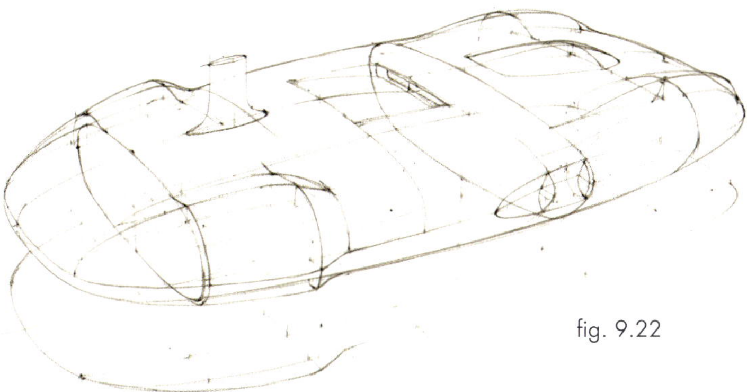

fig. 9.22

This demo walks through each step to render a reflective form in an outdoor environment. There is also a video link. The layering strategies are the ones described earlier on page 179.

Start with a line drawing. The better you understand the sections of the form, the easier it will be to plot the reflections. Remember, plotting the reflections is about predicting where your line of sight will bounce off the reflective surfaces and into the environment.

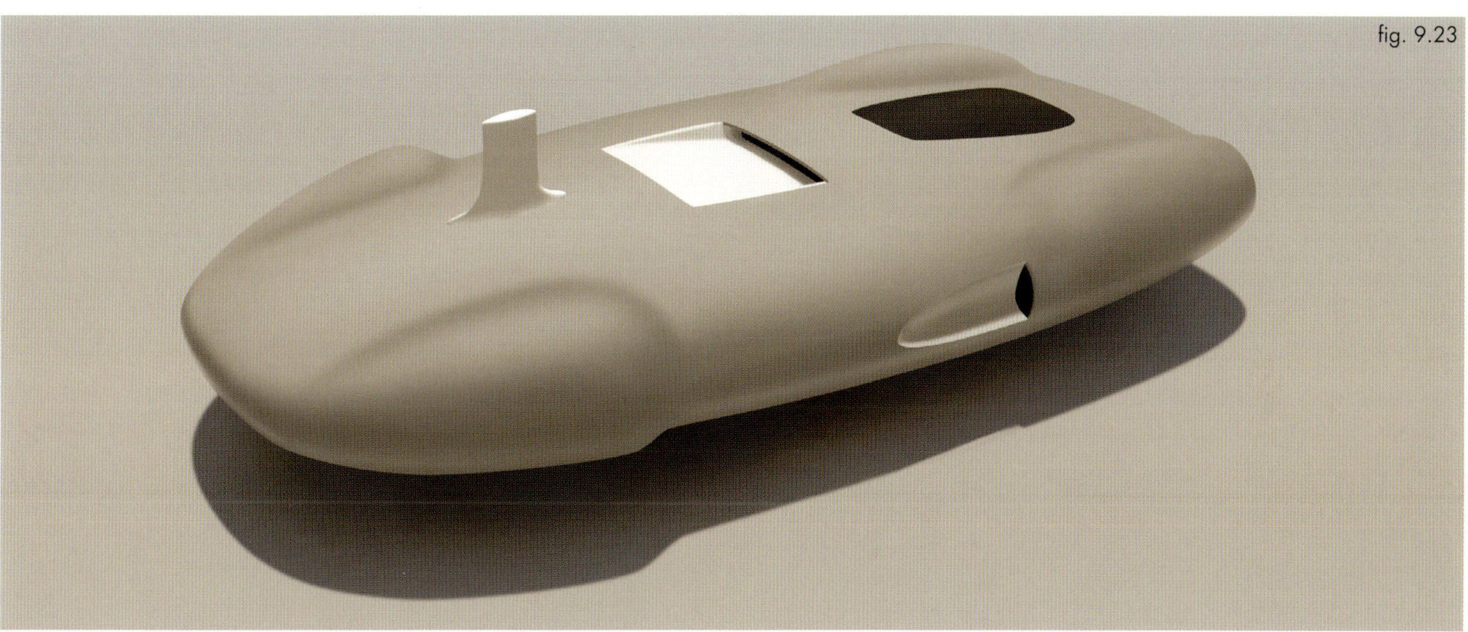

fig. 9.23

Make layers using paths so clean edges will be easy to retain. Decide upon the position of the sun. This form is going to start with a matte surface so the lessons of the first half of the book apply. Build the forms in gray scale (fig. 9.23).

Colorize the matte-gray layer (fig. 9.24). Remember the lesson of how interconnected color and value are. To get this very dark maroon color the values of the gray layer had to be adjusted by lowering the layer's white point.

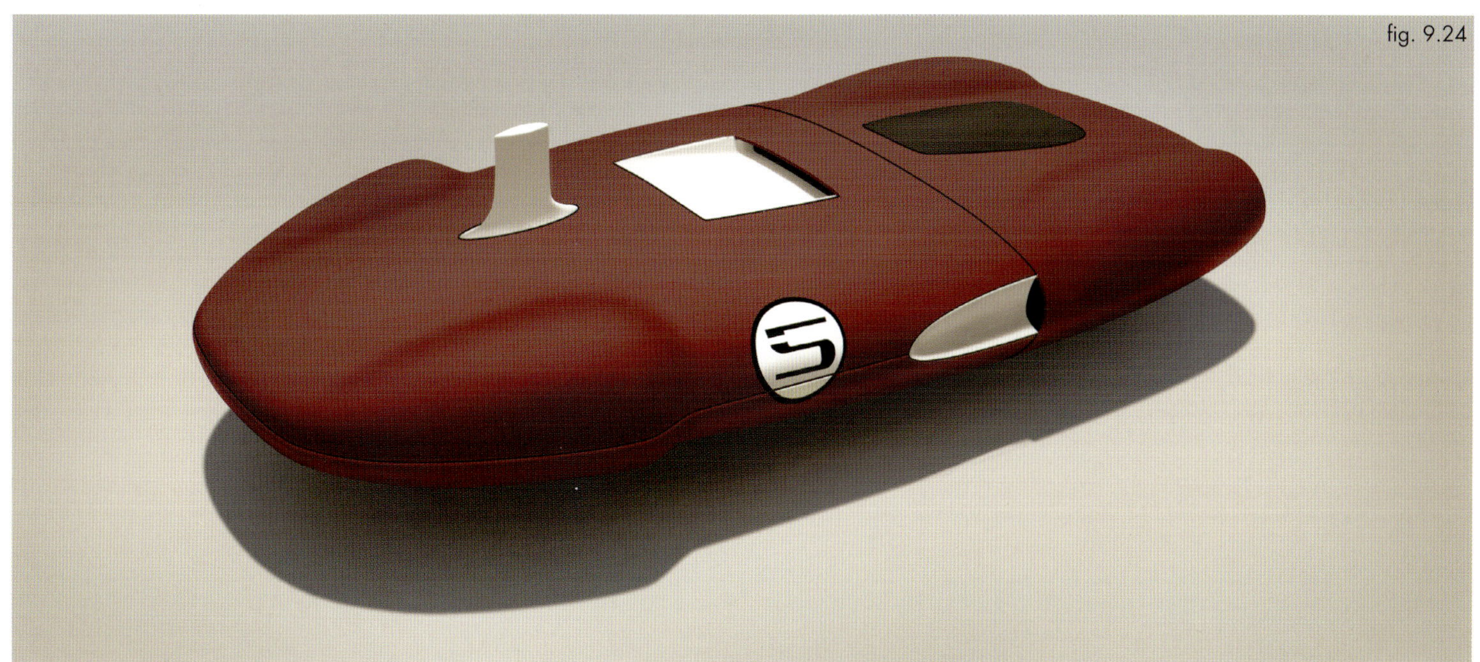

fig. 9.24

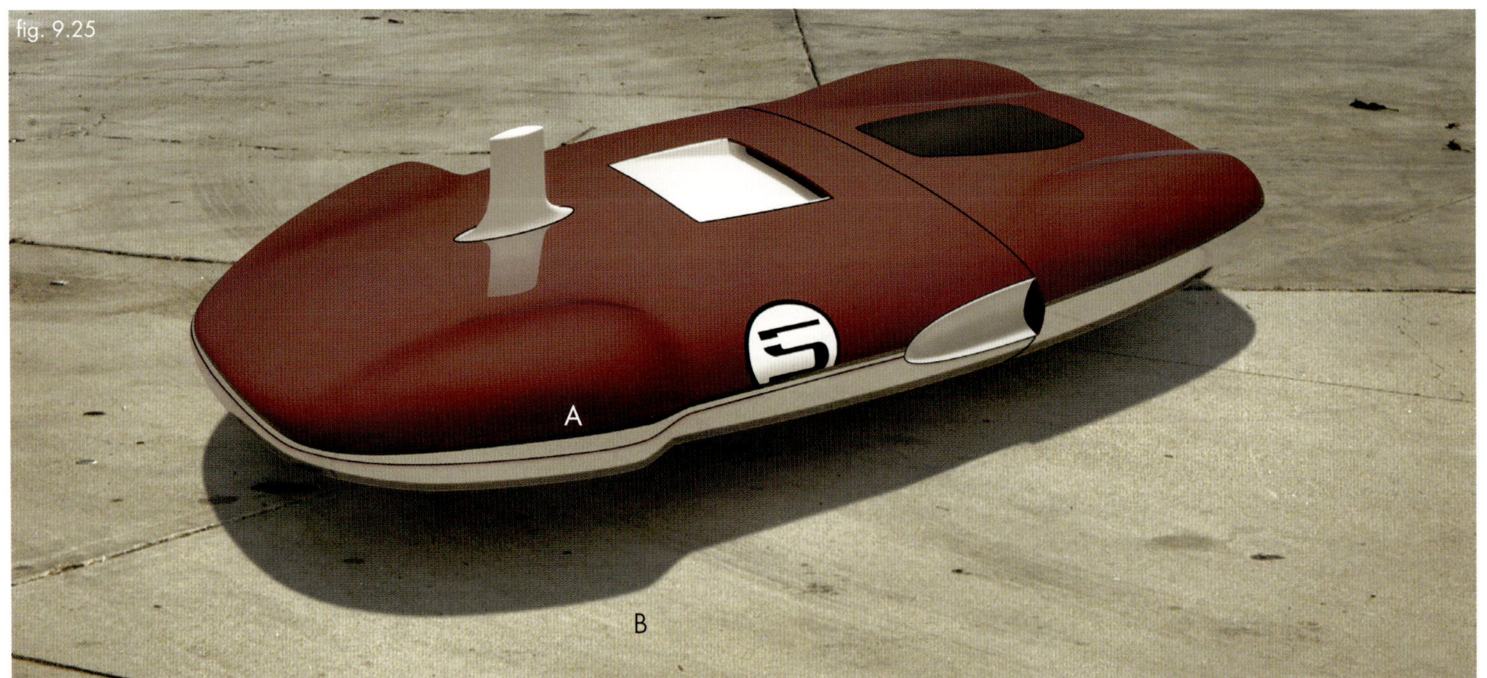

fig. 9.25

The surrounding environment is a simplified one. The sky is clear and the sun is directly overhead. The ground plane is concrete with a distant, dark, low wall surrounding the object (not pictured). Use paths to isolate the edges of the reflection layers. Where the line of sight bounces off the form parallel to the ground plane, render the reflection of the distant wall (A). Below that, the foreground reflects into the form, and the form casts a shadow on the ground (B). Note that a cut line breaks through this reflection.

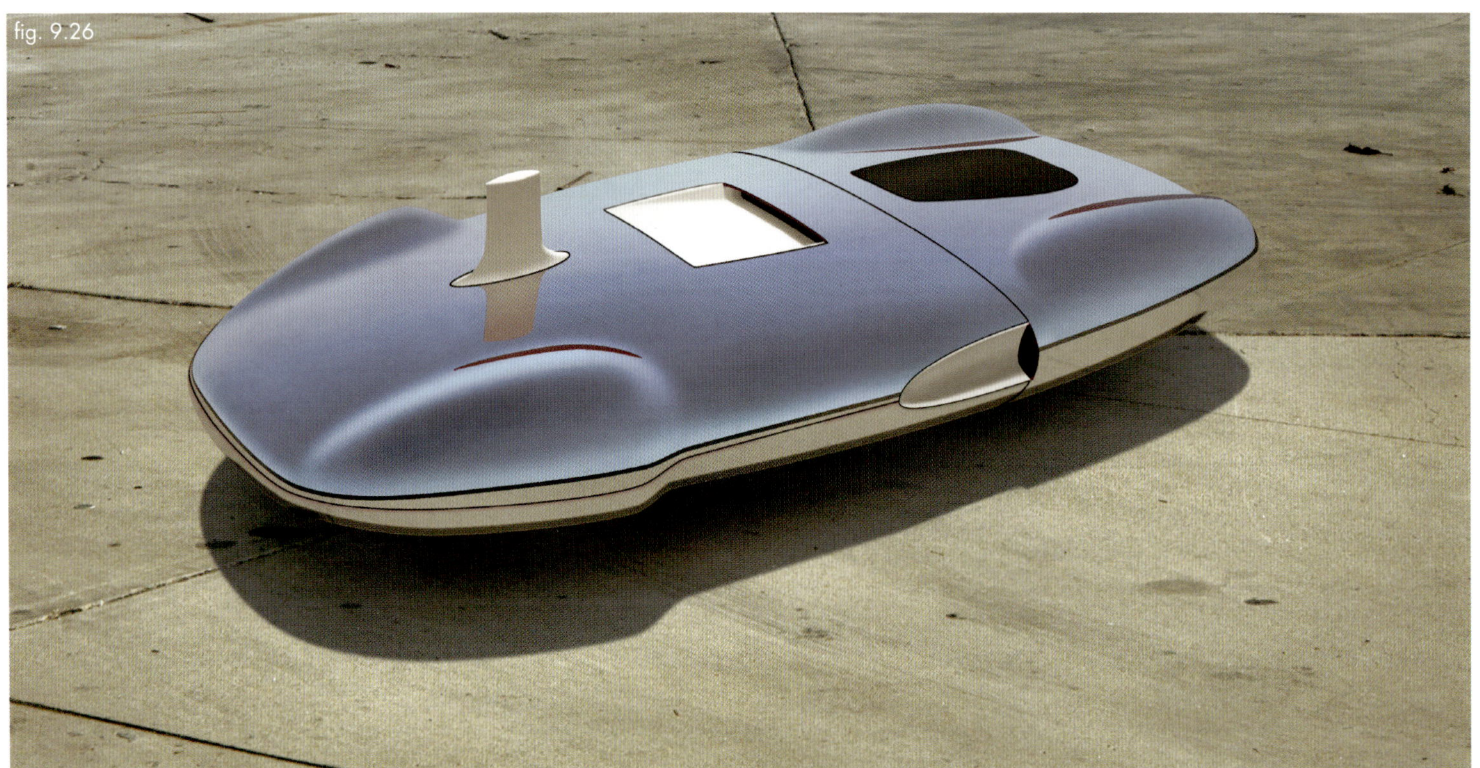
fig. 9.26

The sky reflection comes next. Make educated guesses as to what part of the sky the sight lines bounce off the form towards the sky. This is much harder to render than the ground plane. The ground-plane reflection is just cut and pasted onto the reflection layer and the level adjusted a little darker. The sky has that gradation to it and those colors and values are rendered now. The dark holes that are left out of the sky are where the fender-like forms might reflect other red-painted parts of the object like the puddles and pools observed back on page 181.

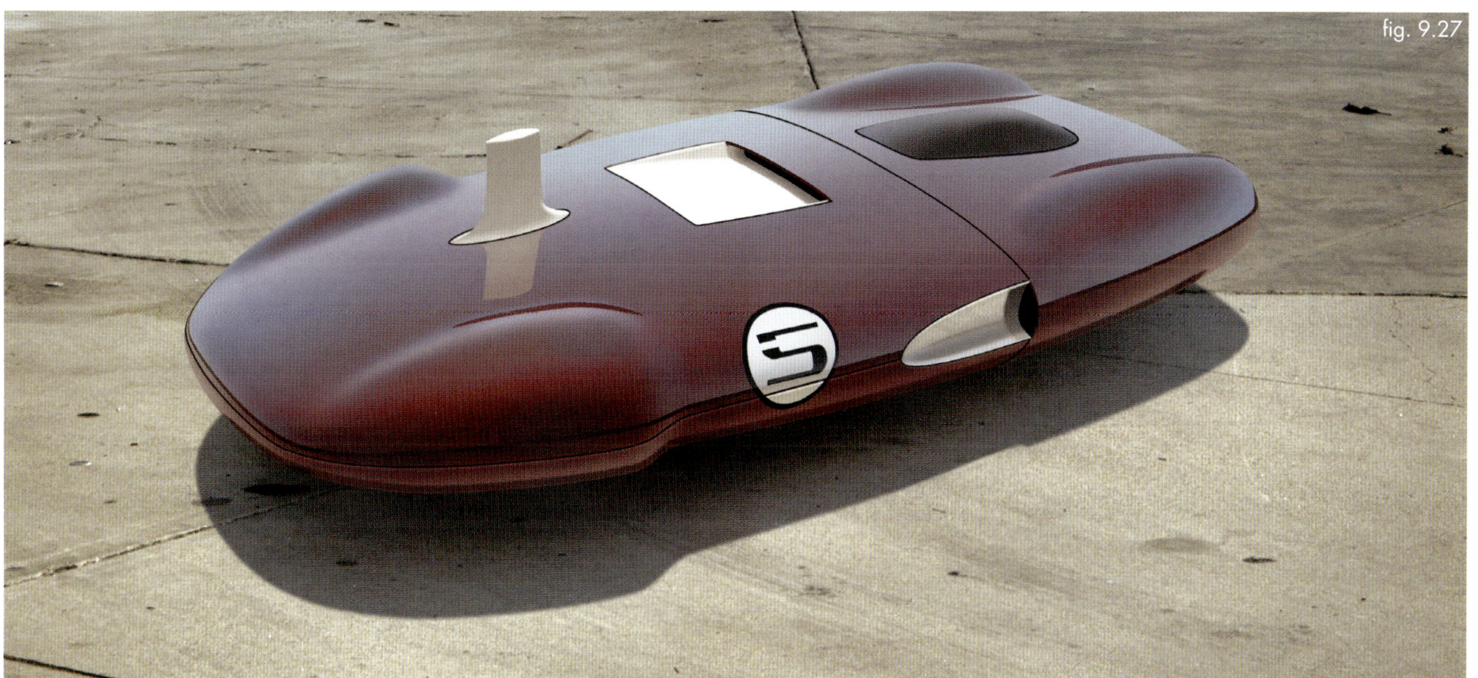

Once the chrome reflection layers are complete, render the layer masks to introduce the Fresnel Effect. Remember that reflections, if rendered on more than one layer, should never overlap each other. All of these reflections, even though they might exist on separate layers for more rendering control, represent just the reflections on the single clear-coat layer. If they overlap and the opacities of the individual layers are adjusted, these overlaps would show through.

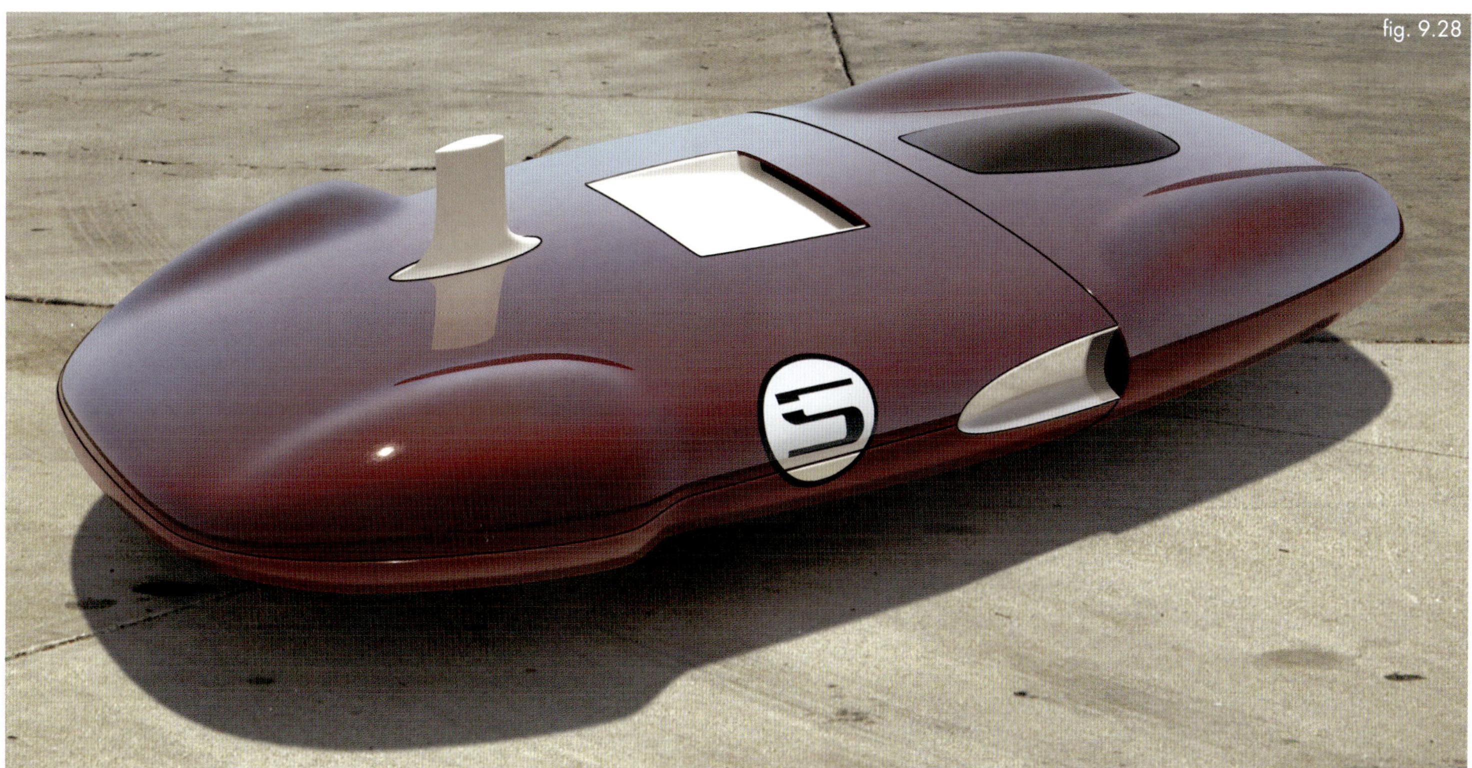

The last step is to add the reflection of the sun. On a clear day the sun does not change the color or value of the sky, except at dawn or dusk. The shape of the sections influences the shape of the sun's reflection. On a spherical form it will be a circle, but in this case the sun's reflection is slightly compressed and tilted by the convex forms of the object. Every step of this rendering was recorded in the video linked to this page.

HOW TO RENDER EYES BY NEVILLE PAGE

Again, creature and concept designer Neville Page shares his approach to rendering, in this case, reflective eyes.

fig. 9.29

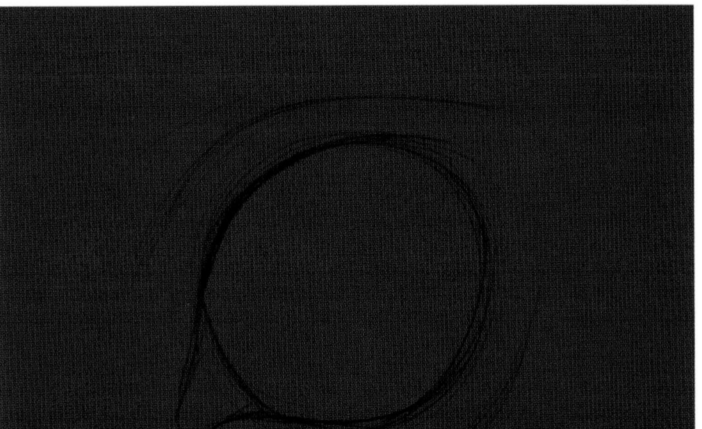

This is going to be a simple eye design. My preference is to work on a darker value background in Photoshop.

fig. 9.30

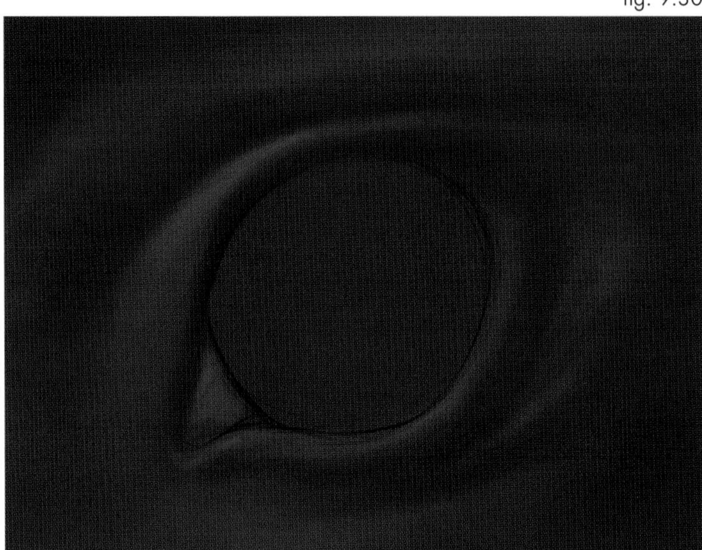

Using simple brushes, block in some form with a little light and dark value.

fig. 9.31

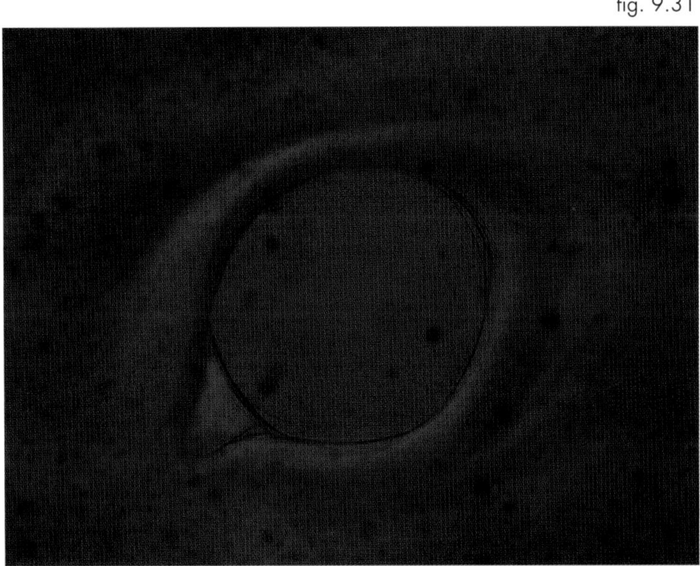

Break up the surface a bit with just a sprinkling of texture.

fig. 9.32

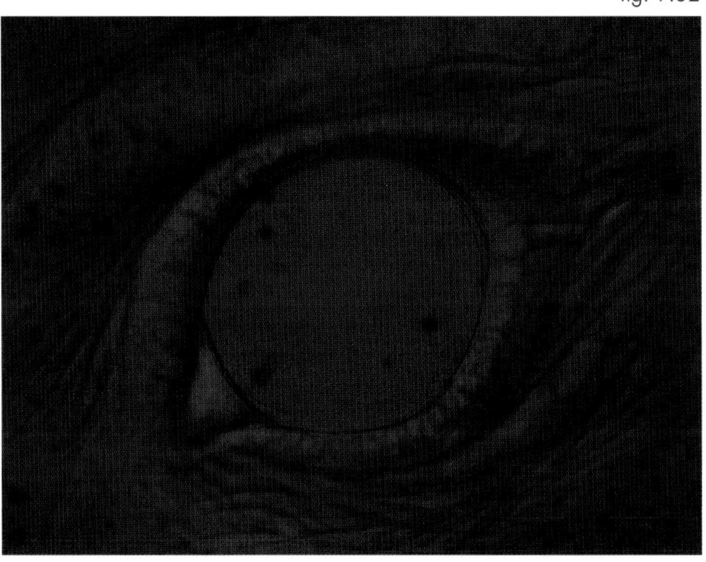

Now, the tedious part, drawing all the wrinkles. Having a good reference is key here.

fig. 9.33

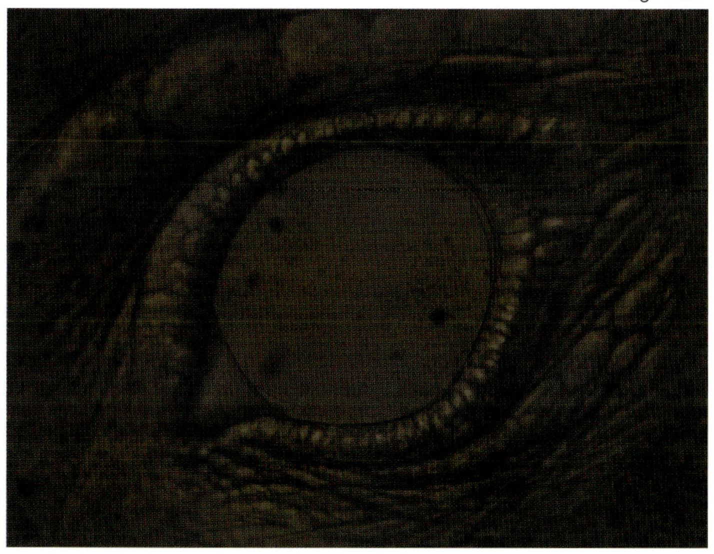

Once defined, start to render each wrinkle and scale. Hours could be spent here, but this is a shorthand version. Just enough to get it to read.

fig. 9.34

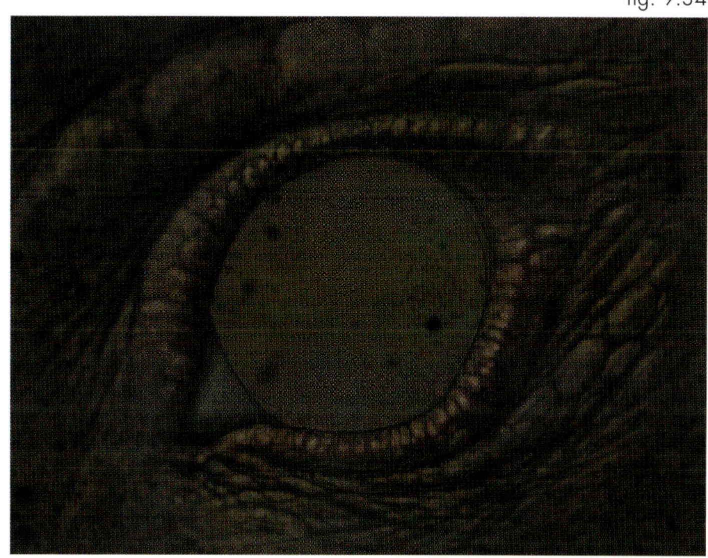

It was too monochromatic so a layer was added to adjust colors.

fig. 9.35

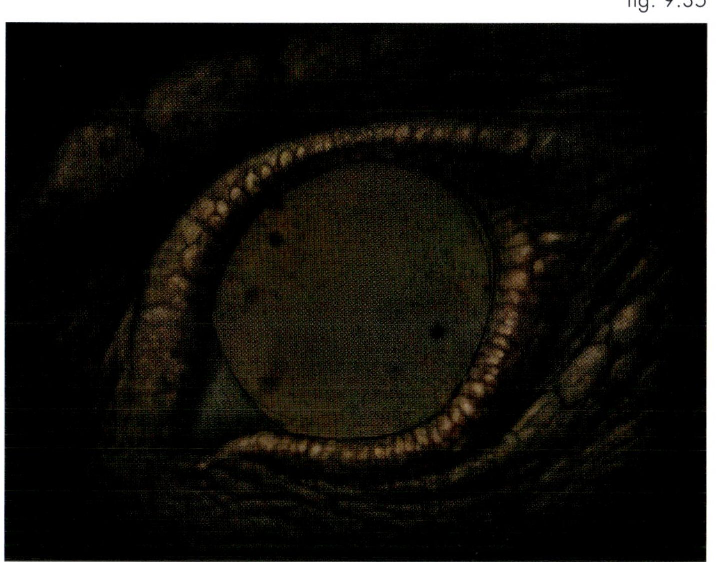

Adjust the contrast and saturation so that there is context for the eyeball.

fig. 9.36

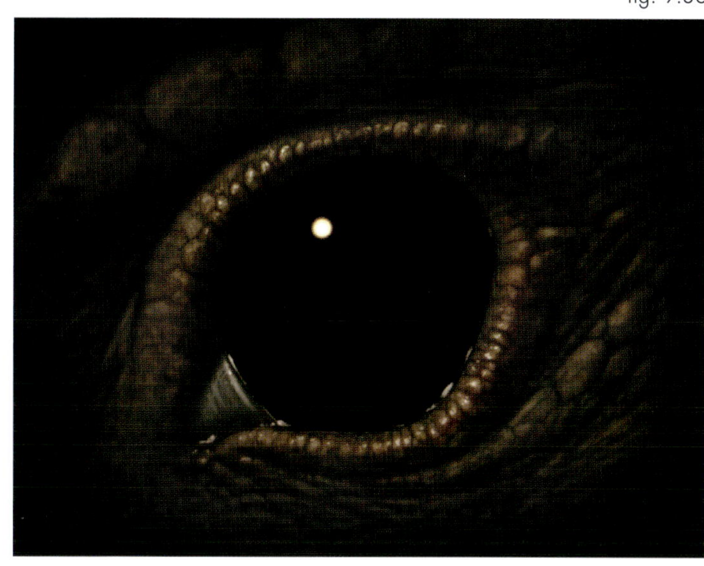

Paint a solid-black area where the eye will go, and then paint in the first reflective highlights.

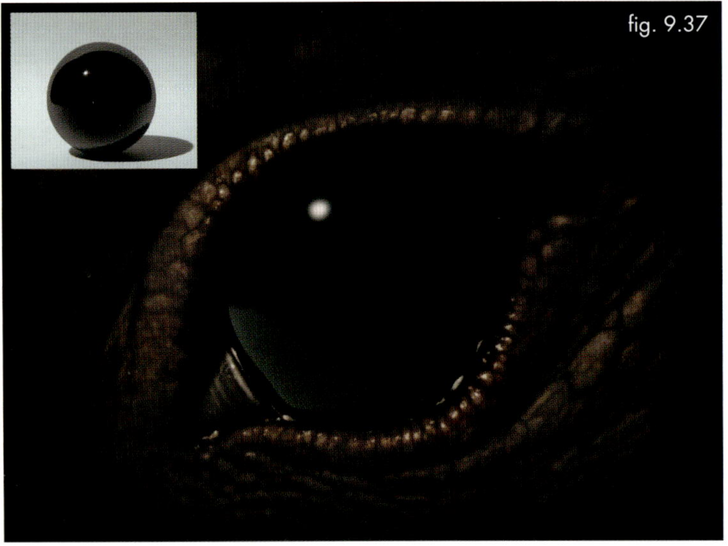
fig. 9.37

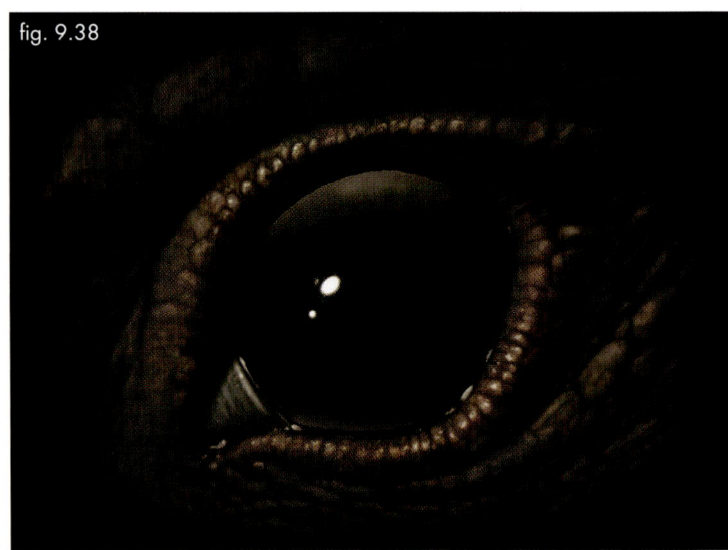
fig. 9.38

The next step is to map out what makes an eye "pop." Reflections! It is critical to understand the environment the character is in, then map it onto the eye. Refer back to Chapter 07. I actually took the black sphere from page 168 and used it as the eyeball.

Considering potential environments, I painted two simple, different scenarios: interior photo studio (fig. 9.38) and exterior environment (fig. 9.39).

Adding either of these simple reflective layers already gives the sense of a complete eye. Now any sclera and iris can be created and placed underneath a reflective layer.

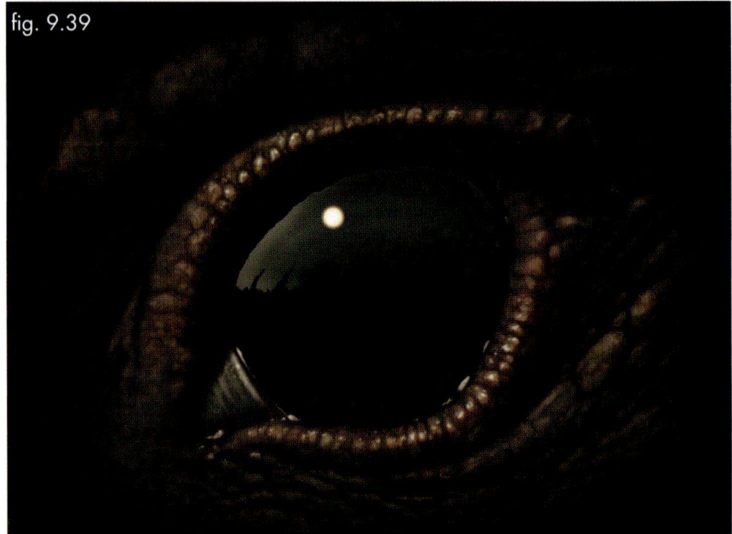
fig. 9.39

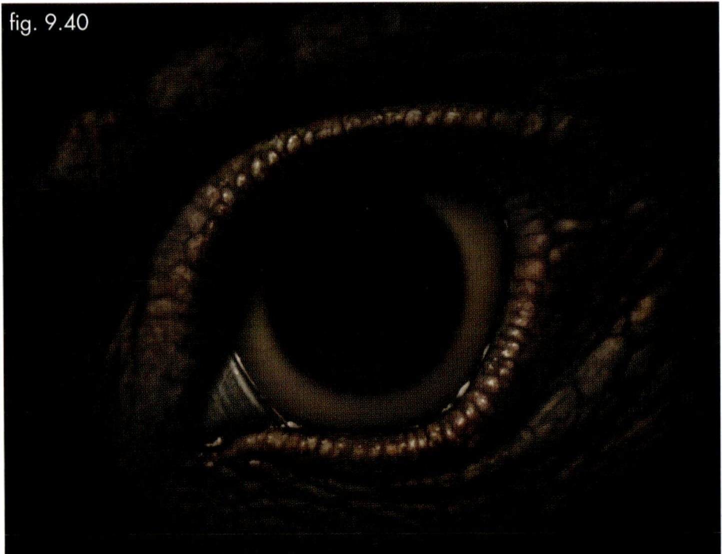
fig. 9.40

To illustrate this, I painted a basic eye. It is a simple render but still generates enough detail and precision to convey a sense of realism.

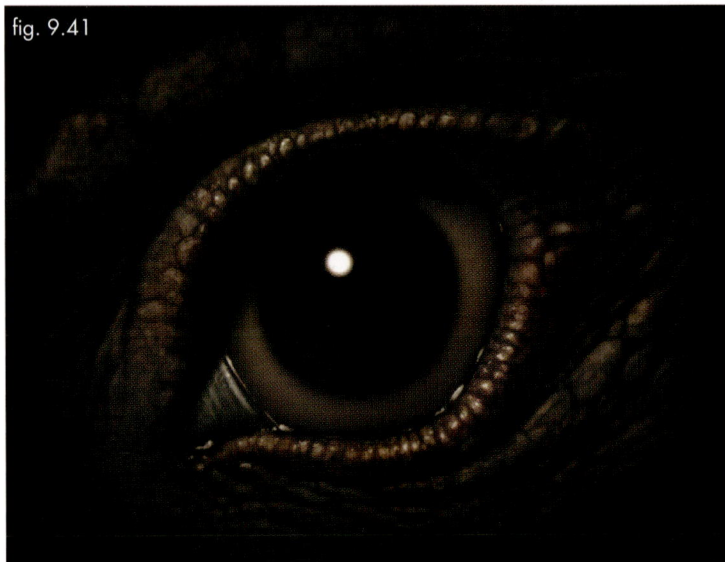
fig. 9.41

Fig. 9.41 uses the basic trick of a singular reflective highlight.

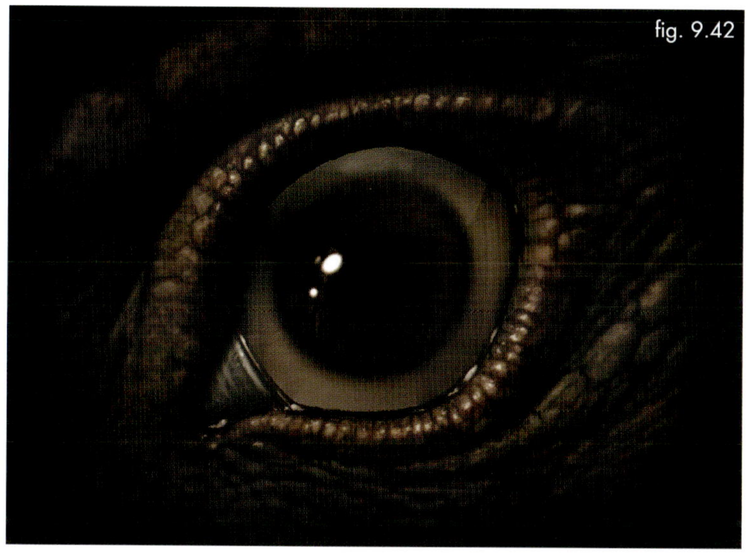

fig. 9.42

Indoor Environment

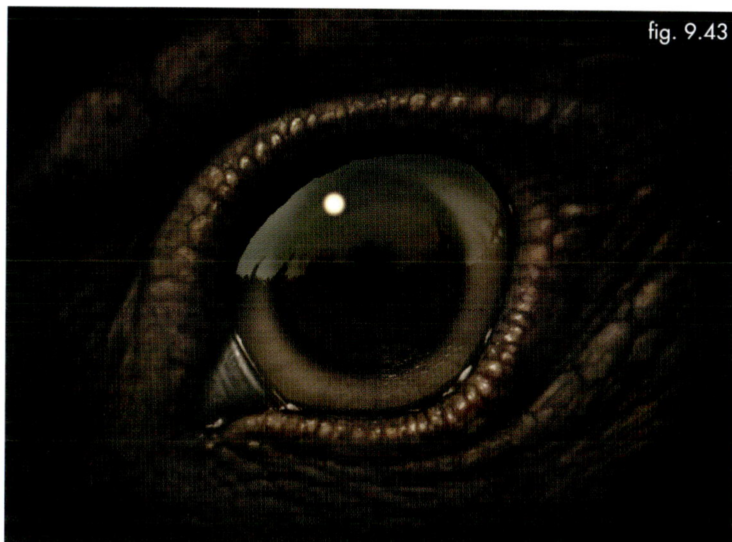

fig. 9.43

Outdoor Environment

Fig. 9.44, below, is the final illustration.

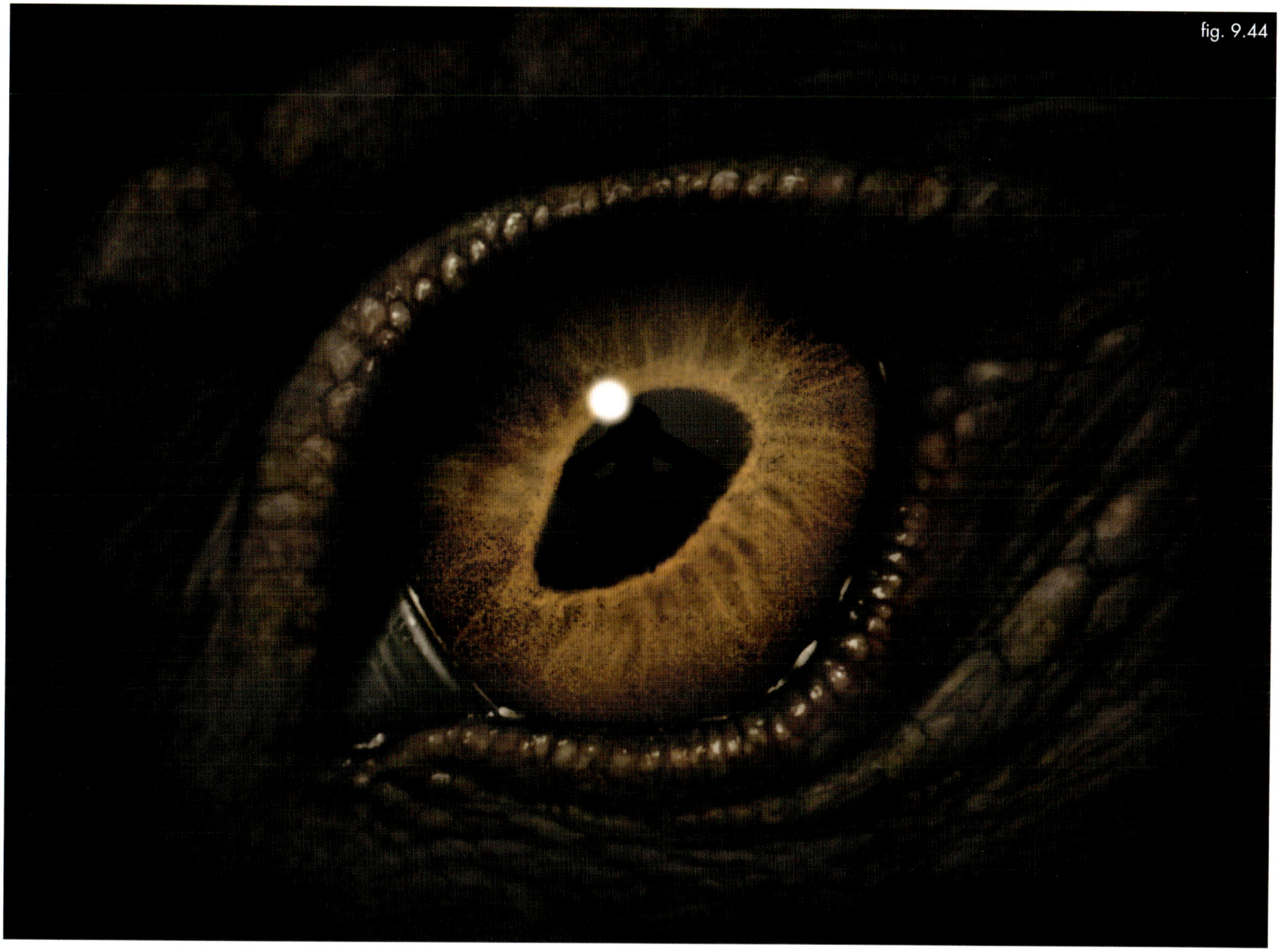

fig. 9.44

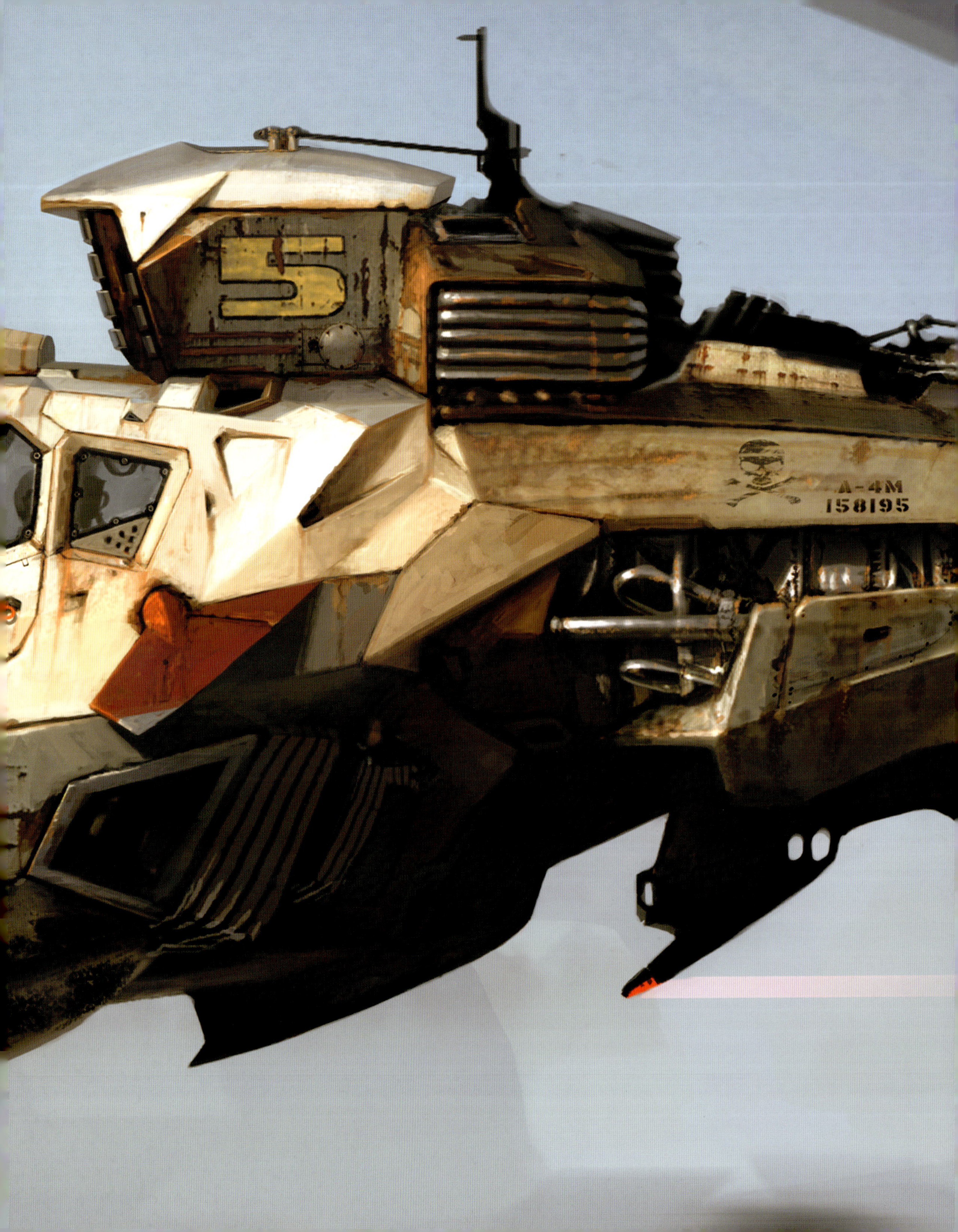
5
A-4M
158195

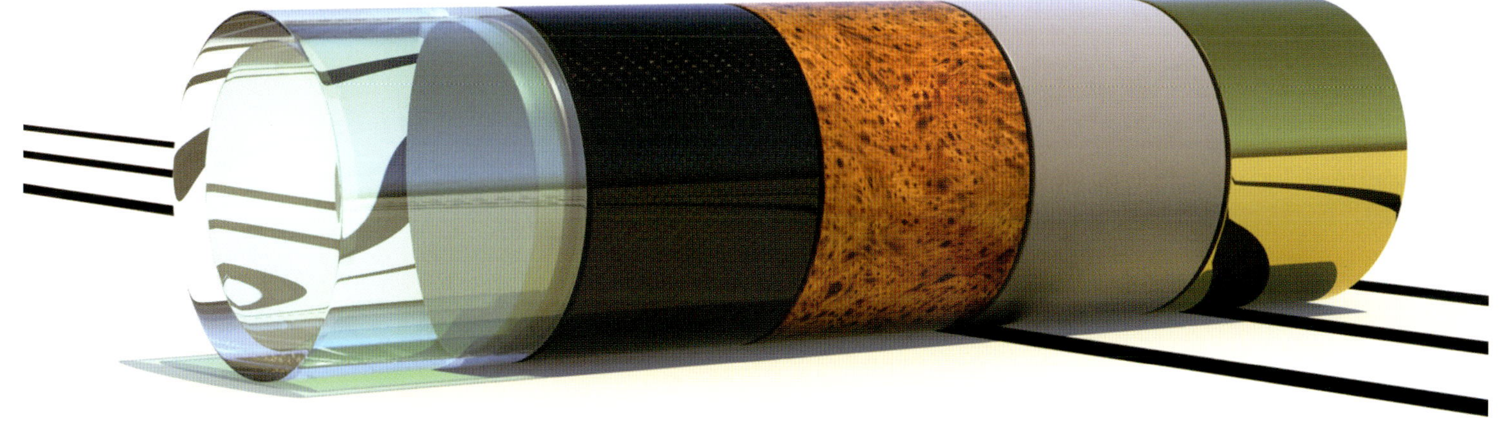

CHAPTER 10

RENDERING SPECIFIC MATERIALS

Now that the fundamentals of rendering reflective surfaces and communicating form with value changes have been explained, these rendering principles based on physics can be applied in varying degrees to indicate any material. The selection of materials featured in this chapter is a nice cross section but by no means an all-encompassing list. The following pages have been designed to help increase awareness of the subtle differences and similarities of some of the more common materials. Sharpening observation skills and applying the knowledge of rendering you now have to understanding these new materials will make your renderings look much richer. The differences in material-indication rendering can be small, so pay special attention to the characteristics of each. Start to build a good photo-reference library. The color variations between polished copper, gold or brass are the types of things you will want to learn to research and then render on your own. Let's get to it!

 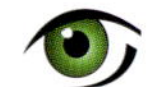

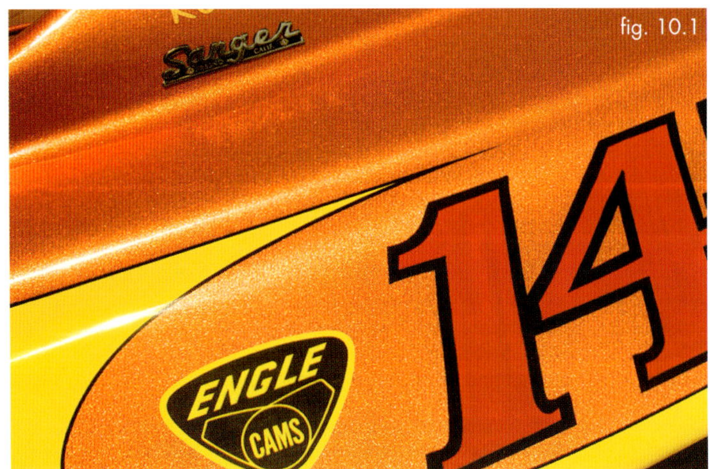

fig. 10.1

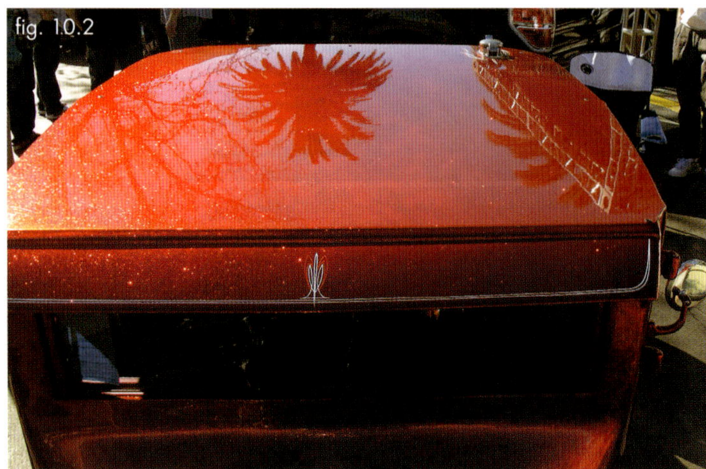
fig. 10.2

Metallic paint differs from and builds upon what has been taught so far, in that it affects the base coat but not the clear coat on top of it. Visually a metallic-painted surface often overrides the strength of the clear coat; it visually trumps the clear coat, which is the reflection of the environment. Here's what happens: A metallic base coat has lots and lots of tiny little mirrors floating in it. When they are sprayed onto a surface they all land on the surface at different angles.

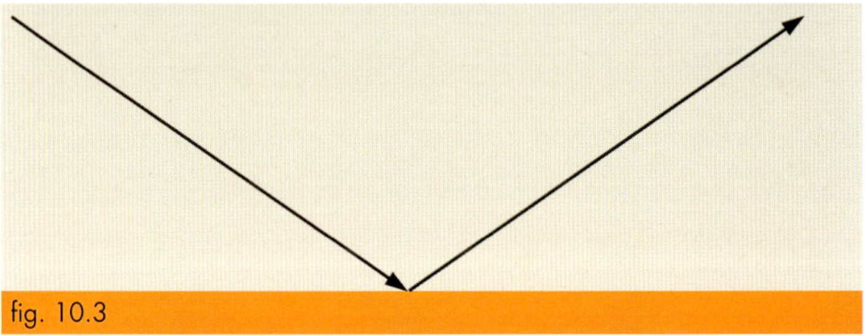
fig. 10.3

Normally, when looking perpendicular at something shiny, your line of site bounces straight back at you. Looking at the surface with a 34° angle of incidence, your line of sight bounces off that spot at the same angle, 34° (fig. 10.3). What happens with the tiny floating mirrors, or metal flakes, is that when you look across that surface, your line of sight bounces off those tiny mirrors in all different directions (fig. 10.4), many of which inevitably bounce the sight lines back to the sun, or other light source, from a wider surface area than would be the case with a glossy, nonmetallic paint. When observing a metallic surface, the reflection of the sun or other light source is intensified compared to a glossy surface.

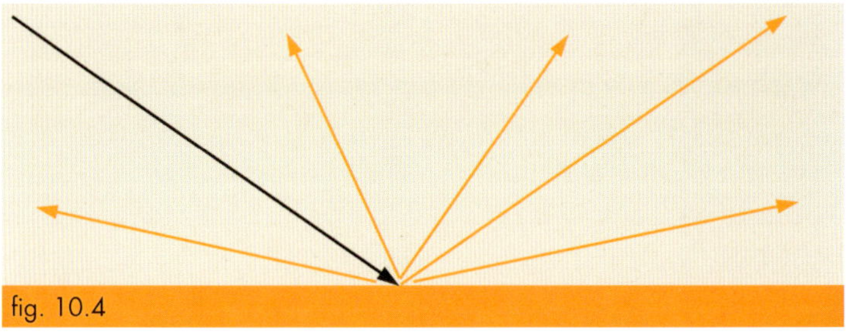
fig. 10.4

Fig. 10.5 is an example of a metallic material: stamped aluminum that has been sandblasted. Notice how brightly it reflects the sun over such an increased area of the surface. The sun's reflection overpowers the reflection of the blue sky. Light wins again. The sandblasting of the metal surface roughens it, which results in the sight lines bouncing all over, much like the metal flakes in the base paint of figs. 10.1 and 10.2, above.

The metallic base layers appear to have little flecks in them. By adding "noise" to a base layer in Photoshop, the same look can be accomplished. However, the image must be printed or projected at a high resolution for the "noise" layer to be visible.

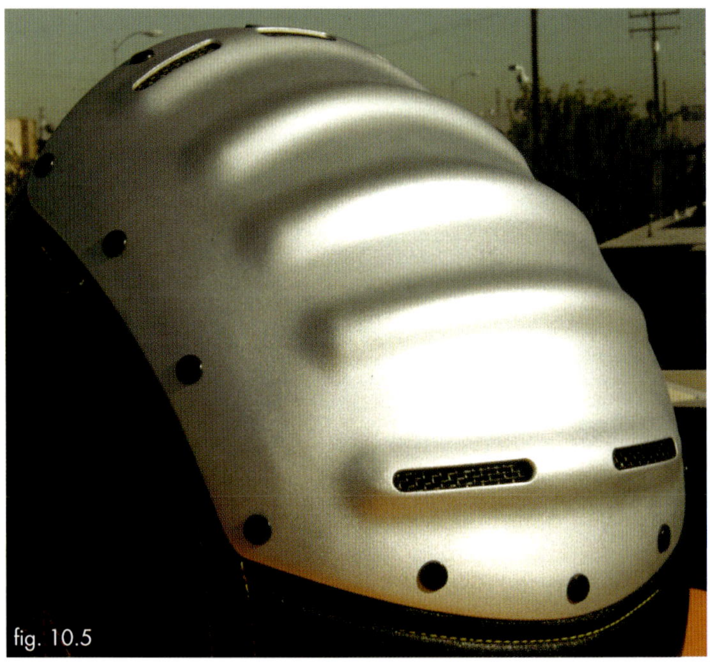
fig. 10.5

To indicate this effect of the metal flakes, first look at the clear coat and wherever it reflects the sun, or other light source, softly expand this reflection of the light across the base-coat layer by "color dodging" it toward the color of the light source. In fig. 10.6, the shoulder of the red car has multiple soft reflections of the spotlights across a much larger surface than the sharp reflections of the lights on the clear coat, which are only small dots. This metallic effect happens independently of the strength of the clear-coat reflections, which are on the layer above it.

The blue car shows what the reflection of the sun looks like on a matte metallic surface. The clear coat has not been sprayed over this base paint yet, so the effect of the metal flake is easier to see. Notice how easy it is to understand the forms of the car with only one light source and no clear-coat reflections, unlike the red car, which is in a more complex indoor setting.

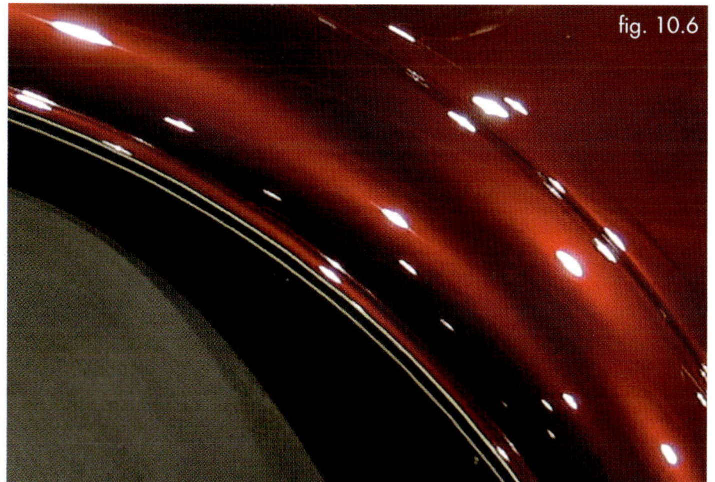
fig. 10.6

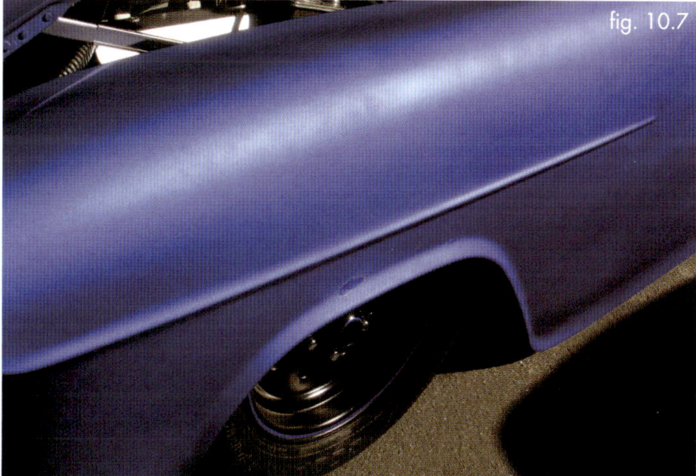
fig. 10.7

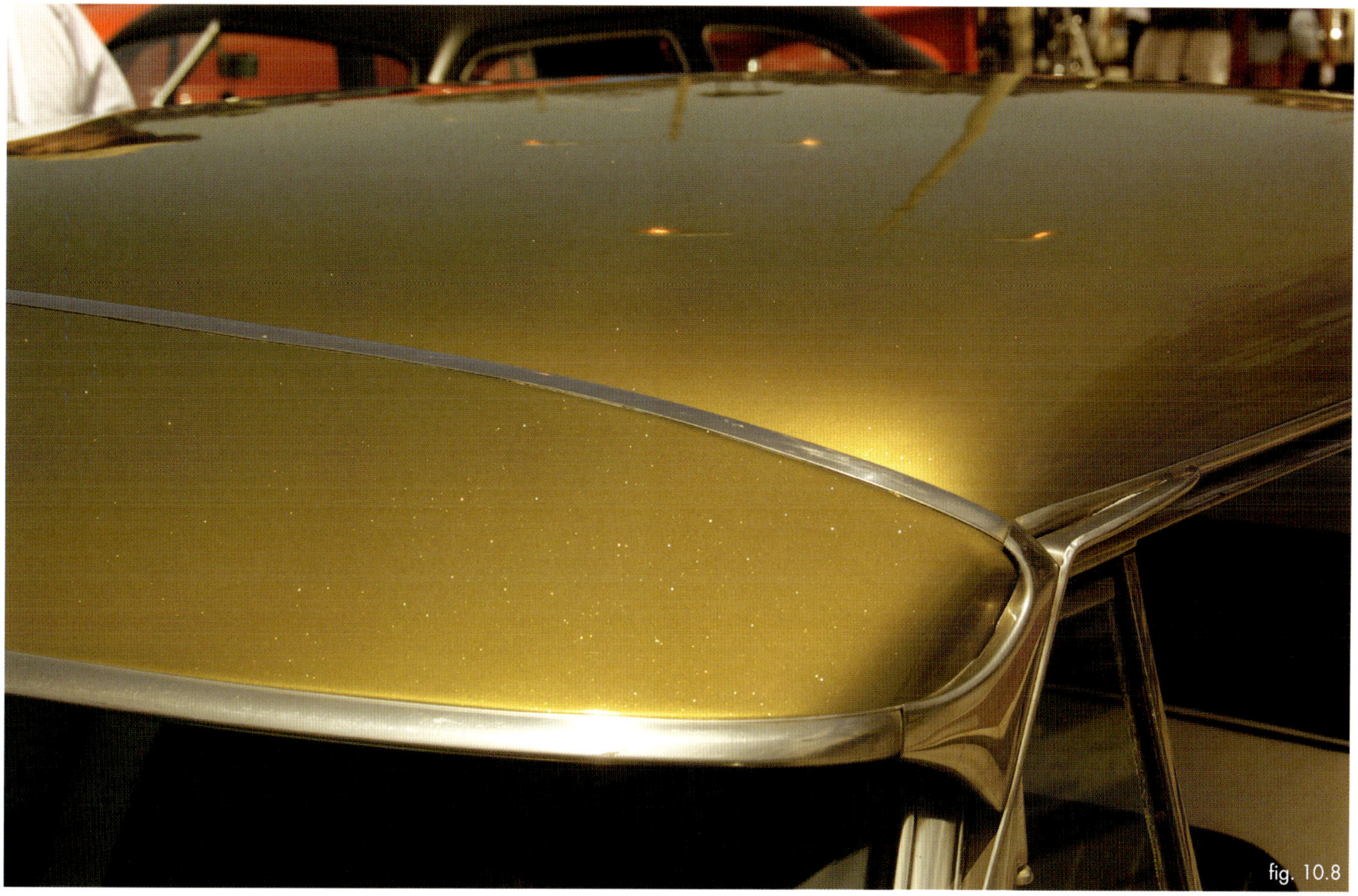

fig. 10.8

METALLIC PAINT VS. GLOSSY PAINT 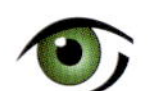

Here are nice examples of metallic paint and how strong the reflection of the light can be, due to the metal flake in the base paint. Contrast this with the glossy painted surfaces on the facing page. A bit of the sun's reflection still stretches here and there across the form, but that is due to scratches in the clear coat and not from any metal flake present in the base paint. It is no surprise that almost all automotive clay scale models are evaluated in a metallic finish. Metallic paint reflects less of the surroundings and more of the light, which in turn increases the value changes across the surface, making it easier to evaluate the forms of the model. If painting a car to show its forms, do not paint it glossy black; paint it metallic silver or metallic gold. Glossy black does not show off the forms well. Just like wearing a black sweater does not show wrinkles and forms as much as a white sweater does.

When rendering reflections in a metallic paint, a fillet reflects the light the same way as a radius. Render them just the same, but remember that the curvature of the sections dictates the width of the reflections of the light source, the core shadows and the distortion of the environment reflection on the clear coat. This kind of stuff is easily forgotten when starting to render. When the fillet is tiny, the reflection of the sun is thin. The softer/flatter the section, the slower the gradation changes. Let the sections of the shape dictate the size of the airbrush used.

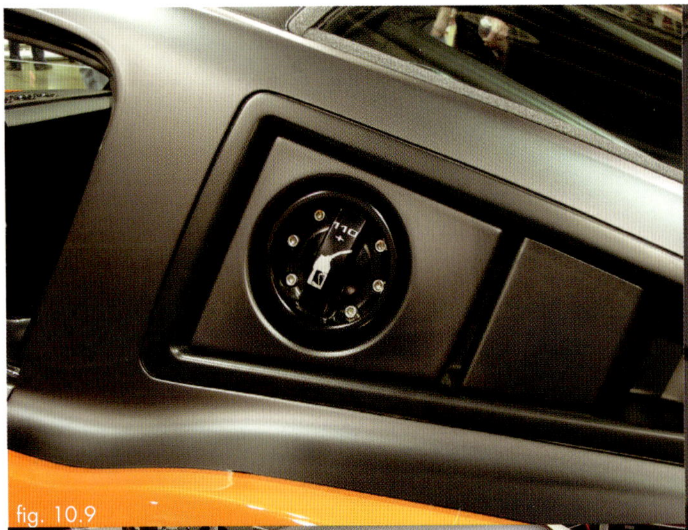

fig. 10.9

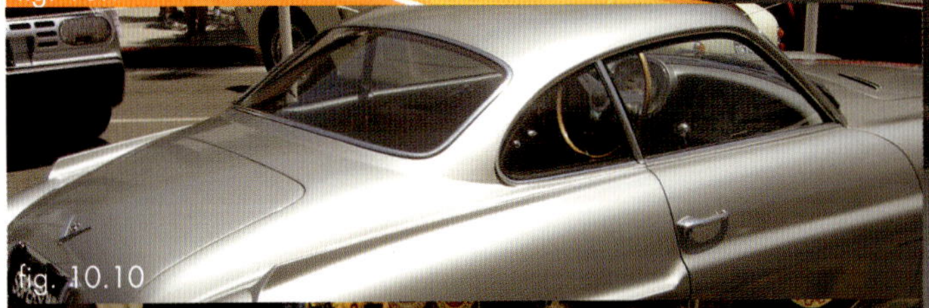

fig. 10.10

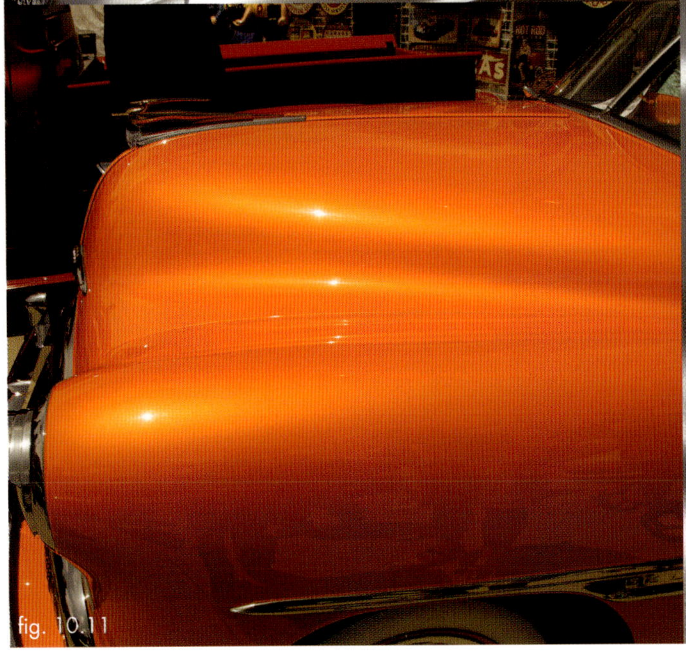

fig. 10.11

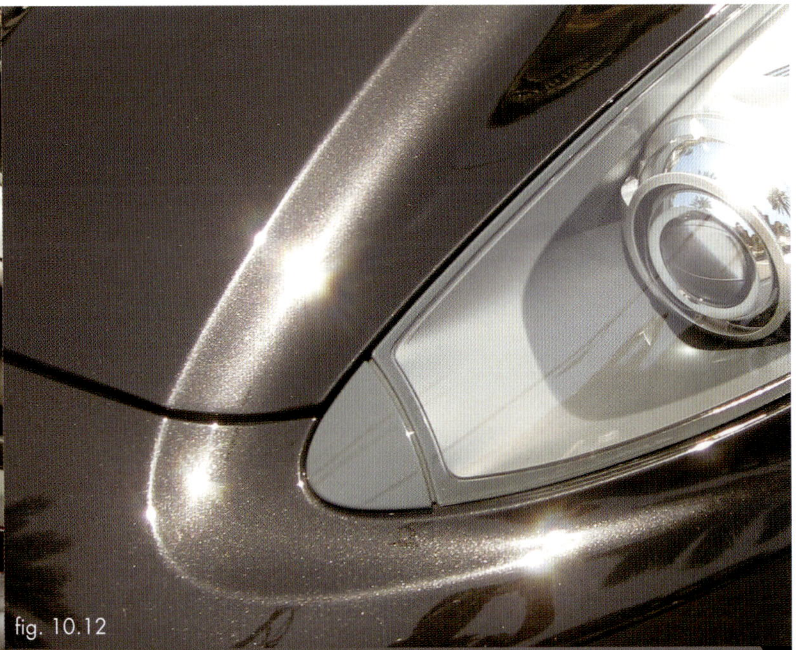

fig. 10.12

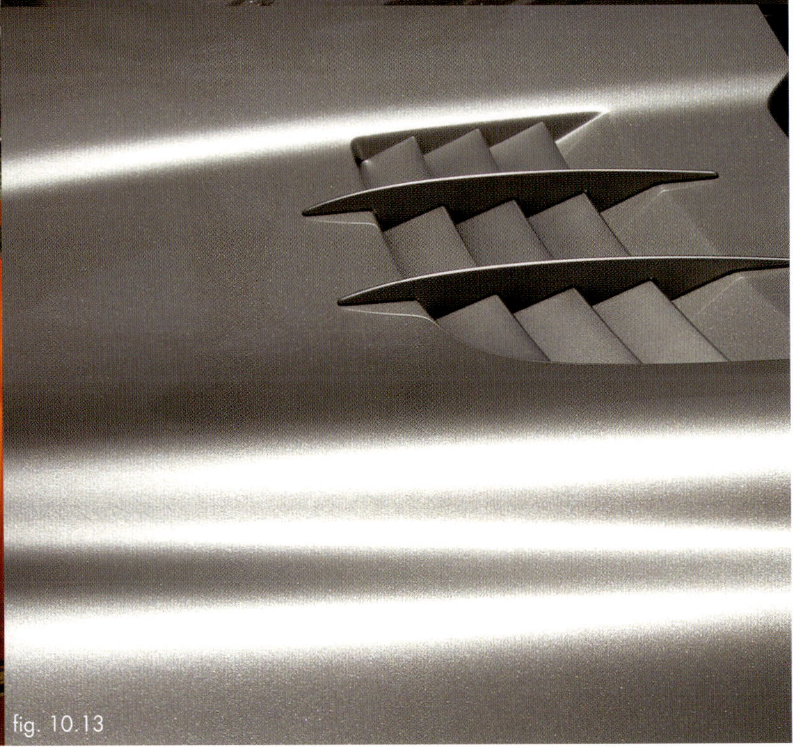

fig. 10.13

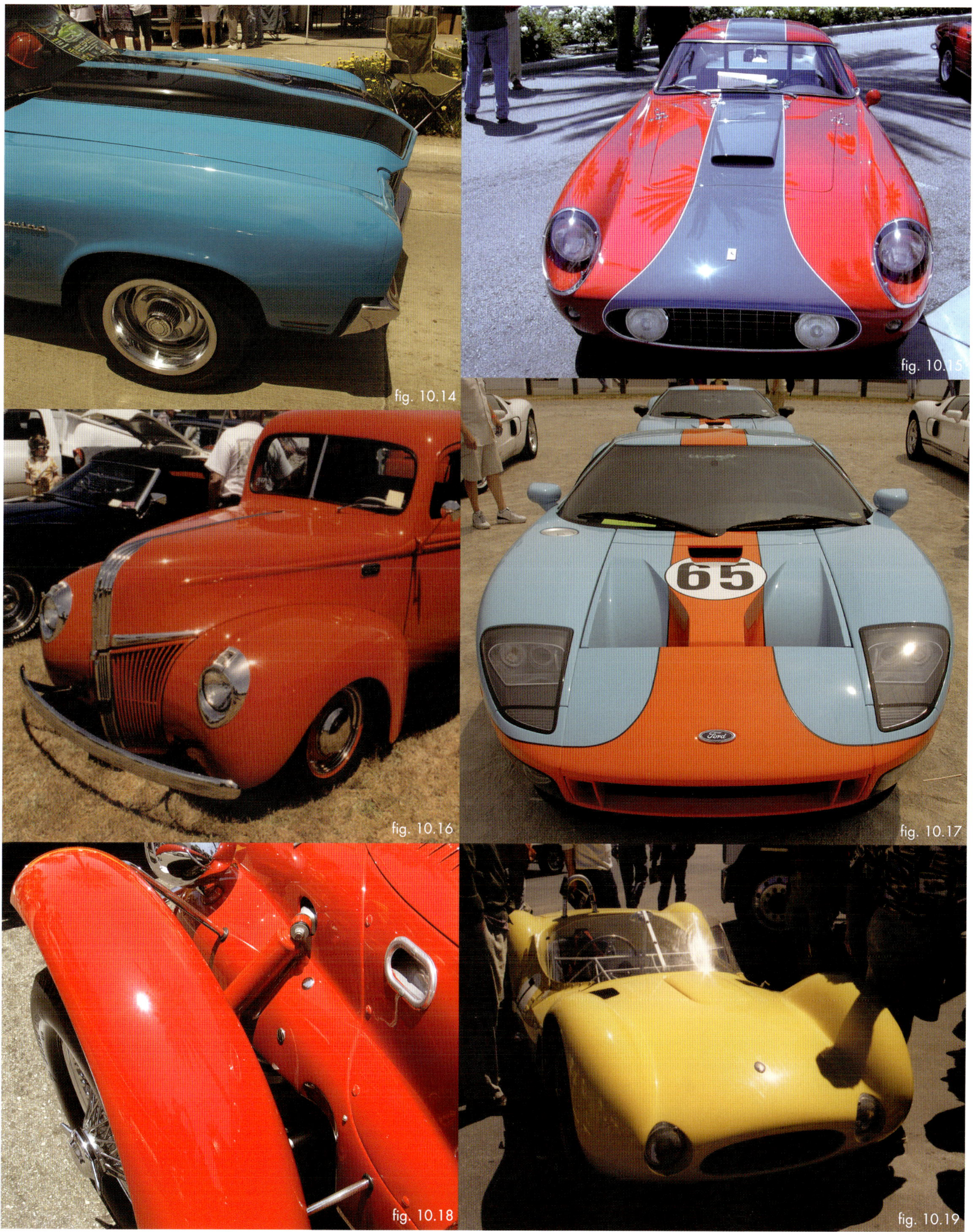
fig. 10.14
fig. 10.15
fig. 10.16
fig. 10.17
fig. 10.18
fig. 10.19

DIGITALLY RENDERING A METALLIC LOOK

The metallic effect happens in the base-paint layer and is characterized by a strong soft reflection of the light source, which increases the overall contrast and range of the value changes across the surfaces of the form. It is not possible to achieve this effect by simply increasing the contrast of the values of the matte-surface base paint. The increase in contrast is caused by a reflection of the light, which may or may not occur in the same position as the passive highlights on the light side of the object. Remember a reflective highlight's position is based on the sight lines bouncing back to the light source (see page 66). Again, the best way to approach rendering this effect is to think about where the sun is reflecting on the clear coat and then color dodge the base paint toward the color of the light source, stretching the sun's reflection across the form of the object, as seen in fig. 10.20.

When using an airbrush set to "color dodge" mode, or by floating a "color dodge" layer above the base layer, it is best to use a low opacity setting for the airbrush, such as 15%, because it does not take much for the effect to be seen. Make a best guess as to how the light source should be softly reflected across the surfaces, then increase the contrast a little. Now add any special metal-flake effects, like those seen in the green paint of the car on page 211, fig. 10.8, and those on page 210, in fig. 10.1. If the finished rendering is going to be in high resolution, a subtle noise filter can be applied to the metallic-paint layer to achieve a little more realism and variation between the metallic paint and the clear coat over it. Use the H2Re app on your smartphone or tablet to watch videos of rendering these metallic effects for both images on this page. From a computer, go to the URL address, which is on page 270 and it will guide you to the video links page on the SRW website, scottrobertsonworkshops.com.

Fig. 10.22, on the facing page, was rendered using these same "color dodge" techniques to create the metallic finish of the silver parts of the suit. By having the dark matte-surface materials right next to the metallic ones, this material variation is made even more pronounced.

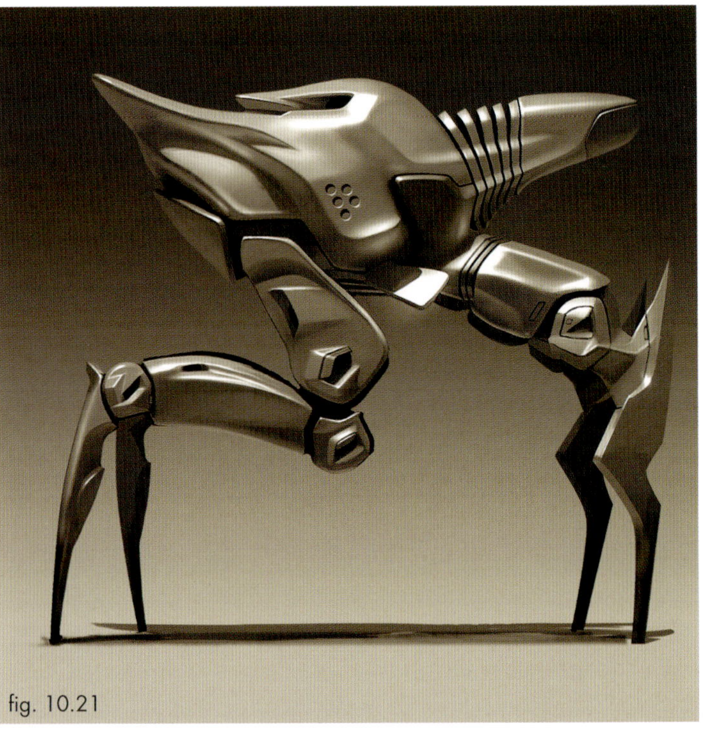

fig. 10.21

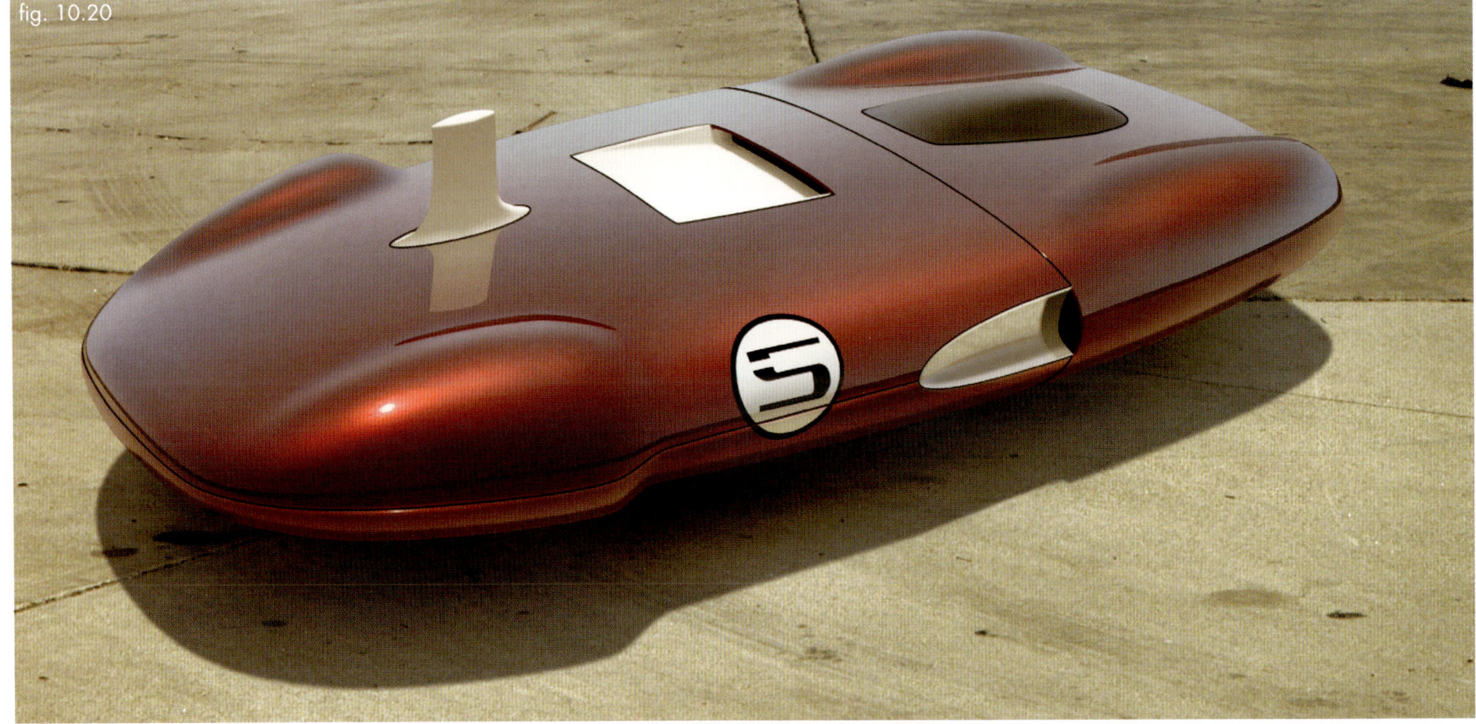

fig. 10.20

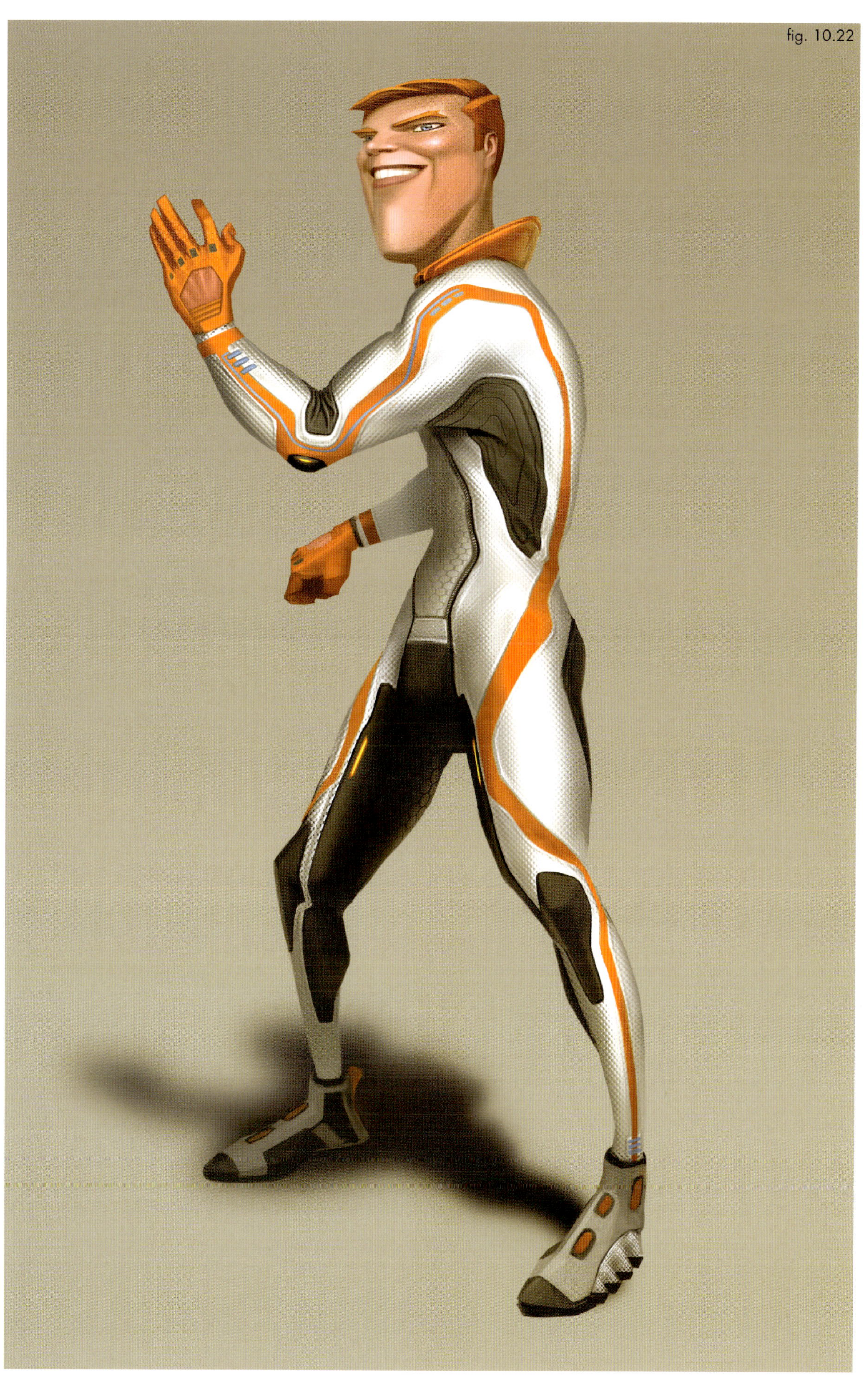
fig. 10.22

TRANSLUCENCE AND SUBSURFACE SCATTERING 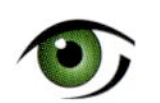

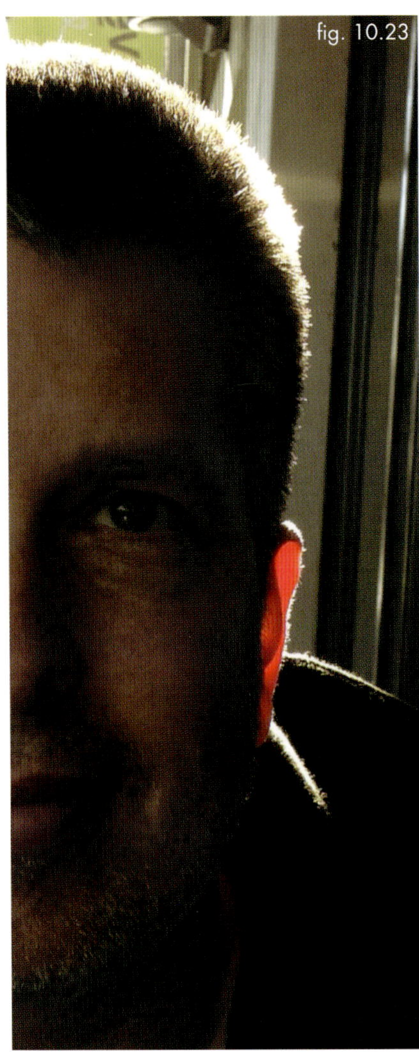
fig. 10.23

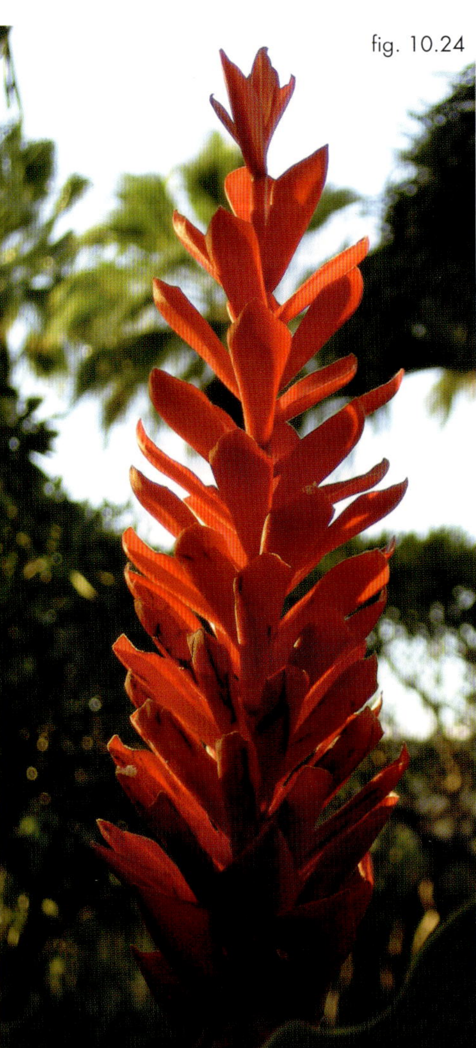
fig. 10.24

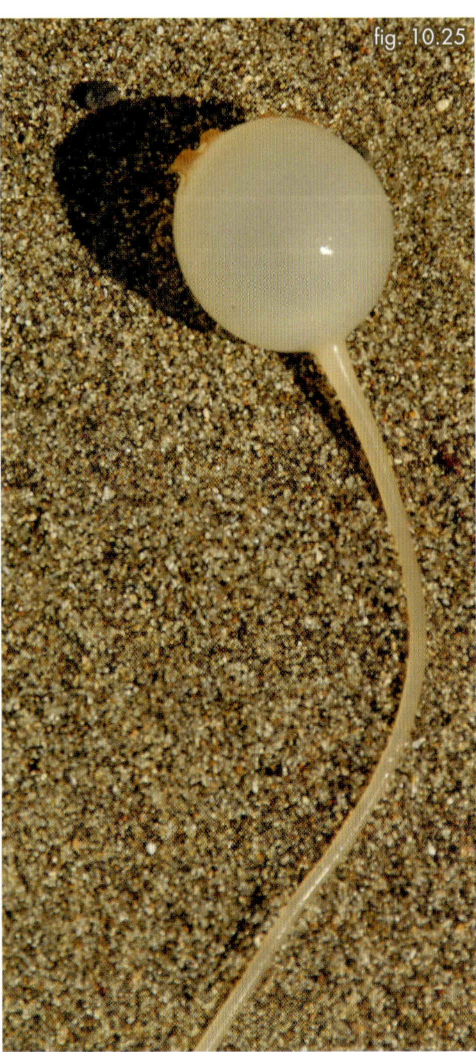
fig. 10.25

Translucent materials are one of the most difficult things to render. To make it easier, work from some good reference either in photographic form or observed in reality. There are many types of translucent materials. Some of the most common are organic materials like skin or plant life, but they can also be manufactured, like wax, or the colored plastic hood ornament in fig. 10.27.

What makes these materials especially tricky to render is that it is hard to predict the behavior of the light rays that illuminate a translucent material. The light rays penetrate this material and then scatter around in different directions within it, creating a glowing effect. This scattering around inside the material is called "subsurface scattering."

When an object is backlit (figs. 10.23 and 10.24), its translucency is usually more obvious. The thinner the cross section of the material, the more pronounced the effect. If an object is more front-lit (fig. 10.25), the light scatters around and will minimize the contrast of the matte-surface value changes. This means rendering a weaker shadow side and core shadow. Depending on the level of the material's translucency, some light might make its way all the way through the material and illuminates the cast shadow, as is fig. 10.27. The color of this cast shadow lighting will match the color of the translucent material.

In fig. 10.28, the cast and core shadows are much lower in contrast than would be expected if this were an opaque material. The organic material is effectively glowing just a little bit from within and reducing the value changes one would expect on the surface.

Since these subsurface-scattering effects can be so varied and dramatic, they will require deeper visual research from nature. Again, using a good reference will make these effects much more believable in your work.

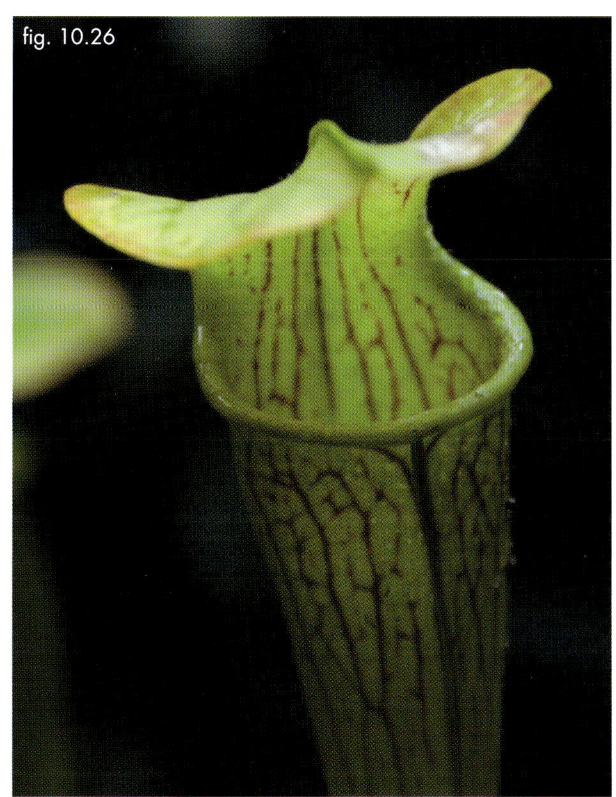
fig. 10.26

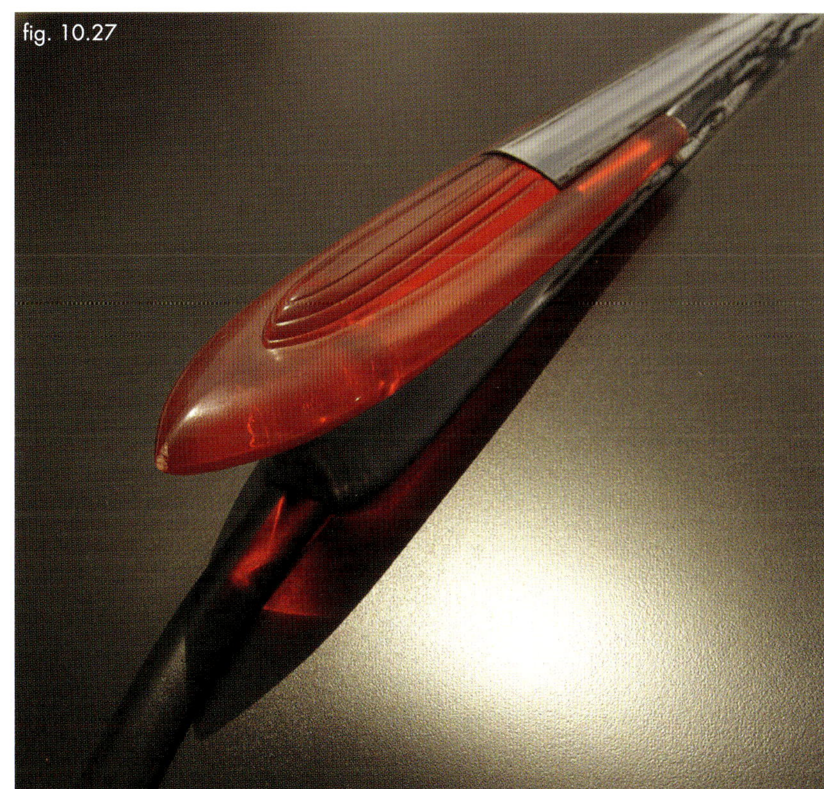
fig. 10.27

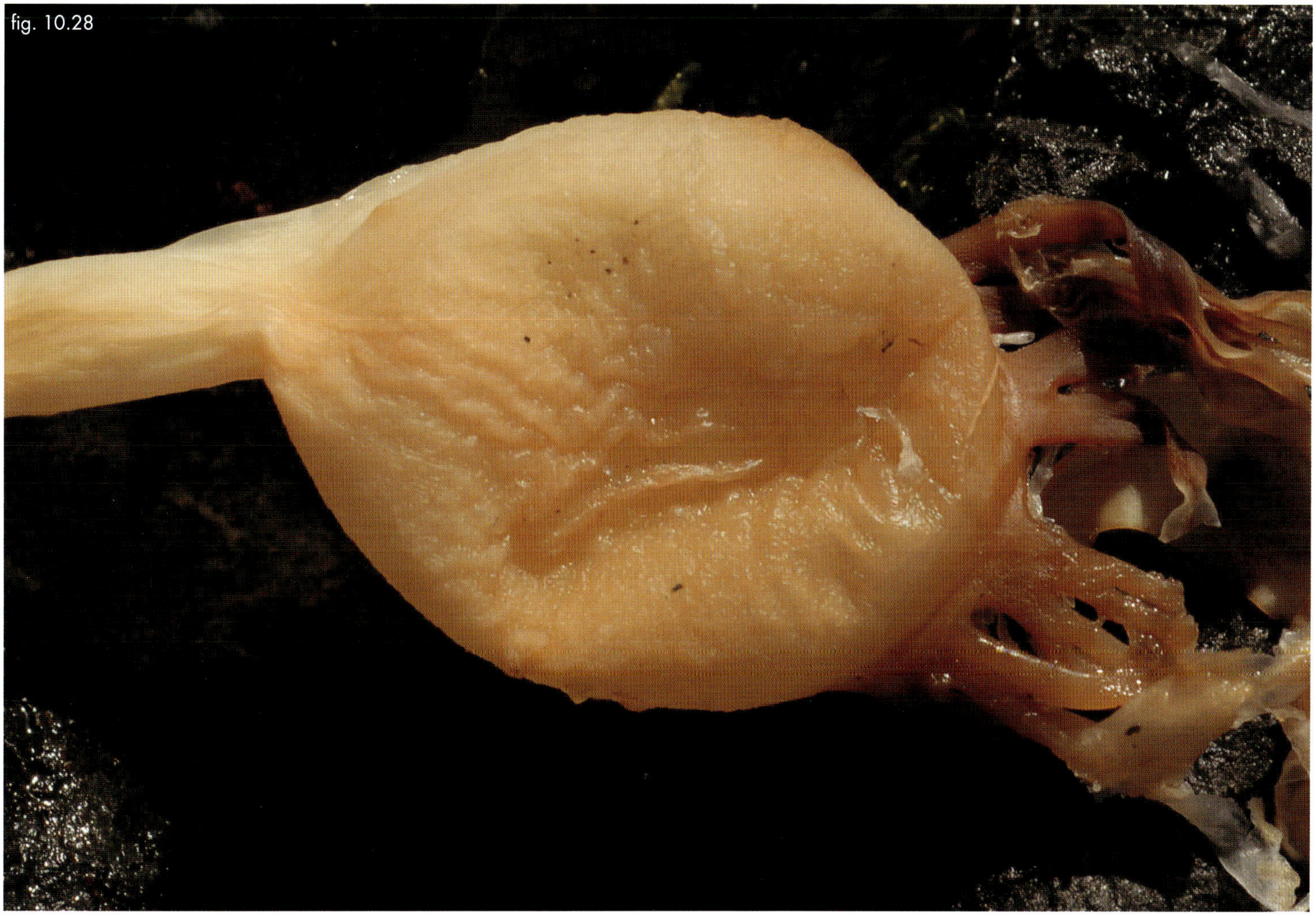

fig. 10.28

GLASS AND PLASTIC

Glass and clear plastic can be rendered in a fairly straightforward manner by following some basic, observable characteristics. Glass is a special material because it is transparent. It is also very shiny, most of the time. This shininess is rendered the same way as a chrome reflective layer over a painted surface.

The most visible part of the reflection is where a light value is being reflected, like the sun or a light sky, creating glare. This glare appears in various levels of opacity on a layer above the glass's color and tinting, and follows the same Fresnel Effect principles as the clear coat. In figs. 10.29, 10.30 and 10.34, a strong Fresnel Effect is observed whereby whatever is behind the glass is least visible wherever the Fresnel Effect is strongest. Conversely, what is behind the glass is most visible wherever the reflection is the weakest. The A side of the glass is less reflective and the B side more reflective. Notice how much more can be seen through the A side (figs. 10.29 and 10.30).

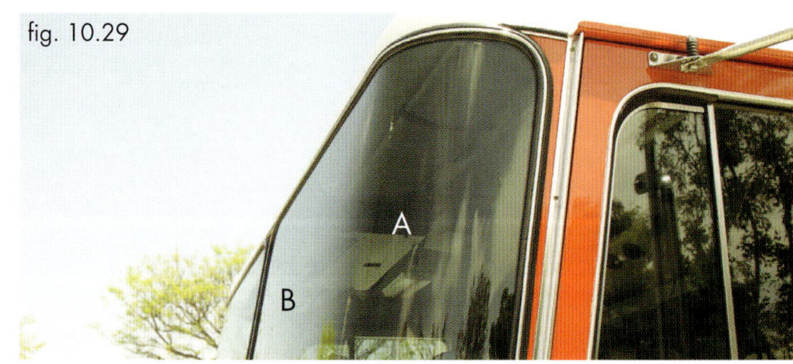

fig. 10.29

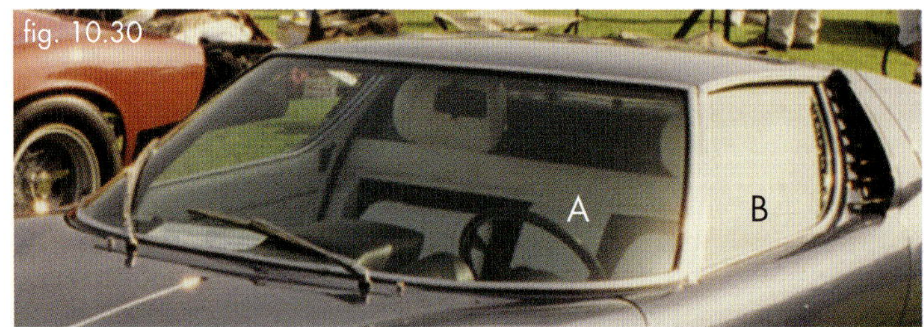

fig. 10.30

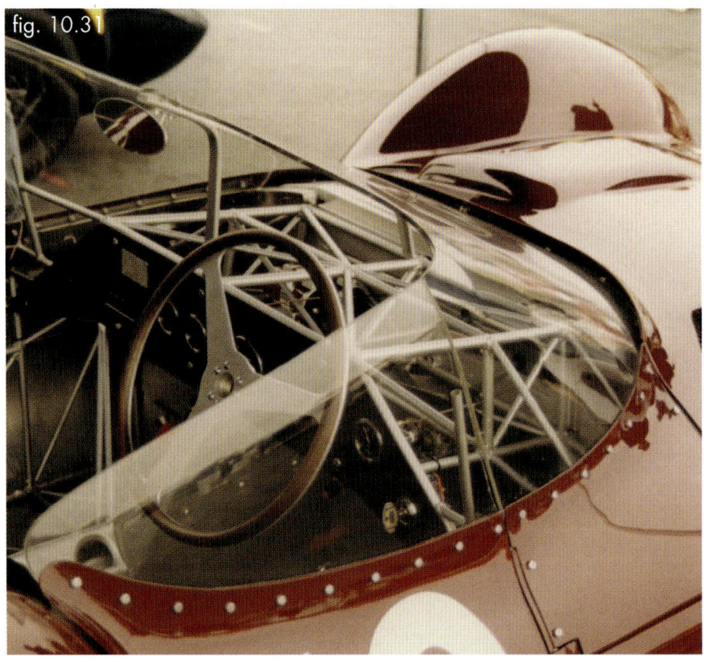
fig. 10.31

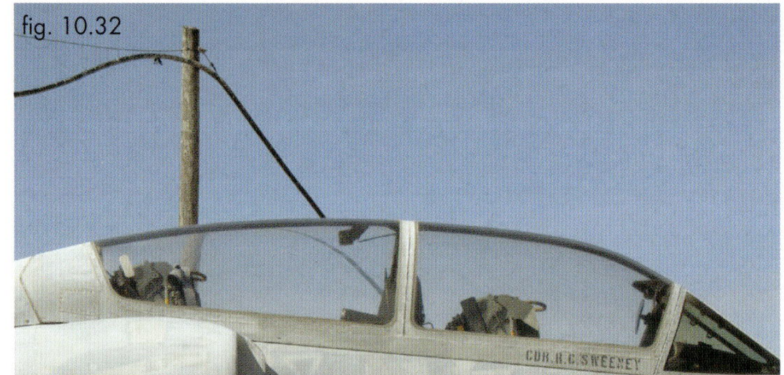
fig. 10.32

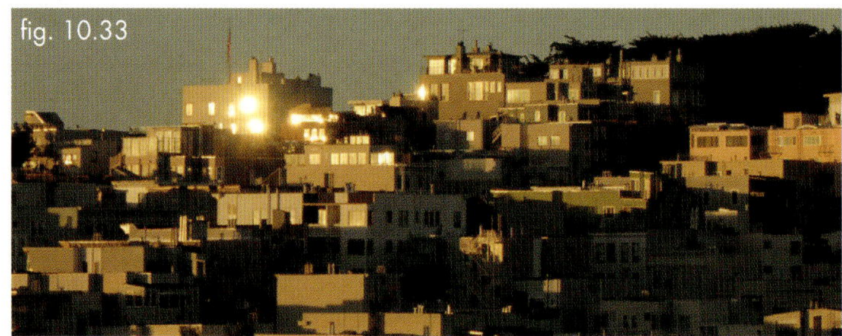
fig. 10.33

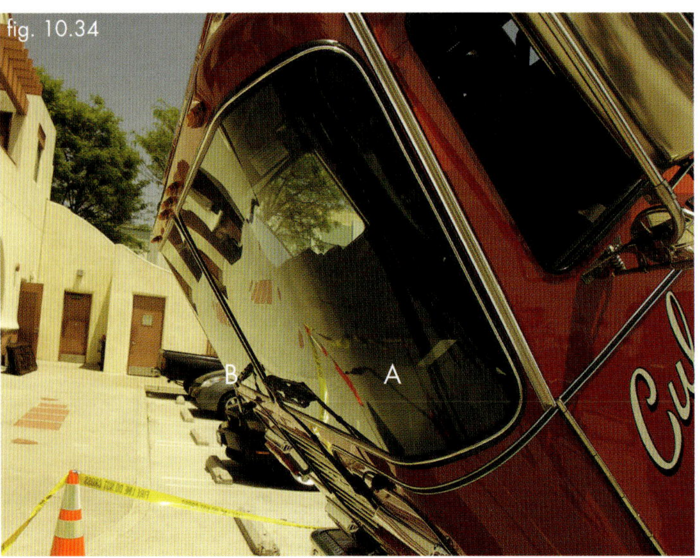

fig. 10.34

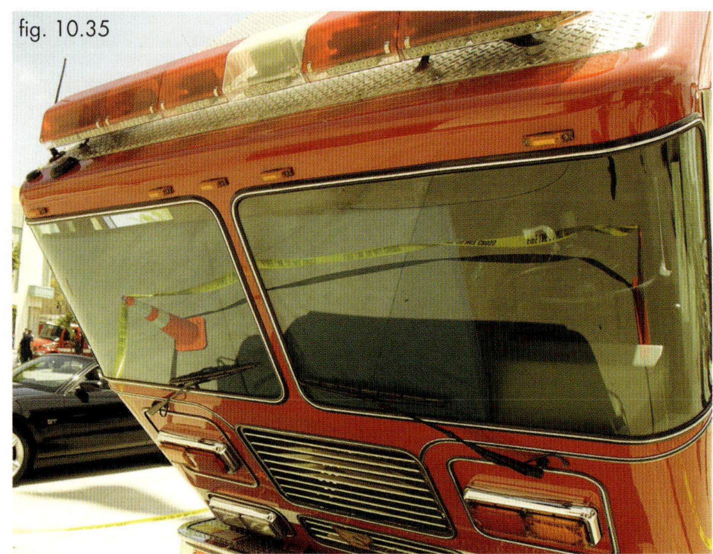
fig. 10.35

A simple way to render glass is to think of it in three layers. Layer 1 is *tint* and can be created by either putting a "multiply" layer in a digital program over what is behind the glass, or if working traditionally, just darkening down everything behind the glass, depending on the desired level of tint. Layer 2 is *color*. The glass can be colorized either by putting a layer set to "overlay" on top of the tint layer, or by shifting what is behind the glass toward the desired glass color. Layer 3 is the *reflection* layer, placed on top of the other layers, to obscure some parts of what is behind the glass. Use the Fresnel Effect to control the strength of the reflection.

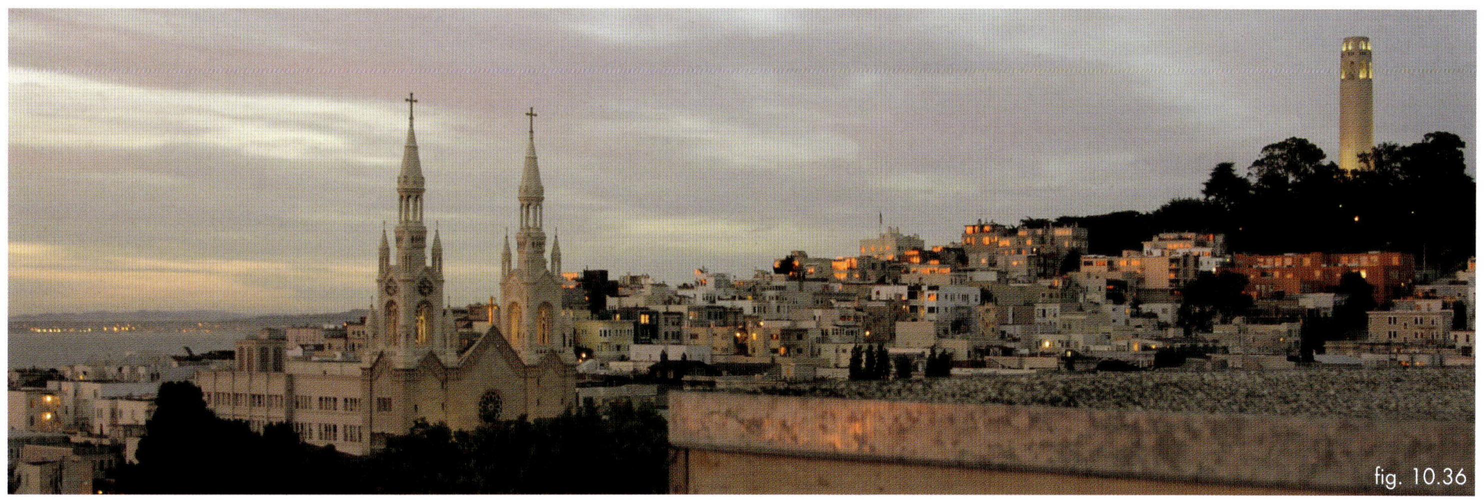
fig. 10.36

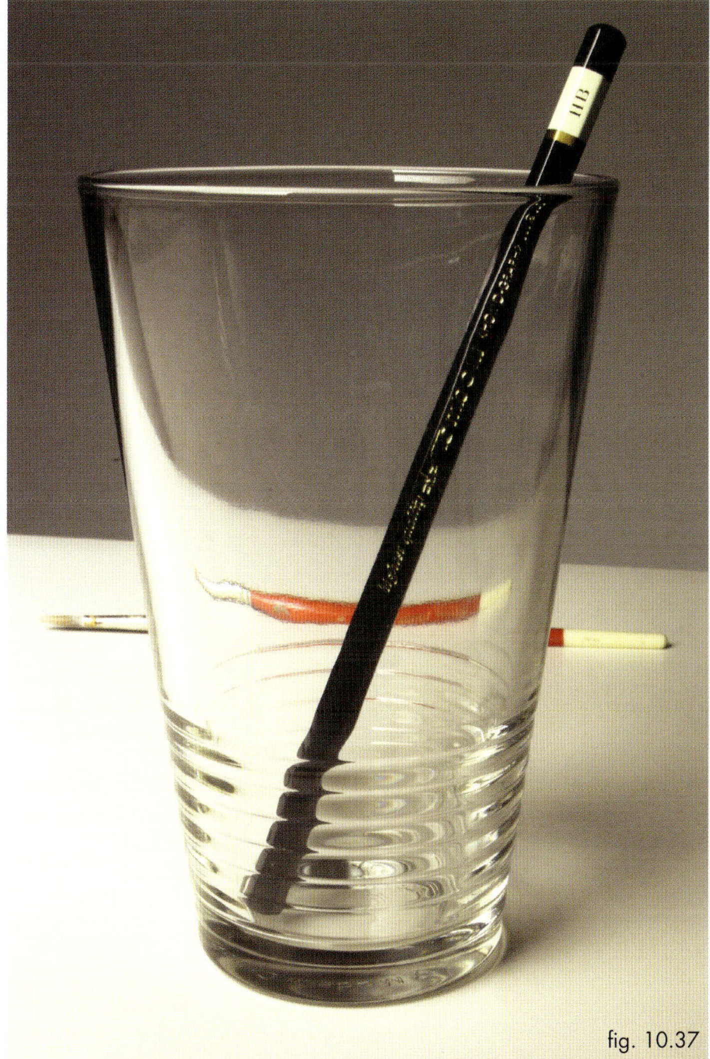

fig. 10.37

Flat planes of glass like windows (figs. 10.33 and 10.36) can very strongly reflect the glare of the sun or whatever light they might be reflecting. The farther away these flat planes are from the viewer, the less the Fresnel Effect is evident. As a shiny object gets farther away, the size of the sun's reflection on it increases over its surface. This might seem counterintuitive, but what is happening is that the angles of incidence of all the various sight lines to a distant object are narrowing; the sight lines are becoming more parallel. This means that more of those sight lines will bounce back to the same spot in the environment around the object. If more of the sight lines are bouncing to the sun than when the object was very near, then the reflection of the sun will appear larger across its surface.

Refraction

Refraction must also be considered. Refraction is the change of direction of a ray of light. This bending makes the object behind the glass appear to be distorted. Figs. 10.32 and 10.37 are nice examples of refraction. In fig. 10.32, the vertical pole and the electrical line are very distorted behind the cockpit's clear canopy. The thicker and less optically correct the glass is, the stronger the refraction will be. Looking at fig. 10.37, on the left side of the glass this refraction is very evident. Notice that the red brush, on the table behind the glass, appears to bend much more toward the sides of the glass. This is because at the very edges of the glass, where the lines of sight are almost tangent to it, there is much more glass to look through than at the middle of the glass. If the glass is very clean then there is no matte-surface value shading to worry about. If it is dirty, then the dirt of the glass will start to show the same matte-surface characteristics as an opaque form.

LUMINESCENCE

Luminescence occurs when the surfaces of objects themselves glow and emit light. There are many forms these lights can take: candlelight, incandescent, L.E.D., neon, fluorescent and even bioluminescent light. Fortunately, all types of light follow some basic laws of physics, which, if understood, will improve how they are rendered.

When a surface glows, it does not receive cast shadows or any form changes due to another light source that might be shining on it. If the glowing light has a color to it (fig. 10.40) then the color will change temperature as it glows more brightly. This means that the lower the strength of the colored light, the more its color is visible. As the colored light grows brighter, the color shifts through the color spectrum toward white at the brightest spot. This can become an enormous topic of study.

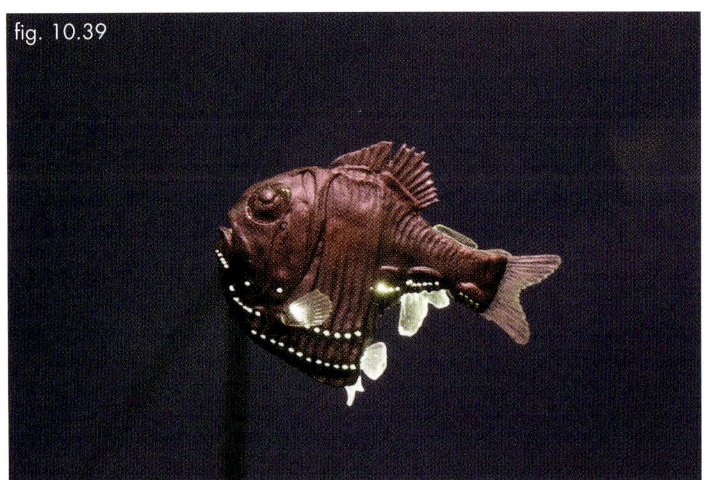
fig. 10.39

These books are great additional resources on the subject of luminescence: *Color and Light: A Guide for the Realist Painter* by James Gurney and *Vision and Art: The Biology of Seeing* by Margaret Livingstone.

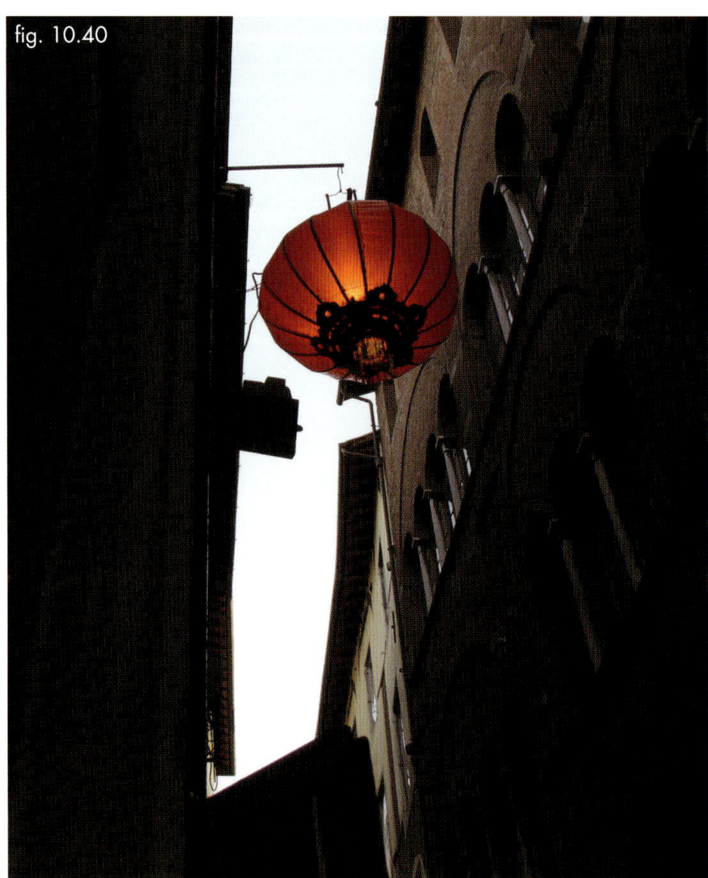
fig. 10.40

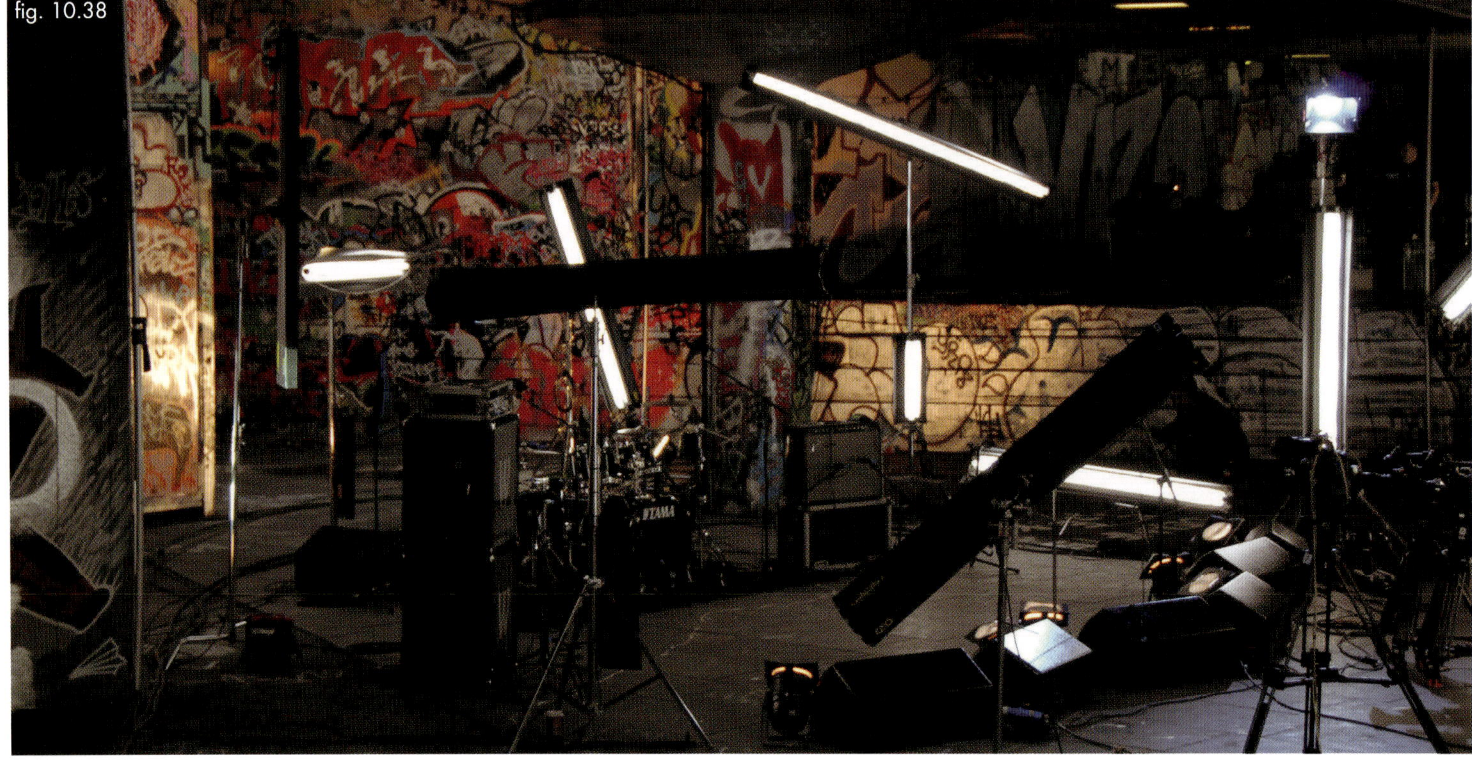

fig. 10.38

One of the best ways to integrate luminescent objects into the whole of a rendering is to make sure the light emitted from them illuminates some of their neighboring surfaces. In fig. 10.41 this can be seen clearly inside the hood of the stoplight, at the upper left of the image. In fig. 10.42 the city lights of Florence, Italy illuminate the underside of the clouds and the streetlamps illuminate the sides of the buildings.

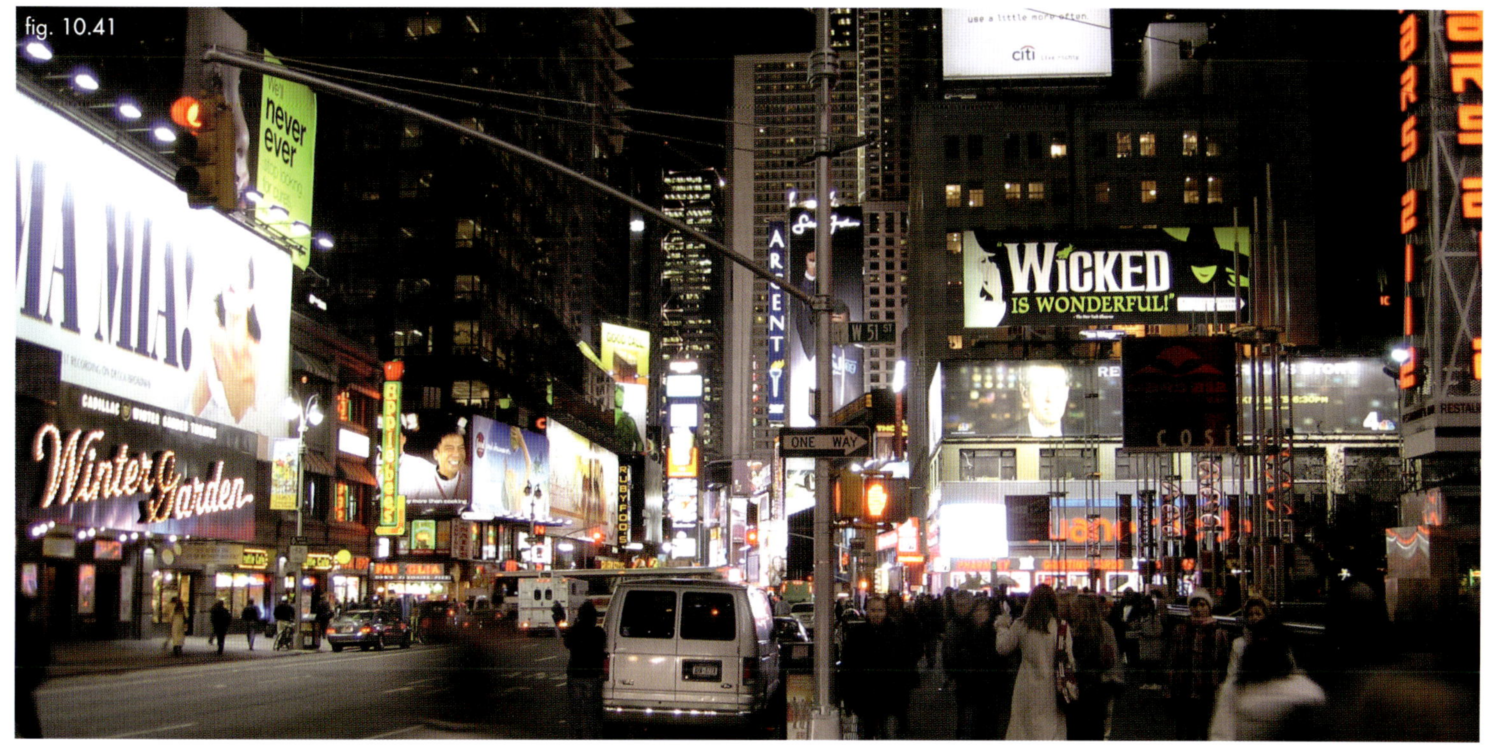

fig. 10.41

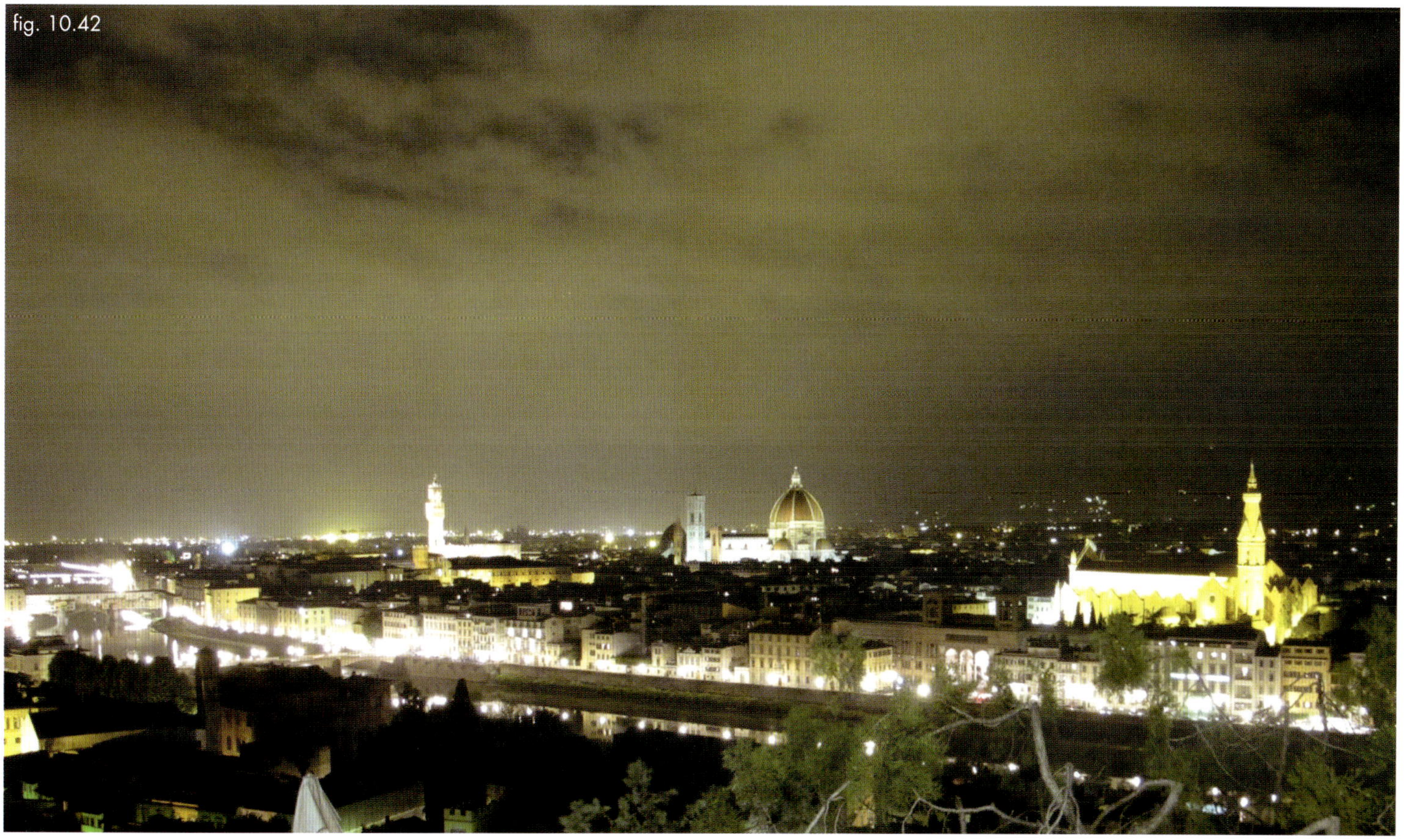

fig. 10.42

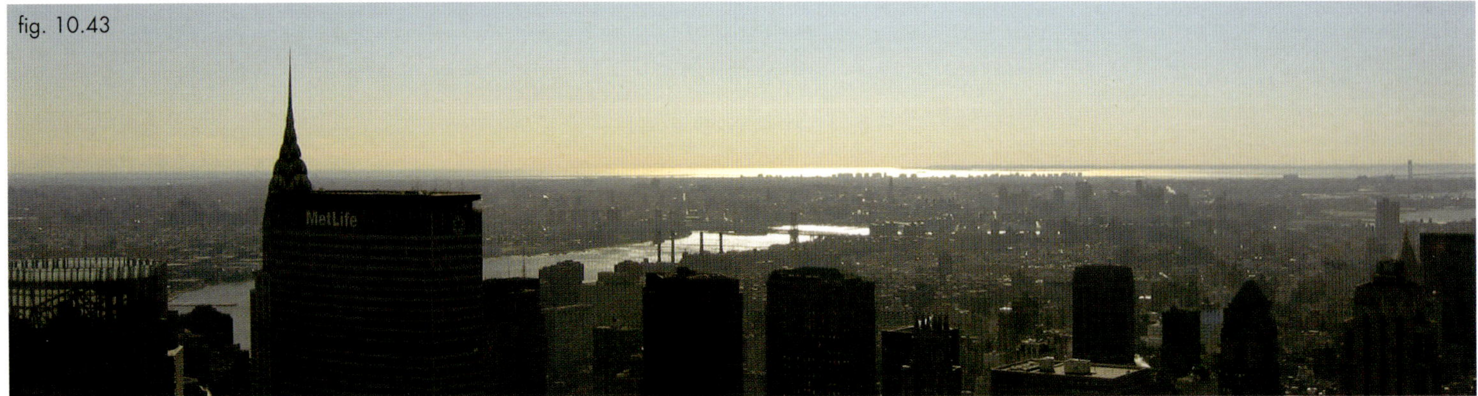

fig. 10.43

Water is pretty interesting because it can look so varied depending on conditions. How clear is the water? Is the wind blowing? What color and value is the sky? Where is the sun? What is reflecting into the water? All of these elements combine to make rendering water a fun and varied subject.

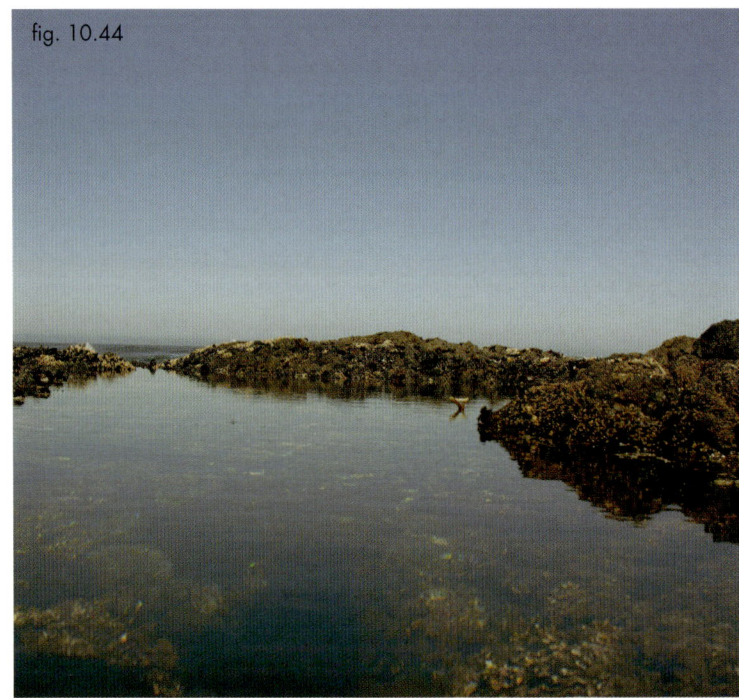
fig. 10.44

In fig. 10.43, look at how strong the reflection of the sun is on the water, all the way out to the horizon. When there are ripples on the surface of the water, the rays of light bounce off those ripples at various angles, making the sun's reflection seem to spread across the water's surface. In other words, the observer's sight lines bounce off the ripples, with many of them bouncing up to the sun. If the waves are rather small, think of rendering the sun's reflection as taking a similar approach to that of rendering a metallic surface. The sun reflection spreads over a larger surface than it would if the water were perfectly still. When water is very smooth (fig. 10.44), it can be rendered like a simple, flat mirror with the Fresnel Effect added to it. The sky being reflected into the water in the foreground is more transparent than the reflection in the distance where

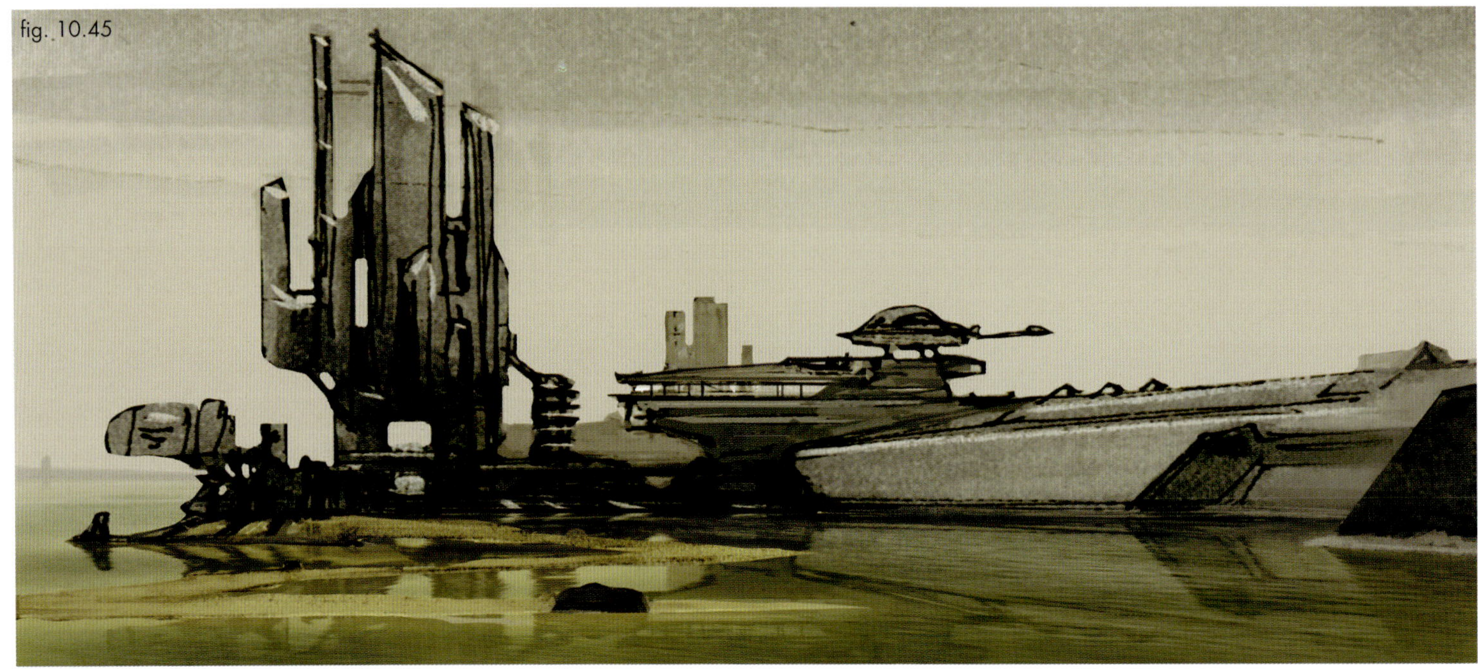
fig. 10.45

the sight lines become more tangent to the water's surface, increasing the reflection's opacity and obscuring the tide-pool details under the surface.

Rendering still or slightly rippled water is a pretty straightforward exercise and all the steps to do it are covered in the video for fig. 10.45. Rendering still water is almost exactly like rendering glass, or a shiny painted surface. Put what is seen beneath the surface of the water on a layer. Put a reflection layer over that. Finally, add a mask layer to create gradations from foreground to background that mimic the Fresnel Effect. This will look very close to reality. All of the variations will exist: either in the base layer where the water's color and clarity are, or in the reflection layer, where the reflection's distortion can be manipulated based on the level of wind or other activity that disrupts the surface.

In figs. 10.46 and 10.47, the ripples are very smooth with just a little left-to-right zigzag distortion of the reflections. In figs. 10.48 and 10.50, the water surface is rougher so the reflections become more vertically distorted. In fig. 10.49, if the water were still, the single-point lights in the distance would just reflect as mirror images of the same shapes in the water, but many more of the ripples are scattering the sight lines back to the same point lights, effectively stretching the reflections vertically across the water.

fig. 10.46

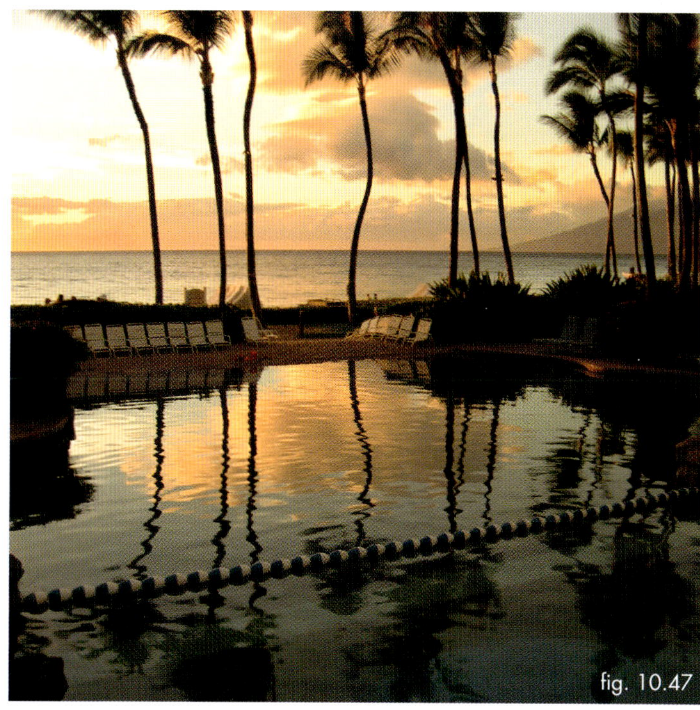

fig. 10.47

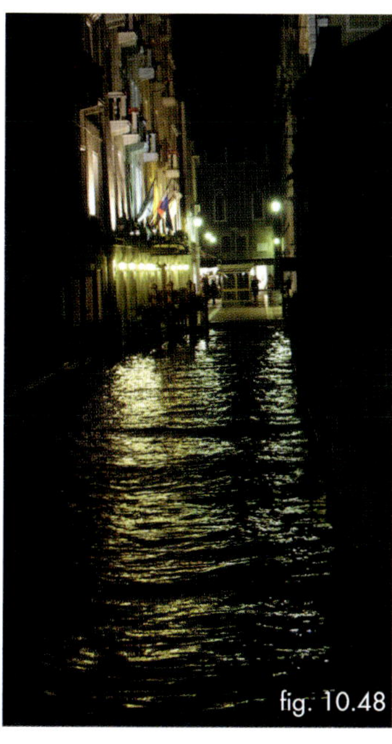

fig. 10.48

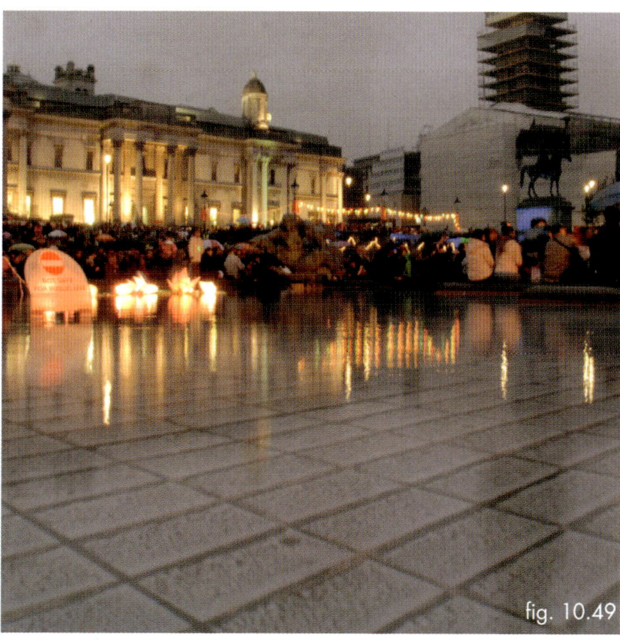

fig. 10.49

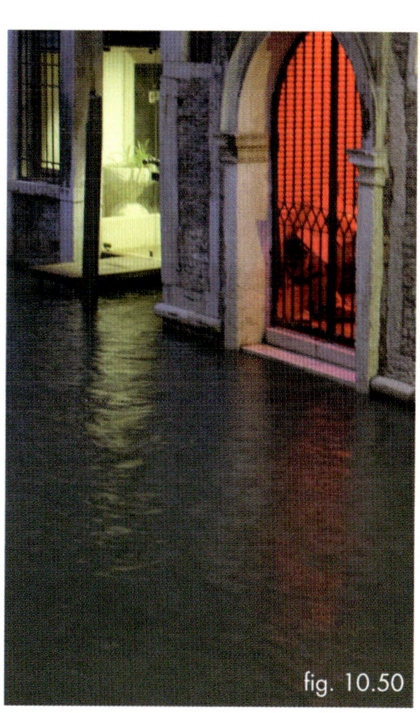

fig. 10.50

WIND ON WATER

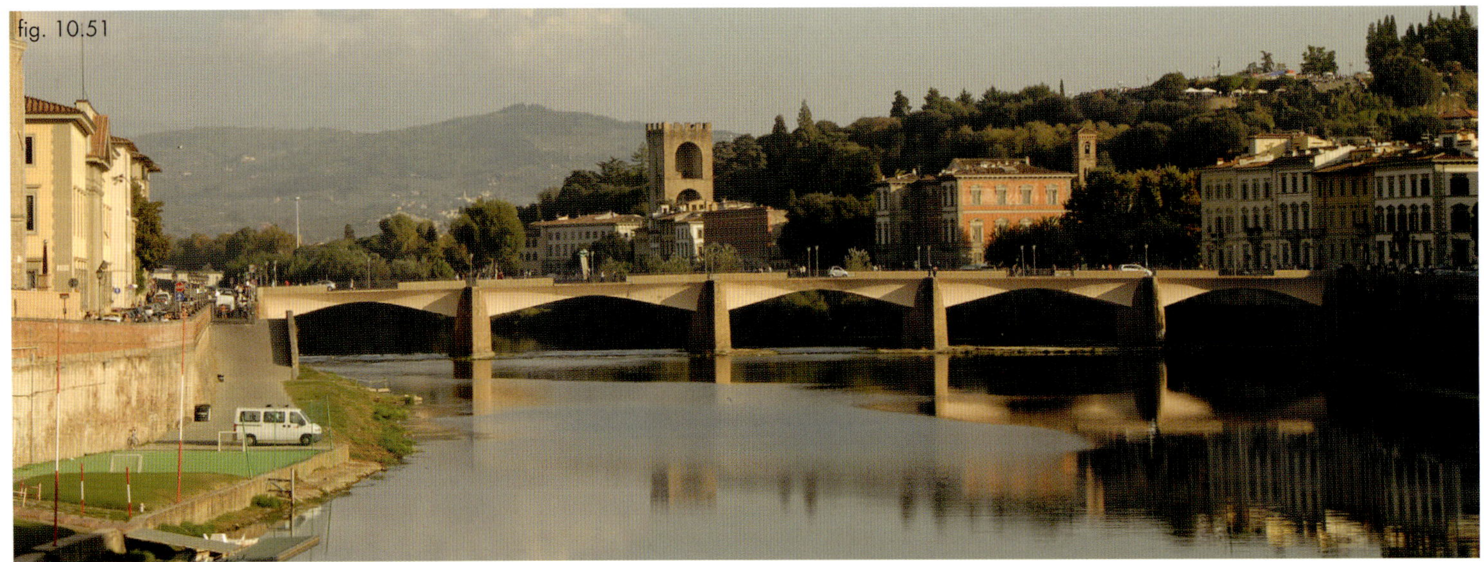
fig. 10.51

As the wind picks up, the water surface can become so rough that very few of the sight lines bounce to a common location in the environment around it. The water will still have an overall color, influenced by the sky it is reflecting, but that reflection will be very diffused with much softer edges. Observe this effect in figs. 10.51 and 10.53. Notice also that there is actually a visible "wind line" where the wind can be seen disturbing the water's surface and softening the reflections.

When the wind really gets blowing, this effect increases and the reflections can basically disappear, as in figs. 10.52 and 10.54. The lack of distinguishable reflections of the buildings or the ship might come as a bit of a surprise after having looked at so many nice reflections throughout this book. These images show how varied and diverse the approach to rendering water can be. Examples of slightly zigzagging reflections, to vertically stretched ones, to almost none at all can be observed in nature. So, when deciding which one or combination to render, apply basic physics, do some visual research and then render away and have fun.

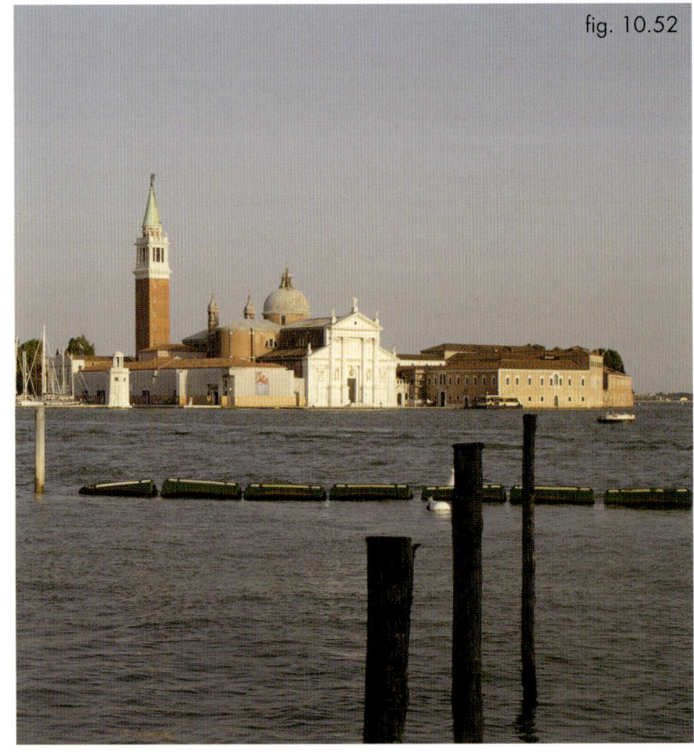
fig. 10.52

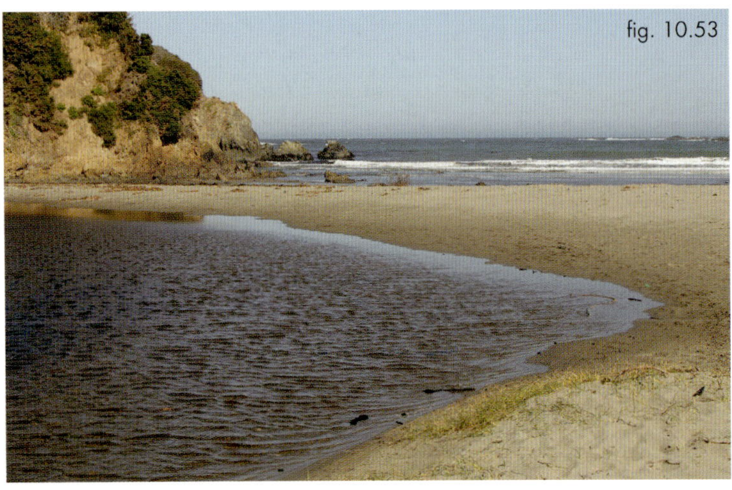
fig. 10.53

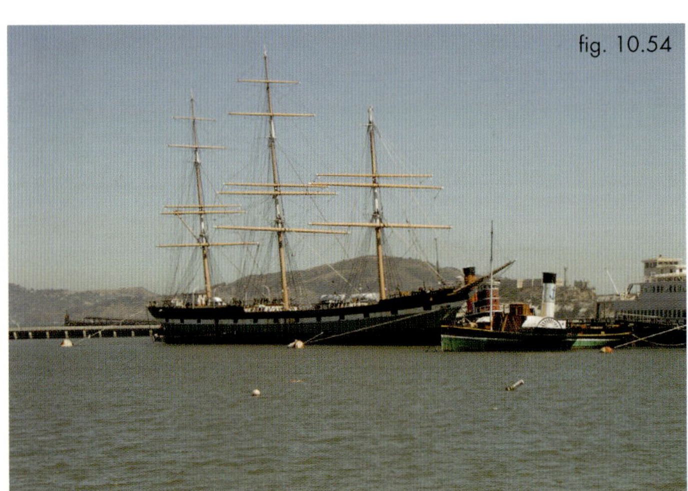
fig. 10.54

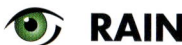

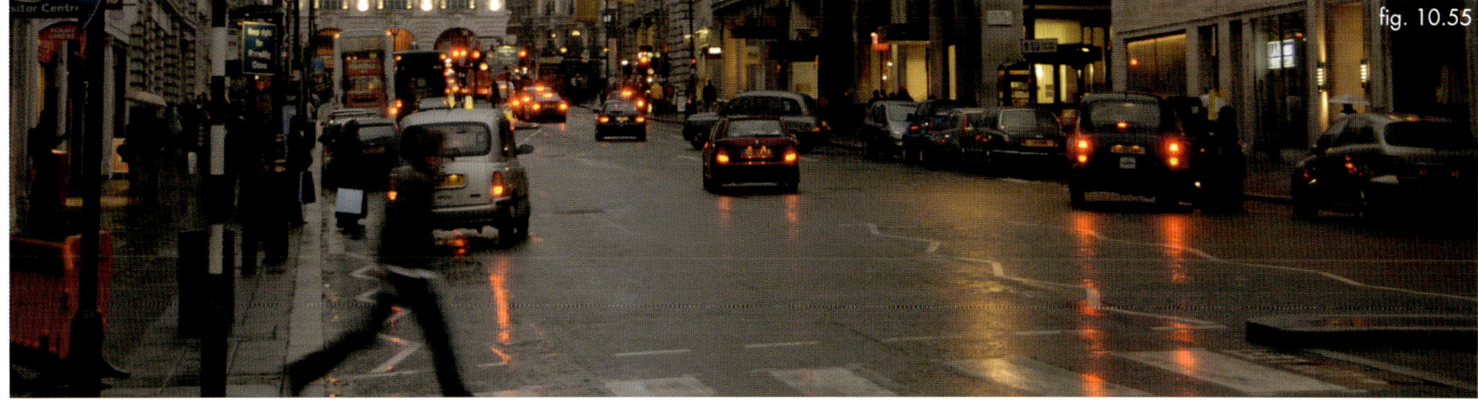
fig. 10.55

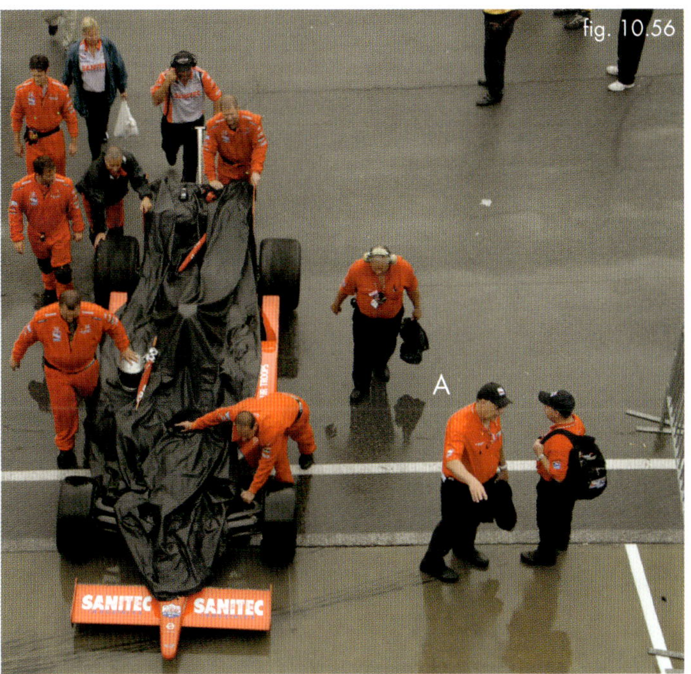

fig. 10.56

When a location is wet, during rainfall and while it dries, think of it as the entire environment being covered by a huge, reflective clear coat. All of the reflective physics discussed so far still apply.

The rendering strategy can be thought of as adding a reflection layer over the entire scene. The way reflections behave in a wet environment varies depending on the roughness of the surfaces.

When a surface is smooth, like point A in fig. 10.56, smaller, crisper reflections occur. That particular reflection is sharp because the water is deeper, or the ground there is smoother than it is just ahead of the man, where his reflection softens and becomes blurry.

Where the ground is rougher, like in figs. 10.55 and 10.58, the reflection of the cars' headlights and taillights stretch across the wet pavement, where the rough texture of the ground acts just like the rougher water in fig. 10.48 on page 223.

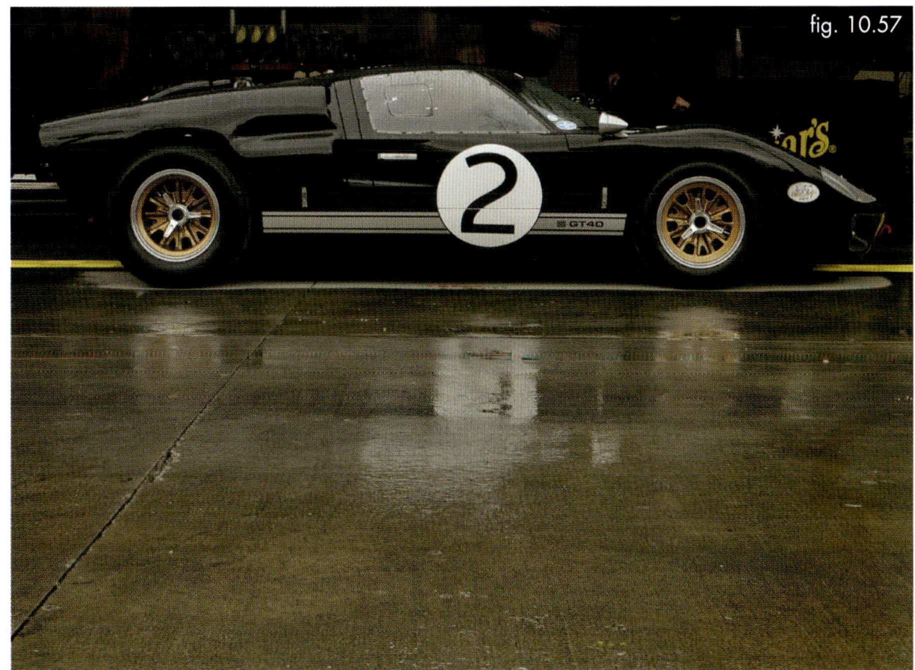

fig. 10.57

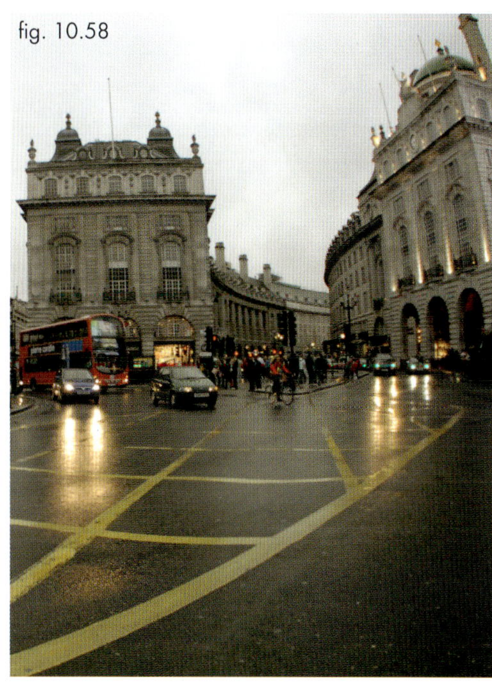
fig. 10.58

SEMIGLOSS 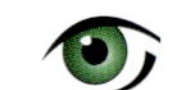

Manufactured semigloss surfaces are basically glossy surfaces with a little surface texture. The amount of texture affects the sharpness of the reflections, just like with water and wet pavement. Figs. 10.59 and 10.60 show two examples of black, semigloss materials. As expected, each car reflects its environment, but the reflections are much softer and more diffused than if the surfaces were smooth and glossy.

To achieve this look in a digital rendering, first render the reflections as if the surface were chrome, then blur the reflections to the desired amount. By adjusting this layer's opacity, making it more or less transparent, the material's reflectivity changes. If the layers were not organized in such a way to allow this, some refinement by hand of the blurring effect will be required.

When rendering traditionally, blur the reflections right from the start. Most of the semigloss effect can be accomplished by rendering only the strongest reflections, where the sight lines are most tangential to the surfaces (fig. 10.61). Note, the blurring of the reflection of the light source creates a kind of soft, metallic look.

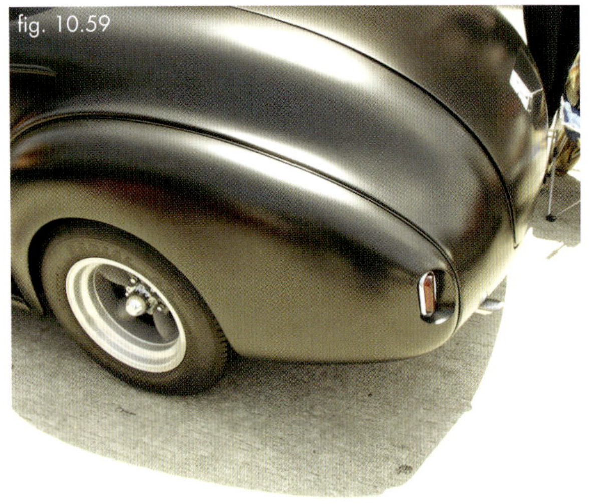
fig. 10.59

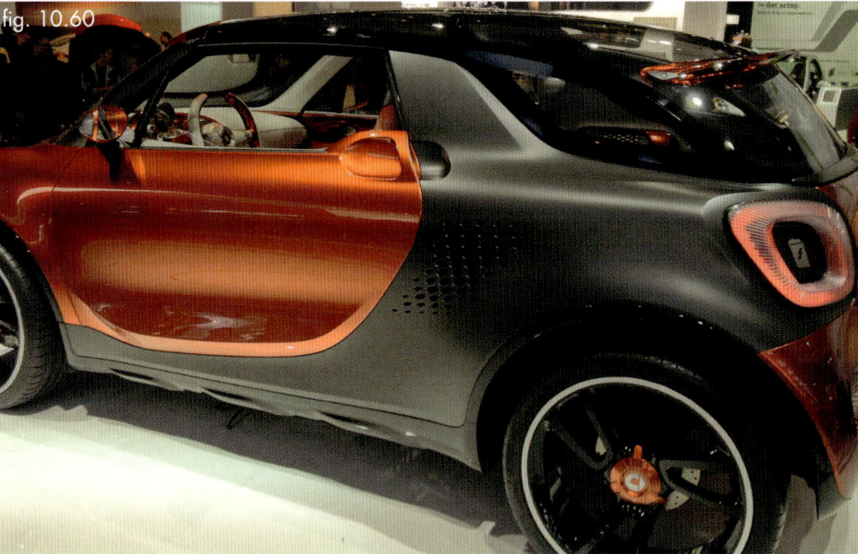
fig. 10.60

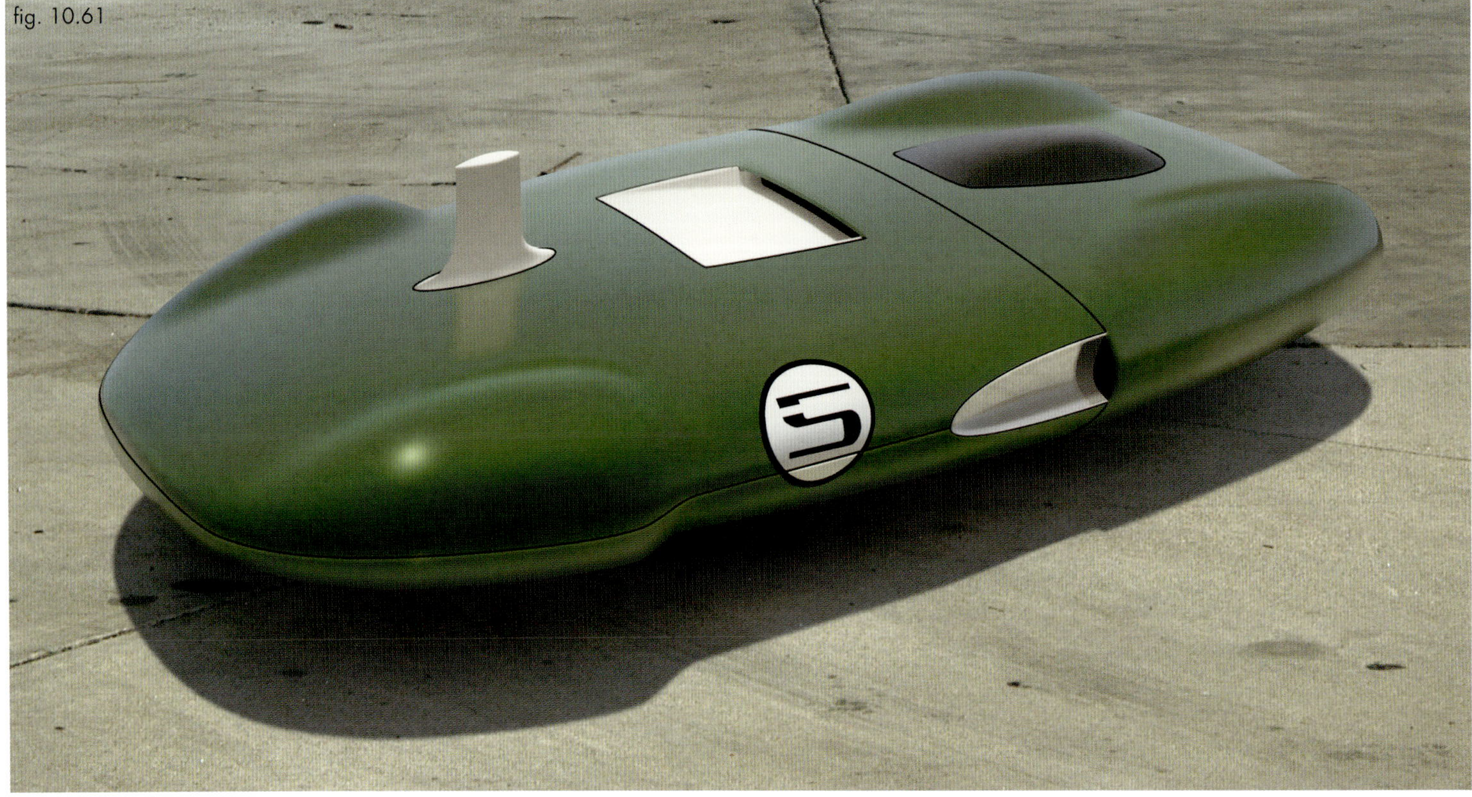
fig. 10.61

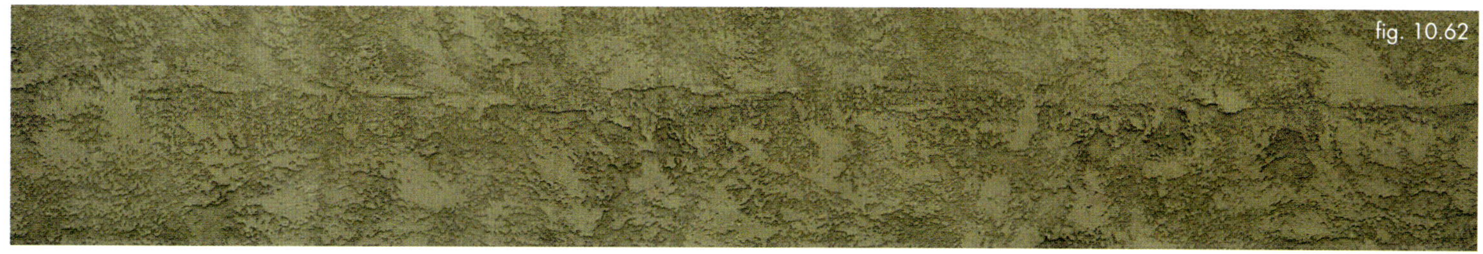
fig. 10.62

Texture changes the reflective qualities of a surface. Conversely, reflectivity affects where the texture of a surface is most visible. Earlier, on page 118, it was shown that the texture of a matte surface is most visible on the Number 2 side. On a round surface, that is just before the core shadow (fig. 10.65). However, on a shiny, textured surface, the texture is most visible where the light source is reflected (figs. 10.63 and 10.64).

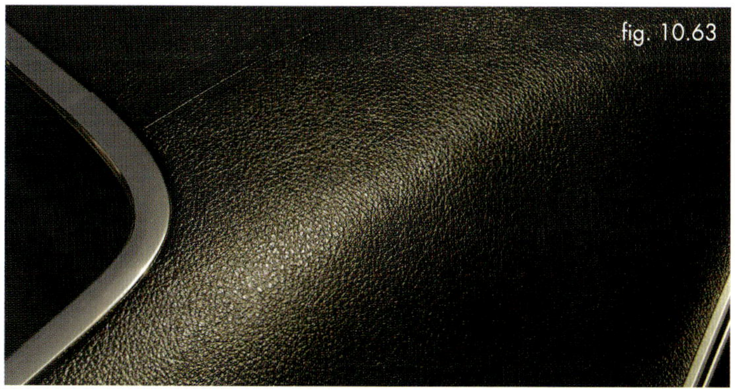

fig. 10.63

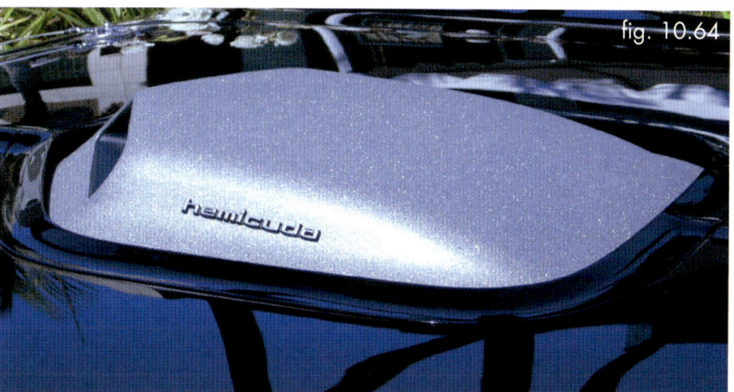

fig. 10.64

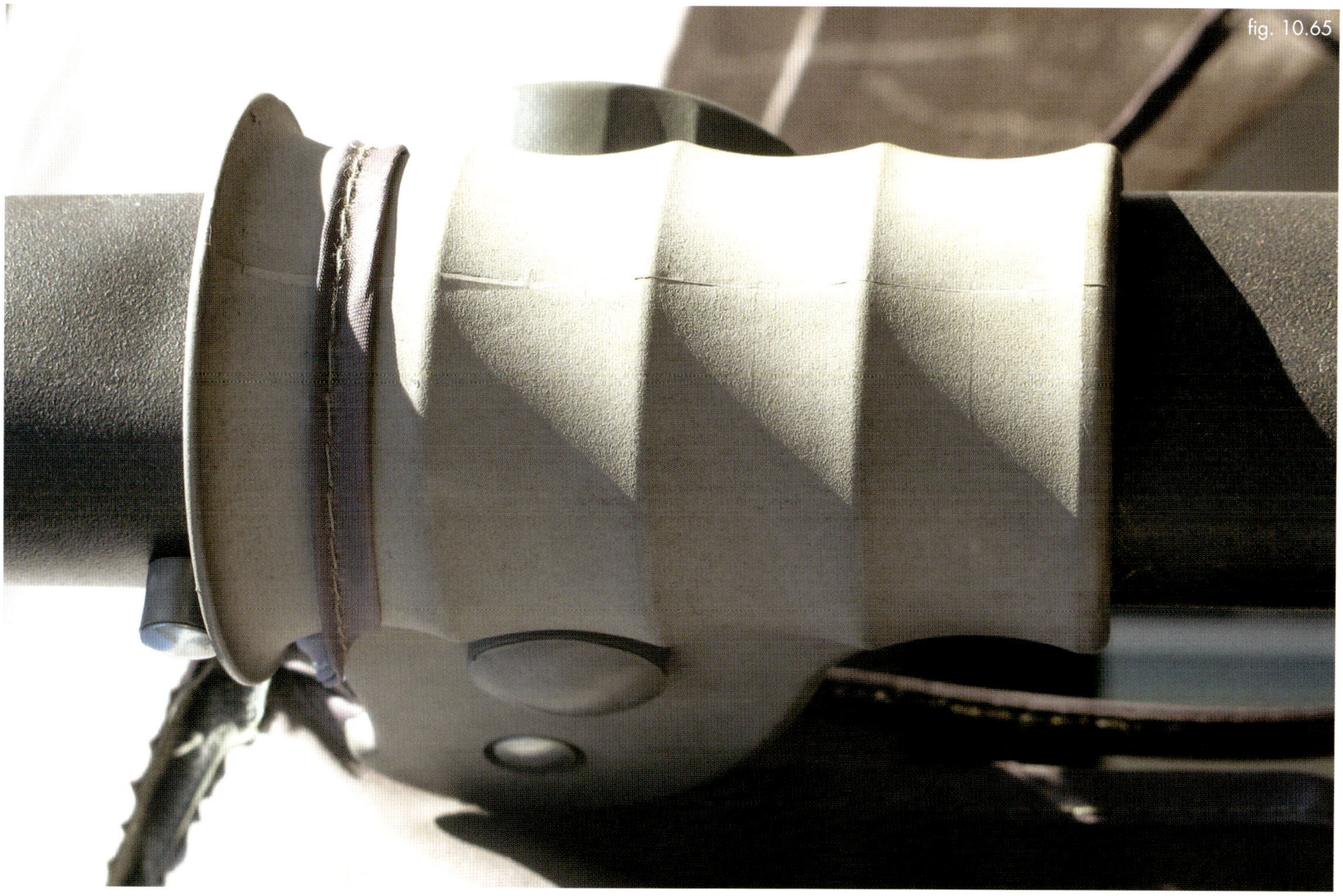

fig. 10.65

CHROME

In Chapter 07, rendering chrome was explained as essentially a clear coat, but there are a few strategies that take it to the next level of realism. One, let the silhouettes of chrome forms disappear and melt a little into their backgrounds. Two, use part lines to help explain the form changes of a surface, as most chrome reflections can get very complex. Three, reflect the colors and values of the environment accurately. If the chrome is in an indoor studio setting where the ceiling is black and there are multiple light sources, do not reflect a blue sky in it! Fig. 10.68 shows how the color of chrome is highly dependent on what is placed around it. Reflections on chrome stay very sharp. There are no shadows to worry about. If there are gradations on the chrome, they are occurring in the environment, not because of the Fresnel Effect. Take time when rendering chrome. There might be entire scenes being reflected. Adding visual complexity to the reflections will make the chrome look more realistic.

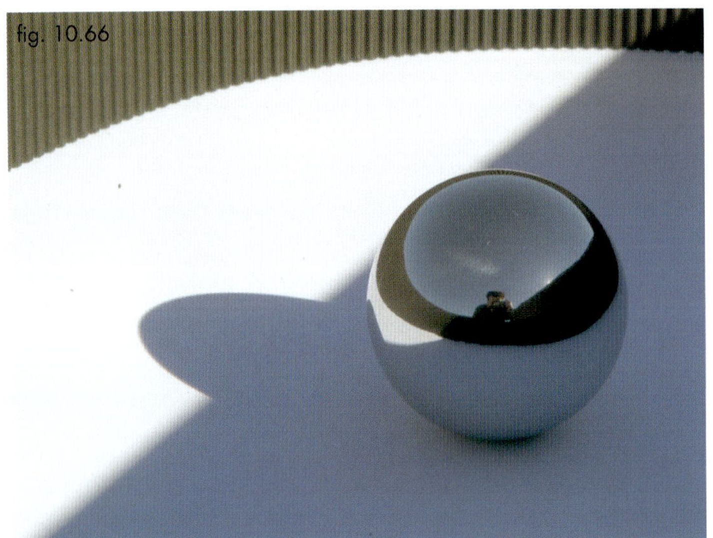
fig. 10.66

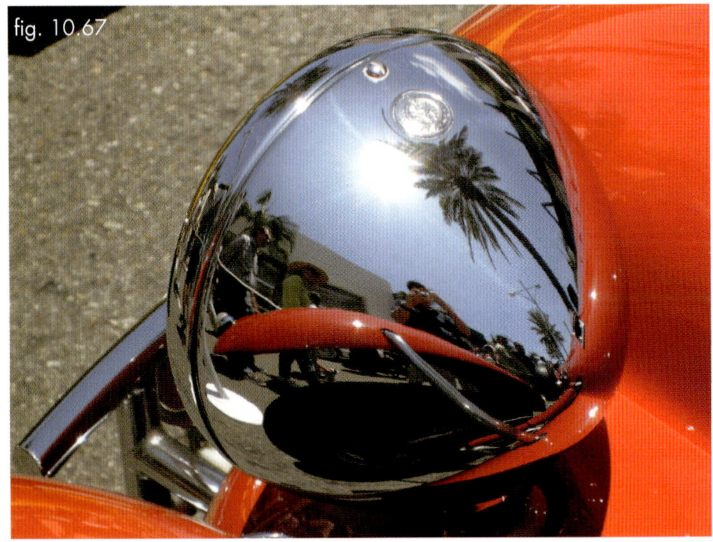
fig. 10.67

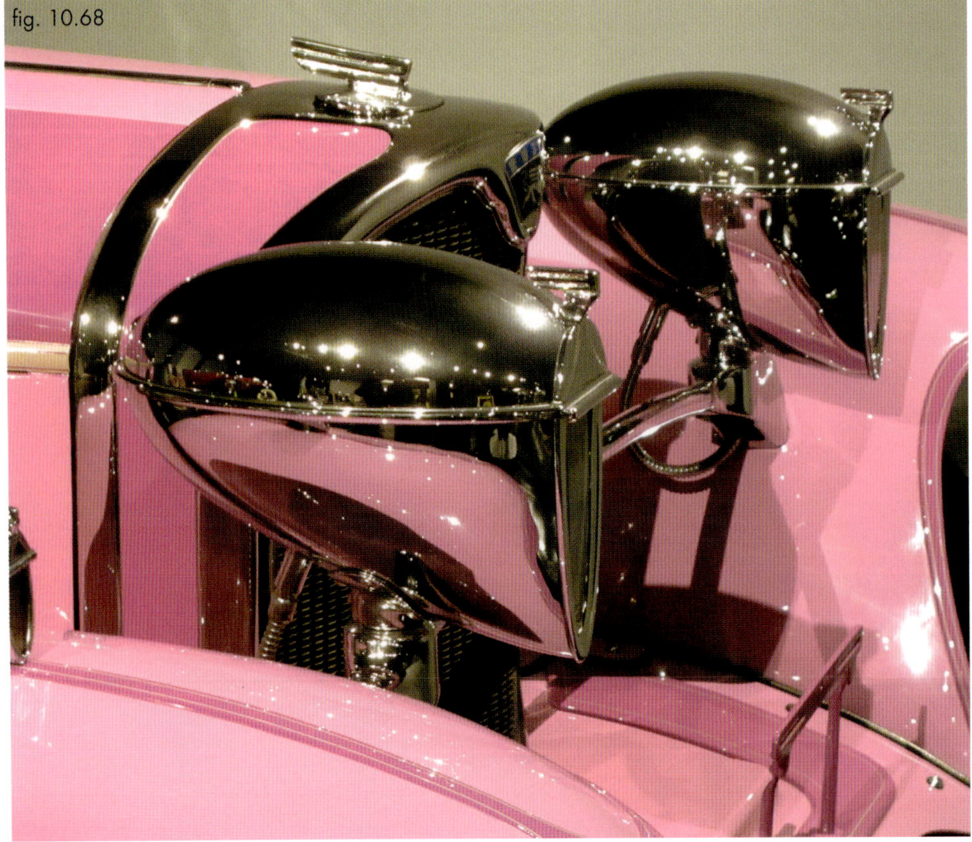
fig. 10.68

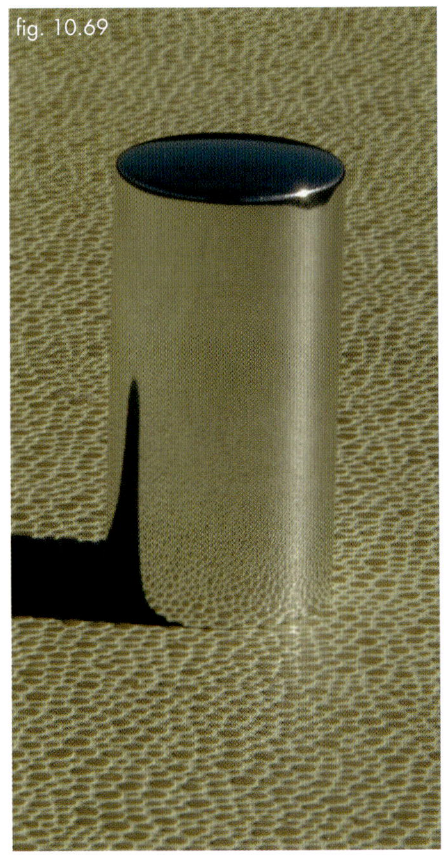
fig. 10.69

Once the base color of any material is correct, the amount of polish or texture becomes the next consideration, informed entirely through the subtleties of how the reflections are rendered. All of the images on this page look like the dark-brown base color of bronze, but the level of polish differs on each. In fig. 10.71, the surfaces are cleaner and more polished than those of fig. 10.70, and the reflections communicate that. On the older bronze (fig. 10.72), only the light seems to be reflected (not the environment) and there is a greenish-gray color occluding all of the creases and crevices of the statue. The way an object is rendered tells a story and reveals its history. When rendering imaginary objects made out of familiar materials like bronze, and placing them in imaginary environments, always make conscious decisions about how to render. The skills from this book can be applied to communicate the story you want to tell.

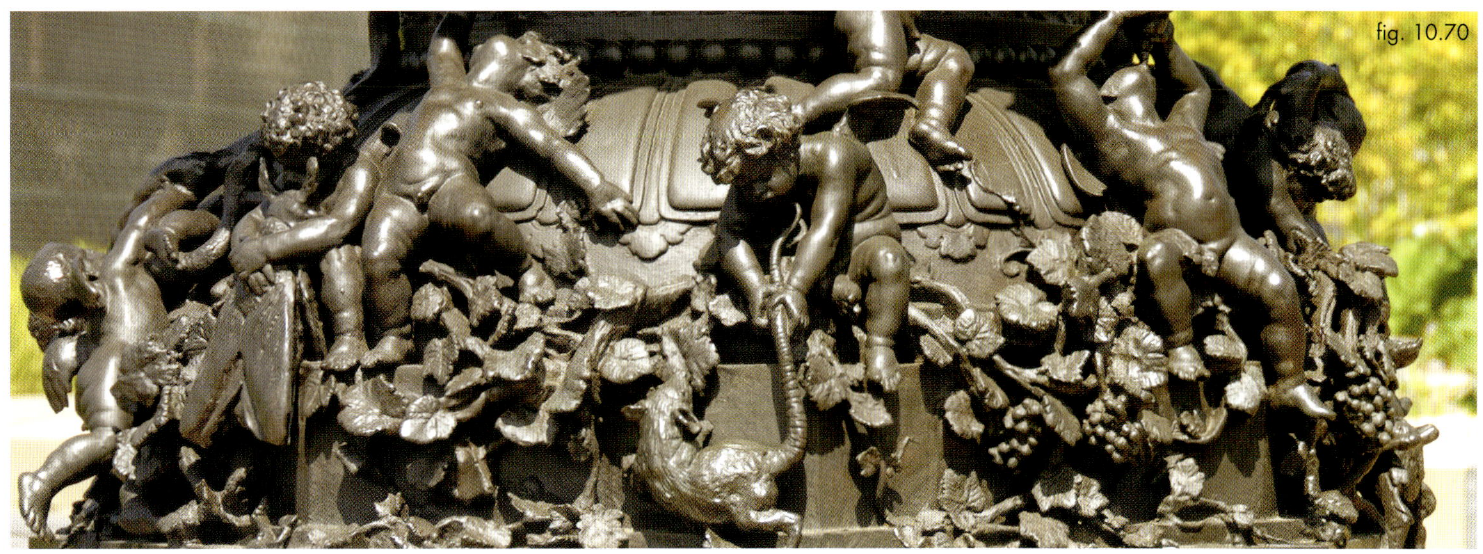
fig. 10.70

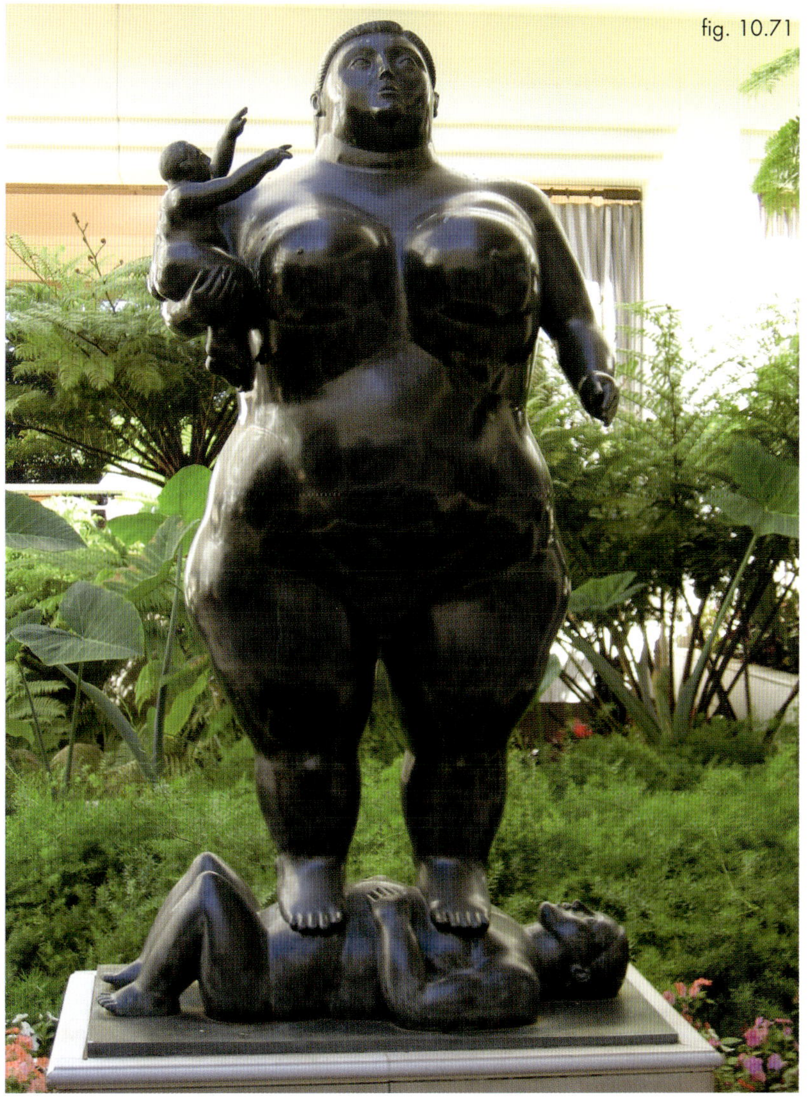
fig. 10.71

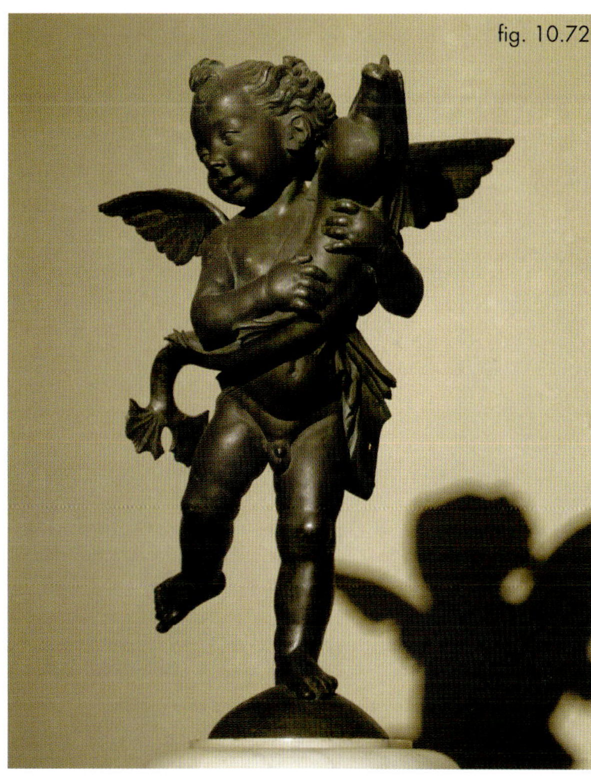
fig. 10.72

BRUSHED METAL

Brushed metal is characterized by small, aligned scratches in the metal that reflect the light, like a series of very small, aligned, cylindrical cross sections. The uneven reflection that results is referred to as "anisotropic." To render this effect, start with a semigloss metallic surface and add aligned scratches that are most visible next to where the light's reflection is strongest. The scratches work a little like a metallic surface, stretching the reflection of the light across more of the surface.

If rendering in Photoshop, try adding a monochromatic noise filter to the surface and then "motion blur" it in the direction of the scratches. Next, "color dodge" it to create the bright reflections of the light source. This subtle change in the scratchy quality of the light source's reflection is what distinguishes this material.

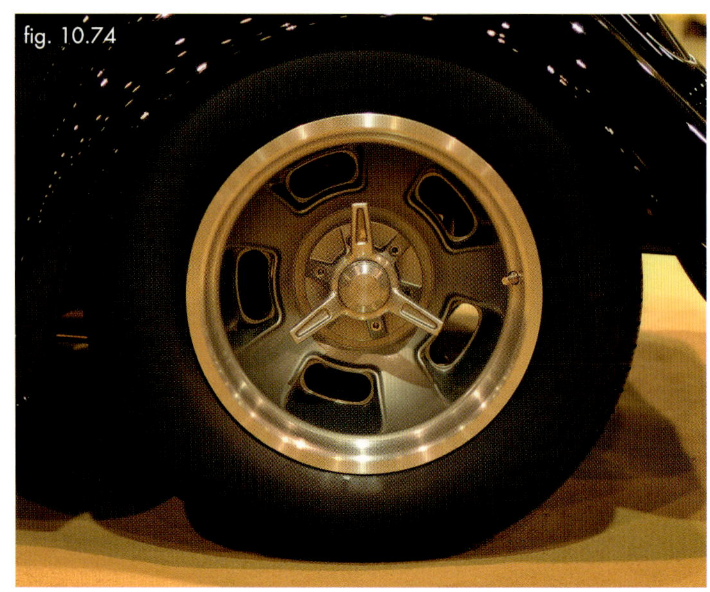
fig. 10.74

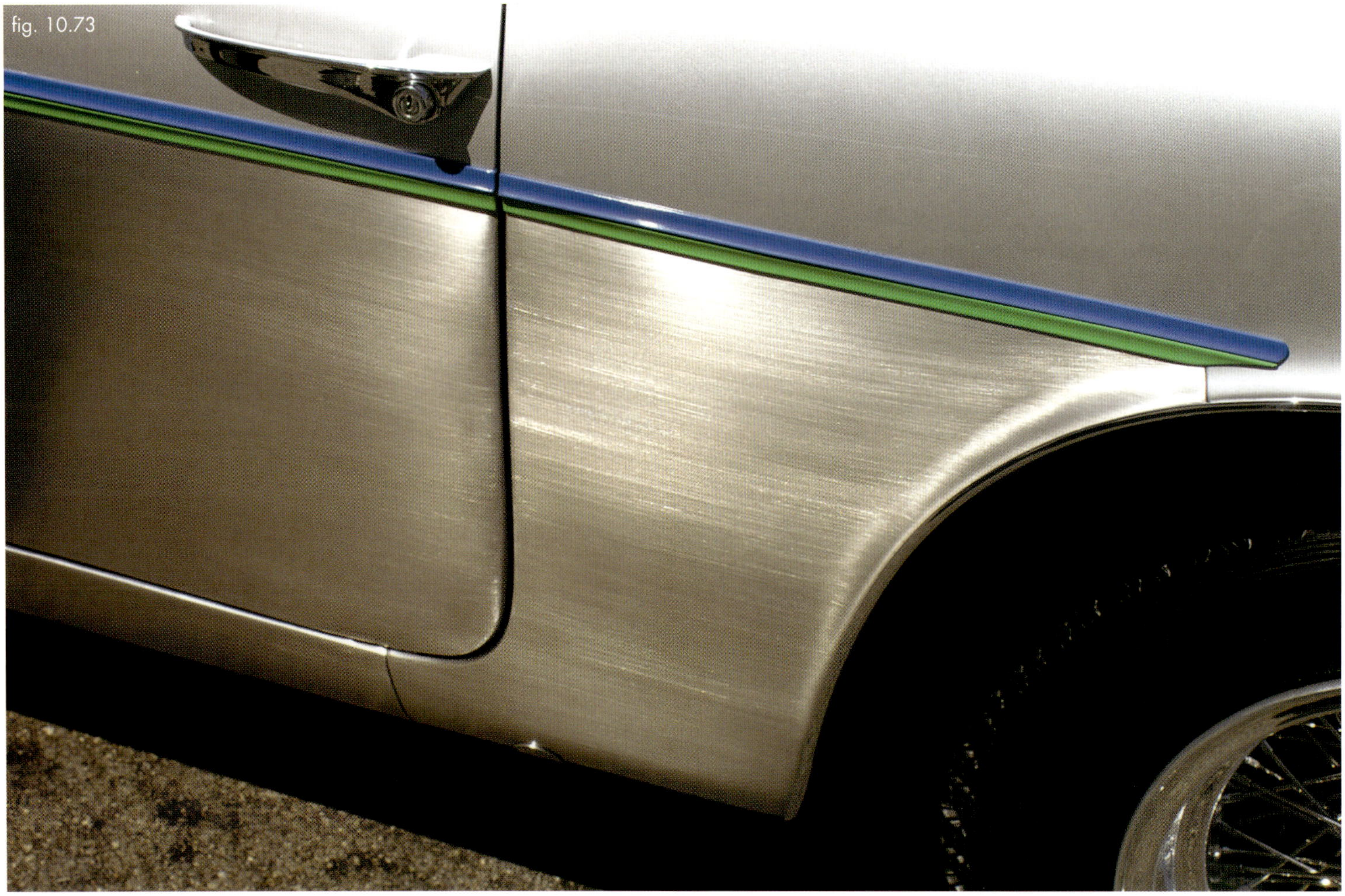

fig. 10.73

Many machined metals exhibit anisotropic qualities in the highlights, due to the machining process. To a lesser extent, these small scratches could be caused by many sources, like polishing with a rough rag or the normal abrasion caused by coming into contact with neighboring surfaces. Notice how the base color of metal can vary from warm to cool. Often, this is the only indication of the metal's type, aside from surface finish. In fig. 10.77, compare the color of the tail pipe (A) with that of the exhaust pipe (B) and gas tank (C). They are slightly different base colors with different levels of reflectivity.

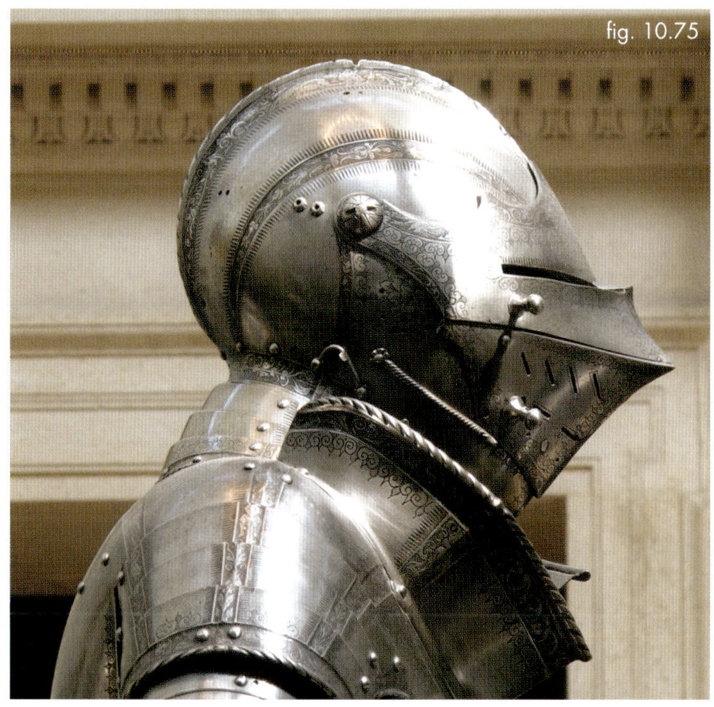

fig. 10.75

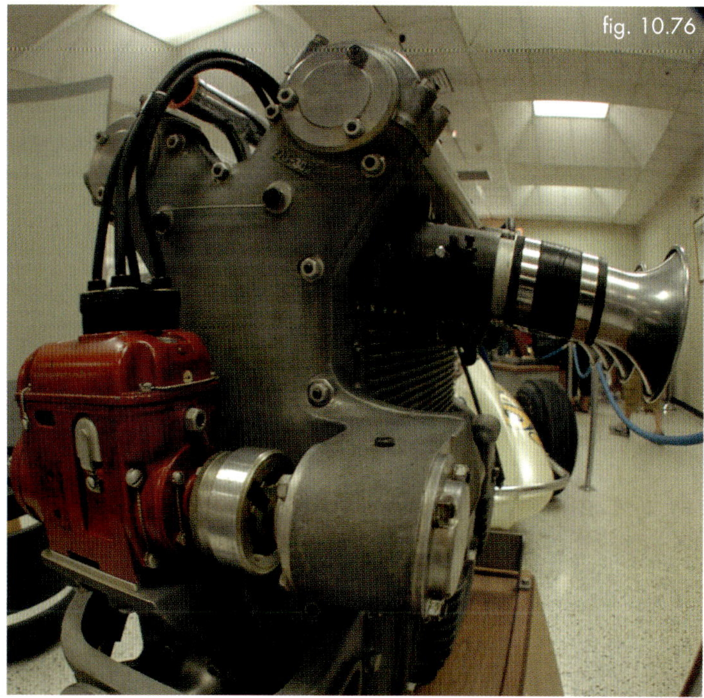

fig. 10.76

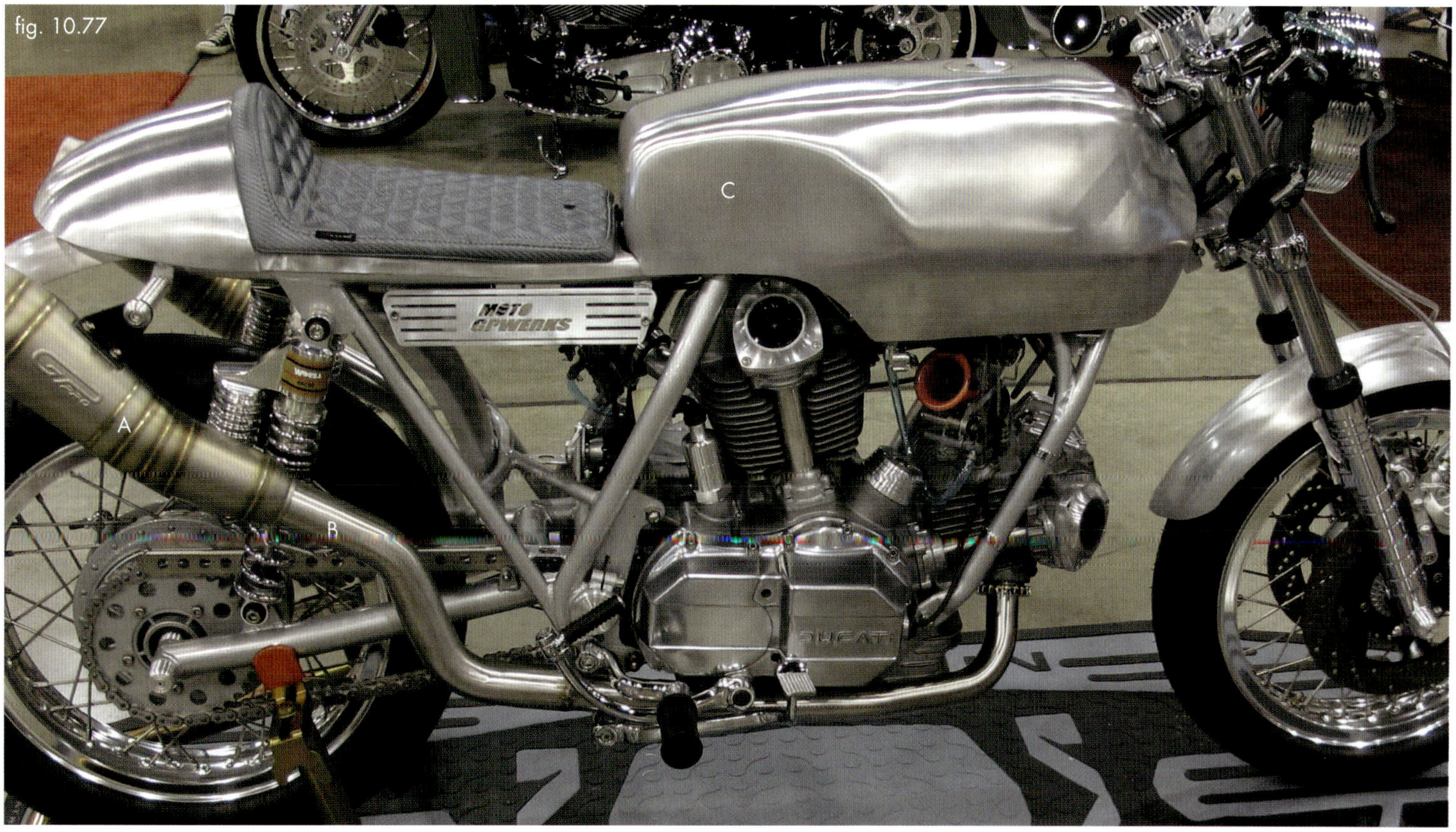

fig. 10.77

ALUMINUM

Rendering aluminum can be as easy as rendering a matte surface, with a little soft reflectivity added. If the aluminum is highly polished, it is almost like rendering chrome. Most of the time, aluminum has a slight semigloss, metallic look to it. Its base color of gray is quite neutral; neither overly warm, nor cool. Unlike clean chrome, be sure to include cast shadows and value range on it, as illustrated by fig. 10.78. This example is quite old and weathered. The older and more weathered aluminum becomes, the less reflective and more matte it appears. Aluminum is a very soft metal, so it scratches easily with regular wear and tear, and therefore tends to be polished often. If the parts are intended to look older, adding overly polished areas and scratches within the highlights of the high-wear zones will give it a nice patina. A lot of these scratches from polishing can be seen in the lighter reflections of the sun, in fig. 10.79.

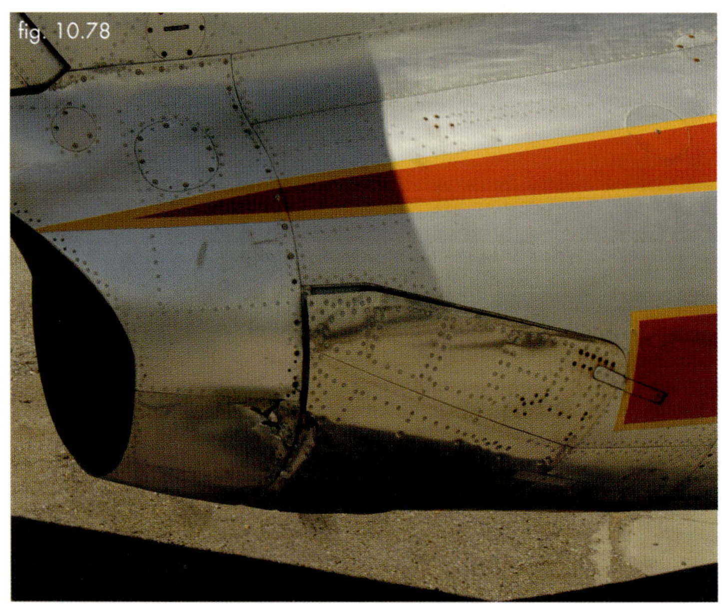
fig. 10.78

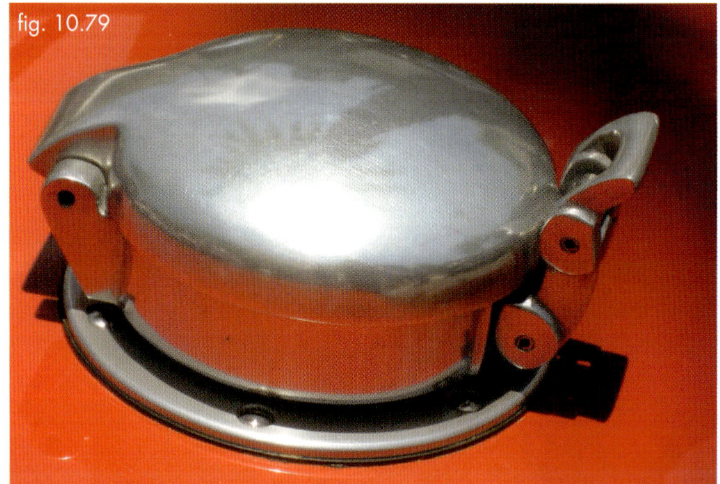
fig. 10.79

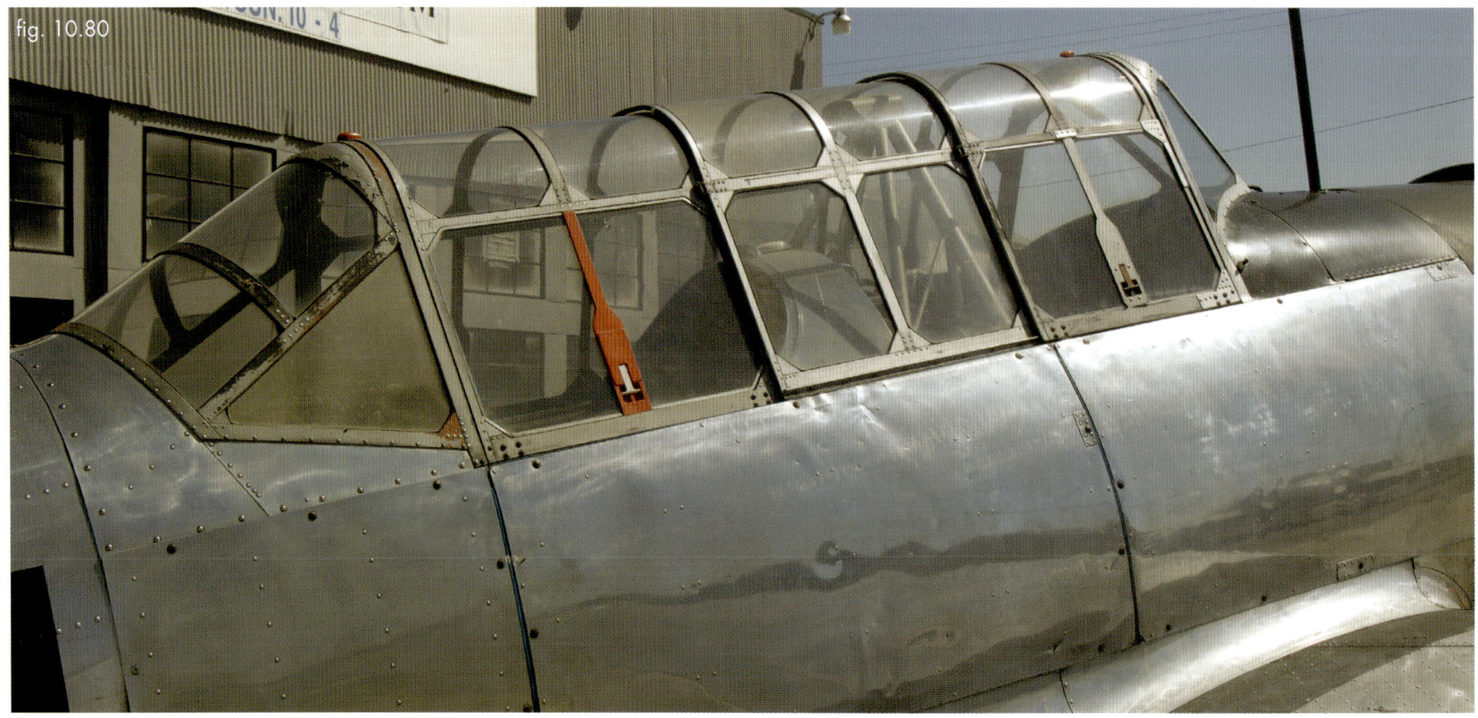
fig. 10.80

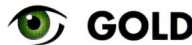

As is the case with most metals, the finish can vary a lot depending on age, polish and the quality of the metal itself. Just looking at this page, there are four very different visual representations of gold. Fig. 10.81 is an example in its more raw form. Fig. 10.83 is a centuries-old cast relief on the door of an Italian building. Its long exposure to the weather, over hundreds of years, has diminished its reflective qualities. Polishing its surface would restore the reflections temporarily, but doing so repeatedly over time would reduce the beautifully sculpted figures to indistinguishable lumps. Fig. 10.84 shows yet another application and presentation of a gold material. The reflective film covering the windows has gone through a metallization manufacturing process to create the golden tint. Rendering this metallic film uses the same process as rendering the polished gold in fig. 10.82.

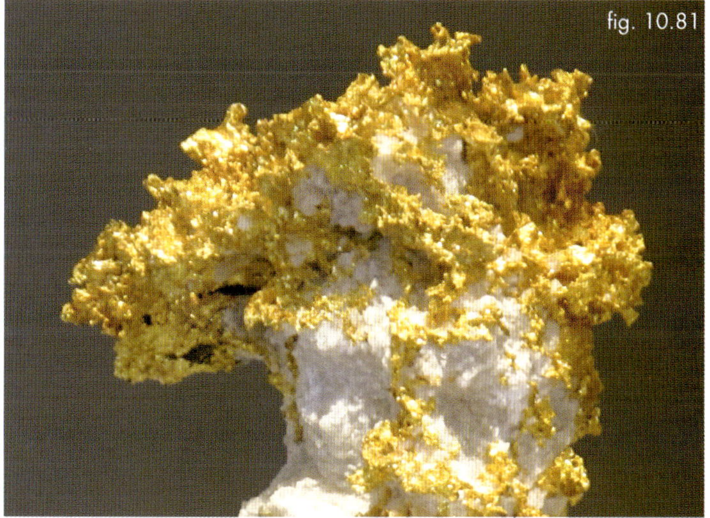
fig. 10.81

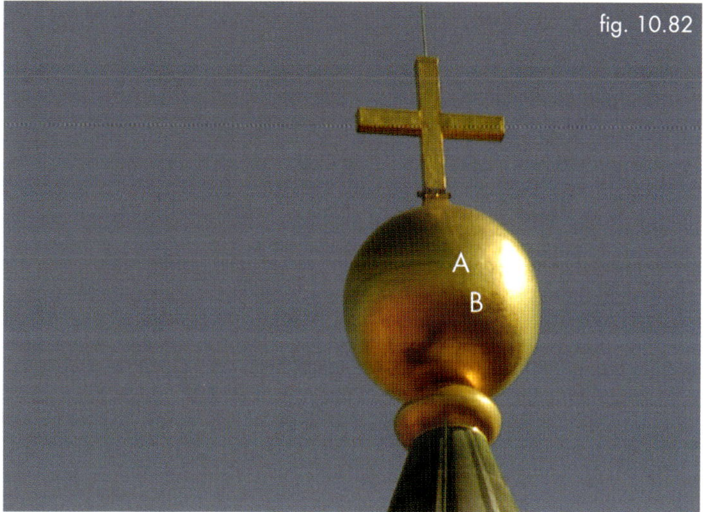

fig. 10.82

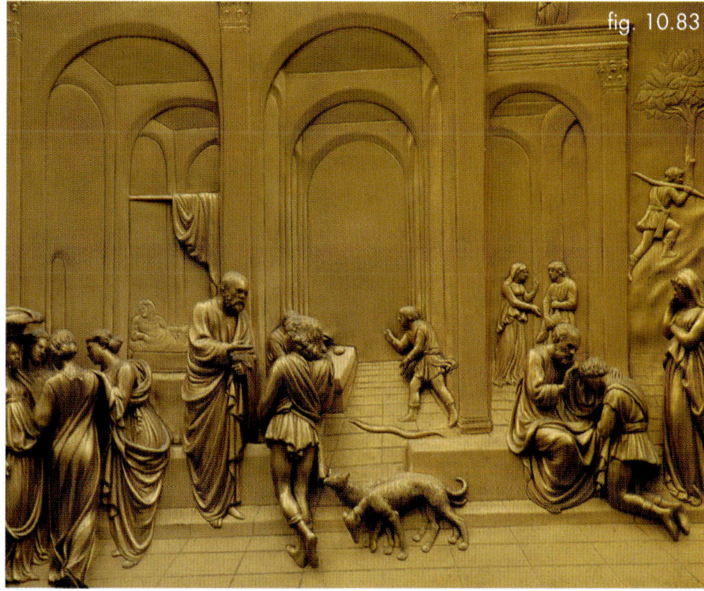
fig. 10.83

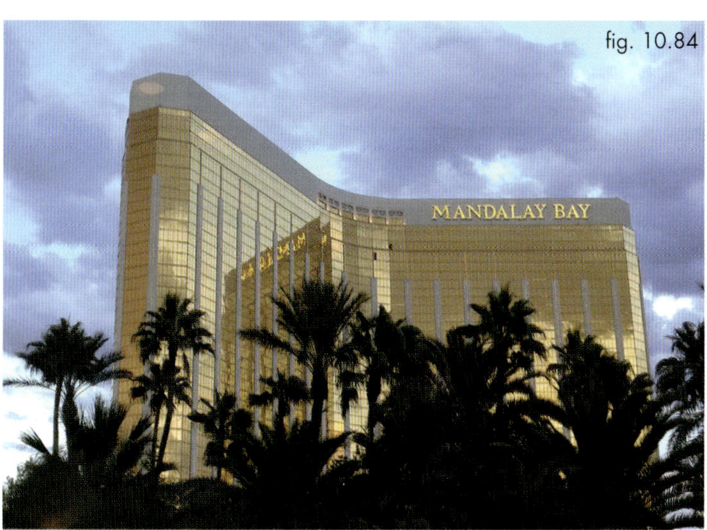

fig. 10.84

When rendering gold, or any material, find a good reference representing its colors and values in context. As evidenced on this page, context can dramatically change the surface appearance of the metal. Remember, when rendering any metallic surface, get the base color correct, and then add the reflections over the top, adjusting opacity to determine its level of polish.

Gold is a bright-yellow color and if color sampling it directly from a photo, pick from somewhere on the Number 2 side where the color is not too greatly influenced by the reflections. Picking point A, in fig. 10.82, would result in the color having too much blue in it and the resulting base color would have a greenish cast. Point B is a better spot to reference. Pick a spot, or average a couple of spots, that provides the most neutral local color of the gold. As the reflections of the environment are added in the reflection layer, the color of the gold will appear warmer or cooler as a result. If working digitally, the addition of a blue-sky reflection over the yellow gold will combine to create green. This happens automatically. Green does not need to be rendered specifically, it is created by the mixing of the two layers.

WOOD

Wood comes in a wide variety of colors, values and levels of reflectivity, but more than anything, it is the grain that makes it instantly recognizable as wood. To render a particular type of wood, it is important to research its unique grain pattern. Searching "wood grain" on the Internet will provide decent images of how the grain looks, depending on how the tree was cut. The more correctly the grain patterns are mapped out across the surfaces of a form, the more photo-real the renderings will look. One of the most important parts of rendering wood grain accurately is to indicate the end grain, if the end of the lumber is visible. Grain does not wrap around a corner like a decal.

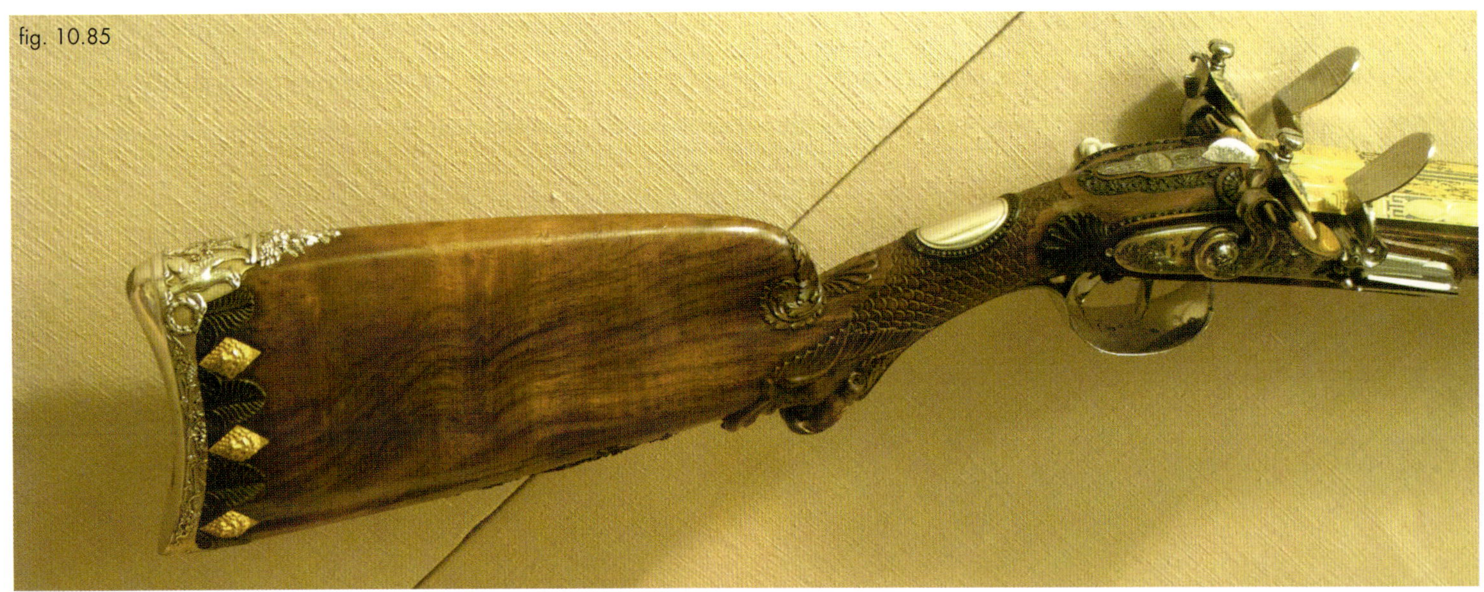

fig. 10.85

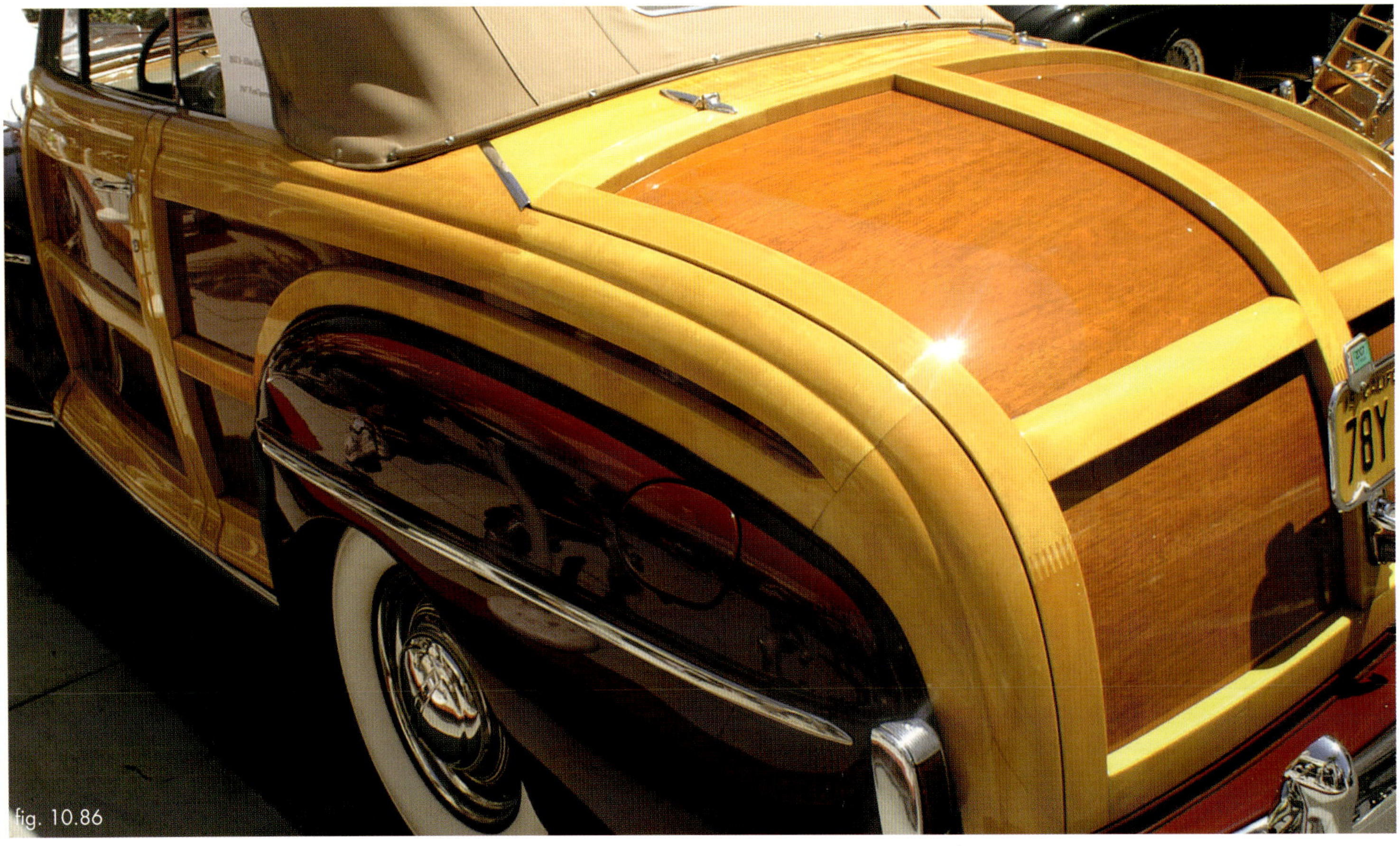

fig. 10.86

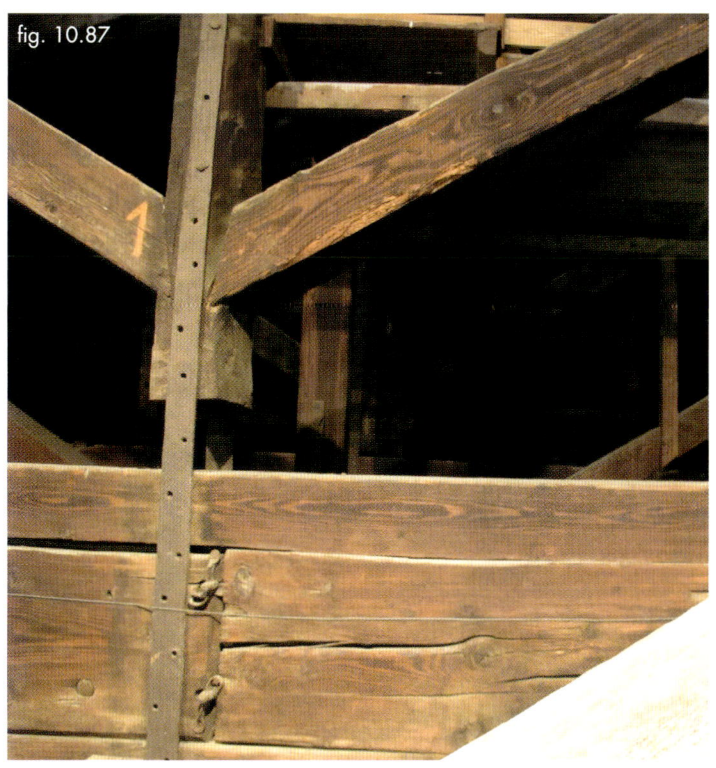
fig. 10.87

Hopefully by now a pattern has emerged in terms of how to render different materials. Figure out the local color, value and unique details of the base material, then render the reflections on a separate layer. Remember, indicating materials is about making them recognizable and familiar enough to the viewers that they understand right away what the materials are. To create the illusion of reality, apply the physics of light, shadow and reflectivity. Knowing how to render a surface to a photo-real level does not mean that is the only illustrative style to use. Even if the art is very abstract, when a quick indication of a reflection, a wooden look or any other material is desired, it will be evident how to render it, even in a stylized manner, using the visual triggers everyone understands.

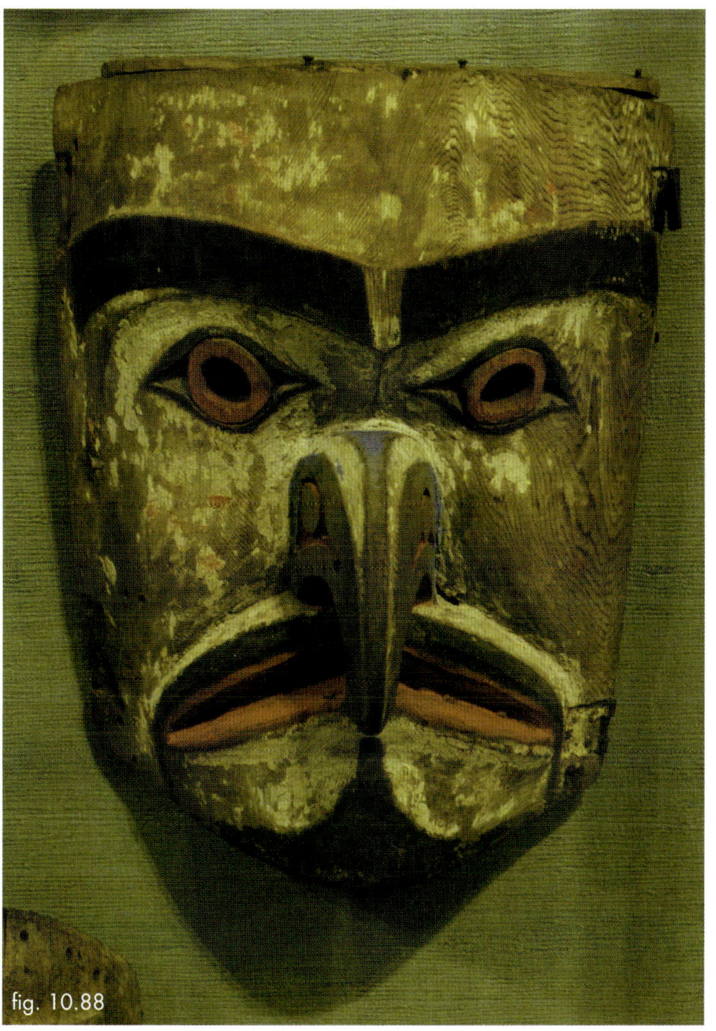
fig. 10.88

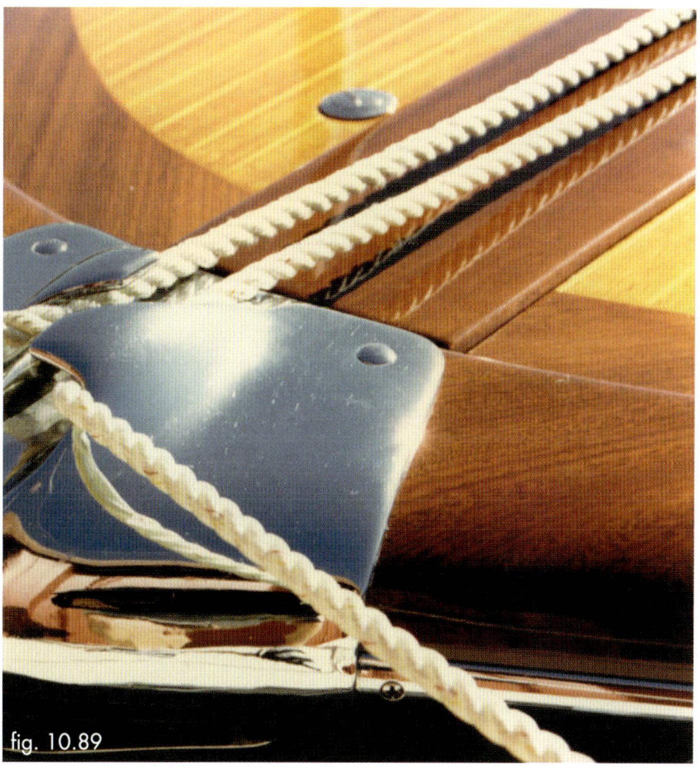
fig. 10.89

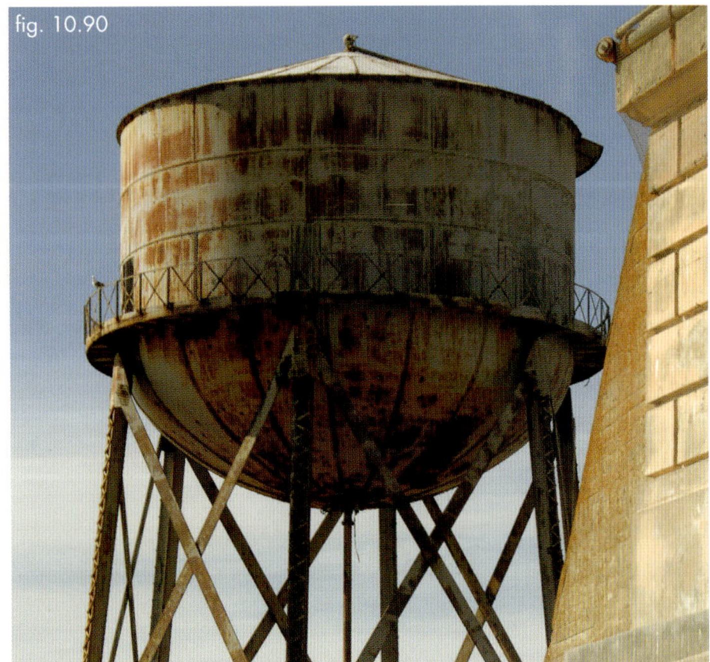
fig. 10.90

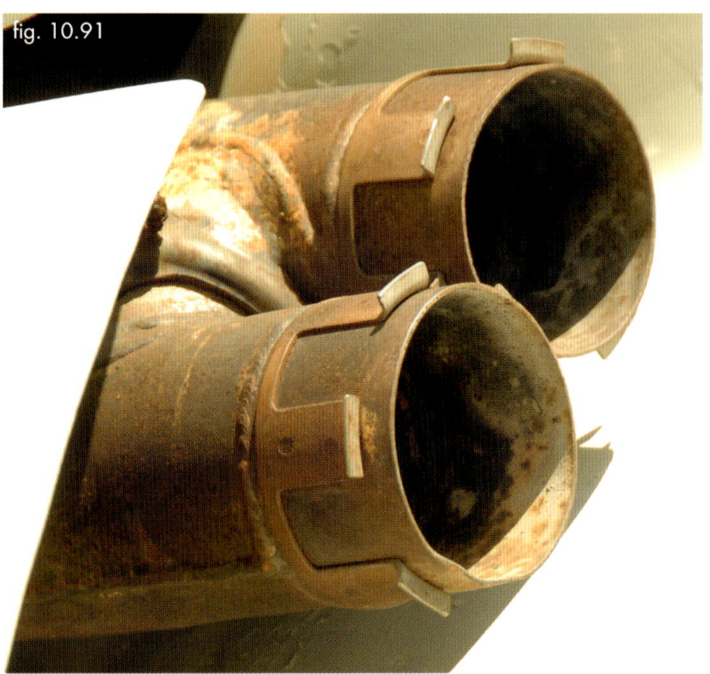
fig. 10.91

Rust is one of my favorite materials to render. When it is added to the steel material of an object, it immediately gives that object a history, and it is fun to imagine its backstory. For example, in fig. 10.92, the rust indicates years of neglect and inclement weather, which caused the rust to drip from the steel fence and stain the concrete wall below it. When placing the oranges and browns of rust, imagine if it rained, where would the water puddle? Eventually, how would the rusty water stain the neighboring surfaces and create decay?

Moisture is the key ingredient in creating rust, so keep this in mind when illustrating or designing the environment of a rusty object. Objects in a very dry, desert have very little rust compared to those in a humid city. Moisture got under the protective finish somehow, so rust is most likely to be found around the seams or part lines of a surface, or in areas where the sun has baked away any protective coating.

The rust effect can be added to an object by turning down the reflectivity of whatever its finish was new, and then rendering the rust as a matte surface over the top. Adding drip lines of rust stains gives it even more realism. Rust usually has a nice range of colors, from yellow to orange to brown. Study some good photo reference to get familiar with the correct colors and values.

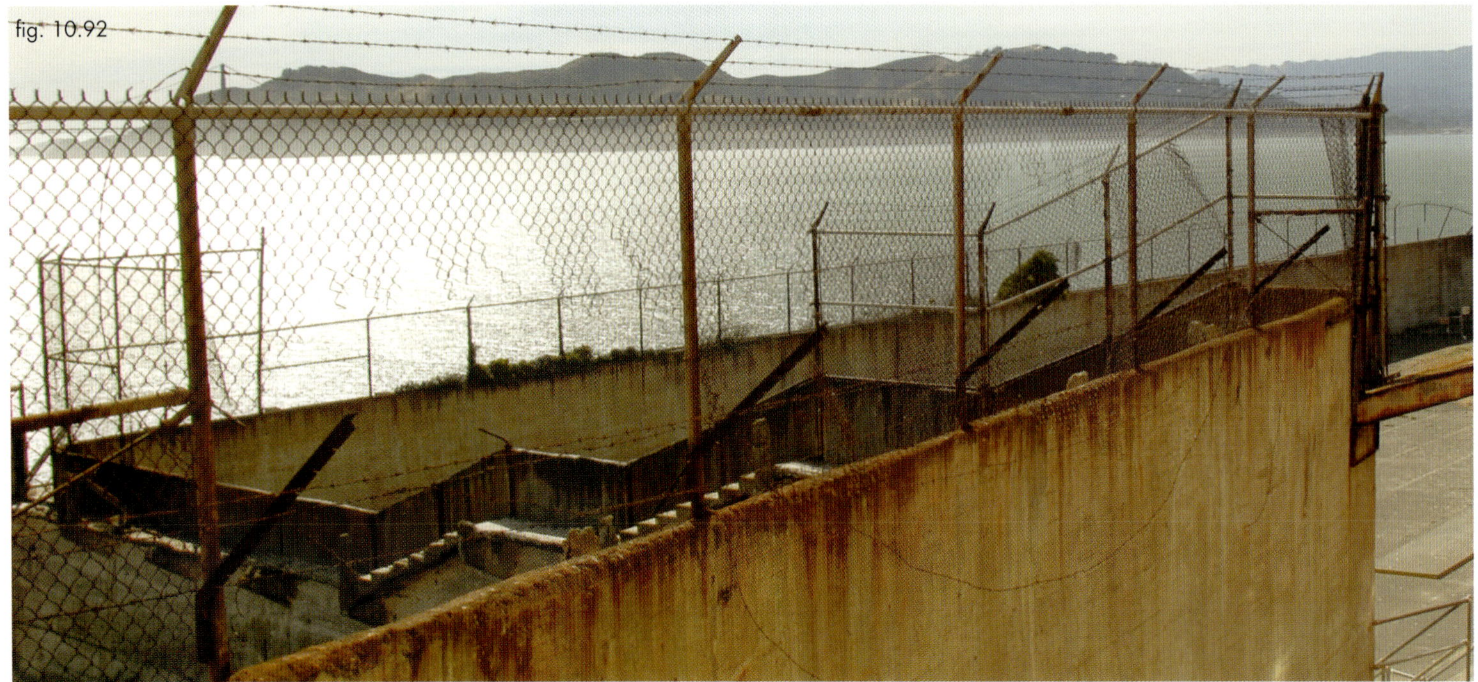
fig. 10.92

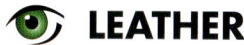

Rendering leather is interesting because it exists in so many textures and levels of reflectivity. But before rendering texture and reflectivity, certain questions must be answered about the design and context of the leather. Is it old, new, thick or thin, and how is it held together or attached to another surface? The leather pouch (fig. 10.93) has seams where the pieces are sewn together and a drawstring to close the top. The seat cover (fig. 10.94) attaches to the car body with snaps. The seams are hidden by piping that needs to be rendered as a small-diameter, flexible cylinder. Depending on the type of leather and how it is used, there might be wrinkles to render as well. These wrinkles can be caused by excessive wear in an area, or just in the way the material wraps around a form. The leather-wrapped steering wheel (fig. 10.95 could be plastic were it not for two visual tip-offs that it is leather. Near point A there are some small wrinkles, as well as some red stitching on the inside of the wheel, both of which indicate this is leather. These visual clues have nothing to do with texture or reflectivity, but everything to do with construction. The leather belt (fig. 10.96) shows some nice wear and stressing where it wraps through the buckle, which help to communicate that the material is leather. As always, working from some good photographic reference is recommended.

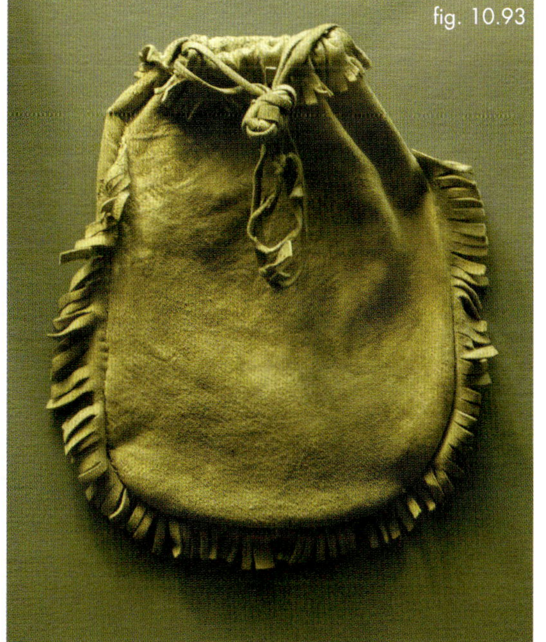
fig. 10.93

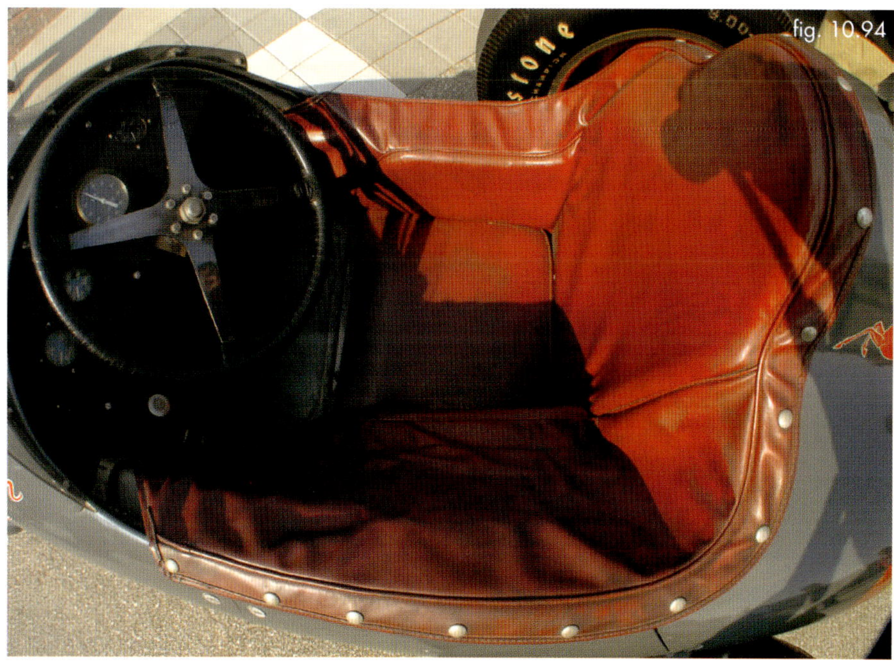
fig. 10.94

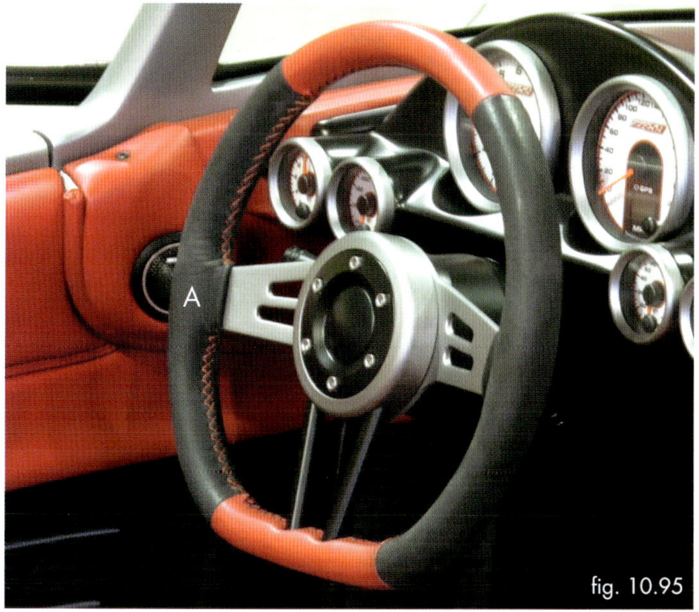

fig. 10.95

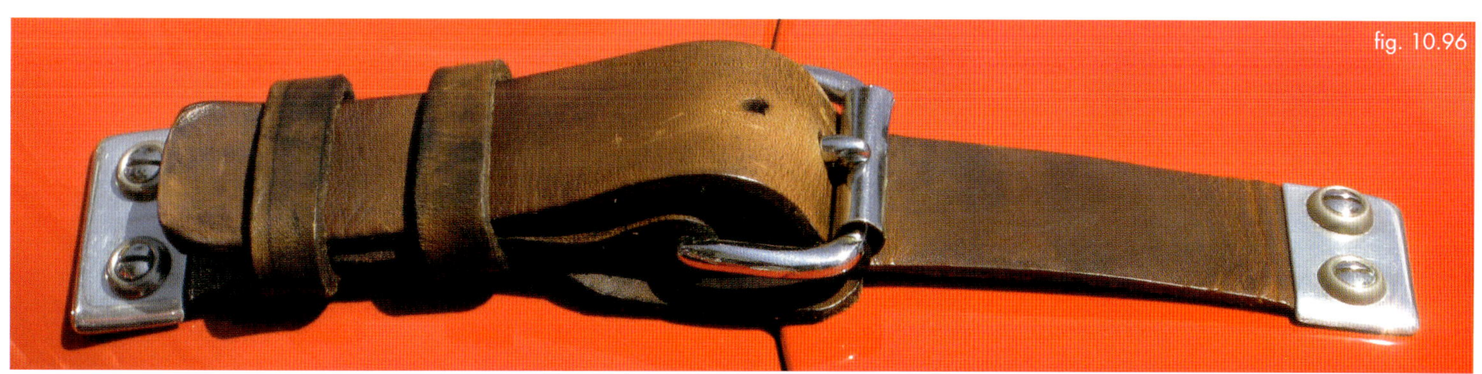
fig. 10.96

CLOTH

Like leather, cloth can have a huge variety of reflective qualities, from very matte surfaces like flannel, to very reflective, metallic finishes like satin (fig. 10.97). Also like leather, it is key to get the construction of the fabric right. Stitches, seams, buttons, snaps and zippers are all great visual cues. Most important, though, is *how the fabric drapes over a form*. In all of the images below, observe how different fabrics wrap around the volumes they cover. Predicting these forms, and accurately rendering them without reference, is a real skill that takes time and practice to acquire. There are entire books dedicated to this single subject and they should be purchased and studied if mastering this skill interests you. For most purposes, it will suffice to photograph or observe someone in the pose needed, and then render the structural form-changes of the fabric from that reference. Remember, it is value, when applied correctly, which communicates the forms of the surfaces and adds realism.

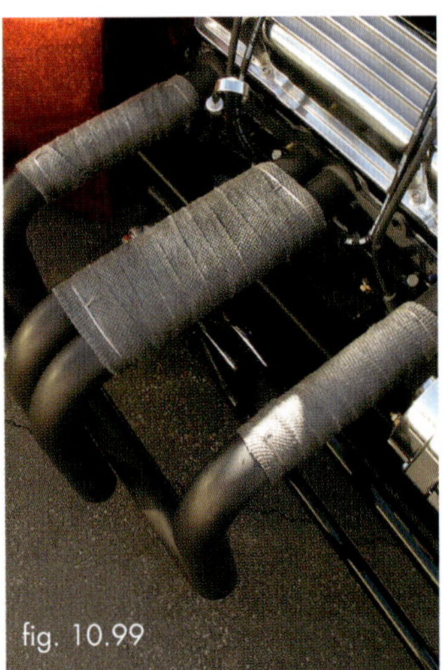

fig. 10.99

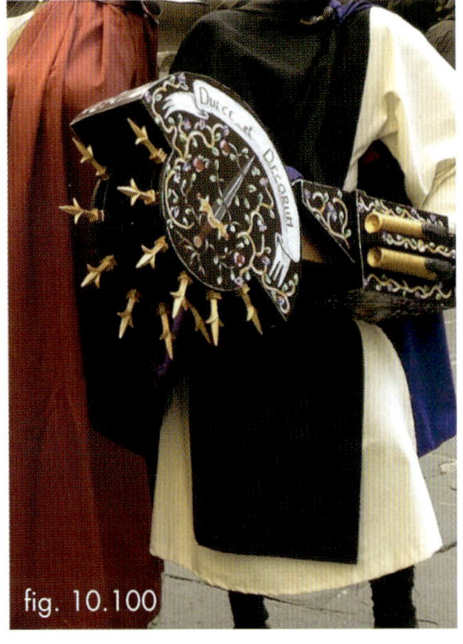

fig. 10.100

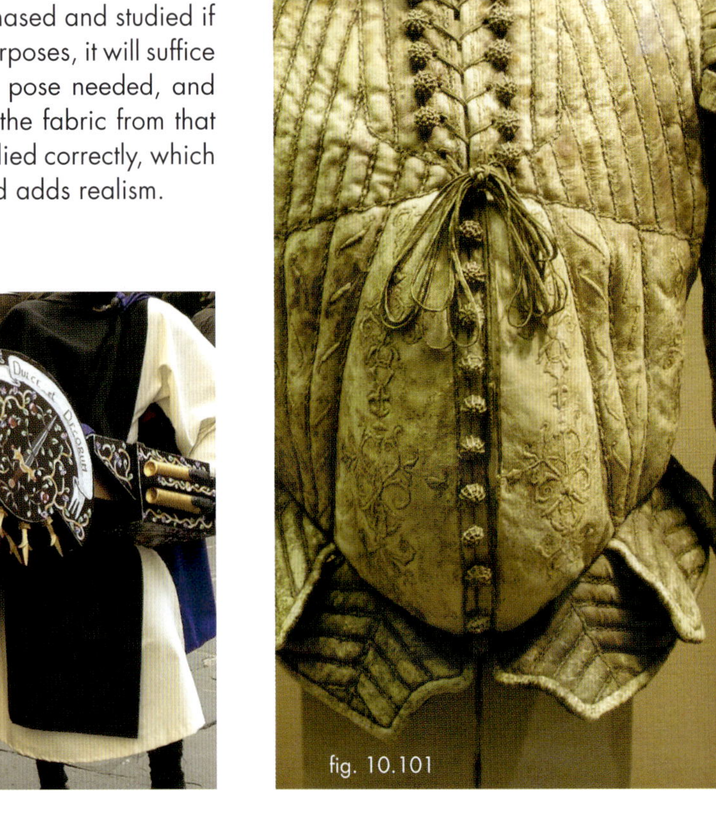

fig. 10.101

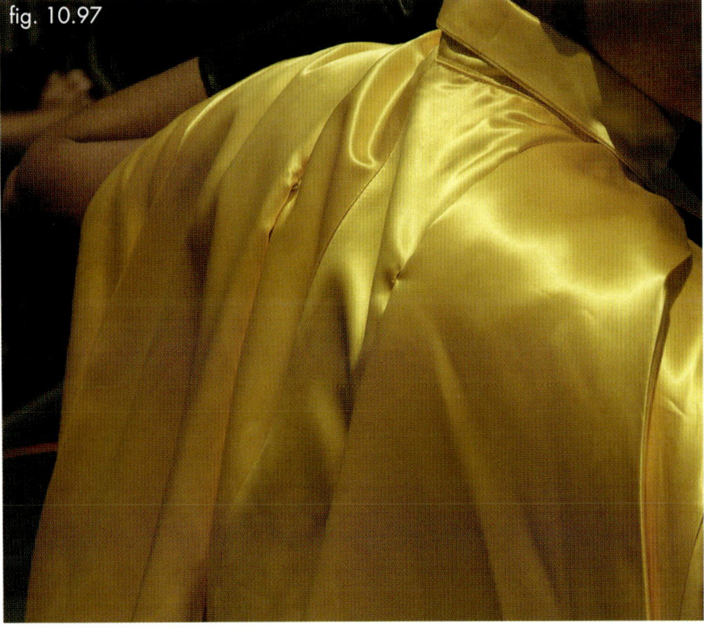

fig. 10.97

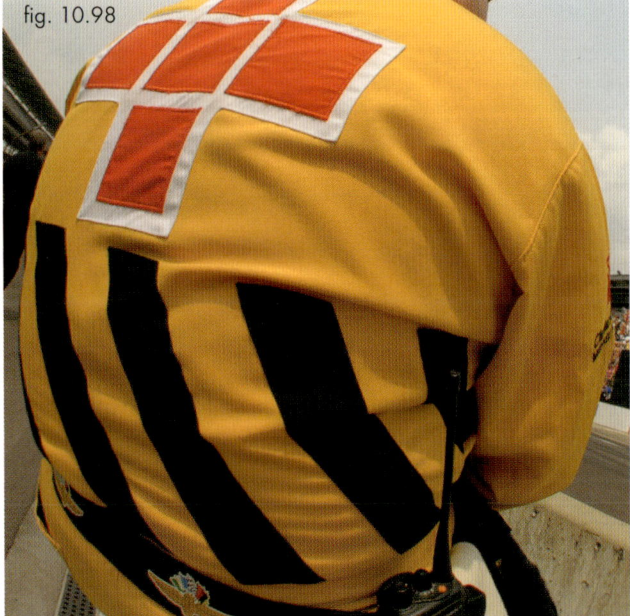

fig. 10.98

Rendering carbon fiber is just like rendering a fabric weave, except that the forms are structurally rigid, so there is no need for stitching or buttons. Before carbon fiber is infused with resin to harden it and hold it in place, the material behaves just like fabric. It can be found in many varieties of weave and color, although the most common color is black. What makes carbon fiber unique is the visible weave in the base material and how this weave reflects the light.

The fiber is usually woven in rows that cross at 90° angles to each other, in a kind of checkerboard pattern. This weave is made up of very small fibers with tubular cross sections, which in turn reflect the light in dramatically different ways due to the 90° alignment of each row in the weave. Usually only one of the directional rows reflects the light source while the other row looks darker. This gives the carbon-fiber weave a lot of visual depth, even though the surface can be very smooth. Rendering this checkerboard pattern, where only one direction of the weave reflects the light, is extremely time-consuming to do exactly right. Typically, just an indication of the carbon fiber is enough to get the point across. In a digital rendering workflow, photograph a flat swatch of carbon-fiber weave and then warp it to look like it is wrapping around the desired form, in perspective. In Photoshop, use the "warp" transform tool. After the weave-image is warped, add the checkered light reflection by "color dodging" it where the light is estimated to be reflected in the clear coat, then add the clear-coat reflections over the top. Observe this effect in fig. 10.104. A nice Fresnel Effect is present in the clear coats of the examples in figs. 10.104 and 10.105. There is a video tutorial for this technique.

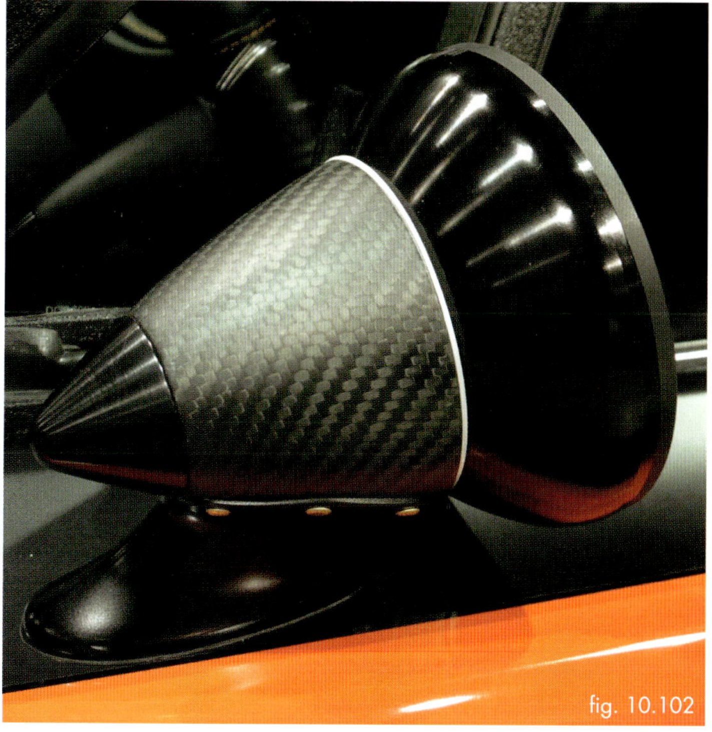
fig. 10.102

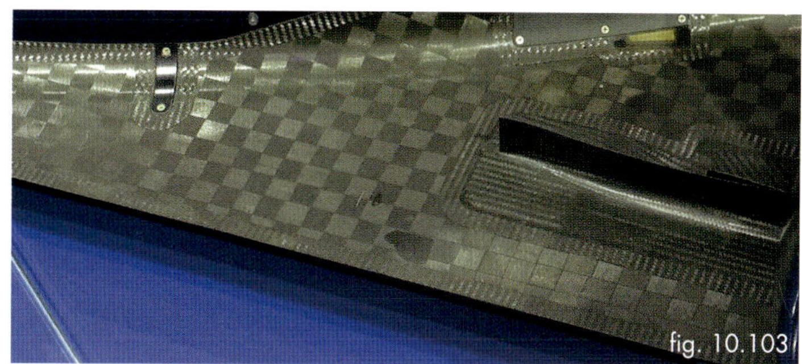
fig. 10.103

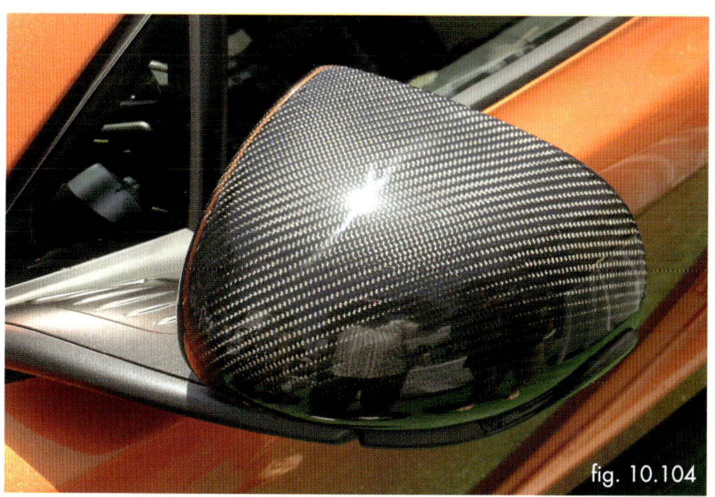
fig. 10.104

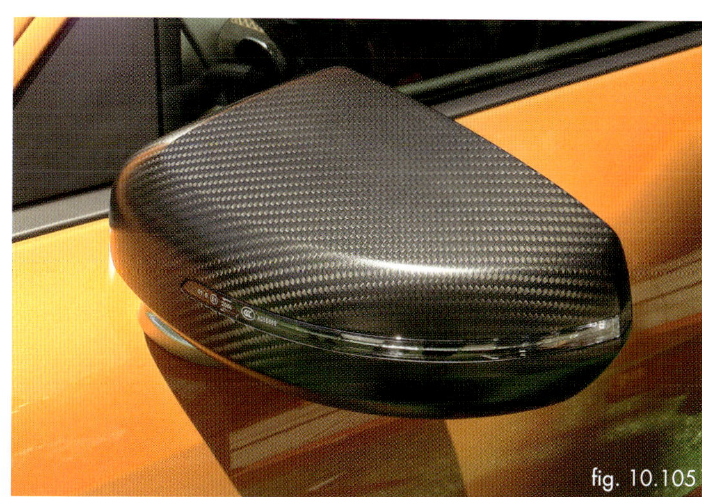
fig. 10.105

WEATHERING 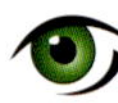 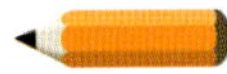

This is where all of the knowledge gained so far in this book can be applied to one rendering. Weathered surfaces can show age through dents, dirt, chipped paint, rust or scratches that reveal multiple layers of paint underneath. Rain-streaked stains, black stains around exhaust pipes or even handwritten notes on a racing tire could fall into this category. Think about weathering in broad terms, not just how the weather itself affects a material, but how over time, interactions with users create opportunities to tell a story about the surfaces. Fig. 10.106 is a nice example of the kind of chipped edges that occur around handles and other high-wear zones. The chipped paint on this old airplane reveals that the surface used to be a different color, or maybe there are several layers of primer under the top layer of paint. This kind of research can be applied to a rendering like fig. 10.107, which shows some chipping of the paint to reveal a primer layer over a metallic base material, and a painted graphic that has been fading and peeling away over time. Watch the video tutorial linked to this page to see this exact layering strategy in Photoshop.

In a small marker sketch (fig. 10.108), an indication of rust and drip lines on the side body of the vehicle can add a lot of visual interest and history to the object. Evidence of a surface's interaction with living things, or exposure to the elements, makes an inanimate object start to feel alive. Viewers of the weathered surface may or may not make a conscious connection to what caused the variation from its original surface finish, but they will sense its history.

It is obvious on the bronze statue (fig. 10.109) that the nose and tops of the claws have been rubbed a lot over a long period of time. Some of the convex surfaces now have a high level of reflectivity, compared to the darker concave forms that show dirt and grime. This simple variation in the surface quality of the object makes one wonder, why is this statue rubbed so much? It must be special. This is all communicated by the weathering of a surface.

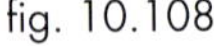

fig. 10.108

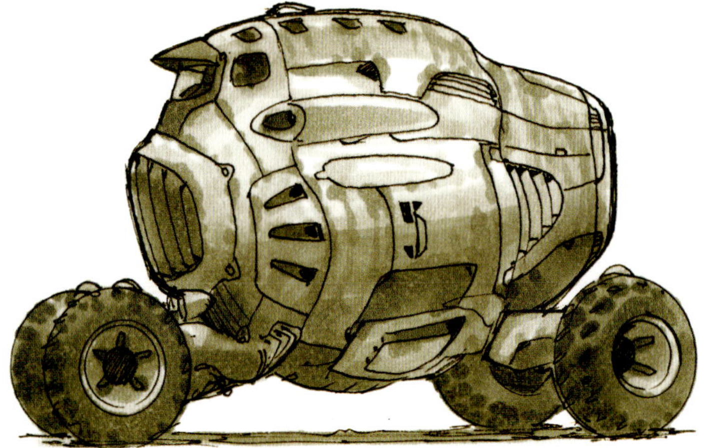

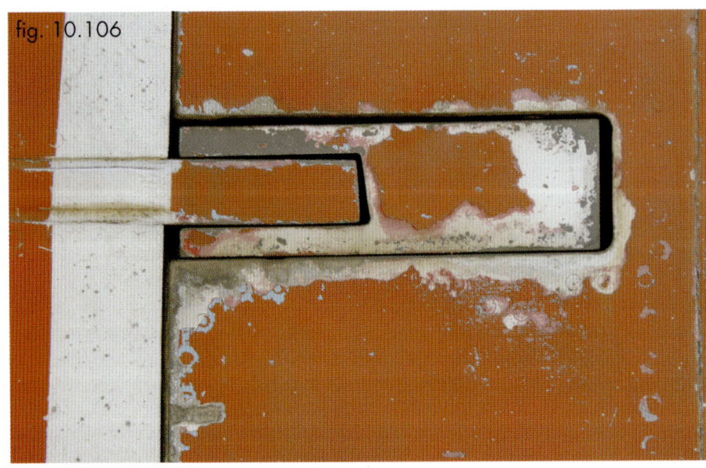

fig. 10.106

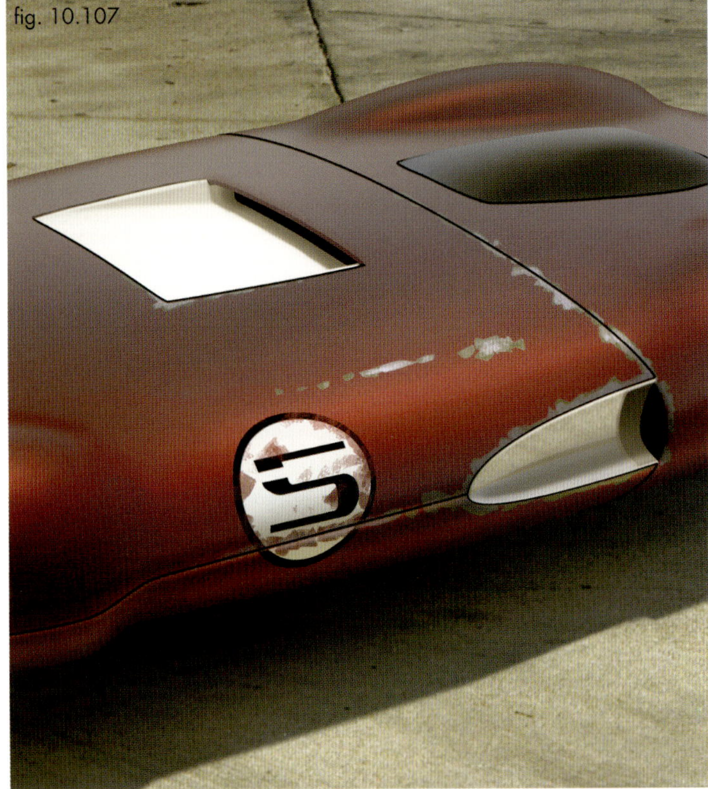

fig. 10.107

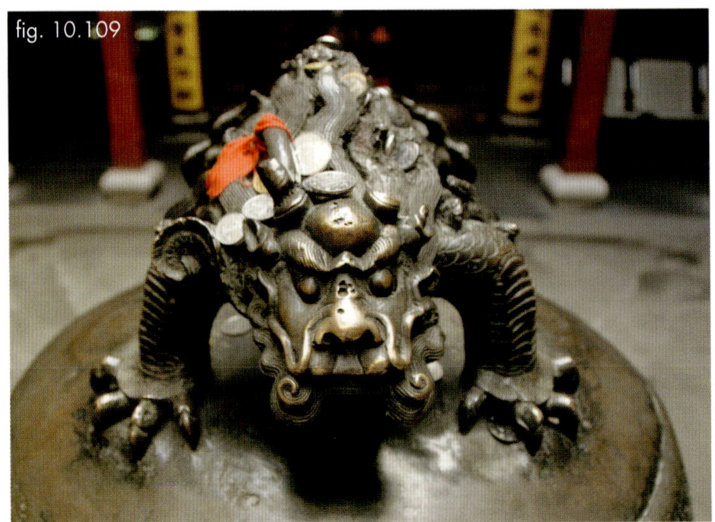

fig. 10.109

fig. 10.110

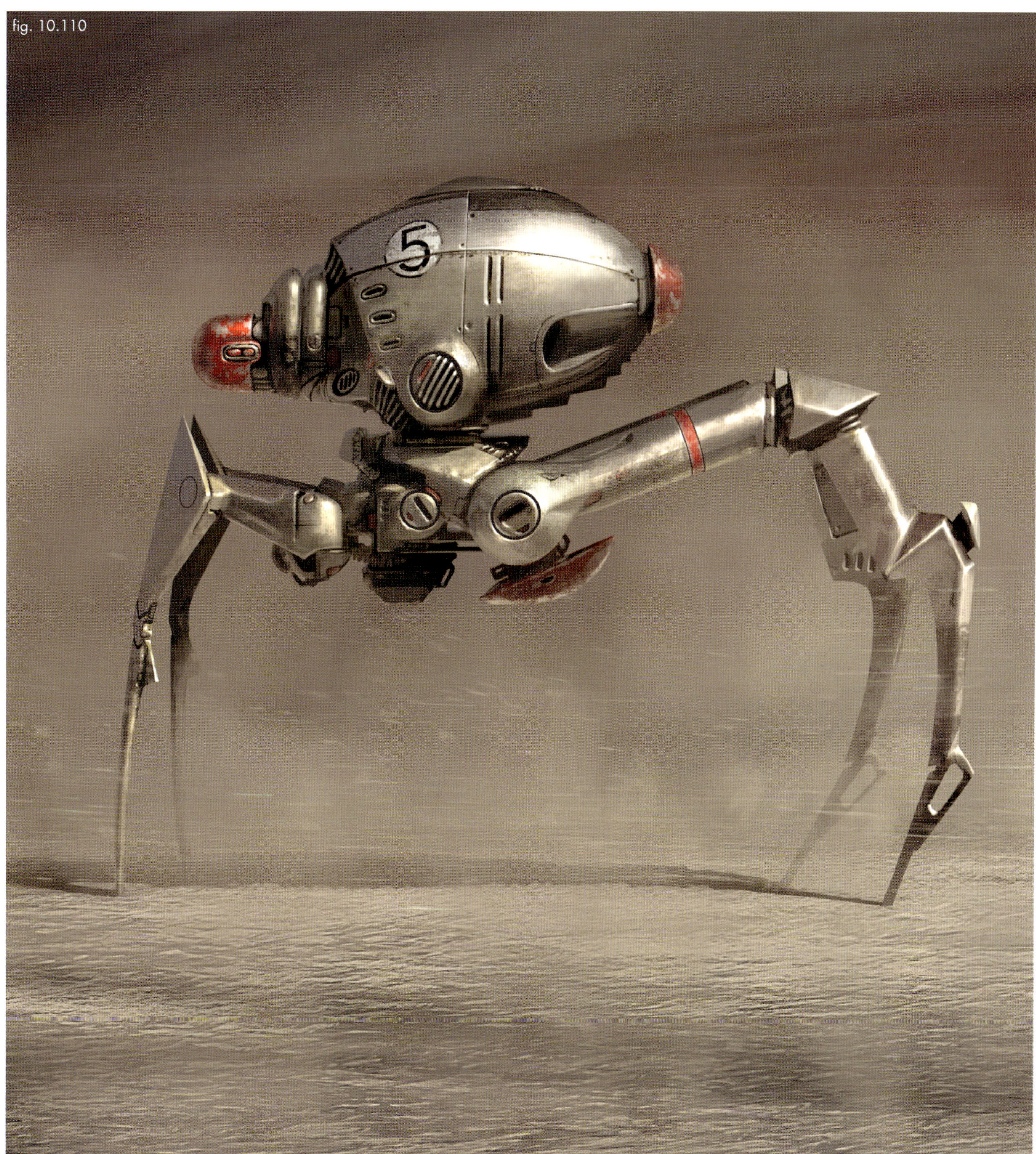

In this rendering, for which there is an entire start-to-finish video tutorial, the dirty metallic surfaces are clearly in a harsh and dusty environment. Rendering the weather in a scene can go a long way to explaining why the materials look the way they do. Dust and debris are blowing hard, hitting the surfaces of the walker in such a way that it becomes clear why the red painted elements are chipping and fading away. Another great way to add life to a scene is by rendering visible wind effects.

CAMERA EFFECTS: MOTION, BLOOM AND GLINTS 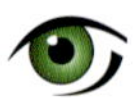

Taking the concept one step further in making a rendering feel more alive is to add a motion-blur effect. This is usually observed in photographs of objects in motion. During the time that a camera's shutter opens and closes to take a picture, objects in motion appear blurry. In fig. 10.111, the spaceship is rendered as though the camera is moving along with the ship so it stays in focus, while the background and falling snow appear blurry. If the camera were in a fixed position and the ship flew by, the ship and snow would be blurry and the background would be in focus. The motion-blur direction of the snow is determined by the direction in which the camera is moving. If the camera were fixed in position, then the snow would be vertically blurred, unless it was very windy. In this example, the camera is moving right to left with the ship, so the snow is blurred horizontally.

All of these spinning wheels are moving relative to the camera's position, which is tracking along next to each vehicle. When reflections on spinning objects are photographed, the spinning elements can seem to disappear and the reflections of the light source can blend together to create fascinating zigzagging patterns.

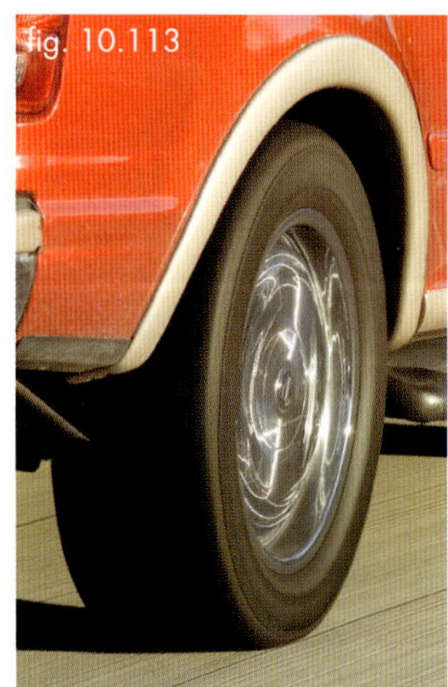
fig. 10.113

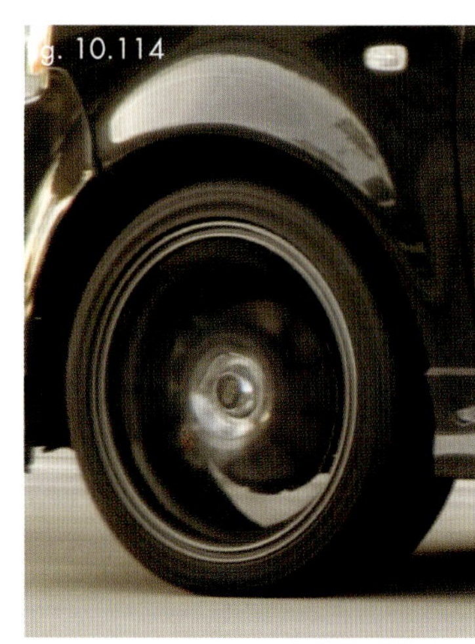
fig. 10.114

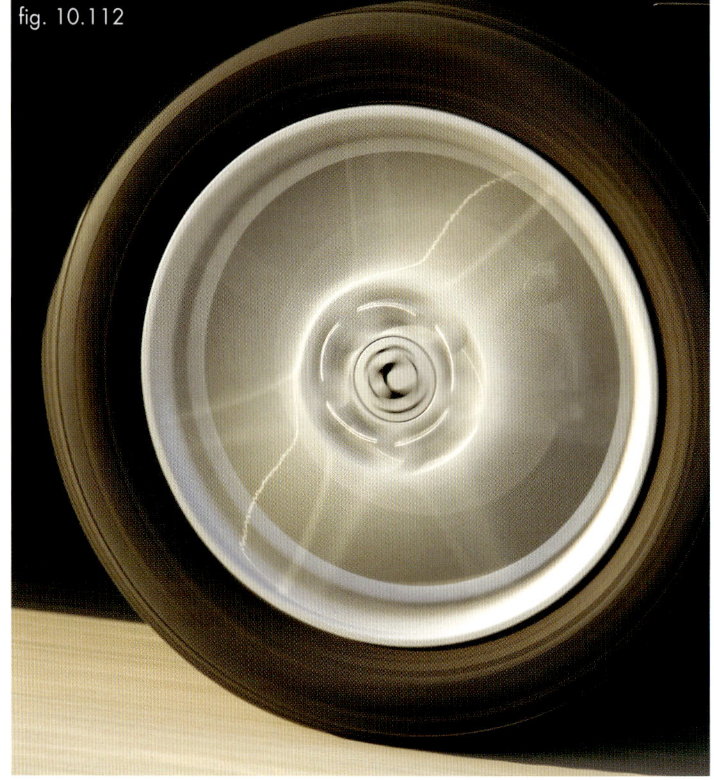
fig. 10.112

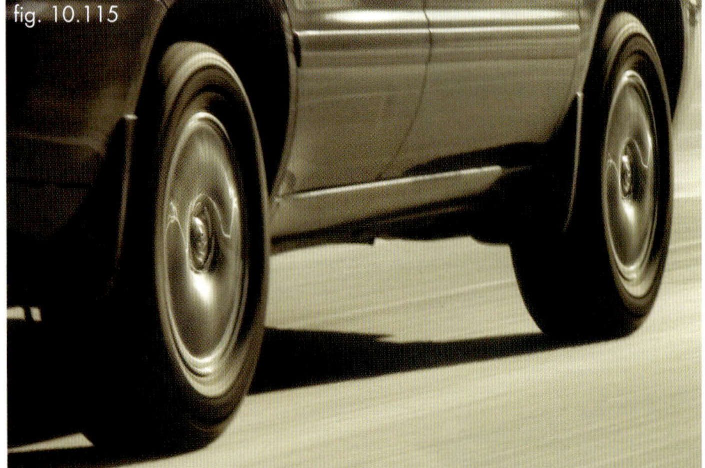
fig. 10.115

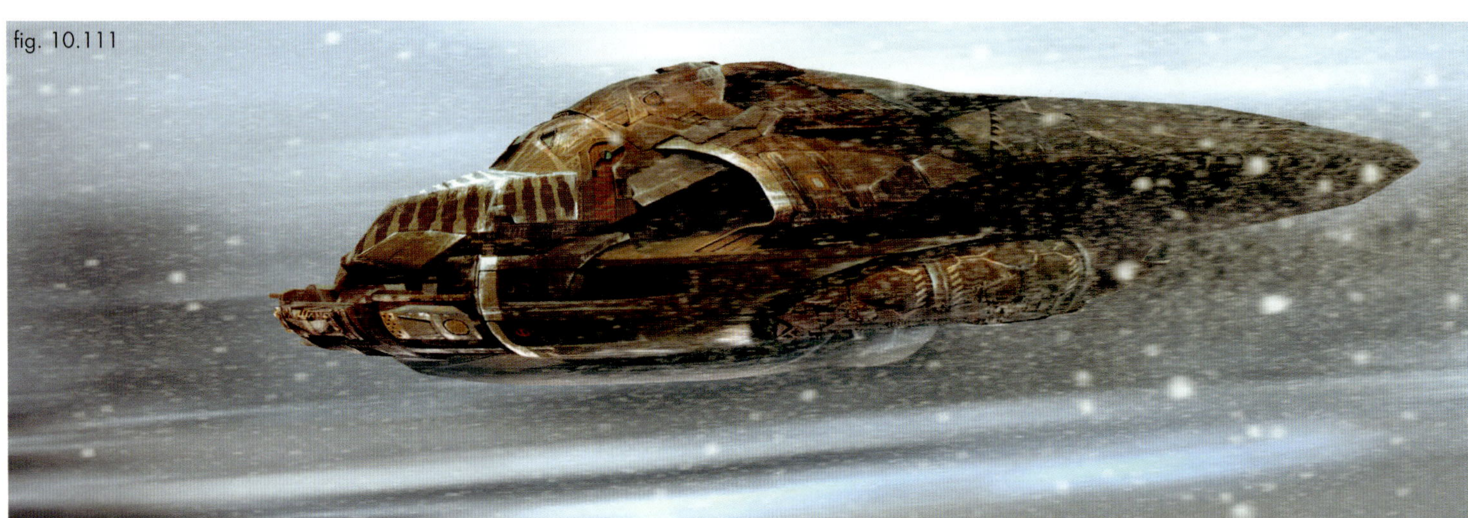
fig. 10.111

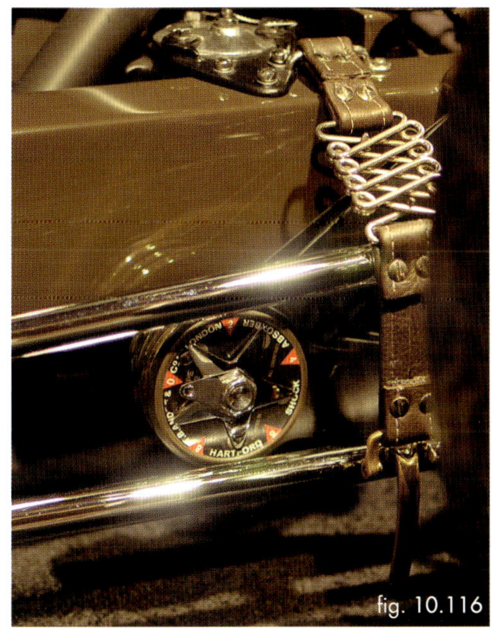
fig. 10.116

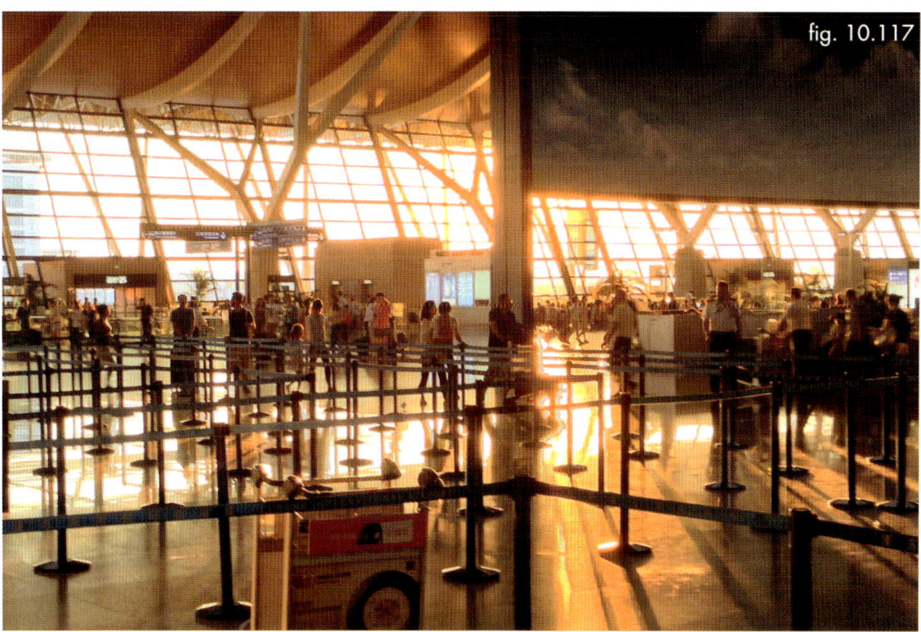
fig. 10.117

Bloom, sometimes referred to as "light bloom," or "glow," is an imaging artifact of real-world cameras. This effect creates fringes or feathers of light extending outward from the center of the bright areas of an image, across the borders of these areas. This effect contributes to the illusion of an extremely bright light, overwhelming the camera or human eye, as in fig. 10.117.

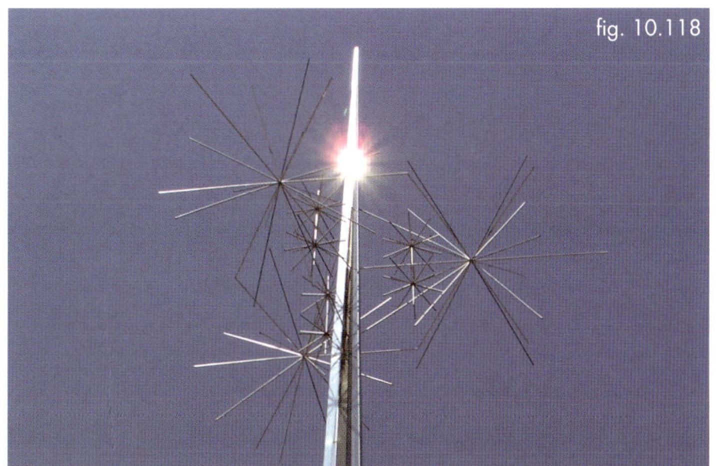
fig. 10.118

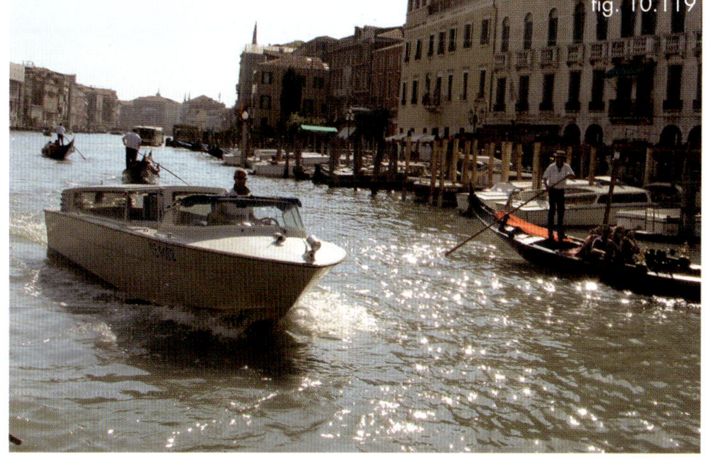
fig. 10.119

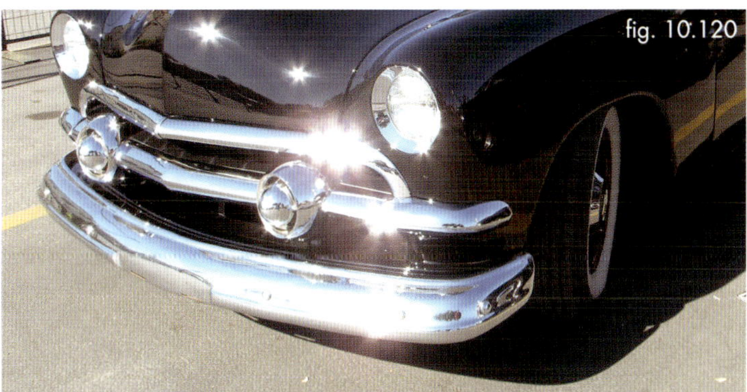
fig. 10.120

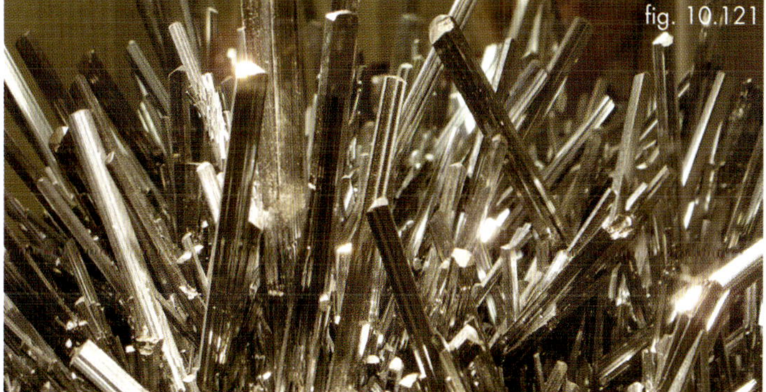
fig. 10.121

Glints are similar to blooms but are usually thought of as the light source being reflected at an angle from a surface, in the form of highlights. The blurring edge is the same as the bloom effect, but it is localized to just around the glint edges.

Glints can be added last to a rendering and can really make the reflective highlights pop.

DEPTH OF FIELD 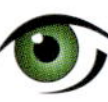 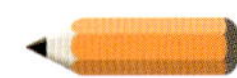

Depth of field is the distance between the farthest and nearest objects in a scene that are both in focus. Even though a camera lens can technically only focus on one depth at a time, the blur before and after this focus depth can be so minimal that the blurring is imperceptible to the human eye.

A large depth of field indicates that objects both near and far appear to be in focus. Artistically, a small, or shallow, depth of field is sometimes preferred to bring the viewer's attention to a particular depth within a scene, where the background and foreground can be blurred and deemphasized.

Fig. 10.123 has the focus set to the head of the lizard, whereby the background beyond that depth grows progressively blurrier. This is the same concept that was applied to the rendering in fig. 10.122.

In fig. 10.124 the reverse is happening. Now the foreground is blurry and background is in sharp focus. In the rendering on the opposite page (fig. 10.126), the same depth-of-field effect was applied. The foreground elements are blurry, while the vehicle in the background is in focus.

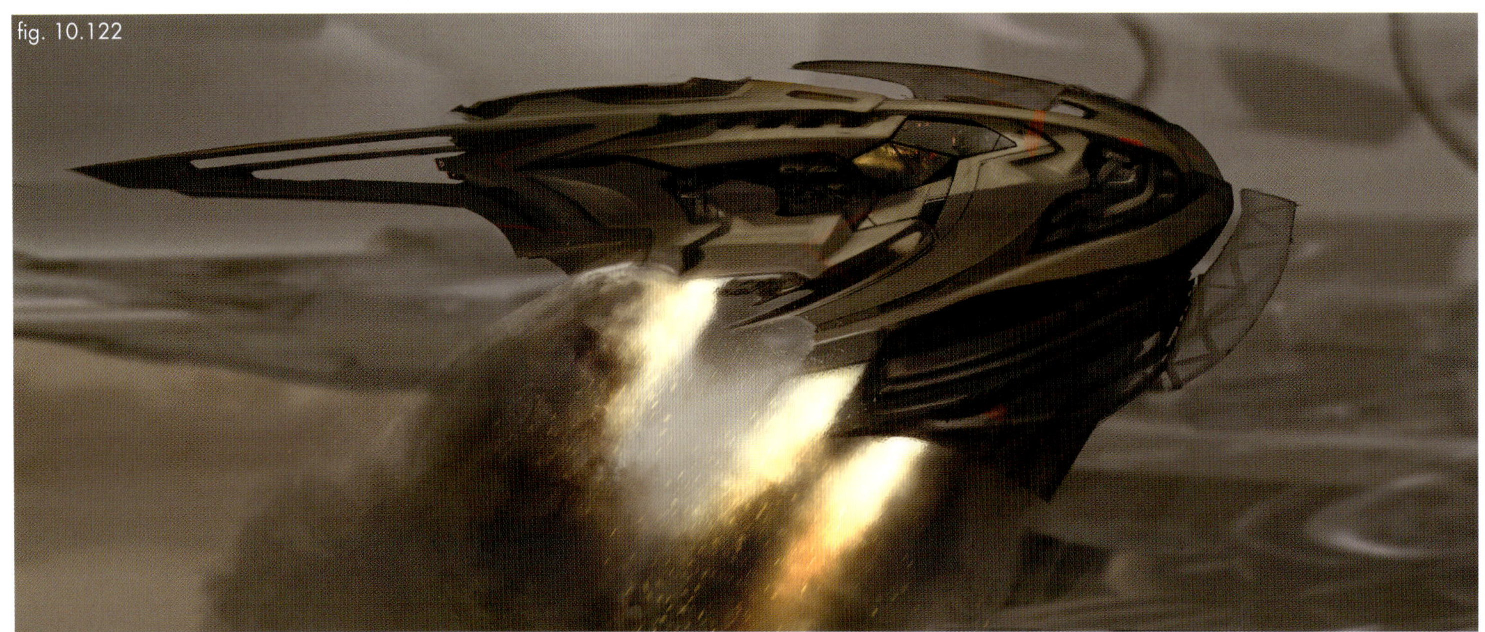
fig. 10.122

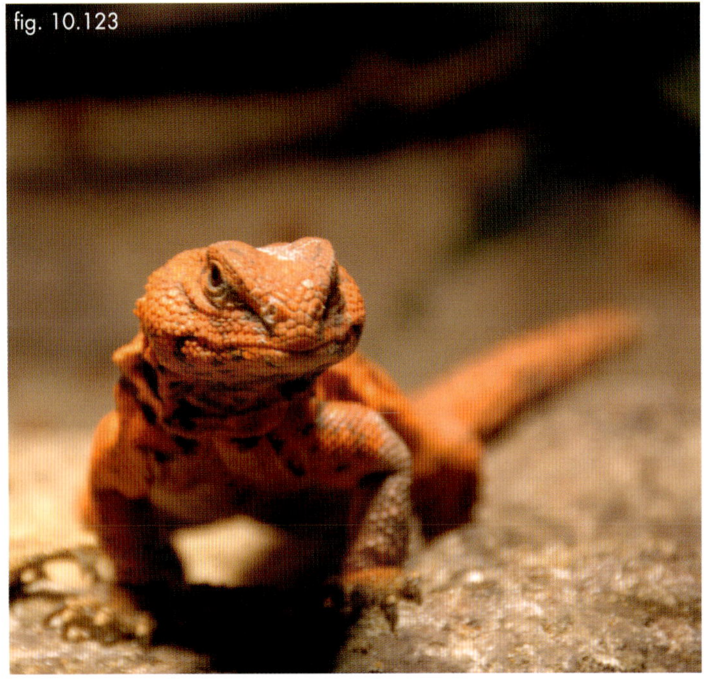
fig. 10.123

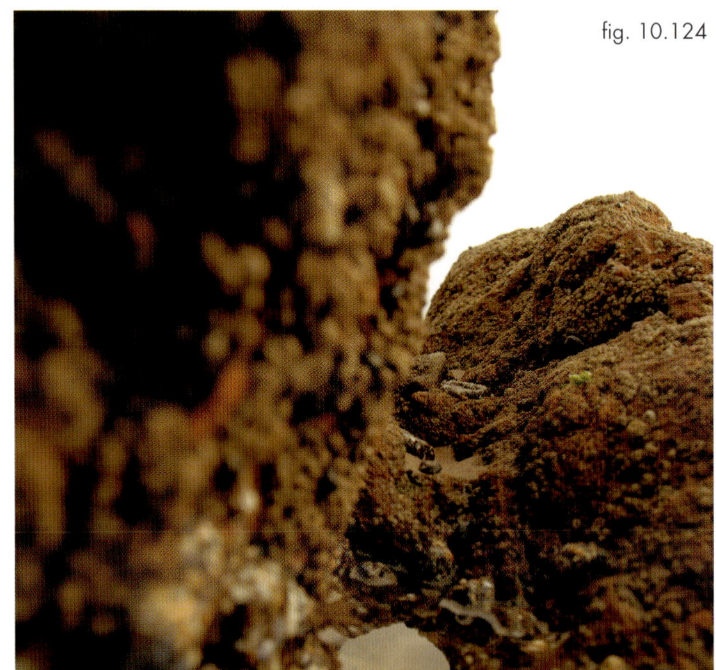
fig. 10.124

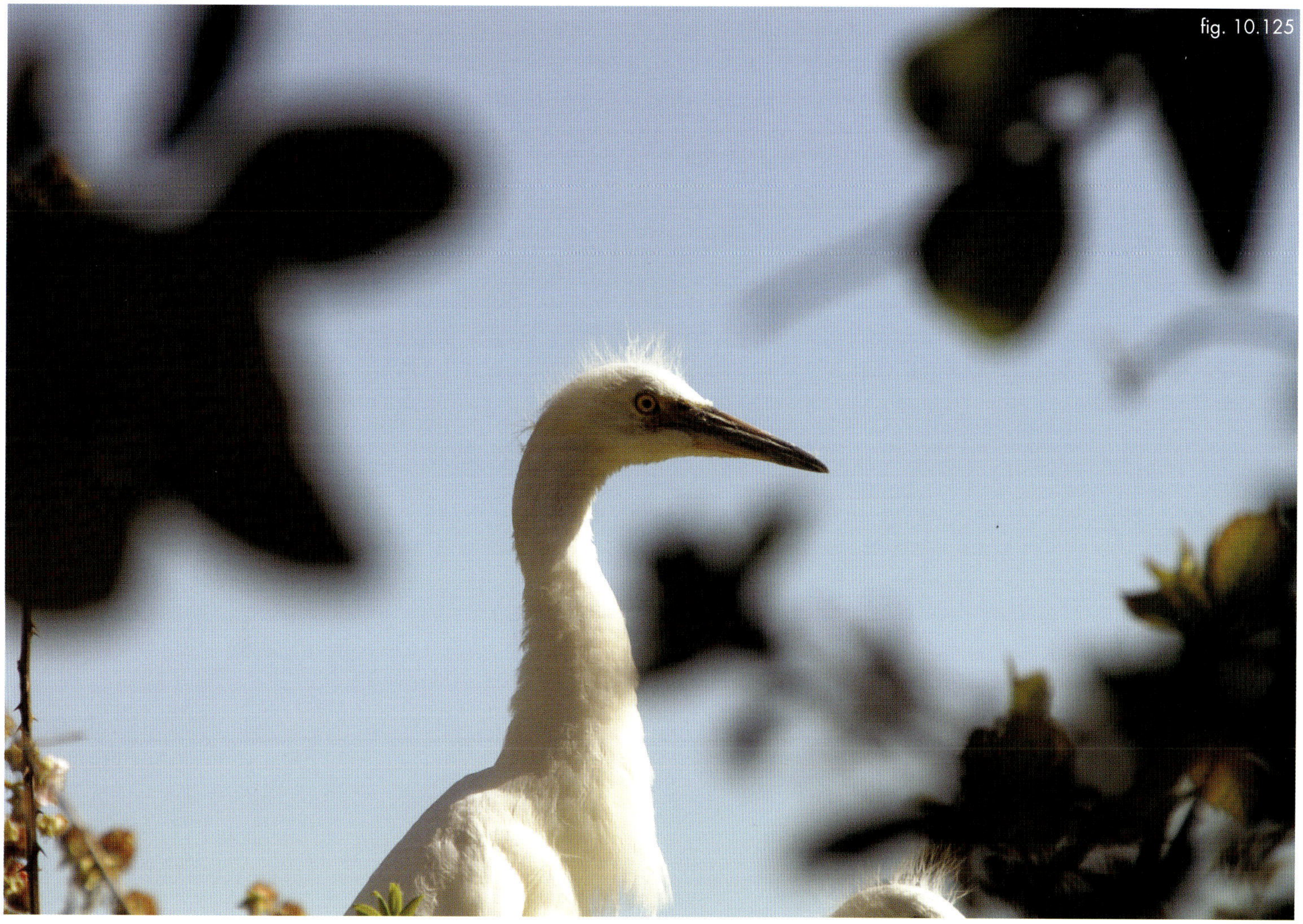

fig. 10.125

A shallow depth of field can be used to help focus the viewer's attention on a particular part of an image. The areas that are crisp and in focus will always command more attention than those that are out of focus. Use this camera effect to bring another level of variation and interest to your work.

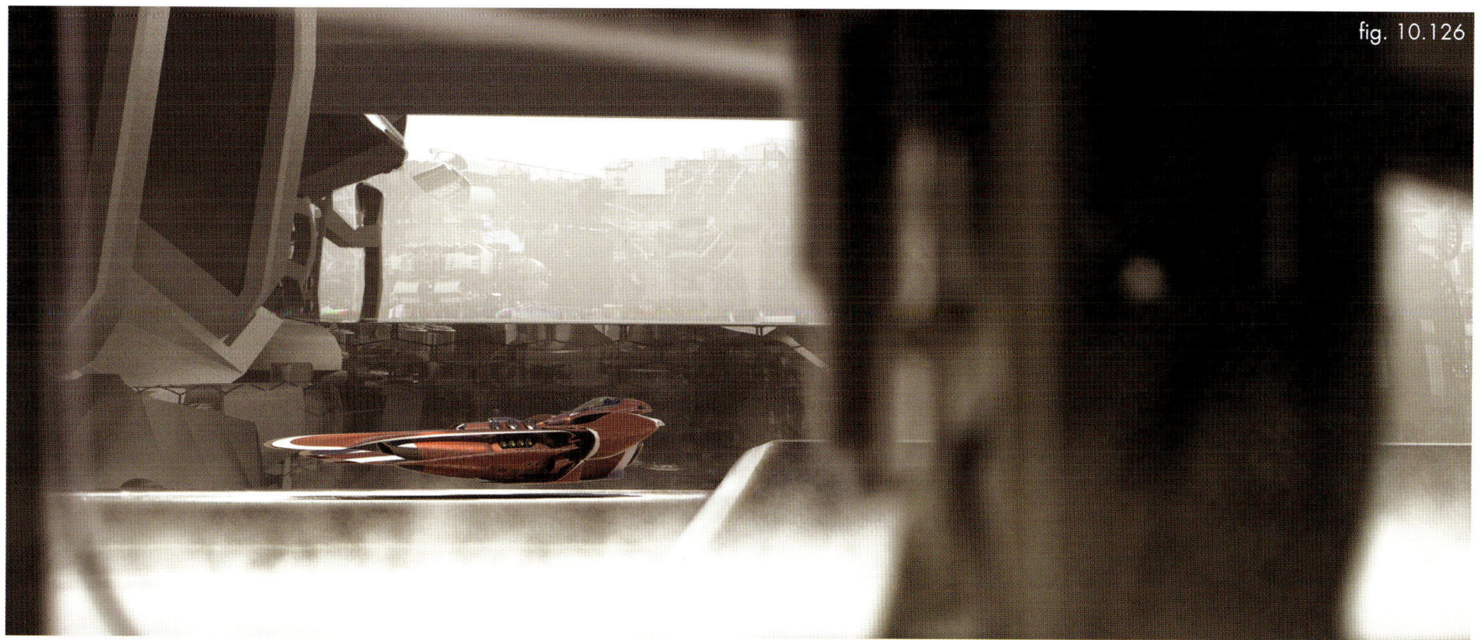

fig. 10.126

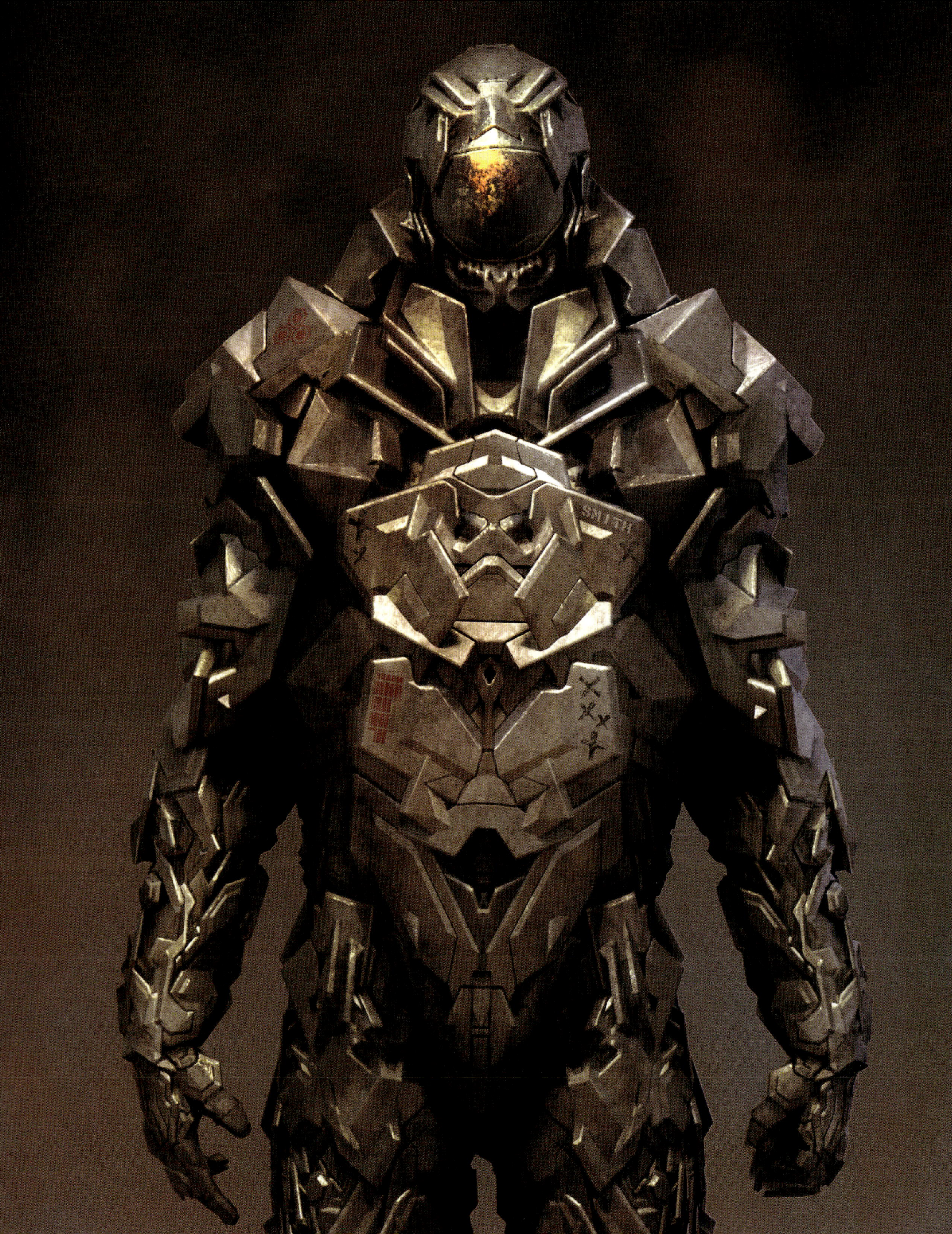
SMITH

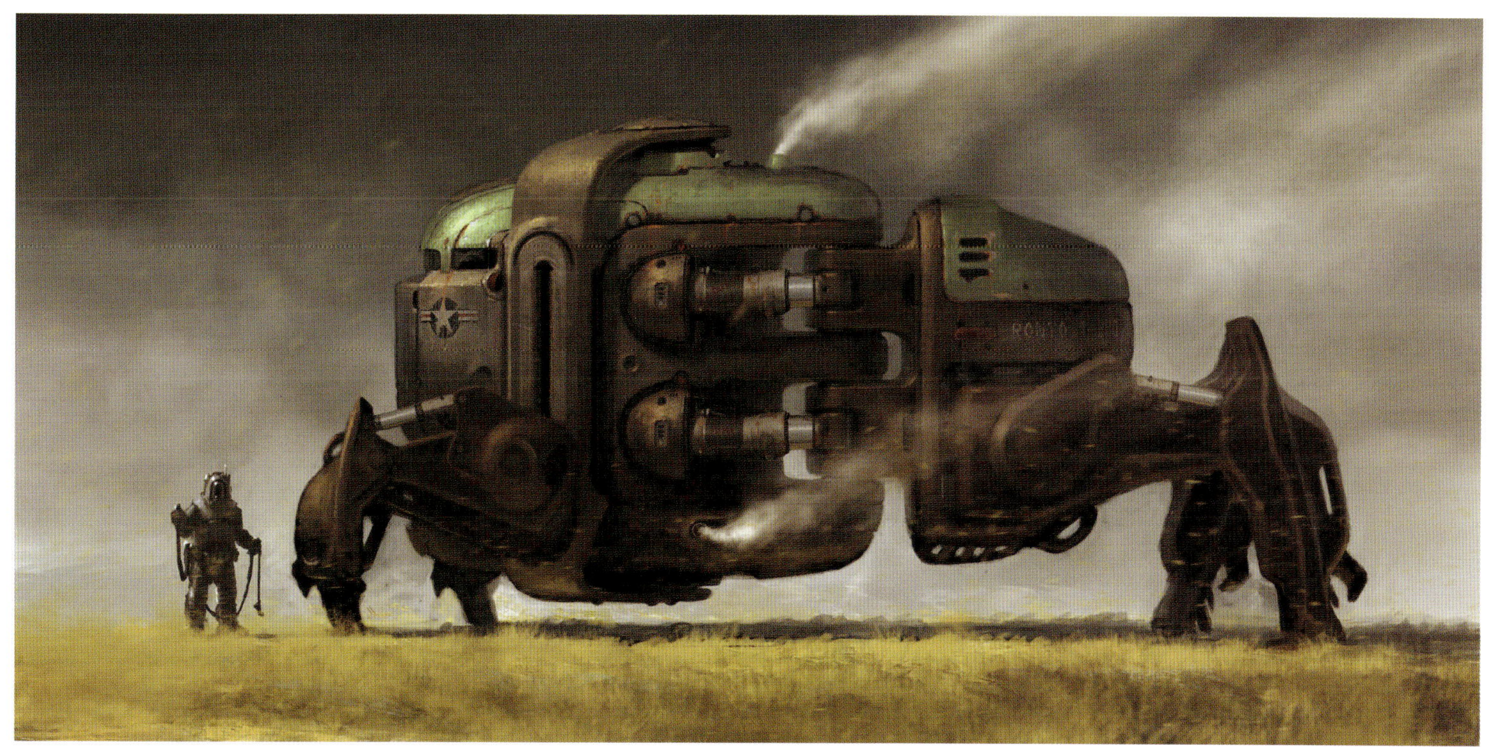

CHAPTER

RENDERING EXAMPLES

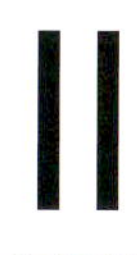

This book is about the fundamentals of rendering. There were simply not enough pages to also cover the advanced workflow techniques of the professional commercial artist and designer. This final chapter provides a taste of the types of advanced techniques that apply the fundamentals learned in this book. Several video tutorials of these more advanced workflows can be found on the *Scott Robertson Design* YouTube channel. Links are provided to those videos. Having gained the knowledge in this book, the videos should be even more educational.

Let's have a look!

FUN WITH VALUES AND WEATHERING

The value rendering in fig. 11.1 was done while on a trip to visit my wife while she was on location editing a movie. Some of my best work has been done while working on my laptop in these locations, totally isolated from the distractions of everyday studio life. This rendering started as a very small sketch, done with Copic marker and ballpoint pen. It was then photographed with an iPhone and opened in Photoshop. The goal of the rendering was to let quite a bit of the line work show through, while focusing on getting the values to describe the surfaces of the vehicle. The first task was to establish a strong 1-2-3 read with a light source that has a fast falloff. After the big volumes were defined, a little metallic effect was added. Finally, some surfaces were rerendered as chrome and a few layers were added to give the vehicle a weathered look. It would be fun to colorize this image and add a couple of human figures for scale and increased interest. Seeing just how far the value scale can be pushed to communicate transitional forms and materials from the imagination is a great exercise and one that not enough young artists and designers challenge themselves to do. Always remember, *value is the key to the communication of the forms.*

fig. 11.1

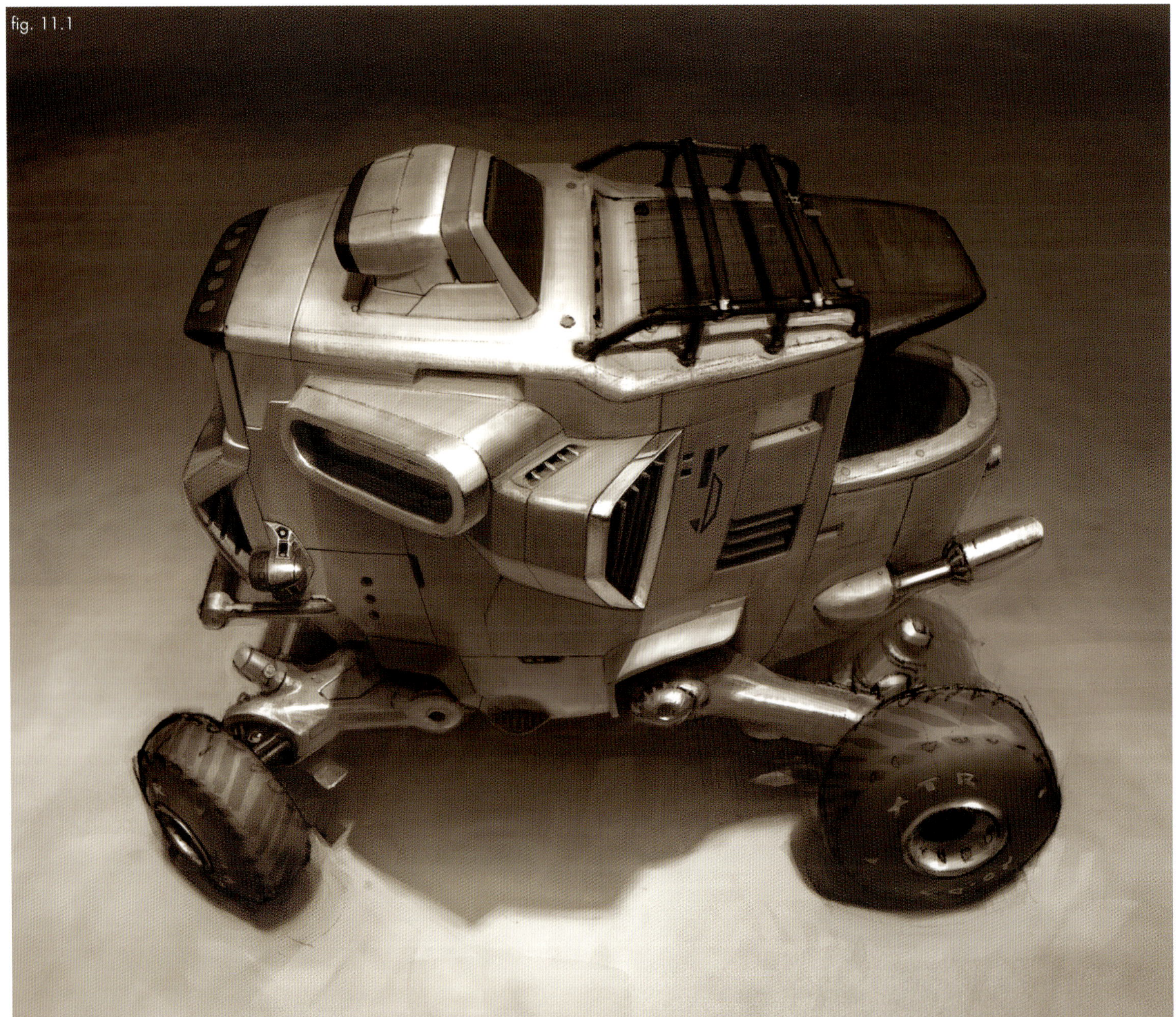

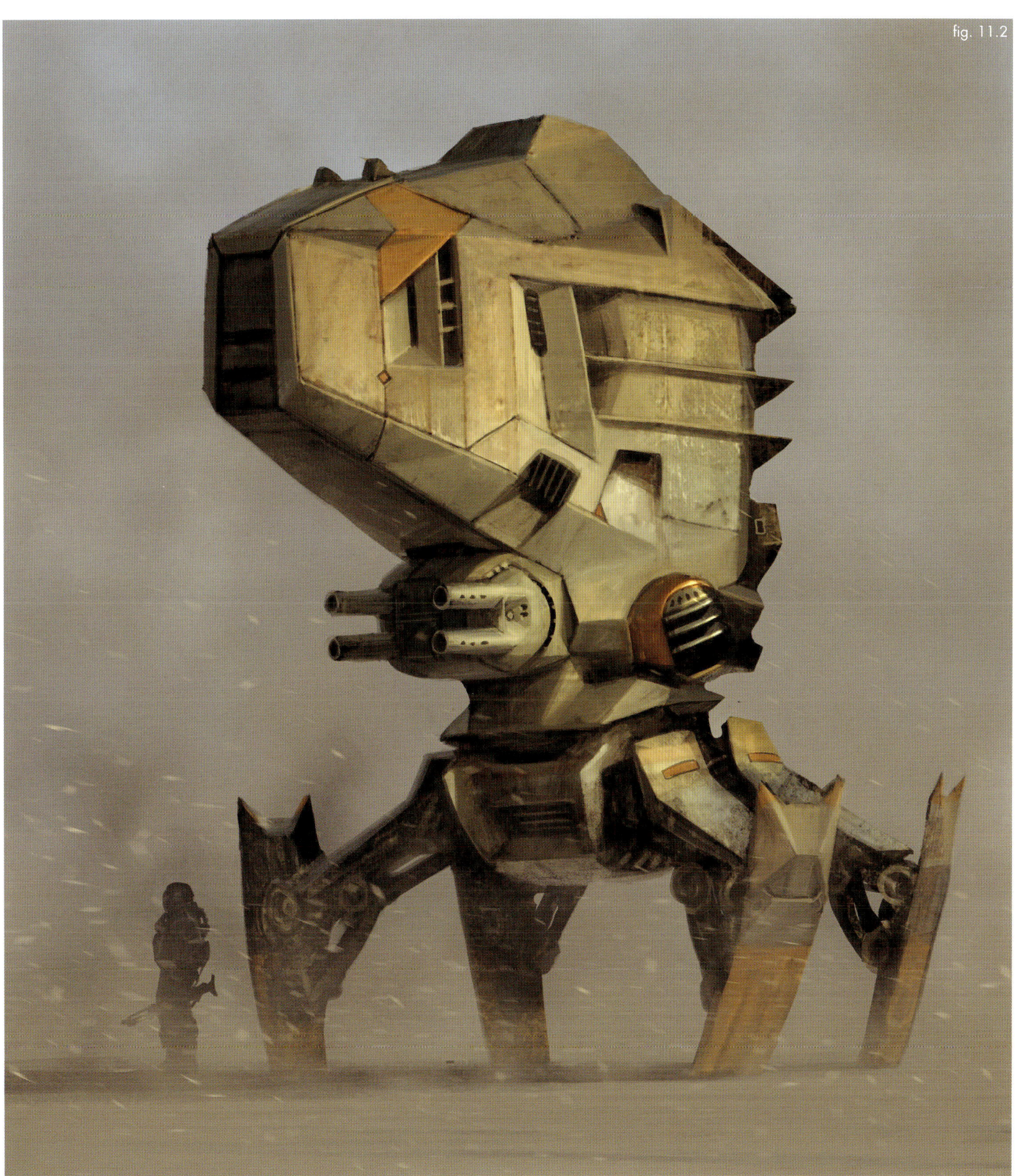

Here is an example from one of the *Scott Robertson Design* video tutorials on YouTube. This rendering started as a small brush-pen sketch with a photo reference cut and pasted over the top. Using different layer modes, "happy accidents" were explored and then the entire mech was rendered on top, using a few layers to refine the color, materials, design, forms and atmosphere. Adding the figure was the last step and it works well to help establish the scale of the mech.

PHOTO BOOTH AS A DESIGN TOOL

Neville Page and I discovered that the program Photo Booth on a Mac could be used to create alien characters by interactively using the mirror filter and the built-in camera. Lately we have switched over to doing these photo studies using iPhones and iPads. In fig. 11.3 the head of the character started as a mirrored Photo Booth snapshot of a crumpled piece of brown paper and the one on the facing page, in fig. 11.4, was the top of a vertical wooden post. Check out the video tutorial that shows this process in action. It basically builds upon the hardwiring that everyone shares, which is always scanning an image for signs of life and seeing "faces" in symmetrical patterns. This technique was refined and put to a lot of use in the making of the book *Alien Race*.

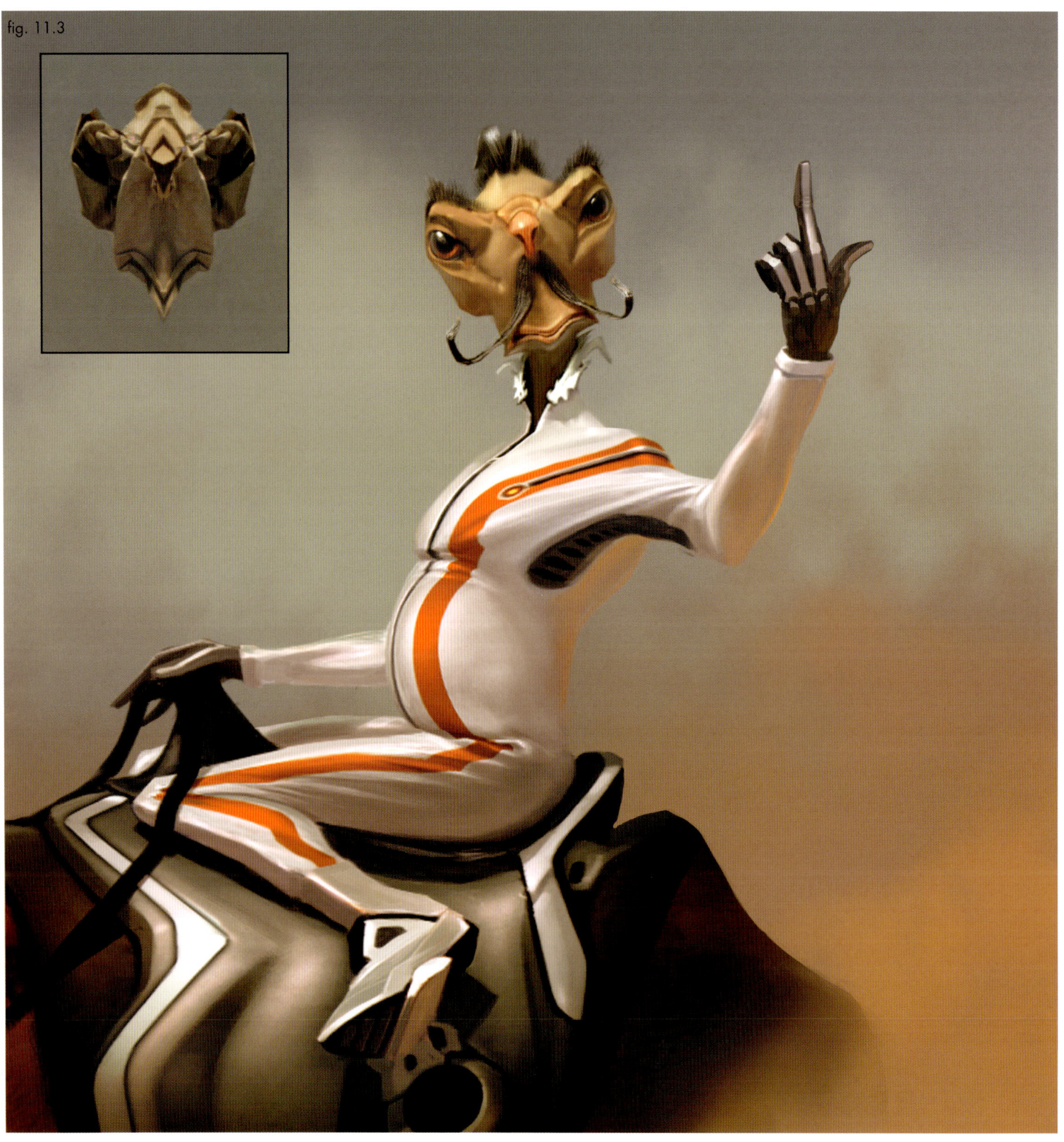
fig. 11.3

fig. 11.4

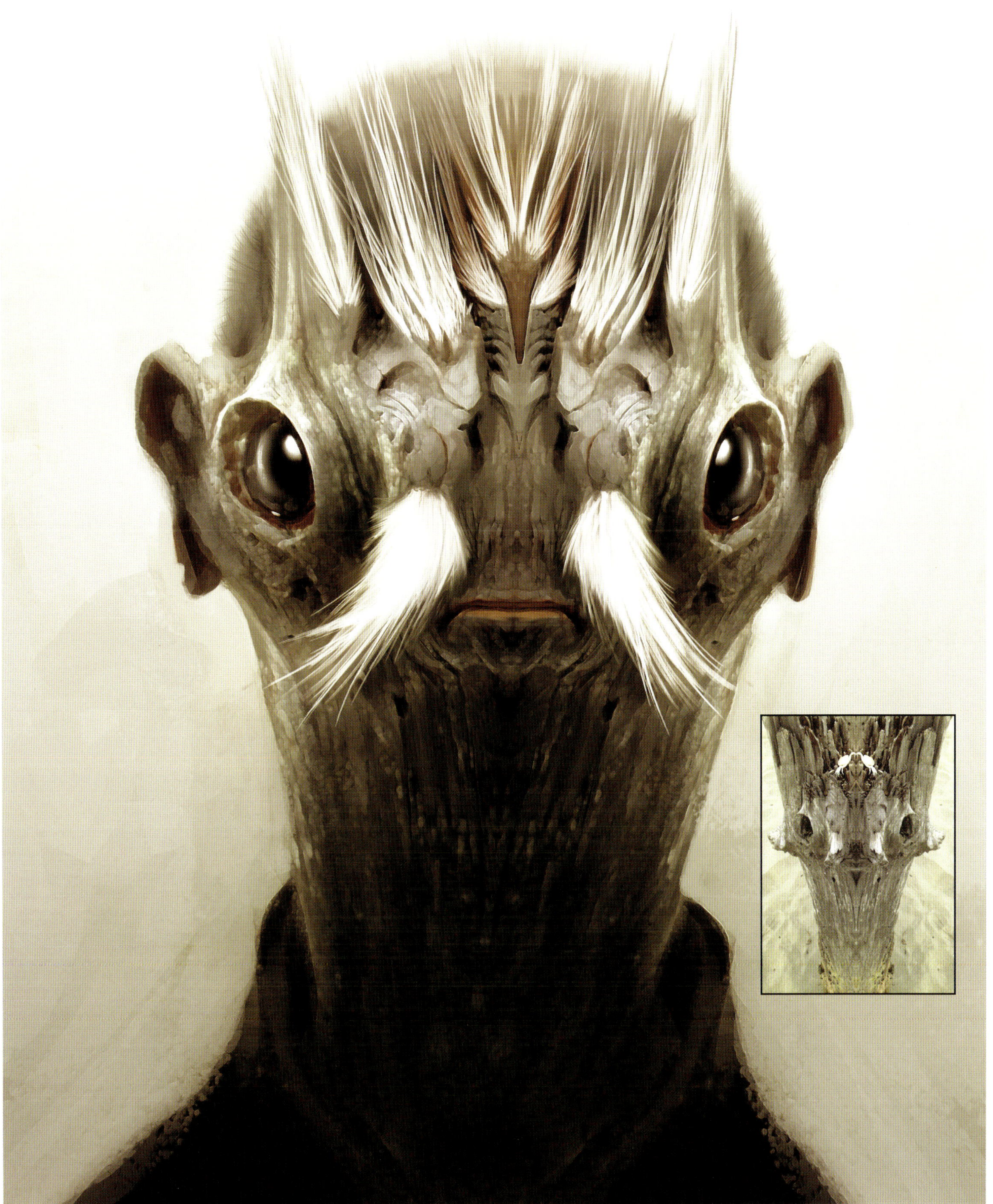

PHOTO PAINT-OVERS

The technique of using photographs to start the digital workflow can be an effective and inspiring way to create imagery. The renderings in figs. 11.5 and 11.6 both began as photographs of dinosaur skeletons. They were then painted over in Photoshop. Once the goal becomes making creative images, as opposed to original pieces of art from a blank piece of paper, tools like photography can become the same as drawing and painting. It is just an auxiliary set of tools to use. None of which can be really well integrated or expanded upon without a strong mastery of the fundamentals.

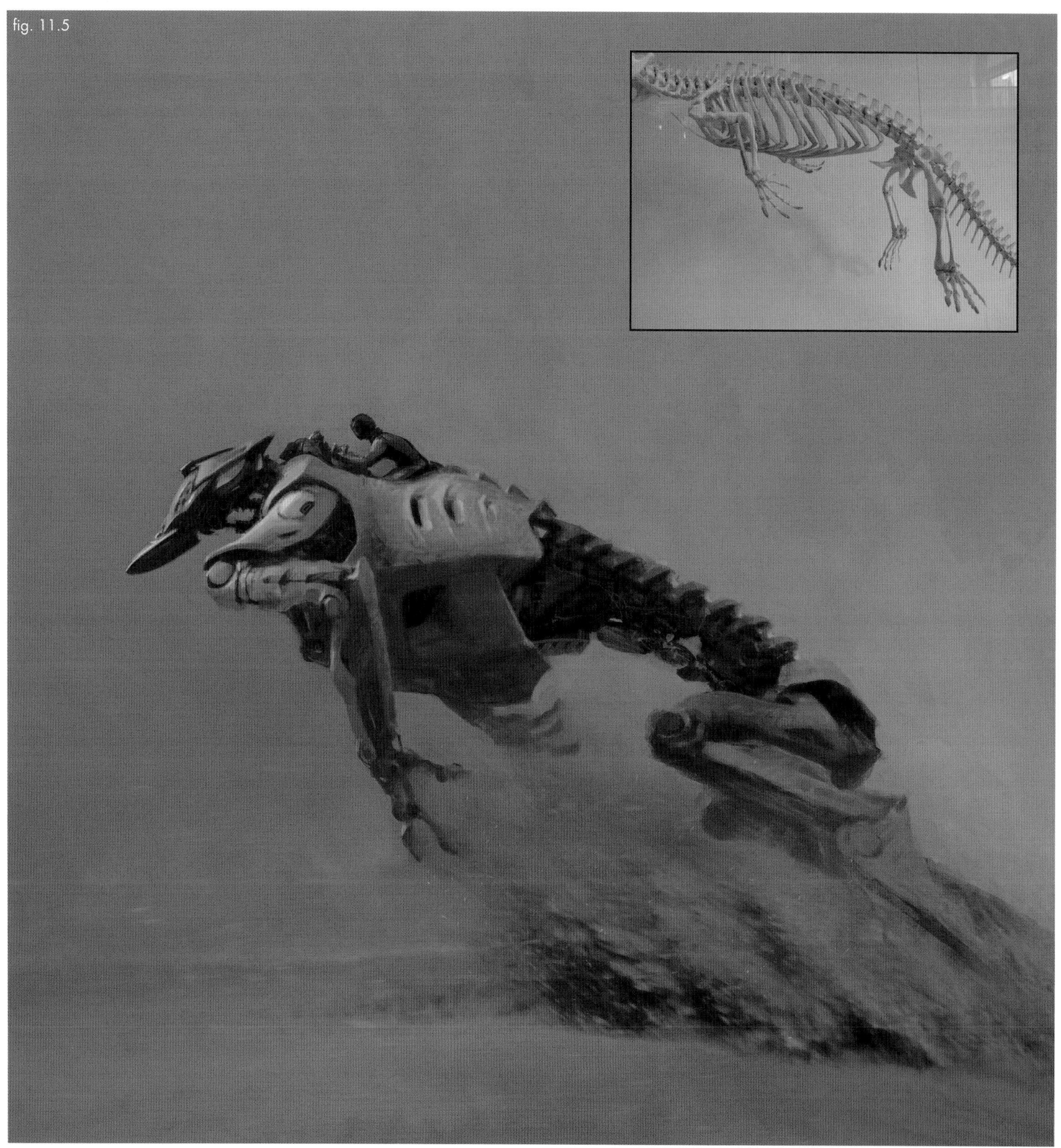
fig. 11.5

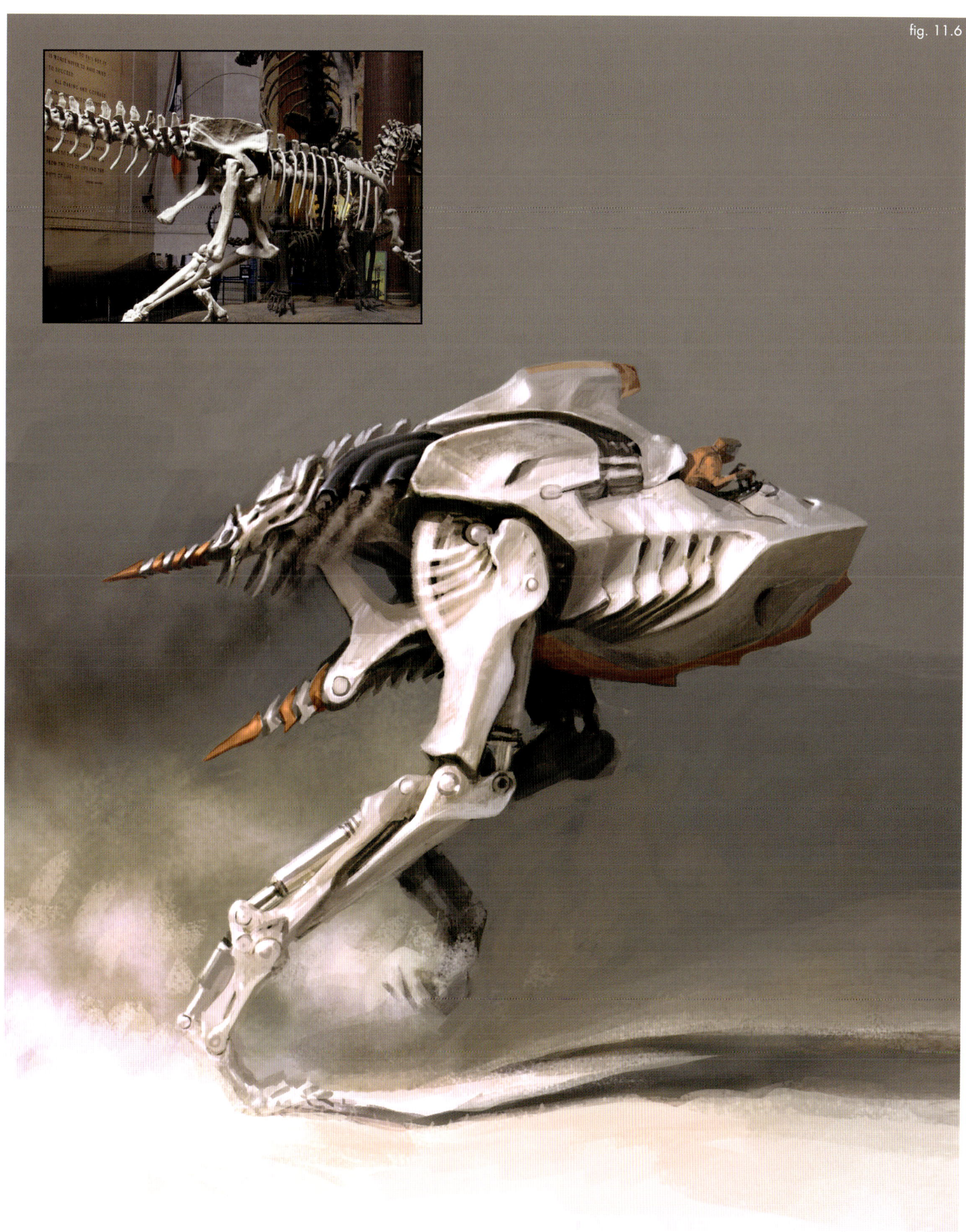

fig. 11.6

RENDERING A SCI-FI SUIT WITH ROBERT SIMONS

Robert Simons was kind enough to share his steps for rendering a sci-fi suited soldier. First, generate loose sketches to establish and explore potential design directions. Do as many of these as it takes until a design is developed that is worthy of the long hours it will take to render it all the way to final-presentation level. The rendering phase will suffer if it is begun prior to loving the concept design.

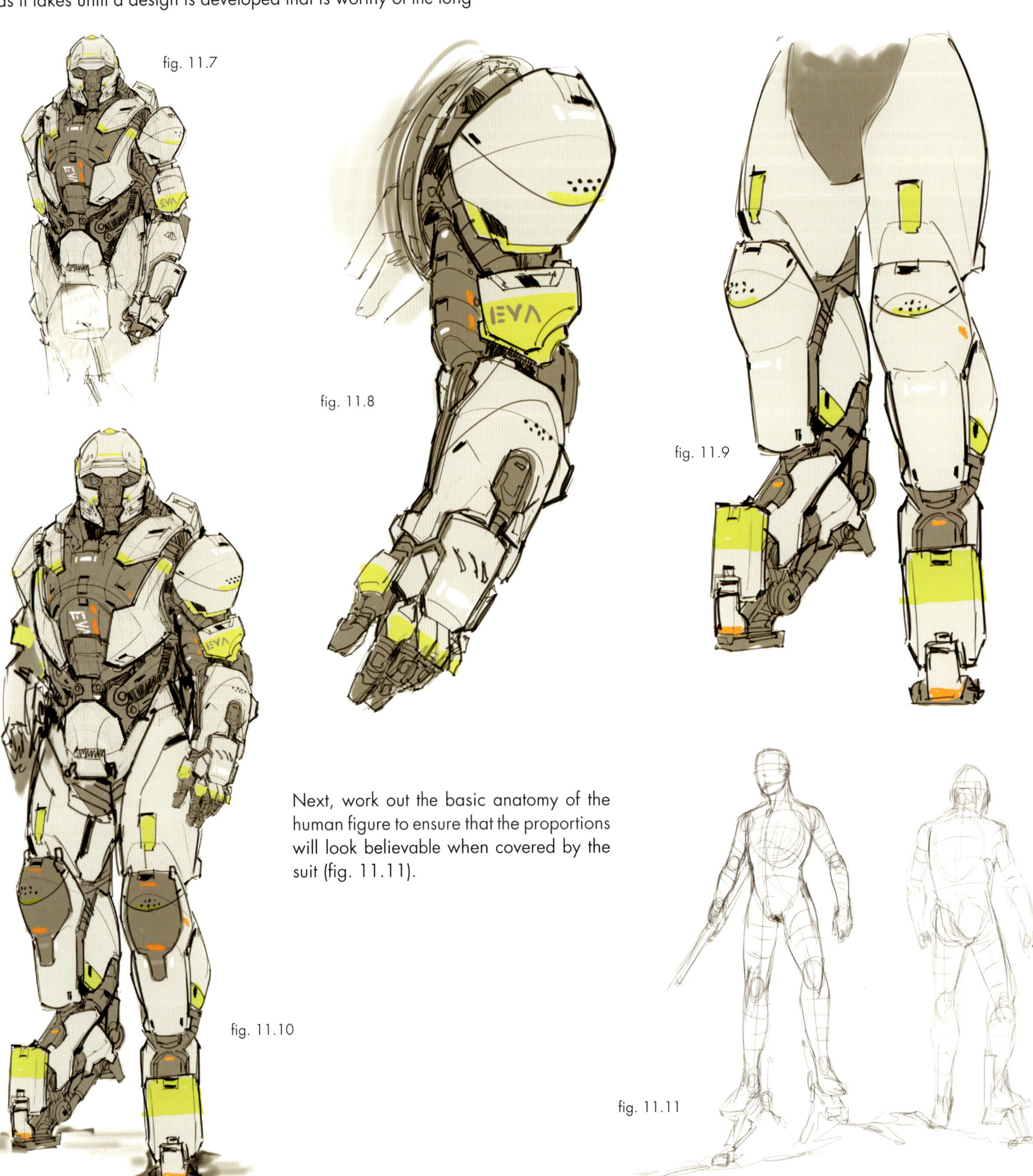

fig. 11.7

fig. 11.8

fig. 11.9

fig. 11.10

Next, work out the basic anatomy of the human figure to ensure that the proportions will look believable when covered by the suit (fig. 11.11).

fig. 11.11

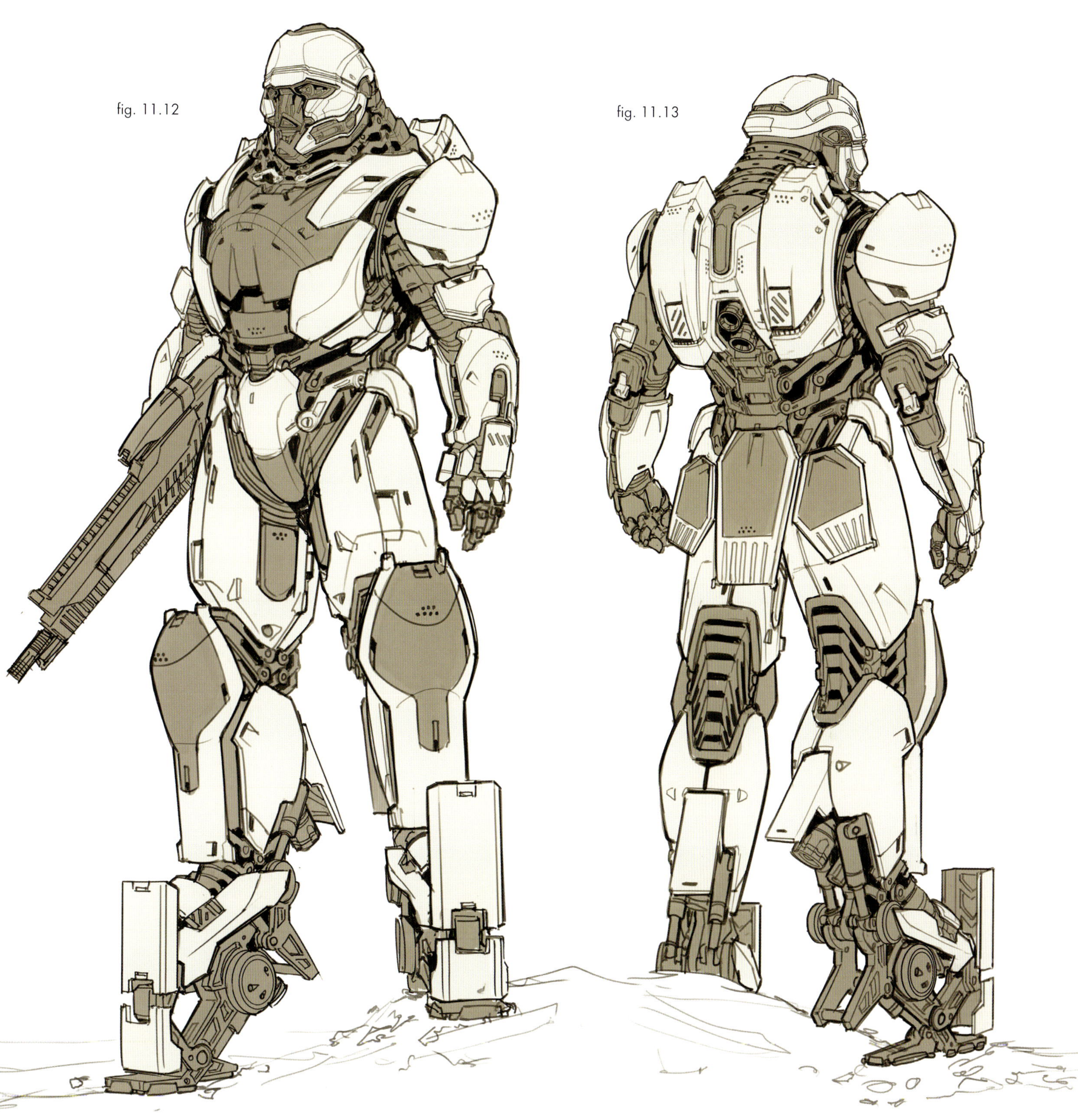

fig. 11.12

fig. 11.13

Now is the time for a final design pass and line drawing before the rendering goes into full gear. Even though this phase is about committing to the line drawing, some of the local values of the component parts of the suit can be blocked out at this point, in gray scale. Taking the time to complete the line drawing effectively serves as a way to pre-visualize the forms, and the final rendering will go much faster as a result. This line drawing was done digitally in SketchBook Pro. Working digitally allows for easier corrections, cleanup and the use of layers on which to make design variations.

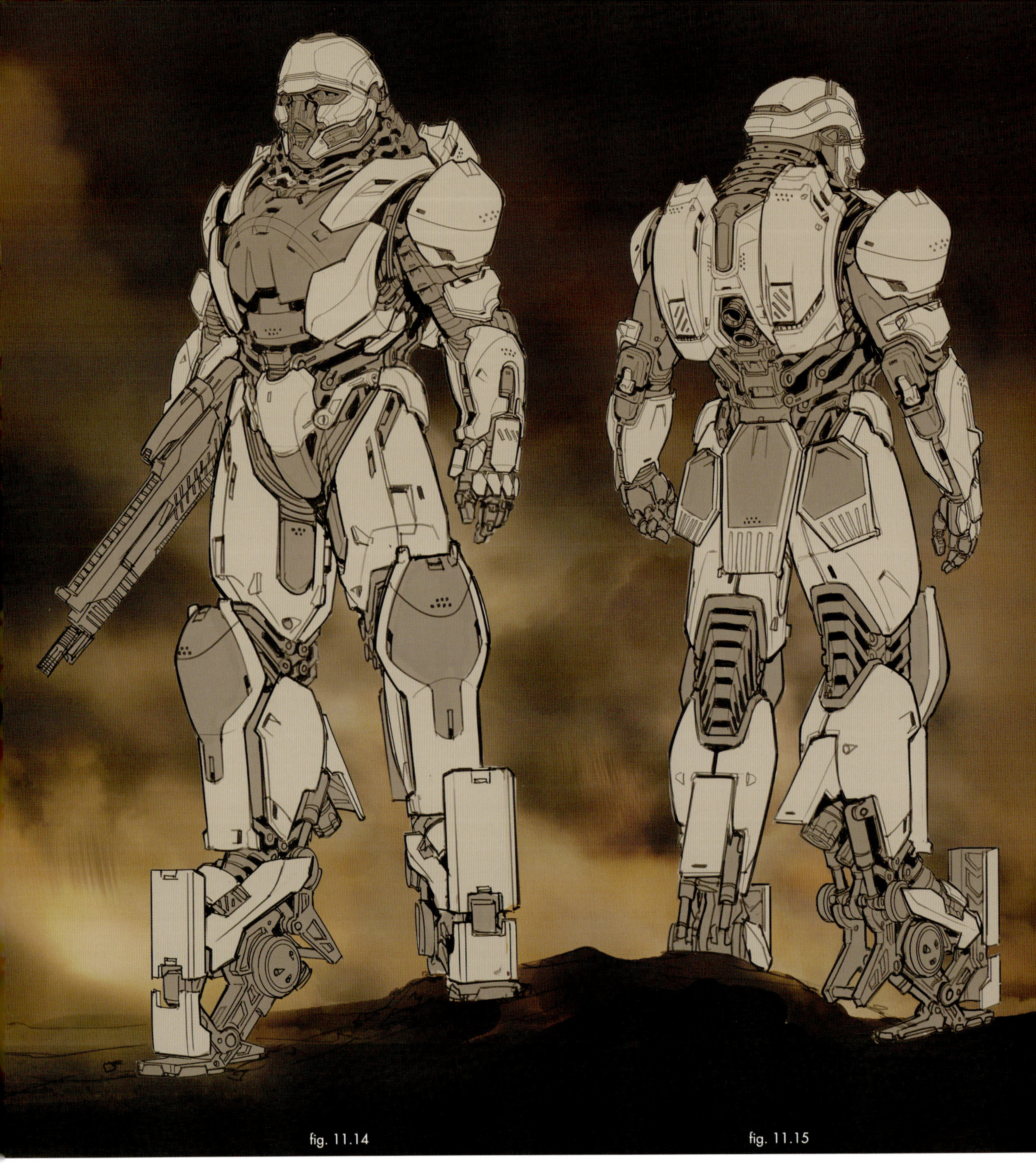

fig. 11.14

fig. 11.15

Since this suit is meant to have a metallic finish, the relative values around it are quite important. Here the background is added and the suit color is warmed up a little to look better in this environment.

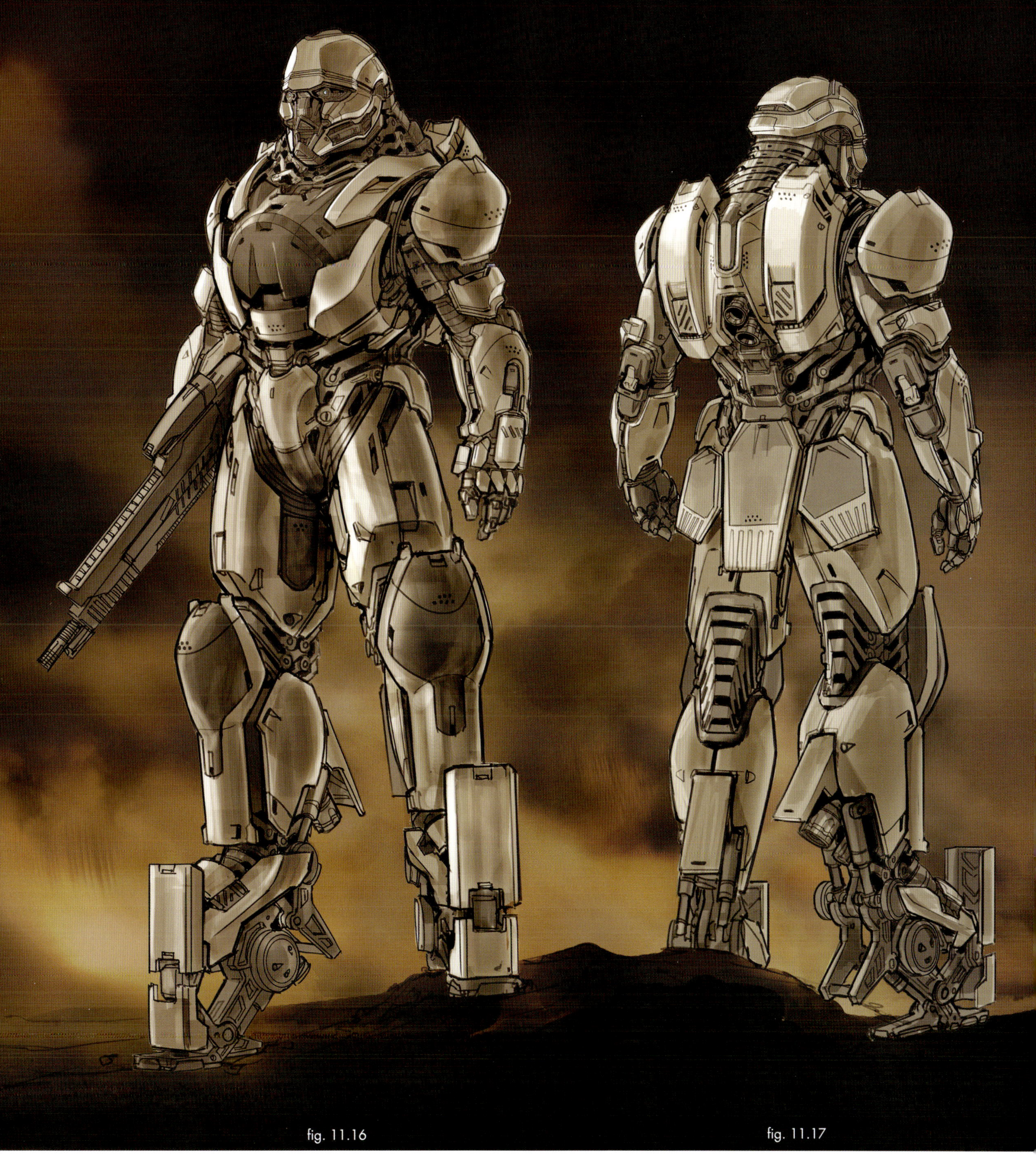

fig. 11.16

fig. 11.17

The primary light source is added from the left, and the forms are loosely blocked out. Next, a secondary, weaker light source is added from the right to provide rim light. Once the basic values are established, the only thing left to do is to refine the surfaces and keep rendering on top of the suit's line-drawing layers until the line work is totally hidden. As the line work disappears, the surfaces look more realistic.

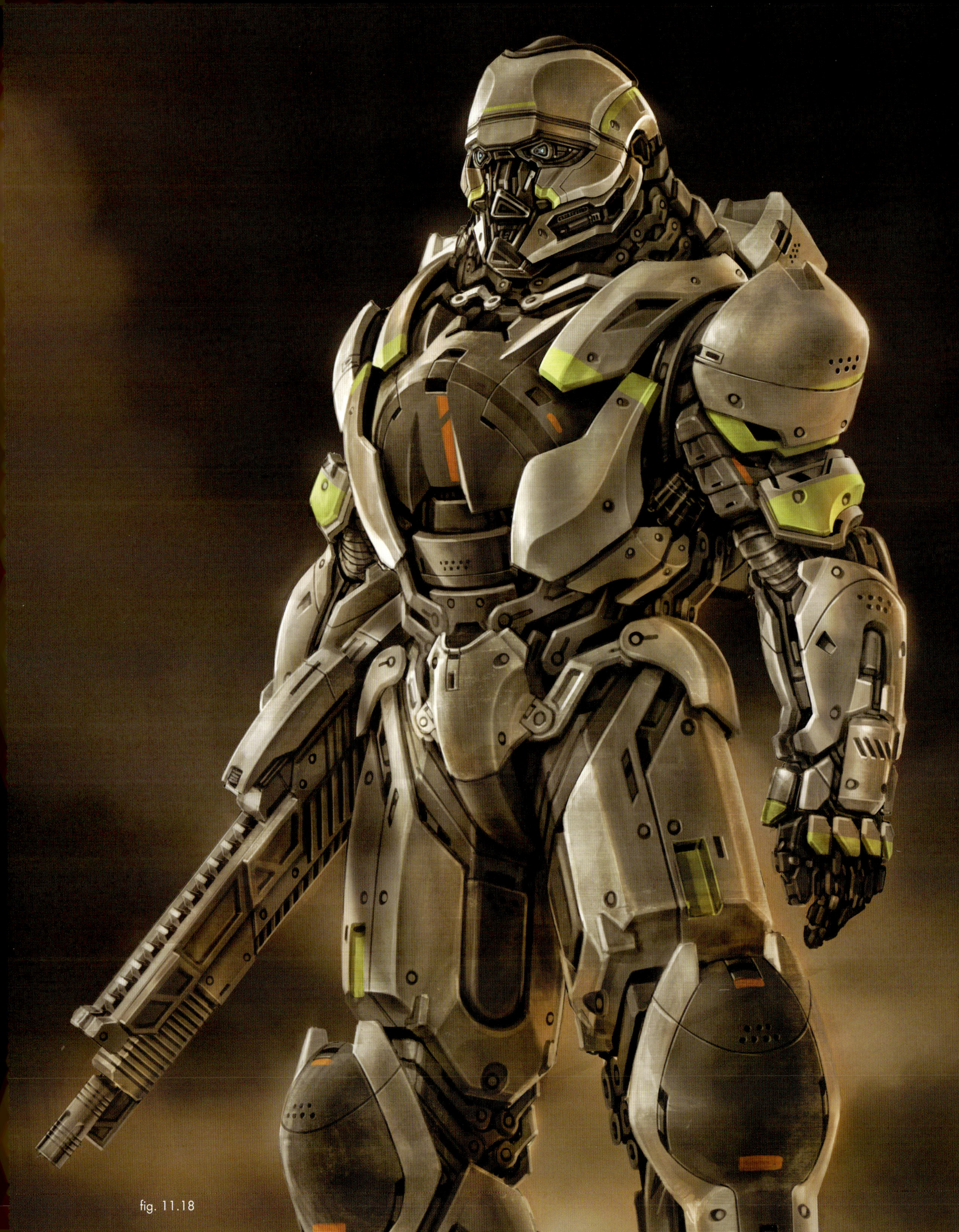

fig. 11.18

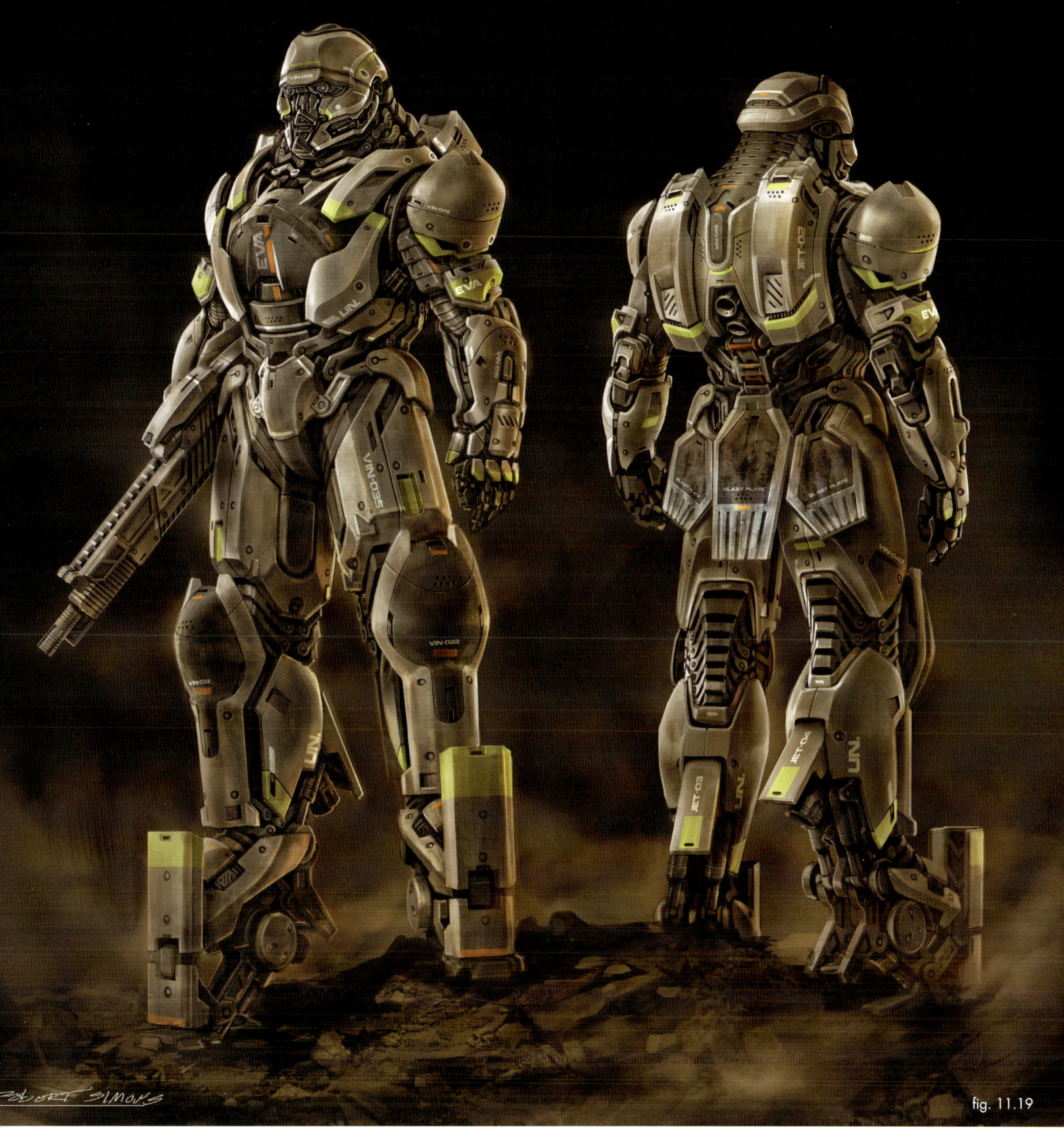

fig. 11.19

In the detail shot to the left, a few photo textures have been laid over the top of the suit layers to help the material look more like brushed metal. This is a very common rendering technique in the concept art world, where getting the point across quickly is more important than perfect accuracy. The job is not to produce art for art's sake, but to communicate design concepts clearly.

Above, final touches to material weathering and color accents were made, completing the rendering in Photoshop.

To see more of Robert's work, visit:
www.robertdraws.com and
www.gadget-bot.com.

2D RENDERING OVER 3D RENDERINGS

Years ago, when 2D digital rendering programs like SketchBook Pro and Photoshop were new, most professional artists and designers fully embraced these programs that added new capabilities to their tool kits. Now 3D modeling and rendering is quickly being adopted by historically 2D-only creative people. My personal favorite modeling and rendering program is MODO, by The Foundry. The real power these new tools bring to 2D artists and designers is that complex materials and views that were too labor intensive to alter in 2D, can now be more easily explored. Lighting and composition studies can be created quickly and different camera lenses and positions can be tried. A big caveat: All of the new tools are worthless without a strong understanding of the fundamentals. Without it, 3D scenes cannot be extended, modified or have new objects added to them properly. Without mastery of basic 2D rendering skills, artists will be slaves to technology and severely limited by the particular 3D program that is learned. Should time be taken to learn a 3D program? Yes. Should time be taken to learn to draw and render with traditional and digital media? Absolutely.

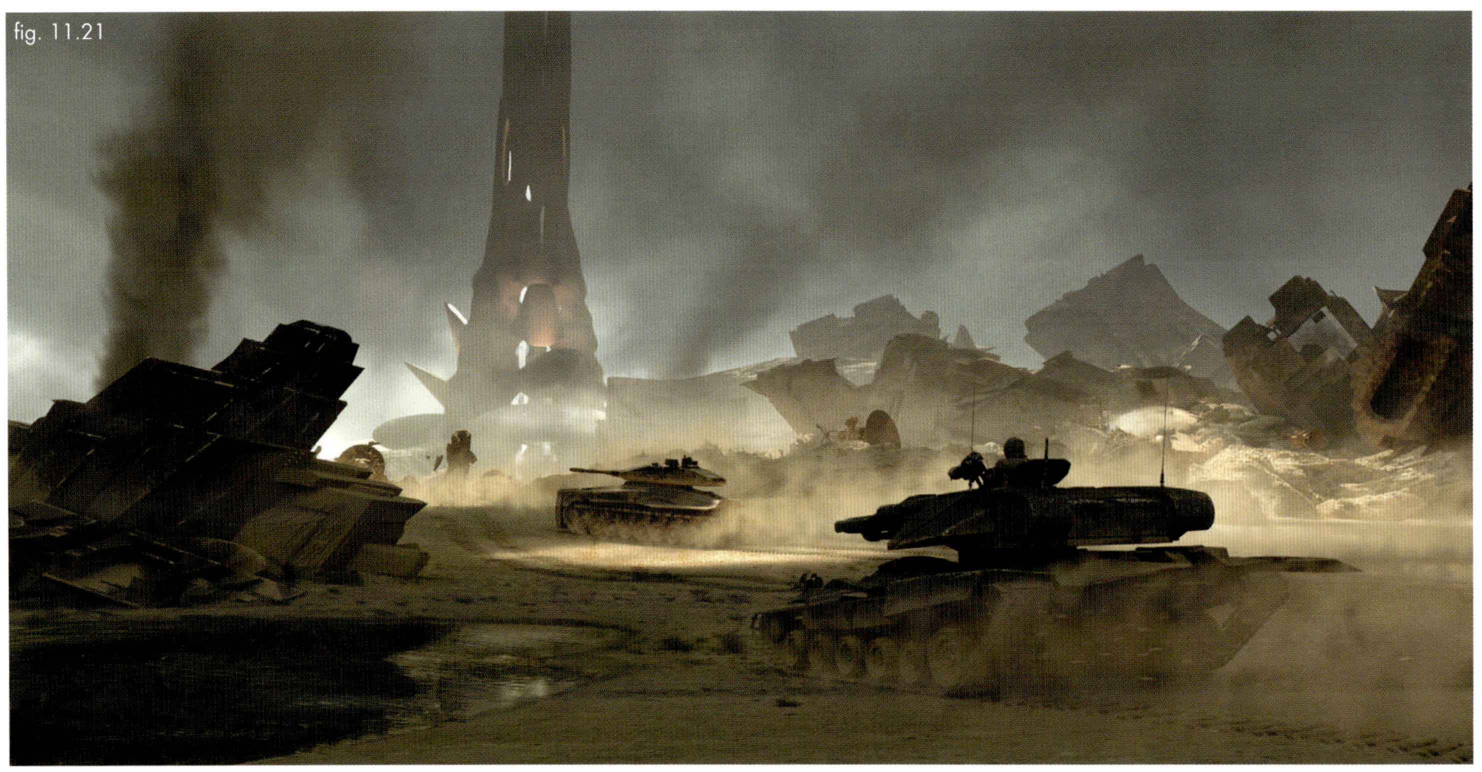
fig. 11.21

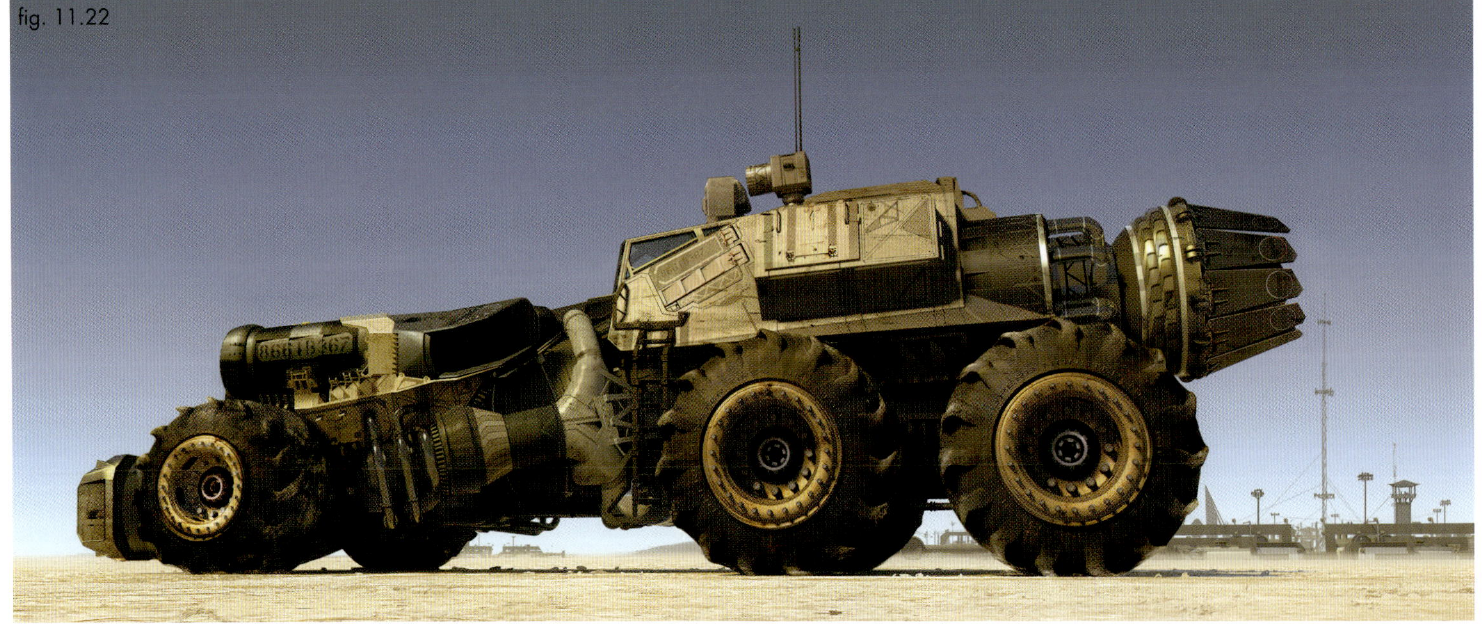
fig. 11.22

Fig. 11.22 shows an image from the book *DRIVE*. I modeled the vehicle in 3D, using MODO, then Danny Gardner rendered over it in 2D, using Photoshop. All of the great detailing and design refinement would have taken at least a week to do to this level in a 3D program, but by using the two programs together it took only one afternoon.

Three-dimensional tools vastly accelerated the production of fig. 11.23. The space-debris field was populated with several purchased 3D models and then a base rendering was output. Lighting, form refinements and the space dust were all rendered in 2D.

Fig. 11.24 also started as a 3D rendering and then the atmospheric effects and motion blur were added in Photoshop. This practice of using fundamental 2D-rendering skills with 3D-generated base images is a very practical and professional approach to creating imagery.

This concludes *How To Render*. You are now armed with the knowledge to create your own images that tell your own stories and clearly illustrate your designs.

Stay inspired.

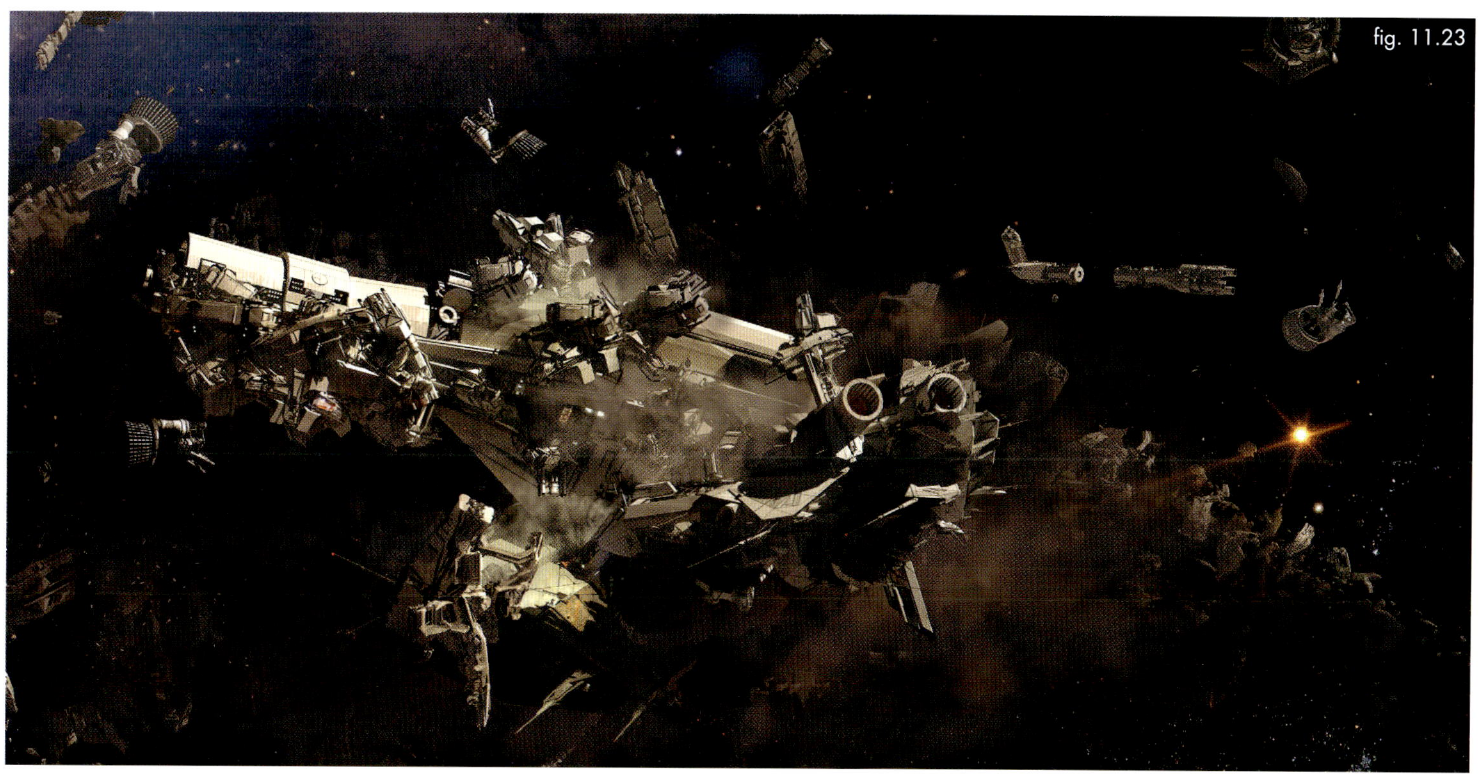
fig. 11.23

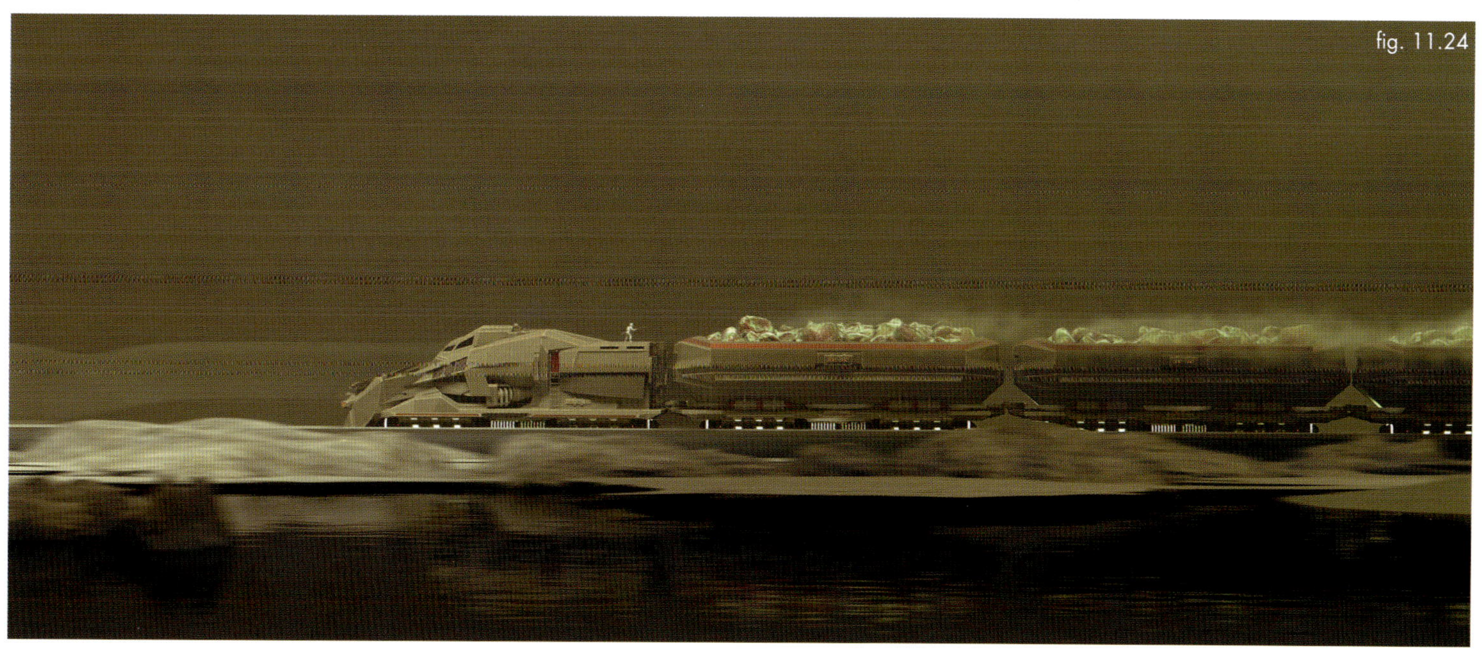
fig. 11.24

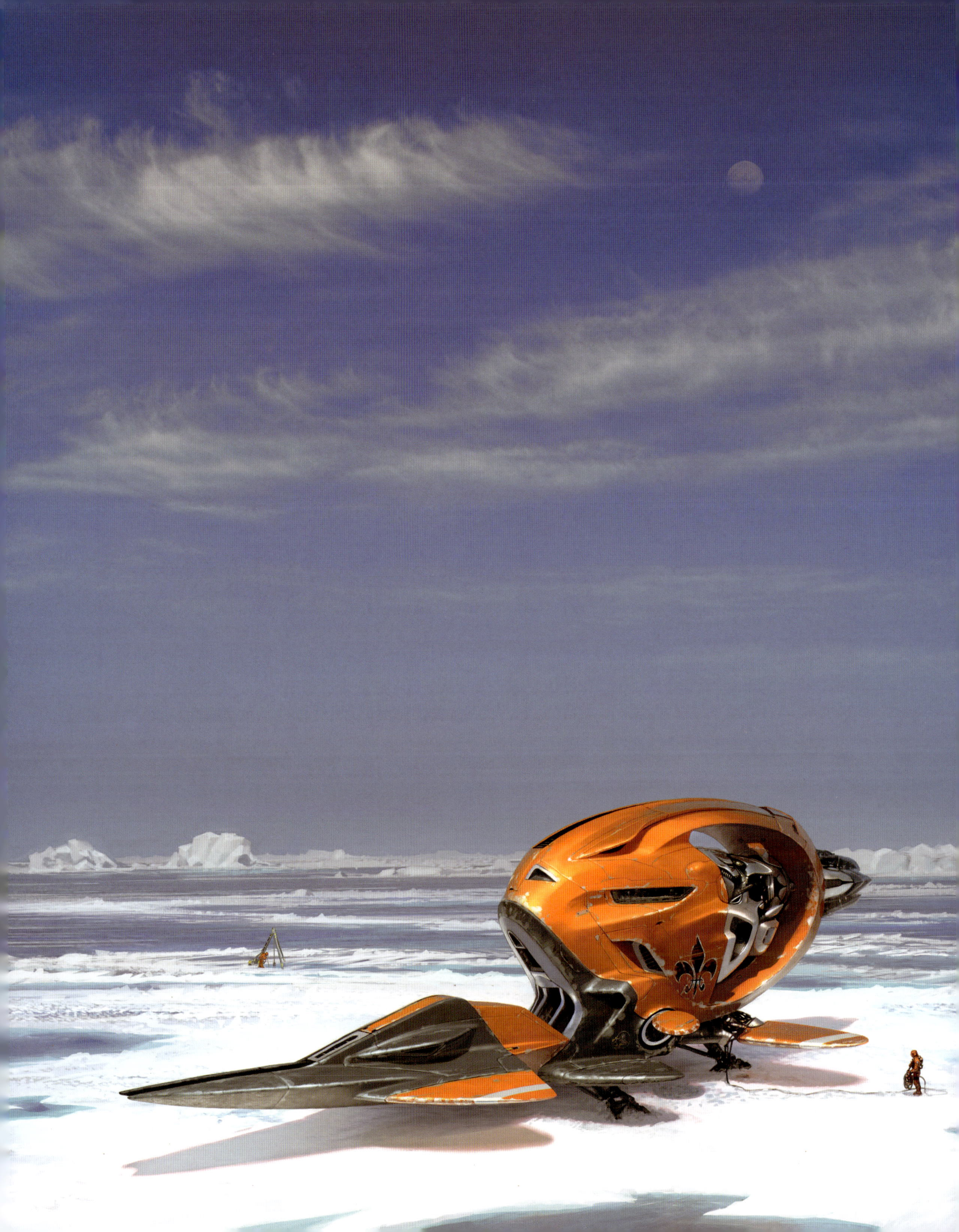

GLOSSARY

1-2-3 read - The strong shifts in value between the three visible sides of an object.

angle of incidence - The measurable angle created when the line of sight intersects a surface.

anisotropic - Exhibits properties with different values when measured in different directions.

artificial light - Any light source that is not sunlight.

atmospheric perspective - A technique of rendering depth or distance in painting by modifying the tone and distinctness of objects perceived as receding from the picture plane, especially by reducing contrasts of light and dark. Also called aerial perspective.

auxiliary vanishing point - That point toward which receding parallel lines appear to converge for secondary elements of an object or a scene, such as a ramp or a pitched roof.

bloom - An imaging artifact of real-world cameras. This effect creates fringes or feathers of light extending outward from the center of the bright areas of an image, across the borders of these areas. Sometimes referred to as light bloom or glow.

cast shadow - The shadow created by an object.

contour line - A line that curves over an object's surface and reveals the item's surface characteristics.

convergence - As parallel lines recede into the distance, they appear to merge at a single point at a person's eye level (also known as the horizon line).

core shadow - The core shadow is the dark band of value gradation that starts at the terminator and gets lighter onto the shadow side of the object as the reflected light increases in strength.

cut line - The necessary clearance gap between two adjacent body panels, such as between a door and the side body of a vehicle. Also referred to as a panel line, part line or shut line.

degree (of ellipse) - The line-of-sight angle at which the plane, defined by a circle in perspective, is viewed.

depth of field - The distance between the farthest and nearest objects in a scene that are both in focus.

diffuse light - Casts a shadow with a soft edge.

direct light - Casts a shadow with a hard edge.

edge lighting - Light shining from behind an object that illuminates just an edge. Also known as rim lighting.

ellipse - A circle in perspective.

equal in–equal out - Referring to how the line of sight's angle at which it hits a surface is exactly the same as the angle at which it bounces away from the surface.

falloff - The rate at which a light's strength diminishes as the distance from its source increases. Also known as decay.

fillet - An additive volume, usually with the cross section of a circle, that blends two intersecting volumes together.

form change - Occurs when rays of light illuminate a volume that has variation to its surface, and the light intersects these varying surface changes at different angles. Also known as value change.

Fresnel Effect - The phenomenon of the strength of the reflection of an environment changing and growing stronger as the surface rolls away from the line of sight.

glint - A light effect produced at an angle that has a light bloom localized at a narrow edge of a reflective surface.

glow - An imaging artifact of real-world cameras that creates fringes or feathers of light extending outward from the center of the bright areas of an image. This effect contributes to the illusion of an extremely bright light, overwhelming the camera or human eye.

GLOSSARY

gradation - A small change from one shade, tone or color to another.

gray scale - An image composed exclusively of shades of gray.

ground plane - The theoretical horizontal plane receding from the picture plane to the horizon line.

half-light - When an object, or a scene, is only half in the light, creating more visual interest and naturally creating a focal point.

halfway to black - An observation method that determines the shadow value of matte objects.

happy accident - When something unexpectedly good comes from what would otherwise be considered a mishap.

hard light - Refers to when all of the light rays are very aligned in a singular direction.

horizon line - A horizontal line across a picture. Its placement defines the viewer's eye level.

iris - The colored part of the eye.

light decay - See 'falloff'

light direction - A line that eminates from a light source.

light plane - Defined by the direction of the light and the direction of the shadow.

line of sight - A straight line extending from the fovea centralis of the eye to an object on which the eye is focused.

linear perspective - A mathematical system for representing three-dimensional objects and space on a two-dimensional surface by means of intersecting lines that are drawn vertically and horizontally and that radiate from one point (one-point perspective), two points (two-point perspective) or several points on a horizon line as perceived by a viewer imagined in an arbitrarily fixed position.

luminescence - Occurs when the surfaces of objects themselves glow and emit light.

light ray angle of incidence - The angle formed by light striking a surface and a line perpendicular to the surface at the point of impact.

local light - Light that does not originate from the sun.

matte surfaces - Having a dull or lusterless surface that partially diffuses reflected light.

minor axis - The line that divides an ellipse in half across its narrow dimension. The minor axis is always perpendicular to the surface on which the ellipse lies.

MODO - 3D modeling and rendering software made by Luxology (www. luxology.com).

negative sunlight - When objects are front-lit.

occlusion - One surface hiding another surface from view.

occlusion shadow - The darkest part of the cast shadow due to a reduction of ambient or reflected light.

orthographic view - A single view of an object onto a drawing surface with no perspective convergence. Also called draft view.

overhand grip - Holding a drawing instrument on its side.

overlay - A sheet of transparent paper placed over a photograph or other artwork for making revisions. This can also describe a digital layering property that mimics this effect.

parallel - Lines or planes that extend in the same direction, everywhere equidistant, and not meeting.

passive highlight - The most brightly lit area of a surface, which occurs where the light rays are most perpendicular to it.

perspective - A technique of depicting volumes and spatial relationships on a flat surface.

perspective grid - A network of lines drawn to represent the perspective of a systematic network of lines on the ground or on X-Y-Z planes.

picture plane - The surface on which images are recorded. Imagine the picture plane being a plate of glass that is perpendicular to the Line of Sight.

photo-real - A style of rendering that emphasizes lifelike recreation of objects, light effects, proportions and textures.

Photoshop - 2D Digital rendering software designed by Adobe Systems (www.adobe.com).

planar volume - A volume with a flat surface.

plane reflectors - Surfaces used to manipulate the way the primary light source reflects into chosen areas of a scene. Also called board reflectors or bounce cards.

pools or puddles - Reflections of the environment or other parts of the car itself that tend to be like floating islands due to reflection flipping.

positive sunlight - When objects are backlit.

reference point - A mark set at a specific location in a drawing so as to permit accurate perspective drawing.

reflected light - Light that reflects or bounces off of one surface and illuminates another surface. Commonly known as "bounce" or "fill" light.

reflection flipping - When the line of sight bounces off a shiny, concave surface.

reflective highlight - The reflection of the light source on the surface of the object.

refraction - The change of direction of a ray of light.

sclera - The white outer layer of the eyeball.

section lines - Parallel lines that curve over an object's surface in a vertical or horizontal manner (or both) and reveal the item's surface characteristics.

shadow direction - A line originating from a point called the shadow origin.

shadow origin - Created when objects under a light source, transfer their height to a vertical line.

shadow side - The area that is not exposed to direct light, but usually receives reflected or ambient light.

SketchBook Pro - 2D modeling and rendering software, developed by Autodesk (www.autodesk.com).

soft light - Refers to when a surface is illuminated from many different points of light, also called diffuse light.

stick - In perspective drawing, any line segment.

subsurface scattering - When light rays penetrate material and then scatter around in different directions within it, creating a glowing effect.

terminator - The transition between the light and shadow sides of an object.

true value - The value of the physical color of the object, regardless of lighting conditions.

underlay - An image or drawing, often of a perspective grid, laid underneath a piece of paper to be the foundation for the overlay drawing.

value - The relative lightness or darkness of a color.

value change - Differences in relative lightness or darkness.

value gradation - To pass imperceptibly from one shade of gray, to another across a plane or surface.

value range - Describes which values, from lightest to darkest, are involved with rendering a particular object.

vanishing point - That point toward which receding parallel lines appear to converge.

writing grip - Traditional way of holding a writing instrument.

X-Y-Z form - An object that posesses height, width and depth that can be given lighting, shadow and cut lines that utilize the perspective and dimensions determined by the positioning of the object within the x-y-z axes.

INDEX

S

T

V

W

X

HOW TO DRAW: DRAWING AND SKETCHING OBJECTS AND ENVIRONMENTS FROM YOUR IMAGINATION

How To Draw is for artists, architects and designers. It is useful to the novice, the student and the professional. You will learn how to draw any object or environment from your imagination, starting with the most basic perspective drawing skills.

Early chapters explain how to draw accurate perspective grids and ellipses that in later chapters provide the foundation for more complex forms. The research and design processes used to generate visual concepts are demonstrated, making it much easier for you to draw things never-before-seen!

Best of all, more than 25 pages can be scanned via a smartphone or tablet using the H2Dr app, which link to video tutorials for that section of the book!

With a combined 26 years of teaching experience, Scott Robertson and Thomas Bertling bring you the lessons and techniques they have used to help thousands of their students become professional artists and designers.

This book is indispensable for anyone who wants to learn, or teaches others, how to draw.

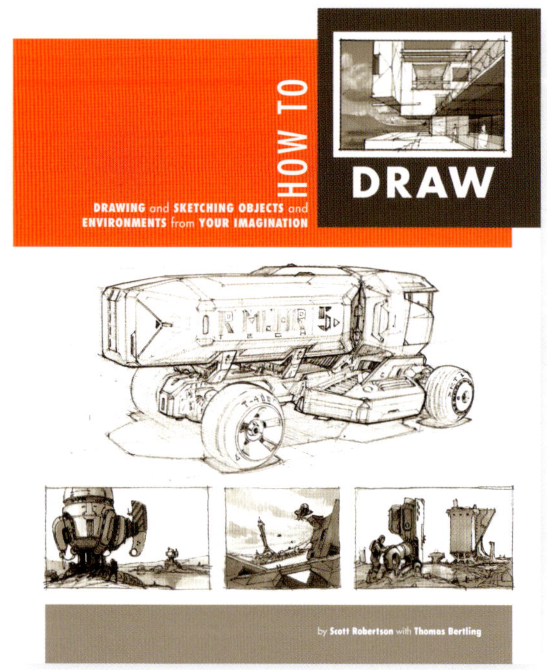

Paperback: 978-1-933492-73-5
Hardcover: 978-1-933492-75-9

SCOTT ROBERTSON DESIGN: YOUTUBE CHANNEL

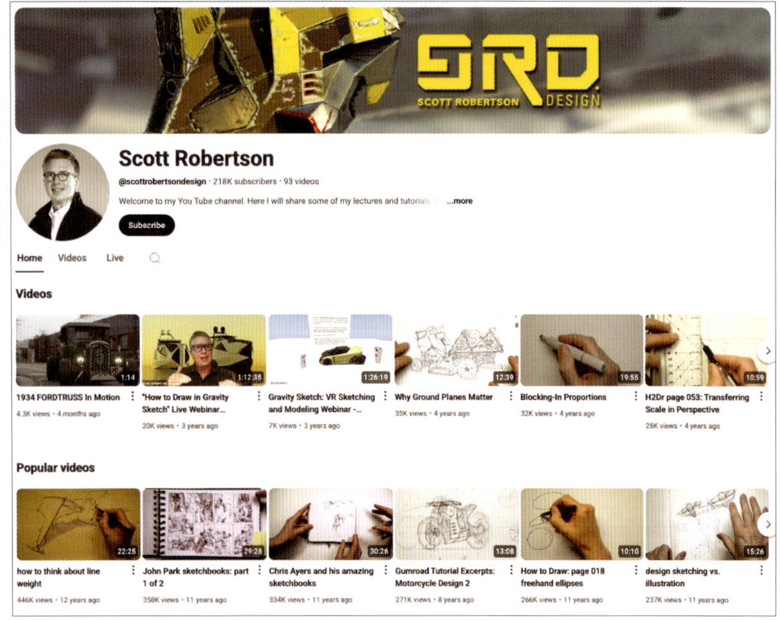

A great FREE educational resource is Scott's YouTube channel,
http://www.youtube.com/scottrobertsondesign

Find plenty of educational tutorials related to drawing, rendering and design. New videos are posted periodically.

MORE FROM SCOTT ROBERTSON

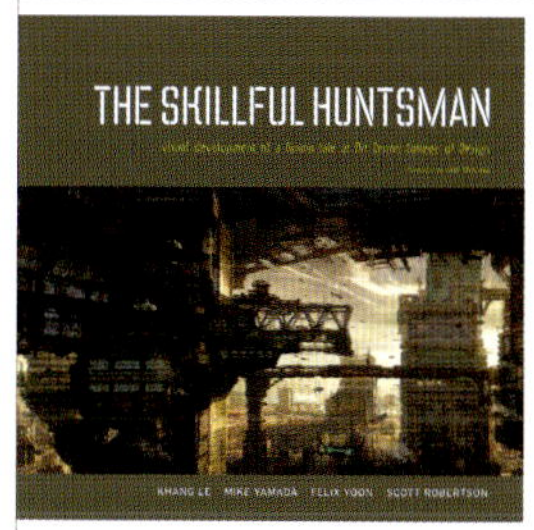

The Skillful Huntsman is a documentation of the creative thought process in designing for entertainment through the work of three talented artists. Guided by Scott Robertson, the artists create original design solutions for the environment, characters, props and much more.

Softcover: 978-0-972667-64-7

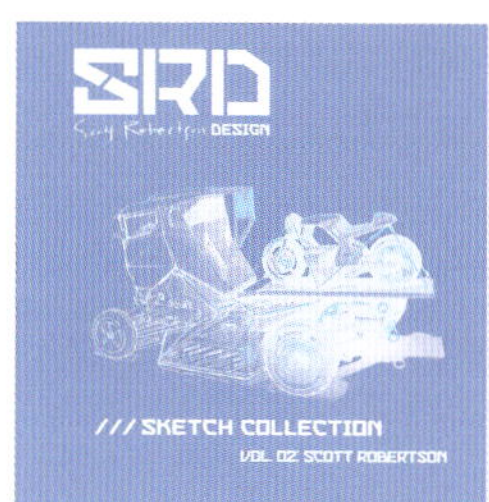

Scott Robertson returns with his much anticipated *SRD Sketch Collection vol. 2*, clocking in at a whopping 288 drawings that are sure to amaze and inspire.

Hardcover: 978-1-624650-37-6

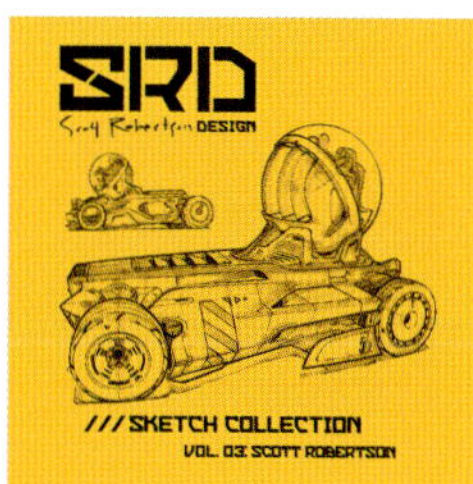

After a brief hiatus from publishing, the acclaimed designer and best-selling author of *How To Draw* and *How To Render* is back with *SRD Sketch Collection vol. 3* to remind the industry why he continues to be a force, with a wide, imaginative range of vehicles represented in the book.

Hardcover: 978-1-624650-97-0

SCOTT ROBERTSON DESIGN: GUMROAD

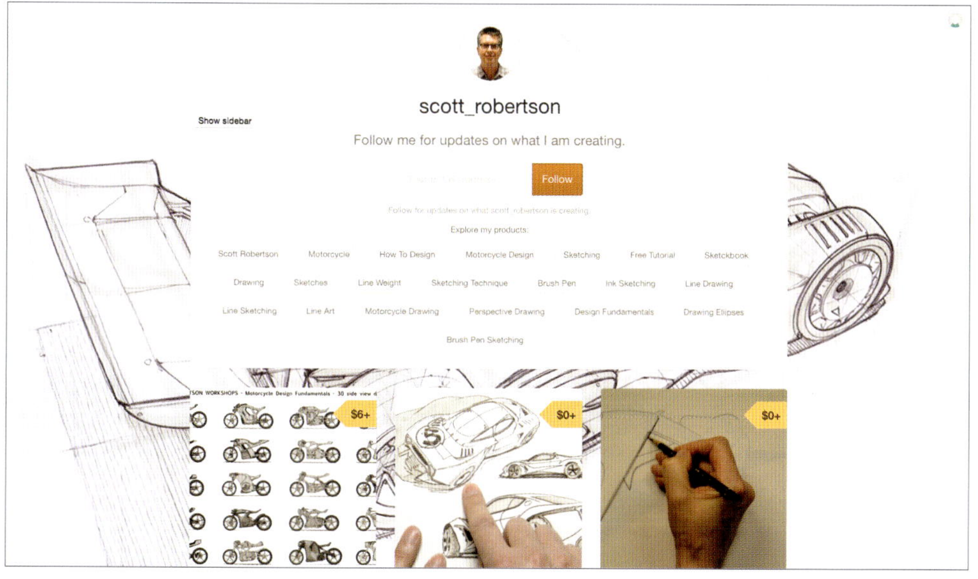

http://www.gumroad.com/scott_robertson

VIDEO LINKS LIST

To download the bonus image files and view all of the educational videos marked by the play button in this book, just type in the URL listed below or scan the QR Code to the right. Enter the username and password, **howtorender**, when prompted.

Be sure to check back from time to time for updates!

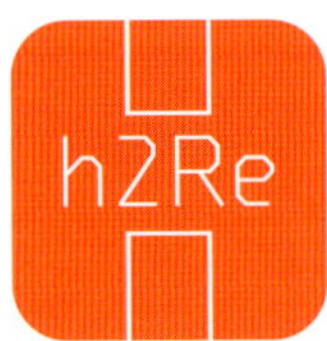

http://scottrobertsonworkshops.com/h2re/linklist

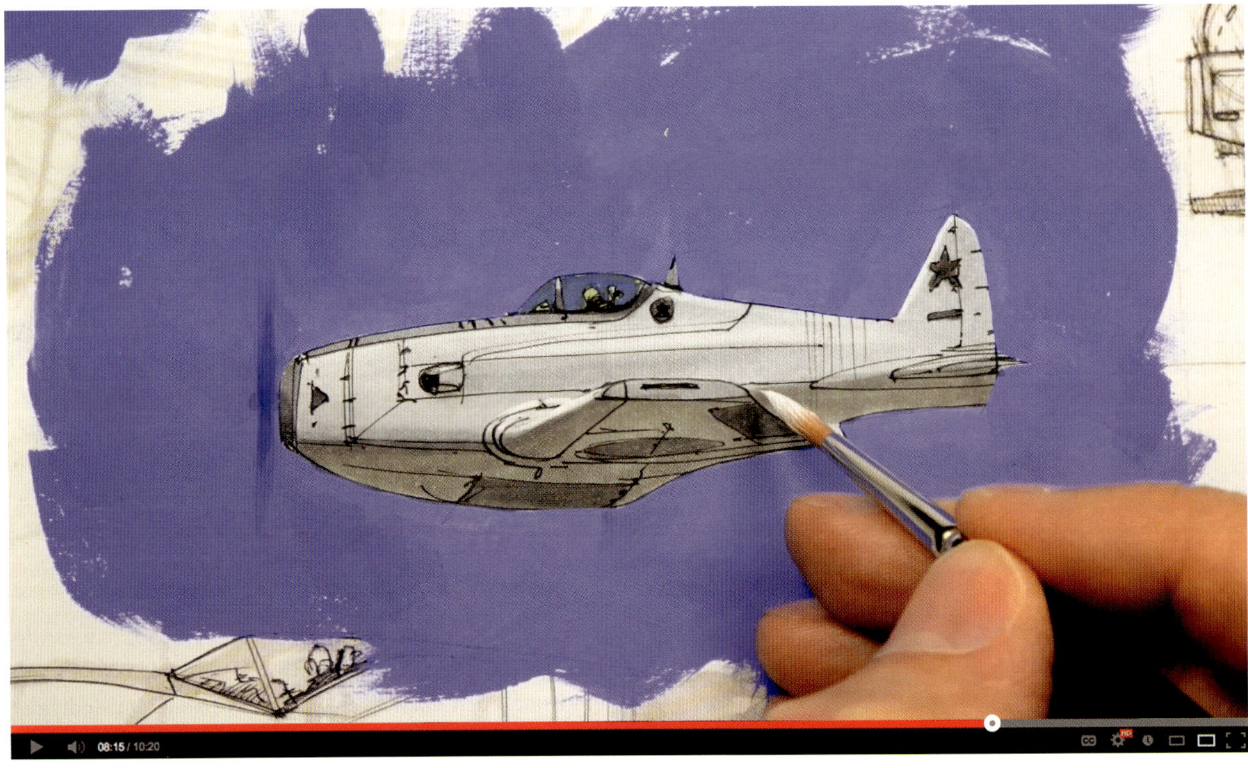

ABOUT THE AUTHORS

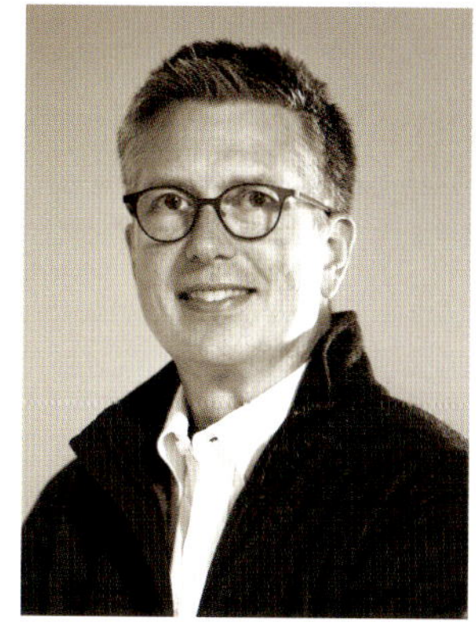

Scott Robertson
Former Chair of Entertainment Design: ArtCenter College of Design
Designer / Author / Educator / Co-Producer

With over 18 years of experience teaching and creating curriculum on how to design, draw, and render at the highest college level, Scott brings unique and unquestioned expertise to the presentation and communication of the subject of this book. He has authored or co-authored 11 books on design and concept art. In addition to books, he has co-produced over 40 educational DVDs with The Gnomon Workshop of which 9 feature his own lectures. At ArtCenter College of Design, Scott formerly chaired the Entertainment Design department that he helped to create.

In addition to teaching, Scott has worked on a wide variety of projects ranging from vehicle and alien designs for the Hot Wheels animated series *Battle Force Five,* to theme park attractions such as the *Men in Black* ride in Orlando, Florida for Universal Studios. Some of his clients have included the BMW subsidiary Design-works/USA, Bell Sports, Giro, Mattel Toys, Spin Master Toys, Patagonia, the feature film *Minority Report,* Nike, Rockstar Games, Sony Online Entertainment, Sony Computer Entertainment of America, Buena Vista Games, THQ, and Fiat, to name just a few.

To see more of Scott Robertson's personal and professional work, please visit **www.drawthrough.com** and his blog at **www.drawthrough.blogspot.com**.

Scott can also be followed online at:
Facebook: www.facebook.com/scott.robertson.005
Instagram: scoro5
X: @scoro5

contact email: **scott@drawthrough.com**

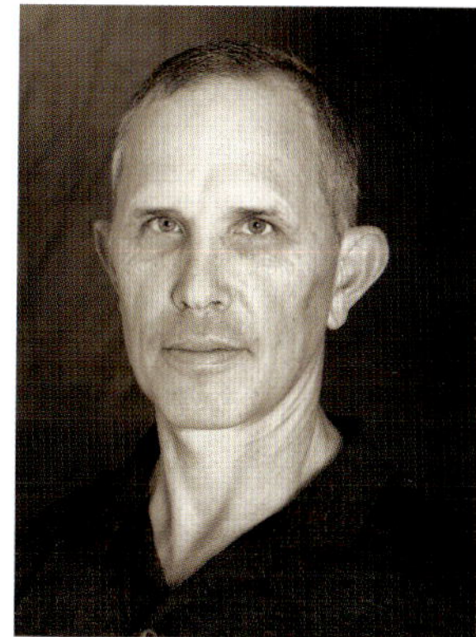

Thomas Bertling
Former Director and Interim Chair of Entertainment Design: ArtCenter College of Design
Designer / Engineer / Educator

Thomas Bertling brings an exceptional blend of practical industrial design experience and pedagogical knowledge to his work. His extensive portfolio includes design projects for a wide array of major clients, spanning diverse fields from medical technology improvement to clean energy development, reflecting a commitment to sustainable innovation. This hands-on expertise ensures that his design guidance is deeply grounded in real-world application and engineering discipline.

Known as a rigorous, caring, and inspiring mentor, Thomas has taught all levels of perspective drawing and design concept development – helping both university students and industry professionals build solid foundational design skills. His learner-centered approach emphasizes clarity in expressing individual creative vision while building deep understanding of core concepts.

Previously, Thomas was instrumental in his leadership of the department of Entertainment Design at ArtCenter College of Design, serving as the founding Director and later as interim Chair, where he developed and taught courses and trained faculty. He now dedicates his pedagogical efforts to curriculum development for K–12 education, aiming to instill strong design thinking and communication skills early in learning.

To see more of Thomas Bertling's personal and professional work, please visit **www.thomasworks.com**.

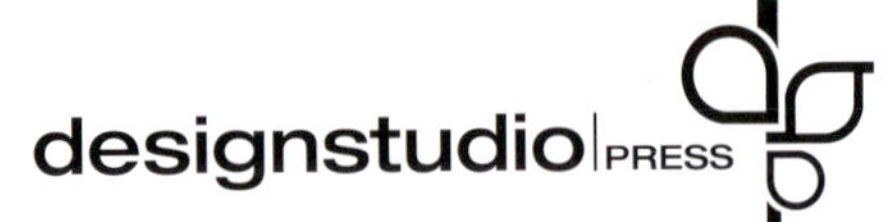

SPECIAL THANKS

Special thanks to my wife Melissa and the Design Studio Press creative team for all the help and support during the creation of this book, and to my mentors who taught me. Lastly, thanks to you for supporting me through your ongoing interest in my books; you keep me inspired to do more!

—Scott Robertson

Special thanks to my parents, Josef and Sabine, who helped me pursue the crazy dream of becoming a designer; to Scott Robertson for being the teacher who made me a teacher; to all my incredible students who inspire me to keep learning & growing, and to my wife Erika & son Lukas who have always believed in me – thank you for your patience, support, and love.

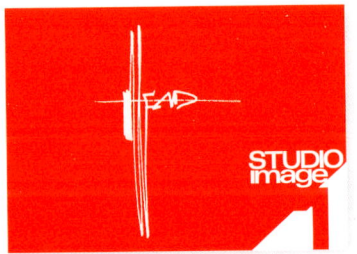

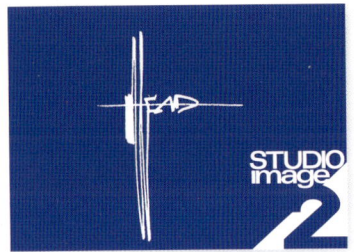

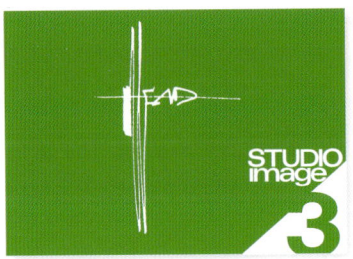

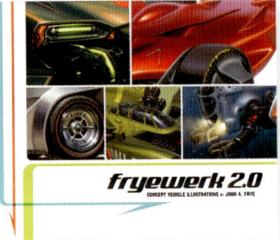

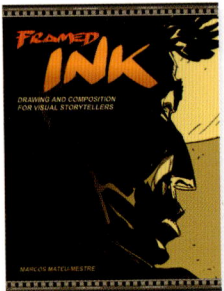

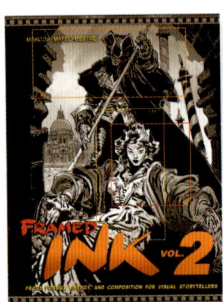

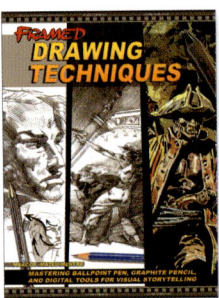

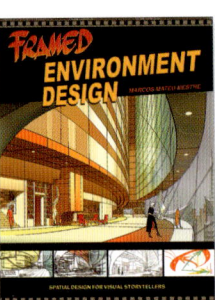

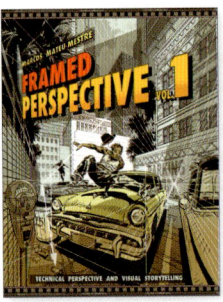

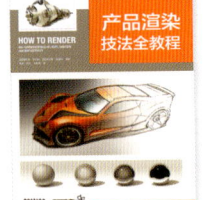

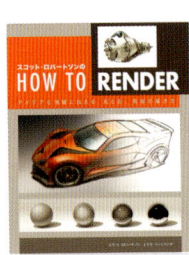

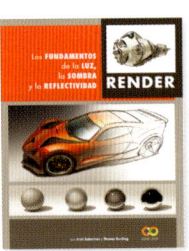

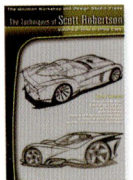

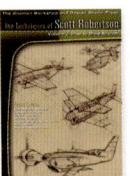

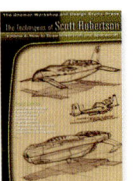

—Thomas Bertling

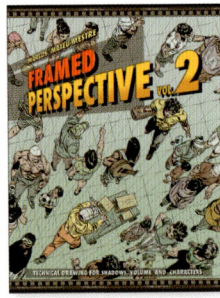

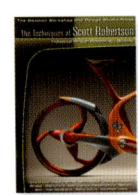

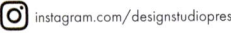

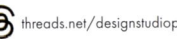

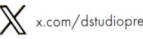